AF714054

Electrical

Level Two

Trainee Guide
2005 *NEC*® Revision

PEARSON
Prentice Hall
Upper Saddle River, New Jersey
Columbus, Ohio

contren®
Learning Series

nccer

National Center for Construction Education and Research
President: Don Whyte
Director of Product Development: Daniele Dixon
Electrical Project Manager: Daniele Dixon
Production Manager: Jessica Martin
Product Maintenance Supervisor: Debie Ness
Desktop Publishers: Laura Parker and Debie Ness

Writing and development services provided by Topaz Publications, Liverpool, New York.

Pearson Education, Inc.
Product Manager: Lori Cowen
Production Editor: Stephen C. Robb
Design Coordinator: Karrie M. Converse-Jones
Text Designer: Kristina D. Holmes
Cover Designer: Kristina D. Holmes
Copy Editor: Sheryl Rose
Scanning Coordinator: Karen L. Bretz
Scanning Technician: Janet Portisch
Production Manager: Pat Tonneman
Marketing Manager: Mark Marsden

This book was set in Palatino and Helvetica by Carlisle Communications, Ltd. It was printed and bound by Courier Kendallville, Inc. The cover was printed by Phoenix Color Corp.

This information is general in nature and intended for training purposes only. Actual performance of activities described in this manual requires compliance with all applicable operating, service, maintenance, and safety procedures under the direction of qualified personnel. References in this manual to patented or proprietary devices do not constitute a recommendation of their use.

Copyright © 2005, 2003, 2000, 1998, 1993 by the National Center for Construction Education and Research (NCCER), Gainesville, FL 32614-1104 and published by Pearson Education, Inc., Upper Saddle River, NJ 07458. All rights reserved. Printed in the United States of America. This publication is protected by copyright, and permission should be obtained from NCCER prior to any prohibited reproduction, storage in a retrieval system, or transmission in any form or by any means, electronic, mechanical, photocopying, recording, or likewise. For information regarding permission(s), write to: NCCER Product Development, P.O. Box 141104, Gainesville, FL 32614-1104.

Pearson Prentice Hall™ is a trademark of Pearson Education, Inc.
Pearson® is a registered trademark of Pearson plc
Prentice Hall® is a registered trademark of Pearson Education, Inc.

Pearson Education Ltd.
Pearson Education Singapore Pte. Ltd.
Pearson Education Canada, Ltd.
Pearson Education—Japan
Pearson Education Australia Pty. Limited
Pearson Education North Asia Ltd.
Pearson Educación de Mexico, S.A. de C.V.
Pearson Education Malaysia Pte. Ltd.

10 9 8 7 6 5 4
ISBN 0-13-168448-5

Preface

TO THE TRAINEE

Electricity powers the applications that make our daily lives more productive and efficient. Thirst for electricity has led to vast job opportunities in the electrical field. Electricians constitute one of the largest construction occupations in the U.S., and they are among the highest-paid workers in the construction industry. According to the U.S. Bureau of Labor Statistics, job opportunities for electricians are expected to be excellent as the demand for skilled craftspeople is projected to outpace the supply of trained electricians.

Electricians install electrical systems in structures. They install wiring and other electrical components, such as circuit breaker panels, switches, and light fixtures. Electricians follow blueprints, the *National Electrical Code®*, and state and local codes. They use specialized tools and testing equipment, such as ammeters, ohmmeters, and voltmeters. Electricians learn their trade through craft and apprenticeship programs. These programs provide classroom instruction and on-the-job training with experienced electricians.

We wish you success as you embark on your second year of training in the electrical craft and hope that you'll continue your training beyond this textbook. There are more than a half-million people employed in electrical work in the U.S., and as most of them can tell you, there are many opportunities awaiting those with the skills and desire to move forward in the construction industry.

NEW WITH *ELECTRICAL LEVEL TWO*

NCCER and Prentice Hall are pleased to present the fifth edition of *Electrical Level Two.* This edition presents a new design and has been fully updated to meet the *2005 National Electrical Code®*. The opening page of each of the twelve modules in this textbook features photos of the construction project Award of Excellence winners from Associated Builders and Contractors, Inc. Check out these awesome projects and see where a career as an electrician could take you.

We invite you to visit the NCCER website at www.nccer.org for the latest releases, training information, newsletter, and much more. You can also reference the Contren® product catalog online at www.crafttraining.com. Your feedback is welcome. You may email your comments to curriculum@nccer.org or send general comments and inquiries to info@nccer.org.

CONTREN® LEARNING SERIES

The National Center for Construction Education and Research (NCCER) is a not-for-profit 501(c)(3) education foundation established in 1995 by the world's largest and most progressive construction companies and national construction associations. It was founded to address the severe workforce shortage facing the industry and to develop a standardized training process and curricula. Today, NCCER is supported by hundreds of leading construction and maintenance companies, manufacturers, and national associations. The Contren® Learning Series was developed by NCCER in partnership with Prentice Hall, the world's largest educational publisher.

Some features of NCCER's Contren® Learning Series are as follows:

- An industry-proven record of success
- Curricula developed by the industry for the industry
- National standardization providing portability of learned job skills and educational credits
- Compliance with Apprenticeship, Training, Employer, and Labor Services (ATELS) requirements for related classroom training (*CFR 29:29*)
- Well-illustrated, up-to-date, and practical information

NCCER also maintains a National Registry that provides transcripts, certificates, and wallet cards to individuals who have successfully completed modules of NCCER's Contren® Learning Series. *Training programs must be delivered by an NCCER Accredited Training Sponsor in order to receive these credentials.*

Special Features of This Book

In an effort to provide a comprehensive user-friendly training resource, we have incorporated many different features for your use. Whether you are a visual or hands-on learner, this book will provide you with the proper tools to get started in the electrical industry.

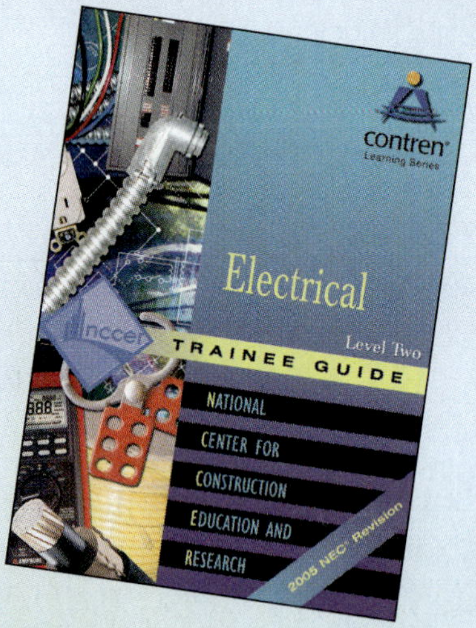

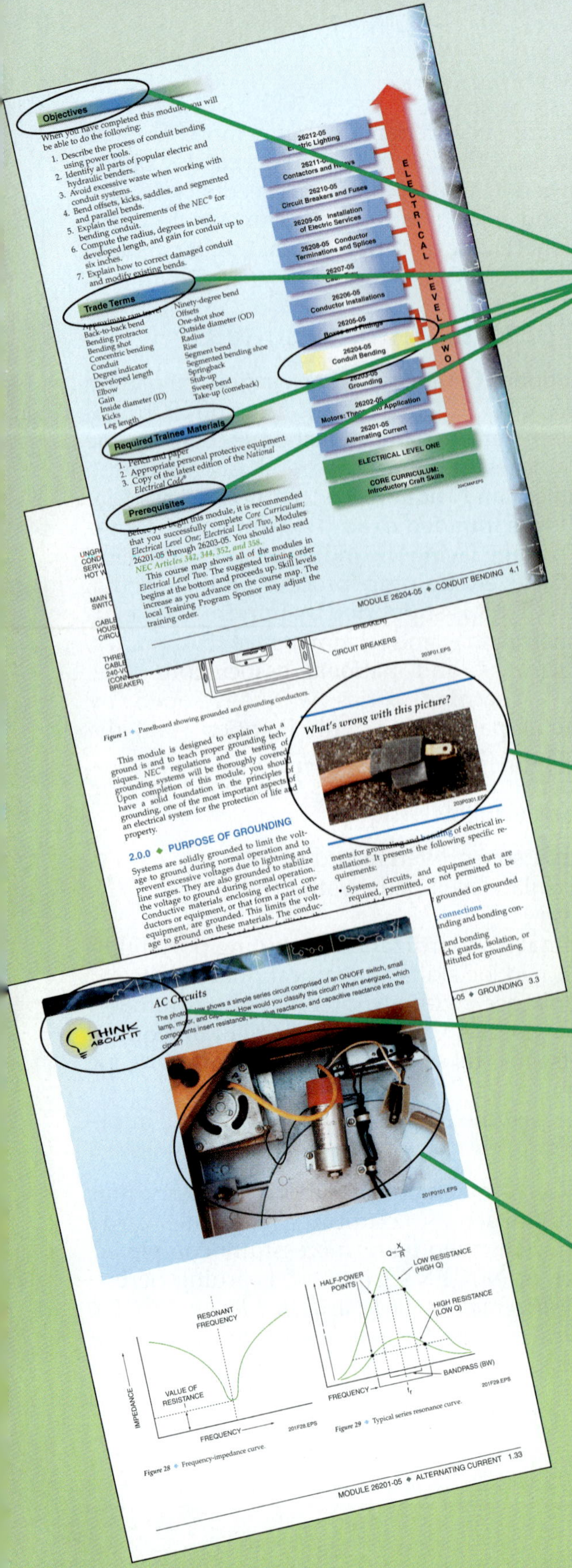

Introduction Page

This page is found at the beginning of each module and lists the Objectives, Trade Terms, Required Trainee Materials, Prerequisites, and Course Map for that module. The Objectives list the skills and knowledge you will need in order to complete the module successfully. The list of Trade Terms identifies important terms you will need to know by the end of the module. Required Trainee Materials list the materials and supplies needed for the module. The Prerequisites for the module are listed and illustrated in the Course Map. The Course Map also gives a visual overview of the entire course and a suggested learning sequence for you to follow.

What's wrong with this picture?

What's wrong with this picture? features include photos of actual code violations for identification and encourage you to approach each installation with a critical eye.

Think About It

Think About It features use "What if?" questions to help you apply theory to real-world experiences and put your ideas into action.

Color Illustrations and Photographs

Full-color illustrations and photographs are used throughout each module to provide vivid detail. These figures highlight important concepts from the text and provide clarity for complex instructions. Each figure is denoted in the text in *italic type* for easy reference.

Inside Track

Inside Track features provide a head start for those entering the electrical field by presenting technical tips and professional practices from master electricians in a variety of disciplines. Inside Tracks often include real-life scenarios similar to those you might encounter on the job site.

Trade Terms

Each module presents a list of Trade Terms that are discussed within the text, and defined in the Glossary at the end of the module. These terms are denoted in the text with **blue bold type** upon their first occurrence. To make searches for key information easier, a comprehensive Glossary of Trade Terms from all modules is found at the back of this book.

Case History

Case History features emphasize the importance of safety by citing examples of the costly (and often devastating) consequences of ignoring *National Electrical Code®* or OSHA regulations.

Notes, Cautions, and Warnings

Safety features are set off from the main text in highlighted boxes and organized into three categories based on the potential danger of the issue being addressed. Notes simply provide additional information on the topic area. Cautions alert you of a danger that does not present potential injury but may cause damage to equipment. Warnings stress a potentially dangerous situation that may cause injury to you or a co-worker.

Step-by-Step Instructions

Step-by-step instructions are used throughout to guide you through technical procedures and tasks from start to finish. These steps show you not only how to perform a task but how to do it safely and efficiently.

Review Questions

Review Questions are provided to reinforce the knowledge you have gained. This makes them a useful tool for measuring what you have learned.

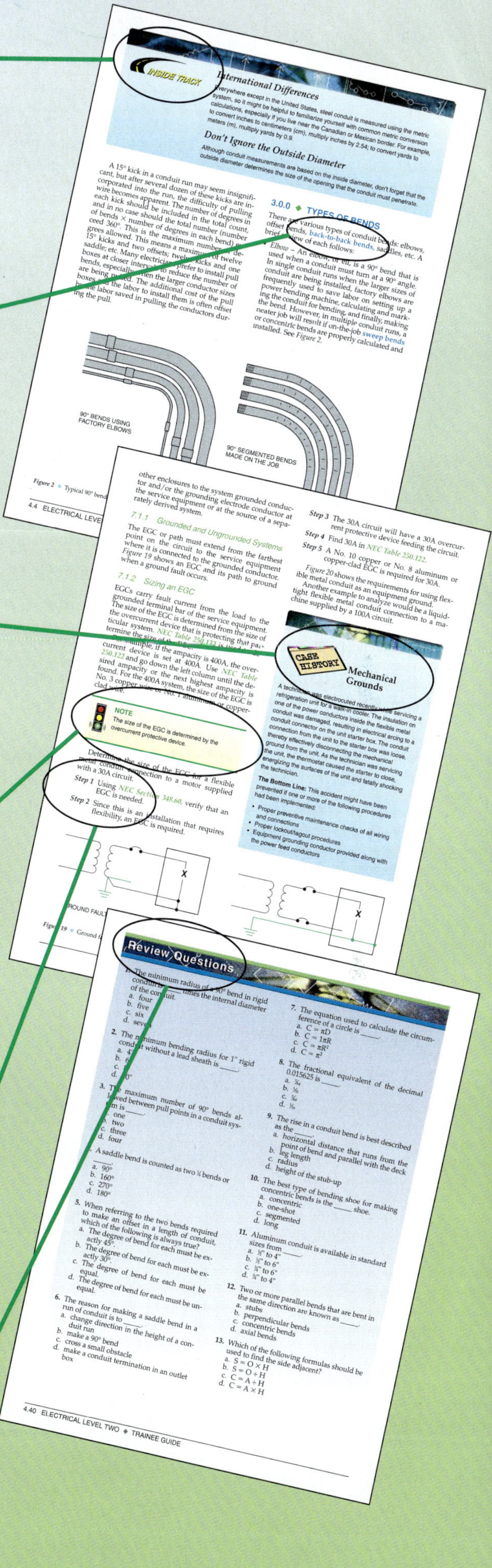

Contren® Curricula

NCCER's training programs comprise more than 40 construction, maintenance, and pipeline areas and include skills assessments, safety training, and management education.

Boilermaking
Carpentry
Carpentry, Residential
Cabinetmaking
Concrete Finishing
Construction Craft Laborer
Construction Technology
Core Curriculum: Introductory Craft Skills
Currículum Básico
Electrical
Electrical, Residential
Electrical Topics, Advanced
Electronic Systems Technician
Exploring Careers in Construction
Fundamentals of Mechanical and Electrical Mathematics
Heating, Ventilating, and Air Conditioning
Heavy Equipment Operations
Highway/Heavy Construction
Instrumentation
Insulating
Ironworking
Maintenance, Industrial
Masonry
Millwright
Mobile Crane Operations
Painting
Painting, Industrial
Pipefitting
Pipelayer
Plumbing
Scaffolding
Sheet Metal
Site Layout
Sprinkler Fitting
Welding

Pipeline

Control Center Operations, Liquid
Corrosion Control
Electrical and Instrumentation
Field Operations, Liquid
Field Operations, Gas
Maintenance
Mechanical

Safety

Field Safety
Orientación de Seguridad
Safety Orientation
Safety Technology

Management

Introductory Skills for the Crew Leader
Project Management
Project Supervision

Acknowledgments

This curriculum was revised as a result of the farsightedness and leadership of the following sponsors:

ABC of Iowa
ABC Ohio Valley
ABC Southwest Ohio
ABC Southwest Pelican Chapter
All Star Electric
Beacon Electric Company
Cuyahoga Valley Career Center
Duck Creek Engineering
Lamphear Electric
Leading Edge Electrical Services
Madison Comprehensive High School
Pumba Electric, LLC
Rust Constructors, Inc.
TIC, The Industrial Company
Tri-City Electrical Contractors, Inc.
Trident Technical College
Washington CITC
Wisconsin ABC

This curriculum would not exist were it not for the dedication and unselfish energy of those volunteers who served on the Authoring Team. A sincere thanks is extended to the following:

John S. Autrey
Rick Centers
Clarence "Ed" Cockrell
Tim Dean
Gary Edgington
Tim Ely
E.L. Jarrell
Dan Lamphear
Leonard R. "Skip" Layne
L.J. LeBlanc
Neil Matthes
Jim Mitchem
Christine Porter
Mike Powers
Wayne Stratton
Irene Ward

A final note: This book is the result of a collaborative effort involving the production, editorial, and development staff at Prentice Hall and the National Center for Construction Education and Research. Thanks to all of the dedicated people involved in the many stages of this project.

NCCER PARTNERING ASSOCIATIONS

American Fire Sprinkler Association
American Petroleum Institute
American Society for Training & Development
American Welding Society
Associated Builders & Contractors, Inc.
Associated General Contractors of America
Association for Career and Technical Education
Carolinas AGC, Inc.
Citizens Development Corps
Construction Industry Institute
Construction Users Roundtable
Design-Build Institute of America
Electronic Systems Industry Consortium
Merit Contractors Association of Canada
Metal Building Manufacturers Association
National Association of Minority Contractors
National Association of State Supervisors for Trade and Industrial Education
National Association of Women in Construction
National Insulation Association
National Ready Mixed Concrete Association
National Systems Contractors Association
National Technical Honor Society
National Utility Contractors Association
North American Crane Bureau
North American Technician Excellence
Painting & Decorating Contractors of America
Portland Cement Association
SkillsUSA
Steel Erectors Association of America
Texas Gulf Coast Chapter ABC
U.S. Army Corps of Engineers
University of Florida
Women Construction Owners & Executives, USA
Youth Training and Development Consortium

Contents

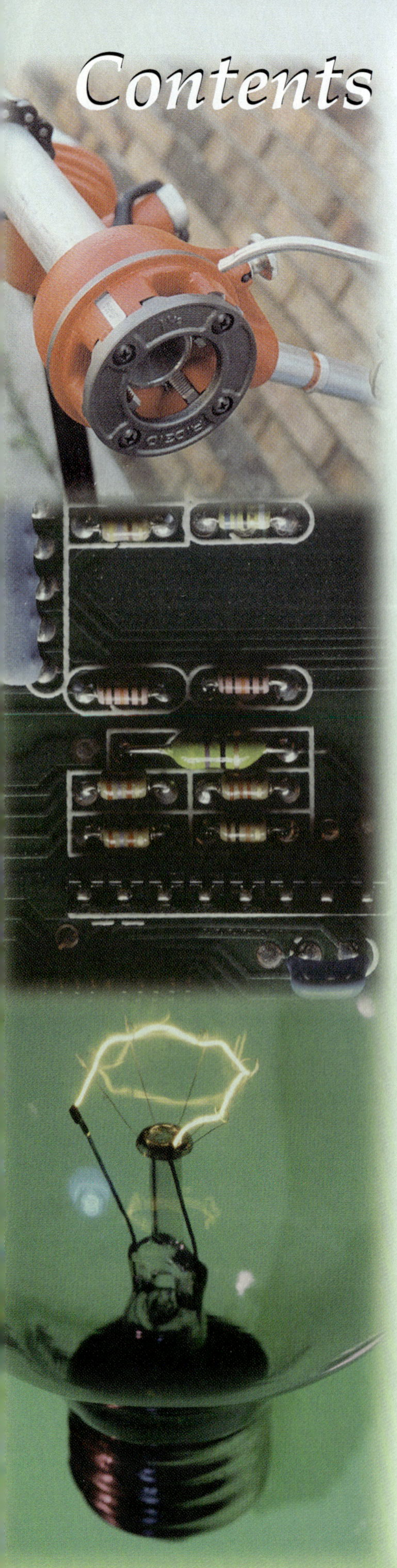

LEGEND
GFCI PROTECTED OUTLET

Alternating Current
26201-05

Inn at the Ballpark
Houston, Texas
Renovation $10–99 Million Award Winner
HOAR Construction, LLC

26201-05
Alternating Current

Topics to be presented in this module include:

Overview

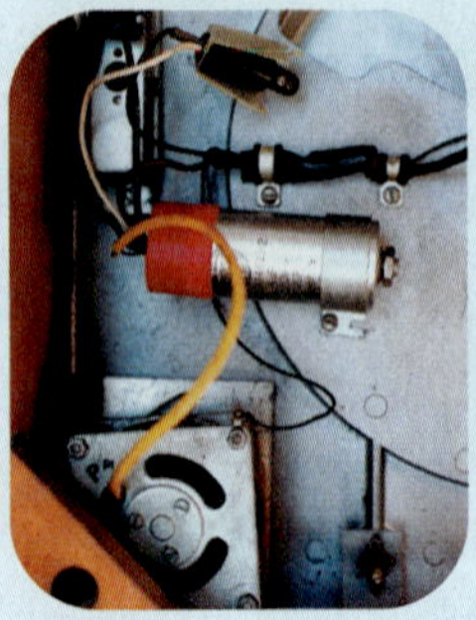

The vast majority of your work as an electrician will be with alternating current (AC) circuits, because AC provides the power to operate the appliances and machines in our homes and businesses. Although direct current (DC) was used in the original electrical systems developed by Thomas Edison, the dynamic nature of AC makes it much easier to generate and distribute than DC. On the other hand, circuit analysis is more difficult because, unlike DC, AC does not act the same way in components such as coils and capacitors as it does in resistive components. When circuits contain these AC components, Ohm's law cannot be directly applied. Special formulas must be used to determine the circuit resistance.

Because alternating current is in a constant state of change with regard to polarities and values, its dynamics make it very useful. Induction-dependent components such as transformers, motors, and coils rely on the changing polarity and voltage values of alternating current in order to function.

Note: *National Electrical Code*® and *NEC*® are registered trademarks of the National Fire Protection Association, Inc., Quincy, MA 02269. All *National Electrical Code*® and *NEC*® references in this module refer to the 2005 edition of the *National Electrical Code*®.

Objectives

When you have completed this module, you will be able to do the following:

1. Calculate the peak and effective voltage or current values for an AC waveform.
2. Calculate the phase relationship between two AC waveforms.
3. Describe the voltage and current phase relationship in a resistive AC circuit.
4. Describe the voltage and current transients that occur in an inductive circuit.
5. Define inductive reactance and state how it is affected by frequency.
6. Describe the voltage and current transients that occur in a capacitive circuit.
7. Define capacitive reactance and state how it is affected by frequency.
8. Explain the relationship between voltage and current in the following types of AC circuits:
   - RL circuit
   - RC circuit
   - LC circuit
   - RLC circuit
9. Describe the effect that resonant frequency has on impedance and current flow in a series or parallel resonant circuit.
10. Define bandwidth and describe how it is affected by resistance in a series or parallel resonant circuit.
11. Explain the following terms as they relate to AC circuits:
    - True power
    - Apparent power
    - Reactive power
    - Power factor
12. Explain basic transformer action.

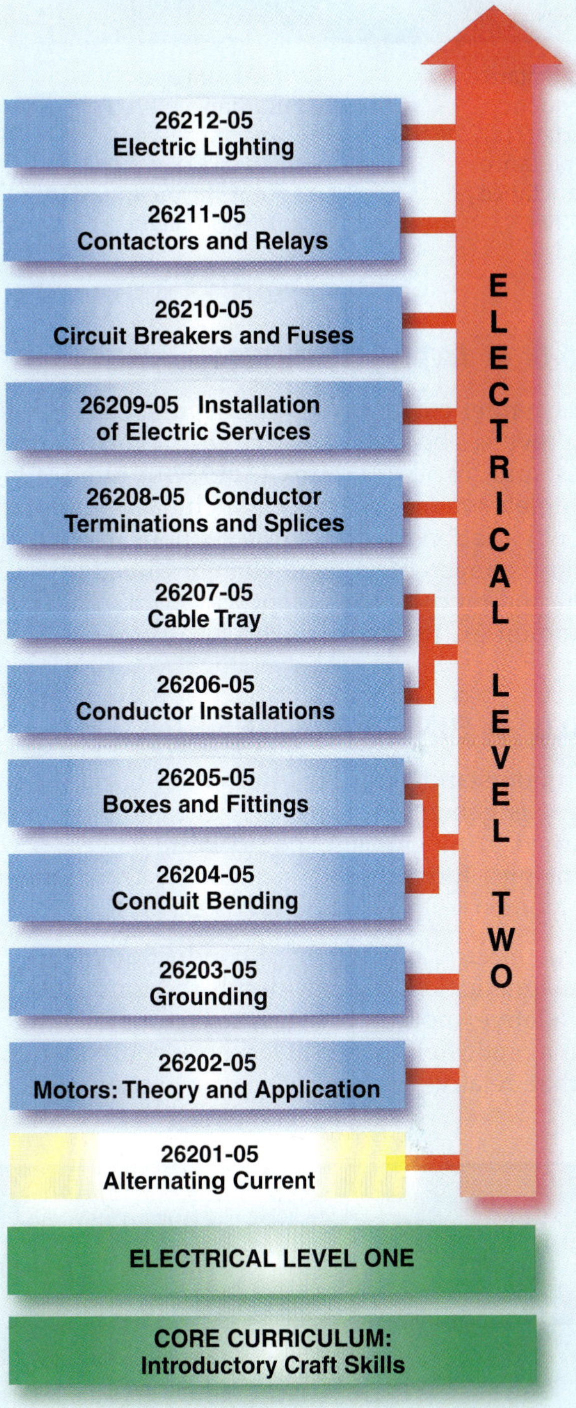

201CMAP.EPS

Prerequisites

Before you begin this module, it is recommended that you successfully complete *Core Curriculum* and *Electrical Level One.*

This course map shows all of the modules in *Electrical Level Two.* The suggested training order begins at the bottom and proceeds up. Skill levels increase as you advance on the course map. The local Training Program Sponsor may adjust the training order.

Required Trainee Materials

1. Pencil and paper
2. Appropriate personal protective equipment
3. Copy of the latest edition of the *National Electrical Code®*

Trade Terms

Capacitance	Peak voltage
Frequency	Radian
Hertz (Hz)	Reactance
Impedance	Resonance
Inductance	Root-mean-square (rms)
Micro	Self-inductance

1.0.0 ◆ INTRODUCTION

Alternating current (AC) and its associated voltage reverses between positive and negative polarities and varies in amplitude with time. One complete waveform or cycle includes a complete set of variations, with two alternations in polarity. Many sources of voltage change direction with time and produce a resultant waveform. The most common AC waveform is the sine wave.

2.0.0 ◆ SINE WAVE GENERATION

To understand how the alternating current sine wave is generated, some of the basic principles learned in magnetism should be reviewed. Two principles form the basis of all electromagnetic phenomena:

- An electric current in a conductor creates a magnetic field that surrounds the conductor.
- Relative motion between a conductor and a magnetic field, when at least one component of that relative motion is in a direction that is perpendicular to the direction of the field, creates a voltage in the conductor.

Figure 1 shows how these principles are applied to generate an AC waveform in a simple one-loop rotary generator. The conductor loop rotates through the magnetic field to generate the induced AC voltage across its open terminals. The magnetic flux shown here is vertical.

There are several factors affecting the magnitude of voltage developed by a conductor through a magnetic field. They are the strength of the magnetic field, the length of the conductor, and the rate at which the conductor cuts directly across or perpendicular to the magnetic field.

Assuming that the strength of the magnetic field and the length of the conductor making the loop are both constant, the voltage produced will vary depending on the rate at which the loop cuts directly across the magnetic field.

The rate at which the conductor cuts the magnetic field depends on two things: the speed of the generator in revolutions per minute (rpm) and the angle at which the conductor is traveling through the field. If the generator is operated at a constant rpm, the voltage produced at any moment will depend on the angle at which the conductor is cutting the field at that instant.

In *Figure 2*, the magnetic field is shown as parallel lines called lines of flux. These lines always go from the north to south poles in a generator. The motion of the conductor is shown by the large arrow.

Assuming the speed of the conductor is constant, as the angle between the flux and the

INSIDE TRACK

Why Do Power Companies Generate and Distribute AC Power Instead of DC Power?

The transformer is the key. Power plants generate and distribute AC power because it permits the use of transformers, which makes power delivery more economical. Transformers used at generation plants step the AC voltage up, which decreases the current. Decreased current allows smaller-sized wires to be used for the power transmission lines. Smaller wire is less expensive and easier to support over the long distances that the power must travel from the generation plant to remotely located substations. At the substations, transformers are again used to step AC voltages back down to a level suitable for distribution to homes and businesses.

There is no such thing as a DC transformer. This means DC power would have to be transmitted at low voltages and high currents over very large-sized wires, making the process very uneconomical. When DC is required for special applications, the AC voltage may be converted to DC voltage by using rectifiers, which make the change electrically, or by using AC motor-DC generator sets, which make the change mechanically.

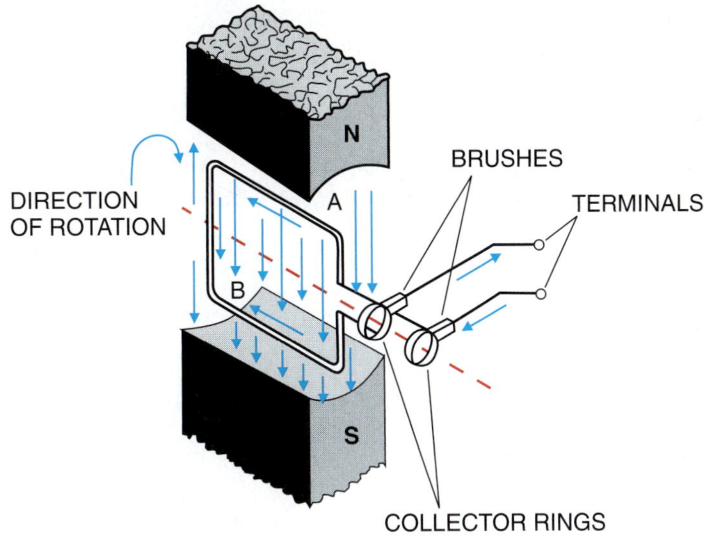

Figure 1 ◆ Conductor moving across a magnetic field.

201F01.EPS

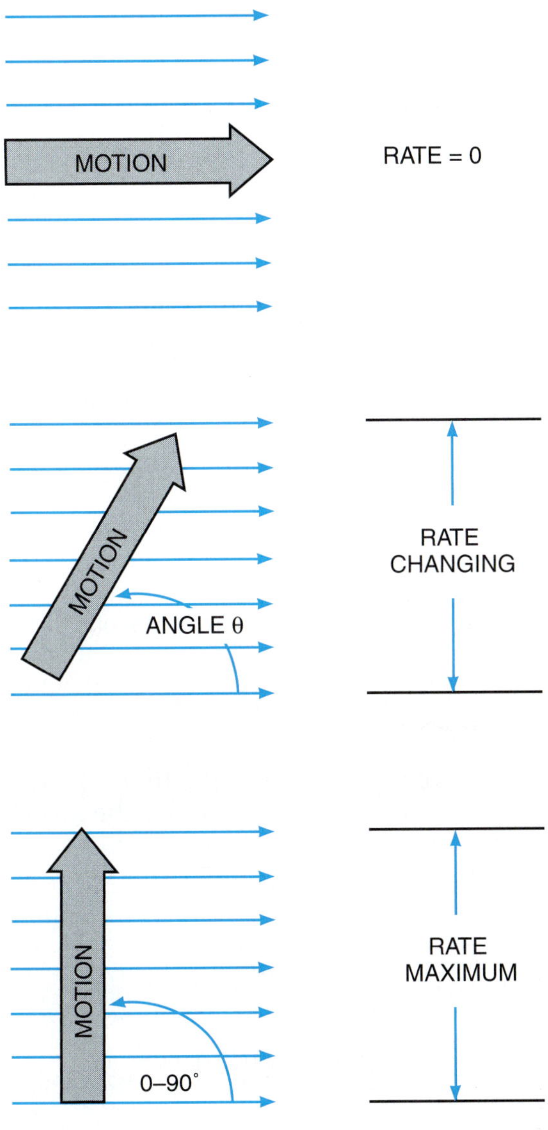

201F02.EPS

Figure 2 ◆ Angle versus rate of cutting lines of flux.

conductor motion increases, the number of flux lines cut in a given time (the rate) increases. When the conductor is moving parallel to the lines of flux (angle of 0°), it is not cutting any of them, and the voltage will be zero.

The angle between the lines of flux and the motion of the conductor is called θ (theta). The magnitude of the voltage produced will be proportional to the sine of the angle. Sine is a trigonometric function. Each angle has a sine value that never changes.

The sine of 0° is 0. It increases to a maximum of 1 at 90°. From 90° to 180°, the sine decreases back to 0. From 180° to 270°, the sine decreases to −1. Then from 270° to 360° (back to 0°), the sine increases to its original 0.

Because voltage is proportional to the sine of the angle, as the loop goes 360° around the circle the voltage will increase from 0 to its maximum at 90°, back to 0 at 180°, down to its maximum negative value at 270°, and back up to 0 at 360°, as shown in *Figure 3*.

Notice that at 180° the polarity reverses. This is because the conductor has turned completely around and is now cutting the lines of flux in the opposite direction. This can be shown using the left-hand rule for generators. The curve shown in *Figure 3* is called a sine wave because its shape is generated by the trigonometric function sine. The value of voltage at any point along the sine wave can be calculated if the angle and the maximum obtainable voltage (E_{max}) are known.

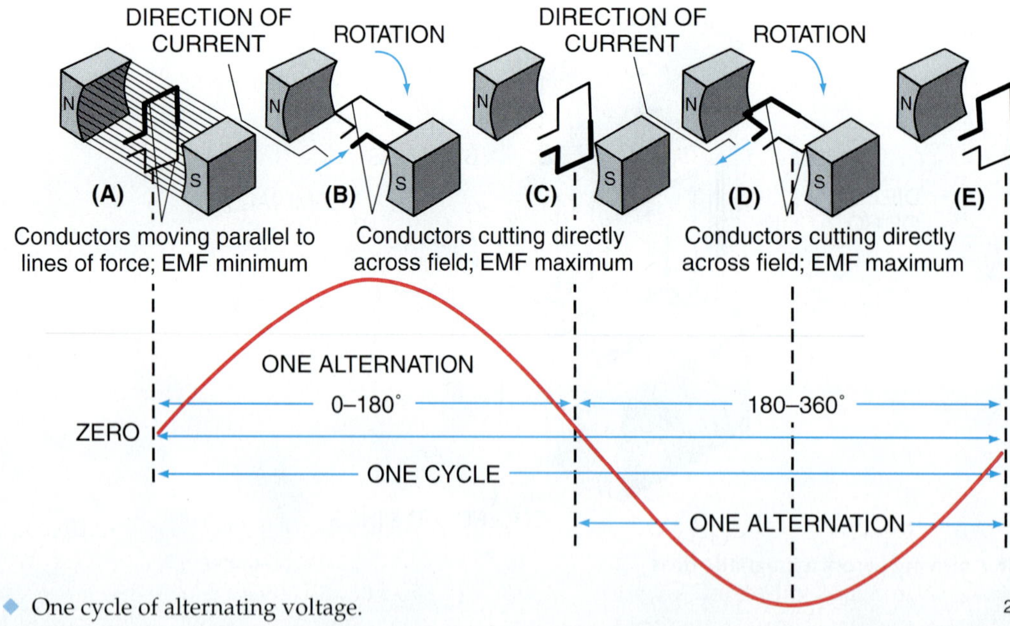

Figure 3 ◆ One cycle of alternating voltage.

201F03.EPS

The formula used is:

$$E = E_{max} \text{ sine } \theta$$

Where:

E = voltage induced

E_{max} = maximum induced voltage

θ = angle at which the voltage is induced

Using the above formula, the values of voltage anywhere along the sine wave in *Figure 3* can be calculated. Sine values can be found using either a scientific calculator or trigonometric tables. With an E_{max} of 10 volts (V), the following values are calculated as examples:

θ = 0°, sine = 0	θ = 45°, sine = 0.707
$E = E_{max}\text{sine } \theta$	$E = E_{max}\text{sine } \theta$
E = (10V)(0)	E = (10V)(0.707)
E = 0V	E = 7.07V
θ = 90°, sine = 1.0	θ = 135°, sine = 0.707
$E = E_{max}\text{sine } \theta$	$E = E_{max}\text{sine } \theta$
E = (10V)(1.0)	E = (10V)(0.707)
E = 10V	E = 7.07V
θ = 180°, sine = 0	θ = 225°, sine = −0.707
$E = E_{max}\text{sine } \theta$	$E = E_{max}\text{sine } \theta$
E = (10V)(0)	E = (10V)(−0.707)
E = 0V	E = −7.07V
θ = 270°, sine = −1.0	θ = 315°, sine = −0.707
$E = E_{max}\text{sine } \theta$	$E = E_{max}\text{sine } \theta$
E = (10V)(−1.0)	E = (10V)(−0.707)
E = −10V	E = −7.07V

3.0.0 ◆ SINE WAVE TERMINOLOGY

3.1.0 Frequency

The **frequency** of a waveform is the number of times per second an identical pattern repeats itself. Each time the waveform changes from zero to a peak value and back to zero is called an alternation. Two alternations form one cycle. The number of cycles per second is the frequency. The unit of frequency is **hertz (Hz).** One hertz equals one cycle per second (cps).

For example, let us determine the frequency of the waveform shown in *Figure 4*.

In one-half second, the basic sine wave is repeated five times. Therefore, the frequency (f) is:

$$f = \frac{5 \text{ cycles}}{0.5 \text{ second}} = 10 \text{ cycles per second (Hz)}$$

3.1.1 Period

The period of a waveform is the time (t) required to complete one cycle. The period is the inverse of frequency:

$$t = \frac{1}{f}$$

Where:

t = period (seconds)

f = frequency (Hz or cps)

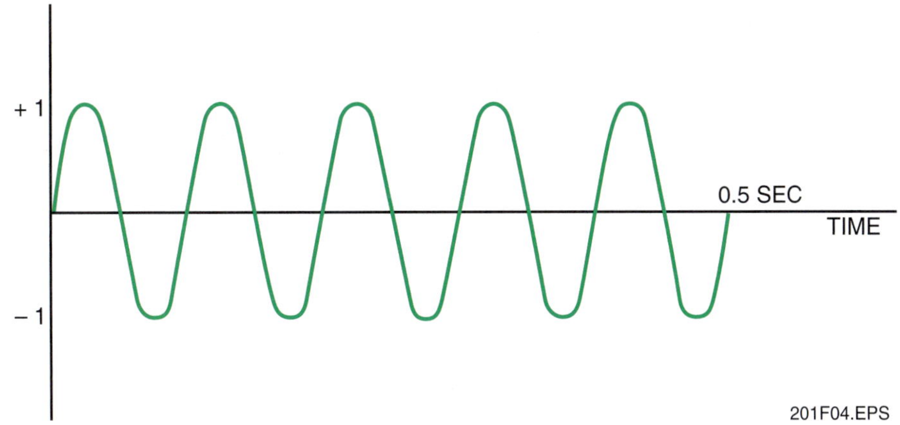

Figure 4 ◆ Frequency measurement.

For example, let us determine the period of the waveform in *Figure 4*. If there are five cycles in one-half second, then the frequency for one cycle is 10 cps ($0.5 \div 5 = 10$). Therefore, the period is:

$$t = \frac{1}{cps}$$

$$t = \frac{1}{10} = 0.1 \text{ second}$$

3.2.0 Wavelength

The wavelength or λ (lambda) is the distance traveled by a waveform during one period. Since electricity travels at the speed of light (186,000 miles/second or 300,000,000 meters/second), the wavelength of electrical waveforms equals the product of the period and the speed of light (c):

$$\lambda = tc$$

or:

$$\lambda = \frac{c}{f}$$

Where:

λ = wavelength (meters)
t = period (seconds)
c = speed of light (meters/second)
f = frequency (Hz or cps)

3.3.0 Peak Value

The peak value is the maximum value of voltage (V_M) or current (I_M). For example, specifying that a sine wave has a **peak voltage** of 170V applies to either the positive or the negative peak. To include both peak amplitudes, the peak-to-peak (p–p) value may be specified. In the above example, the peak-to-peak value is 340V, double the peak value of 170V, because the positive and negative peaks are symmetrical. However, the two opposite peak values cannot occur at the same time. Furthermore, in some waveforms the two peaks are not equal. The positive peak value and peak-to-peak value of a sine wave are shown in *Figure 5*.

3.4.0 Average Value

The average value is calculated from all the values in a sine wave for one alternation or half cycle. The half cycle is used for the average because over a full cycle the average value is zero, which is useless for comparison purposes. If the sine values for all angles up to 180° in one alternation are added and then divided by the number of values, this average equals 0.637.

Since the peak value of the sine is 1 and the average equals 0.637, the average value can be calculated as follows:

$$\text{Average value} = 0.637 \times \text{peak value}$$

Frequency

The frequency of the utility power generated in the United States is normally 60Hz. In some European countries and elsewhere, utility power is often generated at a frequency of 50Hz. Which of these frequencies (60Hz or 50Hz) has the shortest period?

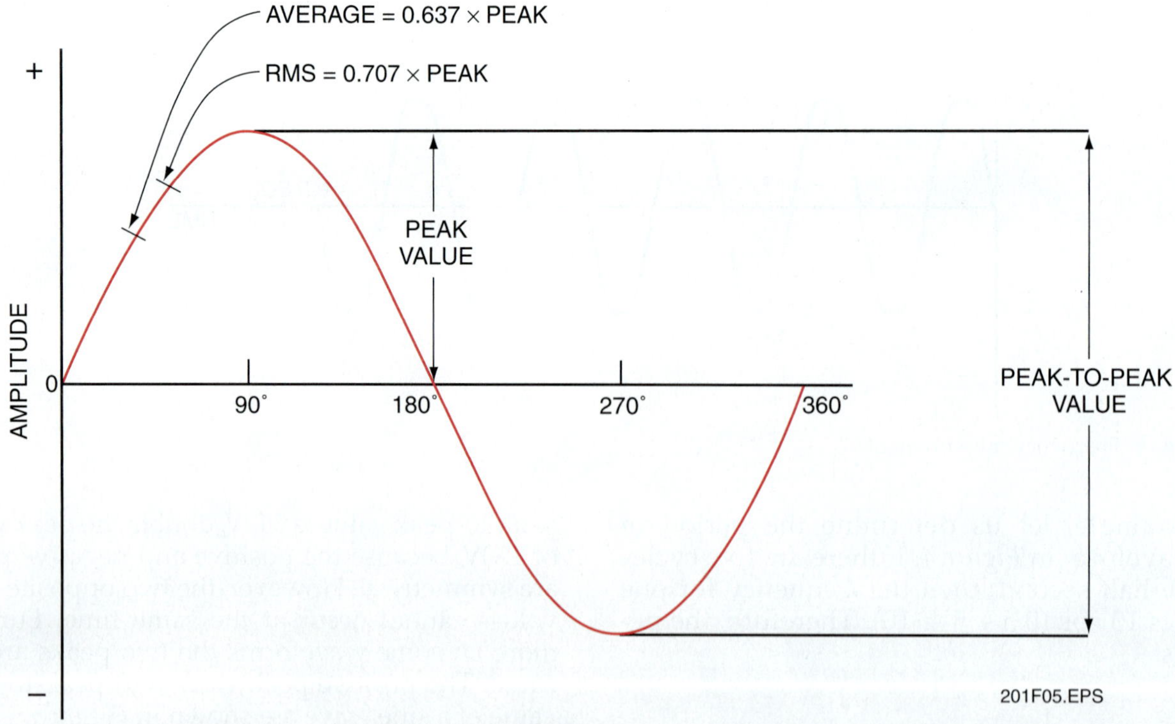

Figure 5 ◆ Amplitude values for a sine wave.

For example, with a peak of 170V, the average value is 0.637 × 170V, which equals approximately 108V. *Figure 5* shows where the average value would fall on a sine wave.

3.5.0 Root-Mean-Square or Effective Value

Meters used in AC circuits indicate a value called the effective value. The effective value is the value of the AC current or voltage wave that indicates the same energy transfer as an equivalent direct current (DC) or voltage.

The direct comparison between DC and AC is in the heating effect of the two currents. Heat produced by current is a function of current amplitude only and is independent of current direction. Thus, heat is produced by both alternations of the AC wave, although the current changes direction during each alternation.

In a DC circuit, the current maintains a steady amplitude. Therefore, the heat produced is steady and is equal to I^2R. In an AC circuit, the current is continuously changing; periodically high, periodically low, and periodically zero. To produce the same amount of heat from AC as from an equivalent amount of DC, the instantaneous value of the AC must at times exceed the DC value.

By averaging the heating effects of all the instantaneous values during one cycle of alternating current, it is possible to find the average heat produced by the AC current during the cycle. The amount of DC required to produce that heat will be equal to the effective value of the AC.

The most common method of specifying the amount of a sine wave of voltage or current is by stating its value at 45°, which is 70.7% of the peak. This is its **root-mean-square (rms)** value. Therefore:

$$\text{Value of rms} = 0.707 \times \text{peak value}$$

For example, with a peak of 170V, the rms value is 0.707 × 170, or approximately 120V. This is the voltage of the commercial AC power line, which is always given in rms value.

4.0.0 ◆ AC PHASE RELATIONSHIPS

In AC systems, phase is involved in two ways: the location of a point on a voltage or current wave with respect to the starting point of the wave or with respect to some corresponding point on the same wave. In the case of two waves of the same frequency, it is the time at which an event of one takes place with respect to a similar event of the other.

Often, the event is the starting of the waves at zero or the points at which the waves reach their maximum values. When two waves are compared in this manner, there is a phase lead or lag of one

INSIDE TRACK

Left-Hand Rule for Generators

Hand rules for generators and motors give direction to the basic principles of induction. For a generator, if you move a conductor through a magnetic field made up of flux lines, you will induce an EMF, which drives current through a conductor. The left-hand rule for generators will help you determine which direction the current will flow in the conductor. It states that if you hold the thumb, first, and middle fingers of the left hand at right angles to one another with the first finger pointing in the flux direction (from the north pole to the south pole), and the thumb pointing in the direction of motion of the conductor, the middle finger will point in the direction of the induced voltage (EMF). The polarity of the EMF determines the direction in which current will flow as a result of this induced EMF. The left-hand rule for generators is also called Fleming's first rule.

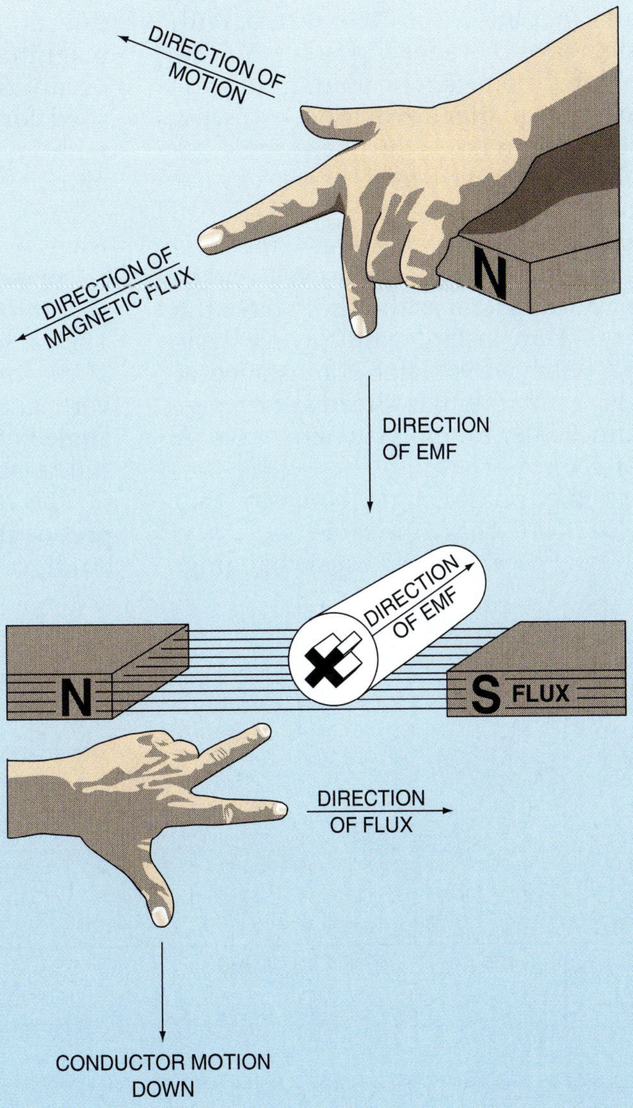

201UA0101.EPS

with respect to the other unless they are alternating in unison, in which case they are said to be in phase.

4.1.0 Phase Angle

Suppose that a generator started its cycle at 90° where maximum voltage output is produced instead of starting at the point of zero output. The two output voltage waves are shown in *Figure 6*. Each is the same waveform of alternating voltage, but wave B starts at the maximum value while wave A starts at zero. The complete cycle of wave B through 360° takes it back to the maximum value from which it started.

Wave A starts and finishes its cycle at zero. With respect to time, wave B is ahead of wave A in its values of generated voltage. The amount it leads in time equals one quarter revolution, which is 90°. This angular difference is the phase angle between waves B and A. Wave B leads wave A by the phase angle of 90°.

The 90° phase angle between waves B and A is maintained throughout the complete cycle and in all successive cycles as long as they both have the same frequency. At any instant in time, wave B has the value that A will have 90° later. For instance, at 180°, wave A is at zero, but B is already at its negative maximum value, the point where wave A will be later at 270°.

To compare the phase angle between two waves, both waves must have the same frequency. Otherwise, the relative phase keeps changing. Both waves must also have sine wave variations, because this is the only kind of waveform that is measured in angular units of time. The amplitudes can be different for the two waves. The phases of two voltages, two currents, or a current with a voltage can be compared.

4.2.0 Phase Angle Diagrams

To compare AC phases, it is much more convenient to use vector diagrams corresponding to the voltage and current waveforms, as shown in *Figure 6*. V_A and V_B represent the vector quantities corresponding to the generator voltage.

A vector is a quantity that has magnitude and direction. The length of the arrow indicates the magnitude of the alternating voltage in rms, peak, or any AC value as long as the same measure is used for all the vectors. The angle of the arrow with respect to the horizontal axis indicates the phase angle.

In *Figure 6*, the vector V_A represents the voltage wave A, with a phase angle of 0°. This angle can be considered as the plane of the loop in the rotary generator where it starts with zero output voltage. The vector V_B is vertical to show the phase angle of 90° for this voltage wave, corresponding to the vertical generator loop at the start of its cycle. The angle between the two vectors is the phase angle.

The symbol for a phase angle is θ (theta). In *Figure 7*, $\theta = 0°$. *Figure 7* shows the waveforms and phasor diagram of two waves that are in phase but have different amplitudes.

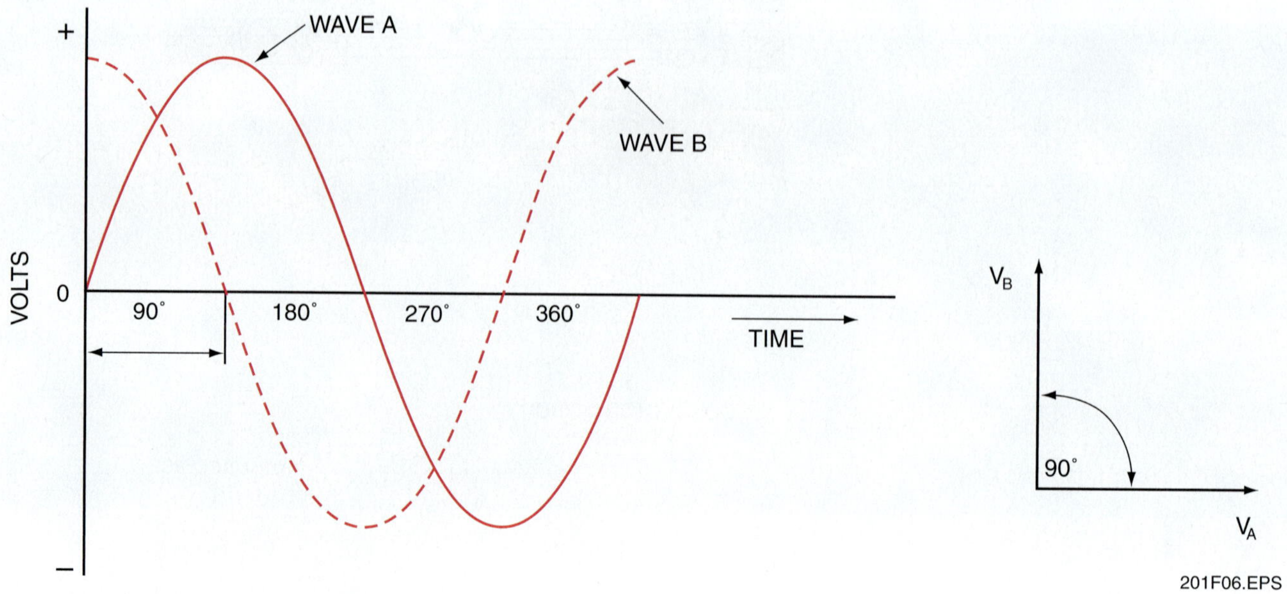

Figure 6 ◆ Voltage waveforms 90° out of phase.

INSIDE TRACK

RMS Amplitude

The root-mean-square (rms) value, also called the *effective value,* is the value assigned to an alternating voltage or current that results in the same power dissipation in a given resistance as DC voltage or current of the same numerical value. This is illustrated below. As shown, 120 (peak) VAC will not produce the same light (350 lumens versus 500 lumens) as 120VDC from a 60W lamp. In order to produce the same light (500 lumens), 120V rms must be applied to the lamp. This requires that the applied sinusoidal AC waveform have a peak voltage of about 170V (170V × 0.707V = 120V).

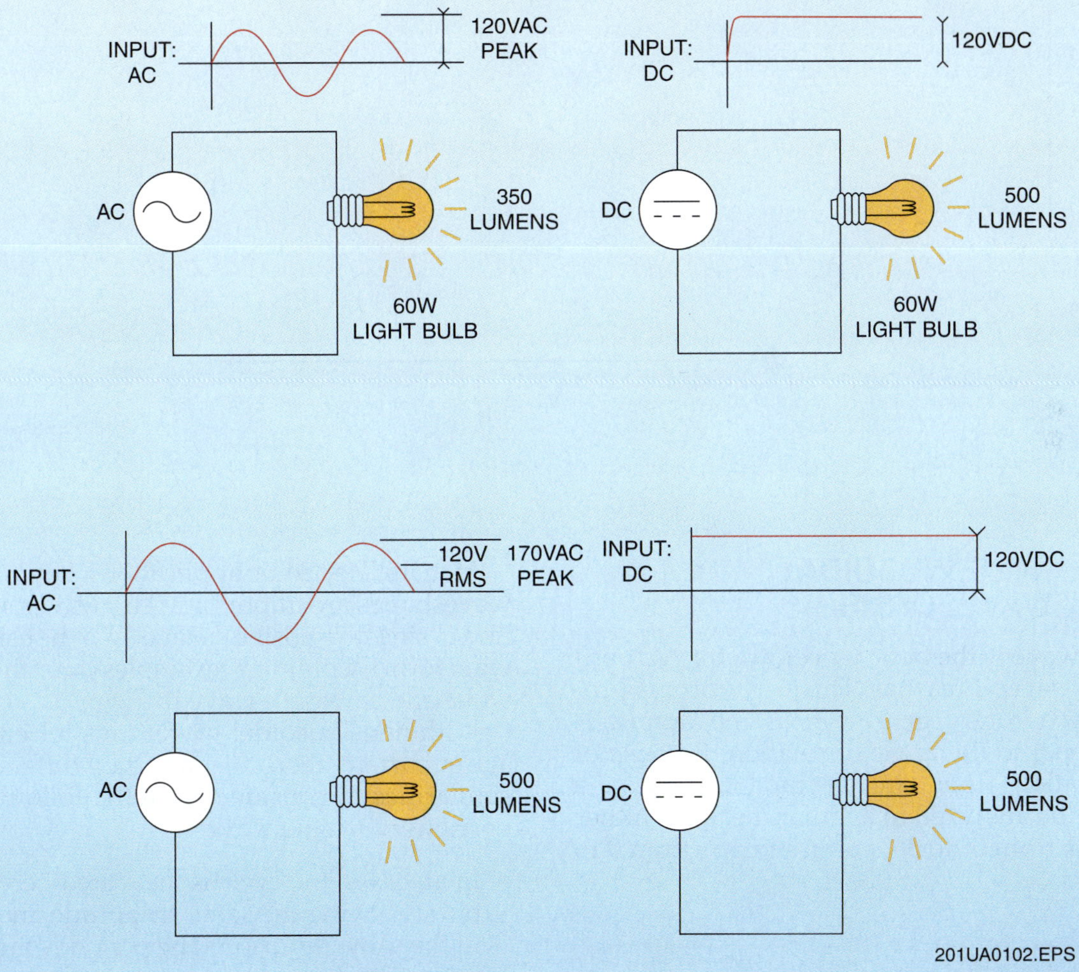

201UA0102.EPS

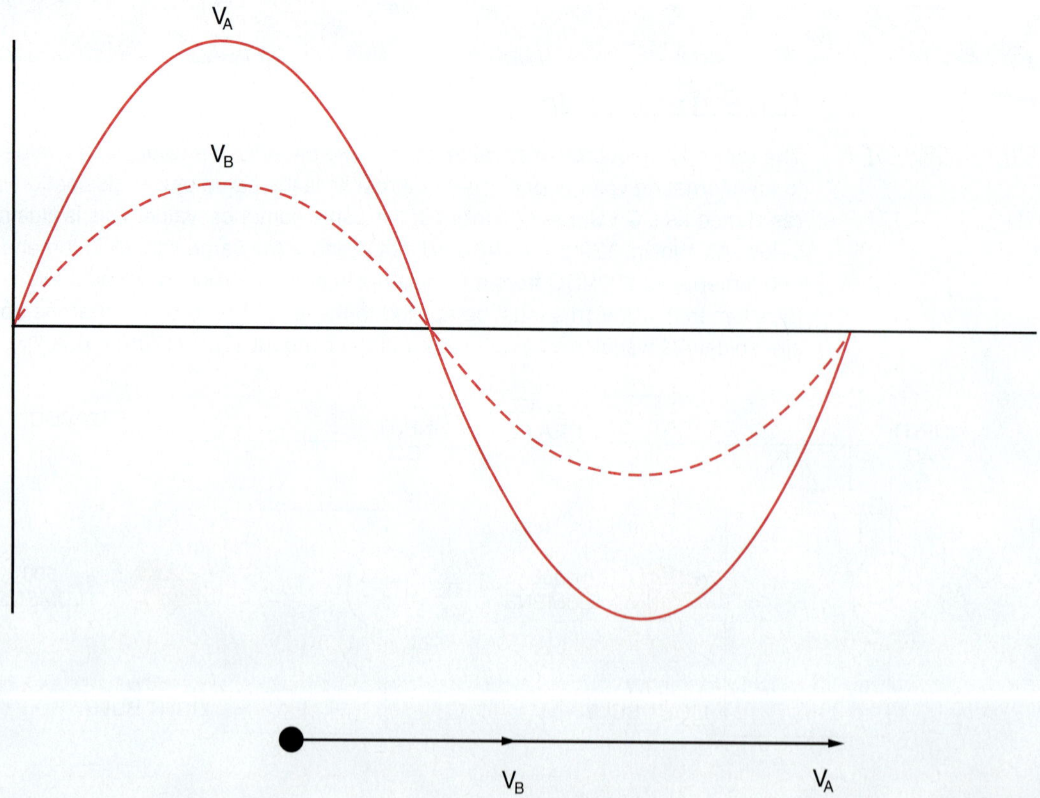

201F07.EPS

Figure 7 ◆ Waves in phase.

5.0.0 ◆ NONSINUSOIDAL WAVEFORMS

The sine wave is the basic waveform for AC variations for several reasons. This waveform is produced by a rotary generator, as the output is proportional to the angle of rotation. Because of its derivation from circular motion, any sine wave can be analyzed in angular measure, either in degrees from 0° to 360° or in **radians** from 0 to 2π radians.

THINK ABOUT IT

Phase Angles

Why is the phase angle 90° in *Figure 6* and 0° in *Figure 7?* Why is the vector diagram in *Figure 7* shown as a straight line?

In many electronic applications, however, other waveshapes are important. Any waveform that is not a sine (or cosine) wave is a nonsinusoidal waveform. Common examples are the square wave and sawtooth wave in *Figure 8.*

With nonsinusoidal waveforms for either voltage or current, there are important differences and similarities to consider. Note the following comparisons with sine waves:

- In all cases, the cycle is measured between two points having the same amplitude and varying in the same direction. The period is the time for one cycle.
- Peak amplitude is measured from the zero axis to the maximum positive or negative value. However, peak-to-peak amplitude is better for measuring nonsinusoidal waveshapes because they can have asymmetrical peaks, as with the rectangular wave in *Figure 8.*
- The rms value 0.707 of peak applies only to sine waves, as this factor is derived from the sine values in the angular measure used only for the sine waveform.
- Phase angles apply only to sine waves, as angular measure is used only for sine waves. Note that the phase angle is indicated only on the sine wave of *Figure 8.*

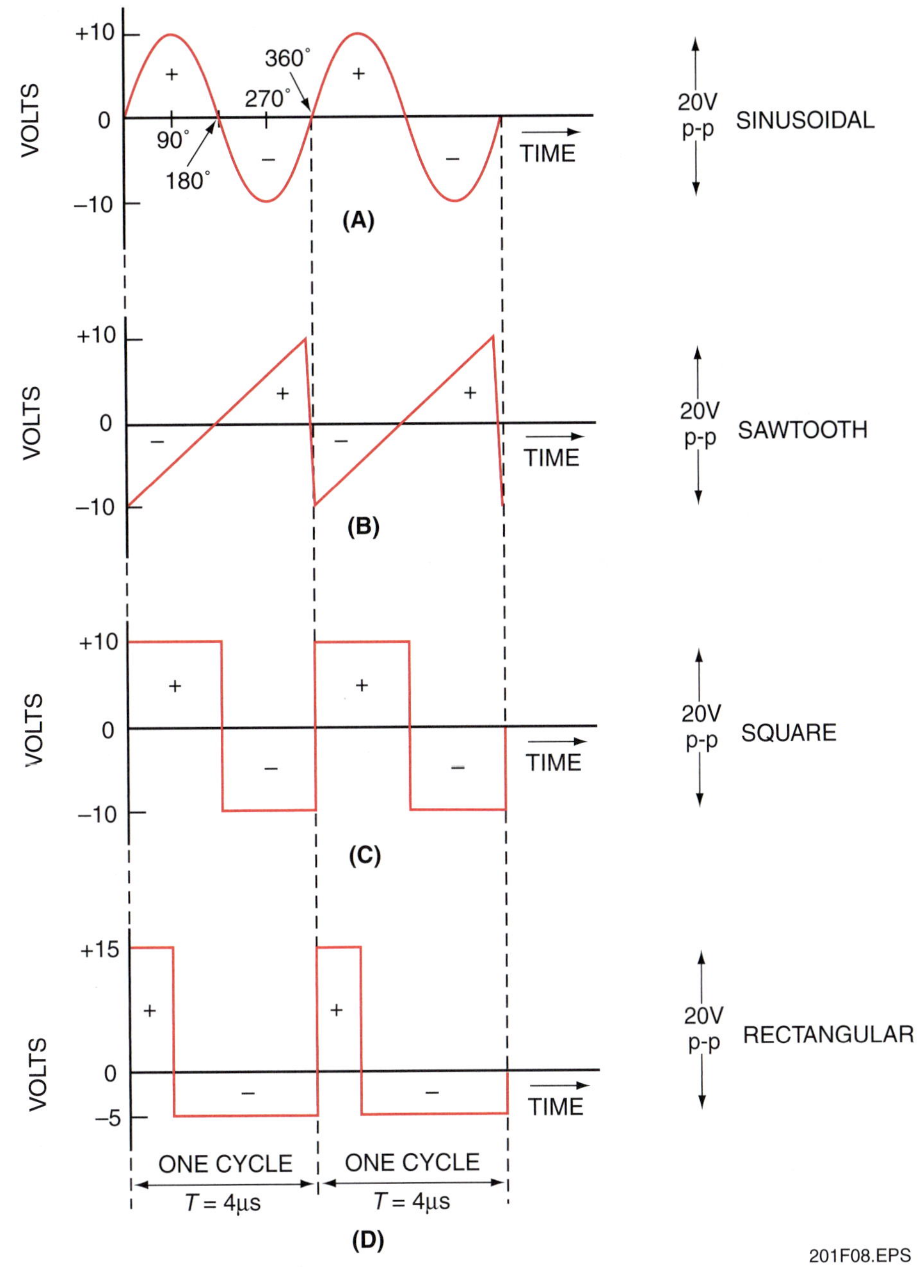

Figure 8 ◆ AC waveforms.

6.0.0 ◆ RESISTANCE IN AC CIRCUITS

An AC circuit has an AC voltage source. Note the circular symbol with the sine wave inside it shown in *Figure 9*. It is used for any source of sine wave alternating voltage. This voltage connected across an external load resistance produces alternating current of the same waveform, frequency, and phase as the applied voltage.

According to Ohm's law, current (I) equals voltage (E) divided by resistance (R). When E is an rms value, I is also an rms value. For any instantaneous value of E during the cycle, the value of I is for the corresponding instant of time.

In an AC circuit with only resistance, the current variations are in phase with the applied voltage, as shown in *Figure 9*. This in-phase relationship between E and I means that such an AC circuit can be analyzed by the same methods used for DC circuits since there is no phase angle to consider. Components that have only resistance

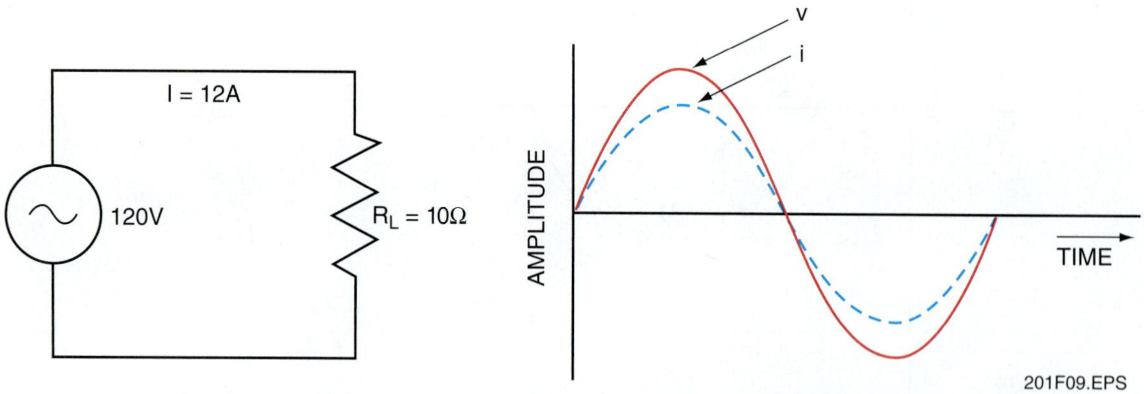

Figure 9 ◆ Resistive AC circuit.

What's wrong with this picture?

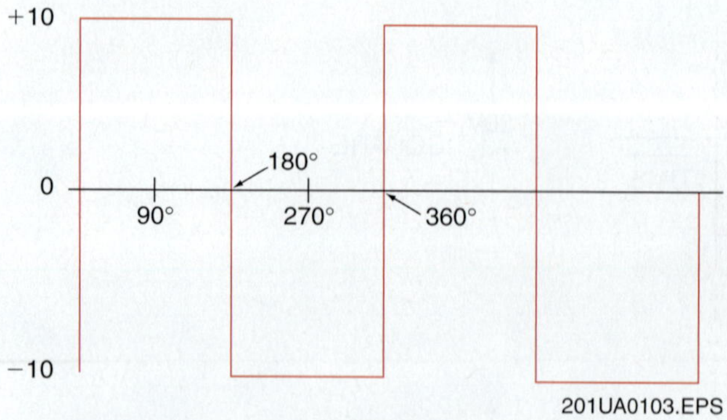

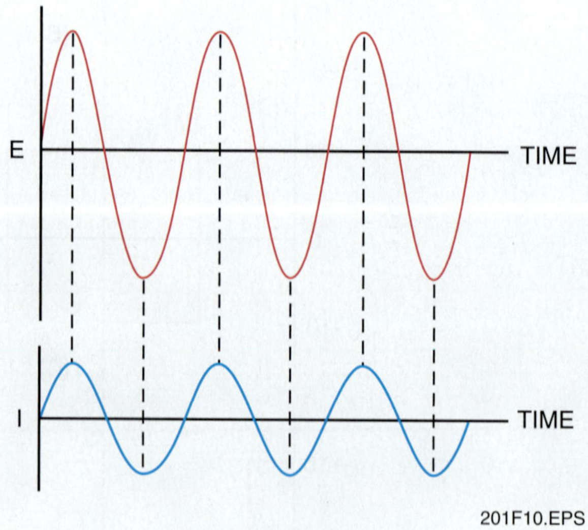

Figure 10 ◆ Voltage and current in a resistive AC circuit.

include resistors, the filaments for incandescent light bulbs, and vacuum tube heaters.

In purely resistive AC circuits, the voltage, current, and resistance are related by Ohm's law because the voltage and current are in phase.

$$I = \frac{E}{R}$$

Unless otherwise noted, the calculations in AC circuits are generally in rms values. For example, in *Figure 9*, the 120V applied across the 10Ω resistance R_L produces an rms current of 12A. This is determined as follows:

$$I = \frac{E}{R_L} = \frac{120V}{10\Omega} = 12A$$

Furthermore, the rms power (true power) dissipation is I^2R or:

$$P = (12A)^2 \times 10\Omega = 1{,}440W$$

Figure 10 shows the relationship between voltage and current in purely resistive AC circuits. The voltage and current are in phase, their cycles begin and end at the same time, and their peaks occur at the same time.

The value of the voltage shown in *Figure 10* depends on the applied voltage to the circuit. The value of the current depends on the applied voltage and the amount of resistance. If resistance is changed, it will affect only the magnitude of the current.

The total resistance in any AC circuit, whether it is a series, parallel, or series-parallel circuit, is calculated using the same rules that were learned and applied to DC circuits with resistance. Power computations are discussed later in this module.

7.0.0 ◆ INDUCTANCE IN AC CIRCUITS

Inductance is the characteristic of an electrical circuit that opposes the change of current flow. It is the result of the expanding and collapsing field caused by the changing current. This moving flux cuts across the conductor that is providing the current, producing induced voltage in the wire itself. Furthermore, any other conductor in the field, whether carrying current or not, is also cut by the varying flux and has induced voltage. This induced current opposes the current flow that generated it.

In DC circuits, a change must be initiated in the circuit to cause inductance. The current must change to provide motion of the flux. A steady DC of 10A cannot produce any induced voltage as long as the current value is constant. A current of 1A changing to 2A does induce voltage. Also, the faster the current changes, the higher the induced voltage becomes, because when the flux moves at a higher speed it can induce more voltage.

However, in an AC circuit the current is continuously changing and producing induced voltage. Lower frequencies of AC require more inductance to produce the same amount of induced voltage as a higher frequency current. The current can have any waveform as long as the amplitude is changing.

The ability of a conductor to induce voltage in itself when the current changes is its **self-inductance** or simply inductance. The symbol for inductance is L and its unit is the henry (H). One henry is the amount of inductance that allows one volt to be induced when the current changes at the rate of one ampere per second.

7.1.0 Factors Affecting Inductance

An inductor is a coil of wire that may be wound on a core of metal or paper, or it may be self-supporting. It may consist of turns of wire placed side by side to form a layer of wire over the core or coil form. The inductance of a coil or inductor depends on its physical construction. Some of the factors affecting inductance are:

- *Number of turns* – The greater the number of turns, the greater the inductance. In addition, the spacing of the turns on a coil also affects inductance. A coil that has widely-spaced turns has a lower inductance than one that has the same number of more closely-spaced turns. The reason for this higher inductance is that the closely-wound turns produce a more concentrated magnetic field, causing the coil to exhibit a greater inductance.
- *Coil diameter* – The inductance increases directly as the cross-sectional area of the coil increases.
- *Length of the coil* – When the length of the coil is decreased, the turn spacing is decreased, increasing the inductance of the coil.
- *Core material* – The core of the coil can be either a magnetic material (such as iron) or a non-magnetic material (such as paper or air). Coils wound on a magnetic core produce a stronger magnetic field than those with non-magnetic cores, giving them higher values of inductance. Air-core coils are used where small values of inductance are required.
- *Winding the coil in layers* – The more layers used to form a coil, the greater the effect the magnetic field has on the conductor. Layering a coil can increase the inductance.

Factors affecting the inductance of a coil can be seen in *Figure 11*.

7.2.0 Voltage and Current in an Inductive AC Circuit

The self-induced voltage across an inductance L is produced by a change in current with respect to time ($\Delta i/\Delta t$) and can be stated as:

$$V_L = L\frac{\Delta i}{\Delta t}$$

Where:

Δ = change
V_L = volts
L = henrys
$\Delta i/\Delta t$ = amperes per second

This gives the voltage in terms of how much magnetic flux is cut per second. When the magnetic flux associated with the current varies the same as I, this formula gives the same results for calculating induced voltage. Remember that the induced voltage across the coil is actually the result of inducing electrons to move in the conductor, so there is also an induced current.

For example, what is the self-induced voltage V_L across a 4h inductance produced by a current change of 12A per second?

$$V_L = L\frac{\Delta i}{\Delta t}$$

$$V_L = 4h \times \frac{12A}{1}$$

$$V_L = 4 \times 12$$

$$V_L = 48V$$

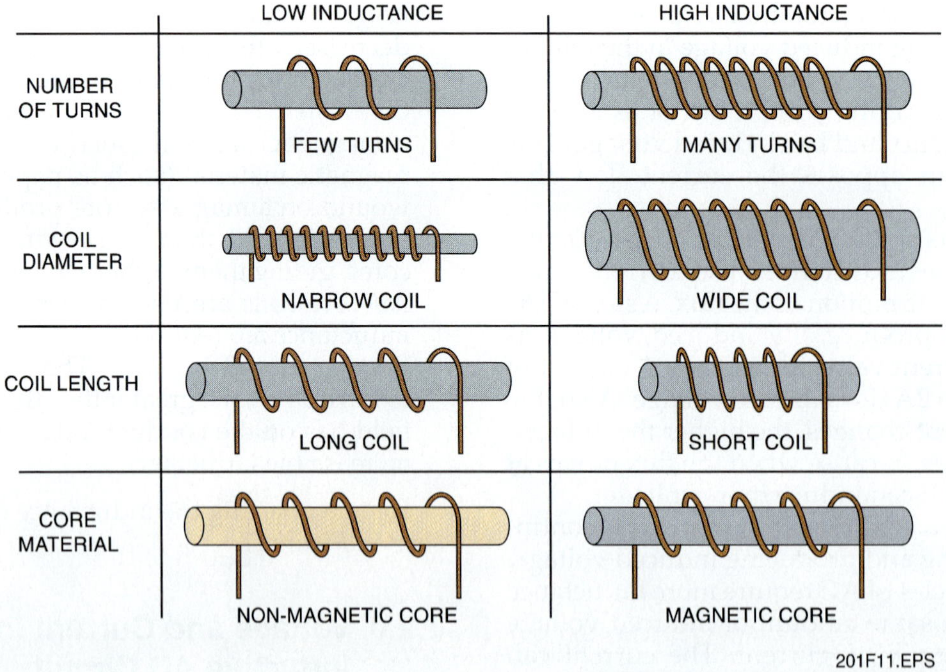

Figure 11 ◆ Factors affecting the inductance of a coil.

THINK ABOUT IT

Inductance in an AC Circuit

Can you name three commonly used electrical devices that insert inductance into an AC circuit?

The current through a 200 microhenry (µh) inductor changes from 0 to 200 milliamps (mA) in 2 microseconds (µsec). (The prefix **micro** means one-millionth.) What is the V_L?

$$V_L = L\frac{\Delta i}{\Delta t}$$

$$V_L = (200 \times 10^{-6})\frac{200 \times 10^{-3}}{2 \times 10^{-6}}$$

$$V_L = 20V$$

The induced voltage is an actual voltage that can be measured, although V_L is produced only while the current is changing. When $\Delta i/\Delta t$ is present for only a short time, V_L is in the form of a voltage pulse. With a sine wave current that is always changing, V_L is a sinusoidal voltage that is 90° out of phase with I_L.

The current that flows in an inductor is induced by the changing magnetic field that surrounds the inductor. This changing magnetic field is produced by an AC voltage source that is applied to the inductor. The magnitude and polarity of the induced current depend on the field strength, direction, and rate at which the field cuts the inductor windings. The overall effect is that the current is out of phase and lags the applied voltage by 90°.

At 270° in *Figure 12*, the applied electromotive force (EMF) is zero, but it is increasing in the positive direction at its greatest rate of change. Likewise, electron flow due to the applied EMF is also increasing at its greatest rate. As the electron flow increases, it produces a magnetic field that is building with it. The lines of flux cut the conductor as they move outward from it with the expanding field.

As the lines of flux cut the conductor, they induce a current into it. The induced current is at its maximum value because the lines of flux are expanding outward through the conductor at their greatest rate. The direction of the induced current is in opposition to the force that generated it. Therefore, at 270° the applied voltage is zero and

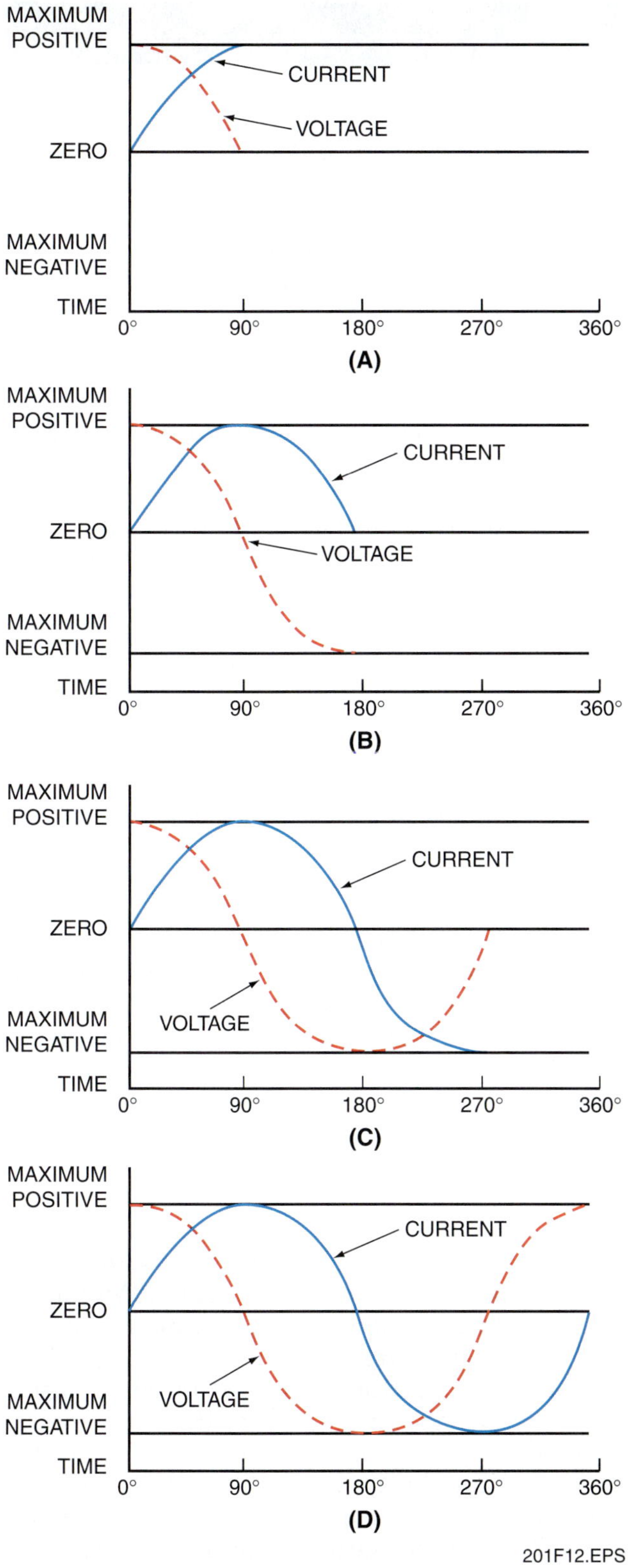

Figure 12 ◆ Inductor voltage and current relationship.

is increasing to a positive value, while the current is at its maximum negative value.

At 0° in *Figure 12*, the applied voltage is at its maximum positive value, but its rate of change is zero. Therefore, the field it produces is no longer expanding and is not cutting the conductor. Because there is no relative motion between the field and conductor, no current is induced. Therefore, at 0° voltage is at its maximum positive value, while current is zero.

At 90° in *Figure 12*, voltage is once again zero, but this time it is decreasing toward negative at its greatest rate of change. Because the applied voltage is decreasing, the magnetic field is collapsing inward on the conductor. This has the effect of reversing the direction of motion between the field and conductor that existed at 0°.

Therefore, the current will flow in a direction opposite of what it was at 0°. Also, because the applied voltage is decreasing at its greatest rate, the field is collapsing at its greatest rate. This causes the flux to cut the conductor at the greatest rate, causing the induced current magnitude to be maximum. At 90°, the applied voltage is zero decreasing toward negative, while the current is maximum positive.

At 180° in *Figure 12*, the applied voltage is at its maximum negative value, but just as at 0°, its rate of change is zero. At 180°, therefore, current will be zero. This explanation shows that the voltage peaks positive first, then 90° later the current peaks positive. Current thus lags the applied voltage in an inductor by 90°. This can easily be remembered using the phrase "ELI the ICE man." ELI represents voltage (E), inductance (L), and current (I). In an inductor, the voltage leads the current just like the letter E leads or comes before the letter I. The word ICE will be explained in the section on **capacitance.**

7.3.0 Inductive Reactance

The opposing force that an inductor presents to the flow of alternating current cannot be called resistance since it is not the result of friction within a conductor. The name given to this force is inductive **reactance** because it is the reaction of the inductor to alternating current. Inductive reactance is measured in ohms and its symbol is X_L.

Remember that the induced voltage in a conductor is proportional to the rate at which magnetic lines of force cut the conductor. The greater the rate or higher the frequency, the greater the counter-electromotive force (CEMF). Also, the induced voltage increases with an increase in inductance; the more turns, the greater the CEMF. Reactance then increases with an increase of frequency and with an increase in inductance. The formula for inductive reactance is as follows:

$$X_L = 2\pi fL$$

INSIDE TRACK

ELI in ELI the ICE Man

Remembering the phrase "ELI" as in "ELI the ICE man" is an easy way to remember the phase relationships that always exist between voltage and current in an inductive circuit. An inductive circuit is a circuit where there is more inductive reactance than capacitive reactance. The L in ELI indicates inductance. The E (voltage) is stated before the I (current) in ELI, meaning that the voltage leads the current in an inductive circuit.

Where:

X_L = inductive reactance in ohms
2π = a constant in which the Greek letter pi (π) represents 3.14 and $2 \times$ pi = 6.28
f = frequency of the alternating current in hertz
L = inductance in henrys

For example, if f is equal to 60Hz and L is equal to 20h, find X_L:

$$X_L = 2\pi fL$$
$$X_L = 6.28 \times 60Hz \times 20h$$
$$X_L = 7{,}536\Omega$$

Once calculated, the value of X_L is used like resistance in a form of Ohm's law:

$$I = \frac{E}{X_L}$$

Where:

I = effective current (amps)
E = effective voltage (volts)
X_L = inductive reactance (ohms)

Unlike a resistor, there is no power dissipation in an ideal inductor. An inductor limits current, but it uses no net energy since the energy required to build up the field in the inductor is given back to the circuit when the field collapses.

8.0.0 ◆ CAPACITANCE

A capacitor is a device that stores an electric charge in a dielectric material. Capacitance is the ability to store a charge. In storing a charge, a capacitor opposes a change in voltage. *Figure 13* shows a simple capacitor in a circuit, schematic representations of two types of capacitors, and a photo of common capacitors.

Figure 14(A) shows a capacitor in a DC circuit. When voltage is applied, the capacitor begins to charge, as shown in *Figure 14(B)*. The charging continues until the potential difference across the capacitor is equal to the applied voltage. This charging current is transient or temporary since it flows only until the capacitor is charged to the applied voltage. Then there is no current in the circuit. *Figure 14(C)* shows this with the voltage across the capacitor equal to the battery voltage or 10V.

The capacitor can be discharged by connecting a conducting path across the dielectric. The stored charge across the dielectric provides the potential difference to produce a discharge current, as shown in *Figure 14(D)*. Once the capacitor is completely discharged, the voltage across it equals zero, and there is no discharge current.

In a capacitive circuit, the charge and discharge current must always be in opposite directions. Current flows in one direction to charge the capacitor and in the opposite direction when the capacitor is allowed to discharge.

Current will flow in a capacitive circuit with AC voltage applied because of the capacitor charge and discharge current. There is no current through the dielectric, which is an insulator. While the capacitor is being charged by increasing applied voltage, the charging current flows in one direction to the plates. While the capacitor is discharging as the applied voltage decreases, the discharge current flows in the reverse direction. With alternating voltage applied, the capacitor alternately charges and discharges.

First, the capacitor is charged in one polarity, and then it discharges; next, the capacitor is charged in the opposite polarity, and then it discharges again. The cycles of charge and discharge current provide alternating current in the circuit at the same frequency as the applied voltage. The amount of capacitance in the circuit will determine how much current is allowed to flow.

Capacitance is measured in farads (F), where one farad is the capacitance when one coulomb is stored in the dielectric with a potential difference of one volt. Smaller values are measured in microfarads (μF). A small capacitance will allow less charge and discharge current to flow than a larger

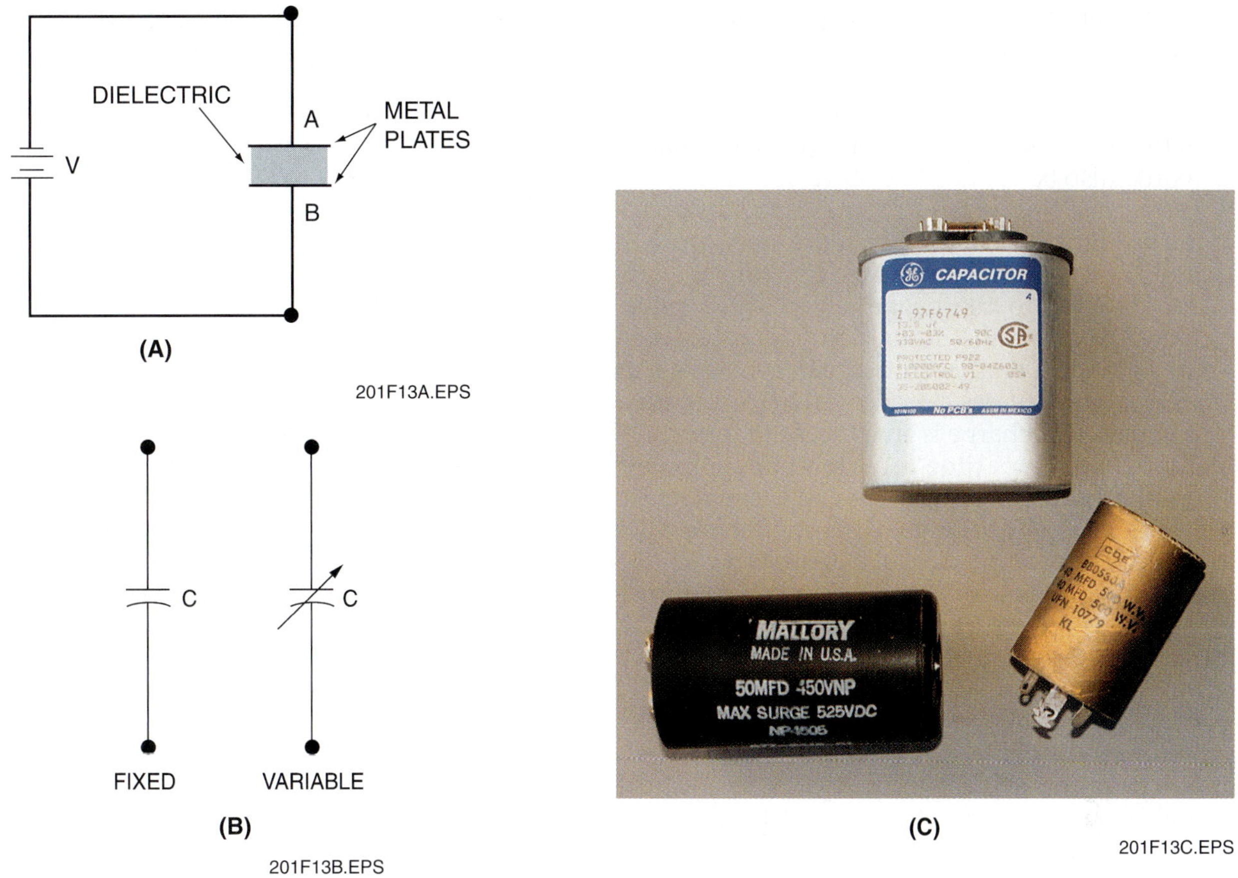

Figure 13 ◆ Capacitors.

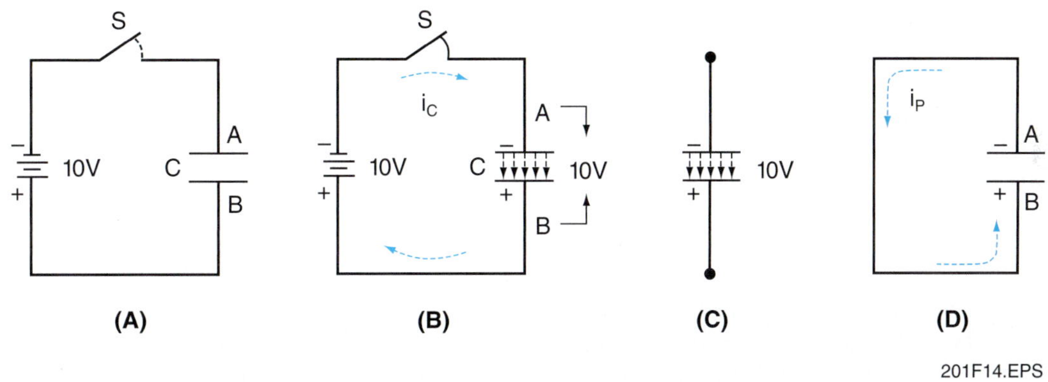

Figure 14 ◆ Charging and discharging a capacitor.

capacitance. The smaller capacitor has more opposition to alternating current, because less current flows with the same applied voltage.

In summary, capacitance exhibits the following characteristics:

- DC is blocked by a capacitor. Once charged, no current will flow in the circuit.
- AC flows in a capacitive circuit with AC voltage applied.
- A smaller capacitance allows less current.

8.1.0 Factors Affecting Capacitance

A capacitor consists of two conductors separated by an insulating material called a dielectric. There are many types and sizes of capacitors with different dielectric materials. The capacitance of a capacitor is determined by three factors:

- *Area of the plates* – The initial charge displacement on a set of capacitor plates is related to the number of free electrons in each plate. Larger plates will produce a greater capacitance than smaller ones. Therefore, the capacitance of a capacitor varies directly with the area of the plates. For example, if the area of the plates is doubled, the capacitance is doubled. If the size of the plates is reduced by 50%, the capacitance would also be reduced by 50%.

- *Distance between plates* – As two capacitor plates are brought closer together, more electrons will move away from the positively charged plate and move into the negatively charged plate. This is because the mutual attraction between the opposite charges on the plates increases as we move the plates closer together. This added movement of charge is an increase in the capacitance of the capacitor. In a capacitor composed of two plates of equal area, the capacitance varies inversely with the distance between the plates. For example, if the distance between the plates is decreased by one-half, the capacitance will be doubled. If the distance between the plates is doubled, the capacitance would be one-half as great.
- *Dielectric permittivity* – Another factor that determines the value of capacitance is the permittivity of the dielectric. The dielectric is the material between the capacitor plates in which the electric field appears. Relative permittivity expresses the ratio of the electric field strength in a dielectric to that in a vacuum. Permittivity has nothing to do with the dielectric strength of the medium or the breakdown voltage. An insulating material that will withstand a higher applied voltage than some other substance does not always have a higher dielectric permittivity. Many insulating materials have a greater dielectric permittivity than air. For a given applied voltage, a greater attraction exists between the opposite charges on the capacitor plates, and an electric field can be set up more easily than when the dielectric is air. The capacitance of the capacitor is increased when the permittivity of the dielectric is increased if all the other parameters remain unchanged.

8.2.0 Calculating Equivalent Capacitance

Connecting capacitors in parallel is equivalent to adding the plate areas. Therefore, the total capacitance is the sum of the individual capacitances, as illustrated in *Figure 15*.

A 10µF capacitor in parallel with a 5µF capacitor, for example, provides a 15µF capacitance for the parallel combination. The voltage is the same across the parallel capacitors. Note that adding parallel capacitance is opposite to the case of inductances in parallel and resistances in parallel.

Connecting capacitances in series is equivalent to increasing the thickness of the dielectric. Therefore, the combined capacitance is less than the smallest individual value. The combined equivalent capacitance is calculated by the reciprocal formula, as shown in *Figure 16*.

Capacitors connected in series are combined like resistors in parallel. Any of the shortcut calculations for the reciprocal formula apply. For example, the combined capacitance of two equal capacitances of 10µF in series is 5µF.

Capacitors are used in series to provide a higher voltage breakdown rating for the combina-

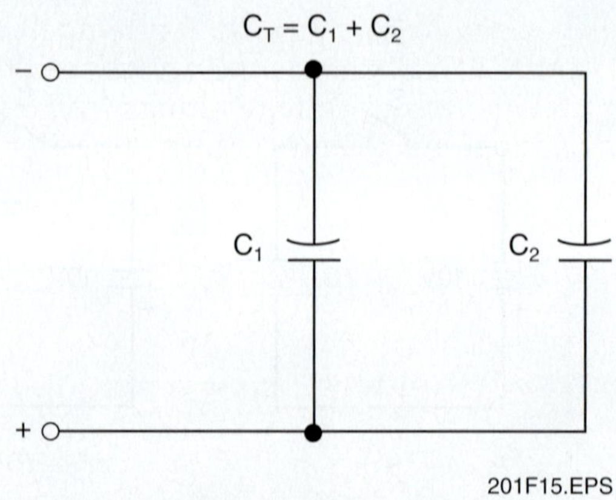

Figure 15 ◆ Capacitors in parallel.

Capacitance

INSIDE TRACK

The concept of capacitance, like many electrical quantities, is often hard to visualize or understand. A comparison with a balloon may help to make this concept clearer. Electrical capacitance has a charging effect similar to blowing up a balloon and holding it closed. The expansion capacity of the balloon can be changed by changing the thickness of the balloon walls. A balloon with thick walls will expand less (have less capacity) than one with thin walls. This is like a small 10µF capacitor that has less capacity and will charge less than a larger 100µF capacitor.

$$C_T = \frac{1}{\frac{1}{C_1} + \frac{1}{C_2}}$$

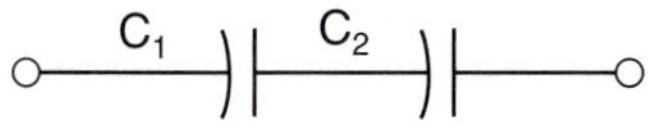

201F16.EPS

Figure 16 ◆ Capacitors in series.

tion. For instance, each of three equal capacitances in series has one-third the applied voltage.

In series, the voltage across each capacitor is inversely proportional to its capacitance. The smaller capacitance has the larger proportion of the applied voltage. The reason is that the series capacitances all have the same charge because they are in one current path. With equal charge, a smaller capacitance has a greater potential difference.

8.3.0 Capacitor Specifications

This specifies the maximum potential difference that can be applied across the plates without puncturing the dielectric.

8.3.1 Voltage Rating

Usually, the voltage rating is for temperatures up to about 60°C. High temperatures result in a lower voltage rating. Voltage ratings for general-purpose paper, mica, and ceramic capacitors are typically 200V to 500V. Ceramic capacitors with ratings of 1 to 5kV are also available.

Electrolytic capacitors are commonly used in 25V, 150V, and 450V ratings. In addition, 6V and 10V electrolytic capacitors are often used in transistor circuits. For applications where a lower voltage rating is permissible, more capacitance can be obtained in a smaller physical size.

The potential difference across the capacitor depends on the applied voltage and is not necessarily equal to the voltage rating. A voltage rating higher than the potential difference applied across the capacitor provides a safety factor for long life in service. With electrolytic capacitors, however, the actual capacitor voltage should be close to the rated voltage to produce the oxide film that provides the specified capacitance.

The voltage ratings are for applied DC voltage. The breakdown rating is lower for AC voltage because of the internal heat produced by continuous charge and discharge.

8.3.2 Leak Resistance

Consider a capacitor charged by a DC voltage source. After the charging voltage is removed, a perfect capacitor would keep its charge indefinitely. After a long period of time, however, the charge will be neutralized by a small leakage current through the dielectric and across the insulated case between terminals, because there is no perfect insulator. For paper, ceramic, and mica capacitors, the leakage current is very slight, or inversely, the leakage resistance is very high. For paper, ceramic, or mica capacitors, R_l is 100MΩ or more. However, electrolytic capacitors may have a leakage resistance of 0.5MΩ or less.

8.4.0 Voltage and Current in a Capacitive AC Circuit

In a capacitive circuit driven by an AC voltage source, the voltage is continuously changing. Thus, the charge on the capacitor is also continuously changing. The four parts of *Figure 17* show the variation of the alternating voltage and current in a capacitive circuit for each quarter of one cycle.

The solid line represents the voltage across the capacitor, and the dotted line represents the current. The line running through the center is the zero or reference point for both the voltage and the current. The bottom line marks off the time of the cycle in terms of electric degrees. Assume that the AC voltage has been acting on the capacitor for some time before the time represented by the starting point of the sine wave.

At the beginning of the first quarter-cycle (0° to 90°), the voltage has just passed through zero and

THINK ABOUT IT

Capacitance

Suppose you had a motor with a bad 30µF starting capacitor and no 30µF direct replacement capacitor was available. As a temporary measure, you are authorized to substitute two equal-value capacitors in its place. What size capacitors (µF) should be used if you are connecting them in parallel?

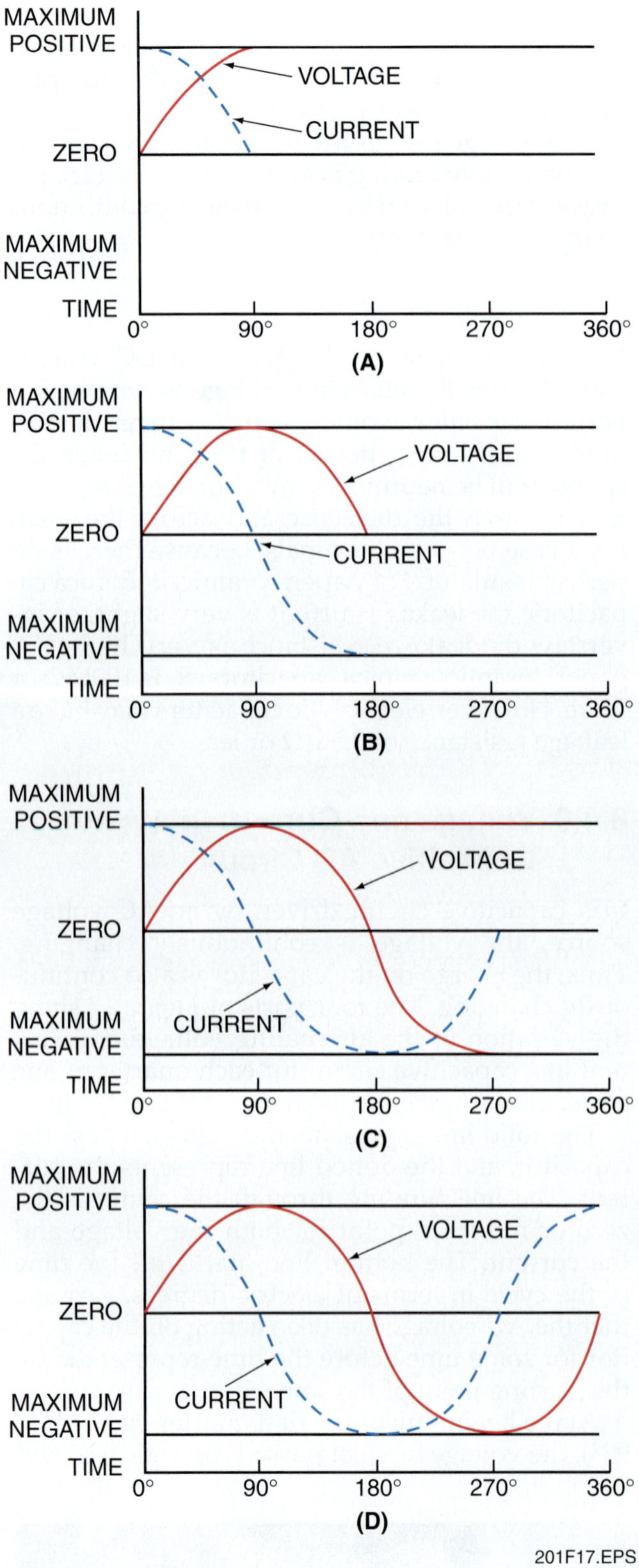

Figure 17 ◆ Voltage and current in a capacitive AC circuit.

is increasing in the positive direction. Since the zero point is the steepest part of the sine wave, the voltage is changing at its greatest rate.

The charge on a capacitor varies directly with the voltage; therefore, the charge on the capacitor is also changing at its greatest rate at the beginning of the first quarter-cycle. In other words, the greatest number of electrons are moving off one plate and onto the other plate. Thus, the capacitor current is at its maximum value.

As the voltage proceeds toward maximum at 90°, its rate of change becomes lower and lower, making the current decrease toward zero. At 90°, the voltage across the capacitor is maximum, the capacitor is fully charged, and there is no further movement of electrons from plate to plate. That is why the current at 90° is zero.

At the end of the first quarter-cycle, the alternating voltage stops increasing in the positive direction and starts to decrease. It is still a positive voltage; but to the capacitor, the decrease in voltage means that the plate that has just accumulated an excess of electrons must lose some electrons. The current flow must reverse its direction. The second part of the figure shows the current curve to be below the zero line (negative current direction) during the second quarter-cycle (90° to 180°).

At 180°, the voltage has dropped to zero. This means that, for a brief instant, the electrons are equally distributed between the two plates; the current is maximum because the rate of change of voltage is maximum.

Just after 180°, the voltage has reversed polarity and starts building to its maximum negative peak, which is reached at the end of the third quarter-cycle (180° to 270°). During the third quarter-cycle, the rate of voltage change gradually decreases as the charge builds to a maximum at 270°. At this point, the capacitor is fully charged and carries the full impressed voltage. Because the capacitor is fully charged, there is no further exchange of electrons and the current flow is zero at this point. The conditions are exactly the same as at the end of the first quarter-cycle (90°), but the polarity is reversed.

Just after 270°, the impressed voltage once again starts to decrease, and the capacitor must lose electrons from the negative plate. It must discharge, starting at a minimum rate of flow and rising to a maximum. This discharging action continues through the last quarter-cycle (270° to 360°) until the impressed voltage has reached zero. The beginning of the entire cycle is 360°, and everything starts over again.

In *Figure 17,* note that the current always arrives at a certain point in the cycle 90° ahead of the voltage because of the charging and discharging action. This voltage-current phase relationship in a capacitive circuit is exactly opposite to that in an inductive circuit. The current through a capacitor leads the voltage across the capacitor by 90°. A convenient way to remember this is the phrase "ELI the ICE man" (ELI refers to inductors, as previously explained). ICE pertains to capacitors as follows:

I = current

C = capacitor

E = voltage

In capacitors (C), current (I) leads voltage (E) by 90°.

It is important to realize that the current and voltage are both going through their individual cycles at the same time during the period the AC voltage is impressed. The current does not go through part of its cycle (charging or discharging) and then stop and wait for the voltage to catch up. The amplitude and polarity of the voltage and the amplitude and direction of the current are continually changing.

Their positions, with respect to each other and to the zero line at any electrical instant or any degree between 0° and 360°, can be seen by reading upward from the time-degree line. The current swing from the positive peak at 0° to the negative peak at 180° is not a measure of the number of electrons or the charge on the plates. It is a picture of the direction and strength of the current in relation to the polarity and strength of the voltage appearing across the plates.

8.5.0 Capacitive Reactance

Capacitors offer a very real opposition to current flow. This opposition arises from the fact that, at a given voltage and frequency, the number of electrons that go back and forth from plate to plate is limited by the storage ability or the capacitance of the capacitor. As the capacitance is increased, a greater number of electrons changes plates every cycle. Since current is a measure of the number of electrons passing a given point in a given time, the current is increased.

Increasing the frequency will also decrease the opposition offered by a capacitor. This occurs because the number of electrons that the capacitor is capable of handling at a given voltage will change plates more often. As a result, more electrons will pass a given point in a given time (greater current flow). The opposition that a capacitor offers to AC is therefore inversely proportional to frequency and capacitance. This opposition is called capacitive reactance. Capacitive reactance decreases with increasing frequency or, for a given frequency, the capacitive reactance decreases with increasing capacitance. The symbol for capacitive reactance is X_C. The formula is:

$$X_C = \frac{1}{2\pi fC}$$

Where:

X_C = capacitive reactance in ohms

f = frequency in hertz

C = capacitance in farads

$2\pi = 6.28\ (2 \times 3.14)$

For example, what is the capacitive reactance of a 0.05µF capacitor in a circuit whose frequency is 1 megahertz?

$$X_C = \frac{1}{2\pi fC} - \frac{1}{(6.28)(10^6\text{ hertz})(5 \times 10^{-8}\text{ farads})}$$

$$X_C = \frac{1}{3.14 \times 10^{-1}} = \frac{1}{0.314} = 3.18\text{ ohms}$$

The capacitive reactance of a 0.05µF capacitor operated at a frequency of 1 megahertz is 3.18 ohms. Suppose this same capacitor is operated at a lower frequency of 1,500 hertz instead of 1 megahertz. What is the capacitive reactance now? Substituting where 1,500 = 1.5×10^3 hertz:

$$X_C = \frac{1}{2\pi fC} = \frac{1}{(6.28)(1.5 \times 10^3\text{ hertz})(5 \times 10^{-8}\text{ farads})}$$

$$X_C = \frac{1}{4.71 \times 10^{-4}} = 2{,}123\text{ ohms}$$

Note a very interesting point from these two examples. As frequency is decreased from 1 megahertz to 1,500 hertz, the capacitive reactance increases from 3.18 ohms to 2,123 ohms. Capacitive reactance increases as the frequency decreases.

INSIDE TRACK

ICE in ELI the ICE Man

Remembering the phrase "ICE" as in "ELI the ICE man" is an easy way to remember the phrase relationships that always exist between voltage and current in a capacitive circuit. A capacitive circuit is a circuit in which there is more capacitive reactance than inductive reactance. This is indicated by the C in ICE. The I (current) is stated before the E (voltage) in ICE, meaning that the current leads the voltage in a capacitive circuit.

THINK ABOUT IT

Frequency and Capacitive Reactance

A variable capacitor is used in the tuner of an AM radio to tune the radio to the desired station. Will its capacitive reactance value be higher or lower when it is tuned to the low end of the frequency band (550kHz) than it would be when tuned to the high end of the band (1,440kHz)?

9.0.0 ◆ LC AND RLC CIRCUITS

AC circuits often contain inductors, capacitors, and/or resistors connected in series or parallel combinations. When this is done, it is important to determine the resulting phase relationship between the applied voltage and the current in the circuit. The simplest method of combining factors that have different phase relationships is vector addition with the trigonometric functions. Each quantity is represented as a vector, and the resultant vector and phase angle are then calculated.

In purely resistive circuits, the voltage and current are in phase. In inductive circuits, the voltage leads the current by 90°. In capacitive circuits, the current leads the voltage by 90°. *Figure 18* shows the phase relationships of these components used in AC circuits. Recall that these characteristics are summarized by the phrase "ELI the ICE man."

ELI = *E* Leads *I* (inductive)

ICE = *I* Capacitive (leads) *E*

The **impedance** Z of a circuit is defined as the total opposition to current flow. The magnitude of the impedance Z is given by the following equation in a series circuit:

$$Z = \sqrt{R^2 + X^2}$$

Where:

Z = impedance (ohms)

R = resistance (ohms)

X = net reactance (ohms)

The current through a resistance is always in phase with the voltage applied to it; thus resistance is shown along the 0° axis. The voltage across an inductor leads the current by 90°; thus inductive reactance is shown along the 90° axis. The voltage across a capacitor lags the current by 90°; thus capacitive reactance is shown along the −90° axis. The net reactance is the difference between the inductive reactance and the capacitive reactance:

$$X = X_L - X_C$$

Where:

X = net reactance (ohms)

X_L = inductive reactance (ohms)

X_C = capacitive reactance (ohms)

The impedance Z is the vector sum of the resistance R and the net reactance X. The angle, called the phase angle, gives the phase relationship between the applied voltage and current.

9.1.0 RL Circuits

RL circuits combine resistors and inductors in a series, parallel, or series-parallel configuration. In a pure inductive circuit, the current lags the voltage by an angle of 90°. In a circuit containing both resistance and inductance, the current will lag the voltage by some angle between zero and 90°.

9.1.1 Series RL Circuit

Figure 19 shows a series RL circuit. Since it is a series circuit, the current is the same in all portions of the loop. Using the values shown, the circuit will be analyzed for unknown values such as X_L, Z, I, E_L, and E_R.

The solution would be worked as follows:

Step 1 Compute the value of X_L.

$X_L = 2\pi fL$

$X_L = 6.28 \times 100 \times 4 = 2{,}512$ ohms

Step 2 Draw vectors R and X_L as shown in *Figure 19*. R is drawn horizontally because the circuit current and voltage across R are in phase. It therefore becomes the reference line from which other angles are measured. X_L is drawn upward at 90° from R because voltage across X_L leads circuit current through R.

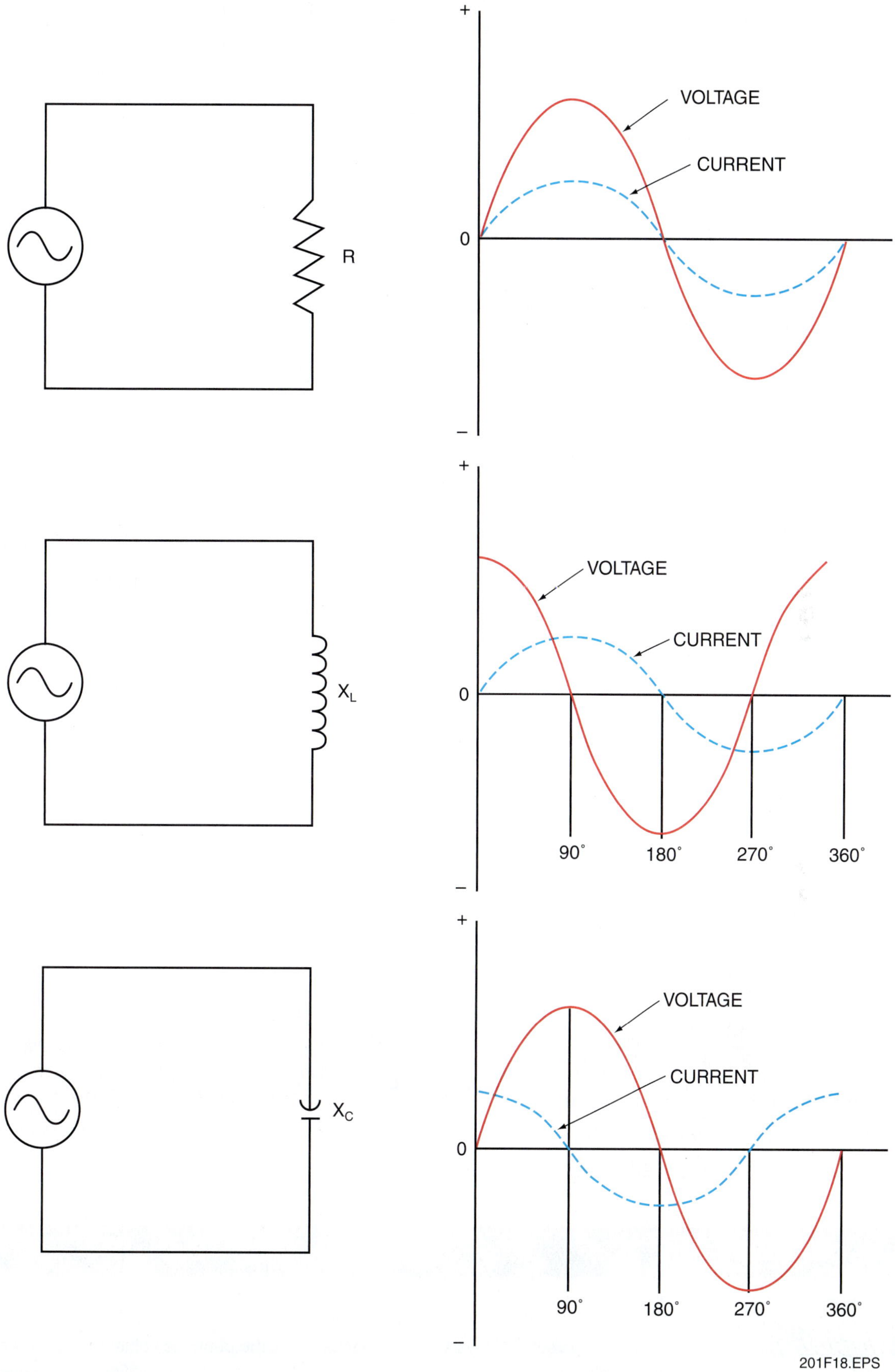

Figure 18 ◆ Summary of AC circuit phase relationships.

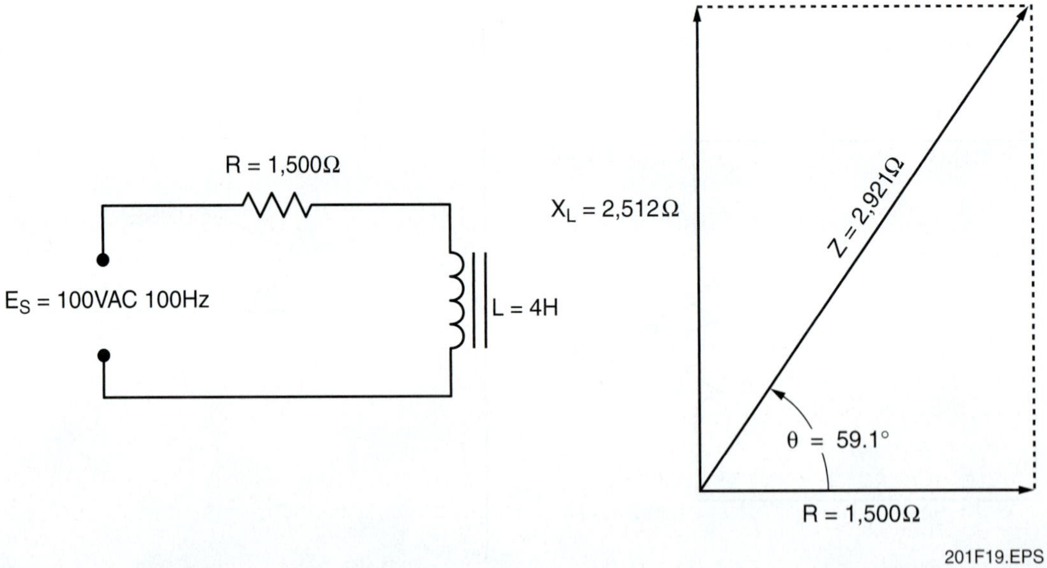

Figure 19 ◆ Series RL circuit and vector diagram.

Step 3 Compute the value of circuit impedance Z, which is equal to the vector sum of X_L and R.

$$\tan = \frac{X_L}{R} = \frac{2{,}512}{1{,}500} = 1.67$$

$$\arctan 1.67 = 59.1°$$

$$\cos 59.1° = 0.5135$$

Find Z using the cosine function:

$$\cos = \frac{R}{Z}$$

$$Z = \frac{R}{\cos}$$

$$Z = \frac{1{,}500}{0.5135} = 2{,}921 \text{ ohms}$$

Step 4 Compute the circuit current using Ohm's law for AC circuits.

$$I = \frac{E}{Z} = \frac{100\text{V}}{2.921\Omega} = 0.034\text{A}$$

Step 5 Compute voltage drops in the circuit.

$$E_L = IX_L = 0.034 \times 2{,}512 = 85 \text{ volts}$$

$$E_R = IR = 0.034 \times 1{,}500 = 51 \text{ volts}$$

Note that the voltage drops across the resistor and inductor do not equal the supply voltage because they must be added vectorially. This would be done as follows (because of rounding, numbers are not exact):

$$\tan = \frac{E_L}{E_R} = \frac{85}{51} = 1.67$$

$$\arctan 1.67 = 59.1°$$

$$\cos = \frac{E_R}{E_Z} \quad E_Z = \frac{E_R}{\cos}$$

$$E_Z = \frac{51}{0.5135} = \text{approx. } 100\text{V} = E_S$$

In this inductive circuit, the current lags the applied voltage by an angle equal to 59.1°.

NOTE

Since we are dealing with right triangles, we could also use the Pythagorean theorem (discussed later) to find this answer.

INSIDE TRACK

Vector Analysis

When using vector analysis, the horizontal line is the in-phase value and the vertical line pointing up represents the leading value. The vertical line pointing down represents the lagging value.

Figure 20 shows another series RL circuit, its associated waveforms, and vector diagrams. This circuit is used to summarize the characteristics of a series RL circuit:

- The current I flows through all the series components.
- The voltage across X_L, labeled V_L, can be considered an IX_L voltage drop, just as V_R is used for an IR voltage drop.
- The current I through X_L must lag V_L by 90°, as this is the angle between current through an inductance and its self-induced voltage.
- The current I through R and its IR voltage drop have the same phase. There is no reactance to sine wave current in any resistance. Therefore, I and IR have the same phase, or this phase angle is 0°.
- V_T is the vector sum of the two out-of-phase voltages V_R and V_L.
- Circuit current I lags V_T by the phase angle.
- Circuit impedance is the vector sum of R and X_L.

In a series circuit, the higher the value of X_L compared with R, the more inductive the circuit is. This means there is more voltage drop across the inductive reactance, and the phase angle increases toward 90°. The series current lags the applied generator voltage.

Several combinations of X_L and R in series are listed in *Table 1* with their resultant impedance and phase angle. Note that a ratio of 10:1 or more for X_L/R means that the circuit is practically all inductive. The phase angle of 84.3° is only slightly less than 90° for the ratio of 10:1, and the total impedance Z is approximately equal to X_L. The voltage drop across X_L in the series circuit will be equal to the applied voltage, with almost none across R.

Table 1 Series R and X_L Combinations

R (Ω)	X_L (Ω)	Z (Ω) (Approx.)	Phase Angle (θ) (°)
1	10	$\sqrt{101} = 10$	84.3°
10	10	$\sqrt{200} = 14$	45°
10	1	$\sqrt{101} = 10$	5.7°

At the opposite extreme, when R is 10 times as large as X_L, the series circuit is mainly resistive. The phase angle of 5.7° means the current has almost the same phase as the applied voltage, the total impedance Z is approximately equal to R, and the voltage drop across R is practically equal to the applied voltage, with almost none across X_L.

9.1.2 Parallel RL Circuit

In a parallel RL circuit, the resistance and inductance are connected in parallel across a voltage source. Such a circuit thus has a resistive branch and an inductive branch.

The 90° phase angle must be considered for each of the branch currents, instead of voltage

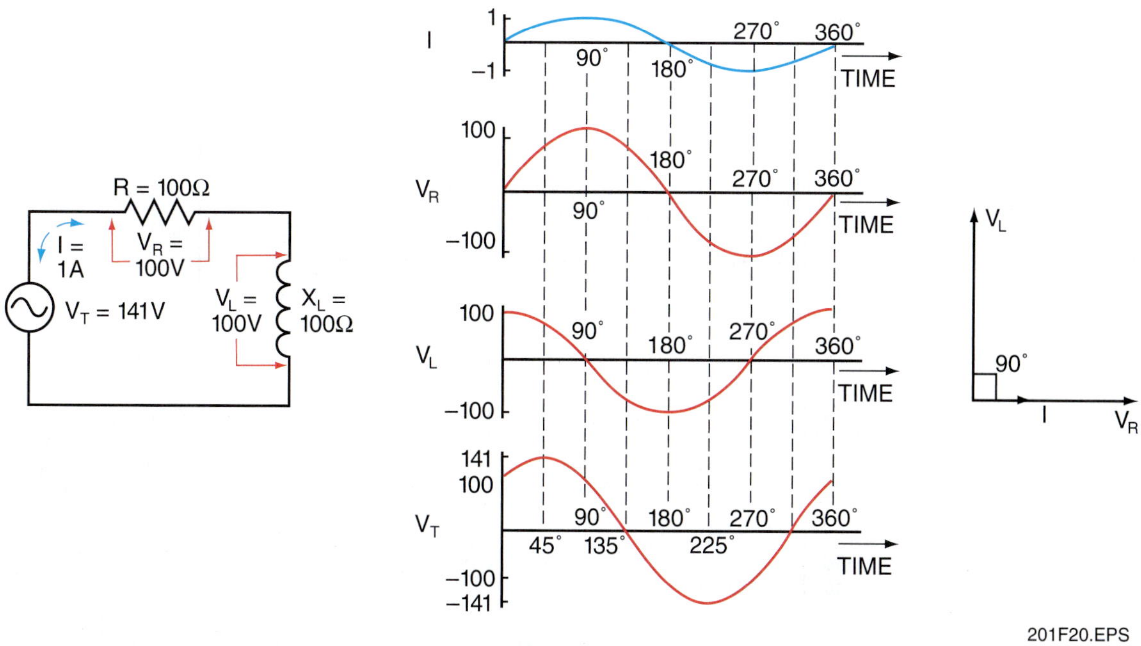

Figure 20 ◆ Series RL circuit with waveforms and vector diagram.

drops in a series circuit. Remember that any series circuit has different voltage drops, but one common current. A parallel circuit has different branch currents, but one common voltage.

In the parallel circuit in *Figure 21*, the applied voltage V_A is the same across X_L, R, and the generator, since they are all in parallel. There cannot be any phase difference between these voltages. Each branch, however, has its individual current. For the resistive branch $I_R = V_A/R$; in the inductive branch $I_L = V_A/X_L$.

The resistive branch current I_R has the same phase as the generator voltage V_A. The inductive branch current I_L lags V_A, however, because the current in an inductance lags the voltage across it by 90°.

The total line current, therefore, consists of I_R and I_L, which are 90° out of phase with each other. The phasor sum of I_R and I_L equals the total line current I_T. These phase relations are shown by the waveforms and vectors in *Figure 21*. I_T will lag V_A by some phase angle that results from the vector addition of I_R and I_L.

The impedance of a parallel RL circuit is the total opposition to current flow by the R of the resistive branch and the X_L of the inductive branch. Since X_L and R are vector quantities, they must be added vectorially.

If the line current and the applied voltage are known, Z can also be calculated by the equation:

$$Z = \frac{V_A}{I_{Line}}$$

The Z of a parallel RL circuit is always less than the R or X_L of any one branch. The branch of a parallel RL circuit that offers the most opposition to current flow has the lesser effect on the phase angle of the current.

Several combinations of X_L and R in parallel are listed in *Table 2*. When X_L is 10 times R, the parallel circuit is practically resistive because there is little inductive current in the line. The small value of I_L results from the high X_L. The total impedance of the parallel circuit is approximately equal to the resistance then, since the high value of X_L in a parallel branch has little effect. The phase angle of −5.7° is practically 0° because almost all the line current is resistive.

As X_L becomes smaller, it provides more inductive current in the main line. When X_L is 1⁄10R, practically all the line current is the I_L component. Then, the parallel circuit is practically all inductive, with a total impedance practically equal to X_L. The phase angle of −84.3° is almost −90° because the line current is mostly inductive. Note that these conditions are opposite from the case of X_L and R in series.

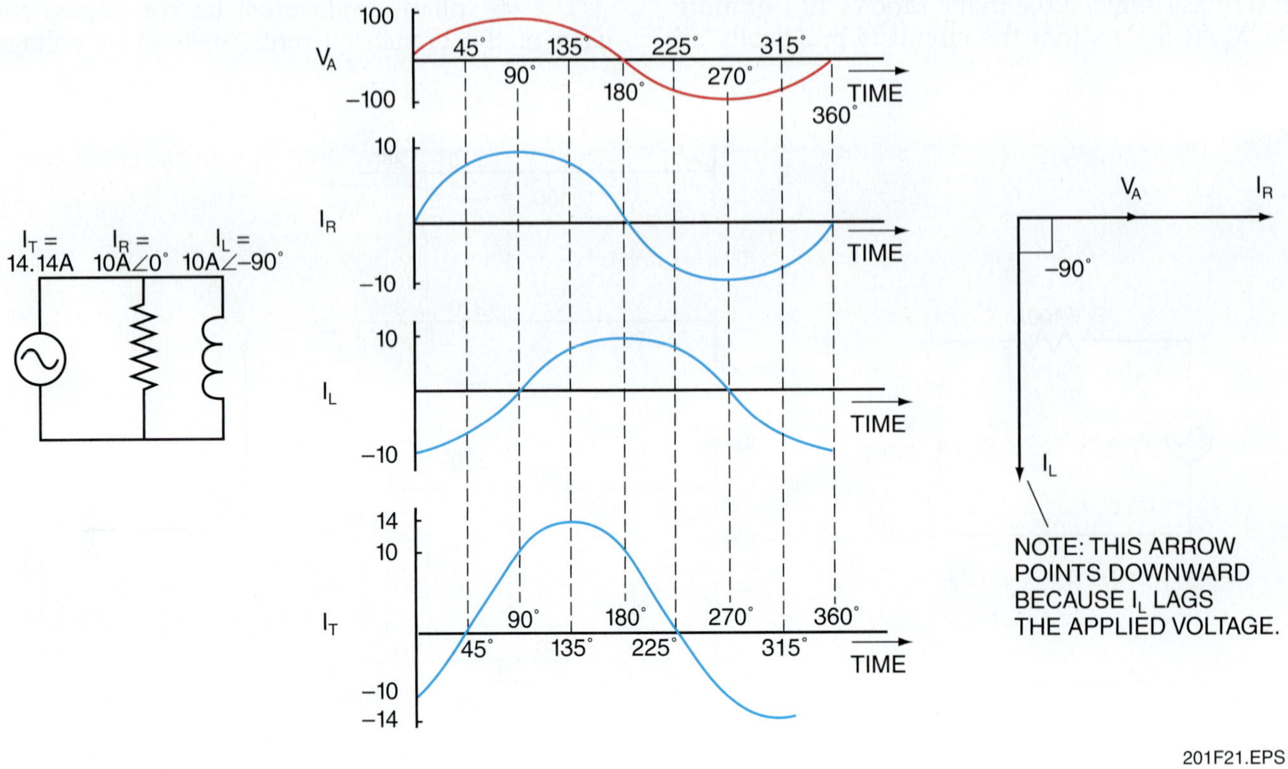

Figure 21 ◆ Parallel RL circuit with waveforms and vector diagram.

Table 2 Parallel R and X_L Combinations

R (Ω)	X_L (Ω)	I_R (A)	I_L (A)	I_T (A) (Approx.)	$Z_T = V_A/I_T$ (Ω)	Phase Angle (θ), (°)
1	10	10	1	$\sqrt{101} = 10$	1	−5.7°
10	10	1	1	$\sqrt{2} = 1.4$	7.07	−45°
10	1	1	10	$\sqrt{101} = 10$	1	−84.3°

9.2.0 RC Circuits

In a circuit containing resistance only, the current and voltage are in phase. In a circuit of pure capacitance, the current leads the voltage by an angle of 90°. In a circuit that has both resistance and capacitance, the current will lead the voltage by some angle between 0° and 90°.

9.2.1 Series RC Circuit

Figure 22(A) shows a series RC circuit with resistance R in series with capacitive reactance X_C. Current I is the same in X_C and R since they are in series. Each has its own series voltage drop, equal to IR for the resistance and IX_C for the reactance.

In *Figure 22(B)*, the phasor is shown horizontal as the reference phase, because I is the same throughout the series circuit. The resistive voltage drop IR has the same phase as I. The capacitor voltage IX_C must be 90° clockwise from I and IR, as the capacitive voltage lags. Note that the IX_C phasor is downward, exactly opposite from an IX_L phasor, because of the opposite phase angle.

If the capacitive reactance alone is considered, its voltage drop lags the series current I by 90°. The IR voltage has the same phase as I, however, because resistance provides no phase shift. Therefore, R and X_C combined in series must be added by vectors because they are 90° out of phase with each other, as shown in *Figure 22(C)*.

As with inductive reactance, θ (theta) is the phase angle between the generator voltage and its series current. As shown in *Figure 22(B)* and *Figure 22(C)*, θ can be calculated from the voltage or impedance triangle.

With series X_C the phase angle is negative, clockwise from the zero reference angle of I because the X_C voltage lags its current. To indicate the negative phase angle, this 90° phasor points downward from the horizontal reference, instead of upward as with the series inductive reactance.

In series, the higher the X_C compared with R, the more capacitive the circuit. There is more voltage drop across the capacitive reactance, and the phase angle increases toward −90°. The series X_C always makes the current lead the applied voltage. With all X_C and no R, the entire applied voltage is across X_C and equals −90°. Several combinations of X_C and R in series are listed in *Table 3*.

9.2.2 Parallel RC Circuit

In a parallel RC circuit, as shown in *Figure 23(A)*, a capacitive branch as well as a resistive branch are connected across a voltage source. The current that leaves the voltage source divides among the branches, so there are different currents in each branch. The current is therefore not a common quantity, as it is in the series RC circuit.

In a parallel RC circuit, the applied voltage is directly across each branch. Therefore, the branch

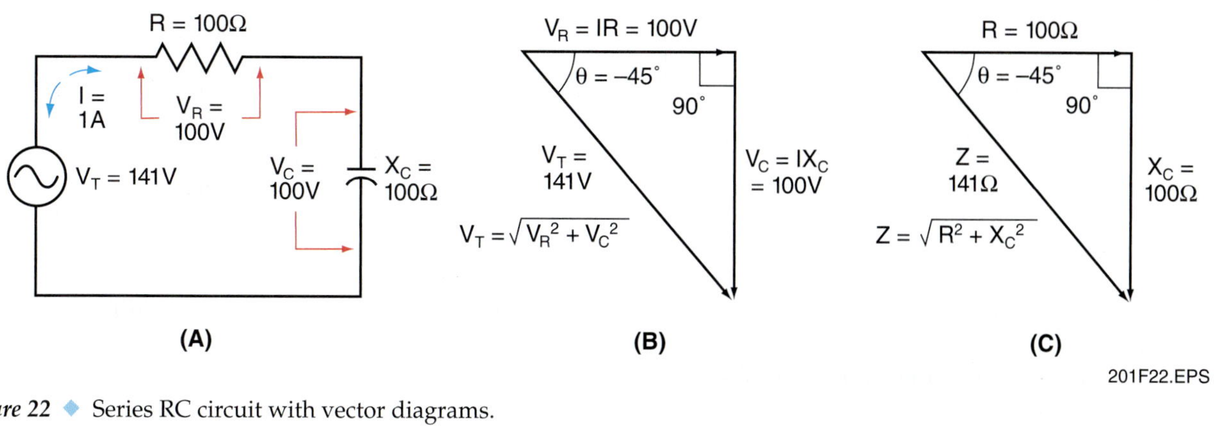

Figure 22 ◆ Series RC circuit with vector diagrams.

Table 3 Series R and X_C Combinations

R (Ω)	X_C (Ω)	Z (Ω) (Approx.)	Phase Angle $(\theta)_Z$ (°)
1	10	$\sqrt{101} = 10$	84.3°
10	10	$\sqrt{200} = 14$	45°
10	1	$\sqrt{101} = 10$	5.7°

voltages are equal in value to the applied voltage and all voltages are in phase. Since the voltage is common throughout the parallel RC circuit, it serves as the common quantity in any vector representation of parallel RC circuits. This means the reference vector will have the same phase relationship or direction as the circuit voltage. Note in *Figure 23(B)* that V_A and I_R are both shown as the 0° reference.

Current within an individual branch of an RC parallel circuit is dependent on the voltage across the branch and on the R or X_C contained in the branch. The current in the resistive branch is in phase with the branch voltage, which is the applied voltage. The current in the capacitive branch leads V_A by 90°. Since the branch voltages are the same, I_C leads I_R by 90°, as shown in *Figure 23(B)*. Since the branch currents are out of phase, they have to be added vectorially to find the line current.

The phase angle, θ, is 45° because R and X_C are equal, resulting in equal branch currents. The phase angle is between the total current I_T and the generator voltage V_A. However, the phase of V_A is the same as the phase of I_R. Therefore, θ is also between I_T and I_R.

The impedance of a parallel RC circuit represents the total opposition to current flow offered by the resistance and capacitive reactance of the circuit. The equation for calculating the impedance of a parallel RC circuit is:

$$Z = \frac{RX_C}{\sqrt{I_R^2 + I_C^2}} \text{ or } Z = \frac{V_A}{I_T}$$

For the example shown in *Figure 23*, Z is:

$$Z = \frac{V_A}{I_T} = \frac{100}{14.14A} = 7.07\Omega$$

This is the opposition in ohms across the generator. This Z of 7.07Ω is equal to the resistance of 10Ω in parallel with the reactance of 10Ω. Notice that the impedance of equal values of R and X_C is not one-half, but equals 70.7% of either one.

When X_C is high relative to R, the parallel circuit is practically resistive because there is little leading capacitive current in the main line. The small value of I_C results from the high reactance of shunt X_C. The total impedance of the parallel circuit is approximately equal to the resistance, since the high value of X_C in a parallel branch has little effect.

As X_C becomes smaller, it provides more leading capacitive current in the main line. When X_C is very small relative to R, practically all the line current is the I_C component. The parallel circuit is practically all capacitive, with a total impedance practically equal to X_C.

The characteristics of different circuit arrangements are shown in *Table 4*.

9.3.0 LC Circuits

An LC circuit consists of an inductance and a capacitance connected in series or in parallel with a voltage source. There is no resistor physically in an LC circuit, but every circuit contains some resistance. Since the circuit resistance of the wiring

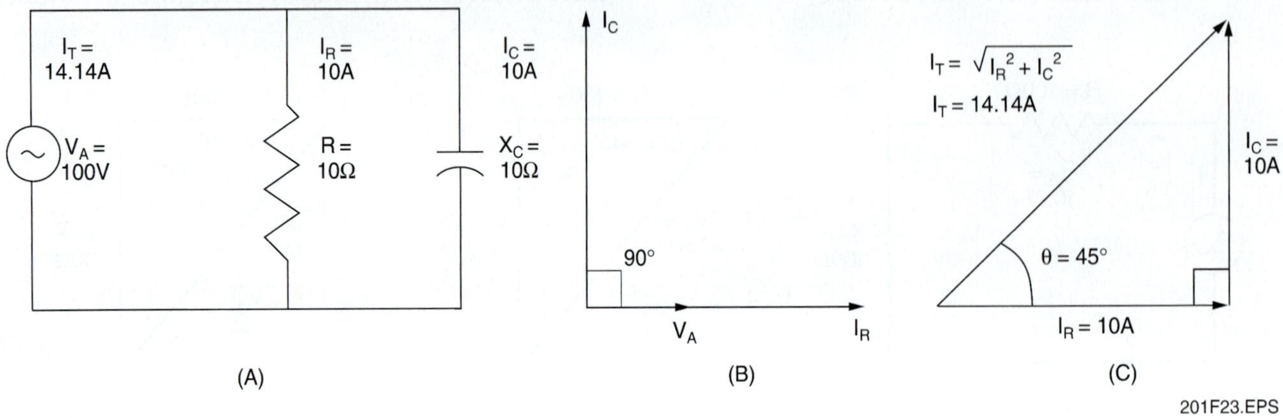

Figure 23 ◆ Parallel RC circuit with vector diagrams.

Table 4 Parallel R and X_C Combinations

R (Ω)	X_C (Ω)	I_R (A)	I_C (A)	I_T (A) (Approx.)	Z_T (Ω) (Approx.)	Phase Angle (θ) (°)
1	10	10	1	$\sqrt{101} = 10$	1	5.7°
10	10	1	1	$\sqrt{2} = 1.4$	7.07	45°
10	1	1	10	$\sqrt{101} = 10$	1	84.3°

and voltage source is usually so small, it has little or no effect on circuit operation.

In a circuit with both X_L and X_C, the opposite phase angles enable one to cancel the effect of the other. For X_L and X_C in series, the net reactance is the difference between the two series reactances, resulting in less reactance than either one. In parallel circuits, the I_L and I_C branch currents cancel. The net line current is then the difference between the two branch currents, resulting in less total line current than either branch current.

9.3.1 Series LC Circuit

As in all series circuits, the current in a series LC circuit is the same at all points. Therefore, the current in the inductor is the same as, and in phase with, the current in the capacitor. Because of this, on the vector diagram for a series LC circuit, the direction of the current vector is the reference or in the 0° direction, as shown in *Figure 24*.

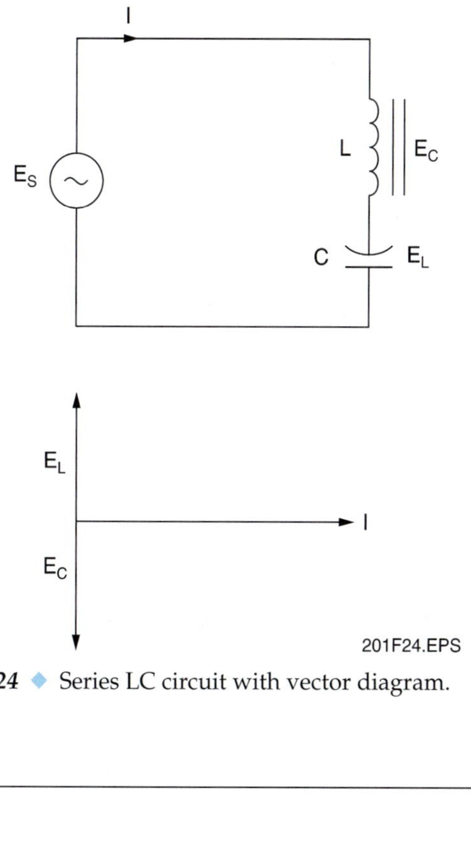

Figure 24 ◆ Series LC circuit with vector diagram.

When there is current flow in a series LC circuit, the voltage drops across the inductor and capacitor depend on the circuit current and the values of X_L and X_C. The voltage drop across the inductor leads the circuit current by 90°, and the voltage drop across the capacitor lags the circuit current by 90°. Using Kirchhoff's voltage law, the source voltage equals the sum of the voltage drops across the inductor and capacitor, with respect to the polarity of each.

Since the current through both is the same, the voltage across the inductor leads that across the capacitor by 180°. The method used to add the two voltage vectors is to subtract the smaller vector from the larger, and assign the resultant the direction of the larger. When applied to a series LC circuit, this means the applied voltage is equal to the difference between the voltage drops (E_L and E_C), with the phase angle between the applied voltage (E_T) and the circuit current determined by the larger voltage drop.

In a series LC circuit, one or both of the voltage drops are always greater than the applied voltage. Remember that although one or both of the voltage drops are greater than the applied voltage, they are 180° out of phase. One of them effectively cancels a portion of the other so that the total voltage drop is always equal to the applied voltage.

Recall that X_L is 180° out of phase with X_C. The impedance is then the vector sum of the two reactances. The reactances are 180° apart, so their vector sum is found by subtracting the smaller one from the larger.

Unlike RL and RC circuits, the impedance in an LC circuit is either purely inductive or purely capacitive.

9.3.2 Parallel LC Circuit

In a parallel LC circuit there is an inductance and a capacitance connected in parallel across a voltage source. *Figure 25* shows a parallel LC circuit with its vector diagram.

As in any parallel circuit, the voltage across the branches is the same as the applied voltage. Since

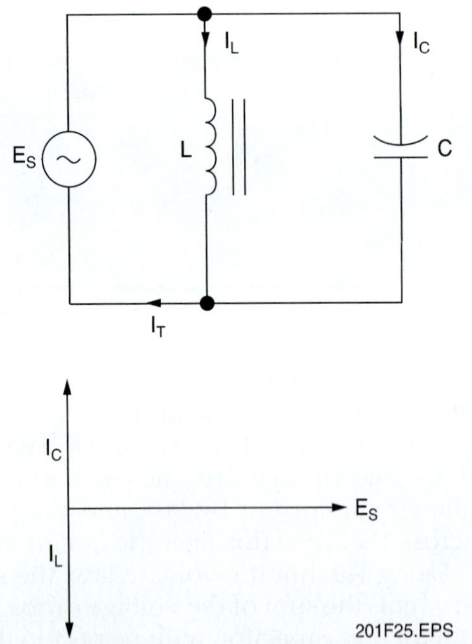

Figure 25 ◆ Parallel LC circuit with vector diagram.

they are actually the same voltage, the branch voltages and applied voltage are in phase. Because of this, the voltage is used as the 0° phase reference and the phases of the other circuit quantities are expressed in relation to the voltage.

The currents in the branches of a parallel LC circuit are both out of phase with the circuit voltage. The current in the inductive branch (I_L) lags the voltage by 90°, while the current in the capacitive branch (I_C) leads the voltage by 90°. Since the voltage is the same for both branches, currents I_L and I_C are therefore 180° out of phase. The amplitudes of the branch currents depend on the value of the reactance in the respective branches.

With the branch currents being 180° out of phase, the line current is equal to their vector sum. This vector addition is done by subtracting the smaller branch current from the larger.

The line current for a parallel LC circuit, therefore, has the phase characteristics of the larger branch current. Thus, if the inductive branch current is the larger, the line current is inductive and lags the applied voltage by 90°; if the capacitive branch current is the larger, the line current is capacitive, and leads the applied voltage by 90°.

The line current in a parallel LC circuit is always less than one of the branch currents and sometimes less than both. The reason that the line current is less than the branch currents is because the two branch currents are 180° out of phase. As a result of the phase difference, some cancellation takes place between the two currents when they combine to produce the line current. The impedance of a parallel LC circuit can be found using the following equations:

$$Z = \frac{X_L \times X_C}{X_L - X_C} \text{ (for } X_L \text{ larger than } X_C\text{)}$$

or:

$$Z = \frac{X_L \times X_C}{X_C - X_L} \text{ (for } X_C \text{ larger than } X_L\text{)}$$

When using these equations, the impedance will have the phase characteristics of the smaller reactance.

9.4.0 RLC Circuits

9.4.1 Series RLC Circuit

Circuits in which the inductance, capacitance, and resistance are all connected in series are called series RLC circuits. The fundamental properties of series RLC circuits are similar to those for series LC circuits. The differences are caused by the effects of the resistance. Any practical series LC circuit contains some resistance. When the resistance is very small compared to the circuit reactance, it has almost no effect on the circuit and can be considered as zero. When the resistance is appreciable, though, it has a significant effect on the circuit operation and therefore must be considered in any circuit analysis. In a series RLC circuit, the same current flows through each component. The phase relationships between the voltage drops are the same as they were in series RC, RL, and LC circuits. The voltage drops across the inductance and capacitance are 180° out of phase. With current the same throughout the circuit as a reference, the inductive voltage drop (E_L) leads the resistive voltage drop (E_R) by 90°, and the capacitive voltage drop (E_C) lags the resistive voltage drop by 90°.

Figure 26 shows a series RLC circuit and the vector diagram used to determine the applied voltage. The vector sum of the three voltage drops is equal to the applied voltage. However, to calculate this vector sum, a combination of the methods learned for LC, RL, and RC circuits must be used. First, calculate the combined voltage drop of the two reactances. This value is designated E_X and is found as in pure LC circuits by subtracting the smaller reactive voltage drop from the larger. This is shown in *Figure 26* as E_X. The result of this calculation is the net reactive voltage drop and is either inductive or capacitive, depending on which of the individual voltage drops is larger. In *Figure 26*, the net reactive voltage drop is inductive since $E_L > E_C$. Once the net reactive voltage drop is known, it is added vectorially to the voltage drop across the resistance.

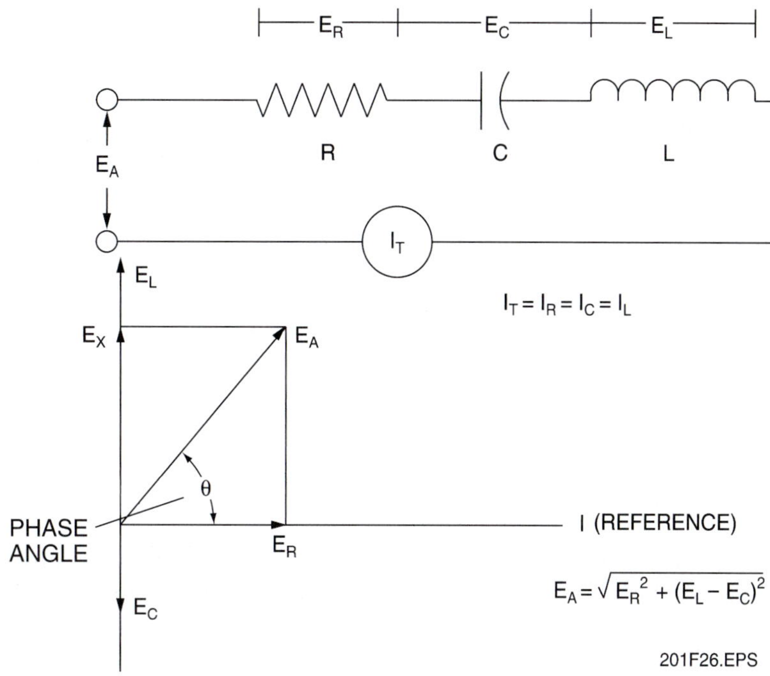

Figure 26 ◆ Series RLC circuit and vector diagram.

The angle between the applied voltage E_A and the voltage across the resistance E_R is the same as the phase angle between E_A and the circuit current. The reason for this is that E_R and I are in phase.

The impedance of a series RLC circuit is the vector sum of the inductive reactance, the capacitive reactance, and the resistance. This is done using the same method as for voltage drop calculations.

When X_L is greater than X_C, the net reactance is inductive, and the circuit acts essentially as an RL circuit. Similarly, when X_C is greater than X_L, the net reactance is capacitive, and the circuit acts as an RC circuit.

The same current flows in every part of a series RLC circuit. The current always leads the voltage across the capacitance by 90° and is in phase with the voltage across the resistance. The phase relationship between the current and the applied voltage, however, depends on the circuit impedance. If the impedance is inductive (X_L greater than X_C), the current is inductive and lags the applied voltage by some phase angle less than 90°. If the impedance is capacitive (X_C greater than X_L), the current is capacitive, and leads the applied voltage by some phase angle also less than 90°. The angle of the lead or lag is determined by the relative values of the net reactance and the resistance.

The greater the value of X or the smaller the value of R, the larger the phase angle, and the more reactive (or less resistive) the current. Similarly, the smaller the value of X or the larger the value of R, the more resistive (or less reactive) the current. If either R or X is 10 or more times greater than the other, the circuit will essentially act as though it is purely resistive or reactive, as the case may be.

9.4.2 Series Resonance

Recall that:

$$X_L = 2\pi fL$$

and:

$$X_C = \frac{1}{2\pi fC}$$

With X_L and X_C being frequency sensitive, any change in frequency will affect the operating characteristics of any reactive circuit.

The effects of frequency variations on the input voltage of a series RLC circuit are shown in *Figure 27* and summarized as follows:

- As frequency is increased, X_L will become larger and X_C will become smaller. As a result, the circuit becomes even more inductive, θ increases, and the voltage across L will increase.
- As frequency decreases, X_L will become smaller and X_C larger, and θ will decrease toward zero.
- At a certain frequency, X_L will equal X_C. This is referred to as **resonance.**
- A further decrease in frequency will make X_C larger than X_L. The circuit will become capacitive and will increase in a negative direction.

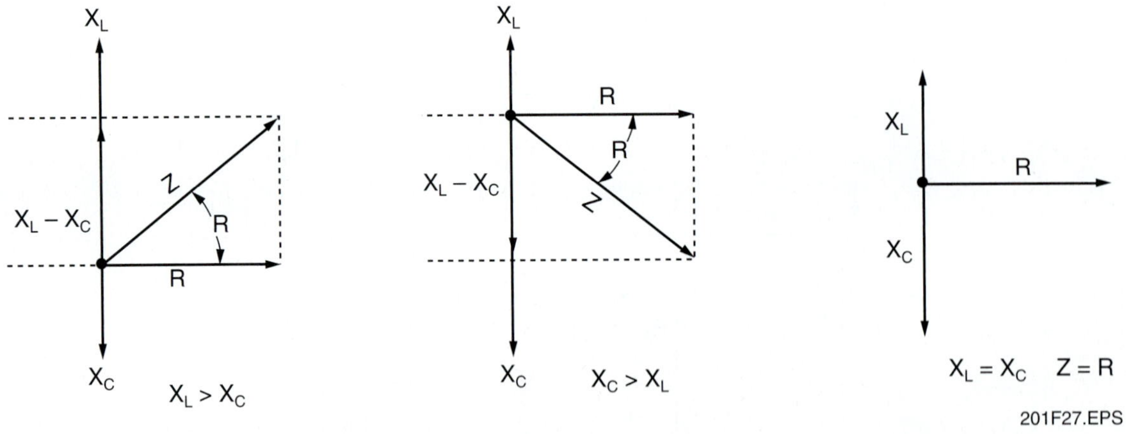

Figure 27 ◆ Effects of frequency variations on an RLC circuit.

To determine the resonant frequency at which $X_L = X_C$ requires only a bit of mathematics:

$$X_L = X_C$$

$$X_L = 2\pi fL$$

$$X_C = \frac{1}{2\pi fC}$$

Therefore, at resonance:

$$2\pi fL = \frac{1}{2\pi fC}$$

or:

$$f_r = \text{resonant frequency} = \frac{1}{2\pi\sqrt{LC}}$$

Figure 27 shows a vector diagram of the resistance and reactance at a frequency when $X_L = X_C$, or resonance. Since X_L and X_C are equal and 180° out of phase, the algebraic sum of X_L and X_C is zero. The only opposition to current flow in the circuit is R. At series resonance:

$$Z = R$$

At frequencies above resonance, X_L is greater than X_C. The circuit becomes inductive and Z increases. At frequencies below resonance, X_C is greater than X_L. The circuit becomes capacitive and Z increases. The point of lowest impedance of the circuit is at resonance. These characteristics are shown in *Figure 28*.

The impedance of a series RLC circuit is minimum at resonance, so the current must therefore be maximum. Both above and below the resonant frequency circuit impedance increases, which means that current decreases. The further the frequency is from the resonant frequency, the greater the impedance, and the smaller the current becomes. At any frequency, the current can be calculated from Ohm's law for AC circuits using the equation I = E/Z. Since at the resonant frequency the impedance equals the resistance, the equation for current at resonance becomes I = E/R.

The letter Q is used to designate the quality of a tuned circuit. It is an indication of its maximum response as well as its ability to respond within a band of frequencies.

To secure maximum currents and response, the resistance must be kept at a low value. At resonance, R is the only resistance in the circuit. The Q of a circuit is the relationship of the reactance of the circuit to its resistance:

$$Q = \frac{X_L}{R} = \frac{X_C}{R}$$

The bandwidth or bandpass of a tuned circuit is defined as those frequency limits above and below resonant frequency where the response of the circuit will drop to 0.707 of its peak response. If current or voltage drops to 0.707 of its peak value, the power drops to 50%.

Bandwidth is frequency above and below resonance where power drops to one-half of its peak value. These are called the half-power points.

If the frequency of an RLC circuit is varied and the values of current at the different frequencies are plotted on a graph, the result is a curve known as the resonance curve of the circuit, as shown in *Figure 29*.

Actually, *Figure 29* shows two curves: one with a high resistance that results in a lower current flow, low Q, and wide bandwidth; and the second with a low resistance that results in high current flow at resonance, high Q, and low bandwidth.

9.4.3 Parallel RLC Circuit

A parallel RLC circuit is basically a parallel LC circuit with an added parallel branch of resistance. The solution of a parallel circuit involves

AC Circuits

THINK ABOUT IT

The photo below shows a simple series circuit comprised of an ON/OFF switch, small lamp, motor, and capacitor. How would you classify this circuit? When energized, which components insert resistance, inductive reactance, and capacitive reactance into the circuit?

201P0101.EPS

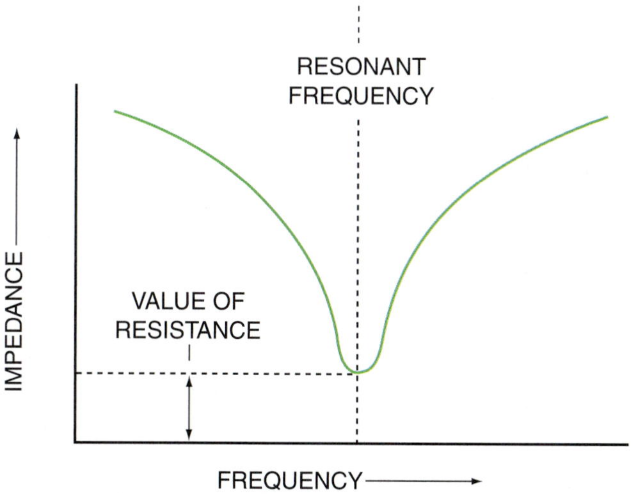

201F28.EPS

Figure 28 ◆ Frequency-impedance curve.

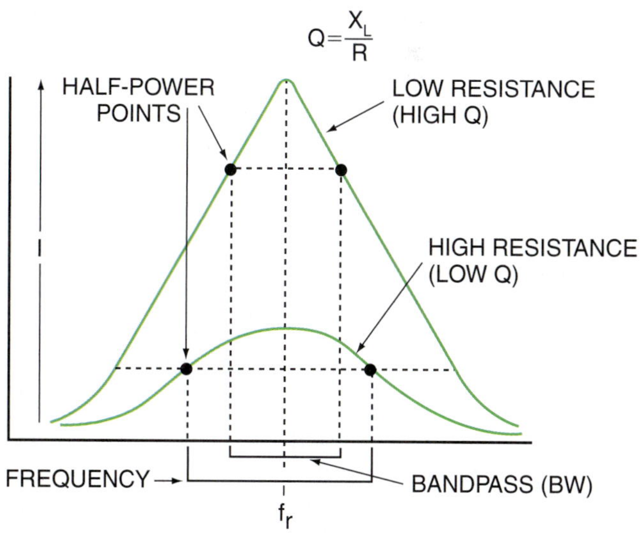

201F29.EPS

Figure 29 ◆ Typical series resonance curve.

the solution of a parallel LC circuit, and then the solution of either a parallel RL circuit or a parallel RC circuit. The reason for this is that a parallel combination of L and C appears to the source as a pure L or a pure C. So by solving the LC portion of a parallel RLC circuit first, the circuit is reduced to an equivalent RL or RC circuit.

The distribution of the voltage in a parallel RLC circuit is no different from what it is in a parallel LC circuit, or in any parallel circuit. The branch voltages are all equal and in phase, since they are the same as the applied voltage. The resistance is simply another branch across which the applied voltage appears. Because the voltages throughout the circuit are the same, the applied voltage is again used as the θ phase reference.

Figure 30 shows the current relationship in a parallel RLC circuit.

The three branch currents in a parallel RLC circuit are an inductive current I_L, a capacitive current I_C, and a resistive current I_R. Each is independent of the others, and depends only on the applied voltage and the branch resistance or reactance.

The three branch currents all have different phases with respect to the branch voltages. I_L lags the voltage by 90°, I_C leads the voltage by 90°, and I_R is in phase with the voltage. Since the voltages are the same, I_L and I_C are 180° out of phase with each other, and both are 90° out of phase with I_R. Because I_R is in phase with the voltage, it has the same zero-reference direction as the voltage. So I_C leads I_R by 90°, and I_L lags I_R by 90°.

The line current (I_T), or total current, is the vector sum of the three branch currents, and can be calculated by adding I_L, I_C, and I_R vectorially. Whether the line current leads or lags the applied voltage depends on which of the reactive branch currents (I_L or I_C) is the larger. If I_L is larger, I_T lags the applied voltage. If I_C is larger, I_T leads the applied voltage.

To determine the impedance of a parallel RLC circuit, first determine the net reactance X of the inductive and capacitive branches. Then use X to determine the impedance Z, the same as in a parallel RL or RC circuit.

Whenever Z is inductive, the line current will lag the applied voltage. Similarly, when Z is capacitive, the line current will lead the applied voltage.

9.4.4 Parallel Resonance

A parallel resonant circuit is a circuit in which the voltage source is in parallel with L and C. The characteristics of parallel resonance are quite different from those of series resonance. However, the frequency at which parallel resonance takes place is identical to the frequency at which series resonance takes place. Therefore, parallel resonance uses the same formula as series resonance:

$$f_r = \frac{1}{2\pi\sqrt{LC}}$$

The properties of a parallel resonant circuit are based on the action that takes place between the parallel inductance and capacitance, which is often called a tank circuit because it has the ability to store electrical energy.

The action of a tank circuit is basically one of interchange of energy between the inductance and capacitance. If a voltage is momentarily applied

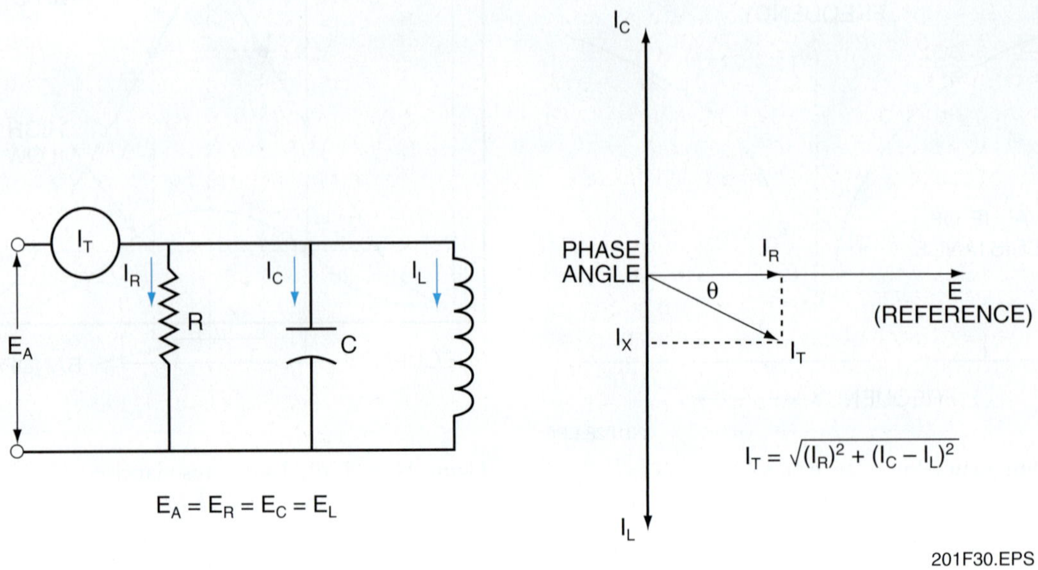

Figure 30 ◆ Parallel RLC circuit and vector diagram.

across the tank circuit, C charges to this voltage. When the applied voltage is removed, C discharges through L, and a magnetic field is built up around L by the discharge current. When C has discharged, the field around L collapses, and in doing so induces a current that is in the same direction as the current that created the field. This current, therefore, charges C in the opposite direction. When the field around L has collapsed, C again discharges, but this time in the direction opposite to before. The discharge current again causes a magnetic field around L, which, when it collapses, charges C in the same direction in which it was initially charged.

This interchange of energy and the circulating current it produces would continue indefinitely, producing a series of sine waves, if this were an ideal tank circuit with no resistance. However, since some resistance is always present, the circulating current gradually diminishes as the resistance dissipates the energy in the circuit in the form of heat. This causes the sine wave current to be damped out. If a voltage were again momentarily applied across the circuit, the interchange of energy and accompanying circulating current would begin again.

At resonance, X_L equals X_C, so the two currents I_L and I_C are also equal. Because the two currents in a parallel LC circuit are 180° out of phase, the line current, which is their vector sum, must be zero. Thus, the only current is the circulating current in the tank circuit. No line current flows, therefore the circuit has infinite impedance as far as the voltage source is concerned.

These two conditions of zero line current and infinite impedance are characteristic of ideal parallel resonant circuits at resonance. In practical circuits that contain some resistance, the theoretical conditions of zero line current and infinite impedance are not realized. Instead, practical parallel resonant circuits have minimum line current and maximum impedance at resonance. This is the exact opposite of series resonant circuits, which have maximum current and minimum impedance at resonance.

In the ideal parallel resonant circuit at resonance, the branch currents I_L and I_C are equal, so the line current is zero and the circuit impedance is infinite. Above and below the resonant frequency, one of the reactances X_L or X_C is larger than the other. The two branch currents are therefore unequal, and the line current, which equals their vector sum (or arithmetic difference), has some value greater than zero. Since line current flows, the circuit impedance is no longer infinite. The further the frequency is from the resonant frequency, the greater the difference between the values of the reactances. As a result, the line current is larger and the circuit impedance is smaller.

The principal effect of the resistance in a parallel resonant circuit is that it causes the current in the inductive branch to lag the applied voltage by a phase angle of less than 90°, instead of exactly 90° as in the case of the ideal circuit. As a result, the two branch currents are not 180° out of phase. For simplicity, resonance can still be considered as occurring when X_L equals X_C, but now when the two branch currents are added vectorially, their sum is not zero. This means that at resonance, some line current flows. Since there is line current, the impedance cannot be infinite, as it is in the ideal circuit. Thus at resonance, practical parallel resonant circuits have minimum line current and maximum resistance, instead of zero line current and infinite impedance, as do ideal circuits.

For parallel resonance, Q also measures the quality of a circuit. In parallel resonant circuits, Q depends on circuit resistance. The Q of a parallel resonant circuit is defined as:

$$Q = \frac{X_L}{R} \text{ or } \frac{X_C}{R}$$

Recognize this as the same equation used for the Q of a series resonant circuit. As a result, resistance has the same effect on the Q of a parallel resonant circuit as it does on a series resonant circuit. The lower the resistance, the higher the Q of the circuit and the narrower its bandpass. Conversely, the greater the resistance, the lower the Q and the wider the bandpass.

Recall that for every series resonant circuit there is a range of frequencies above and below the resonant frequency at which, for practical purposes, the circuit can be considered as being at resonance. This range of frequencies is called the bandwidth, and consists of all the frequencies at which the circuit current was 0.707 or more times its value at resonance. Parallel resonant circuits also have a bandwidth, but it is defined in terms of the frequency-vs.-impedance curve, and consists of all the frequencies that produce a circuit impedance 0.707 or more times the impedance at resonance. *Figure 31* shows the bandpass or bandwidth as all the frequencies between F_1 and F_2.

Circuit resistance affects the width and steepness of the frequency-impedance curve. Therefore, resistance affects the circuit bandpass. A low resistive circuit causes a steep curve and narrow bandpass. A high resistive circuit causes a flatter frequency-impedance curve and therefore, a wide bandpass.

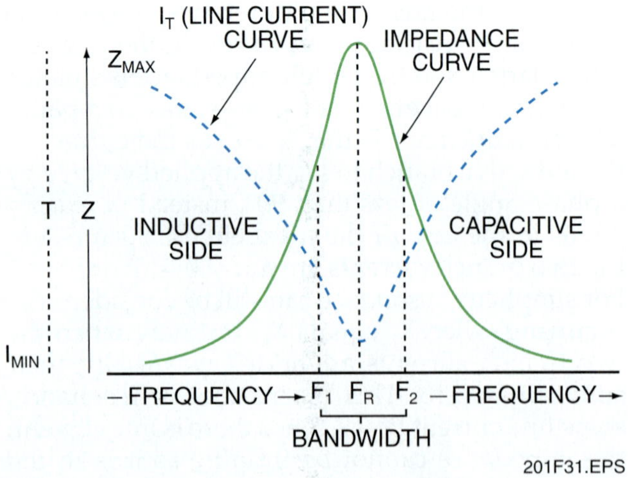

Figure 31 ◆ Tuned parallel circuit curves.

10.0.0 ◆ POWER IN AC CIRCUITS

In DC circuits, the power consumed is the sum of all the I^2R heating in the resistors. It is also equal to the power produced by the source, which is the product of the source voltage and current. In AC circuits containing only resistors, the above relationship also holds true.

10.1.0 True Power

The power consumed by resistance is called true power and is measured in units of watts. True power is the product of the resistor current squared and the resistance:

$$P_T = I^2R$$

This formula applies because current and voltage have the same phase across a resistance. To find the corresponding value of power as a product of voltage and current, this product must be multiplied by the cosine of the phase angle θ:

$$P_T = I^2R \text{ or } P_T = EI \cos \theta$$

Where E and I are in rms values to calculate the true power in watts, multiplying I by the cosine of the phase angle provides the resistive component for true power equal to I^2R.

For example, a series RL circuit has 2A through a 100Ω resistor in series with the X_L of 173Ω. Therefore:

$$P_T = I^2R$$
$$P_T = 4 \times 100$$
$$P_T = 400W$$

Furthermore, in this circuit the phase angle is 60° with a cosine of 0.5. The applied voltage is 400V. Therefore:

$$P_T = EI \cos \theta$$
$$P_T = 400 \times 2 \times 0.5$$
$$P_T = 400W$$

In both cases, the true power is the same (400W) because this is the amount of power supplied by the generator and dissipated in the resistance. Either formula can be used for calculating the true power.

10.2.0 Apparent Power

In ideal AC circuits containing resistors, capacitors, and inductors, the only mechanism for power consumption is $I^2_{eff}R$ heating in the resistors. Inductors and capacitors consume no power. The only function of inductors and capacitors is to store and release energy. However, because of the

INSIDE TRACK

The Importance of Resonance

A common use of RCL circuits is in resonant filter applications. Because RCL circuits provide a means for frequency selectivity, resonance in electrical circuits is very important to many types of electrical and electronic systems. Series and/or parallel RCL resonant filters are widely used to filter out or reduce the effects of unwanted harmonic frequencies and/or unwanted electromagnetic interference (EMI) signals present in power distribution circuits. Resonant filters are also used to increase the efficiency or power factor of a circuit or system.

In communications, the ability of a radio or television receiver to select a certain frequency transmitted by a certain station, and eliminate frequencies from other stations, is made possible by the use of series and/or parallel RCL resonant circuits.

phase shifts that are introduced by these elements, the power consumed by the resistors is not equal to the product of the source voltage and current. The product of the source voltage and current is called apparent power and has units of volt-amperes (VA).

The apparent power is the product of the source voltage and the total current. Therefore, apparent power is actual power delivered by the source. The formula for apparent power is:

$$P_A = (E_A)(I)$$

Figure 32 shows a series RL circuit and its associated vector diagram.

This circuit is used to calculate the apparent power and compare it to the circuit's true power:

$$P_A = (E_A)(I) \qquad P_T = EI\cos\theta$$

$$P_A = (400V)(2A) \qquad \theta = \frac{R}{X_L} = \frac{100}{173} = 60°$$

$$P_A = 800VA \qquad P_T = (400V)(2A)(\cos 60°)$$

$$P_T = (400V)(2A)(0.5)$$

$$P_T = 400W$$

Note that the apparent power formula is the product of EI alone without considering the cosine of the phase angle.

10.3.0 Reactive Power

Reactive power is that portion of the apparent power that is caused by inductors and capacitors in the circuit. Inductance and capacitance are always present in real AC circuits. No work is performed by reactive power; the power is stored in the inductors and capacitors, then returned to the circuit. Therefore, reactive power is always 90° out of phase with true power. The units for reactive power are volt-amperes-reactive (VARs).

In general, for any phase angle θ between E and I, multiplying EI by sine θ gives the vertical component at 90° for the value of the VARs. In *Figure 32*, the value of sine 60° is 800 × 0.866 = 692.8 VARs.

Note that the factor sine θ for the VARs gives the vertical or reactive component of the apparent power EI. However, multiplying EI by cosine θ as the power factor gives the horizontal or resistive component for the real power.

10.4.0 Power Factor

Because it indicates the resistive component, cosine θ is the power factor (pf) of the circuit, converting the EI product to real power. For series circuits, use the formula:

$$pf = \cos\theta = \frac{R}{Z}$$

For parallel circuits, use the formula:

$$pf = \cos\theta = \frac{I_R}{I_T}$$

In *Figure 32* as an example of a series circuit, R and Z are used for the calculations:

$$pf = \cos\theta = \frac{R}{Z} = \frac{100\Omega}{200\Omega} = 0.5$$

The power factor is not an angular measure but a numerical ratio with a value between 0 and 1, equal to the cosine of the phase angle. With all resistance and zero reactance, R and Z are the same for a series circuit of I_R and I_T and are the same for a parallel circuit. The ratio is 1. Therefore, unity power factor means a resistive circuit. At the opposite extreme, all reactance with zero resistance makes the power factor zero, meaning that the circuit is all reactive.

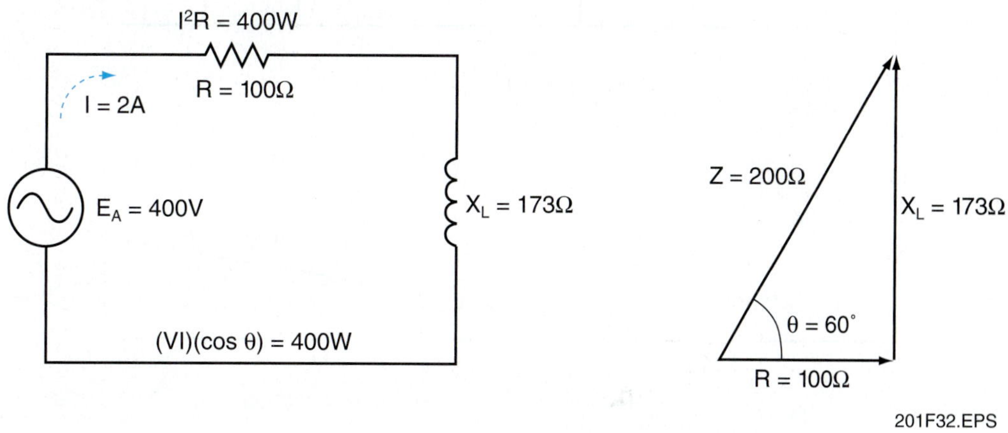

Figure 32 ◆ Power calculations in an AC circuit.

The power factor gives the relationship between apparent power and true power. The power factor can thus be defined as the ratio of true power to apparent power:

$$pf = \frac{P_T}{P_A}$$

For example, calculate the power factor of the circuit shown in *Figure 33.*

The true power is the product of the resistor current squared and the resistance:

$$P_T = I^2R$$
$$P_T = 10A^2 \times 10\Omega$$
$$P_T = 1{,}000W$$

The apparent power is the product of the source voltage and total current:

$$P_A = (I_T)(E)$$
$$P_A = 10.2A \times 100V$$
$$P_A = 1{,}020VA$$

Calculating total current:

$$I_T = \sqrt{I_R^2 + (I_C - I_L)^2} = \sqrt{10A^2 + (4A - 2A)^2}$$
$$I_T = 10.2A$$

The power factor is the ratio of true power to apparent power:

$$pf = \frac{P_T}{P_A}$$
$$pf = \frac{1{,}000}{1{,}020}$$
$$pf = 0.98$$

As illustrated in the previous example, the power factor is determined by the system load. If the load contained only resistance, the apparent power would equal the true power and the power factor would be at its maximum value of one. Purely resistive circuits have a power factor of unity or one. If the load is more inductive than capacitive, the apparent power will lag the true power and the power factor will be lagging. If the load is more capacitive than inductive, the apparent power will lead the true power and the power factor will be leading. If there is any reactive load on the system, the apparent power will be greater than the true power and the power factor will be less than one.

10.5.0 Power Triangle

The phase relationships among the three types of AC power are easily visualized on the power triangle shown in *Figure 34*. The true power (W) is the horizontal leg, the apparent power (VA) is the

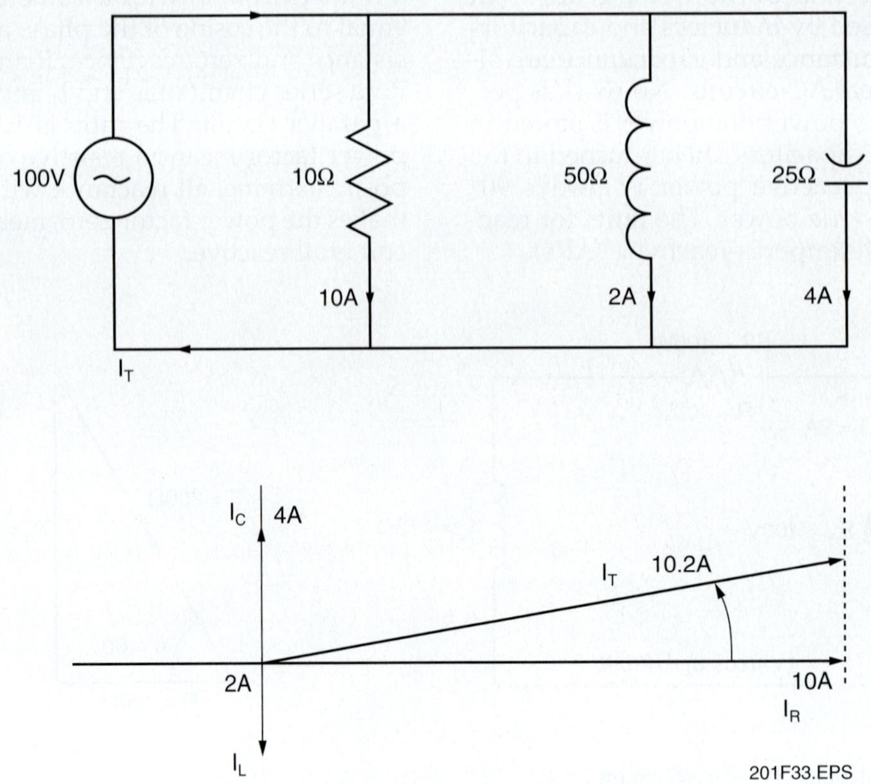

Figure 33 ◆ RLC circuit calculation.

$$P_A = \sqrt{(P_T^2) + (P_{RX}^2)}$$

$$P_T = \sqrt{(P_A^2) - (P_{RX}^2)}$$

$$P_{RX} = \sqrt{(P_A^2) - (P_T^2)}$$

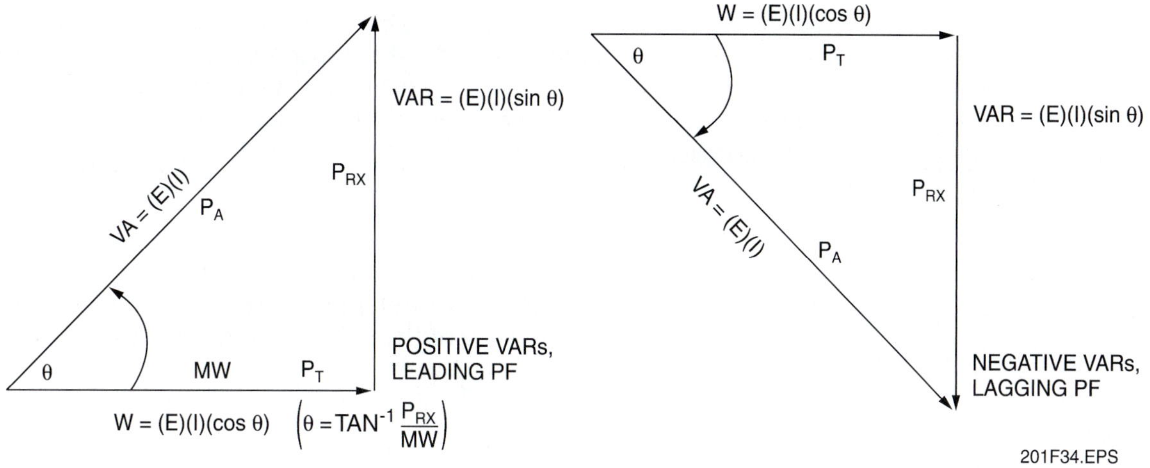

Figure 34 ◆ Power triangle.

hypotenuse, and the cosine of the phase angle between them is the power factor. The vertical leg of the triangle is the reactive power and has units of volt-amperes-reactive (VARs).

As illustrated on the power triangle (*Figure 34*), the apparent power will always be greater than the true power or reactive power. Also, the apparent power is the result of the vector addition of true and reactive power. The power magnitude relationships shown in *Figure 34* can be derived from the Pythagorean theorem for right triangles:

$$c^2 = a^2 + b^2$$

Therefore, c also equals the square root of $a^2 + b^2$, as shown below:

$$c = \sqrt{a^2 + b^2}$$

11.0.0 ◆ TRANSFORMERS

A transformer is a device that transfers electrical energy from one circuit to another by electromagnetic induction (transformer action). The electrical energy is always transferred without a change in frequency, but may involve changes in the magnitudes of voltage and current. Because a transformer works on the principle of electromagnetic induction, it must be used with an input source voltage that varies in amplitude.

THINK ABOUT IT

Power Factor

In power distribution circuits, it is desirable to achieve a power factor approaching a value of 1 in order to obtain the most efficient transfer of power. In AC circuits where there are large inductive loads such as in motors and transformers, the power factor can be considerably less than 1. For example, in a highly inductive motor circuit, if the voltage is 120V, the current is 12A, and the current lags the voltage by 60°, the power factor is 0.5 or 50% (cosine of 60° = 0.5). The apparent power is 1,440VA (120V × 12A), but the true power is only 720W [120V × (0.5 × 12A) = 720W]. This is a very inefficient circuit. What would you do to this circuit in order to achieve a circuit having a power factor as close to 1 as possible?

Transformers

INSIDE TRACK

Transformers are essential to all electrical systems and all types of electronic equipment. They are especially essential to the operation of AC high-voltage power distribution systems. Transformers are used to both step up voltage and step down voltage throughout the distribution process. For example, a typical power generation plant might generate AC power at 13,800V, step it up to 230,000V for distribution over long transmission lines, step it down to 13,800V again at substations located at different points for local distribution, and finally step it down again to 240V and 120V for lighting and local power use.

11.1.0 Transformer Construction

Figure 35 shows the basic components of a transformer. In its most basic form, a transformer consists of:

- A primary coil or winding
- A secondary coil or winding
- A core that supports the coils or windings

A simple transformer action is shown in *Figure 36*. The primary winding is connected to a 60Hz AC voltage source. The magnetic field or flux builds up (expands) and collapses (contracts) around the primary winding. The expanding and contracting magnetic field around the primary winding cuts the secondary winding and induces an alternating voltage into the winding. This voltage causes AC to flow through the load. The voltage may be stepped up or down depending on the design of the primary and secondary windings.

11.1.1 Core Characteristics

Commonly used core materials are air, soft iron, and steel. Each of these materials is suitable for particular applications and unsuitable for others. Generally, air-core transformers are used when the voltage source has a high frequency (above 20kHz). Iron-core transformers are usually used when the source frequency is low (below 20kHz). A soft-iron transformer is very useful where the transformer must be physically small yet efficient. The iron-core transformer provides better power transfer than the air-core transformer. Laminated

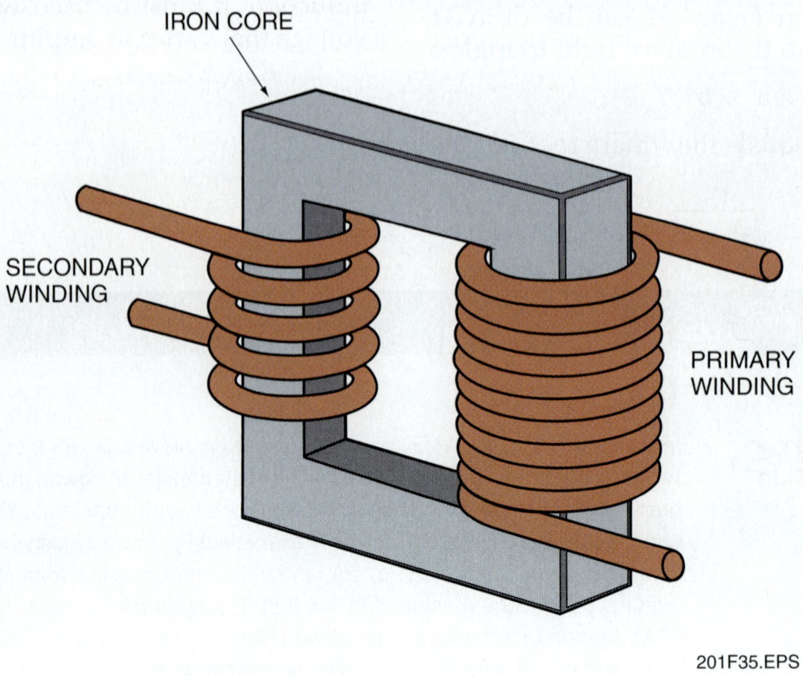

Figure 35 ◆ Basic components of a transformer.

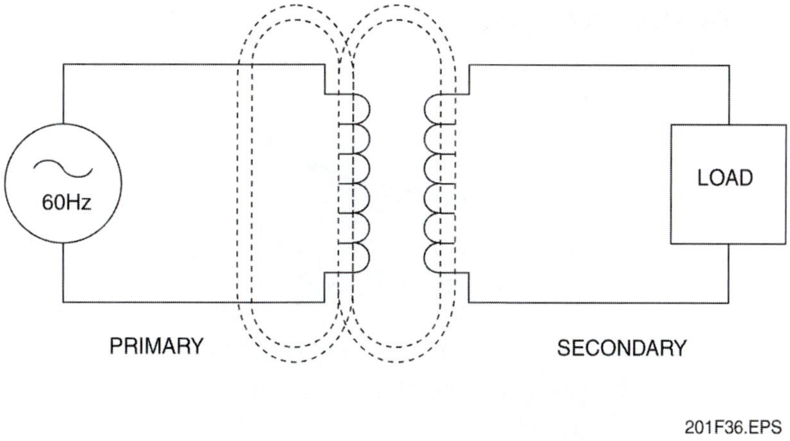

Figure 36 ◆ Transformer action.

sheets of steel are often used in a transformer to reduce one type of power loss known as eddy currents. These are undesirable currents, induced into the core, which circulate around the core. Laminating the core reduces these currents to smaller levels. These steel laminations are insulated with a nonconducting material, such as varnish, and then formed into a core as shown in *Figure 37*. It takes about 50 such laminations to make a core one-inch thick. The most efficient transformer core is one that offers the best path for the most lines of flux, with the least loss in magnetic and electrical energy.

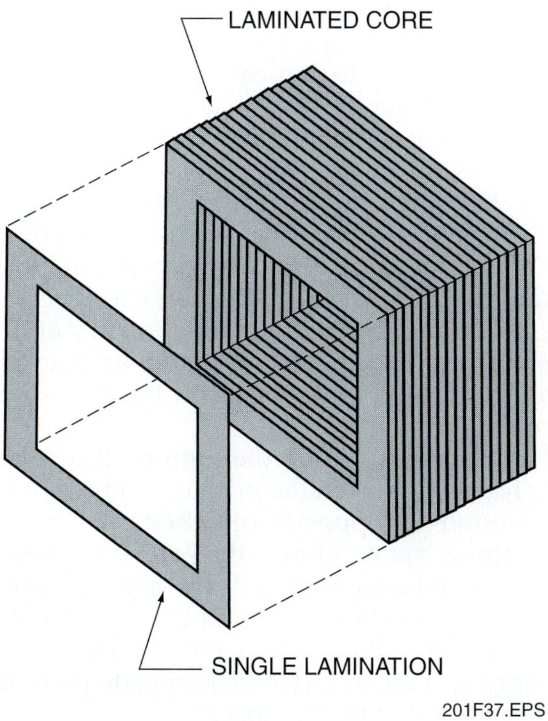

Figure 37 ◆ Steel laminated core.

11.1.2 Transformer Windings

A transformer consists of two coils called windings, which are wrapped around a core. The transformer operates when a source of AC voltage is connected to one of the windings and a load device is connected to the other. The winding that is connected to the source is called the primary winding. The winding that is connected to the load is called the secondary winding. *Figure 38* shows a cutaway view of a typical transformer.

The wire is coated with varnish so that each turn of the winding is insulated from every other turn. In a transformer designed for high-voltage applications, sheets of insulating material such as paper are placed between the layers of windings to provide additional insulation.

When the primary winding is completely wound, it is wrapped in insulating paper or cloth. The secondary winding is then wound on top of the primary winding. After the secondary winding is complete, it too is covered with insulating paper. Next, the iron core is inserted into and around the windings as shown.

Sometimes, terminals may be provided on the enclosure for connections to the windings. The figure shows four leads, two from the primary and two from the secondary. These leads are to be connected to the source and load, respectively.

11.2.0 Operating Characteristics

11.2.1 Energized with No Load

A no-load condition is said to exist when a voltage is applied to the primary, but no load is connected to the secondary. Assume the output of the secondary is connected to a load by a switch that is open. Because of the open switch, there is no current flowing in the secondary winding. With the

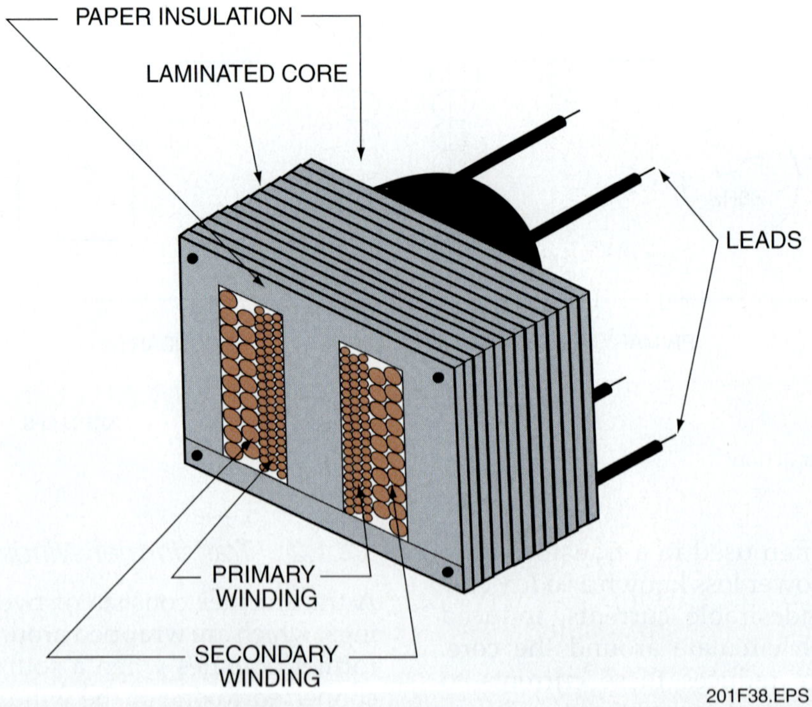

Figure 38 ◆ Cutaway view of a transformer core.

switch open and an AC voltage applied to the primary, there is, however, a very small amount of current, called exciting current, flowing in the primary. Essentially, what this current does is excite the coil of the primary to create a magnetic field. The amount of exciting current is determined by three factors: the amount of voltage applied (E_A); the resistance (R) of the primary coil's wire and core losses; and the X_L, which is dependent on the frequency of the exciting current. These factors are all controlled by transformer design.

This very small amount of exciting current serves two functions:

- Most of the exciting energy is used to support the magnetic field of the primary.
- A small amount of energy is used to overcome the resistance of the wire and core. This is dissipated in the form of heat (power loss).

Exciting current will flow in the primary winding at all times to maintain this magnetic field, but no transfer of energy will take place as long as the secondary circuit is open.

11.2.2 Phase Relationship

The secondary voltage of a simple transformer may be either in phase or out of phase with the primary voltage. This depends on the direction in which the windings are wound and the arrangement of the connection to the external circuit (load). Simply, this means that the two voltages may rise and fall together, or one may rise while the other is falling. Transformers in which the secondary voltage is in phase with the primary are referred to as like-wound transformers, while those in which the voltages are 180° out of phase are called unlike-wound transformers.

Dots are used to indicate points on a transformer schematic symbol that have the same instantaneous polarity (points that are in phase). The use of phase-indicating dots is illustrated in *Figure 39*. In the first part of the figure, both the primary and secondary windings are wound from top to bottom in a clockwise direction, as viewed from above the windings. When constructed in this manner, the top lead of the primary and the top lead of the secondary have the same polarity. This is indicated by the dots on the transformer symbol.

The second part of the figure illustrates a transformer in which the primary and secondary are wound in opposite directions. As viewed from above the windings, the primary is wound in a clockwise direction from top to bottom, while the secondary is wound in a counterclockwise direction. Notice that the top leads of the primary and secondary have opposite polarities. This is indicated by the dots being placed on opposite ends of the transformer symbol. Thus, the

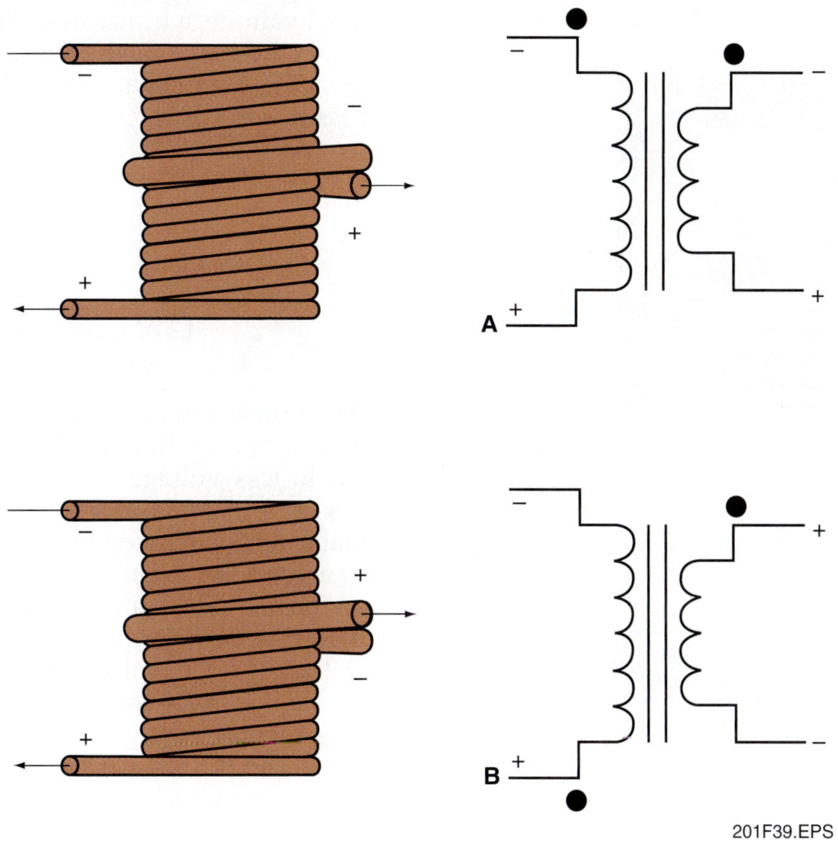

Figure 39 ◆ Transformer winding polarity.

polarity of voltage at the terminals of the transformer secondary depends on the direction in which the secondary is wound with respect to the primary.

11.3.0 Turns and Voltage Ratios

To understand how a transformer can be used to step up or step down voltage, the term turns ratio must be understood. The total voltage induced into the secondary winding of a transformer is determined mainly by the ratio of the number of turns in the primary to the number of turns in the secondary, and by the amount of voltage applied to the primary. Therefore, to set up a formula:

$$\text{Turns ratio} = \frac{\text{Number of turns in the primary}}{\text{Number of turns in the secondary}}$$

The first transformer in *Figure 40* shows a transformer whose primary consists of 10 turns of wire, and whose secondary consists of a single turn of wire. As lines of flux generated by the primary expand and collapse, they cut both the 10 turns of the primary and the single turn of the secondary. Since the length of the wire in the secondary is approximately the same as the length of the wire in each turn of the primary, the EMF induced into the secondary will be the same as the EMF induced into each turn of the primary.

This means that if the voltage applied to the primary winding is 10 volts, the CEMF in the primary is almost 10 volts. Thus, each turn in the primary will have an induced CEMF of approximately $\frac{1}{10}$ of the total applied voltage, or one volt. Since the same flux lines cut the turns in both the secondary and the primary, each turn will have an EMF of one volt induced into it. The first transformer in *Figure 40* has only one turn in the secondary, thus, the EMF across the secondary is one volt.

The second transformer represented in *Figure 40* has a 10-turn primary and a two-turn secondary. Since the flux induces one volt per turn, the total voltage across the secondary is two volts. Notice that the volts per turn are the same for both primary and secondary windings. Since the CEMF in the primary is equal (or almost) to the applied voltage, a proportion may be set up to express the value of the voltage induced in terms of the voltage applied to the primary and the number of turns in each winding. This proportion also

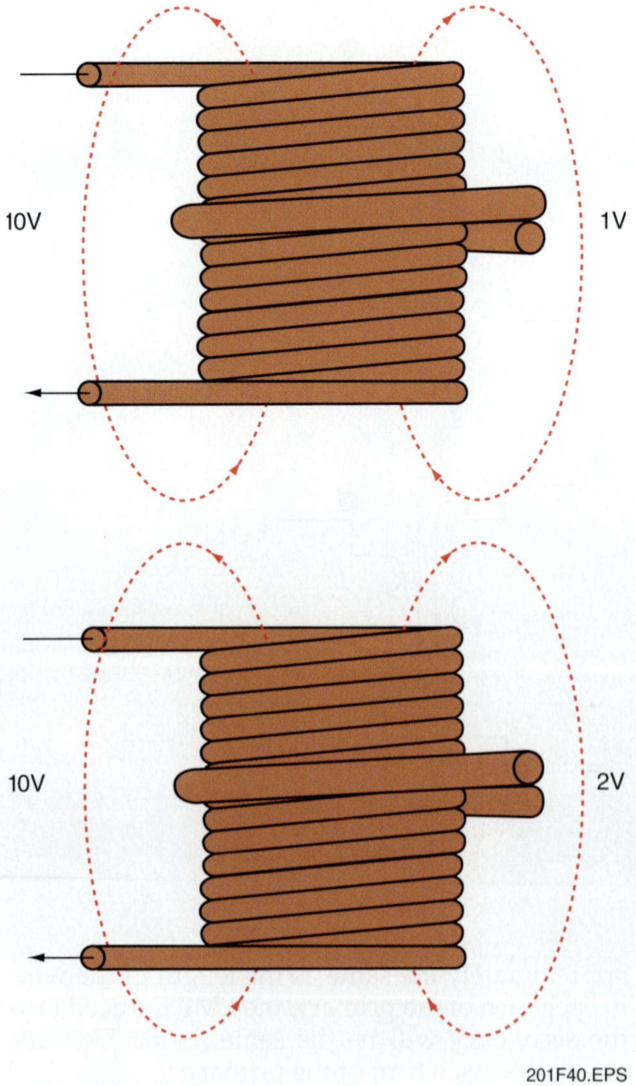

Figure 40 ◆ Transformer turns ratio.

shows the relationship between the number of turns in each winding and the voltage across each winding, and is expressed by the equation:

$$\frac{E_S}{E_P} = \frac{N_S}{N_P}$$

Where:

N_P = number of turns in the primary
E_P = voltage applied to the primary
E_S = voltage induced in the secondary
N_S = number of turns in the secondary

The equation shows that the ratio of secondary voltage to primary voltage is equal to the ratio of secondary turns to primary turns. The equation can be written as:

$$E_P N_S = E_S N_P$$

For example, a transformer has 100 turns in the primary, 50 turns in the secondary, and 120VAC applied to the primary (E_P). What is the voltage across the secondary (E_S)?

$$N_P = 100\text{ turns} \qquad E_P = 120\text{VAC}$$
$$N_S = 50\text{ turns}$$
$$\frac{E_S}{E_P} = \frac{N_S}{N_P} \textit{ or } E_S = \frac{E_P N_S}{N_P}$$
$$E_S = \frac{120\text{ V} \times 50\text{ turns}}{100\text{ turns}} = 60\text{VAC}$$

The transformers in *Figure 40* have fewer turns in the secondary than in the primary. As a result, there is less voltage across the secondary than across the primary. A transformer in which the voltage across the secondary is less than the voltage across the primary is called a step-down transformer. The ratio of a 10-to-1 step-down transformer is written as 10:1.

A transformer that has fewer turns in the primary than in the secondary will produce a greater voltage across the secondary than the voltage applied to the primary. A transformer in which the voltage across the secondary is greater than the voltage applied to the primary is called a step-up transformer. The ratio of a 1-to-4 step-up transformer should be written 1:4. Notice in the two ratios that the value of the primary winding is always stated first.

11.4.0 Types of Transformers

Transformers are widely used to permit the use of trip coils and instruments of moderate current and voltage capacities and to measure the characteristics of high-voltage and high-current circuits. Since secondary voltage and current are directly related to primary voltage and current, measurements can be made under the low-voltage or low-current conditions of the secondary circuit and still determine primary characteristics. Tripping transformers and instrument transformers are examples of this use of transformers.

The primary or secondary coils of a transformer can be tapped to permit multiple input and output voltages. *Figure 41* shows several tapped transformers. The center-tapped transformer is particularly important because it can be used in conjunction with other components to convert an AC input to a DC output.

11.4.1 Isolation Transformer

Isolation transformers are wound so that their primary and secondary voltages are equal. Their

Turns and Voltage Ratios

THINK ABOUT IT

What is the magnitude of the voltage and current supplied by the secondary of the transformer in the circuit shown below?

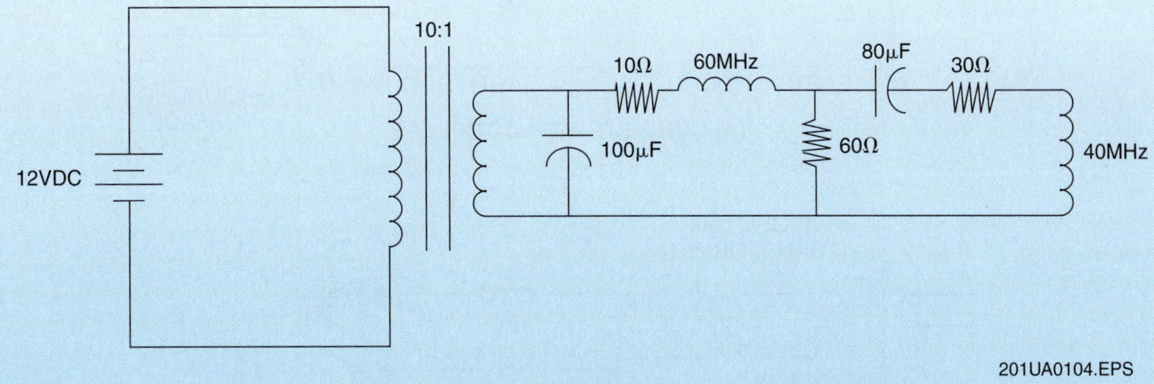

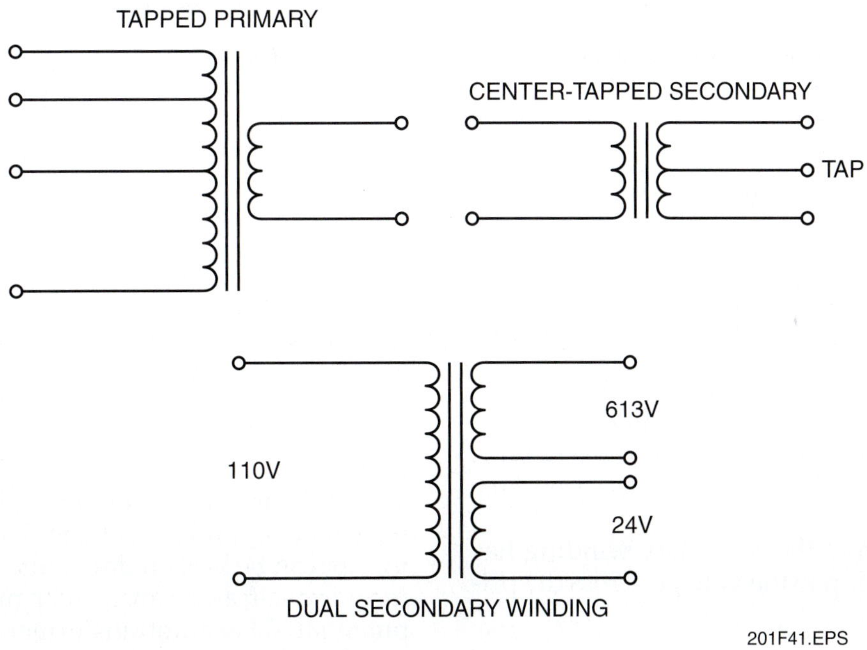

Figure 41 ◆ Tapped transformers.

purpose is to electrically isolate a piece of electrical equipment from the power distribution system.

Many pieces of electronic equipment use the metal chassis on which the components are mounted as part of the circuit (*Figure 42*). Personnel working with this equipment may accidentally come in contact with the chassis, completing the circuit to ground, and receive a shock as shown in *Figure 42(A)*. If the resistances of their body and the ground path are low, the shock can be fatal. Placing an isolation transformer in the circuit as shown in *Figure 42(B)* breaks the ground current path that includes the worker. Current can no longer flow from the power supply through the chassis and worker to ground; however, the equipment is still supplied with the normal operating voltage and current.

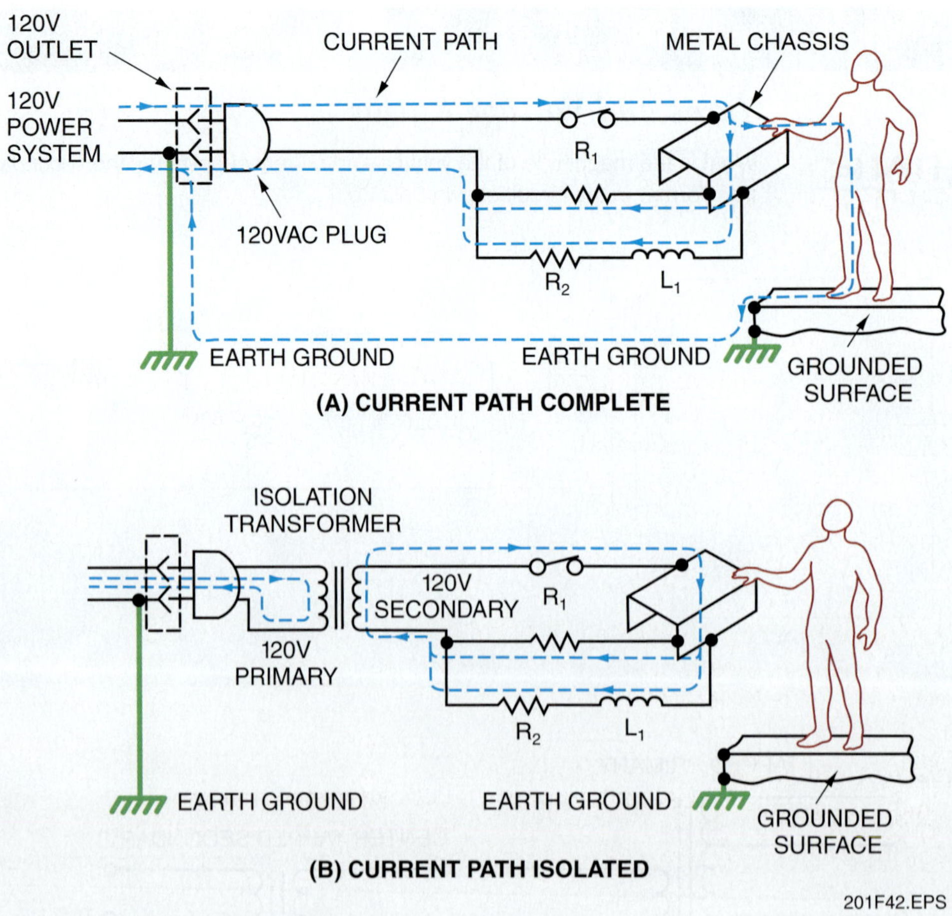

Figure 42 ◆ Importance of an isolation transformer.

11.4.2 Autotransformer

In a transformer, it is not necessary for the primary and secondary to be separate and distinct windings. *Figure 43* is a schematic diagram of what is known as an autotransformer. Note that a single coil of wire is tapped to produce what is electrically both a primary and a secondary winding.

The voltage across the secondary winding has the same relationship to the voltage across the primary that it would have if they were two distinct windings. The movable tap in the secondary is used to select a value of output voltage either higher or lower than E_P, within the range of the transformer. When the tap is at Point A, E_S is less than E_P; when the tap is at Point B, E_S is greater than E_P.

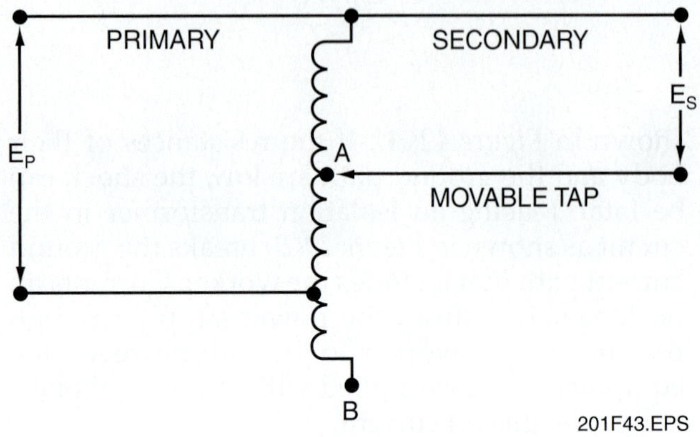

Figure 43 ◆ Autotransformer schematic diagram.

Autotransformers rely on self-induction to induce their secondary voltage. The term autotransformer can be broken down into two words: auto, meaning self; and transformer, meaning to change potential. The autotransformer is made of one winding that acts as both a primary and a secondary winding. It may be used as either a step-up or step-down transformer. Some common uses of autotransformers are as variable AC voltage supplies and fluorescent light ballast transformers, and to reduce the line voltage for various types of low-voltage motor starters.

11.4.3 Current Transformer

A current transformer differs from other transformers in that the primary is a conductor to the load and the secondary is a coil wrapped around the wire to the load. Just as any ammeter is connected in line with a circuit, the current transformer is connected in series with the current to be measured. *Figure 44* is a diagram of a current transformer.

WARNING!
Do not open a current transformer under load.

Since current transformers are series transformers, the usual voltage and current relationships do not apply. Current transformers vary considerably in rated primary current, but are usually designed with ampere-turn ratios such that the secondary delivers five amperes at full primary load.

Current transformers are generally constructed with only a few turns or no turns in the primary. The voltage in the secondary is induced by the changing magnetic field that exists around a single conductor. The secondary is wound on a circular core, and the large conductor that makes up the primary passes through the hole in its center. Because the primary has few or no turns, the secondary must have many turns (providing a high turns ratio) in order to produce a usable voltage. The advantage of this is that you get an output off the secondary proportional to the current flowing through the primary, without an appreciable voltage drop across the primary. This is because the primary voltage equals the current times the impedance. The impedance is kept near zero by using no or very few primary turns. The disadvantage is that you cannot open the secondary circuit with the primary energized. To do so would cause the secondary current to drop rapidly to zero. This would cause the magnetic field generated by the secondary current to collapse rapidly. The rapid collapse of the secondary field through the many turns of the secondary winding would induce a dangerously high voltage in the secondary, creating an equipment and personnel hazard.

Because the output of current transformers is proportional to the current in the primary, they are most often used to power current-sensing meters and relays. This allows the instruments to respond to primary current without having to handle extreme magnitudes of current.

11.4.4 Potential Transformer

The primary of a potential transformer is connected across or in parallel with the voltage to be measured, just as a voltmeter is connected across a circuit. *Figure 45* shows the schematic diagram for a potential transformer.

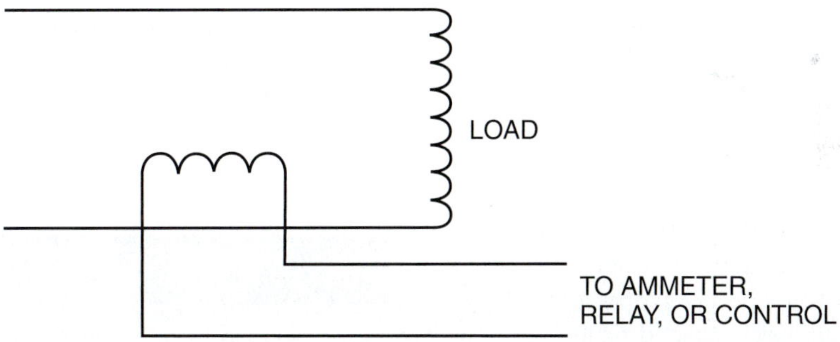

Figure 44 ◆ Current transformer schematic diagram.

INSIDE TRACK

Isolation Transformers

In addition to being used to protect personnel from receiving electrical shocks, shielded isolation transformers are widely used to prevent electrical disturbances on power lines from being transmitted into related load circuits. The shielded isolation transformer has a grounded electrostatic shield between the primary and secondary windings that acts to direct unwanted signals to ground.

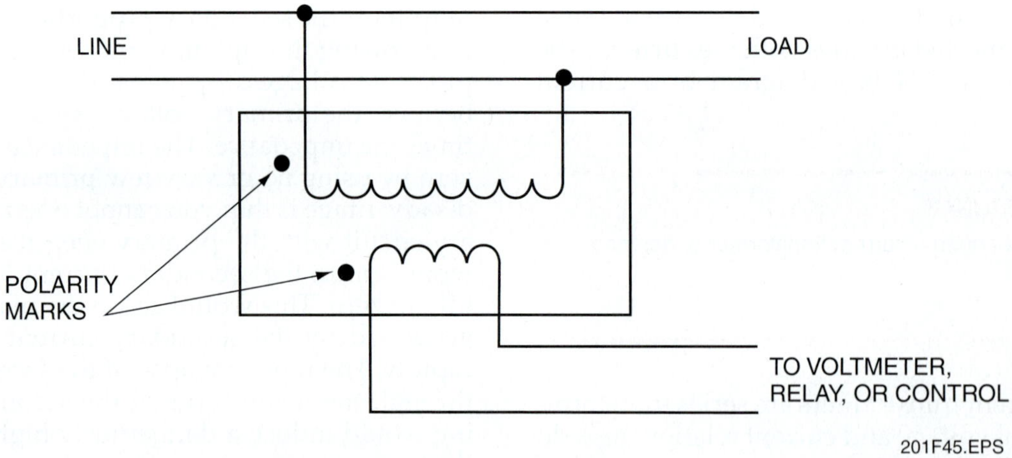

Figure 45 ◆ Potential transformer.

Potential transformers are basically the same as any other single-phase transformer. Although primary voltage ratings vary widely according to the specific application, secondary voltage ratings are usually 120V, a convenient voltage for meters and relays.

Because the output of potential transformers is proportional to the phase-to-phase voltage of the primary, they are often used to power voltage-sensing meters and relays. This allows the instruments to respond to primary voltage while having to handle only 120V. Also, potential transformers are essentially single-phase step-down transformers. Therefore, power to operate low-voltage auxiliary equipment associated with high-voltage switchgear can be supplied off the high-voltage lines that the equipment serves via potential transformers.

CASE HISTORY

AC Power Provides Many Benefits, but Working with It Can Be Dangerous

Working with AC power can be dangerous unless proper safety methods and procedures are followed. The National Institute for Occupational Safety and Health (NIOSH) investigated 224 incidents of electrocutions that resulted in occupational fatalities. One hundred twenty-one of the victims were employed in the construction industry. Two hundred twenty-one of the incidents (99%) involved AC. Of the 221 AC electrocutions, 74 (33%) involved AC voltages less than 600V and 147 (66%) involved 600V or more. Forty of the lower-voltage electrocutions involved 120/240V.

Factors relating to the causes of these electrocutions included the lack of enforcement of existing employer policies including the use of personal protective equipment and the lack of supervisory intervention when existing policies were being violated. Of the 224 victims, 194 (80%) had some type of electrical safety training. Thirty-nine victims had no training at all. It is notable that 100 of the victims had been on the job less than one year.

The Bottom Line: Never assume that you are safe when working at lower voltages. All voltage levels must be considered potentially lethal. Also, safety training does no good if you don't put it into practice every day. Always put safety first.

THINK ABOUT IT

Putting It All Together

A power company's distribution system has capacitor banks that are automatically switched into the system by a temperature switch during hot weather. Why?

INSIDE TRACK

Current Transformers

Although the use of a current transformer completely isolates the secondary and the related ammeter from the high-voltage lines, the secondary of a current transformer should never be left open circuited. To do so may result in dangerously high voltage being induced in the secondary.

Potential Transformers

In addition to being used to step down high voltages for the purpose of safe metering, potential transformers are widely used in all kinds of control devices where the condition of high voltages must be monitored. One such example involves the use of a potential transformer-operated contactor in emergency lighting standby generator circuits. Under normal conditions with utility power applied, the contactor is energized and its normally closed contacts are open. If the power fails, the contactor de-energizes, causing its contacts to close and activating the standby generator circuit.

Review Questions

1. An electric current always produces _____.
   a. mutual inductance
   b. a magnetic field
   c. capacitive reactance
   d. high voltage

2. The number of cycles an alternating electric current undergoes per second is known as _____.
   a. amperage
   b. frequency
   c. voltage
   d. resistance

3. What is the peak voltage in a circuit with an rms voltage of 120VAC?
   a. 117 volts
   b. 120 volts
   c. 150 volts
   d. 170 volts

4. Which of the following conditions exist in a circuit of pure resistance?
   a. The voltage and current are in phase.
   b. The voltage and current are 90° out of phase.
   c. The voltage and current are 120° out of phase.
   d. The voltage and current are 180° out of phase.

5. When the current increases in an AC circuit, what role does inductance play?
   a. It increases the current.
   b. It plays no role at all.
   c. It causes the overcurrent protection to open.
   d. It reduces the current.

6. Which of the following conditions exist in a circuit of pure inductance?
   a. The voltage and current are in phase.
   b. The voltage and current are 90° out of phase.
   c. The voltage and current are 120° out of phase.
   d. The voltage and current are 180° out of phase.

7. The opposition to current flow offered by the capacitance of a circuit is known as _____.
   a. mutual inductance
   b. pure resistance
   c. inductive reactance
   d. capacitive reactance

8. The total opposition to current flow in an AC circuit is known as _____.
   a. resistance
   b. capacitive reactance
   c. inductive reactance
   d. impedance

9. A power factor is not an angular measure, but a numerical ratio with a value between 0 and 1, equal to the _____ of the phase angle.
   a. sine
   b. tangent
   c. cosine
   d. cotangent

10. The two windings of a conventional transformer are known as the _____ windings.
    a. mutual and inductive
    b. high and low voltage
    c. primary and secondary
    d. step-up and step-down

Summary

The process by which current is produced electromagnetically is called induction. As the conductor moves across the magnetic field, it cuts the lines of force, and electrons within the conductor flow, creating an electromotive force (EMF). EMF is also known as voltage. There are three conditions that must exist before a current can be produced in this way:

- There must be a magnetic field through which the conductor can pass.
- There must be a conductor in which the voltage will be produced, and the conductor should be perpendicular to the field.
- There must be motion. Either the magnetic field or the conductor must move.

Several factors control the magnitude of the induced current. Voltage will be increased if:

- The speed with which the conductor cuts through the magnetic field is increased (the faster the conductor cuts through the field, the greater the current pulse)
- The strength of the field is increased (the stronger the field, the greater the current pulse)
- The conductor is wound to form a coil (the voltage increases directly with the number of turns of the coil)

A decrease in voltage occurs as the conductor intersects the magnetic field at an angle less than 90°. The greatest current is produced when the conductor intersects the magnetic field at right angles (perpendicular) to the flux lines.

It should be emphasized that a current may be induced by using the magnetic field of a permanent magnet or the magnetic field of another current-carrying conductor (electromagnet).

The magnetic field is among the reasons why phases of current-carrying conductors should not be separated in a raceway; all phase conductors (including the neutral) should be contained in the same raceway. For example, if phase A is separated from phase B and phase C by a metal enclosure, conduit wall, etc., the magnetic field around the conductors will cut across the conduit, causing the conduit to heat up.

Notes

Trade Terms Introduced in This Module

Capacitance: The storage of electricity in a capacitor; capacitance produces an opposition to voltage change. The unit of measurement for capacitance is the farad (F) or microfarad (μF).

Frequency: The number of cycles an alternating electric current, sound wave, or vibrating object undergoes per second.

Hertz (Hz): A unit of frequency; one hertz equals one cycle per second.

Impedance: The opposition to current flow in an AC circuit; impedance includes resistance (R), capacitive reactance (X_C), and inductive reactance (X_L). Impedance is measured in ohms.

Inductance: The creation of a voltage due to a time-varying current; also, the opposition to current change, causing current changes to lag behind voltage changes. The unit of measure for inductance is the henry (H).

Micro: Prefix designating one-millionth of a unit. For example, one microfarad is one-millionth of a farad.

Peak voltage: The peak value of a sinusoidally varying (cyclical) voltage or current is equal to the root-mean-square (rms) value multiplied by the square root of two (1.414). AC voltages are usually expressed as rms values; that is, 120 volts, 208 volts, 240 volts, 277 volts, 480 volts, etc., are all rms values. The peak voltage, however, differs. For example, the peak value of 120 volts (rms) is actually $120 \times 1.414 = 169.71$ volts.

Radian: An angle at the center of a circle, subtending (opposite to) an arc of the circle that is equal in length to the radius.

Reactance: The imaginary part of impedance. Also, the opposition to alternating current (AC) due to capacitance (X_C) and/or inductance (X_L).

Resonance: A condition reached in an electrical circuit when the inductive reactance neutralizes the capacitance reactance, leaving ohmic resistance as the only opposition to the flow of current.

Root-mean-square (rms): The square root of the average of the square of the function taken throughout the period. The rms value of a sinusoidally varying voltage or current is the effective value of the voltage or current.

Self-inductance: A magnetic field induced in the conductor carrying the current.

Additional Resources

This module is intended to present thorough resources for task training. The following reference works are suggested for further study. These are optional materials for continued education rather than for task training.

Introduction to Electric Circuits, Latest Edition. New York: Prentice Hall.

Principles of Electric Circuits, Latest Edition. New York: Prentice Hall.

CONTREN® LEARNING SERIES — USER FEEDBACK

The NCCER makes every effort to keep these textbooks up-to-date and free of technical errors. We appreciate your help in this process. If you have an idea for improving this textbook, or if you find an error, a typographical mistake, or an inaccuracy in NCCER's *Contren®* textbooks, please write us, using this form or a photocopy. Be sure to include the exact module number, page number, a detailed description, and the correction, if applicable. Your input will be brought to the attention of the Technical Review Committee. Thank you for your assistance.

Instructors – If you found that additional materials were necessary in order to teach this module effectively, please let us know so that we may include them in the Equipment/Materials list in the Annotated Instructor's Guide.

Write: Product Development
National Center for Construction Education and Research
P.O. Box 141104, Gainesville, FL 32614-1104

Fax: 352-334-0932

E-mail: curriculum@nccer.org

Craft ______ Module Name ______

Copyright Date ______ Module Number ______ Page Number(s) ______

Description ______

______

______

______

(Optional) Correction ______

______

______

(Optional) Your Name and Address ______

______

______

Motors: Theory and Application

26202-05

Lake House Spa & Pool Barn
at Lake Austin Spa Resort
Austin, Texas
Commercial $5–10 Million Award Winner
Spaw Glass Contractors, Inc.

26202-05

Motors: Theory and Application

Topics to be presented in this module include:

Overview

Motor windings are electromagnets in which like polarities repel one another and unlike polarities attract each other. The rotor in any motor is the part that turns and is used to drive the load. The stationary windings are typically mounted in a doughnut-like arrangement into which the rotor is installed.

Windings are built into the rotor, and current flows through these windings. Likewise, a current flow is created through the stationary windings. Any time current flows through a conductor, a magnetic field is created around that conductor. Since both windings have current flowing through them at the same time, magnetic fields are generated around both windings. As these magnetic fields interact, they either repel or attract each other. Manipulating the current flow through the rotor and the secondary windings enables the rotor to turn at controlled speeds or in different directions. This is the basic operation of any motor. Motors can be designed to supply high speed, high torque, or both, depending on the demands of the load.

Note: *National Electrical Code*® and *NEC*® are registered trademarks of the National Fire Protection Association, Inc., Quincy, MA 02269. All *National Electrical Code*® and *NEC*® references in this module refer to the 2005 edition of the *National Electrical Code*®.

Objectives

Upon completion of this module, you will be able to do the following:

1. Define the following terms:
   - Ampacity
   - Branch circuit
   - Circuit breaker
   - Controller
   - Duty
   - Equipment
   - Full-load amps
   - Ground fault circuit interrupter
   - Interrupting rating
   - Motor circuit switch
   - Thermal protector
   - NEMA design letter
   - Nonautomatic
   - Overcurrent
   - Overload
   - Power factor
   - Rated full-load speed
   - Rated horsepower
   - Remote control circuit
   - Service factor
   - Thermal cutout
2. Describe the various types of motor enclosures.
3. Describe how the rated voltage of a motor differs from the system voltage.
4. Describe the basic construction and components of a three-phase squirrel cage induction motor.
5. Explain the relationships among speed, frequency, and the number of poles in a three-phase induction motor.
6. Describe how torque is developed in an induction motor.
7. Explain how and why torque varies with rotor reactance and slip.
8. Define percent slip and speed regulation.
9. Explain how the direction of a three-phase motor is reversed.
10. Describe the component parts and operating characteristics of a three-phase wound-rotor induction motor.
11. Describe the component parts and operating characteristics of a three-phase synchronous motor.
12. Define torque, starting current, and armature reaction as they apply to DC motors.
13. Explain how the direction of rotation of a DC motor is changed.
14. Describe the design and characteristics of a DC shunt, series, and compound motor.
15. Describe dual-voltage motors and their applications.
16. Describe the methods for determining various motor connections.
17. Describe general motor protection requirements as delineated in the *NEC*®.

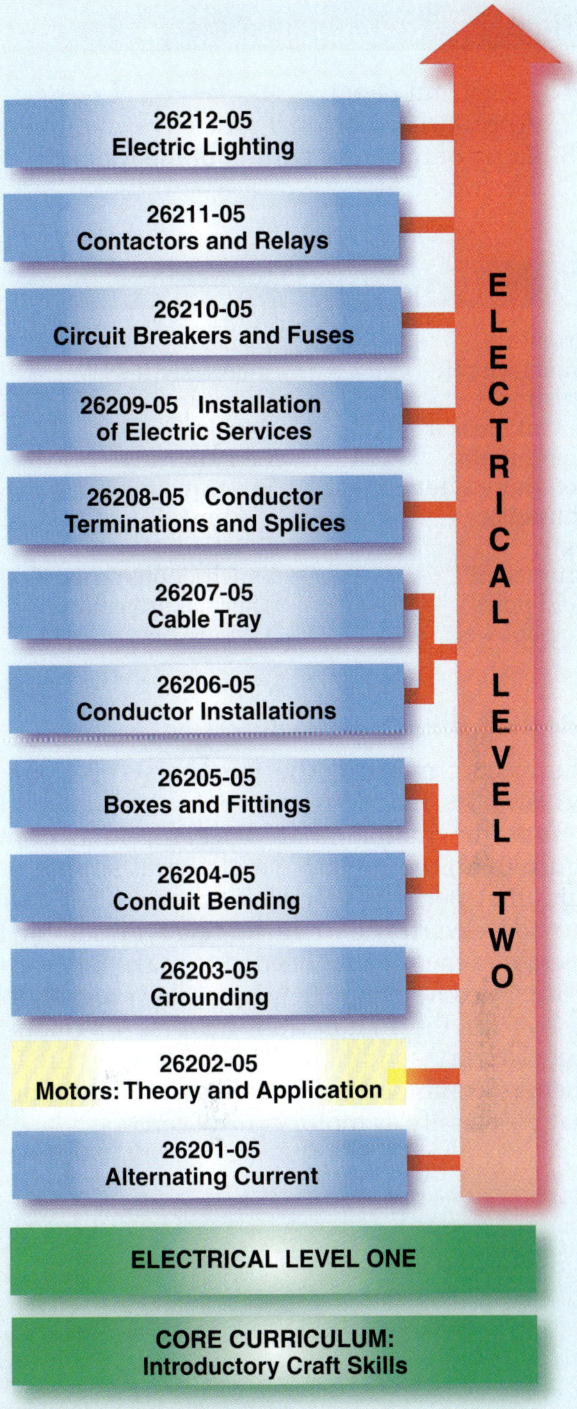

202CMAP.EPS

Prerequisites

Before you begin this module, it is recommended that you successfully complete *Core Curriculum; Electrical Level One;* and *Electrical Level Two*, Module 26201-05.

This course map shows all of the modules in *Electrical Level Two.* The suggested training order begins at the bottom and proceeds up. Skill levels increase as you advance on the course map. The local Training Program Sponsor may adjust the training order.

Required Trainee Materials

1. Pencil and paper
2. Appropriate personal protective equipment
3. Copy of the latest edition of the *National Electrical Code®*

Trade Terms

Armature
Branch circuit
Brush
Circuit breaker
Commutator
Continuous duty
Controller
Duty
Equipment
Field poles
Horsepower
Hours
Intermittent duty
Overcurrent
Overload
Periodic duty
Revolutions per minute (rpm)
Rotation
Synchronous speed
Thermal protector
Varying duty

1.0.0 ◆ INTRODUCTION

The electric motor is the workhorse of modern industry. Its functions are almost unlimited. To control the motors that drive machinery and **equipment,** we must have electrical supply circuits that perform certain functions. They must provide electrical current to cause the motor to operate in the manner needed to make it perform its intended function. They must also provide protection for the motor from adverse mechanical and electrical conditions. These functions are frequently combined within electrical equipment that we classify as motor control centers.

A thorough understanding of the functions of the various components of a motor control center is desirable from both a maintenance and a troubleshooting standpoint. Properly-maintained motor control centers ensure a minimum of downtime for unscheduled repairs, increase productivity, and contribute to a safer working environment.

2.0.0 ◆ DC MOTOR PRINCIPLES

When a bar of magnetic material is given an induced magnetic charge, a field of magnetic force is developed around the bar. We picture this field as consisting of lines of force that exist in the space surrounding the bar. These lines of force appear to leave one end of the bar and extend outside the bar to the opposite end.

The end of the bar magnet where magnetic lines appear to start is called the north pole of the magnet, while the end where these lines reenter the magnet is called the south pole. Actually, the lines extend inside the magnet from the south pole to the north pole, completing a closed loop. This principle is shown in *Figure 1*.

There are several characteristics of these magnetic lines of force that must be remembered when dealing with electric motors. They are:

- Magnetic lines of force are continuous and always form closed loops.
- Magnetic lines of force do not cross.
- Magnetic lines of force with polarities in the same direction repel each other. In other words, a north pole will repel a north pole and a south pole will repel a south pole.

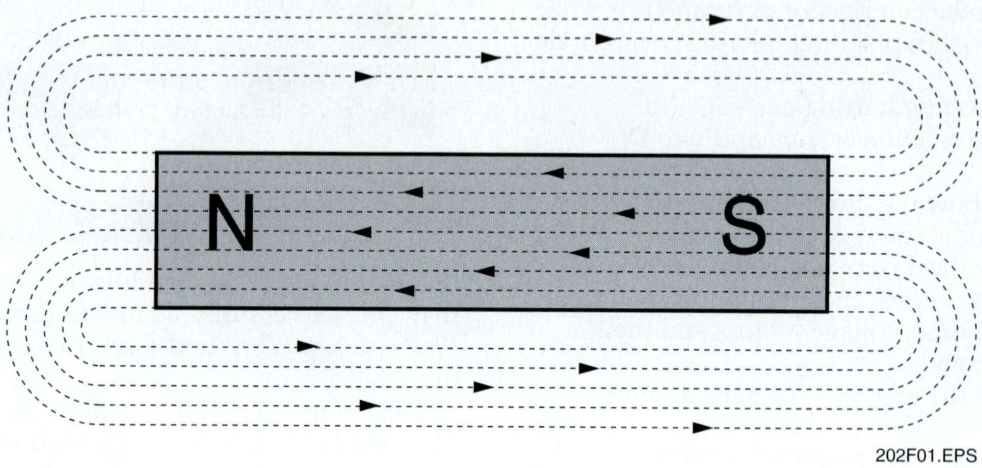

Figure 1 ◆ Magnetic field.

Early Electric Motors

INSIDE TRACK

The first U.S. patent for a motor was issued to Thomas Davenport in 1837. He reported that he used silk from his wife's wedding gown as insulation for the conductors, but despite this sacrifice, his motor was not commercially successful. Practical electric motors, like the practical light bulb, did not appear until the late 19th century.

- Magnetic lines of force having polarities in opposite directions tend to attract each other and combine. In other words, a south pole will be attracted by a north pole and vice versa.
- Magnetic lines of force tend to shorten themselves. Therefore, the magnetic lines of force existing between two unlike poles cause the poles to tend to pull together.
- Magnetic lines of force pass through all known materials, magnetic or nonmagnetic. Some materials provide a much easier path for these lines than others. These materials have high permeability and low reluctance. (Reluctance is discussed later in this module.)

2.1.0 Hand Rules

When a current is passed through the wire, circular lines of force are produced around the wire. These flux lines go in a direction described by the left-hand rule. This rule is shown in *Figure 2*.

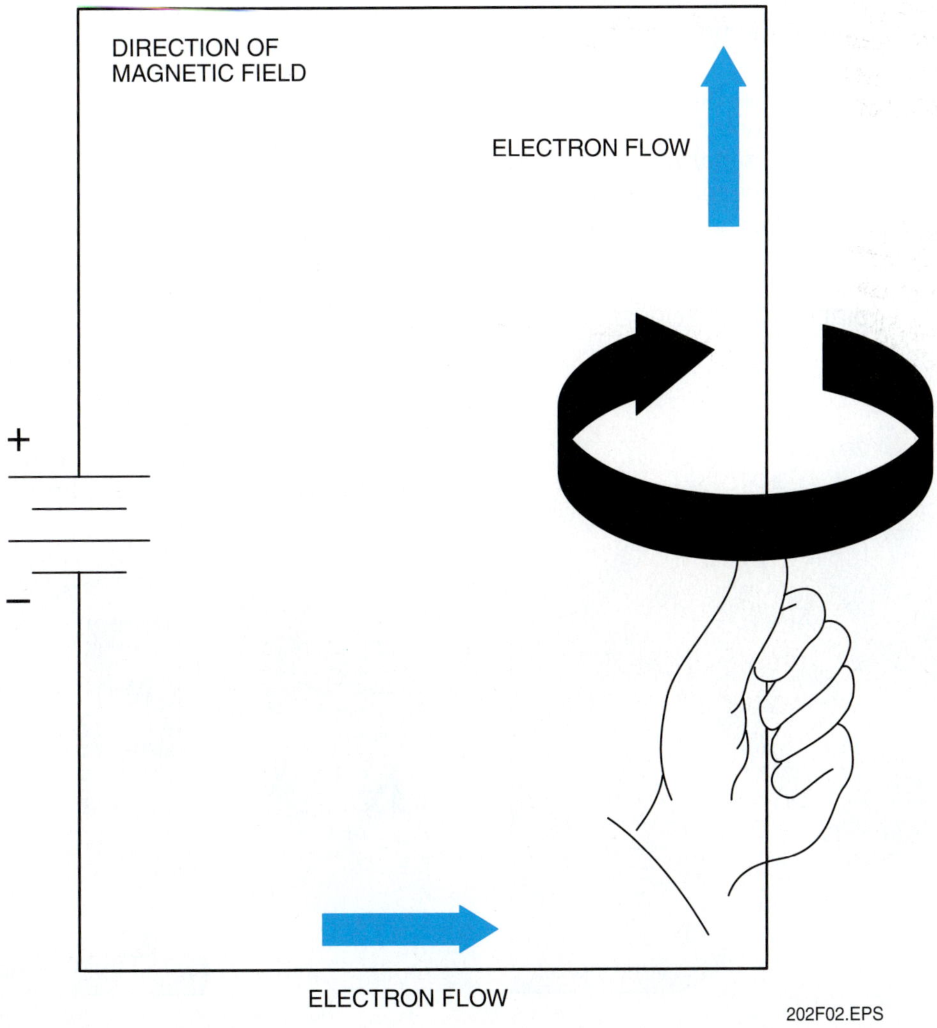

Figure 2 ◆ Left-hand rule for conductors.

The left-hand rule shows the direction of the flux lines around a wire that is carrying current. When the thumb points in the direction of the electron current, the fingers will point in the direction of the magnetic lines of force.

The right-hand rule for motors shows the direction that a current-carrying wire will be moved in a magnetic field (see *Figure 3*). When the forefinger is pointed in the direction of the magnetic field lines, and the center finger is pointed in the direction of the current in the wire, the thumb will point in the direction that the wire will be moved.

2.2.0 DC Motor Components

A DC motor consists of a few major components, each with a specific purpose in the motor's operation.

The **armature** is a movable electromagnet located between the poles of another fixed permanent (field) magnet, as shown in *Figure 4*.

The magnetic field from the armature conductors interacts with the magnetic field from the field magnet. The result of the field interaction is motor action.

Current in a conductor also has its associated magnetic field. When a conductor is placed in another magnetic field from a separate source, the two fields can react to produce motor action. The conductor must be perpendicular to the magnetic field, as illustrated in *Figure 5*. This way, the perpendicular magnetic field of the current is in the same plane as the external magnetic field.

Unless the two fields are in the same plane, they cannot affect each other. In the same plane, however, lines of force in the same direction reinforce to make a stronger field, while lines in the opposite direction cancel and result in a weaker field. The stronger field tends to move the conductor toward the weaker field, as illustrated in *Figure 5*.

These directions are summarized as follows:

- With the conductor at 90°, or perpendicular to the external field, the reaction between the two magnetic fields is at its maximum.
- With the conductor at 0°, or parallel to the external field, there is no effect between them.
- When the conductor rise is at an angle between 0° and 90°, only the perpendicular component is effective.

In motor action, the wire only moves in a straight line, and it stops moving once out of the field, even though current still exists. A practical motor must develop continuous rotary motion. To produce this, a twisting force called torque must be developed.

Torque is produced by mounting a loop in a fixed magnetic field. Current is applied and the flux lines along both sides of the loop interact, causing the loop to act like a lever with a force pushing on its two sides in opposite directions. This is shown in *Figure 6*.

The combined forces result in a turning force or torque, because the rotor or armature is arranged to pivot on its axis. The overall turning force on the armature depends on several factors, including field strength, armature current strength, and the physical construction of the armature, especially the distance from the loop sides to the axis lines.

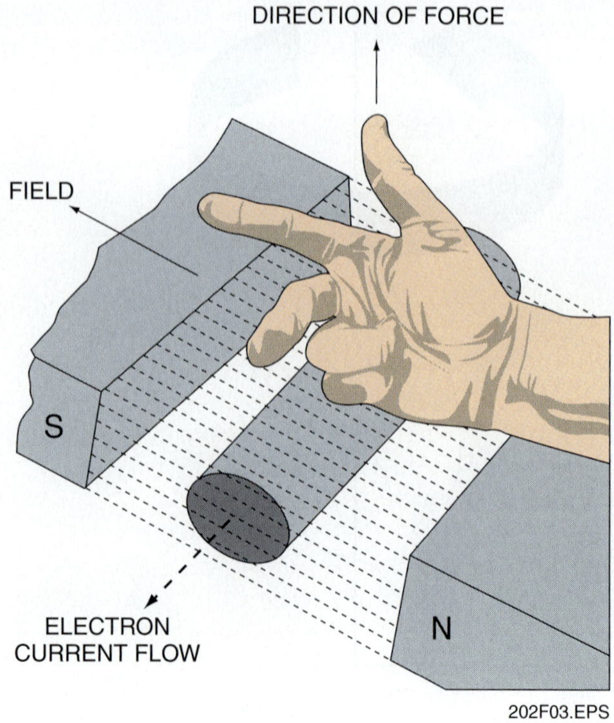

Figure 3 ◆ Right-hand rule for motors.

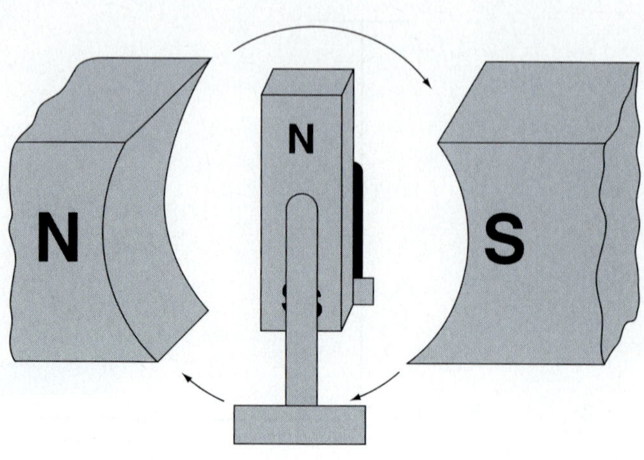

Figure 4 ◆ Basic motor action.

Hand Rules for Motors

THINK ABOUT IT

Use the left-hand conductor rule and the right-hand motor rule to explain the motor action pictured in *Figure 5*.

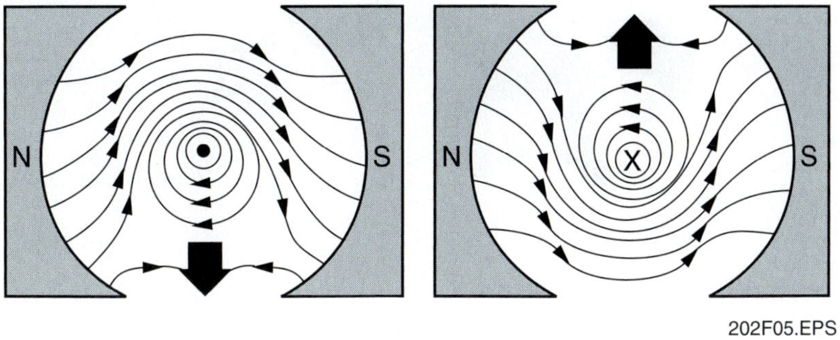

Figure 5 ◆ Motor action.

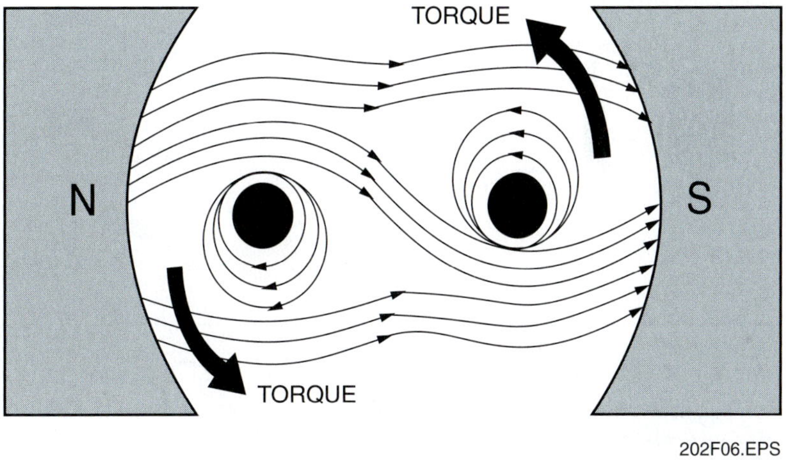

Figure 6 ◆ Torque.

Because of the lever action, the forces on the sides of the armature loop will increase as the loop sides are farther from the axis; therefore, larger armatures will produce greater torques.

In the practical motor, the torque determines the energy available for doing useful work. The greater the torque, the greater the energy. If a motor does not develop enough torque to turn its load, it stalls.

To get continuous **rotation,** the armature must be kept moving in the same direction. This requires reversing the direction of current through the armature for every 180° of revolution. A **commutator** is used to provide this switching action. This is shown in *Figure 7*.

The commutator on a DC motor is a conducting ring that is split into two segments, with each segment connected to an end of the armature loop. Current enters the side of the armature closest to the south pole of the field and leaves the side closest to the north pole of the field. The interaction of the two fields produces a torque, and the armature rotates in that direction.

A **brush** makes contact with each segment of the commutator, providing a connection between the movable commutator and the stationary DC power source. *Figure 8* shows various brushes and commutator connections used in DC motors.

The problem of switching commutator segments in a simple single-loop motor is that when

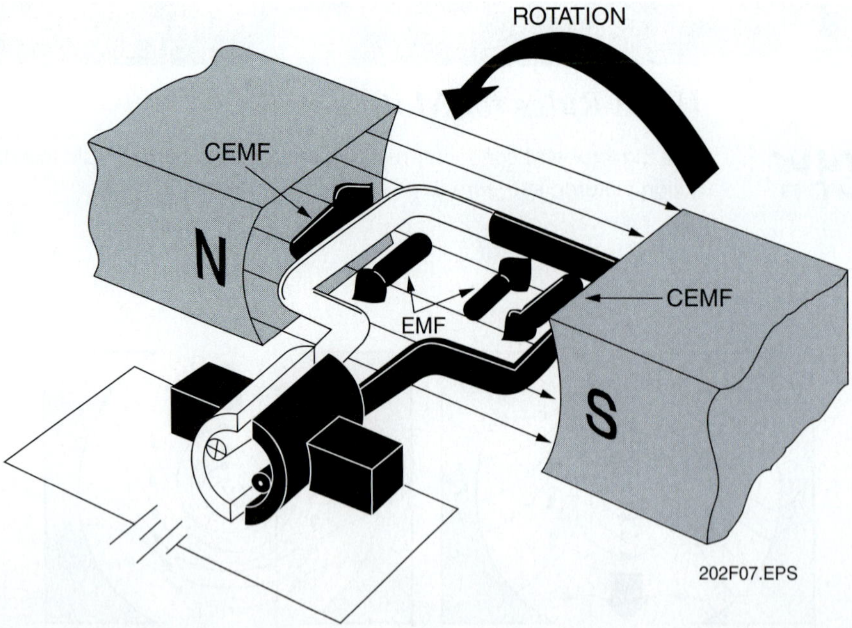

Figure 7 ◆ Single-loop armature DC motor.

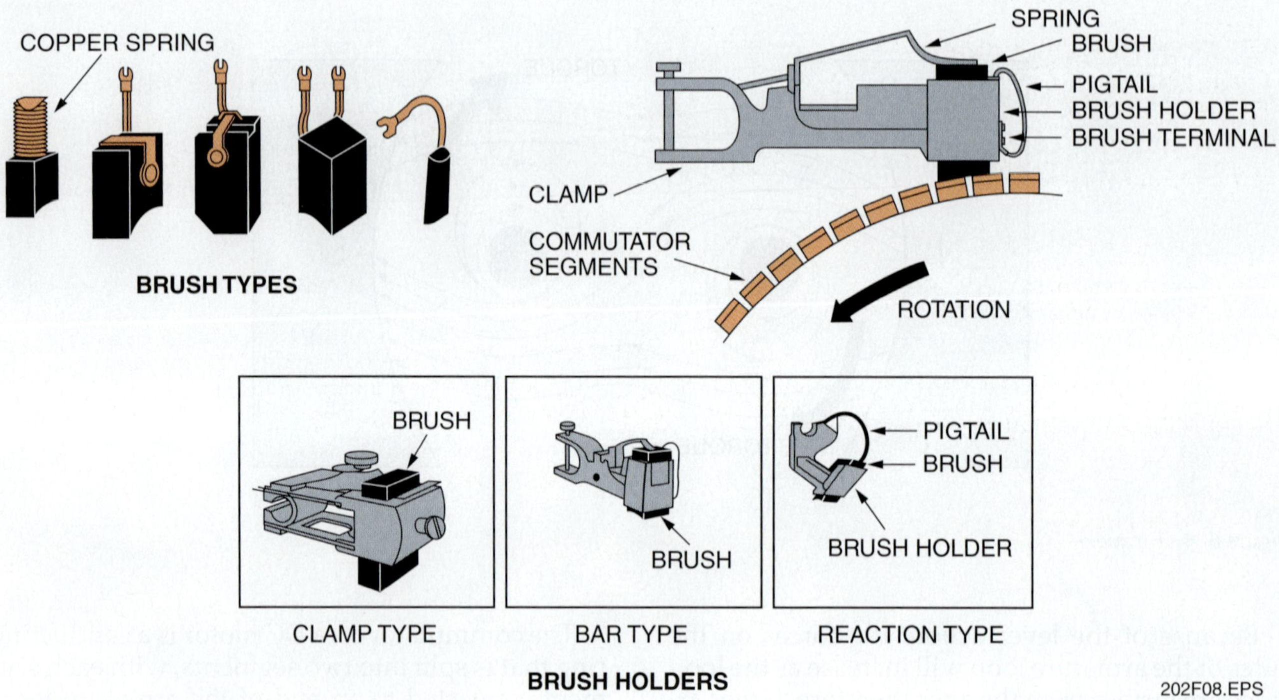

Figure 8 ◆ Brushes, brush rigging, and commutator connections.

the motor stops, there is no way of predicting the position of the armature at rest. If the armature stops in a position where the commutator is in the middle of switching, the motor will not start unless you physically turn the armature.

This problem can be overcome by winding more coils on the armature and by using more commutator segments. This will produce a self-starting motor. The motor shown in *Figure 9* uses three armature coils and three commutator segments. Regardless of where the armature comes to rest, there is always a path for current that will produce torque to rotate the armature.

2.3.0 The Neutral Plane

The armature turns when torque is produced, and torque is produced as long as the fields of the

Understanding the Theory of Torque

THINK ABOUT IT

A motor provides torque (a turning force) similar to that needed to turn a nut. In the illustration shown here, which wrench should be able to turn the nut easier?

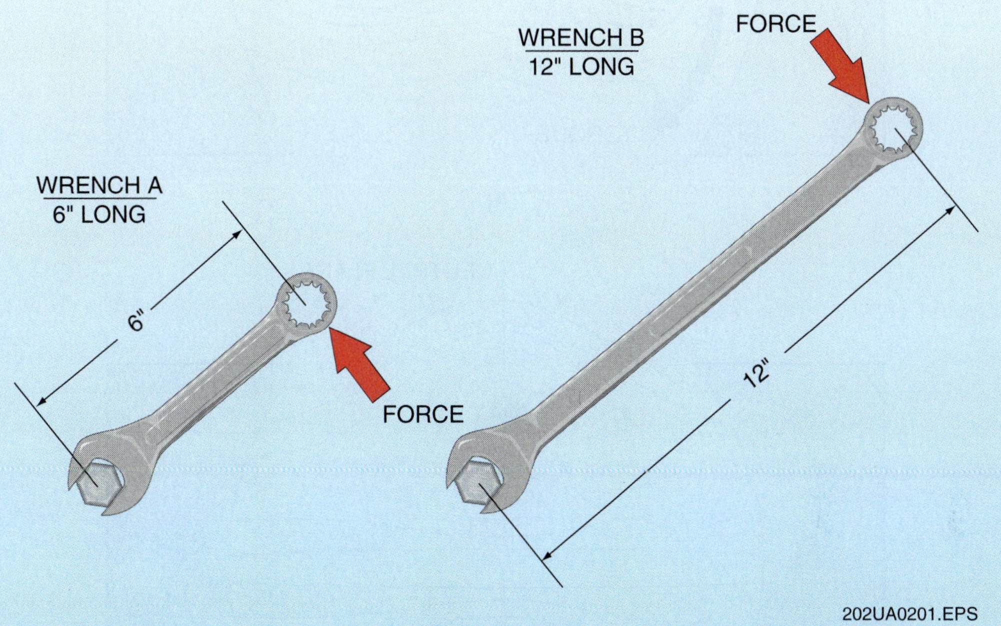

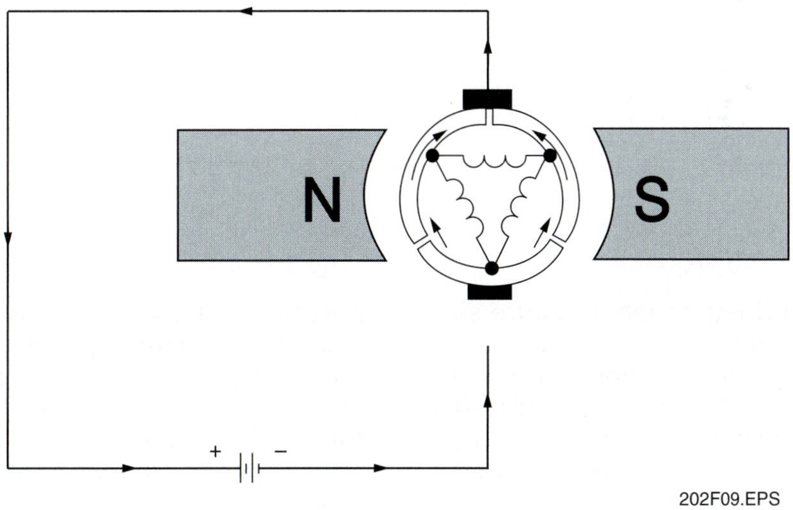

Figure 9 ◆ Self-starting motor.

magnet and armature interact. When the loop reaches a position perpendicular to the field, the interaction of the magnetic fields stops. This position is the neutral plane, shown in *Figure 10*.

In the neutral plane, no torque is produced and the rotation of the armature should stop. However, inertia tends to keep the armature in motion even after the prime moving force is removed; thus the armature tends to rotate past the neutral plane. At the neutral position, the commutator disconnects from the brushes, and once the armature goes past neutral, the sides of the loop reverse positions. The switching action of the commutator maintains the direction of current through the

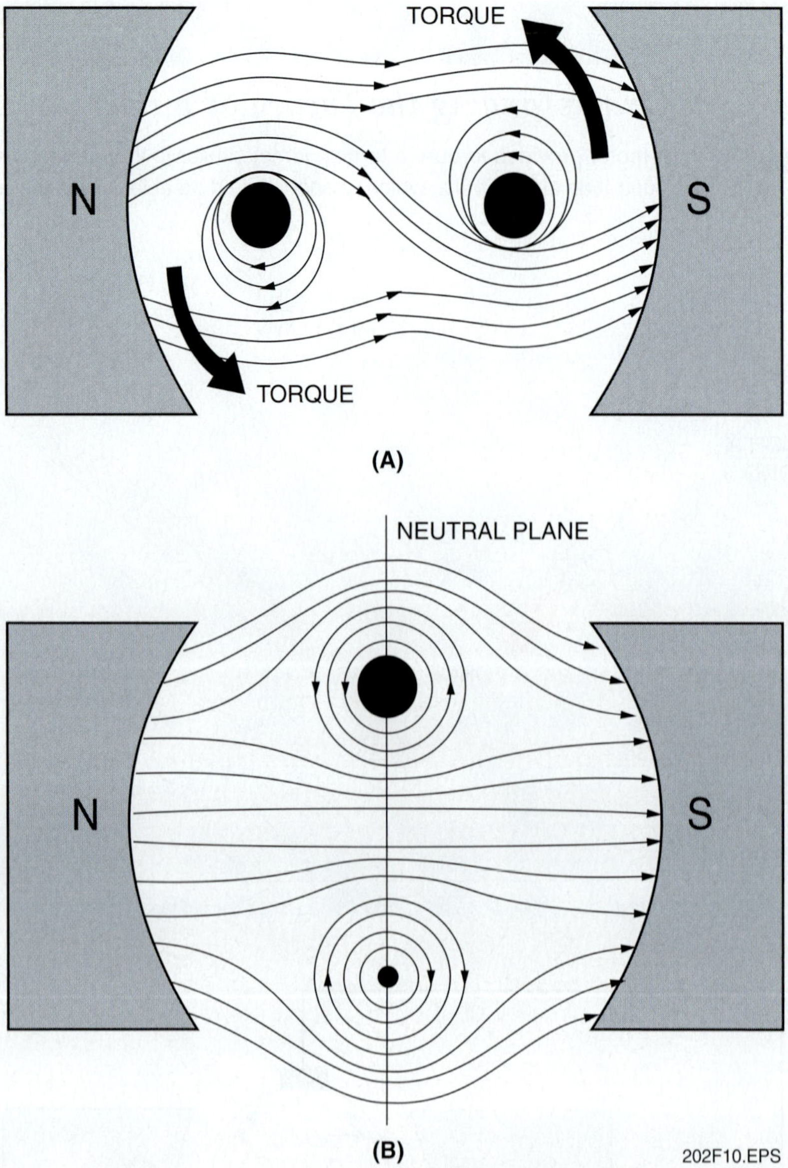

Figure 10 ◆ Neutral plane.

armature. Current still enters the armature side that is closest to the south pole.

Since the magnet's field direction remains the same throughout, the interaction of fields after commutation keeps the torque going in the original direction; thus, continuous rotation is maintained. See *Figure 11*.

Although such an elementary DC motor can be built and operated, it has two serious shortcomings that prevent it from being useful: first, such a motor cannot always start by itself; and second, once started, it operates very irregularly.

When the elementary DC motor runs, its operation is erratic because it produces torque irregularly. Maximum torque is produced only when the plane of the single-loop armature is parallel with the plane of the field. This is the position at right angles to the neutral plane. Once the armature passes this plane of maximum torque, less and less torque is developed until it arrives at the neutral plane again. Inertia carries the armature past the neutral plane and so the motor continues to turn. Its irregularity in producing torque, however, prevents the single-loop elementary DC motor from being used for practical jobs.

2.4.0 Two-Loop DC Motors

The basic DC motor is improved by building the armature with two or more loops. The loops are placed at right angles to each other; when one loop lies in the neutral plane, the other is in the plane of maximum torque. See *Figure 12*.

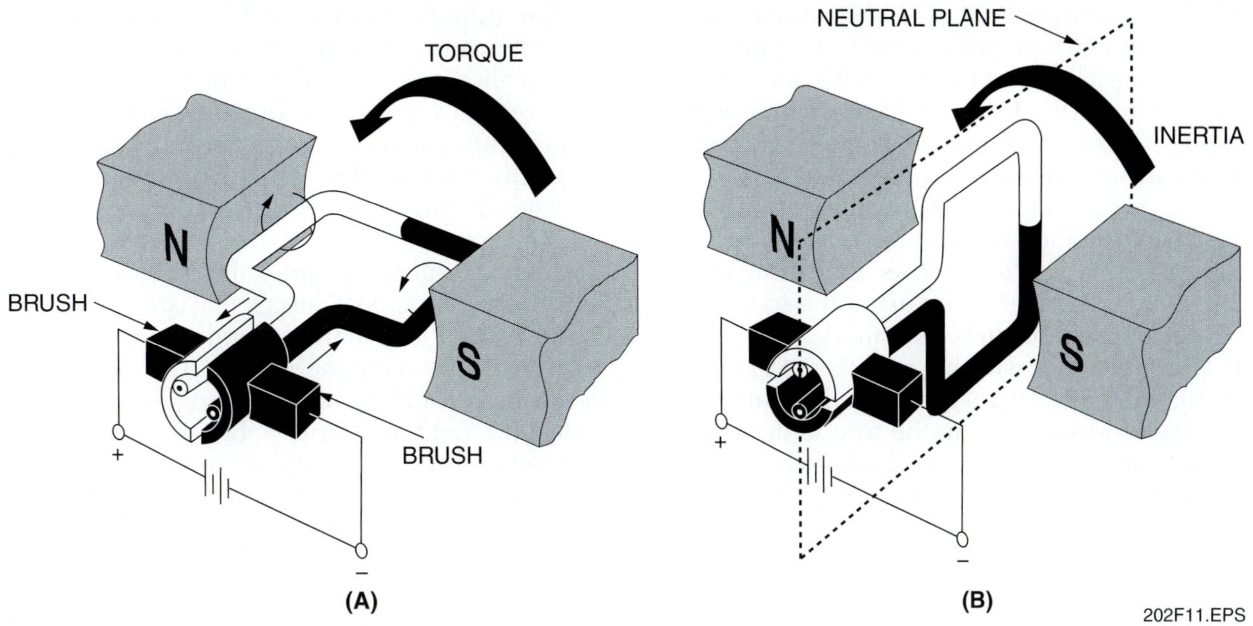

Figure 11 ◆ Neutral plane in a DC motor.

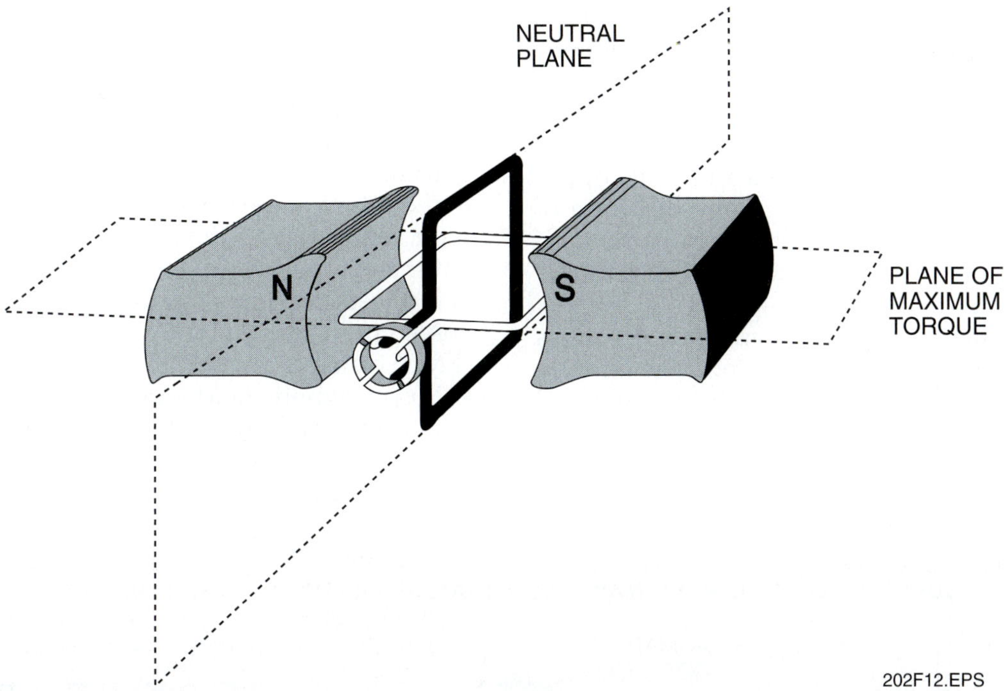

Figure 12 ◆ Two-loop armature DC motor.

THINK ABOUT IT

Two-Loop DC Motors

Using the basic principles of motor action, explain why the two-loop DC motor is self-starting and why its torque is erratic.

In this case, the commutator is split into two pairs or four segments, with one segment associated with each end of each armature loop. This sets up two parallel loop circuits. Only one loop at a time is ever connected if power is supplied through one pair of fixed brushes to one set of ring segments.

In this multi-loop armature, the commutator serves two functions: it maintains current through the armature in the same direction at all times, and it switches power to the armature loop nearing the maximum torque position.

This motor is self-starting because at least one winding will have interaction with the main field. With this two-loop system, the torque developed is steadier and stronger but still somewhat erratic, because only one loop at a time provides the torque that drives the motor.

2.5.0 Armature Reaction

When a motor armature is supplied with current, a magnetic flux is built up around the conductors of the armature windings. Armature reaction is caused by two magnetic fields: the main magnetic field from the field magnets and the magnetic field produced by the armature. These two fields combine to produce a new resultant magnetic field.

The resultant field is distorted and shifts opposite the main field and opposite the direction of armature rotation. This distortion shifts the neutral plane of the motor. See *Figure 13* for an illustration of armature reaction.

The amount of armature reaction determines how far the neutral plane is shifted. The amount of armature reaction depends on the amount and direction of the armature current. The concern over the neutral plane shift occurs because of the need for commutation.

Commutation, or the switching of the armature polarity, must take place at the neutral plane in order to allow the output current from the machine to remain in the same direction without arcing. When commutation takes place anywhere other than the neutral plane, it is like a switch that is opened during high current—it will draw an arc.

This armature reaction can be overcome by installing interpole windings. Interpoles are special electromagnetic pole pieces that are connected in series with the armature winding. The armature current causes a magnetic field to form around the windings. Their action is self-regulating, and the interpole field will apply the proper amount of cancellation field for any set of conditions. For a high armature reaction, the canceling field is strong. For a low armature reaction, the canceling field is weaker. See *Figure 14*.

2.6.0 Counter-Electromotive Force (CEMF)

When a DC motor is in operation, it acts much like a DC generator. A magnetic field is produced by the **field poles,** and a loop of wire in the armature turns and cuts this magnetic field. To understand counter-electromotive force (CEMF), first disregard the fact that external current is being applied to the rotor via the carbon brushes on the commutator segments. As the armature wires rotate and cut the magnetic field of the field poles, a voltage is induced in them similar to that which was discussed in induction motors. This induced voltage (EMF) causes a current to flow in them and a resulting magnetic field is created.

Before analyzing the relative direction between the current induced in the armature windings and the current that caused it in the field poles, first

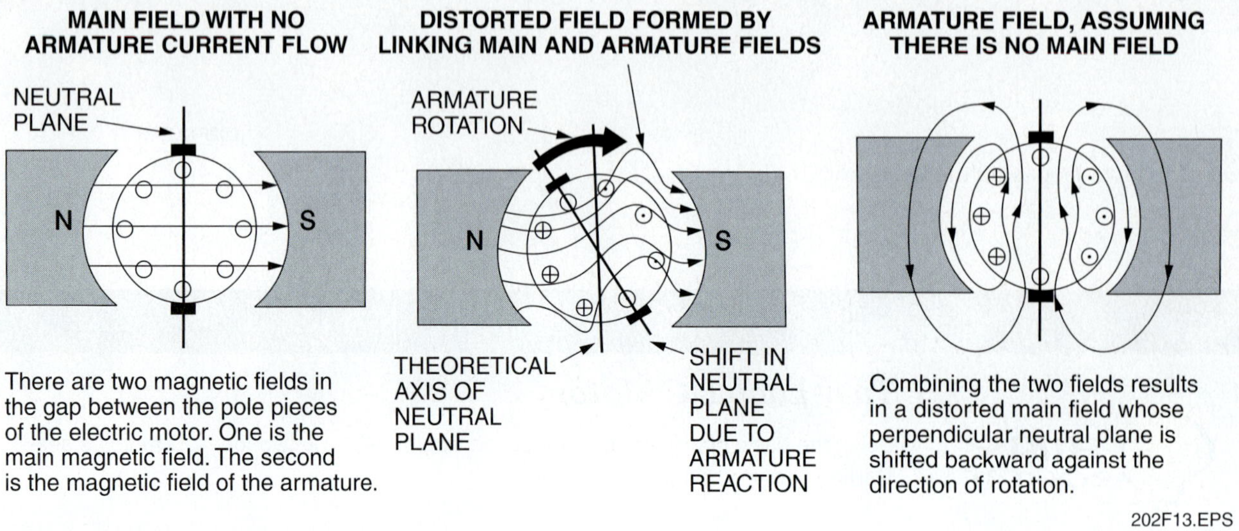

Figure 13 ◆ Armature reaction.

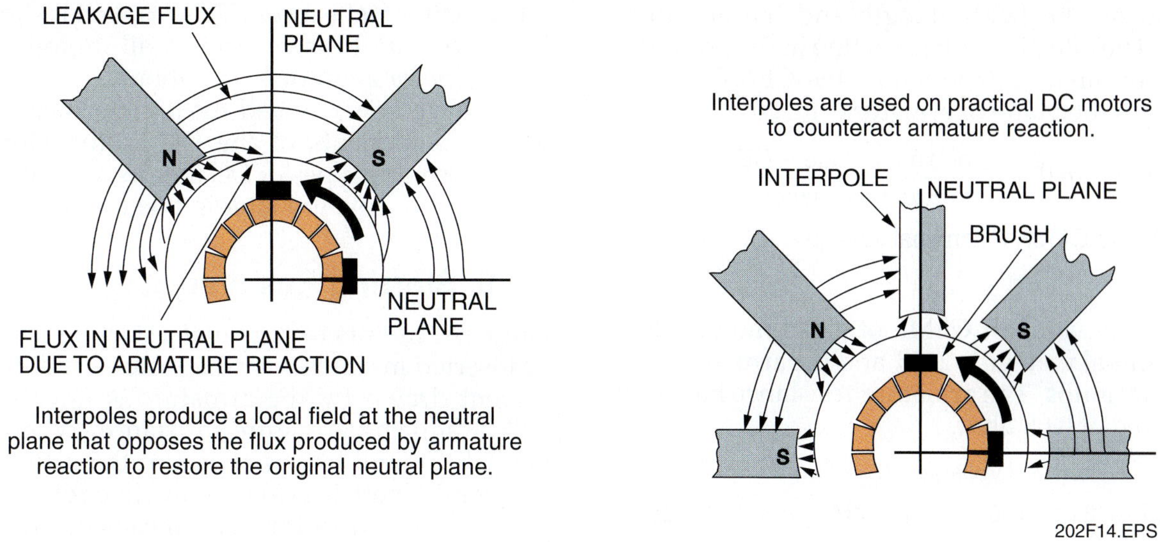

Figure 14 ◆ Interpoles.

remember the left-hand rule. Using your left hand, hold it such that your index finger points in the direction of the magnetic field (north to south) and your thumb points in the direction of rotational force on a given conductor. Your middle finger will now point in the direction of current flow for that conductor. This current would be in opposition to the current that is flowing from the battery. Since this induced voltage and induced current are opposite to those of the battery, they are called CEMF. The two currents are flowing in opposite directions. This would mean that the battery voltage and the CEMF are opposite in polarity. See *Figure 15*.

When first discussing CEMF, we disregarded the fact that external DC was being applied to the armature via the brushes. The induced voltage and resulting current flow was then shown to flow opposite to the externally applied current. This was an oversimplification, since only one current flows. Since the CEMF can never become as large as the external applied voltage, and since they are opposite in polarity, the CEMF works to cancel only a part of the applied voltage. The single current that flows is smaller due to the CEMF.

Since the CEMF of a motor is generated by the action of the armature windings cutting the lines of force set up by the field poles, the value of it will

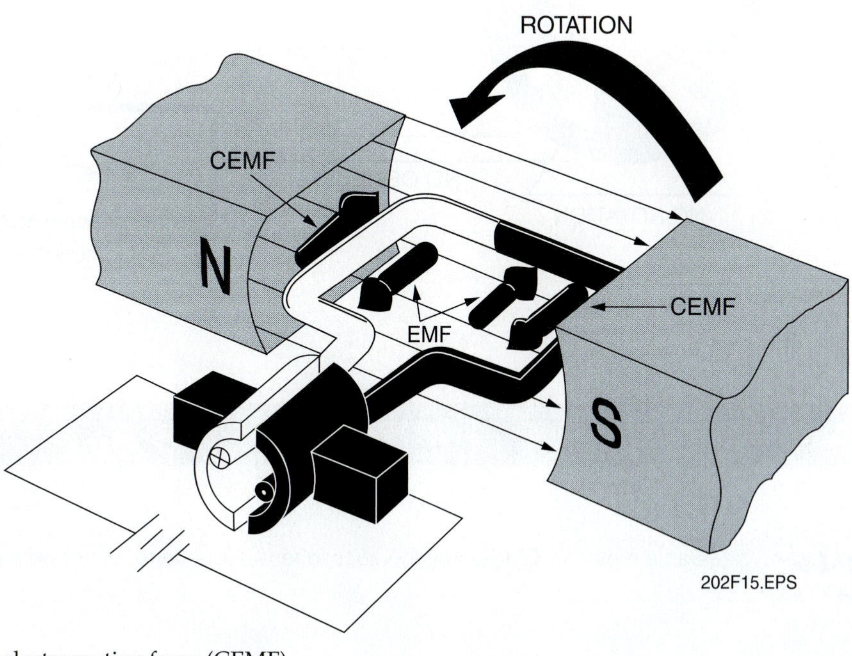

Figure 15 ◆ Counter-electromotive force (CEMF).

depend on the field strength and the armature speed. The effective voltage acting in the armature is the terminal voltage minus the CEMF. Ohm's law gives the value of armature current by:

$$\text{Armature } (I_A) = \frac{\text{terminal voltage} - \text{CEMF}}{\text{armature resistance } (R_A)}$$

$$\text{Where CEMF} = \text{terminal voltage} - (I_A \times R_A)$$

Example:

Find the value of CEMF of a DC motor when the terminal voltage is 240V and the armature current is 60 amps. The armature resistance has been measured at 0.08 ohm.

$$\text{CEMF} = \text{terminal voltage} - (I_A \times R_A)$$

$$\text{CEMF} = 240 - (60 \times 0.08) = 240 - 4.8 = 235.2\text{V}$$

CEMF acts as an automatic current limiter that reduces armature current to a level adequate to drive the motor but not great enough to heat the armature to where it is in danger of burning out. CEMF acts as a load for the DC power supply feeding the motor, so that the low-resistance motor windings do not draw excessive amounts of current.

If we stalled the armature so that no CEMF was produced, we would find that the motor draws so much current it heats up. This reaction is shown in *Figure 16*. CEMF is present in all motors and is necessary for a motor's operation.

We have now covered the basic principles and major components of the DC motor. However, there are many types of DC motors, and several of them will be covered later in this module.

2.7.0 Starting Resistance

Large DC motors require that a starting resistance be inserted in series with the motor armature. The current drawn by the armature is governed by CEMF and the armature resistance. When starting, CEMF will be zero because the rotor is at a standstill. There is also no inductive reactance, as in AC induction motors. This means that the starting current will be abnormally high unless limited by external starting resistance.

Figure 17 shows a shunt motor that is connected directly across a 250V line. The armature resistance is known to be 0.5 ohm. The full-load current of the motor is known to be 25 amps and the shunt field current is one amp. The resulting armature current under full-load conditions would therefore be 24 amps.

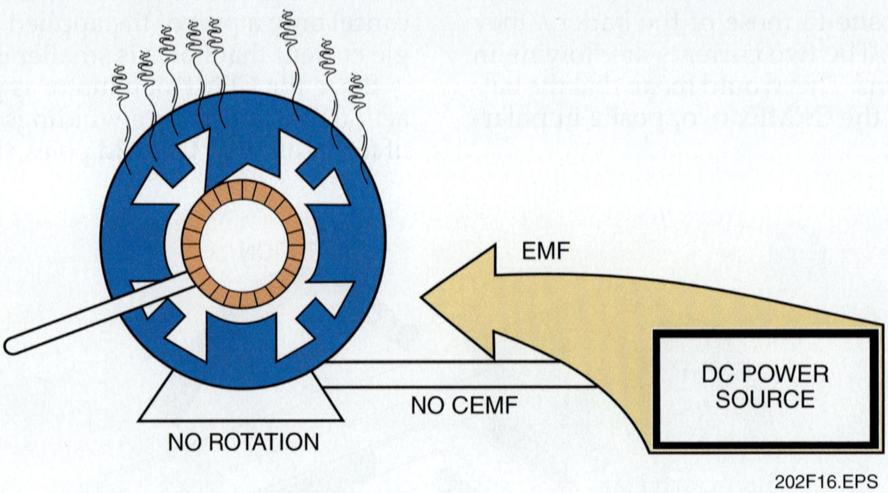

Figure 16 ◆ No CEMF.

THINK ABOUT IT

CEMF

Can a motor's CEMF equal the applied terminal voltage? If not, why not?

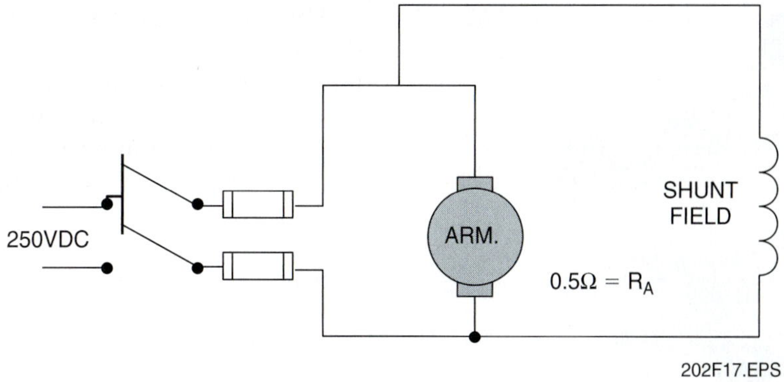

Figure 17 ◆ Shunt motor.

If starting resistance is not used, the value of the armature current (I_A) can be found using the following equation:

$$I_A = \frac{\text{terminal voltage} - \text{CEMF}}{R_A}$$

$$I_A = \frac{250V - 0V}{0.5 \text{ ohm}}$$

$$I_A = 500 \text{ amps}$$

This amount of starting current is too high and may result in excessive torque and heat that may cause damage to the motor. When starting resistance is added in series with the armature, the starting current can be limited to 1.5 times the full-load current value. After starting, this external resistance can be removed from service.

If we want to limit the starting armature current to 1.5 times the full-load value, we can solve for the size of resistance that would be required using the previous equations.

$$R_{starting} = \frac{(\text{terminal voltage} - \text{CEMF}) - (I_A \times R_A)}{I_A}$$

Where:

$$\text{Starting } I_A = 1.5 \times \text{steady state}$$
$$= 1.5 \times 24 \text{ amps}$$
$$= 36 \text{ amps}$$

At the moment of motor start, when the rotor is at a standstill and the CEMF is zero, the series resistance will be:

$$R_{starting} = \frac{(250V - 0V) - (36A)(0.5\Omega)}{36A} = 6.44\Omega$$

To find the wattage required in the starting resistance, take the square of the current multiplied by the resistance, where watt loss is calculated by the I^2R method.

Example:

Find the power developed in both watts and **horsepower** in a DC motor that has a terminal voltage of 240V and an armature current of 60A. The armature resistance is known to be 0.08Ω.

$$\text{CEMF} = V_T - (I_A \times R_A)$$
$$\text{CEMF} = 240 - (60 \times 0.08) = 235.2V$$
$$\text{Power} = EI$$
$$\text{Power} = 235.2V \times 60A = 14{,}112 \text{ watts}$$
$$\text{Horsepower} = \frac{\text{watts}}{746}$$
$$= \frac{14{,}112 \text{ watts}}{746}$$
$$= 18.92\text{hp}$$

3.0.0 ◆ TYPES OF DC MOTORS

There are two basic types of motor connections that are in common use. They are the series motor and the shunt motor. The series motor is so called because the field is connected in series with the armature winding. The shunt motor has the field coils connected in parallel with the armature (rotor) winding. One additional type of motor is a compound motor. This motor has both a series- and a shunt-connected field. *Figure 18* shows a typical DC motor.

3.1.0 Shunt Motors

The field circuit of a shunt motor is connected across the supply line and is in parallel with the armature. A shunt motor connection is shown in *Figure 19*.

When an external load is applied to the shunt motor, it tends to slow down. The slight decrease in

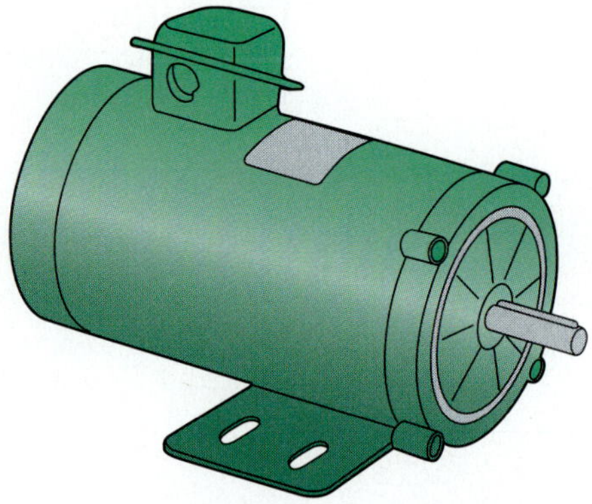

202F18.EPS

Figure 18 ◆ Typical DC motor.

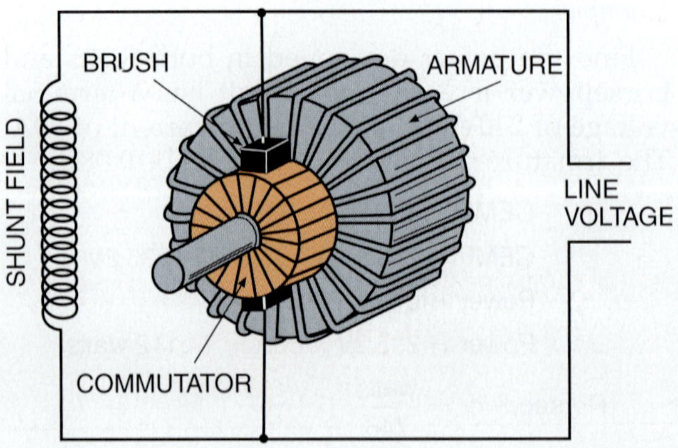

202F19.EPS

Figure 19 ◆ Shunt DC motor.

speed causes a corresponding decrease in CEMF. Since the armature resistance is low, the resulting increases in armature current and torque are relatively large. Therefore, the torque is increased until it matches the opposing torque of the load. The speed of the motor then remains constant at the new value as long as the load is constant.

If the load on the shunt motor is reduced, the motor tends to speed up. The increased speed causes a corresponding increase in CEMF and a relatively large decrease in armature current and torque.

Thus, it may be seen that the amount of current through the armature of a shunt motor depends largely upon the load on the motor. The larger the load, the larger the armature current; the smaller the load, the smaller the armature current. The change in speed causes a change in CEMF and armature current in each case.

The main advantage of a shunt-wound motor is that its speed is fairly constant, changing only a few **revolutions per minute (rpm)** when the amount of load changes. The main disadvantage of this connection is that the motor does not develop much torque when it is first started. If a motor is to be started with a large load, it is generally series connected.

It is important to note that the shunt field circuit of a DC motor should never be opened when the motor is operating, especially when unloaded. This is because an open field may cause the motor to rotate at dangerously high speeds. Large DC shunt motors have a field rheostat with a no-field release feature that disconnects the motor from the power source if the field circuit opens.

3.1.1 Torque

A DC shunt motor has high torque at any rated speed. At startup, a DC shunt motor can develop up to 150% of its normal running torque as long as the resistors in the starting circuit can withstand the heating effect of the current.

3.1.2 Speed Control

DC motors have excellent speed control. To operate the motor above rated speed, a field rheostat is used to reduce the field current and field flux. To operate below rated speed, resistors are used to reduce the armature voltage.

3.1.3 Speed Regulation

The speed regulation of a shunt motor drops from 5% to 10% from no-load to full-load. As a result, a shunt motor is superior to the series DC motor but is inferior to a differential compound-wound DC motor.

INSIDE TRACK

DC Motor Applications

DC motors were developed before AC motors and one of their first uses was in electric trolleys. The DC motor is still widely used in applications that require accurate speed control or high starting torque. For example, in an elevator, the motor must start under a heavy load and accelerate smoothly. It must also stop precisely and reverse direction easily. A DC motor is a good choice for this application.

3.2.0 Series Motors

The field coils of a series motor are connected in series with the armature (*Figure 20*). The value of current through the armature and the field is the same. Hence, if the armature current changes, the field current must also change.

As the motor speeds up, the armature current and field current decrease. With a weaker field, the armature speed will increase still more. The limiting factor on the speed is the load.

If there is no load on the motor, the armature will speed up to such an extent that the windings might be thrown from the slots and the commutator destroyed by the excessive centrifugal forces. For this reason, series motors are seldom belt-connected to their loads. The belt might break, allowing the motor to overspeed and destroy itself. Series motors are usually connected to their loads directly or through gears.

The series motor is used where there is a wide variation in both torque and speed requirements, such as traction equipment, blowers, hoists, cranes, and so forth.

3.2.1 Torque

The DC series motor develops 500% of its full-load torque at starting. Therefore, this type of motor is used in applications where large amounts of starting torque are needed, such as cranes, railway applications, and other equipment with high starting torque demands. With a series motor, any increase in load causes an increase in both the armature current and the field current. Since torque depends on the interaction of these two flux fields, the torque increases as the square of the value of the current increases. Therefore, series motors produce greater torque than shunt motors for the same increase in current. The series motor shows a greater reduction in speed for an equal change in load.

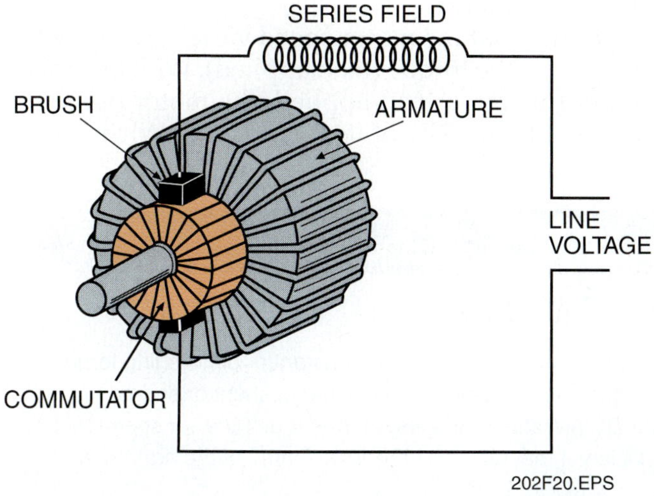

Figure 20 ◆ Series DC motor.

3.2.2 Speed Control and Speed Regulation

The speed control of a series motor is poorer than that of a shunt motor because if the load is reduced, a simultaneous reduction of current occurs in both the armature and field windings, and therefore, there is a greater increase in speed than there would be in a shunt-wound motor.

If the mechanical load were to be disconnected completely from a series motor, the motor would continue to accelerate until the motor armature self-destructed. For this reason, series-wound motors are always permanently connected to their loads.

The speed of a series DC motor is controlled by varying the applied voltage. A series motor **controller** is usually designed to start, stop, reverse, and regulate speed. The direction of rotation of a series motor is changed by reversing either the armature or field winding current flow.

3.3.0 Compound Motors

Compound DC motors are used whenever it is necessary to obtain speed regulation characteristics not obtainable with either the shunt- or series-wound motor. Because many applications require high starting torque and constant speed under load, the compound motor is used. Some industrial applications include drives for elevators, stamping presses, rolling mills, and metal shears. The compound motor has a normal shunt winding and a series winding on each field pole. They may be connected as a long shunt, as shown in *Figure 21(A)*, or a short shunt, as shown in *Figure 21(B)*. When the series winding is connected to aid the shunt winding, the machine is known as a cumulative compound motor. When the series field opposes the shunt field, the machine is known as a differential compound motor.

3.3.1 Torque

The operating characteristics of a cumulative compound-wound motor are a combination of the series motor and the shunt motor. A cumulative compound-wound motor develops high torque for sudden increases in load.

3.3.2 Speed

Unlike the series motor, the cumulative compound-wound motor has definite no-load speeds and will not build up self-destructive speeds if the

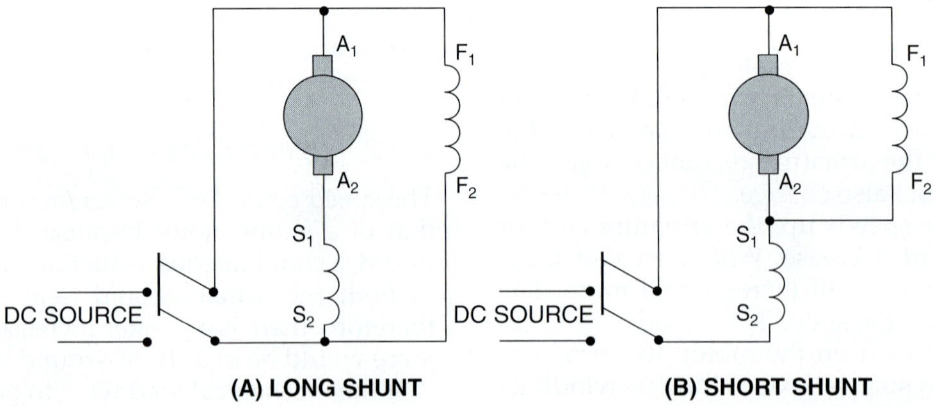

Figure 21 ◆ Long and short shunts.

INSIDE TRACK: Compound DC Motors

The compound motor avoids some of the limitations of the series-wound and shunt-wound motors. The shunt field has a constant current, so the motor will not self-destruct like a series motor. The series winding, on the other hand, provides strong torque.

load is removed. Speed control of a cumulative compound-wound motor can be controlled by inserting resistors in the armature circuit to reduce the applied voltage. When the motor is to be used for installations where the rotation must be reversed frequently, such as in elevators, hoists, and railways, the controller should have voltage dropping resistors and switching arrangements to accomplish reversal.

3.3.3 Speed Regulation

The speed regulation of a cumulative compound-wound motor is inferior to that of a shunt motor and superior to that of a series motor.

3.4.0 Operating Characteristics

Different types of motors have different operating characteristics. Therefore, the proper type of DC motor should be selected when the load to be driven is known. *Figure 22* shows the operating characteristics of a typical DC shunt motor.

Notice that the motor speed is relatively independent of the torque (load applied) from 0 to 150% of the rated capacity of the motor. Such motors find application where relatively constant speed over a wide load range is required. *Figure 23* shows the operating characteristics of a typical DC series motor.

Notice that the motor speed varies greatly with respect to the torque (load applied). With less than half of its rated load applied, the motor operates at more than 150% of its rated speed. When 150%

INSIDE TRACK: Permanent Magnet DC Motors

Many ¼hp to 3hp variable-speed DC motors available for constant or diminishing torque applications use permanent magnets for the field poles instead of shunt or series windings. They employ variable DC armature voltages up to 90V or 180V for speed control. However, they are inefficient if they use only rheostat control of the armature voltage.

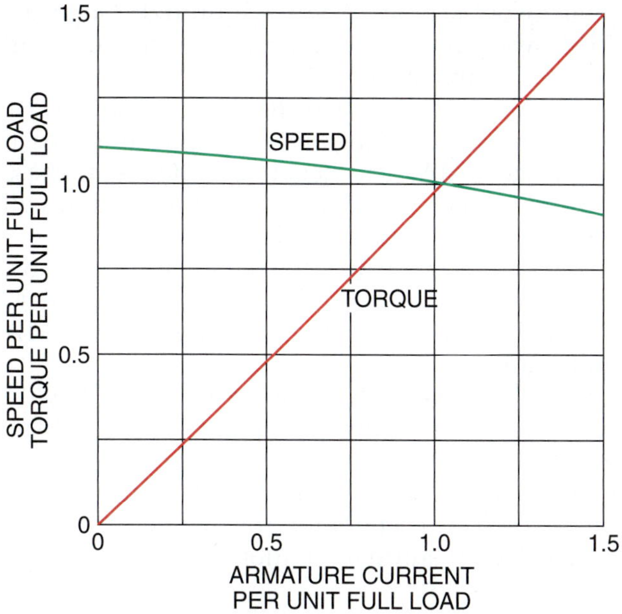

202F22.EPS

Figure 22 ◆ Operating characteristics of a typical DC shunt motor.

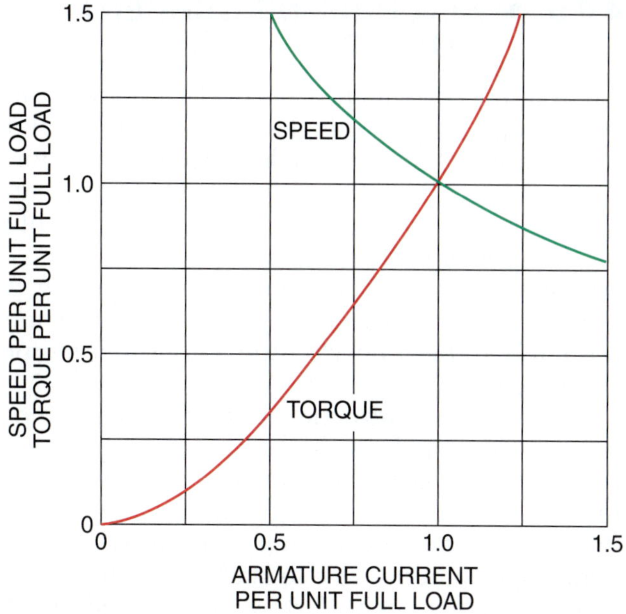

202F23.EPS

Figure 23 ◆ Operating characteristics of a typical series motor.

THINK ABOUT IT

DC Motors

Given the speed and torque characteristics of shunt and DC motors, which one would be better suited to the varying loads of an escalator?

of the rated load is applied to the motor, it drops to 75% of its rated speed. Such motors find application where a constant heavy load exists or where great speed variations are tolerable.

Figure 24 shows the operating characteristics of a DC compound motor. Notice that the motor speed is relatively constant over the operating range. Its speed does vary with the torque somewhat more than the shunt motor, but will not run away or markedly decrease, as with the series motor. Such motors find application where the load is not known exactly or where some speed variation is tolerable with load variation.

3.5.0 Brushless DC Motors

The brushless DC motor was developed to eliminate commutator problems in missiles and spacecraft operating above the Earth's atmosphere. Two general types of brushless motors are in use: the inverter-induction motor and a DC motor with an electronic commutator.

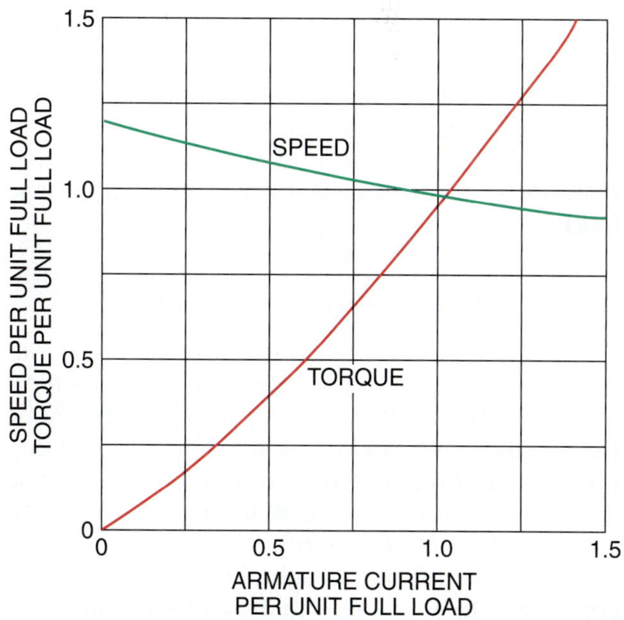

202F24.EPS

Figure 24 ◆ Operating characteristics of a typical DC compound motor.

4.0.0 ◆ ALTERNATING CURRENT MOTORS

Alternating current motors can be divided into two major types: single-phase motors and polyphase motors. The single-phase motor is normally limited to fractional horsepower ratings up to about five horsepower. They are commonly used to power such things as fans, small pumps, appliances, and other devices not requiring a great amount of power. Single-phase motors are not likely to be connected to complicated motor control circuitry.

Polyphase motors make up the majority of motors needed to drive large machinery such as pumps, large fans, and compressors. These motors have several advantages over single-phase motors in that they do not require a separate winding or other device to start the motor. They have relatively high starting torque and good speed regulation for most applications.

There are two classes of polyphase motors: induction and synchronous. The rotor of a synchronous motor revolves at **synchronous speed,** or the speed of the revolving magnetic field in the stator. The rotor of an induction motor revolves at a speed somewhat less than synchronous speed. The differences in rotor speed are due to differences in construction and operation. Both will be discussed in depth after a review of motor theory.

4.1.0 Polyphase Motor Theory

AC motors consist of two parts: the stator, or stationary part; and the rotor, or revolving part. The stator is connected to the incoming three-phase AC power. The rotor in an induction motor is not connected to the power supply, whereas the rotor of a synchronous motor is connected to external power. Both induction and synchronous motors operate on the principle of a rotating magnetic field.

4.1.1 Rotating Fields

This section shows how the stator windings can be connected to a three-phase AC input to create a magnetic field that rotates. Another magnetic field in the rotor can be made to chase it by being attracted and repelled by the stator field. Because the rotor is free to turn, it follows the rotating magnetic field in the stator.

Polyphase AC is brought into the stator and connected to windings that are physically displaced 120° apart. These windings are connected to form north and south magnetic poles, as shown in *Figure 25*. An analysis of the electromagnetic polarity of the poles at points 1 through 7 in *Figure 25* shows how the three-phase AC creates magnetic fields that rotate.

At point 1, the magnetic field in coil (pole) 1–1A is at its maximum. Negative voltages are shown in 1–2A and 3–3A. The negative voltages in these windings create smaller magnetic fields that will tend to aid the field set up in 1–1A.

At point 2, phase 3 creates a maximum negative flux in 3–3A windings. This strong negative field is aided by the weaker magnetic fields in 1–1A and 1–2A.

The three-phase AC input rises and falls with each cycle. Analyzing each point on the voltage graph shows that the resultant magnetic field rotates clockwise. When the three-phase input completes a full cycle at point 7, the magnetic field has completed an entire revolution of 360°.

4.1.2 Rotor Behavior in a Rotating Field

An oversimplification of rotor behavior shows how the magnetic field of the stator influences the rotor. Assume that a simple bar magnet is placed in the center of the stator diagrams shown in *Figure 25*. Also assume that the bar magnet is free to rotate. It has been aligned such that at point 1, its south pole is opposite the large north of the stator field.

Unlike poles attract and like poles repel. As the AC completes a cycle, going from point 1 to point 7, the stator field rotates and pulls the bar magnet with it because of the attraction of unlike poles and the repulsion of like poles. The bar magnet is rotating at the same speed as the revolving flux of the stator. This speed is known as synchronous speed. The synchronous speed of a motor is given by the equation:

$$N = \frac{120f}{P}$$

Where:

N = speed in rpm
f = frequency in cycles per second
P = number of magnetic poles

4.1.3 Induction

Current flowing through a conductor sets up a magnetic field around the length of the conductor. Conversely, a conductor in a magnetic field will produce a current when the magnetic lines of flux cut across the conductor. This action is called induction because there is no physical connection between the magnetic field and the conductor. Current is induced in the conductor.

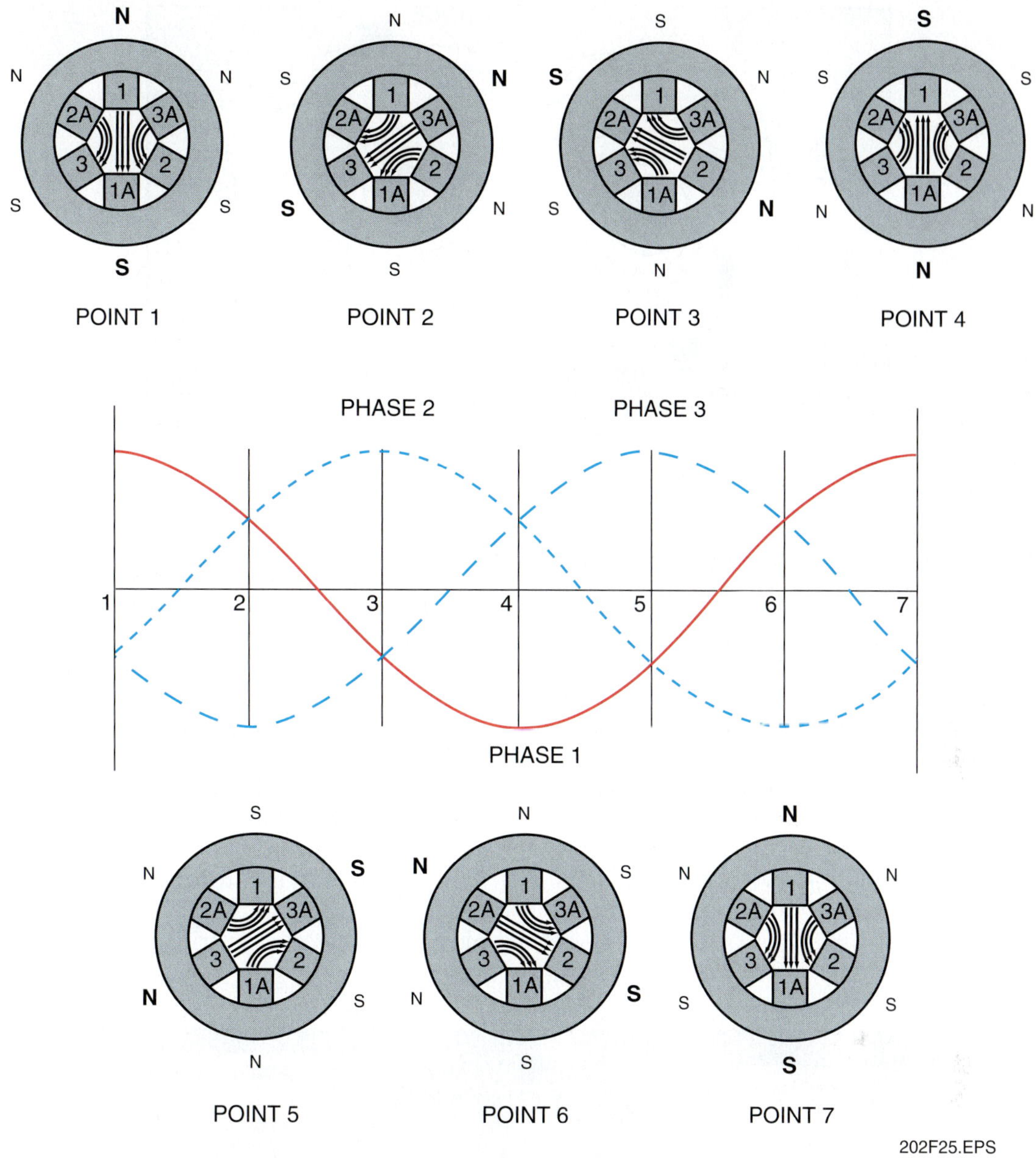

Figure 25 ◆ AC generation.

4.2.0 Three-Phase Induction Motors

In a three-phase induction motor, the driving torque is caused by the reaction of a current-carrying conductor in a magnetic field. In induction motors, the rotor currents are supplied by electromagnetic induction. The stator windings are supplied with three-phase power and produce a rotating magnetic field.

The rotor is not electrically connected to the power supply. The induction motor derives its name from the mutual inductance taking place between the stator and the rotor under operating conditions. The rotating field produced by the stator cuts the rotor conductors, inducing a voltage into the conductors. The induced voltage causes rotor current. This develops motor torque due to the reaction of a current-carrying conductor in a magnetic field. This torque causes the rotor to rotate. This principle is shown in *Figure 26*.

The three-phase (3ϕ) induction motor has a frame, or stationary part, which is the stator. The stator is made of laminated steel rings with slots on the inside circumference. The motor stator windings are the phase windings. They are symmetrically placed on the stator and may be either wye- or delta-connected. Depending on how the stator is wound, it may have two, four, or any even number of poles.

There are two varieties of three-phase induction motors: the squirrel cage rotor motor and the wound rotor motor.

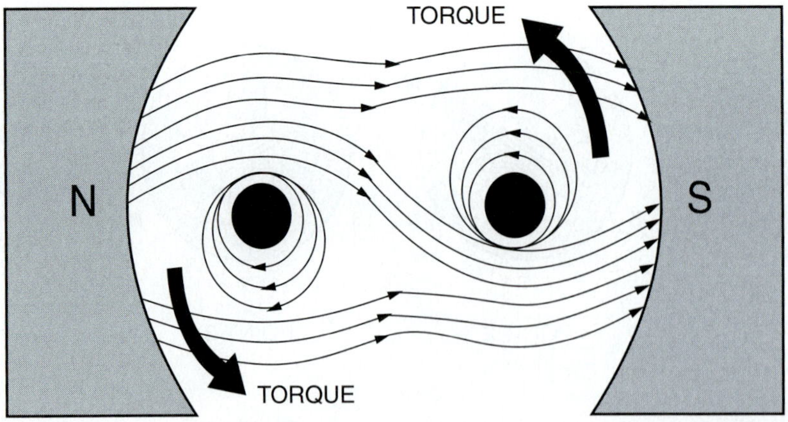

Figure 26 ◆ Producing torque.

4.2.1 Squirrel Cage Induction Motor

The squirrel cage is probably the most popular rotor in use. Three-phase squirrel cage induction motors consist of a stator, a rotor, and two end shields that house the bearings that support the rotor shaft. The frame is usually made of cast steel. The stator core is pressed into the frame. In this rotor, the bars are connected together at the ends by shorting rings made of similar material. The conductor bars carry large currents at low voltages. The bearings can be either sleeve or ball bearings. *Figure 27* shows the main components of an induction motor.

It is not necessary to insulate the bars from the core because the current will follow the path of least resistance and is confined to the cage windings. *Figure 28* shows how a squirrel cage rotor is constructed.

The squirrel cage rotor induction motor has a fixed rotor circuit. The resistance and reactance of the windings are determined when the motor is designed. The standard cage rotor motor is a general-purpose motor. It is used to drive loads that require variable torque at relatively constant speed with high full-load efficiency. Some examples are blowers, centrifugal pumps, and fans.

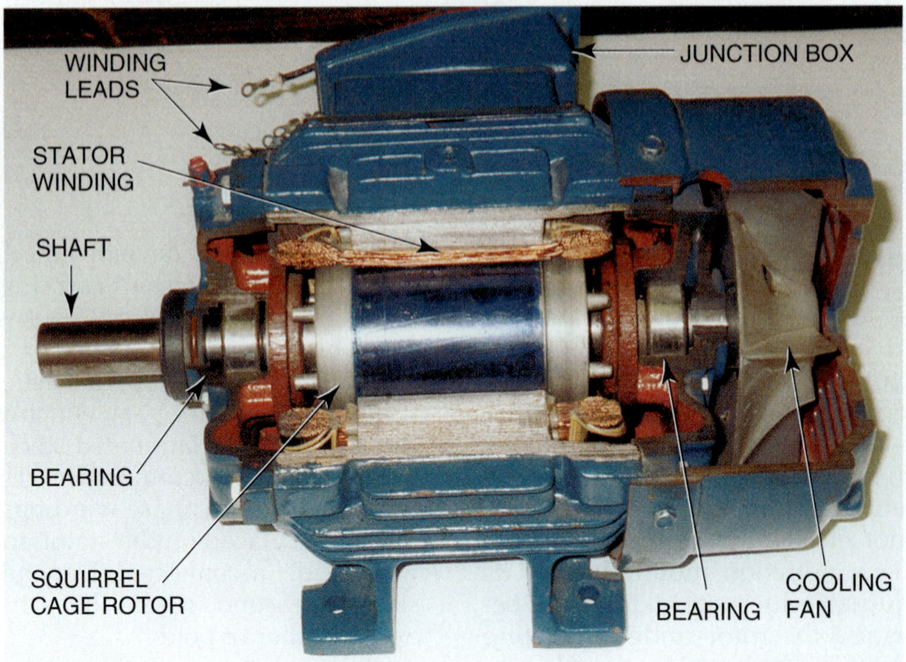

Figure 27 ◆ Main components of an induction motor.

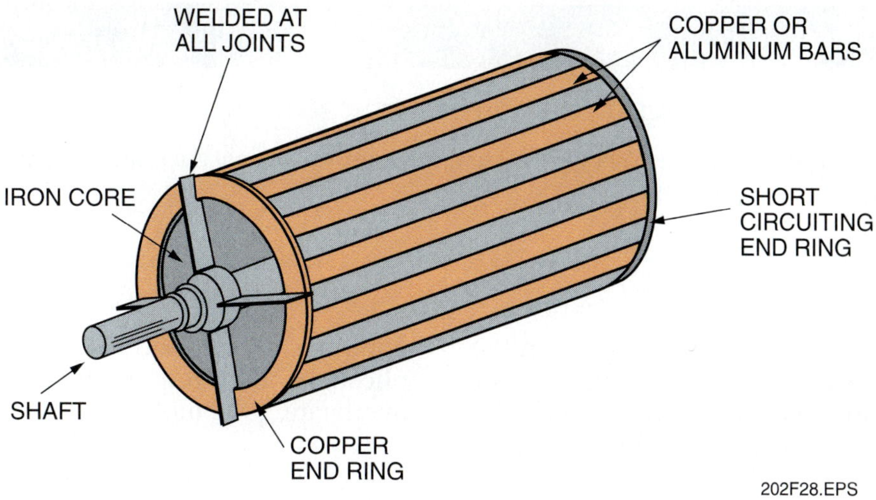

Figure 28 ◆ Squirrel cage rotor.

INSIDE TRACK

Squirrel Cage Motor Applications

Squirrel cage induction motors offer many advantages. For example, maintenance costs are low because these motors have no brushes or slip rings, but work entirely through induction. They also have a high starting torque, so they are useful in common applications such as overhead doors, large compressors, fans, and printing presses.

Due to the absence of any moving electrical contacts, they are suitable for use where they are exposed to flammable dust or gas.

If the load requires special operating characteristics, such as high starting torque, the squirrel cage rotor can be designed to have high resistance bars for a starting circuit and low resistance bars for running operation. A rotor of this type is called a double squirrel cage rotor.

4.2.2 Wound Rotor Induction Motor

A wound rotor (*Figure 29*) has a winding that is similar to the three-phase stator windings. The rotor windings are usually wye-connected with the free ends of the windings connected to three slip rings mounted on the rotor shaft. The slip rings are shown physically mounted on the end of the rotor shaft in *Figure 29*. They are used with brushes to form an electromechanical connection to the rotor.

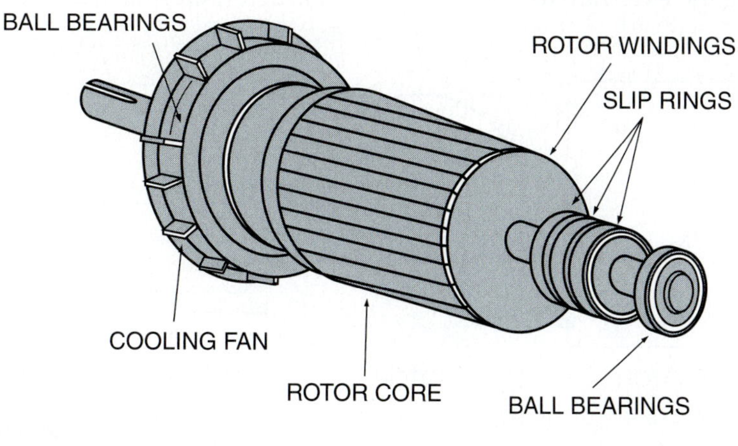

Figure 29 ◆ Wound rotor.

> **THINK ABOUT IT**
>
> ### *Wound Rotor Motors*
>
> When might you want a wound rotor motor to run like a squirrel cage motor at full speed?

Slip rings are contact surfaces mounted on the shaft of a motor or generator to which the rotor windings are connected and against which the brushes ride. The brushes are sliding contacts, usually made of carbon, that make continuous electrical connection to the rotating part of a motor or generator.

The wound rotor motor often uses an external wye-connected resistor connected to the rotor through slip rings. The resistor provides a means of varying the rotor resistance. This can be used when the motor is started to produce a high starting torque. As the motor accelerates, the resistance is reduced. When the motor has reached full speed, the slip rings are short circuited, and the operation is similar to that of a squirrel cage rotor induction motor. A schematic representation of this is shown in *Figure 30*.

The wound rotor induction motor is used when it is necessary to vary the rotor resistance, to limit starting current, or to vary the motor speed. Speed can be varied by as much as 50% to 75%; the greater the resistance inserted in the rotor circuit, the lower the speed will be below synchronous speed. When the motor is operating below full speed, the percent slip is increased and the motor is operating at reduced efficiency and horsepower. When all resistance is cut completely out, the speed is somewhat less than that obtained with squirrel cage rotors. Because the rotor circuit heat generation is largely external to the rotor windings, the wound rotor motor is used for applications that require frequent starts without overheating the motor.

4.2.3 Wound Rotor Speed Control

The insertion of resistance in the rotor circuit not only limits the starting surge of current, but also produces a high starting torque and provides a means of adjusting the speed. If the full resistance of the speed controller is cut into the rotor circuit when the motor is running, the rotor current decreases and the motor slows down. As the rotor speed decreases, more voltage is induced in the rotor windings and more rotor current is developed to create the necessary torque at the reduced rotor speed.

If all the resistance is removed from the rotor circuit, both the current and motor speed will increase. However, the rotor speed will always be less than the synchronous speed of the field developed by the stator windings. Recall that this is also true of the squirrel cage induction motor. The speed of a wound rotor motor can be controlled manually or automatically with timing relays, contactors, and pushbutton speed selection.

The advantages of the wound rotor motor are high starting torque with moderate starting current, smooth acceleration under heavy load, no excessive heating during starting, good running characteristics, and adjustable speed control. The

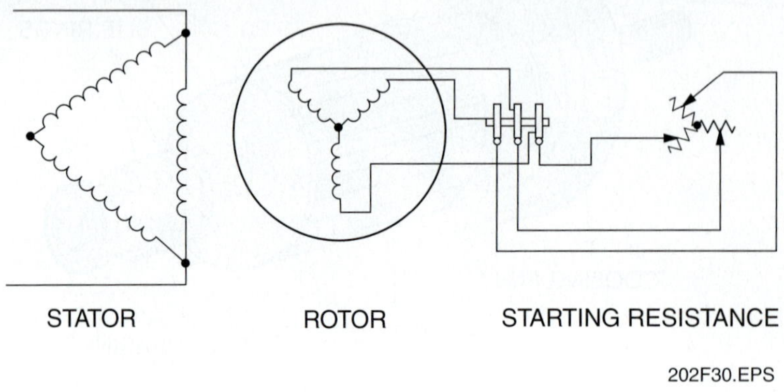

Figure 30 ◆ Wound rotor motor circuit.

INSIDE TRACK

Wound Rotor Motor Applications

Like DC motors, wound rotor motors are used where high inertia loads must be started easily or often. Wound rotor motors have starting torques in the range of 225% of full-load torque. They are used for hoists, hydraulic gates, yard locomotives, and cranes.

chief disadvantage is that both initial and maintenance costs are greater than those of the squirrel cage rotor motor.

4.2.4 Torque

The torque on the rotor of an induction motor tends to turn the rotor in the same direction as the rotating field. If the motor is not driving a load, it will accelerate to nearly the same speed as the rotating field. As the rotor accelerates, the magnitude of the induced voltage in the rotor decreases. This is because the relative motion between the rotating field and the rotor conductors is reduced. It is impossible for an induction motor to operate at synchronous speed because there would be no relative motion between the rotating field and the rotor. Thus, there would be no induced voltage, no rotor current, no rotor magnetic field, and no torque.

4.2.5 Slip

In an induction motor, the rotor always rotates at a speed less than the synchronous speed. The rotor speed is such that sufficient torque is produced to balance the restraining torque caused by motor friction and mechanical load. The difference between the synchronous speed and the rotor speed is known as slip. Slip is expressed mathematically as follows:

$$S = \frac{N - N_R}{N} \times 100\%$$

Where:

S = slip

N = synchronous speed

N_R = rotor speed

To express the quantity as a percent, multiply by 100.

Example:

A four-pole, 208V, 2hp, 60Hz, three-phase induction motor has a no-load speed of 1,790 rpm and a full-load speed of 1,650 rpm.

Find the percent slip for each case below:

- No-load condition
- Full-load condition
- Locked-rotor condition (standstill)

Before any calculations can be made, we must first calculate synchronous speed.

$$N = \frac{120f}{P}$$

$$N = \frac{120 \times 60}{4}$$

$$N = 1{,}800\text{rpm}$$

- At no-load condition:

$$S = \frac{N - N_R}{N} \times 100\%$$

$$S = \frac{1{,}800 - 1{,}790}{1{,}800} \times 100\%$$

$$S = 0.556\%$$

- At full-load condition:

$$S = \frac{N - N_R}{N} \times 100\%$$

$$S = \frac{1{,}800 - 1{,}650}{1{,}800} \times 100\%$$

$$S = 8.33\%$$

- At locked-rotor condition:

$$S = \frac{N - N_R}{N} \times 100\%$$

$$S = \frac{1{,}800 - 0}{1{,}800} \times 100\%$$

$$S = 100\%$$

Figure 31 shows how torque relates to speed over the operating range of a motor. Note that speed is proportional to torque on the left side up to pullout torque. Beyond this point, however, torque decreases as speed increases.

Slip is the difference between the synchronous speed and the actual speed of the rotor in an induction motor. Slip is necessary to permit motor action to occur. Under increasing load, the rotor torque increases. Since percent slip is proportional

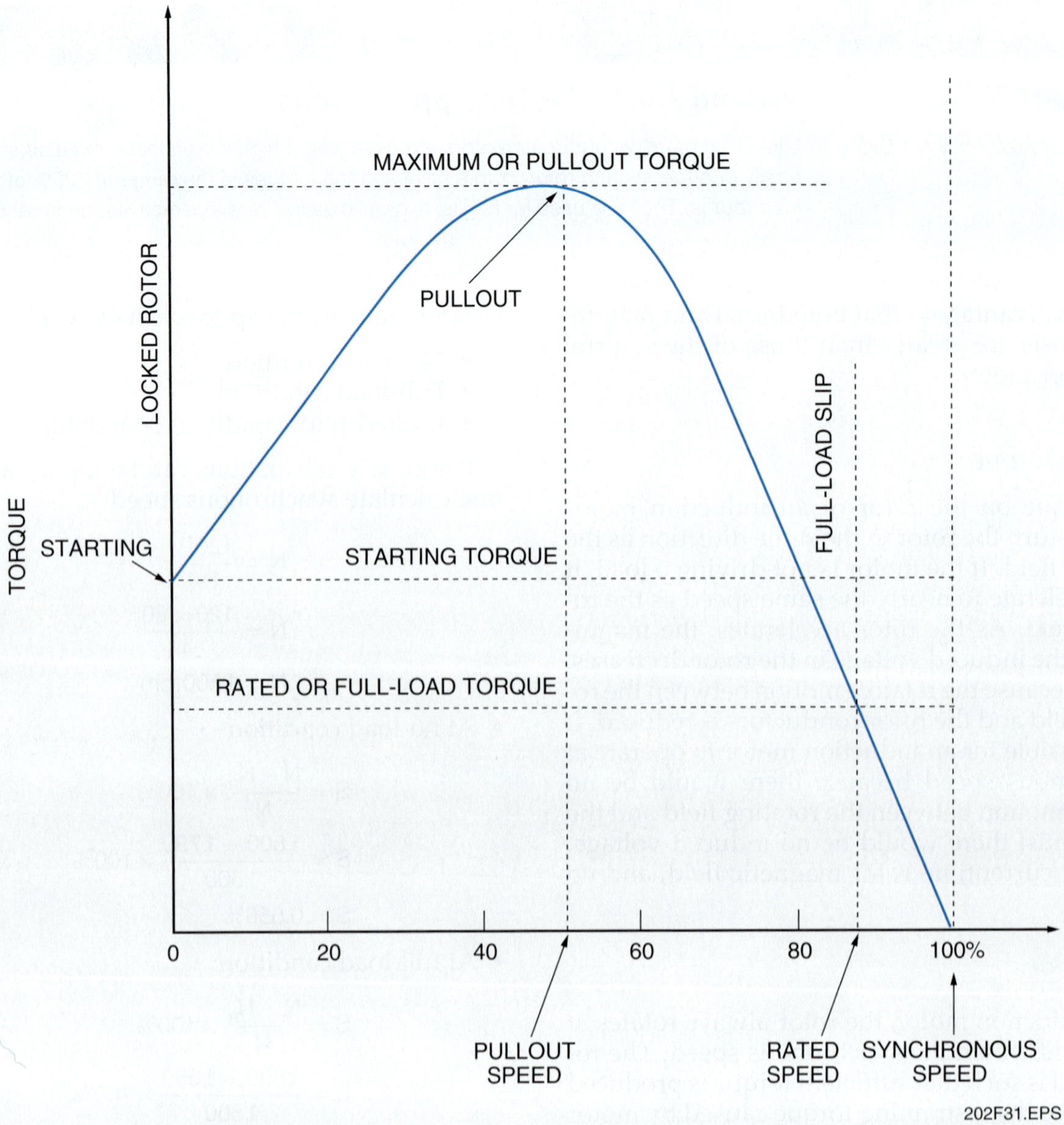

Figure 31 ◆ Typical torque-speed curve.

to torque, the amount of slip will increase. This increase means a higher current draw by the motor due to the greater difference between the rotor and the magnetic field. Motor supply voltages, current, torque, speed, and rotor impedance are closely related. By changing the resistance and reactance of the rotor, the characteristics of the motor can be changed; however, for any one rotor design these characteristics are fixed.

4.2.6 Starting Current

At the moment a three-phase induction motor is started, the current supplied to the motor stator terminals may be as high as six times the motor full-load current. This is because at starting, the rotor is at rest; therefore, the rotating magnetic field of the stator cuts the squirrel cage rotor at the maximum rate, inducing large amounts of EMF in the rotor.

This results in proportionally high currents at the input terminals of the motor as was previously discussed. Because of this high inrush, current starting protection as high as 300% of full-load current must be provided to allow the motor to start and come up to speed.

Because 100% slip exists at the instant the motor is energized (see *Figure 31*), the rotor current lags the rotor EMF by a large angle. This means that the maximum current flow occurs in a rotor conductor at a time after the maximum amount of stator flux has passed by. This results in a high starting current at a low power factor, which results in a low value of starting torque.

As the rotor speeds up, the rotor frequency and reactance decrease, causing the torque to increase up to its maximum value, then decrease to the value needed to carry the load.

4.2.7 Loaded Torque

If a load is now placed on the shaft, the rotor will tend to slow down. As it slows down, more flux lines are cut until enough torque is developed to overcome the load placed on the shaft.

The motor now runs under load at a slower speed than before the load was placed on the shaft. This normal range of operation is shown in the lower right corner of *Figure 31* as the rated or full-load torque.

In this range, the slip will vary from 2% to 10%, depending on the load applied and the motor. Rated slip will occur at the point where 100% rated load is applied. Increased load means increased slip, which means the rotor is now rotating slower. An induction motor is considered to be a constant speed motor. We will now examine how much speed fluctuates from no-load speed to full-load speed.

Example:

A two-pole induction motor has a no-load slip of 2% and a full-load slip of 8%. What are the no-load speed, full-load speed, and percent speed change?

$$\text{Nominal} = \frac{120 \times 60}{2} = 3{,}600 \text{ rpm}$$

$$\text{No-load speed} = \frac{100\% - 2\%}{100\%} \times 3{,}600 \text{ rpm} = 3{,}528 \text{ rpm}$$

$$\text{Full-load speed} = \frac{100\% - 8\%}{100\%} \times 3{,}600 \text{ rpm} = 3{,}312 \text{ rpm}$$

Percent speed change

$$= \frac{\text{no-load speed} - \text{full-load speed}}{\text{no-load speed}} \times 100\%$$

$$= \frac{3{,}528 - 3{,}312}{3{,}528} \times 100\%$$

$$= 6.12\%$$

4.2.8 Overload Condition

If the load is increased above full-rated load, everything happens as stated before to increase torque up to a certain point. *Figure 32* shows typical torque and current curves.

In *Figure 32*, note how the torque climbs as the load is increased.

This will continue as load is increased until the pullout torque point is reached. Beyond this point, the torque decreases and the motor will quickly

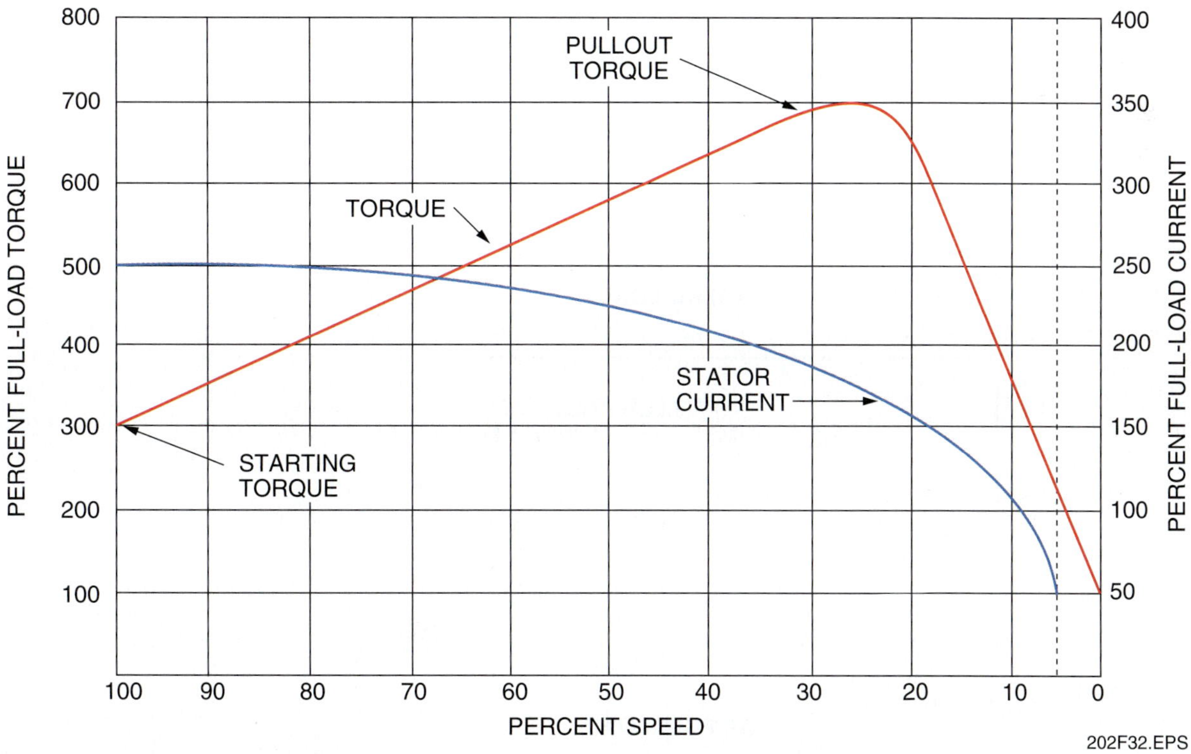

Figure 32 ◆ Torque and current curves.

stall. A typical situation is when a bench circular saw or a lathe stalls on a heavy cut. The machine will slow down as its cutting load is increased until it suddenly stalls and hums or growls loudly. The condition will persist until the load is relieved or a fuse blows or a breaker trips. The motor has simply reached a point where it cannot continue to increase its torque. Any further increase in load will cause a stall.

4.2.9 Power Factor

The power factor of a squirrel cage induction motor is poor at no-load and low-load conditions. At no-load conditions, the power factor can be as low as 15% lagging. However, as load is increased, the power factor increases. At high-rated load, the power factor may be as high as 85% to 90% lagging.

The power factor at no-load speed is low because the magnetizing component of input current is a large part of the total input current of the motor. When the load on the motor is increased, the in-phase current supplied to the motor increases, but the magnetizing component of current remains practically the same. This means that the resultant line current is more nearly in phase with the voltage, and the power factor is improved when the motor is loaded compared with an unloaded motor, which chiefly draws its magnetizing current.

Figure 33 shows the increase in power factor from no-load conditions to full-load conditions. In the no-load diagram, the in-phase current (I_{ENERGY}) is small when compared to the magnetizing current (I_M); thus, the power factor is poor at no-load conditions. In the full-load diagram, the in-phase current has increased, while the magnetizing current remains the same. As a result, the angle of lag of the line current decreases, and the power factor increases.

4.2.10 Speed Control

The speed of a three-phase squirrel cage induction motor depends on the frequency of the applied voltage and the number of poles. As a result, these motors are used in applications where speed remains constant or where it can be controlled by other means such as variable frequency drives.

4.2.11 Reversing Rotation

The direction of rotation of a three-phase induction motor can be readily reversed. The motor will

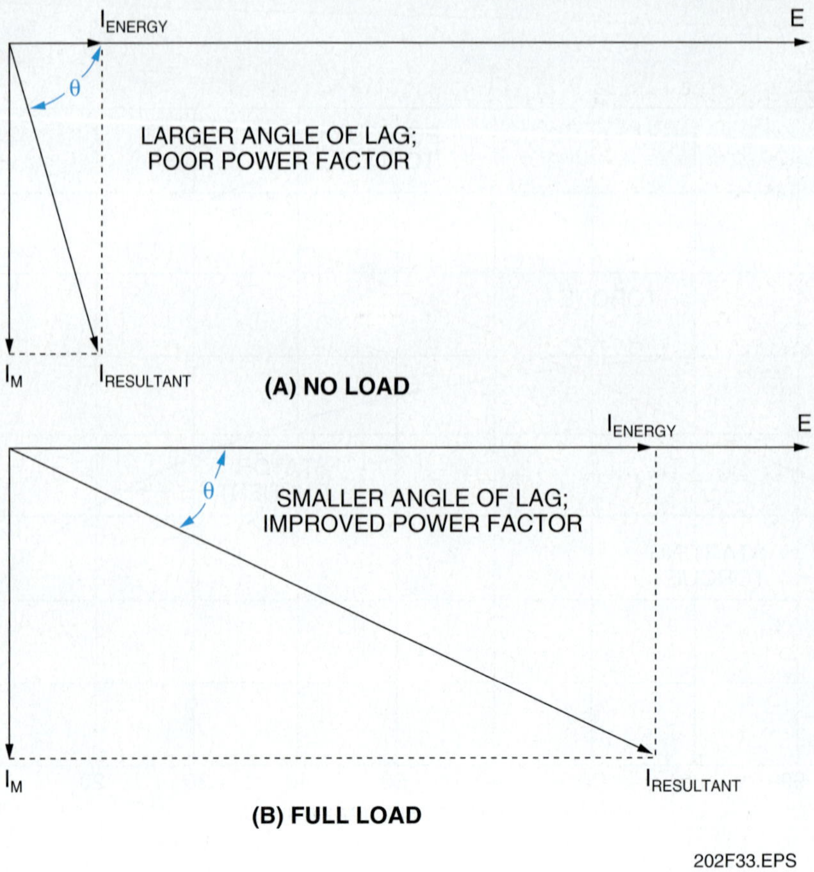

Figure 33 ◆ Power factor versus load for an induction motor.

Inside Track: Variable-Speed Drives

Variable-speed drives, known as VSDs or ASDs (adjustable-speed drives), are powerful electronic devices that are available for virtually any size motor in all types of applications. Of the various types of VSDs available, the most efficient versions for AC motors are VFDs (variable-frequency drives), which control both the frequency and the voltage applied to the motor. By changing the frequency of the rotating stator field, you change the speed of the rotor, thus changing the speed of the motor. VSDs are often used in energy management systems to conserve energy by supporting variable loads such as those that occur in heating, ventilating, and air conditioning systems. In addition to controlling the speed of a motor, VSDs are available to control both motor starting and stopping functions, as well as to provide controlled acceleration and deceleration.

rotate in the opposite direction if any two of the three incoming leads are reversed, as shown in *Figure 34*.

4.3.0 Synchronous Motors

The synchronous motor is a three-phase motor that operates at synchronous speed from no-load conditions to full-load conditions.

4.3.1 Characteristics

This type of motor has a revolving field that is energized from a source separate from the stator winding. The rotor is excited by a DC source. The magnetic field set up by the direct current on the rotor then locks in with the rotating magnetic field of the stator and causes the rotor to revolve at synchronous speed. By changing the magnitude of DC excitation, the power factor of the motor can be changed over a wide variety of power factors from leading to lagging. Because of the unique ability of synchronous motors to change power factors, they are often used as power-factor correctors. They are most often used in applications that require constant speed from no-load conditions to full-load conditions.

4.3.2 Construction

The construction of synchronous motors is essentially the same as the construction of three-phase

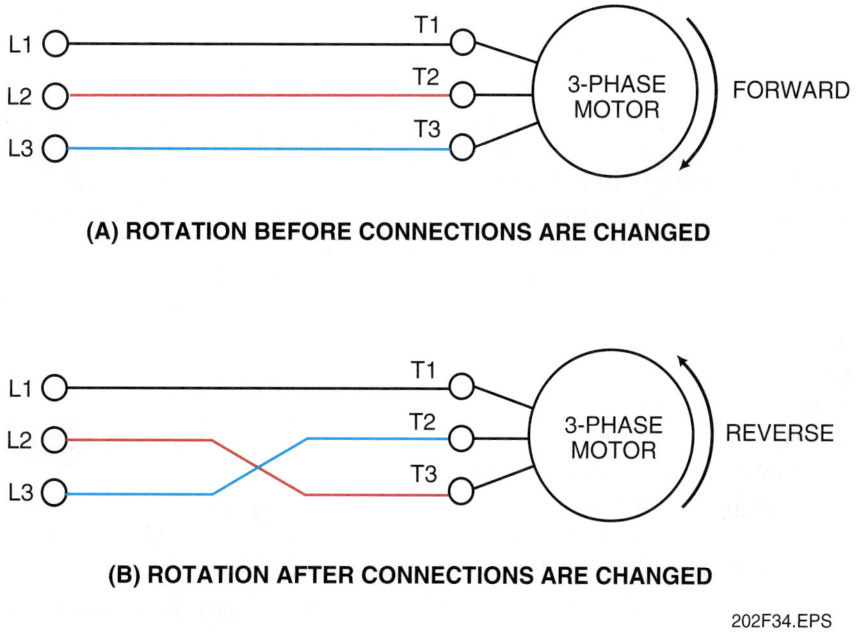

Figure 34 ◆ Three-phase induction motor rotational direction change.

Synchronous Motors

Three-phase synchronous motors can be used in industrial applications to correct the low power factor of a number of induction motors or other inductive devices that are operating at less than their rated load levels. Synchronous motors can accomplish power factor correction while driving their own mechanical loads. Correcting a low power factor created by inductive loads through the use of synchronous motors reduces energy costs by making efficient use of the power supplied to the industrial facility. The use of synchronous motors can eliminate the need for dedicated capacitor banks or switched capacitor banks and the surges caused by them.

generators. They have three stator windings that are 120° apart and a wound rotor that is connected to slip rings where the rotor excitation current is applied.

When three-phase AC is applied to the stator, a revolving magnetic field is created just as it is in induction motors. The rotor is energized with DC, which creates a magnetic field around the rotor. The strong rotating magnetic field of the stator attracts the rotor field. This results in a strong turning force on the rotor shaft.

This is how the synchronous motor works once it is started. However, one of the disadvantages of this type of motor is that it cannot be started just by applying AC to the stator. When AC is applied to the stator, the high-speed rotating magnetic field rushes past the rotor poles so quickly that the rotor does not have a chance to get started. The rotor is locked; it is repelled in one direction and then in another direction. In its purest form, the synchronous motor has no starting torque.

This is more easily understood using *Figure 35*. When the stator and rotor fields are energized, the poles of the rotating field approach the rotor poles of opposite polarity. The attracting force will tend to turn the rotor in a direction opposite the rotating field. As the rotor starts to move in that direction, the rotating field moves past the rotor poles and tends to pull the rotor in the same direction as the rotating field. The result is no starting torque.

To allow this type of motor to start, a squirrel cage winding is added to the rotor to cause it to start like an induction motor. This winding is called an amortisseur winding. The rotor windings are constructed so that definite north and south poles are created and these poles, when excited by DC, will lock in with the revolving field. The rotor windings are wound about the salient field poles, which are connected in series for opposite polarity.

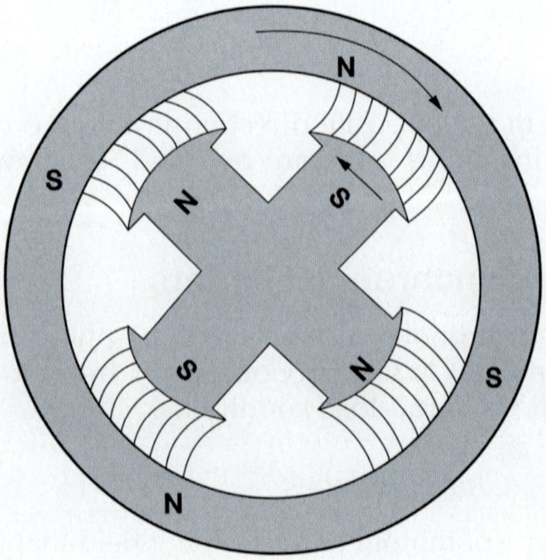

TENDENCY OF ROTOR TO
TURN COUNTERCLOCKWISE

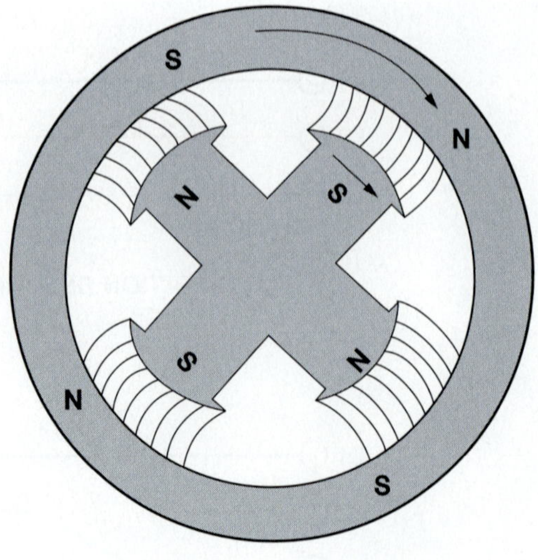

TENDENCY OF ROTOR TO
TURN CLOCKWISE

202F35.EPS

Figure 35 ◆ Synchronous motor operation at start.

The number of field poles must equal the number of stator poles. The rotor field windings are brought out to slip rings that are mounted on the rotor shaft. The field current is supplied through carbon brushes to the field windings. *Figure 36* shows a simplification of a synchronous motor. *Figure 37* shows the construction of the rotor pole assembly.

4.3.3 Principles of Operation

When a synchronous motor is started, current is first applied to the stator windings. Current is induced in the amortisseur winding and the motor starts as an induction motor. The motor then comes up to near-synchronous speed (about 5% to 10% slip). At that point, the field is excited, and the motor, turning at high speed, pulls into synchronism. When this occurs, the rotor is turning at synchronous speed, and the squirrel cage winding will not be generating any current, and therefore will not affect the synchronous motor's operation. The amortisseur windings serve an additional purpose. When the load changes frequently, the motor speed is not steady because the torque angle (discussed later) oscillates (or hunts) back and forth, trying to settle at its required value. This momentary change in speed creates a current due to induction, and there will be torque in the amortisseur winding. This momentary torque serves to dampen or stabilize the oscillating torque angle. That is why amortisseur windings are sometimes referred to as damper windings.

4.3.4 Rotor Field Excitation

The rotor must be excited from an external DC source. *Figure 38* shows a simplified synchronous motor excitation circuit. Notice that the DC field current can be varied by the rheostat; however, this does not change the speed of the motor. It only changes the power factor of the motor stator circuit. If full resistance is applied to the rotor field circuit, then the field strength of the rotor is at its minimum and the power factor is extremely lagging. As the DC field strength is increased, the power factor improves. If current is increased sufficiently, the power factor can be increased to near unity or 100%. This value of field current is referred to as normal excitation. By increasing the rotor field strength further, the power factor decreases but in a leading direction; that is, the stator circuit becomes capacitive and the motor is said to be overexcited. The synchronous motor can be used to counteract the lagging power factor in circuits by adding capacitive reactance to the circuit, thereby bringing the overall power factor closer to unity.

If the rotor DC field windings of a synchronous motor are open when the stator is energized, a high AC voltage will be induced in it because the

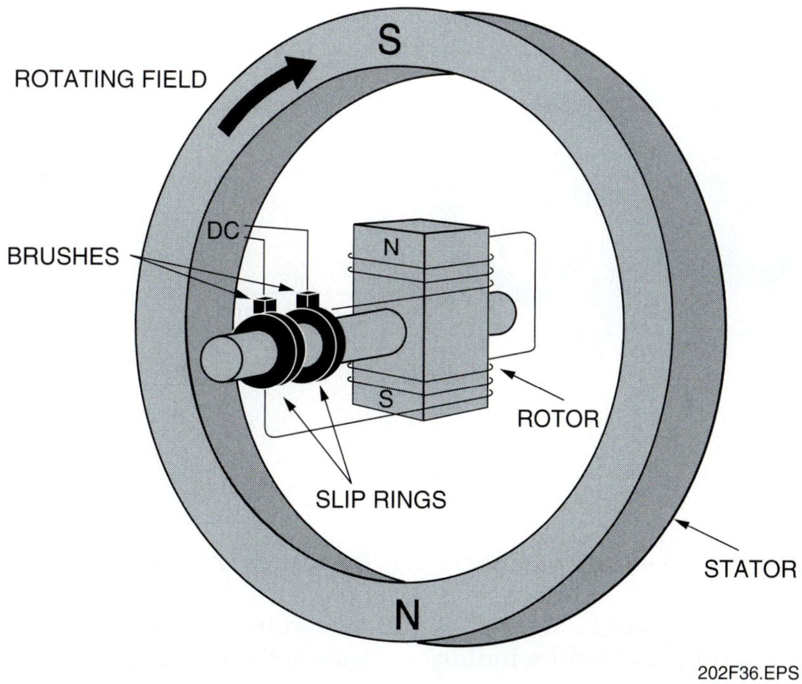

Figure 36 ◆ Simplification of a synchronous motor.

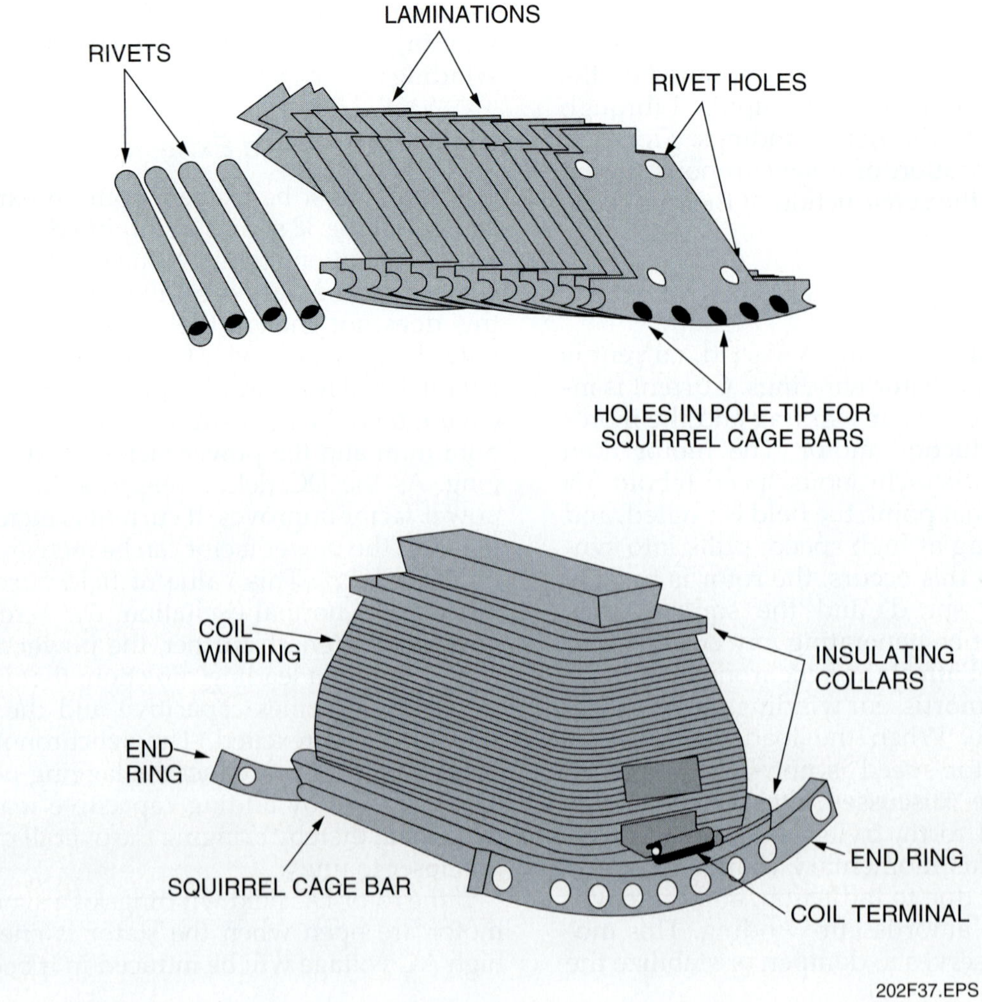

Figure 37 ◆ Pole assembly.

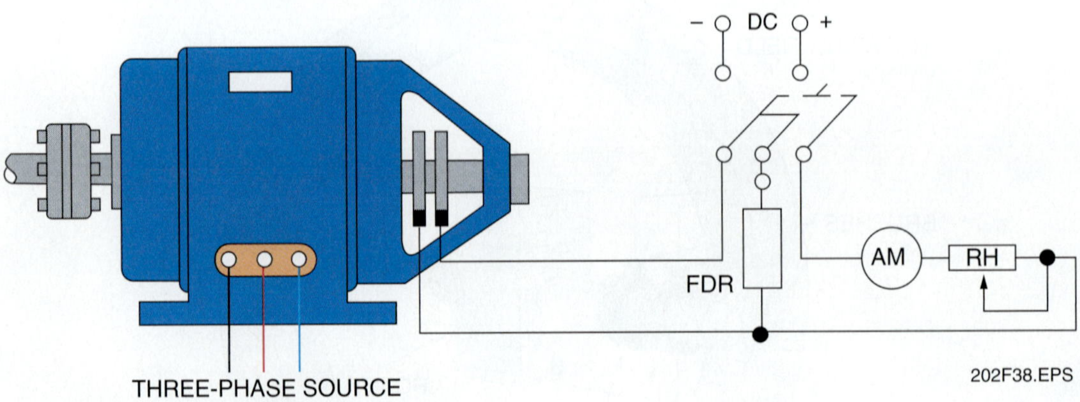

Figure 38 ◆ Simplified synchronous motor excitation circuit.

rotating field sweeps through the large number of turns at synchronous speed.

It is therefore necessary to connect a resistor of low resistance across the rotor DC field winding during the starting period. During the starting period, the DC field winding is disconnected from the source, and the resistor is connected across the field terminals. This permits alternating current to flow in the DC field winding. Because the impedance of the winding is high compared with the inserted external resistance, the internal voltage drop limits the terminal voltage to a safe value.

4.3.5 Synchronous Motor Pullout

When a synchronous motor loses synchronism with the system to which it is connected, it is said to be out of step. This occurs when the following take place singly or in combination:

- Excessive load applied to the shaft
- Supply voltage reduced excessively
- Motor excitation lost or too low

Torque pulsations applied to the shaft of a synchronous motor are also a possible cause of loss of synchronism if the pulsations occur at an unfavorable period relative to the natural frequency of the rotor with respect to the power system.

A prevalent cause of loss of synchronism is a fault occurring on the supply system. Underexcitation of the rotor is also a distinct possibility.

Synchronous motor pullout is significant in that the squirrel cage or amortisseur winding is designed for starting only. They are not as hardy as those found in induction motors. The amortisseur winding will not overheat if the motor starts, accelerates, and reaches synchronous speed within a time interval determined to be normal for the motor. However, the motor must continue to operate at synchronous speed. If the motor operates at a speed less than synchronous, the amortisseur winding may overheat and suffer damage.

Protection against a synchronous motor losing synchronism can be provided by polarized field frequency relays and out-of-step relays as well as various digital methods.

4.3.6 Synchronous Motor Torque Angle

Once the rotor is brought up to high speed (close to synchronous speed) it will lock on to the rotating magnetic field. Under these conditions, a running torque will be developed. The rotor will rotate at synchronous speed in a direction and at a speed determined by synchronous speed.

While the motor is running, the two rotating fields will line up perfectly. The rotor pole will always lag behind the stator pole by some angle. This angle is called the torque angle and is shown in *Figure 39*.

As the load on the shaft increases, the torque angle increases even though the rotor continues to turn at synchronous speed. This behavior continues until the torque angle is approximately 90°. At that point, the motor is developing a maximum torque. Any further increase in load will cause either of the following to occur:

- If the increase in load is momentary or very small, the rotor will slip a pole. In other words, the stator field will lose hold of the rotor and grab onto it again the next time around.

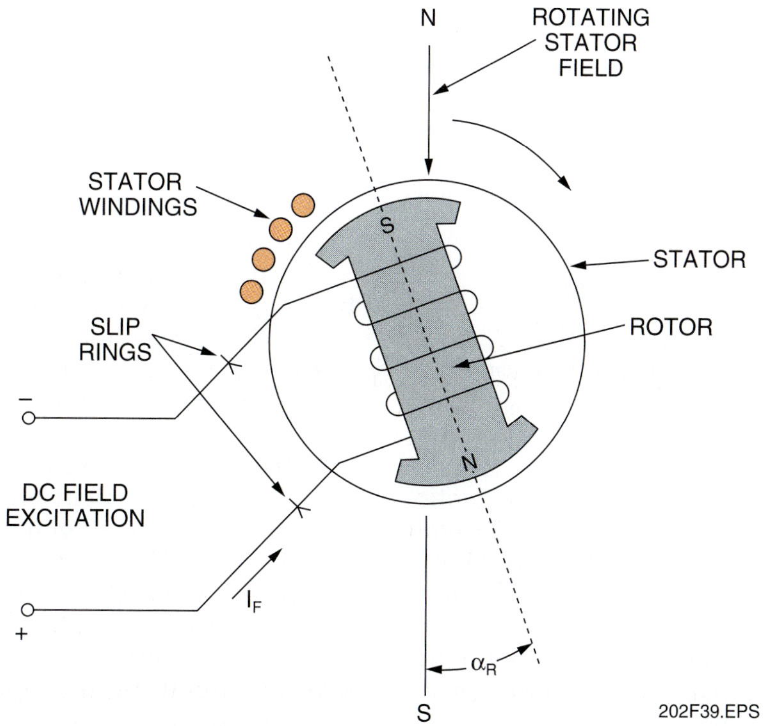

Figure 39 ◆ Torque angle.

- If the increase in load is large enough and is not momentary, the motor will lose synchronism and will either stall or cause the rotor to suffer thermal damage.

In both cases, a noticeable straining sound will be heard.

The synchronous motor should not be used where fluctuations in torque are violent. As a rule, it is also not used in small sizes (under 50hp), because it requires DC excitation. It is more difficult to start than induction motors and falls out of step quite readily when system disturbances occur. Its common applications are in motor generator sets, air compressors, and compressors in refrigerating plants.

4.4.0 Single-Phase AC Motors

Single-phase motors operate on a single-phase power supply. This is important because in the typical home or office and many areas of industrial plants, the only power source available is single-phase AC. Not only do single-phase AC motors eliminate the need for three-phase AC lines, but they are also easier to manufacture in small sizes and are, therefore, less expensive.

Examples of the many applications of single-phase AC motors today are the following: refrigerators, freezers, washers, dryers, power tools, typewriters, copying machines, heating systems, water pumps, computer peripherals, and various small appliances.

There are two basic types of single-phase motors. First, there is the single-phase induction motor. Its theory of operation is similar to that of the three-phase induction motor; hence, it runs at a speed slightly lower than synchronous speed. Second, there is the single-phase synchronous motor.

4.4.1 Single-Phase Induction Motors

Single-phase AC induction motors are extremely popular. Unlike polyphase induction motors, the stator field in the single-phase motor does not rotate. Instead, it simply alternates polarity between poles as the AC voltage changes polarity.

Voltage is induced in the rotor, and a magnetic field is produced around the rotor. This field will always be in opposition to the stator field. However, the interaction between the rotor and stator fields will not produce rotation (see *Figure 40*). Because this force is across the rotor and through the pole pieces, there is no rotary motion, just a push and/or pull along this line.

If the rotor is rotated by some outside force (a twist of your hand, for example), the push-pull along the line is disturbed. Look at the fields shown in *Figure 40* as the motor begins to rotate. At this instant, the south pole on the rotor is being attracted to the left-hand pole. The north rotor pole is being attracted to the right-hand pole. All of this is a result of the rotor being rotated 90° by the outside force.

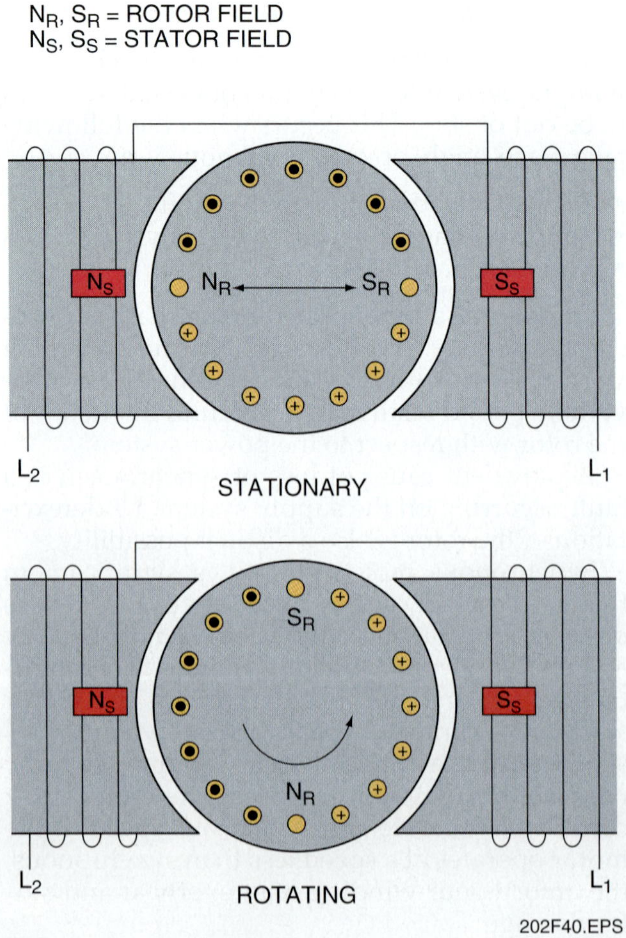

Figure 40 ◆ AC induction motor.

The pull that now exists between the two fields becomes a rotary force, turning the rotor toward magnetic correspondence with the stator. Because the two fields continuously alternate, they will never actually line up and the rotor will continue to turn once started.

Since a single-phase rotor will rotate if it has a rotating magnetic field present, all that remains is to find a means of generating a rotating field at the start. There are a number of practical means for generating a rotating field. All the methods used for single-phase induction motors involve the simulation of a second phase for a starting circuit. In this module, we will discuss the following types of motors: split-phase, capacitor-start, capacitor-run, shaded-pole, and repulsion-start.

Single-Phase Induction Motors

INSIDE TRACK

Outside of large industrial and commercial facilities, single-phase induction motors are the most common type of motor used. While they are initially less expensive than polyphase motors, they are also less efficient and more costly to maintain. They are typically available in small sizes from ⅛hp to 1hp (previously referred to as fractional horsepower sizes) and in sizes up to 10hp.

202P0201.EPS

4.4.2 Split-Phase Induction Motor

The split-phase motor, shown schematically in *Figure 41*, has a stator composed of slotted laminations that contain an auxiliary (starting) winding and a running (main) winding. The axes of these two windings are displaced by an angle of 90 electrical degrees. The starting winding has fewer turns and smaller wire than the running winding and, therefore, has different electrical characteristics. The main winding occupies the lower half of the slots and the starting winding occupies the upper half. The two windings are connected in parallel across the single-phase line supplying the motor. The motor derives its name from the action of the stator during the starting period.

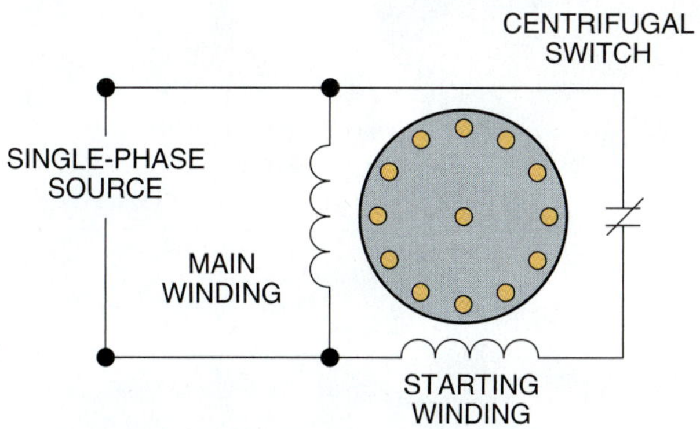

202F41.EPS

Figure 41 ◆ Split-phase motor.

When energized with single-phase AC, the two windings are physically different enough in position and construction to produce a magnetic revolving field that rotates around the stator air gap at synchronous speed. As the rotating field moves around the air gap, it cuts across the rotor conductors and induces a voltage in them. The interaction between the rotor and stator causes the rotor to accelerate in the direction in which the stator field is rotating.

When the rotor has come up to about 75% of synchronous speed, a centrifugally operated switch disconnects the starting winding from the line supply, and the motor continues to run on the main winding alone. As the motor ages, the centrifugal switch contacts pit and corrode. When this happens, they may get stuck in the closed position. To safeguard against the winding burning up, a thermal relay is also used. If the motor draws the high starting current for more than 5 or 10 seconds, the relay will de-energize.

INSIDE TRACK

Modern Split-Phase Induction Motor

Photo (A) shows a centrifugally-actuated start winding switch that is closed when the motor is at rest. The start winding switch is opened and closed by the movement of an actuator disk against a contact lever. The actuator disk is moved back and forth on the rotor shaft by a centrifugal weight assembly. When the motor is at rest, springs retract the weights and cause the actuator disk to move toward the bearing at the end of the shaft. This pushes on the contact lever, closing the switch contacts. After the motor starts and reaches about 75% of its rated speed, the weights swing out against the spring tension and retract the contact disk from the contact lever. This opens the switch contacts, removing the starting winding from the circuit. If the switch does not open after starting, the motor will operate at a reduced speed until it overheats the starting winding and activates the thermal relay. Photo (B) shows the starting winding (green) and the running winding (copper).

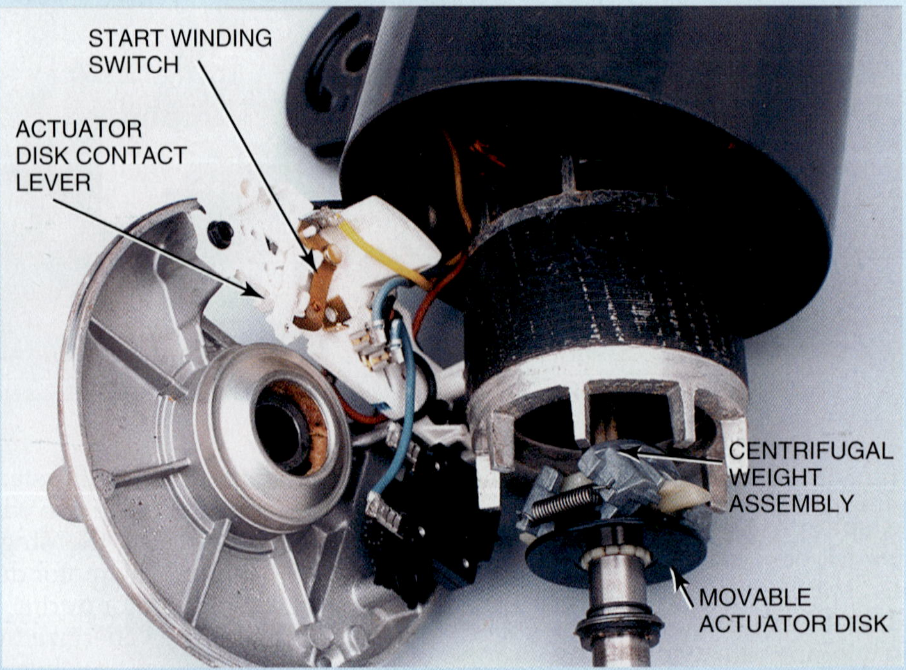

(A) 202P0202.EPS

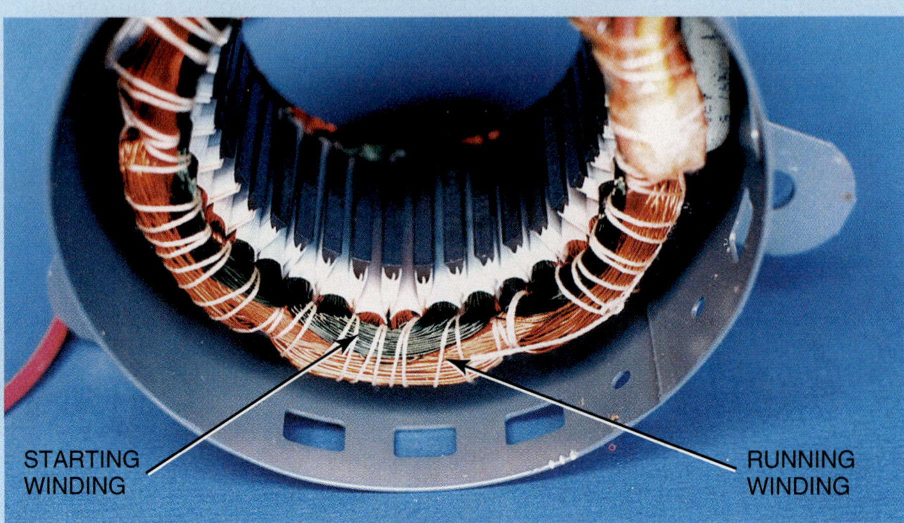

(B) 202P0203.EPS

In a split-phase motor, the starting torque is 150% to 200% of the full-load torque, and the starting current is six to eight times the full-load current. Fractional-horsepower split-phase motors are used in a variety of devices such as washers, oil burners, and ventilating fans. The direction of rotation of the split-phase motor can be reversed by interchanging the starting winding leads.

4.4.3 Capacitor-Type Induction Motor

The capacitor-type motor is a modified form of split-phase motor. A typical capacitor-type motor is shown in *Figure 42*. The capacitor is located on top of the motor.

To develop a larger starting torque than that available with a standard split-phase motor, a capacitor is placed in series with the auxiliary winding of a split-phase motor, as shown in *Figure 43*. This is called a capacitor-start motor. The capacitor tends to create a greater electrical phase separation of the two windings. Also, because the reactance of a capacitor is 180° out of phase with the inductive reactance of the motor windings when they are combined, they yield a lower total impedance. This allows a larger current to produce a greater magnetic field.

The net effect of the capacitor is to give its motor a starting torque of about four times its rated torque. The split-phase motor, on the other hand, produces a starting torque of about one to two times its rated torque. Once the capacitor motor has come up to speed and the starting winding has been disconnected, it will have the same running characteristics as the split-phase motor.

To reverse the direction of rotation of the capacitor-start motor and split-phase motor, the connection of either winding would have to be reversed. Since the starting winding is disconnected at a high speed, this reversal can be accomplished only at standstill or at low speeds when the centrifugal switch is still closed.

The capacitor-start motor is made in sizes from ¼ to 10hp (150W to 7.5kW). The starting capacitor is the dry-type electrolytic capacitor made for AC use. Typical values are from 200 to 600 microfarads (μF). *Figure 44* shows a comparison of torque slip curves for a split-phase and capacitor-start motor. It also shows the typical effect of the starting capacitor.

A variation of the capacitor-start motor is one in which the capacitor and auxiliary winding are not disconnected. The centrifugal switch in *Figure 43* is eliminated, and the auxiliary winding is left in all the time. This motor is called a capacitor-run motor.

202F42.EPS

Figure 42 ◆ Capacitor motor.

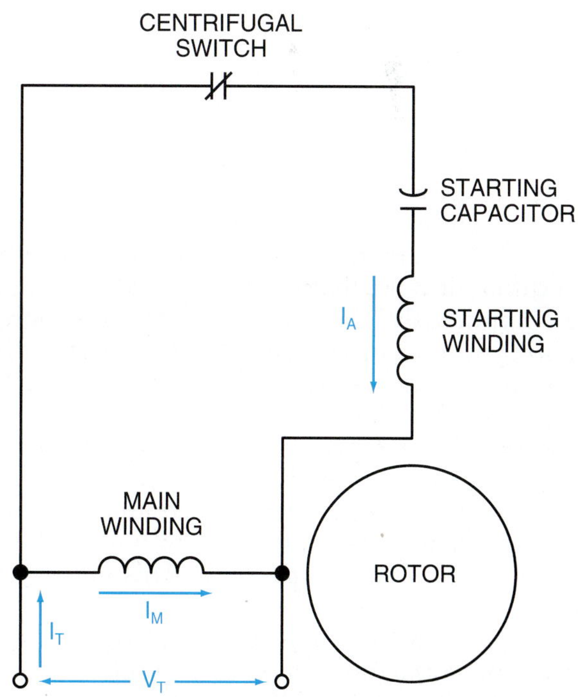

202F43.EPS

Figure 43 ◆ Capacitor-start motor schematic.

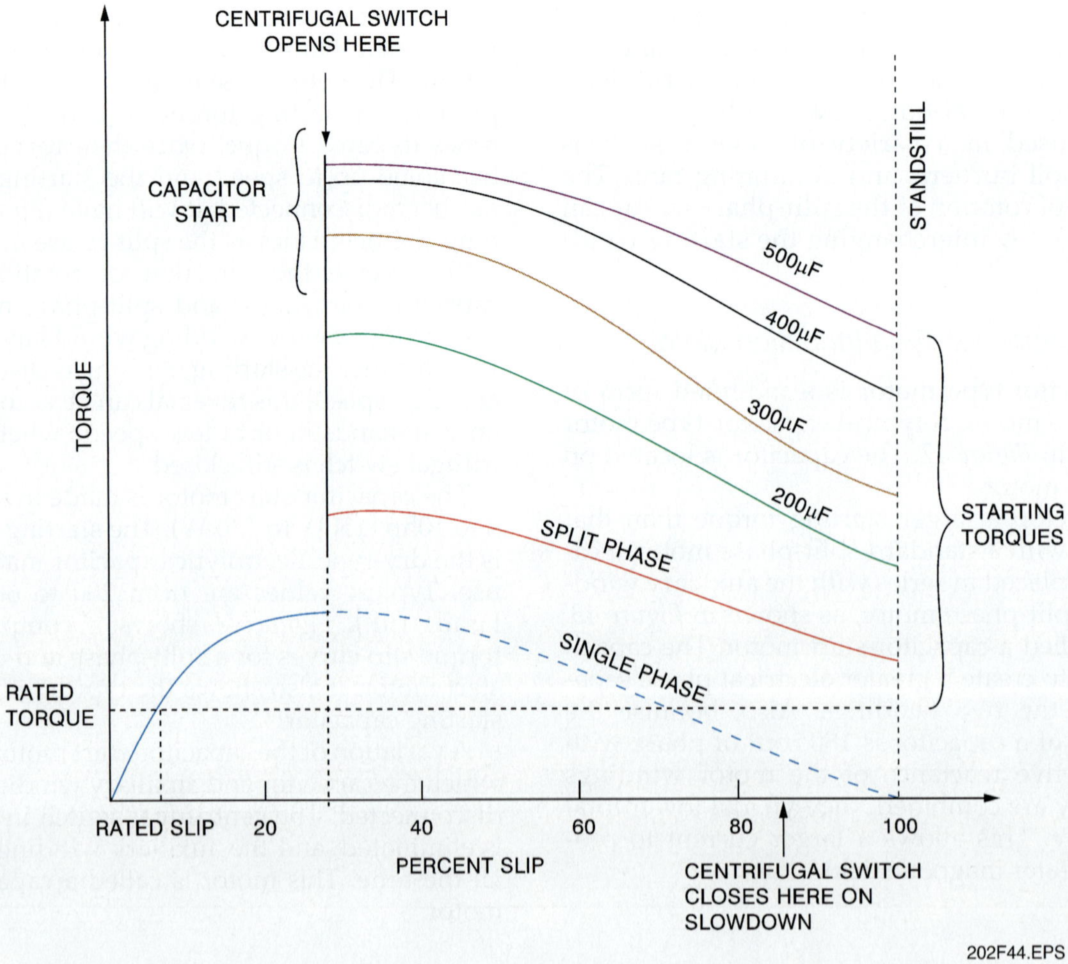

Figure 44 ◆ Torque-slip curves.

The capacity used for running under load is not the same as that needed for starting. Furthermore, the capacitors used for starting cannot be used for continuous operation. Since the capacitor used in this motor is in all the time, it must be of a different type; that is, one capable of operating continuously. The net result is that the motor has improved running characteristics; however, it does not provide a starting torque as large as that of the capacitor-start motor.

Among the improvements are higher efficiency and power factor at rated load, lower line current, and very quiet operation. It should also be pointed out that the start winding must be designed for continuous operation. This makes the motor somewhat more costly.

Another variation is the capacitor-start, capacitor-run motor. This motor combines the useful features of the capacitor-start and the capacitor-run motors by using two different capacitors, as shown in *Figure 45*.

4.4.4 Shaded-Pole Induction Motor

The shaded-pole motor employs a salient-pole stator and a cage rotor. The projecting poles on the stator resemble those of DC machines, except that the entire magnetic circuit is laminated and a portion of each pole is split to accommodate a short circuited copper strap called a shading coil. This motor is generally manufactured in very small sizes and runs up to $\frac{1}{20}$hp. A four-pole motor of this type is illustrated in *Figure 46*.

The shading coils are placed around the leading pole tip, and the main pole winding is concentrated and wound around the entire pole. The four coils that make up the main winding are connected in series across the motor terminals. An inexpensive type of two-pole motor that uses shading coils is illustrated in *Figure 47*.

Referring to *Figure 47*, we see that during part of the cycle when the main pole flux (ϕ_1) is increasing, the shading coil is cut by the flux, and the resulting induced EMF and current in the

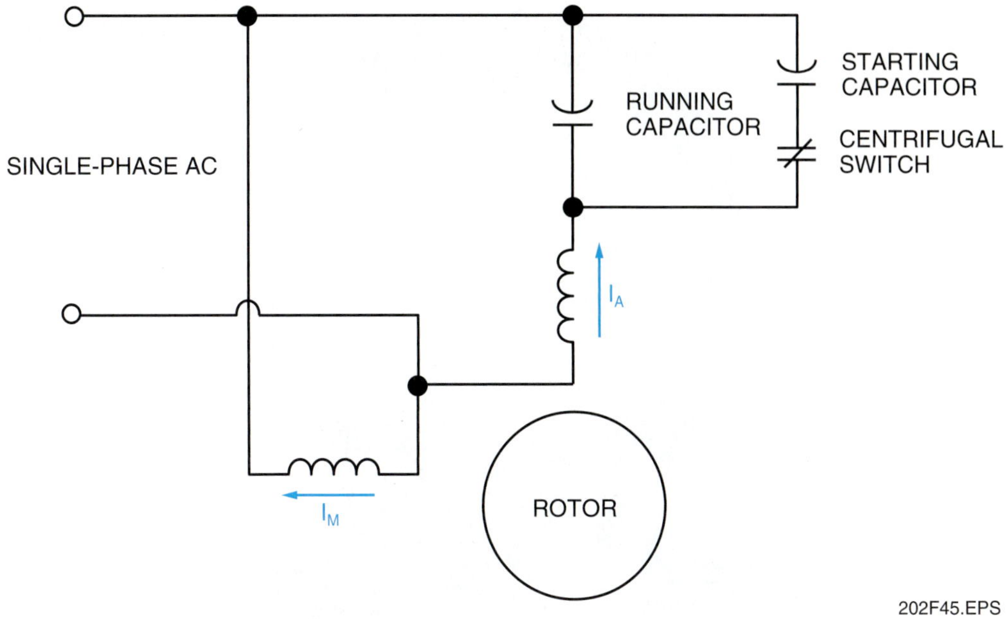

Figure 45 ◆ Capacitor-start, capacitor-run motor schematic.

INSIDE TRACK

Capacitor-Start, Capacitor-Run Motors

These motors run quietly and smoothly and have a high starting torque. Their construction is similar to a split-phase motor in that they use a centrifugal switch to remove only the start capacitor from the starting winding while leaving the run capacitor connected to the winding. These motors have a higher power factor than ordinary split-phase motors.

202P0204.EPS

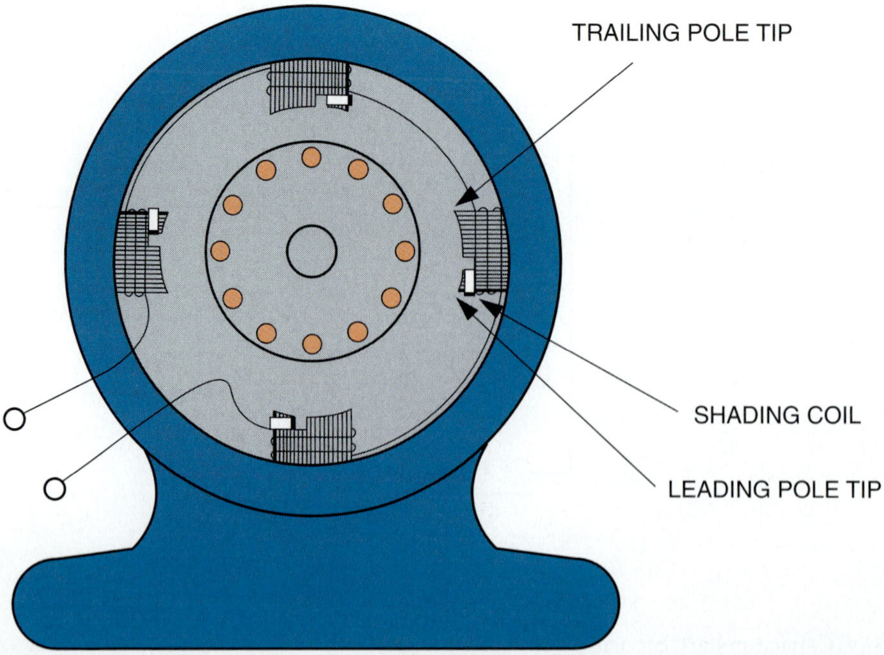

Figure 46 ◆ Four-pole shaded-pole motor.

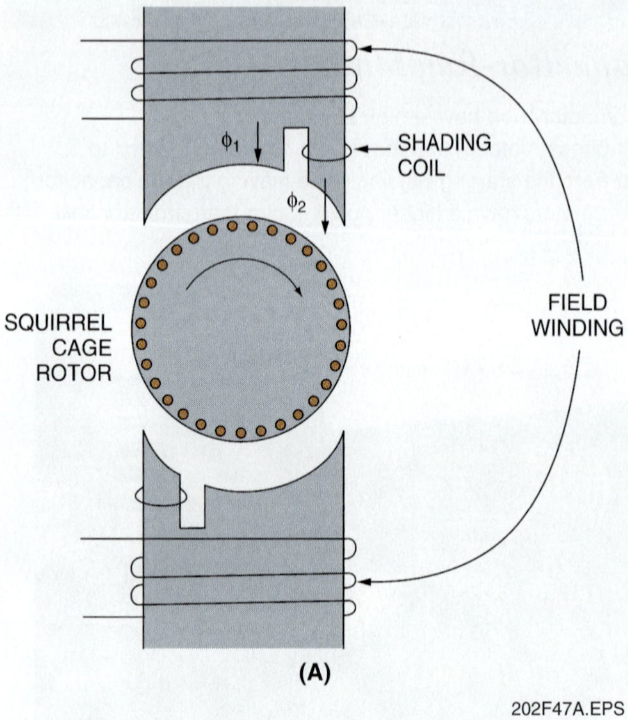

Figure 47 ◆ Two-pole shaded-pole motor.

Shaded-Pole Motors

INSIDE TRACK

The efficiency of this type of motor can be as low as 5%, but this low efficiency is rarely significant because these motors use very little power to begin with ($\frac{1}{20}$hp or less).

shading coil tend to prevent the flux from rising readily through it. Thus, the greater portion of the flux rises in that portion of the pole that is not in the vicinity of the shading coil ($\phi_1 > \phi_2$). When the flux reaches its maximum value, the rate of change of flux is zero, and the voltage and current in the shading coil are also at zero. At this time, the flux is distributed more uniformly over the entire pole face ($\phi_1 = \phi_2$).

As the main flux decreases toward zero, the induced voltage and current in the shading coil reverse their polarity, and the resulting force tends to prevent the flux from collapsing through the iron in the region of the shading coil ($\phi_2 > \phi_1$). The result is that the main flux rises first in the unshaded portion of the pole and later in the shaded portion. This action is equivalent to a sweeping movement of the field across the pole face in the direction of the shaded pole. The cage rotor conductors are cut by this moving field, and the force exerted on them causes the rotor to turn in the direction of the sweeping field.

Most shaded-pole motors have only one edge of the pole split, and therefore, the direction of rotation is not reversible. However, some shaded-pole motors have both leading and trailing pole tips split to accommodate shading coils. The leading pole tip shading coils form one series group, and the trailing pole tip shading coils form another series group. Only the shading coils in one group are simultaneously active, while those in the other group are on an open circuit.

The shaded-pole motor is similar in operating characteristics to the split-phase motor. It has the advantages of simple construction and low cost. It has no sliding electrical contacts and is reliable in operation. However, it has low starting torque, low efficiency, and a high noise level. It is normally used to operate small fans. The shading coil and split pole are also used in timers to make them self-starting.

4.4.5 Single-Phase Synchronous Motor

The single-phase synchronous motor, as its name implies, runs at synchronous speed. It finds use where a constant speed is needed, such as in turntables and clocks. It is started in the same way as any of the single-phase induction motors and therefore has a rotating field. By having a modified rotor, the motor pulls into synchronism and runs at synchronous speed.

5.0.0 ◆ MULTIPLE-SPEED INDUCTION MOTORS

The speed of an induction motor depends on the power supply frequency and the number of pairs of poles used in the motor. Obviously, to alter motor speed it is merely necessary to change one of these two factors. By far the most common method used involves changing the number of poles, generally at some type of external controller.

There are two types of multiple-speed squirrel cage induction motors in common use: the multiple-winding motor and the consequent-pole motor. Both feature poles that may be changed, as required, by shifting key external connections, and in this way they provide for operating the motor at a limited number of different speeds.

5.1.0 Multiple-Winding Motor

In the multiple-winding motor, two or more separate windings are placed in the stator core slots, one over the other, as shown in *Figure 48*. For example, a four-pole winding can be positioned in the core slots and have a two-pole winding placed on top of it. The windings are insulated from each other and arranged so that only one winding at a time can be energized. Switching speeds is normally accomplished by switching contacts that are in the motor controller external to the motor itself.

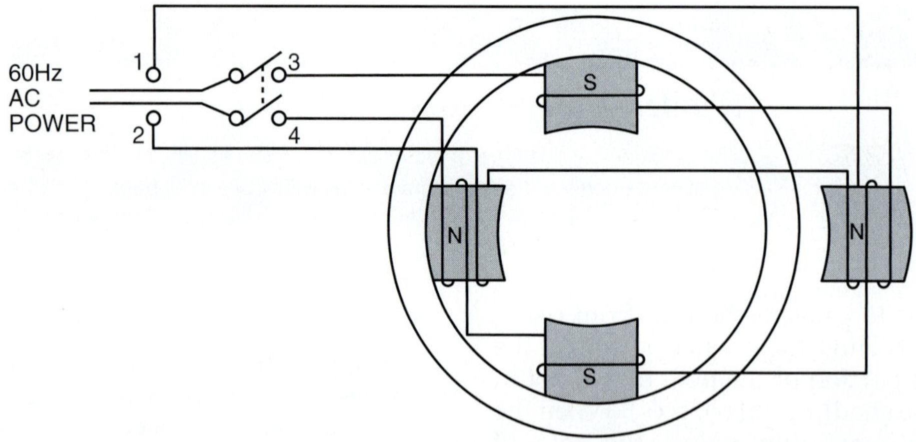

202F48.EPS

Figure 48 ◆ Two-winding, two-speed motor.

INSIDE TRACK

Hysteresis Synchronous Motor

This unusual synchronous motor has an external rotor. These motors have low noise levels, high efficiency, and constant speed, and are used for applications such as phonograph turntables.

202P0205.EPS

5.2.0 Consequent-Pole Motor

In the consequent-pole motor, there are two speeds. The motor is constructed to have a certain number of poles for high-speed operation and then, by a switching action, double this number of poles to give low-speed operation. The switching action is illustrated by the use of the two-phase motor in *Figure 49*. If you trace the wiring in *Figure 49*, you can see how the system is phased so that both magnetic north and south poles are produced at the winding projections. With two-phase power applied to the two-pole motor (two poles per phase), a rotating magnetic field of 3,600 rpm is produced.

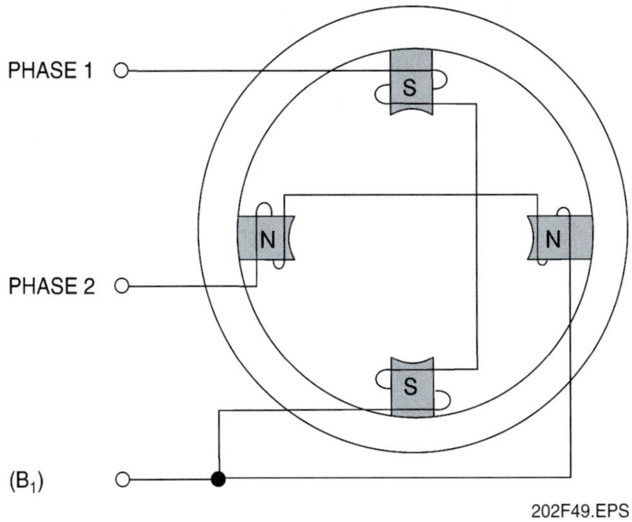

Figure 49 ◆ High-speed consequent-pole motor.

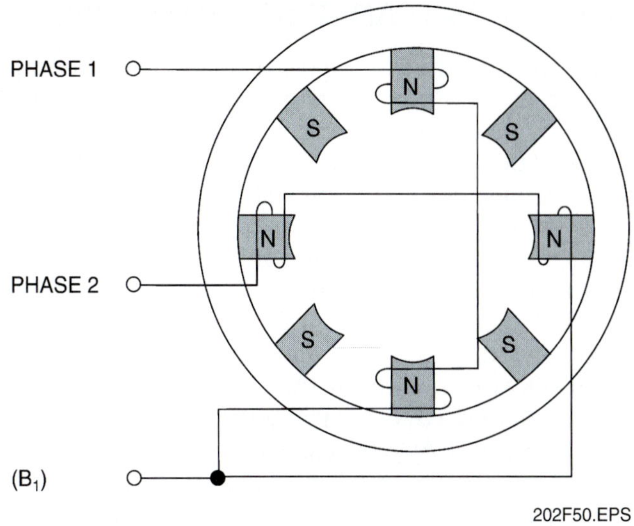

Figure 50 ◆ Low-speed consequent-pole motor.

In *Figure 50*, the connections are changed so that the system is phased to produce four magnetic north poles at the winding projections. Since every north pole must have a south pole, consequent south poles are produced between the projecting north poles as a consequence of having formed north poles. Accordingly, in *Figure 50*, there are twice as many pole groups as in *Figure 49*. Therefore, a four-pole rotating magnetic field of 1,800 rpm is produced.

Figure 51 shows the short-jumpering arrangement of the consequent-pole motor. In this, all windings are in series, and alternate north and south poles are produced. To produce consequent poles, the series connection is replaced with a parallel connection accomplished by the long-jumpering arrangement. By connecting the motor in this manner, four salient monopoles are produced and, as a result, create four opposite consequent poles. In the practical consequent-pole motor, all necessary internal connection rearrangements are accomplished at an external control panel.

Consequent-pole motor characteristics depend on the intended application. In a constant-horsepower motor, torque varies inversely with speed. It is used for driving machine tools.

In a constant-torque motor, horsepower varies directly with speed. It is used to drive pumps and air compressors, as well as in constant-pressure blowers. In a variable-torque, variable-horsepower motor, both torque and horsepower change with changes in speed. This is the type of motor found in household fans and air conditioners.

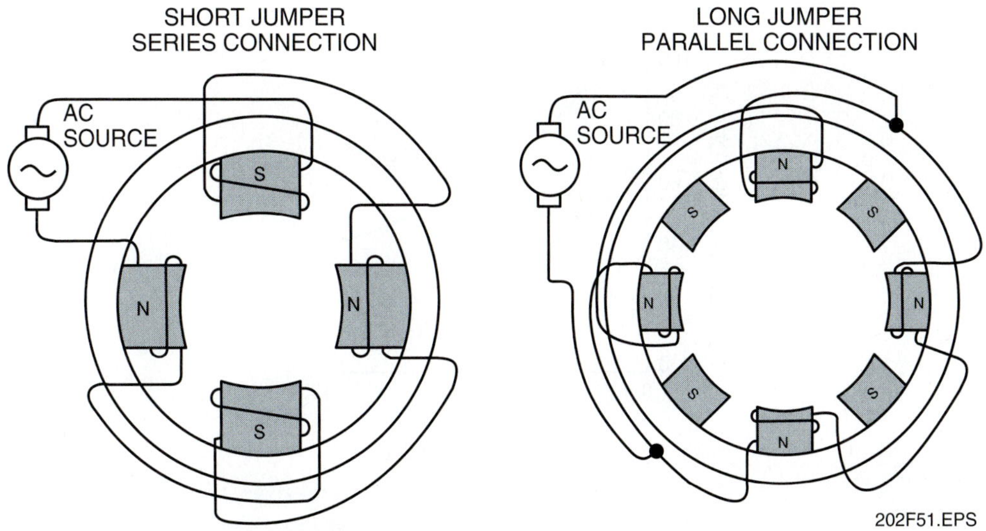

Figure 51 ◆ Single-phase consequent-pole motor.

6.0.0 ◆ VARIABLE-SPEED DRIVES

The use of adjustable speed in industrial equipment is increasing due to the need for better equipment control and for energy savings where partial power is required. AC drives compete with DC drives, eddy current drives, and mechanical and hydraulic systems as methods to control speed. Reliability, cost, and control capabilities are the major factors in system selection.

A drive system includes both the drive controller and the motor being driven. This module focuses specifically on the electronic drive components and covers various types of control for both DC and AC drives. This section provides a basic review of some fundamental principles that are important to understand when starting up, operating, or troubleshooting a variable-speed drive system.

6.1.0 Types of Adjustable Speed Loads

Most drive controllers can be adjusted or modified to optimize performance and provide the most efficient and cost-effective drive, depending on the load characteristics of the application.

It is important to understand the speed and torque characteristics as well as the maximum horsepower requirements for the type of load to be considered. Based on this, either a constant-torque controller or a variable-torque controller is selected. The most common types of loads are shown in *Figure 52*. Note that a load requires the same amount of torque at low speed as at high speed.

For a constant-torque load, the torque remains constant throughout the speed range, and the

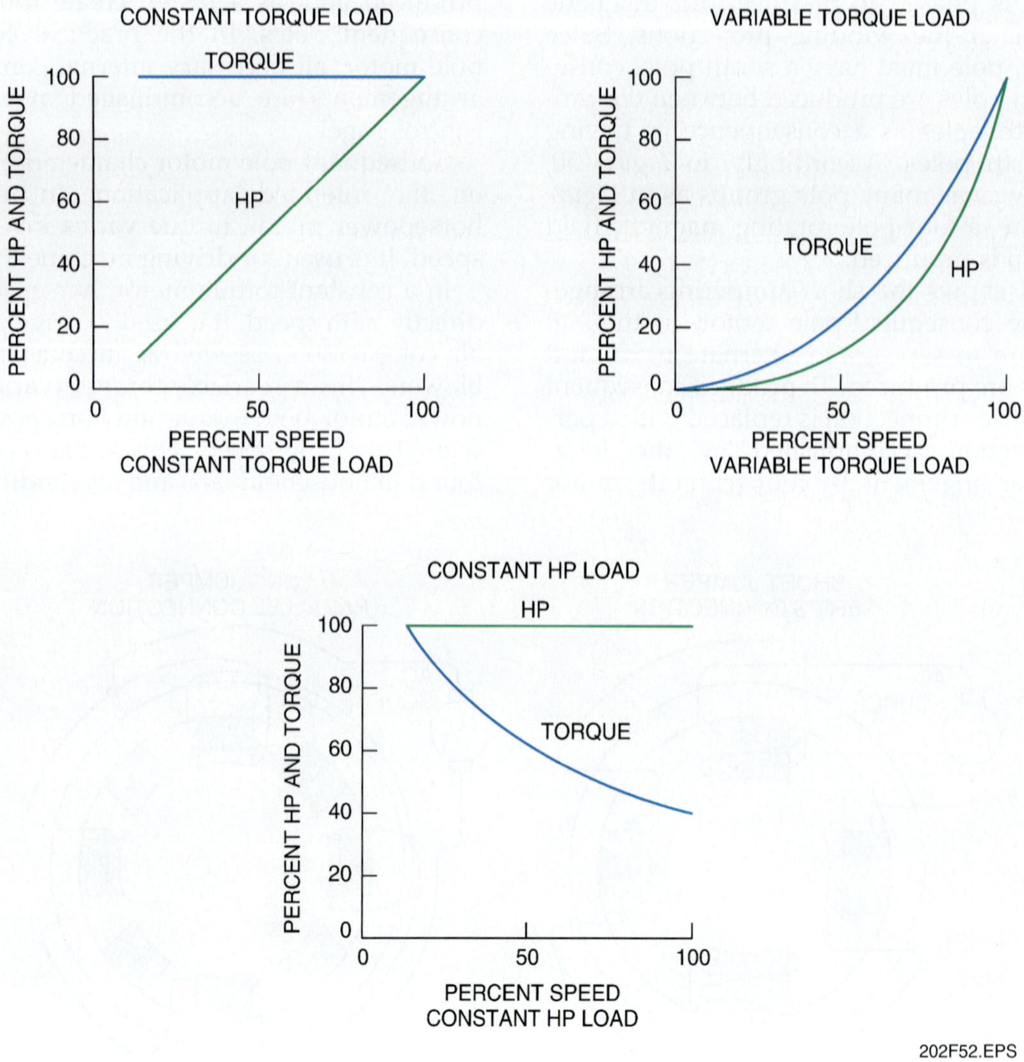

Figure 52 ◆ Types of adjustable speed loads.

horsepower increases and decreases in direct proportion to the speed. This applies to applications such as conveyors, as well as applications in which shock loads, overloads, or high inertia loads are encountered.

A variable-torque load requires much lower torque at low speeds than at high speeds. Horsepower varies approximately as the cube of the speed, and the torque varies approximately as the square of the speed. This applies to applications such as centrifugal fans, pumps, and blowers.

A constant-horsepower load requires high torque at low speeds, low torque at high speeds, and thus constant horsepower at any speed. It applies to applications such as lathes requiring slow speeds for deep cuts and high speeds for finishing. Usually, very high starting torques are required.

6.2.0 Motor Considerations

For industrial applications, motors are required to function at varying torques and speeds, and in forward and reverse directions. Besides operating as a motor, the machine may also function as a brake or a generator for short periods.

The various operating modes for industrial drives are shown in *Figure 53*. Positive and negative speed (rotation) are plotted on the horizontal axis and the torques are plotted on the vertical axis. The four quadrants of operation are labeled 1, 2, 3, and 4.

A machine operating in quadrant 1 has positive torque and speed, which means that they both act in the same direction (in this case, clockwise). A machine in this quadrant is functioning as a motor. It delivers mechanical power to a load. The machine will also act as a motor in quadrant 3, but torque and speed are reversed from quadrant 1 (counterclockwise).

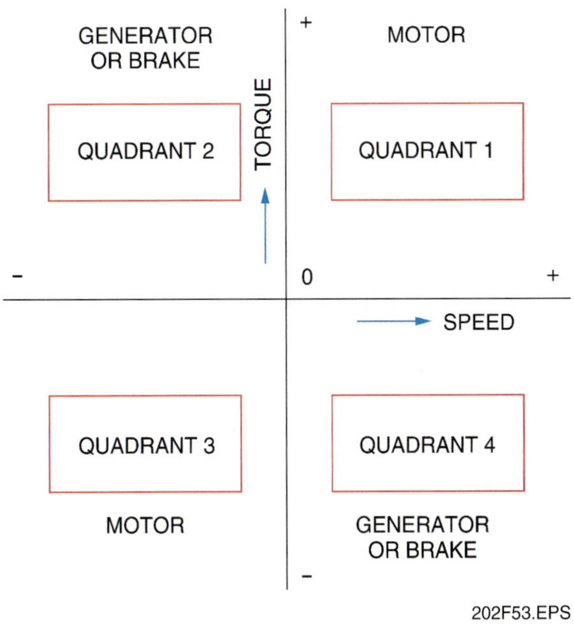

Figure 53 ◆ Electric drive operation in four quadrants.

While operating in quadrant 2, a machine will develop a positive torque and a negative speed. The torque is acting clockwise, and the speed is counterclockwise. In this quadrant, the machine is absorbing mechanical power from the load and functions as a generator. This mechanical power is converted into electric power and is generally transmitted back into the line. The electric power may also be dissipated in an external resistor, which is known as dynamic braking.

Depending on its connections, a machine may also be used as a brake while operating in quadrant 2. Absorbed mechanical power is converted to electric power, then converted into heat. If the machine absorbs electric line power as it is converting mechanical power into electric power, it functions as a brake. Both power inputs are dissipated as heat. Large power drives seldom use the brake mode of operation as it is very inefficient. The circuitry is generally chosen so that the machine will function as a generator when it is operating in quadrant 2. Quadrant 4 operation is identical to quadrant 2 except that speed and torque are reversed.

6.2.1 Typical Torque-Speed Curves

A three-phase motor has a torque-speed curve that is a good example of an electrical machine's behavior as a generator brake. The solid curve in *Figure 54* is the torque-speed curve for a machine acting as a motor in quadrant 1, a brake in quadrant 2, and a generator in quadrant 4.

If the stator leads are reversed, the torque-speed curve is shown by the dotted curve. Now the motor operates as a motor in quadrant 3, a generator in quadrant 2, and a brake in quadrant 4. The machine functions as a brake or a generator in quadrants 2 and 4, but it always runs as a motor in quadrants 1 and 3.

Figure 55 shows the torque-speed curve of a DC shunt motor. Motor, generator, and brake modes are apparent. The dotted curve represents reversed armature leads.

Variable-speed electric drives are designed to vary speed and torque in a smooth and continuous manner so as to satisfy load requirements. Typically, this is accomplished by shifting the torque-speed characteristic back and forth along the horizontal axis. The torque-speed characteristic of the motor is shifted by varying the armature

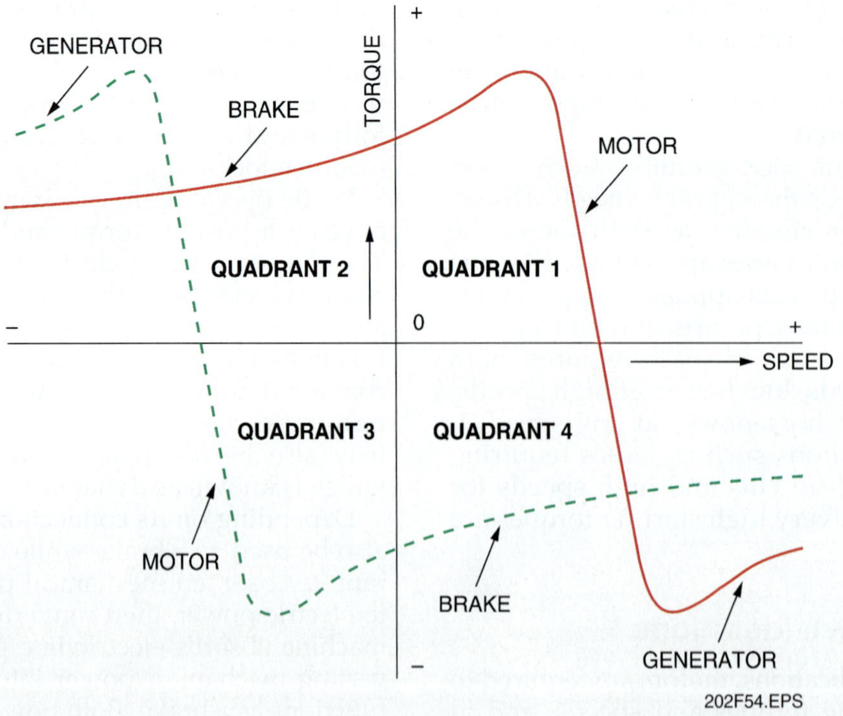

Figure 54 ◆ Four-quadrant operation for a squirrel cage motor.

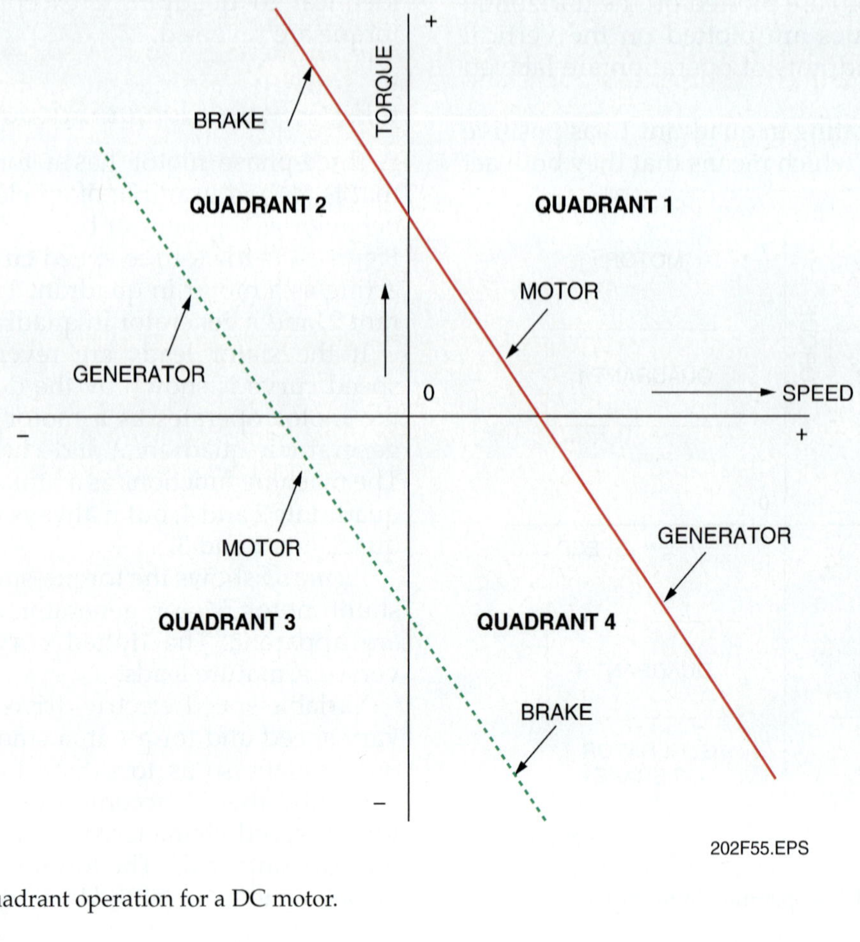

Figure 55 ◆ Four-quadrant operation for a DC motor.

voltage. Also, the curve of an induction motor can be shifted by varying the voltage and frequency applied to the stator.

In describing the various methods of motor control, only the behavior of power circuits will be discussed. The many ways of shaping and controlling triggering pulses will not be covered. They constitute a complex subject that involves sophisticated electronics, logic circuits, integrated circuits, and microprocessors.

6.2.2 Motor Heating

Since a variable-speed drive system includes both the drive controller and the motor, the design engineer should always consider the capabilities of the motor to perform acceptably under the desired operating conditions. One of the factors that you should be aware of is motor heating. When operating a motor at reduced speeds, the ability to dissipate heat is also reduced due to the slower cooling fan speed. This factor should be considered when maintaining the motor, modifying its enclosure or surrounding area, or troubleshooting the drive system.

6.3.0 Motor Speed Control

It is important to understand how DC or AC motor speed can be varied in order to understand how a drive controller accomplishes that task. This section reviews the fundamentals of DC and AC motor speed control.

6.3.1 Varying the Speed of a DC Shunt Motor

A DC shunt motor is shown in *Figure 56(A)*. Basically, there are two ways of varying the running speed of a DC shunt motor:

- Adjusting the voltage (and current) applied to the field winding. As the field voltage is increased, the motor slows down. This method is suggested by *Figure 56(B)*.
- Adjusting the voltage (and current) applied to the armature. As the armature voltage is increased, the motor speeds up. This method is suggested by *Figure 56(C)*.

6.3.2 Field Control

Here is how method 1, adjusting the field voltage, works. As the field voltage is increased, by reducing R_V in *Figure 56(B)*, for example, the field current is increased. This results in a stronger magnetic field, which induces a greater CEMF in the armature winding. The greater CEMF tends to oppose the applied DC voltage and thus reduces the armature current, I_A. Therefore, an increased field current causes the motor to slow down until the induced CEMF has returned to near its normal value.

Going in the other direction, if the field current is reduced, the magnetic field gets weaker. This causes a reduction in CEMF created by the rotating armature winding. The armature current increases, forcing the motor to spin faster, until the CEMF is once again approximately equal to what it was before. The reduction in magnetic field strength is compensated for by an increase in armature speed.

This method of speed control has certain positive features. It can be accomplished by a small, inexpensive rheostat, since the current in the field winding is fairly low. Also, because of the low value of the field current, I_F, the rheostat R_V does not dissipate very much energy. Therefore, this method is energy efficient.

However, there is one major drawback to speed control from the field winding: to increase the speed, you must reduce I_F and weaken the magnetic field, thereby lessening the motor's torque-producing ability. The ability of a motor to create torque depends on two things: the current in the armature conductors and the strength of the magnetic field. As I_F is reduced, the magnetic field is weakened, and the motor's torque-producing ability declines. Unfortunately, it is at this point

INSIDE TRACK

Using a Motor as Both a Generator and a Brake

DC motors that power subway cars are also used for regenerative braking. When driven by the train's momentum, such as when slowing down upon approaching a station, the motor acts as a generator and puts current back into the system. Thus, while the train's motors are creating a significant amount of power from the train's movement, the train makes use of the resistive torque created by the power generation to slow down the train.

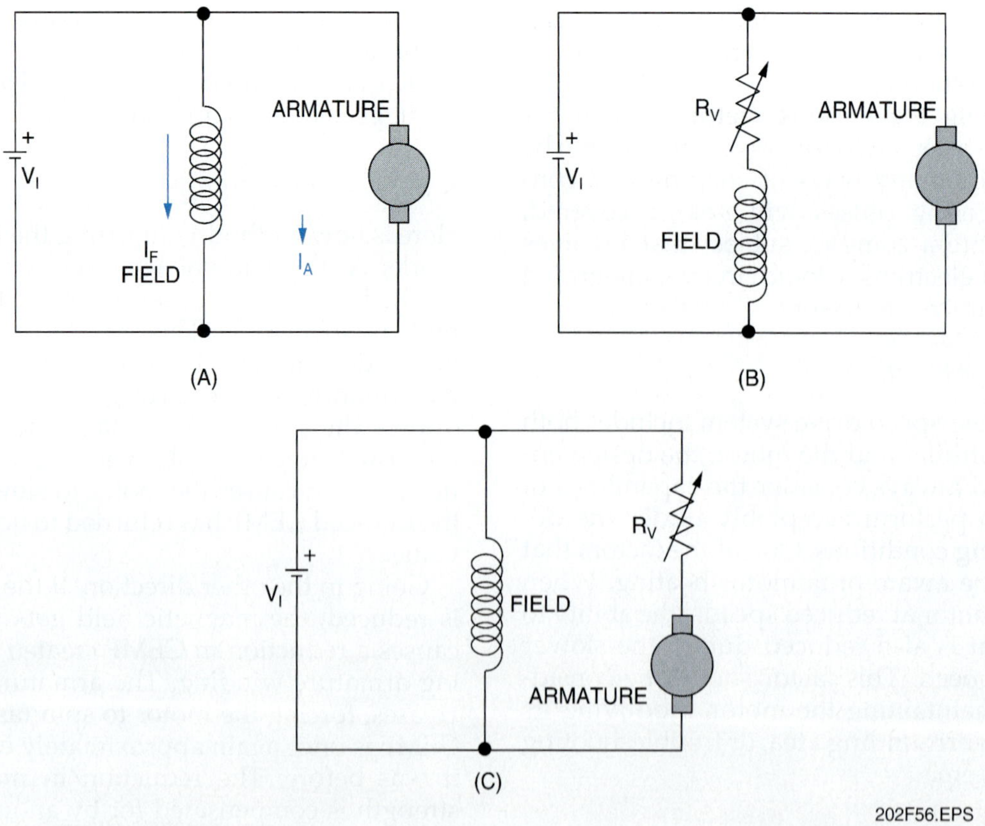

Figure 56 ◆ DC shunt motor schematic.

that the motor needs all the torque-producing ability it can get, since it probably requires greater torque to drive the load at a faster speed.

6.3.3 Armature Control

From the torque-producing point of view, method 2, armature control, is much better. As the armature voltage and current are increased by reducing R_V, the motor starts running faster, which normally requires more torque. The reason for the rise in speed is that the increased armature voltage demands an increased CEMF to limit the increase in armature current to a reasonable amount. The only way the CEMF can increase is for the armature winding to spin faster, since the magnetic field strength is fixed. In this instance, the ingredients are all present for increased torque production, since the magnetic field strength is kept constant and I_A is increased.

The problem with the armature control method of *Figure 56(C)* is that R_V, the rheostat, must handle the armature current, which is relatively large. Therefore, the rheostat must be physically large and expensive, and it will waste a considerable amount of energy.

6.3.4 Varying the Speed of an AC Motor

The principle of speed control for adjustable-frequency drives is based on the following fundamental formula for a standard AC motor:

$$N_s = \frac{120f}{P}$$

Where:

N_s = synchronous speed (rpm)
f = frequency
P = number of poles

The number of poles of a particular motor is set in its design and manufacture.

The adjustable-frequency system controls the frequency (f) applied to the motor. The speed (N_s) of the motor is then proportional to this applied frequency. Control frequency is adjusted by means of a potentiometer or external signal, depending on the application.

The frequency output of the controller is adjustable over its design speed range. Therefore, the speed of the motor is adjustable over this same range. Because an electronic means of generating variable frequencies is being used, the speed range often exceeds the 60 hertz (Hz) rated speed of the motor.

When variable-frequency speed control is employed, the motor supply voltage cannot be allowed to remain at a steady value. The magnitude of the motor voltage must be increased or decreased in proportion to the frequency. That is, the voltage-to-frequency ratio, V/f, must remain approximately constant.

For instance, if the motor has a nameplate rating of 240V at 60Hz, the voltage-to-frequency ratio is 4 (240 ÷ 60 = 4). If the motor is speeded up by adjusting its variable-frequency inverter to 90Hz, the voltage magnitude must be increased to 360V, since 4 × 90 = 360. If the motor is slowed down by adjusting the inverter frequency to 45Hz, the voltage magnitude must be decreased to 180V, since 4 × 45 = 180.

The stator's magnetic field strength must remain constant under all operating conditions. If the stator field strength should happen to rise much above the design value, the motor's core material would go into magnetic saturation. This would effectively lower the core's permeability, thereby inhibiting proper induction of voltage and current in the rotor loops (or bars), thus detracting from the torque-producing capability of the motor. On the other hand, if the stator field strength should happen to fall much below the design value, the weakened magnetic field would simply induce lower values of voltage and current in the rotor loops. This would also detract from the torque-producing ability of the motor.

Therefore, the magnetic field produced by the stator windings must hold a constant rms value, regardless of frequency. The magnetizing current of an induction motor is the current that flows through the stator winding when the rotor is spinning at steady-state speed with no torque load. The magnetizing current for an induction motor is given by Ohm's law:

$$I_{mag} = \frac{V}{X_L}$$

Where:

V = rms value of the applied stator voltage

X_L = inductive reactance of the stator winding

In the equation, X_L does not remain constant as the supply frequency is adjusted; it varies in proportion to the frequency ($X_L = 2\pi fL$). Therefore V must also be varied in proportion to the frequency, so that the Ohm's law division operation yields an unvarying value of magnetizing current.

Alternatively, using $X_L = 2\pi fL$, we can rewrite the equation:

$$I_{mag} = \frac{V}{X_L}$$

$$I_{mag} = \frac{V}{2\pi fL}$$

$$I_{mag} = \frac{1}{2\pi L} \times \frac{V}{f}$$

Since 1 ÷ 2πL is a constant determined by the motor's construction, the magnetizing current is kept constant by maintaining the V/f ratio.

The controller can automatically maintain the required volts/cycle (V/Hz) ratio to the motor at any speed. This provides maximum motor capability throughout the speed range.

The V/Hz setting is typically preset at the factory. However, on many controllers it can be adjusted or changed to fine tune controller operation.

7.0.0 ◆ MOTOR ENCLOSURES

Motors are usually designed with covers over the moving parts. These covers, called enclosures, are classified by NEMA (National Electrical Manufacturers Association) according to the degree of environmental protection provided and the method of cooling. If the cover has openings, the motor is classified as an open motor; if the enclosure is complete, the motor is classified as an enclosed motor. Each of these types of motors has many modifications. *Table 1* lists the various types for both open and totally enclosed motors.

Table 1 Motor Enclosures

Open	Totally Enclosed
General purpose	Nonventilated
Drip-proof	Fan-cooled
Splash-proof	Fan-cooled guarded
Guarded	Explosion-proof
Semi-guarded	Dust- and ignition-proof
Drip-proof guarded	Pipe-ventilated
Externally ventilated	Water-cooled
Pipe-ventilated	Water-to-air-cooled
Weather-protected (Type I & Type II)	
Encapsulated windings	
Sealed windings	

The different standard types as explained and defined by NEMA are as follows:

- *General purpose* – Has ventilating openings which permit the passage of external cooling air over and around the windings of the machine.
- *Drip-proof* – Ventilating openings are so constructed that successful operation is not interfered with when drops of liquid or solid particles strike or enter the enclosure at any angle from 0° to 15° downward from the vertical.
- *Splash-proof* – Ventilating openings are so constructed that successful operation is not interfered with when drops of liquid or solid particles strike or enter the enclosure at any angle not greater than 100° downward from the vertical.
- *Guarded* – Openings giving direct access to live metal or rotating parts (except smooth surfaces) are limited in size by the structural parts or by screens, baffles, grills, expanded metal, or other means to prevent accidental contact with hazardous parts.
- *Semi-guarded* – Some of the ventilating openings, usually in the top half, are guarded as in the case of a guarded machine, but the others are left open.
- *Drip-proof guarded* – This type of machine has ventilating openings as in a guarded machine.
- *Externally ventilated* – Designating a machine that is ventilated by a separate motor-driven blower mounted on the machine enclosure. Mechanical protection may be as defined above. This machine is sometimes known as a blower-ventilated or force-ventilated machine.
- *Pipe-ventilated* – Openings for the admission of ventilating air are so arranged that inlet ducts or pipes can be connected to them.
- *Weather-protected* – Type I: Ventilation passages are so designed as to minimize the entrance of rain, snow, and airborne particles to the electrical parts. Type II: In addition to the enclosure described for a Type I machine, ventilating passages at both intake and discharge are so arranged that high-velocity air and airborne particles blown into the machine by storms or high winds can be discharged without entering the internal ventilating passages leading directly to the electric parts.
- *Encapsulated windings* – An AC squirrel cage machine having random windings filled with an insulating resin, which also forms a protective coating.
- *Sealed windings* – An AC squirrel cage machine making use of form-wound coils and an insulation system that, through the use of materials, processes, or a combination of materials and processes, results in a sealing of the windings and connections against contaminants.

INSIDE TRACK

Self-Cooling Motors

Conventional squirrel cage fan-cooled motors may overheat when operated at reduced speeds. Many manufacturers now offer inverter duty-rated motors with increased self-cooling capability.

7.1.0 Totally Enclosed

- *Nonventilated* – Not equipped for cooling by means external to the enclosing parts.
- *Fan-cooled* – Equipped for exterior cooling by means of a fan or fans that are integral with the machine but external to the enclosing parts.
- *Fan-cooled guarded* – All openings giving direct access to the fan are limited in size by design of the structural parts or by screens, grills, expanded metal, etc. to prevent accidental contact with the fan.
- *Explosion-proof* – Designed and constructed to withstand an explosion of a specified gas or vapor that may occur within it and to prevent the ignition of the specified gas or vapor surrounding the machine by sparks, flashes, or explosions of the specified gas or vapor that may occur within the machine casing.
- *Dust- and ignition-proof* – Designed and constructed in a manner that will exclude ignitable amounts of dust or amounts which might affect performance or rating, and that will not permit arcs, sparks, or heat otherwise generated or liberated inside the enclosure to cause ignition of exterior accumulations or atmospheric suspensions of a specific dust on or in the vicinity of the enclosure.
- *Pipe-ventilated* – Openings are so arranged that when inlet and outlet ducts or pipes are connected to them there is no free exchange of the internal air and the air outside the case.
- *Water-cooled* – Cooled by circulating water, with the water or water conductors coming in direct contact with the machine parts.
- *Water-to-air-cooled* – Cooled by circulating air, which in turn is cooled by circulating water.

7.2.0 Open Motor

The most common type of motor is the open motor. It has ventilating openings that permit the passage of external cooling air over and around its windings. If these are limited in size and shape, the motor is called a protected motor, since it is protected from any large pieces of material that may somehow enter the motor, thus damaging its internal parts. A protected motor also prevents a person from touching the rotating or electrically-energized parts of the motor. Drip-proof and splash-proof motors are constructed such that drops of liquid cannot enter the motor.

7.3.0 Enclosed Motor

The totally enclosed motor is designed to prevent the free exchange of air between the inside and outside of the actual motor housing. It is used where hostile environmental conditions and the motor application require maximum protection of the internal parts of the motor.

8.0.0 ◆ NEMA FRAME DESIGNATIONS

Frame sizes were developed by NEMA to ensure interchangeability of motors among manufacturers. They appear on motor nameplates to give information about the machine's physical dimensions. Key dimensions are:

- Distance from motor feet to shaft centerline
- Bolt-hole center-to-center distance between front and back feet
- Exposed shaft distance from shaft end to shaft shoulder

Tables are available to correlate frame size to dimensions. The system for designating the frames of motors and generators consists of a series of numbers in combination with letters.

8.1.0 Small Machines

The frame number for small machines is the D dimension in inches multiplied by 16. The following letters shall immediately follow the frame number to denote variations:

B – Carbonator pump motors

C – Type C face-mounting motors

G – Gasoline pump motors

H – A frame having an F dimension larger than that of the same frame without the suffix H

J – Jet pump motors

K – Sump pump motors

M – Oil burner motors

N – Oil burner motors

Y – Special mounting dimensions (must obtain dimensional diagram from manufacturer)

Z – All mounting dimensions are standard except the shaft extension

8.2.0 Medium Machines

The system for numbering frames of medium machines is as follows:

- The first two digits of the frame number are equal to four times the D dimension in inches. (If this product is not a whole number, the first two digits of the frame number shall be the next higher whole number.)
- The third and, when required, the fourth digit of the frame number are obtained from the value of 2F in inches.

Figure 57 shows a typical end view of a foot-mounted machine. The many different dimensions can be found on a dimension sheet for that machine (*Figure 58*). The NEMA frame designation will provide information relating to both the D and 2F dimensions.

Table 2 may be used to determine the D dimension and 2F dimension for medium-size motors. The D dimension is the distance from the centerline of the shaft to the bottom of the feet. The 2F dimension is the distance between the centerlines of the mounting holes in the feet or base of the machine.

Medium machines also use letters that denote variations. These letters follow the frame number. Since there are many more varieties of medium-size machines, the letter relates to the different aspects of mounting and shaft orientation.

For example, to understand the NEMA frame designation, we will take a typical motor frame designation and determine the D and 2F dimensions. Then we will use a different set of dimensions to determine the frame designation number.

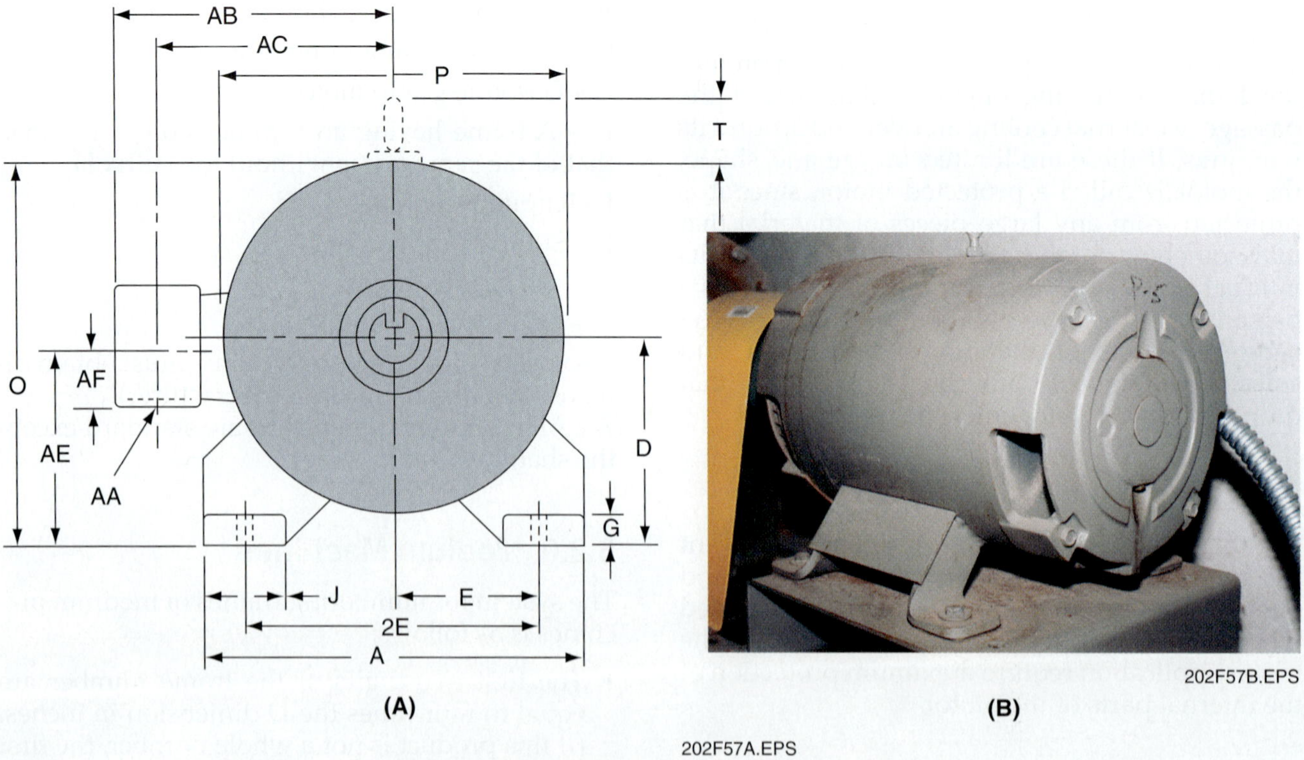

Figure 57 ◆ End view of a foot-mounted motor.

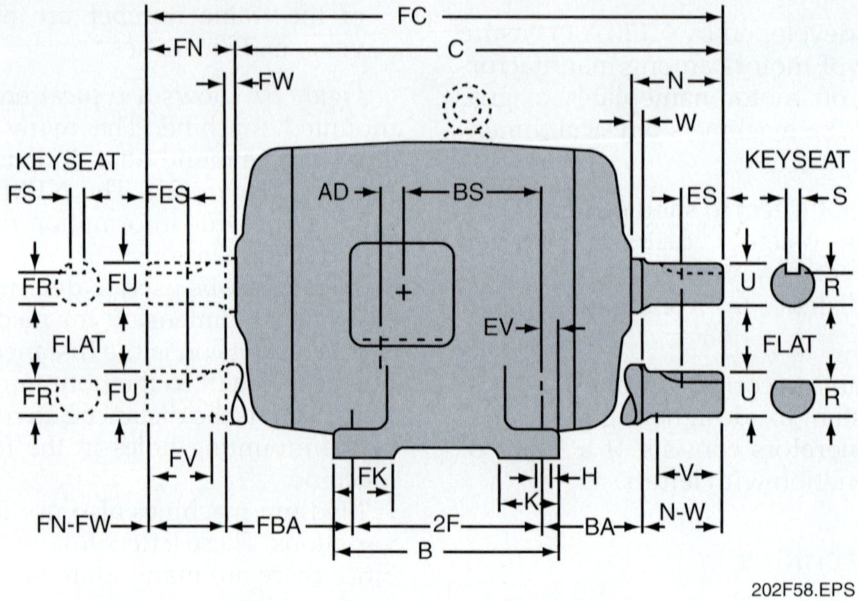

Figure 58 ◆ Lettering of dimension sheets for foot-mounted machines (side view).

Table 2 Frame Dimension Chart

Frame Number Series	D	Third/Fourth Digit in Frame Number						
		1	2	3	4	5	6	7
		2F Dimensions						
140	3.50	3.00	3.50	4.00	4.50	5.00	5.50	6.25
160	4.00	3.50	4.00	4.50	5.00	5.50	6.25	7.00
180	4.50	4.00	4.50	5.00	5.50	6.25	7.00	8.00
200	4.50	4.50	5.00	5.50	6.50	7.00	8.00	9.00
210	5.00	4.50	5.00	5.50	6.50	7.00	8.00	9.00
220	5.50	5.00	5.50	6.25	6.75	7.50	9.00	10.00
250	6.25	5.50	6.25	7.00	8.25	9.00	10.00	11.00
280	7.00	6.25	7.00	8.00	9.50	10.00	11.00	12.50
320	8.00	7.00	8.00	9.00	10.50	11.00	12.00	14.00
360	9.00	8.00	9.00	10.00	11.25	12.25	14.00	16.00
400	10.00	9.00	10.00	11.00	12.25	13.75	16.00	18.00
440	11.00	10.00	11.00	12.50	14.50	16.50	18.00	20.00
500	12.50	11.00	12.50	14.00	16.00	18.00	20.00	22.00
580	14.50	12.50	14.00	16.00	18.00	20.00	22.00	25.00
680	17.00	16.00	18.00	20.00	22.00	25.00	28.00	32.00

Frame Number Series	D	Third/Fourth Digit in Frame Number							
		8	9	10	11	12	13	14	15
		2F Dimensions							
140	3.50	7.00	8.00	9.00	10.00	11.00	12.50	14.00	16.00
160	4.00	8.00	9.00	10.00	11.00	12.50	14.00	16.00	18.00
180	4.50	9.00	10.00	11.00	12.50	14.00	16.00	18.00	20.00
200	5.00	10.00	11.00	...	...	...	...	...	...
210	5.25	10.00	11.00	12.50	14.00	16.00	18.00	20.00	22.00
220	5.50	11.00	12.50	...	...	...	...	...	...
250	6.25	12.50	14.00	16.00	18.00	20.00	22.00	25.00	28.00
280	7.00	14.00	16.00	18.00	20.00	22.00	25.00	28.00	32.00
320	8.00	16.00	18.00	20.00	22.00	25.00	28.00	32.00	36.00
360	9.00	18.00	20.00	22.00	25.00	28.00	32.00	36.00	40.00
400	10.00	20.00	22.00	25.00	28.00	32.00	36.00	40.00	45.00
440	11.00	22.00	25.00	28.00	32.00	36.00	40.00	45.00	50.00
500	12.50	25.00	28.00	32.00	36.00	40.00	45.00	50.00	56.00
580	14.50	28.00	32.00	36.00	40.00	45.00	50.00	56.00	63.00
680	17.00	36.00	40.00	45.00	50.00	56.00	63.00	71.00	80.00

Example 1:

A typical medium-size frame number is a 256T. Since we know this is a medium frame, we divide the first two digits by 4:

$$25 \div 4 = 6.25$$

Therefore, the D dimension is 6.25 inches.

To determine the 2F dimension, we use *Table 2* and the third digit in the frame number. The third digit is 6 and the frame is a 250 series. Using the table, the 2F dimension is 10 inches. The T in the frame number is included as part of a frame designation for which standard dimensions have been established.

Medium-size frames can have multiple letters that denote a variety of different applications and arrangements. A 256AT has the same dimensions, with the A added to denote an industrial DC machine.

Example 2:

A frame has a D dimension of 3.5 inches and a 2F dimension of 4 inches, and all standard dimensions have been established. Multiplying the

D dimension by 4 will give the first two digits of the frame designation:

$$3.5 \times 4 = 14$$

Using the table, a 140 frame series and a 2F dimension of 4 inches provides a third digit of 3. Since it is a standard dimension frame, the letter will be the suffix. This frame has a designation of 143T.

Full-load torque (rather than horsepower) determines the frame size required to house the motor. Thus, a motor developing a large amount of horsepower at high speed will have the same frame size as a machine developing less horsepower at a slower speed.

More compact design, better ventilation, and insulation systems with higher temperature ratings have enabled manufacturers to house motors in increasingly smaller frame sizes. NEMA re-rates occurred in 1952 and 1964. Motors manufactured before 1952 are generally referred to as pre-U-frame motors. Those manufactured between 1952 and 1964 are called U-frame motors; those manufactured since 1964 are called T-frame motors.

9.0.0 ◆ MOTOR RATINGS AND NAMEPLATE DATA

The ratings of an electric motor include:

- Voltage
- Full-load current
- Speed
- Number of phases and frequency
- Full-load horsepower
- Service classification

Except for full-load horsepower and service classification, these are self-explanatory units. The horsepower rating that is stamped on the motor nameplate by the manufacturer is the horsepower load the motor will carry without damaging any part of the motor.

Electric motor service classification depends on the type of service for which the motor is designed. A motor will usually fall into one of two classifications. General-purpose motors are those motors designed for use without restriction to a particular application. They meet certain specifications as standardized by NEMA. A definite-purpose motor is one that is designed in standard ratings and with standard operating characteristics for use under service conditions other than usual or for use on a particular type of application. A special-purpose motor is one with special operating characteristics or special mechanical construction, or both, that is designed for a particular application and that does not meet the definition of a general-purpose or a definite-purpose motor.

The most common machine rating is the **continuous duty** rating defining the output (in kilowatts for DC generators, kilovolt-amperes at a specified power factor for AC generators, and horsepower for motors) that can be carried indefinitely without exceeding established limitations. For **intermittent duty, periodic duty,** or **varying duty,** a machine may be given a short-time rating defining the load that can be carried for a specific time. Standard periods for short-time ratings are 5, 15, 30, and 60 minutes. Speeds, voltages, and frequencies are also specified in ratings, and provision is made for possible variations in voltage and frequency.

For example, motors must operate successfully at voltages 10% above and below rated voltage and, for AC motors, at frequencies 5% above and below rated frequency; the combined variation of voltage and frequency may not exceed 10%. Other performance conditions are so established that reasonable short-time **overloads** can be carried. Thus, the user of a motor can expect to be able to apply an overload of 25% for a short time at 90% of normal voltage with an ample margin of safety.

9.1.0 Nameplate Data

The *NEC Section 430.7* has specified information that must be listed on a motor nameplate based on its type. Requirements can also be found in *NEMA Standards MG-1* and *MG-2*. Required information plus additional information is shown on the nameplate in *Figure 59*.

9.1.1 Rated Voltage

Power plant induction motors are designed to operate with a balanced three-phase voltage source applied at the terminals. The rated voltage on the nameplate is usually lower than the voltage of the electrical system. For example, a 460V motor is designed to operate in a 480V system. Here, an assumption is made by motor manufacturers that there will be a voltage drop of 20V from the transformer down to the motor terminals (see *Table 3*). The rated or nameplate voltage is the voltage at which the motor will operate most effectively. When other than rated voltage is applied, performance will change and motor life may be reduced.

Many three-phase motors have two voltages listed on the nameplate. For example, 230/460V means the motor can be connected for either 230V or 460V operation. In these cases, a connection diagram is usually found on the nameplate, as shown in *Figure 60*. These diagrams refer to low-voltage and high-voltage connections.

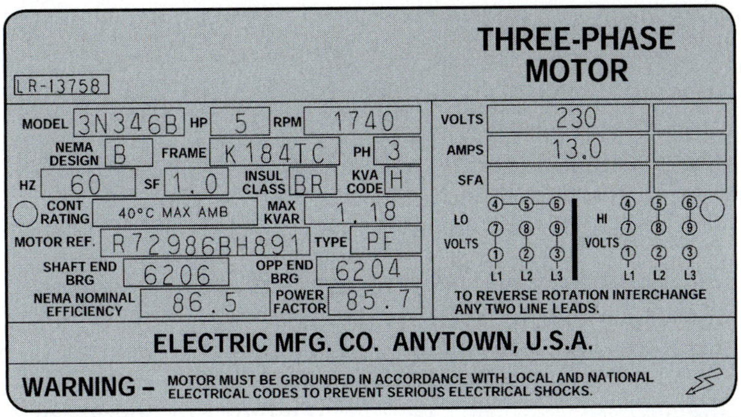

202F59.EPS

Figure 59 ◆ Nameplate data.

9.1.2 Full-Load Amps (FLA)

The FLA rating appearing on the nameplate indicates the current the motor will draw at nameplate horsepower, frequency, and voltage. Most manufacturers test to determine this value on a periodic basis during production, ensuring reasonable accuracy. The *NEC*® requires that the rated full-load current be the basis for determining the proper sizing of cable, overload protective devices, and other **overcurrent** protection in the motor circuit. Since many motors can be connected for one of two voltage ratings, they have two FLA ratings.

The FLA ratings are guaranteed if the induction motor is operating at full-load conditions and the applied voltage and frequency are the same as stated on the nameplate. When voltage or frequency are not the same, however, the current drawn by the motor at full-load conditions will be different from the nameplate indication (see *Table 4*). It is possible to damage a motor operated below its rated voltage or frequency, since the current the motor draws at full-load conditions increases in both cases. If the overload protective device is not sized properly, motor life may be shortened by this overcurrent condition.

Table 3 Induction Motor Voltages

System Voltage	Rated Voltage
216	208
240	230
480	460
600	575
2,400	2,300
4,160	4,000
4,800	4,600
6,900 and 7,200	6,600
13,200 and 13,800	13,200

Table 4 Motor Operation

Mode	Full-Load Current
110% of rated volts	7% decrease
90% of rated volts	11% increase
105% of rated frequency	5%–6% decrease
95% of rated frequency	5%–6% increase

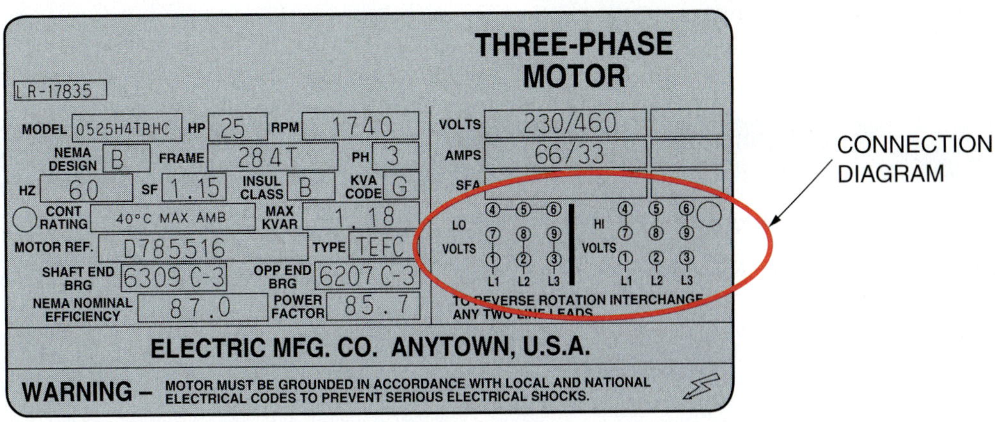

202F60.EPS

Figure 60 ◆ High-voltage and low-voltage connection diagrams shown on motor nameplate.

9.1.3 Rated Full-Load Speed

The rated full-load speed is the value indicated in rpm on the nameplate. It is the speed at which the shaft will turn at the nameplate horsepower when supplied with power at the nameplate voltage and frequency. If the driven load is less than the nameplate horsepower, the shaft will turn faster than full-load speed.

If the motor is operating unloaded, the shaft will turn very close to synchronous speed. For example, with a full-load speed of 1,750 rpm, it can be inferred that the motor's synchronous speed is 1,800 rpm. The machine will operate from close to 1,800 rpm down to 1,750 rpm, from no-load to full-load conditions.

Common synchronous speeds are 3,600, 1,800, 1,200, 900, and 600 rpm. Synchronous speed is rarely found on the motor nameplate unless the machine has been retrofitted and has not yet been tested for new full-load speed.

9.1.4 Rated Horsepower

An induction motor is really a torque generator. It delivers a needed torque to a driven machine at a certain speed. Thus:

$$\text{Horsepower} = \frac{\text{load torque in ft-lbs} \times \text{rpm}}{5{,}250}$$

For induction motors that are built to NEMA standards, the ratings will range from ½hp to 400hp, with 24 categories in all. If horsepower requirements fall between any two ratings, the larger motor size should be selected.

Remember, an induction motor will try to deliver any amount of horsepower the load requires. If properly sized, most motors operate at something less than the motor nameplate horsepower. Standard motors are designed to operate at nameplate values from sea level up to an altitude of 3,300 feet if the ambient temperature does not exceed 104°F (40°C). Above this altitude, the nameplate horsepower no longer applies.

NEMA standards provide a method for determining the proper temperature rise, or the new maximum ambient temperature, at higher elevations. However, the standards do not provide a direct method for deriving the horsepower. Several methods are available to estimate true motor horsepower output.

9.1.5 Duty or Time Rating

All polyphase induction motors have either a **duty** or a time rating, which is the elapsed time the motor can operate at nameplate horsepower without shortening its life. The time rating of a motor is determined by operating the machine at full-load conditions and measuring the time it takes for the windings to heat up to the temperature rating of the insulation.

Standard time ratings are 5, 15, 30, and 60 minutes, and continuous or 24 **hours.** A motor with a time (or duty) rating other than continuous is a smaller motor that is given a higher horsepower rating for a shorter period of time, thus reducing size and cost. In power plant uses, most motor ratings are continuous at 104°F (40°C).

9.1.6 NEMA Design Letters

The NEMA design letter defines the starting torque characteristics of an induction motor. It is one of the most important pieces of information on the nameplate; unfortunately, when a motor is replaced, the NEMA design letter is usually ignored, often leading to misapplication of the new machine. For fans or centrifugal pumps, starting torque requirements increase with the square of the change in speed. For mixers or loaded conveyor belts, however, starting torque requirements change very little with speed.

THINK ABOUT IT

Motor Horsepower and Speed

Can you replace a fan motor rated at a specific horsepower and speed with a motor rated at the same horsepower but a higher speed to increase airflow? If not, why not?

To account for these differences, NEMA has formulated design letters A, B, C, D, and F. The difference between motors with these letters is mainly in the design of the rotor, although there are also a few external differences. Design A and B motors are intended to drive conventional loads such as fans, blowers, and centrifugal pumps. About 80% of industrial motors are NEMA Design B.

Generally, the starting current is about five to seven times the rated full-load current. From *Table 5*, it can be seen that for larger motors, the starting current can be very significant, and across-the-line starting of larger motors could result in objectionable line-voltage dips. These voltage dips could result in other control equipment dropping out on low voltage and could even cause lights to dim.

Table 5 Current for 220V, 60-Cycle Squirrel Cage Motors

HP	Rated Full-Load Current	Starting (Maximum) Current	
		Classes B, C, D	Class F
½	2.0	12	—
1	3.5	24	—
1½	5.0	35	—
2	6.5	45	—
3	9	60	—
5	15	90	—
7½	22	120	—
10	27	150	—
15	40	220	—
20	52	290	—
25	64	365	—
30	78	435	270
40	104	580	360
50	125	725	450
60	150	870	540
75	185	1,085	675
100	246	1,450	900
125	310	1,815	1,125
150	360	2,170	1,350
200	480	2,900	1,800

9.1.7 Insulation Class

The electrical insulation system in a motor determines the machine's ultimate life span more than any other component. By some estimates, over 60% of all motors brought to repair shops are there because of premature failure of the insulation system.

The insulation class is a NEMA designation that identifies the class of material used to insulate the windings. Four letters designate the four classifications. They are A, B, F, and H. The insulation class defines the temperature that the insulation can be subjected to without suffering damage. The insulation class is shown on the nameplate in *Figure 61*.

Class A is now obsolete insofar as industrial motors are concerned. Class A was once the most common classification for motor insulation, especially for small motors. Class A comprises materials or combinations of materials such as cotton or paper, when suitably impregnated or coated, or other materials capable of operation at the temperature rise assigned for Class A insulation for the particular machine.

Class B is the predominant class of insulation used in motor manufacturing and rewinding

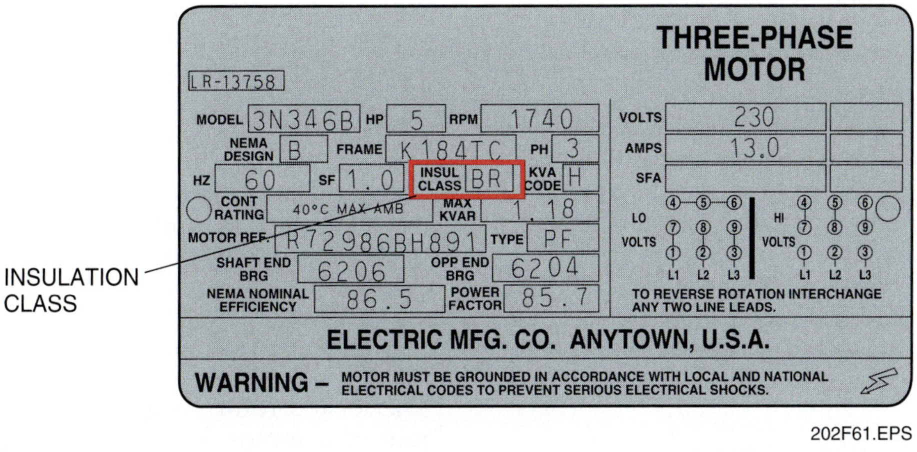

Figure 61 ◆ Nameplate showing insulation class.

today. This class is the basic standard of the industry. It includes materials such as mica, glass fiber, polyester, and aramid laminates, etc., with suitable bonding substances, or other materials, not necessarily inorganic, capable of operation at the temperature rise assigned for Class B insulation for the particular machine. (The insulation class may be designated more specifically by the use of additional letters, such as the BR shown in *Figure 61*.)

Class F incorporates materials that are similar to those in Class B but are capable of operation at the temperature rise assigned for Class F for the particular machine.

Class H insulation systems comprise materials or combinations of materials such as silicone elastomer, mica, glass fiber, polyester, and aramid laminates, etc., with suitable bonding substances such as silicone resins, or other materials capable of operation at the temperature rise assigned for Class H insulation for the particular machine.

When replacing motors, ensure that the insulation class is equal to or better than that of the motor removed from service.

9.1.8 Nominal Rated Voltage

Nominal rated voltage is defined as the voltage rating at which the motor is designed to operate.

9.1.9 Minimum Starting Voltage

Minimum starting voltage may be defined as the lowest voltage at which a motor will start without drawing an excessive/trip current.

9.1.10 Frequency

Frequency is given for AC motors in hertz or cycles per second. Standard frequencies for AC motors are 50Hz and 60Hz. Alternating current in the U.S. is 60Hz.

9.1.11 Service Factor

The service factor is a multiplier for the nameplate horsepower rating that determines the amount of overload the motor can withstand. This extra horsepower is available if the motor is already operating at rated voltage and frequency and is in an environment that does not exceed the ambient temperature rating. The most common service factor appearing on a motor nameplate is 1.15.

9.1.12 NEMA Code Letters

The high current draw of the motor during the first moments of startup is called the in-rush or locked-rotor current. It can be derived from the kVA code letter on the motor nameplate. The letter corresponds to the kilovolt-amperes per rated horsepower (kVA/hp) required during the first moments of motor startup. *Table 6* provides the kVA/hp value for each kVA code—a letter from A to V, excluding I, O, and Q. The locked-rotor current is required when sizing fuses or determining a **circuit breaker** setting in an induction motor circuit.

9.1.13 Bearings

Polyphase induction motors require either anti-friction or sleeve bearings. Anti-friction bearings are standard in medium (integral) horsepower motor sizes through 125hp/1,800 rpm. They are optional in 150 to 600hp/1,800 rpm sizes. Sleeve bearings are standard in 500hp/3,600 rpm and larger sizes.

Since radial loads are higher at the drive end of the motor, the drive-end bearing has a higher load rating than the bearing at the opposite end. A typical nameplate (*Figure 62*) might depict both bearing duties as:

- Shaft end brg: 6,206
- Opp end brg: 6,204

Bearing internal clearances are: C1 and C2 (smaller-than-normal clearance); standard clearance (normal); and C3, C4, and C5 (larger-than-normal clearance). Electric motors usually require a C3 internal clearance. Some bearing manufacturers have a different designation for motor bearings that have a larger-than-normal internal clearance.

9.1.14 Rated Amperage

Rated amperage may be defined as the full-load current required to produce full-rated horsepower at the motor's rated voltage and frequency. *Figure 62* shows the amperage for a typical motor.

9.1.15 Rated Horsepower

Horsepower is a rating used to specify the capacity of an electric motor to produce mechanical power to drive a specific piece of equipment (see *Figure 62*).

9.1.16 Locked-Rotor Current

The locked-rotor current is the steady-state current of a motor with the rotor locked and with rated voltage applied at rated frequency. NEMA has designated a set of code letters (*NEC Section 430.7*) to define locked-rotor kilovolt-amperes

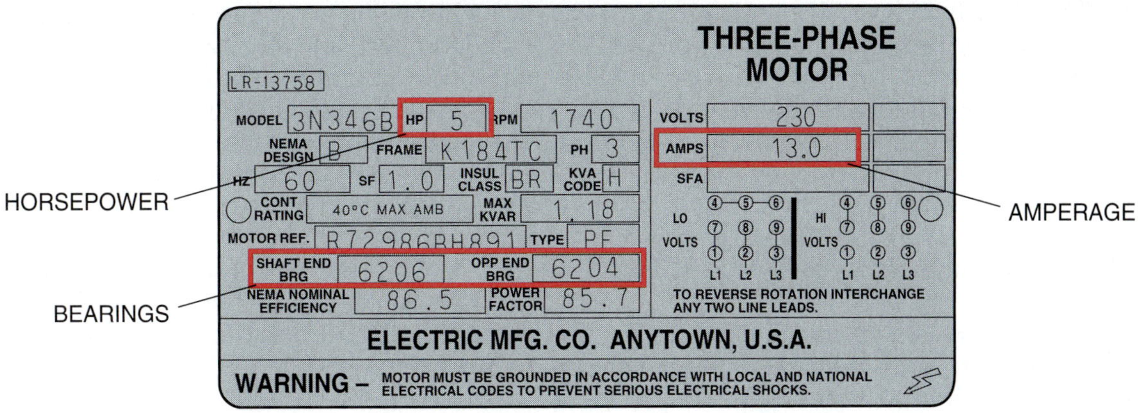

202F62.EPS

Figure 62 ◆ Nameplate showing bearings, horsepower, and amperage.

(kVA) per horsepower (*Table 6*). This code letter appears on the nameplate of all AC squirrel cage induction motors. The kVA rating is an indication of the current draw and, indirectly, the impedance of the locked rotor.

The current drawn by the motor under stall conditions can be calculated using the values given in *Table 6*. The current drawn by the motor under stalled conditions must be considered when selecting the motor protection and starting package and in coordination with the power system protective devices.

Table 6 Locked-Rotor Code Letters

Code Letter	kVA Per Horsepower with Locked Rotor
A	0–3.14
B	3.15–3.54
C	3.55–3.99
D	4.0–4.49
E	4.5–4.99
F	5.0–5.59
G	5.6–6.29
H	6.3–7.09
J	7.1–7.99
K	8.0–8.99
L	9.0–9.99
M	10.0–11.19
N	11.2–12.49
P	12.5–13.99
R	14.0–15.99
S	16.0–17.99
T	18.0–19.99
U	20.0–22.39
V	22.4–AND UP

Reprinted with permission from NFPA 70, the *National Electrical Code*®. Copyright © 2004, National Fire Protection Association, Quincy, MA 02269. This reprinted material is not the complete and official position of the National Fire Protection Association on the referenced subject, which is represented only by the standard in its entirety.

9.1.17 Starting Current

The total instantaneous starting current comprises the locked-rotor current plus the transient in-rush that flows until the motor magnetic circuit stabilizes.

9.1.18 Temperature Rise

The temperature rise may be defined as the measure of the heat produced by the operation of the motor. Several conditions contribute to temperature rise. Examples include running current, hysteresis losses, and friction of rotating parts.

9.1.19 Power Factor

The power factor (pf) is the ratio of active power of an alternating or pulsating current (when measured with a wattmeter) to the apparent power indicated by an ammeter and voltmeter. It is also referred to as the phase factor. The power factor is the measure of the system or equipment efficiency.

9.2.0 Motor Protection

Fuses are normally used for motor overload protection. When used, fuses must be provided in each ungrounded conductor and also the grounded conductor of a three-wire, three-phase AC system with one conductor grounded. When non-fuse overload protective devices are used (*NEC Section 430.37*), follow the guidelines as stated in *Table 7*. Note that each motor winding must be individually protected against short circuits and ground faults.

In general, when providing overcurrent protection for motor circuits against overcurrents due to grounds and short circuits (*NEC Section 430.52),* follow the guidance listed in *Table 8*. *Table 8* provides time delay and instantaneous trip values for various types of motors as a percentage of full-load current.

When sizing the overload protection for continuous duty motors, refer to *NEC Section 430.32*. Motors that have a service factor of 1.15 and/or a maximum temperature rise of 40°C shall be provided with overload protection limited to 125% of the full-load current of the motor (this also applies to the secondary circuit of a wound rotor motor). All other motors shall be limited to 115% of the full-load current. Motors that are rated at less than 1hp have exceptions to this rule. Refer to *NEC Article 430* for these exceptions.

If the desired overload ratings are not available when sizing overload protection for the motor, then use the next highest available overload rating. This is allowed provided that 140% of full-load current is not exceeded by motors that have a service factor of 1.15 and/or a maximum temperature rise of 40°C. For all other motors, this maximum value would be 130% rather than 115% as stated earlier.

Additionally, in certain situations, motors with installed overloads rated as discussed previously may not start or be able to carry system load. In these instances, it is permissible to increase the overload settings to the respective 140%/130% values. When motor starting is still a problem, the overload protective device may be shorted out during the equipment startup sequence provided that the circuit breaker or fuse protecting the branch is not set at greater than 400% of the full-load current value and the motor does not have an automatic starter.

Overload protection for adjustable speed drives is based on the rated input to the power conversion equipment. If overload protection is supplied with the equipment, then no further overload protection is required. The rating of this disconnecting means shall be no less than 115% of the power conversion equipment rated input current and it shall be physically located in the incoming line. Overload protection, if not shunted, should allow a sufficient time delay for the motor to start and accelerate.

9.2.1 Thermal Protectors

Thermal protectors that are integral with the motor are often used to protect the motor from overloads and starting failures. All motors with a voltage rating greater than 600V must have a thermal

Table 7 Minimum Number of Overload Units

Type of Motor	Supply System	Number and Location of Overload Units (such as trip coils or relays)
Single-phase AC or DC	Two-wire, single-phase AC or DC, ungrounded	One in either conductor
Single-phase AC or DC	Two-wire, single-phase AC or DC, one grounded conductor	One in ungrounded conductor
Single-phase AC or DC	Three-wire, single-phase AC or DC, grounded neutral	One in either ungrounded conductor
Single-phase AC	Any three-phase supply	One in ungrounded conductor
Two-phase AC	Three-wire, two-phase AC, ungrounded	Two, one in each phase
Two-phase AC	Three-wire, two-phase AC, one grounded conductor	Two, one in each ungrounded conductor
Two-phase AC	Four-wire, two-phase AC, grounded or ungrounded	Two, one per phase in ungrounded conductors
Two-phase AC	Five-wire, two-phase AC, grounded neutral or ungrounded	Two, one per phase in any ungrounded phase wire
Three-phase AC	Any three-phase supply	Three, one in each phase*

*Exception: An overload unit in each phase shall not be required where overload protection is provided by other approved means.

Reprinted with permission from NFPA 70, the *National Electrical Code®*. Copyright © 2004, National Fire Protection Association, Quincy, MA 02269. This reprinted material is not the complete and official position of the National Fire Protection Association on the referenced subject, which is represented only by the standard in its entirety.

Table 8 Motor Protection

Type of Motor	Percent of Full-Load Current: Nontime Delay Fuse	Dual Element (Time Delay) Fuse**	Instantaneous Trip Breaker	Inverse Time Breaker*
Single-phase motors	300	175	800	250
AC polyphase motors other than wound-rotor:				
Squirrel cage:				
Other than Design B, energy efficient	300	175	800	250
Design B, energy-efficient	300	175	1,100	250
Synchronous†	300	175	800	250
Wound rotor	150	150	800	150
Direct-current (constant voltage)	150	150	250	150

For certain exceptions to the values specified, see ***NEC Section 430.54.***

*The values given in the last column also cover the ratings of nonadjustable inverse time types of circuit breakers that may be modified per ***NEC Section 430.52.***

**The values in the Nontime Delay Fuse column apply to time delay Class CC fuses.

†Synchronous motors of the low-torque, low-speed type (usually 450 rpm or lower), such as are used to drive reciprocating compressors, pumps, etc., that start unloaded, do not require a fuse rating or circuit breaker setting in excess of 200% of the full-load current.

Reprinted with permission from NFPA 70, the *National Electrical Code®*. Copyright © 2004, National Fire Protection Association, Quincy, MA 02269. This reprinted material is not the complete and official position of the National Fire Protection Association on the referenced subject, which is represented only by the standard in its entirety.

protector and its overload must not have an automatic reset feature. They shall trip no higher than the following percentage of full-load current:

- Motor full-load current not exceeding 9 amps—170%
- Motor full-load current between 9.1 and 20 amps—156%
- Motor full-load current greater than 20.1 amps—140%

This requirement is based on the maximum full-load motor current as listed in the tables provided in *NEC Article 430.*

Motors that are rated 1,500hp include a device that is set to de-energize the motor once the actual temperature rise of the motor equals the rated temperature rise of the motor insulation. Thermal protectors are usually sized and installed by the motor manufacturer.

9.2.2 Branch Considerations

According to *NEC Section 430.22,* when a single motor used in a continuous duty application is supplied from a **branch circuit,** the ampacity of the branch circuit must be not less than 125% of the motor full-load current as determined by *NEC Section 430.6(A)(1).* If a multiple-speed motor is used, then the ampacity shall be based on the highest of the full-load current ratings on the motor nameplate. Where motors have unusual duty cycle requirements, use the requirements listed in *Table 9*, as referenced in *NEC Section 430.22(E)*.

Per *NEC Section 430.24*, when sizing conductors supplying several motors, the capacity shall not be less than 125% of the largest motor plus the sum of the full-load current ratings of all other motors in the group. Values for the full-load amps are taken from *NEC Tables 430.247 through 430.250.* Several motors or loads are permitted to be provided for on one branch circuit if:

- The system voltage is <600 volts.
- The branch protective device protects the smallest installed motor.
- All motors are 1hp, <20A (15A) on 120V (600V) circuits where each motor draws <6A, overloads are installed on the motor, and short circuit current and ground fault current do not exceed the branch circuit rating.
- It is part of a factory-listed assembly.

In instances where taps are used, short circuit current and ground fault current protection may not be required for the taps used. This is true provided that the tap used has the same ampacity as the branch circuit it is connected to. Additionally, the tap cannot be longer than 25' and it must also be physically protected from damage.

Table 9 Duty Cycle Service				
	Percentages of Nameplate Current Rating			
Classification of Service	**5-Minute Rated Motor**	**15-Minute Rated Motor**	**30- and 60-Minute Rated Motor**	**Continuous Rated Motor**
Short-Time Duty				
Operating valves, raising or lowering rolls, etc.	110	120	150	—
Intermittent Duty				
Freight and passenger elevators, tool heads, pumps, drawbridges, turntables, etc. For arc welders, see *NEC Section 630.11*	85	85	90	140
Periodic Duty				
Rolls, ore- and coal-handling machines, etc.	85	90	95	140
Varying Duty	110	120	150	200

Any motor application shall be considered as continuous duty unless the nature of the apparatus it drives is such that the motor will not operate continuously with load under any condition of use.

Reprinted with permission from NFPA 70, the *National Electrical Code®*. Copyright © 2004, National Fire Protection Association, Quincy, MA 02269. This reprinted material is not the complete and official position of the National Fire Protection Association on the referenced subject, which is represented only by the standard in its entirety.

10.0.0 ◆ CONNECTIONS AND TERMINAL MARKINGS FOR AC MOTORS

The markings on the external leads of an induction motor are sometimes missing or illegible, and proper identification must be made before the motor can be connected to the line. This section describes the procedures for identifying leads in either a wye-connected or delta-connected, three-phase, nine-lead motor.

The required materials for this procedure are:

- Appropriate personal protective equipment
- 12V battery such as an automotive battery
- Analog meter with a large scale and low range (digital meters may not clearly capture the voltage kick)
- Test leads and jumpers
- Momentary contact, normally open (N.O.) pushbutton switch
- Labels to mark leads as they are identified
- Three-phase, nine-lead induction motor

Before starting, identify whether the motor to be tagged is wye-connected or delta-connected. Compare the two connection diagrams in *Figure 63* and *Figure 64*. You will see that both types of motors have nine leads and six coils. In a wye-connected motor, three coils are connected together and three coils are isolated. A delta-connected motor has three sets of two coils connected together. Using an ohmmeter, insulate all motor leads from one another and check for continuity between each lead. Start by placing one probe on one lead and check through the remaining leads. As you identify which leads show

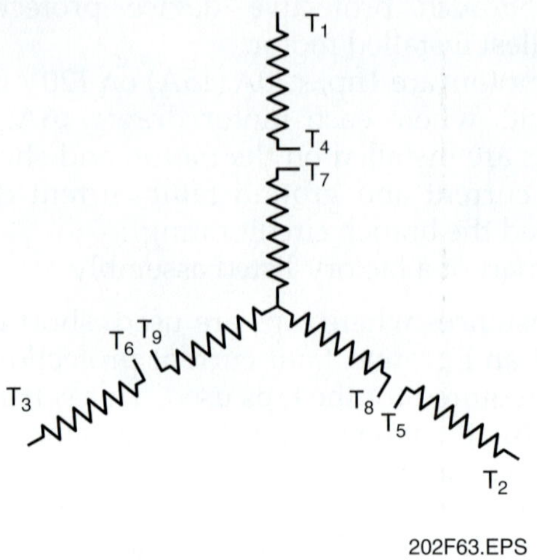

Figure 63 ◆ Dual-voltage, three-phase wye connection.

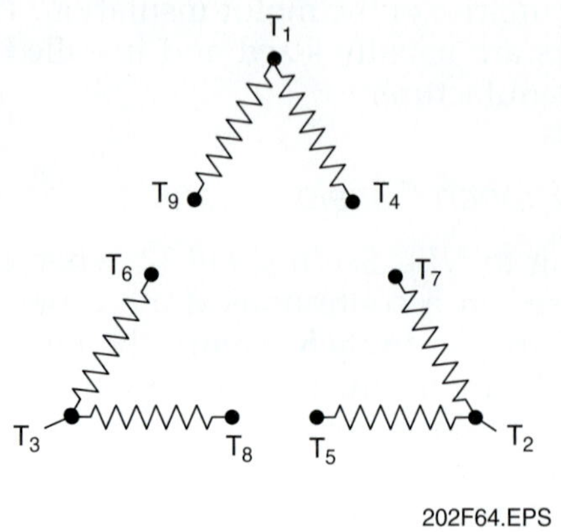

Figure 64 ◆ Dual-voltage, three-phase delta connection.

continuity, group them together. Continue this procedure with each lead until all leads are grouped. When you have completed this procedure, you should have either three wires in one group and three sets of two wires grouped together, which would indicate a wye-connected motor, or three sets of three wires grouped together, indicating a delta-connected motor.

> **NOTE**
> If the leads are partially tied together, mark or identify them in a way that will allow you to reconnect them after the testing procedure is completed.

10.1.0 Identifying the Terminals of Wye-Connected Motors

Figure 65 shows the coil arrangement for a three-phase, wye-connected motor. To identify the terminals of a wye-connected motor, proceed as follows:

Step 1 Taking the group of three common leads, arbitrarily identify them as leads 7, 8, and 9.

Step 2 Using the diagram in *Figure 66* as a guide, connect the positive lead from the battery to lead 7. Connect the lead from the battery negative through the N.O. switch to leads 8 and 9 simultaneously.

Step 3 Connect one of the three remaining lead pairs to the voltmeter terminals.

Step 4 Close the N.O. switch while observing the DC voltmeter. Use the lowest scale practical without over-ranging the meter. If the meter deflection is upward, note the voltage reading. Observe that the reading occurs only on the initial energization of the windings and then decays. Note the peak reading only and ignore the deflection in the opposite direction that occurs when the switch is opened. If the meter initially deflects downward, reverse the test lead connections.

Step 5 Continue with the remaining two lead pairs. The pair with the highest voltage reading is the winding associated with lead 7. The lead with positive polarity is identified as lead 4 and the lead with negative polarity is lead 1.

Step 6 Repeat Step 3, but apply the positive lead of the battery to lead 8 and the negative lead of the battery to leads 7 and 9. The positive lead of the pair with the highest voltage is identified as lead 5 and the lead with negative polarity is lead 2.

Step 7 Repeat Step 3, but apply the positive lead of the battery to lead 9 and the negative lead of the battery to leads 7 and 8. The positive lead of the pair with the highest voltage is marked lead 6 and the negative lead is lead 3.

Step 8 To confirm that all leads are correctly identified, connect the motor to the circuit. Be sure to observe proper connection procedures for the applied voltage. Once connected to the circuit, start the motor and take current readings on all three lines. If the motor starts correctly and the current readings are approximately equal, the procedure was a success.

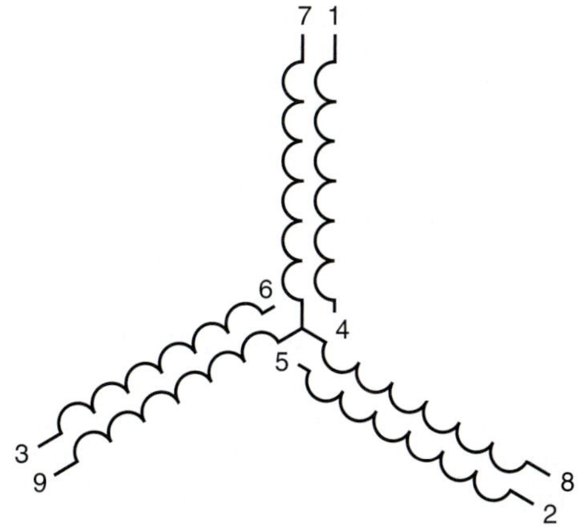

Figure 65 ◆ Coil arrangement in a wye-connected motor.

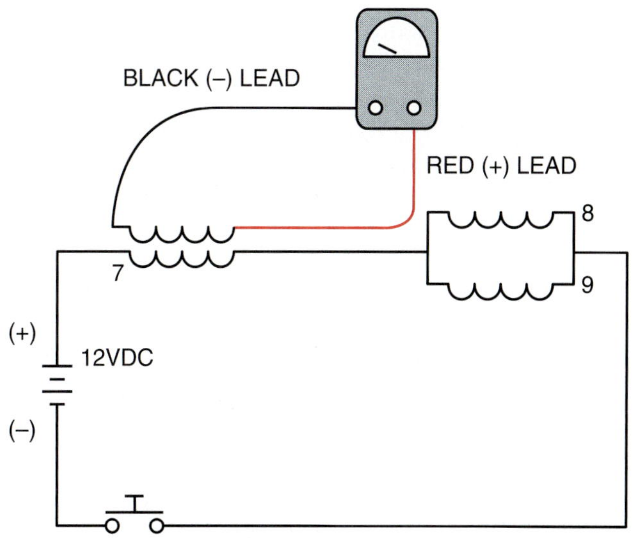

Figure 66 ◆ Battery hookup for wye-connected motor lead identification.

Motor Lead Identification

THINK ABOUT IT

Refer to the procedure for identifying the leads in a wye-connected motor and answer the following questions:

- In Step 2, what would happen if 8 and 9 were connected one at a time?
- In Step 4, why does the voltage reading decay and later deflect downward?
- In Step 5, how much larger is the high voltage reading than the lower readings? Why?

10.2.0 Identifying the Terminals of Delta-Connected Motors

A delta-connected motor has three sets of three leads. *Figure 67* shows how the coils are arranged in a delta-connected, three-phase motor. In this figure, the coils that are side by side are actually wound on the same poles on the motor. As discussed previously, this will allow some transformer interaction between adjacent coils. To identify the terminals of a delta-connected motor, proceed as follows:

Step 1 Using an ohmmeter on a low scale, measure the resistance between each of the three leads in one group. When performing this measurement, you should see that the resistance between two of the leads is about twice that between either of those two and the third. The lead that shows the least resistance to the other two will be lead 1. Refer to all three wires in this set as set 1.

Step 2 Repeat Step 1 with the second set to identify the lead with the least resistance as lead 2. Refer to all three wires in this set as set 2.

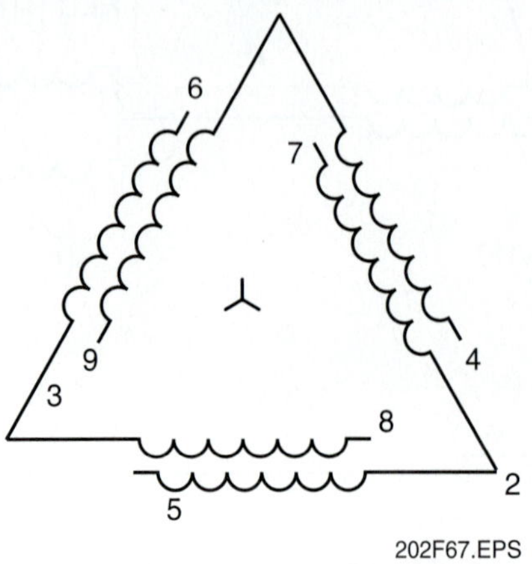

202F67.EPS

Figure 67 ◆ Coil arrangement in a delta-connected motor.

Step 3 Repeat Step 1 with the final set of leads to identify the lead with the least resistance as lead 3. Refer to all three wires in this set as set 3.

Step 4 Using the diagram in *Figure 68* as a guide, connect lead 1 to the positive terminal of a DC voltage source and one of the remaining leads in that set (set 1) to the negative terminal. Attach the red lead of the voltmeter to lead 2 and the black lead to one of the two unknown leads in set 2. Press the pushbutton and observe the meter needle. If the correct leads have been selected, a voltage will be induced into this coil. If not, connect the second lead to the voltmeter and repeat the test. If there is still no induced voltage, disconnect the unknown lead in set 1 from the DC voltage and connect the remaining unknown lead to the negative source. Repeat the test until the leads with an induced voltage have been identified. Once these leads are located, identify the lead connected to the negative terminal of the DC voltage source as lead 4 and the lead connected to the negative voltmeter probe as lead 7. Identify the other lead in set 2 as lead 5.

Step 5 The remaining lead in set 1 will be lead 9. Leaving lead 1 on the positive DC terminal, connect the negative terminal to lead 9. Attach the red lead of the voltmeter to lead 3 and the black lead to one of the two unknown leads in set 3. Press the pushbutton and observe the meter needle. If the correct leads have been selected, a voltage will be induced into this coil. If not, connect the second lead to the voltmeter and repeat the test. Once these leads are located, identify the lead connected to the negative voltmeter probe as lead 6.

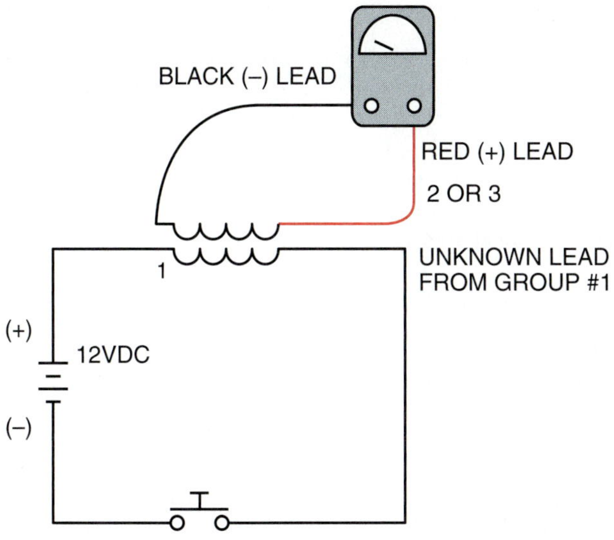

Figure 68 ◆ Battery hookup for delta-connected motor lead identification.

Step 6 The remaining lead in set 3 is lead 8. To verify this, connect lead 3 to the positive terminal of the DC voltage source and lead 8 to the negative terminal. Attach the red lead of the voltmeter to lead 2 and the black lead to lead 5. Press the pushbutton and observe the meter needle. If the results are correct, a voltage will be induced into this coil, resulting in meter needle deflection.

Step 7 To confirm that all leads are correctly identified, connect the motor to the circuit. Be sure to observe proper connection procedures for the applied voltage. Once connected to the circuit, start the motor and take current readings on all three lines. If the motor starts correctly and the current readings are approximately equal, the procedure was a success.

11.0.0 ◆ *NEC®* REQUIREMENTS

NEC Article 430 covers the application and installation of motors, motor circuits, and motor control connections, including conductors, short circuit and ground fault protection, starters, disconnects, and overload protection.

NEC Article 440 contains provisions for motor-driven equipment and for branch circuits and controllers for HVAC equipment.

All motors must be installed in a location that allows adequate ventilation to cool the motors. Furthermore, the motors should be located so that maintenance, troubleshooting, and repairs can be readily performed. Such work could consist of lubricating the motor bearings or perhaps replacing worn brushes. Testing the motor for open circuits and ground faults is also necessary from time to time.

When motors must be installed in locations where combustible material, dust, or similar material may be present, special precautions must be taken in selecting and installing the motors.

Any exposed live parts of motors operating at 50V or more between terminals must be guarded; that is, they must be installed in a room, enclosure, or location so as to allow access only by qualified persons (electrical maintenance personnel). If such a room, enclosure, or location is not feasible, an alternative is to elevate the motors not less than 8' above the floor. In all cases, adequate space must be provided around motors with exposed live parts, even when properly grounded, to allow for maintenance, troubleshooting, and repairs.

The chart in *Table 10* summarizes *NEC®* installation rules.

A summary of *NEC Article 430* is shown in *Figure 69*. Detailed information may be found in the *NEC®* under the articles or sections indicated.

Table 10 Summary of *NEC*® Requirements for Motor Installations

Application	Requirement	*NEC*® Reference
Location	Motors must be installed in areas with adequate ventilation. They must also be arranged so that sufficient work space is provided for replacement and maintenance.	*NEC Section 430.14(A)*
	Open motors must be located or protected so that sparks cannot reach combustible materials.	*NEC Section 430.14(B)*
	In locations where dust or flying material will collect on or in motors in such quantities as to seriously interfere with the ventilation or cooling of motors and thereby cause dangerous temperatures, suitable types of enclosed motors that will not overheat under the prevailing conditions must be used.	*NEC Section 430.16*
Disconnecting means	A motor disconnecting means must be within sight from the controller location (with exceptions) and disconnect both the motor and controller. The disconnect must be readily accessible and clearly indicate the OFF/ON positions (open/closed).	*NEC Article 430, Part IX* *NEC Section 430.104*
	Motor control circuits require a disconnecting means to disconnect them from all supply sources.	*NEC Section 430.74*
	The disconnecting means must be as specified in the code.	*NEC Section 430.109*
Wiring methods	Flexible connections such as Type AC cable, Greenfield, flexible metal tubing, etc., are standard for motor connections.	*NEC Articles 300 and 430*
Motor control circuits	All conductors of a remote motor control circuit outside of the control device must be installed in a raceway or otherwise protected. The circuit must be wired so that an accidental ground in the control device will not start the motor.	*NEC Section 430.73*
Guards	Exposed live parts of motors and controllers operating at 50 volts or more must be guarded by installation in a room, enclosure, or other location so as to allow access by only qualified persons, or elevated 8 feet or more above the floor.	*NEC Section 430.232*
Adjustable speed drive systems	Requirements for adjustable speed drives and their motors.	*NEC Article 430, Part X*
Motors operating over 600 volts	Special installation rules apply to motors operating at over 600 volts.	*NEC Article 430, Part XI*
Controller grounding	Motor controllers must have their enclosures grounded.	*NEC Section 430.244*

INSIDE TRACK

Motor Connections

On dual-voltage and/or multi-speed motors, always check the wiring connection diagrams given on the motor nameplate to wire the motor for the correct voltage and/or speed.

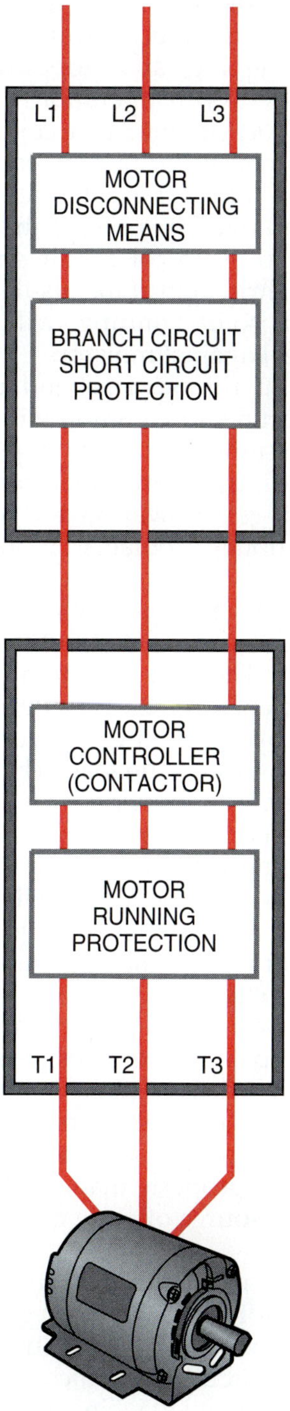

NEC Article 430, Part IX Sections 430.101 through 430.113

Disconnects motor and controllers from circuit.

1. Continuous rating of 115% or more of motor FLC. Also see ***NEC Article 430, Part II.***
2. Disconnecting means shall be as listed in ***NEC Section 430.109.***
3. Must be located in sight of motor location and driven machinery. The controller disconnecting means can serve as the disconnecting means if the controller disconnect is located in sight of the motor location and driven machinery.

NEC Article 430, Part IV Sections 430.51 through 430.58

Protects branch circuit from short circuits or grounds.

1. Must carry starting current of motor.
2. Rating must not exceed values in ***NEC Table 430.52*** unless not sufficient to carry starting current of motor.
3. Values of branch circuit protective devices shall in no case exceed exceptions listed in ***NEC Section 430.52.***

NEC Article 430, Part VII Sections 430.81 through 430.91

Used to start and stop motors.

1. Must have current rating of 100% or more of motor FLC.
2. Must be able to interrupt LRC.
3. Must be rated as specified in ***NEC Section 430.83.***

NEC Article 430, Part III Sections 430.31 through 430.44

Protects motor and controller against excessive heat due to motor overload.

1. Must trip at following percent or less of motor FLC for continuous motors rated more than one horsepower.
   a) 125% FLC for motors with a marked service factor of not less than 1.15 or a marked temperature rise of not over 40°C.
   b) 115% FLC for all others. (See the *NEC®* for other types of protection.)
2. Three thermal units required for any three-phase AC motor.
3. Must allow motor to start.
4. Select size from FLC on motor nameplate.

NEC Article 430, Part II Sections 430.21 through 430.29

Specifies the sizes of conductors capable of carrying the motor current without overheating.

1. To determine the ampacity of conductors, switches, branch circuit overcurrent devices, etc., the full-load current values given in ***NEC Tables 430.247 thru 430.250*** shall be used instead of the actual current rating marked on the motor nameplate. ***(See NEC Section 430.6.)***
2. According to ***NEC Section 430.22,*** branch circuit conductors supplying a single motor used in a continuous duty application shall have an ampacity of not less than 125% of motor FLC, as determined by ***NEC Section 430.6(A)(1).***

202F69.EPS

Figure 69 ◆ Summary of requirements for motors, motor circuits, and controllers.

12.0.0 ◆ TROUBLESHOOTING

The useful life of an induction motor depends largely on the condition of its insulation. In general, the insulation should be suitable for the operating requirements.

12.1.0 Stator Windings

The stator (stationary) windings appear to be so simple and rugged as to cause one to frequently overlook the necessity for certain maintenance procedures. However, a glance into the average motor repair shop will make it apparent that the induction motor stator is a vulnerable piece of equipment. Most of the work going on will be involved with replacing or repairing stator windings.

Stator troubles can usually be traced to one or more of the following causes:

- Worn bearings
- Moisture
- Overloading
- Operating single phase
- Poor insulation

Dust and dirt are usually contributing factors. Some forms of dust are highly conductive and contribute materially to insulation breakdown. The effect of dust on the motor temperature through restriction of ventilation is another reason for keeping the machine clean, either by periodically blowing it out with compressed air or by dismantling and cleaning it. The compressed air must be dry and throttled down to a low pressure that will not endanger the insulation.

Moisture – One of the most subtle enemies of motor insulation is moisture. Needless to say, motor insulation must be kept reasonably dry, although many applications make this practically impossible unless the motor is totally enclosed or otherwise protected from the direct effects of moisture. If operated in a damp place, a special moisture-resistant coating should be applied to the windings.

Dipping and baking – The life of a winding depends on keeping it in its original (or new) condition for as long as possible. In a new machine, the winding is snug in the slots and the insulation is fresh and flexible, being newly treated with varnish and therefore resistant to the deteriorating effects of moisture and other foreign matter. This condition is best maintained by periodic cleaning, followed by varnish and oven treatments.

One condition that frequently hastens winding failure is movement of the coils because of vibration during operation. After insulation dries out, it loses its flexibility and the mechanical stresses caused by starting, plugging, and stopping, as well as the natural stresses in operation under load, will precipitate short circuits in the coils and possibly failures from coil to ground, usually at the point where the coil leaves the slot. The effect of periodic varnish and oven treatments that are properly carried out so as to fill all air spaces caused by drying and shrinkage of the insulation, thereby maintaining a solid winding, will also provide an effective seal against moisture.

Rotor windings – The rotors of wound rotor motors have many features in common with the stators; therefore, the same comments apply to the care of rotor windings as are given for the care of stator windings. However, the rotor introduces some additional problems because it is a rotating element.

Most wound rotors have a three-phase winding, and are, therefore, susceptible to trouble from single-phase operation. The first symptom of an open-rotor circuit is lack of torque, with a decrease in speed accompanied by a growling noise, or perhaps a complete failure to start the load. The first place to look for an open secondary circuit is in the resistance bank or the control circuit external to the rotor. Short circuiting the rotor circuit at the slip rings and then operating the motor will usually determine whether the trouble is in the control circuit or in the rotor itself. It may be one of the stud connections to the slip rings.

If the rotor is wave wound with the windings made up of copper strap coils with clips connecting the top and bottom halves of the coil, inspect these end connections for possible signs of heating, which would be an indication of a partially open circuit. Faulty or improperly made end connections are a common source of open circuits in rotor windings.

A ground in a rotor circuit will not affect the performance of the motor unless another ground should also develop, which might cause the equivalent of a short circuit, in which case it would have the effect of unbalancing the rotor electrically. In addition to reduced torque, another symptom of this condition might be excessive vibration of the motor. There might also be sparking and uneven wear of the collector rings.

Another method of checking for short circuits in the rotor windings is to raise the brushes off the slip rings and energize the stator. If the rotor winding is free from short circuits, it should have little or no tendency to rotate, even when disconnected from the load. If it does show evidence of considerable torque or the tendency to come up to speed, the rotor should be removed and the winding opened and examined for a fault. In making this test, note that some rotors having a wide tooth design may show a tendency to rotate even though the windings are in good condition.

Still another check that can be made when the rotor is in place and the stator is energized (also with the brushes raised) is to check the voltage across the rings to see if they are balanced. When making this check, be sure that any inequality in voltage measurements is not due to the relative position of the rotor and stator phases. To avoid inaccurate measurements, the rotor should be moved to several positions during the voltage test.

12.2.0 Squirrel Cage Rotors

Squirrel cage rotors are more rugged and in general require less maintenance than wound rotors, but may have trouble because of open circuits or high resistance points in the rotor circuit. The symptoms of such conditions are generally the same as with wound rotor motors: slowing down under load and reduced starting torque. Such conditions can usually be detected by looking for evidence of heating at the end ring connections, which is particularly noticeable when shutting down after operating under load.

In brazed rotors, any fractures in the rotor bars will usually be found either at the point of connection to the end ring or at the point where the bar leaves the laminations. Discolored rotor bars are also evidence of excessive heating.

Brazing broken bars or replacing bars should only be done by a qualified person. Considerable technique is required for this kind of work, and it is recommended that the manufacturer's nearest district sales office be consulted before attempting such repairs in the shop or plant, unless an experienced operator is available.

With die-case rotors, look for cracks or other imperfections that may have developed in the end rings. A faulty die-case rotor can rarely be effectively repaired and should be replaced if defective.

12.3.0 The Air Gap

A small air gap is characteristic of the induction motor. The size of the air gap has an important bearing on the power factor of the motor. Doing anything to affect it, such as grinding the rotor laminations or filing the stator teeth, results in increased magnetizing current with a resultant lower power factor.

Good maintenance procedures call for periodically checking the air gap with a feeler gauge to ensure against a worn sleeve bearing that might permit the rotor to rub the laminations. (A very light rub can produce enough heat to destroy the coil insulation.) Gap measurements should be made on the shaft end of the motor.

On large machines, it is desirable to keep a record of these checks. Four measurements should be taken approximately 90° apart, with one of these points being the load side; that is, the point on the rotor periphery that corresponds with the load side of the bearing.

A comparison of the new measurements with previously-recorded measurements will permit the early detection of bearing wear.

12.4.0 Overloading and Single-Phase Operation

Often, a motor of adequate capacity that was properly applied in the original application is later found to be overloaded or otherwise unsuited for the job. This usually happens because of one or more of the following:

- More severe duty imposed on the motor
- A change in equipment
- A change in equipment parts
- A change in operating time

Connecting measuring instruments to the motor circuit will quickly disclose the reason for motor overheating, failure to start the load, or other abnormal symptoms.

Control circuits for many older systems were not provided with relay protection, and single-phase operation of polyphase induction motors on such circuits has frequently been responsible for motor burnout. Usually this has resulted from one of the fuses blowing while the motor is up to speed and under load. Under such conditions, the portion of winding that remains in the circuit will endeavor to carry the load until it fails due to overheating.

The effect of increasing the load on the motor beyond its rated capacity is simply to increase the operating temperature, which shortens the life of the insulation. Momentary overloads usually do no damage; consequently, the tendency is to use the thermal type of overload protection in present-day controls. Obviously, the ideal place to measure the thermal effect of overload is on the motor itself.

The polyphase induction motor is the simplest and most foolproof piece of rotating electrical apparatus. The largest single cause of winding failures is probably the rotor rubbing the stator iron, usually because of a worn or failed bearing.

13.0.0 ◆ TESTING ELECTRIC MOTORS

A simple test light or continuity tester may be used to test for an open motor coil. With all power circuits shut off, connect the continuity test leads across each motor coil in turn. If the coil is operational, the light will glow or the dial of the ohmmeter will swing to full scale.

To test for a grounded coil, connect one of the test leads to the motor frame and the other lead to one of the field coil wires. If the light glows or the ohmmeter dial swings toward zero, the coil is grounded.

To prepare for an insulation resistance test, first take the equipment or circuit to be tested out of service. Check between the equipment terminals and ground using a voltmeter (at the proper range setting) to be sure there is no voltage present. If possible, disconnect all leads to the motor being tested. When a motor or circuit is not completely isolated, make sure you are aware of all the components that will be tested when the megohmmeter (megger) is connected. Should an interconnected circuit be overlooked, the instrument readings may be lower than expected.

The testing of wiring can be performed on all types of systems if two rules are kept in mind:

- Be sure all wiring is de-energized.
- Know what wiring is included in the test and make a record of it.

When a distribution panel is present, check the entire system to ground by attaching one megger

INSIDE TRACK

Testing with a Megohmmeter

While multimeters or ohmmeters can be used to check for open or short circuits in a motor, they can't be used to test the insulation resistance of a motor because the resistance to be measured is in the megohm range. A megohmmeter (megger) is designed to apply a high voltage to the motor to check the insulation resistance under load. Because all motor insulation degrades over time, many industrial/commercial facilities conduct a periodic maintenance program that includes checking the insulation resistance of motors on a regular schedule so that the rate of degradation can be used to predict motor failures before they occur.

202P0206.EPS

Motor Maintenance and Repair

INSIDE TRACK

Modern industrial culture has developed a "use it up, throw it away, and buy it new" mentality. Motors are often poorly maintained and quickly replaced. However, most motors are simple in construction and relatively easy to maintain and repair. Dirt and metallic particles are common causes of motor failure.

If your facility has a number of expensive motors in use, make sure that a preventive maintenance schedule is in place. Many facilities do not have such procedures. If there is no maintenance schedule, create one.

lead to the dead (de-energized) post of the open main power switch and the other lead to a grounded portion of the system such as the panel housing or an incoming conduit.

Individual circuits are tested to ground by opening distribution panel switches, fuses, or circuit breakers and testing each circuit in turn.

Once all power has been disconnected, disconnect the motor from the line, either by using the switch or by disconnecting the wiring at the motor terminals. If the switch is used, remember that the insulation resistance of the connecting wire, switch panel, and contacts will all be measured at the same time. Connect the positive megger lead to one of the motor lines and the negative test lead to the frame of the motor. If insulation resistance minimums have been established, the reading can be checked against them. Always check with your supervisor before using a megger. Since every brand of megger operates somewhat differently, always refer to the operating manual or seek assistance from your supervisor as to the proper operation of the instrument.

14.0.0 ◆ MOTOR INSTALLATION

The best motors on the market will operate improperly if they are installed incorrectly. Therefore, all personnel involved with the installation of electric motors should understand the procedures for installing the various types of motors that will be used.

WARNING!

When a motor is received at the job site, always refer to the manufacturer's instructions and follow them to the letter. Failure to do so could result in serious injury or death. Install and ground according to *NEC*® requirements and good practices. Consult qualified personnel with any questions or problems.

Keep the following in mind when installing new motors:

- *Uncrating* – Once the motor has been carefully uncrated, check to see if any damage has occurred during handling. Be sure that the motor shaft and armature turn freely. This is also a good time to check to determine if the motor has been exposed to dirt, grease, grit, or excessive moisture during shipment or storage. Motors in storage should have shafts turned over once each month to redistribute grease in the bearings. The measure of insulation resistance is a good dampness test. Clean the motor of any dirt or grit.

WARNING!

Never start a motor that has been wet until it has been completely dried and thoroughly tested.

- *Lifting* – Eyebolts or lifting lugs on motors are intended only for lifting the motor and factory motor-mounted standard accessories. These lifting provisions should never be used when lifting or handling the motor when the motor is attached to other equipment as a single unit. The eyebolt lifting capacity rating is based on a lifting alignment coincident with the eyebolt centerline. The eyebolt capacity reduces as deviation from this alignment increases.
- *Guards* – Rotating parts such as pulleys, couplings, external fans, and shaft extensions must be permanently guarded against accidental contact with clothing or body extremities.
- *Requirements* – All motors must be installed, protected, and fused in accordance with *NEC Article 430.* For general information on grounding, refer to *NEC Article 250* and *NEC Article 430, Part XIII*.
- *Thermal protector information* – The motor nameplate may or may not be stamped to indicate thermal protection.

INSIDE TRACK

Installing Motors

The shaft of this air conditioner motor must be aligned precisely with the shaft of the driven device. Note the micrometer attached to the motor and the load to achieve exact alignment.

202P0207.EPS

THINK ABOUT IT

Putting It All Together

Count the motors in your home. The typical home may easily have over 30 motors (including electronic equipment and tools). A century ago, a typical home might have had none. Determine what types of motors you have. Are they AC or DC? What identifying information can you determine by examining each motor nameplate?

Review Questions

1. The name of the motor part that rotates during operation is the _____.
   a. stator
   b. shunt/capacitor
   c. brushes/commutator
   d. armature/rotor

2. The electrical energy required to produce one horsepower of mechanical energy is _____.
   a. 476W
   b. 647W
   c. 746W
   d. 864W

3. The principal reason for developing the brushless DC motor was to _____.
   a. eliminate commutator problems
   b. improve efficiency
   c. increase horsepower
   d. improve airplanes

4. The name of the stationary motor part that produces the magnetic field during operation is the _____.
   a. stator
   b. shunt/capacitor
   c. brushes/commutator
   d. armature/rotor

5. The name of the most popular rotor in use is the _____ rotor.
   a. wound
   b. squirrel cage
   c. multiphase
   d. induction

6. The starting torque characteristics of an induction motor are defined by _____.
   a. rotor resistance
   b. starting current
   c. load
   d. stator resistance

7. The speed of an induction motor depends on the power supply frequency and _____.
   a. current
   b. voltage
   c. size
   d. number of pairs of poles

8. The NEMA frame designation for gasoline pump motors is _____.
   a. B
   b. C
   c. G
   d. H

9. The letters FLA on a motor nameplate stand for _____.
   a. fused last application
   b. fast lower arm
   c. full-load amps
   d. full-load armature

10. To determine the horsepower rating of a motor that was already installed and in place, you would _____.
    a. multiply the rated voltage times the FLA
    b. divide the nominal efficiency by the power factor
    c. check the nameplate data
    d. use a torque wrench and horsepower data

11. The multiplier to the nameplate horsepower rating is called the _____.
    a. maximum kVAR
    b. service factor
    c. power factor
    d. NEMA nominal efficiency

12. A _____ is used to protect the motor from overloads and starting failures.
    a. fuse
    b. circuit breaker
    c. centrifugal switch
    d. thermal protector

13. The code requirements for motor disconnects are covered in _____.
    a. *NEC Article 430, Part VIII*
    b. *NEC Article 430, Part IX*
    c. *NEC Article 430, Part X*
    d. *NEC Article 430, Part XI*

Review Questions

14. One of the five causes of stator problems in motors is _____.
    a. no motor protection device
    b. worn bearings
    c. torque
    d. no air circulation

15. Before starting a motor that has been wet, _____.
    a. clean it with alcohol
    b. thoroughly dry it
    c. rotate it by hand
    d. heat it

Summary

This module discussed AC and DC motor theory, construction, and various motor types and applications. This discussion included torque, speed, and speed regulations as well as the fundamental concepts associated with variable speed drive systems. Motor enclosures were described, including open and totally enclosed motors and motor frame designations. Discussions of horsepower and calculation of load under various conditions were included. *NEC*® requirements and basic troubleshooting techniques were also discussed.

Notes

Trade Terms Introduced in This Module

Armature: The rotating windings of a DC motor.

Branch circuit: The circuit conductors between the final overcurrent device protecting the circuit and the outlet(s).

Brush: A conductor between the stationary and rotating parts of a machine. It is usually made of carbon.

Circuit breaker: A device designed to open and close a circuit by nonautomatic means and to open the circuit automatically on a predetermined overcurrent without injury to itself when properly applied within its rating.

Commutator: A device used on electric motors or generators to maintain a unidirectional current.

Continuous duty: Operation at a substantially constant load for an indefinitely long time.

Controller: A device that serves to govern, in some predetermined manner, the electric power delivered to the apparatus to which it is connected.

Duty: Describes the length of operation. There are four designations for circuit duty: continuous, periodic, intermittent, and varying.

Equipment: A general term including material, fittings, devices, appliances, fixtures, apparatus, and the like used as a part of, or in connection with, an electrical installation.

Field poles: The stationary portion of a DC motor that produces the magnetic field.

Horsepower: The rated output capacity of the motor. It is based on breakdown torque, which is the maximum torque a motor will develop without an abrupt drop in speed.

Hours: The duty cycle of a motor. Most fractional horsepower motors are marked continuous for around-the-clock operation at the nameplate rating in the rated ambient conditions. Motors marked one-half are for ½-hour ratings, and those marked one are for 1-hour ratings.

Intermittent duty: Operation for alternate intervals of (1) load and no load; or (2) load and rest; or (3) load, no load, and rest.

Overcurrent: Any current in excess of the rated current of equipment or the ampacity of a conductor. It may result from an overload, short circuit, or ground fault.

Overload: Operation of equipment in excess of the normal, full-load rating, or of a conductor in excess of rated ampacity, which, after a sufficient length of time, will cause damage or dangerous overheating. A fault, such as a short circuit or ground fault, is not an overload.

Periodic duty: Intermittent operation at a substantially constant load for a short and definitely specified time.

Revolutions per minute (rpm): The approximate full-load speed at the rated power line frequency. The speed of a motor is determined by the number of poles in the winding. A four-pole, 60Hz motor runs at an approximate speed of 1,725 rpm. A six-pole, 60Hz motor runs at an approximate speed of 1,140 rpm.

Rotation: For single-phase motors, the standard rotation, unless otherwise noted, is counterclockwise facing the lead or opposite shaft end. All motors can be reconnected at the terminal board for opposite rotation unless otherwise indicated.

Synchronous speed: When the speed of the rotor is equal to the speed of the stator. The speed is determined by multiplying 120 times the frequency divided by the number of poles.

Thermal protector: A protective device for assembly as an integral part of a motor or motor compressor that, when properly applied, protects the motor against dangerous overheating due to overload or failure to start.

Varying duty: Operation at varying loads and/or intervals of time.

Additional Resources

This module is intended to present thorough resources for task training. The following reference works are suggested for further study. These are optional materials for continued education rather than for task training.

American Electricians' Handbook, Latest Edition. New York: Croft and Summers, McGraw-Hill.

National Electrical Code® Handbook, Latest Edition. Quincy, MA: National Fire Protection Association.

CONTREN® LEARNING SERIES — USER FEEDBACK

The NCCER makes every effort to keep these textbooks up-to-date and free of technical errors. We appreciate your help in this process. If you have an idea for improving this textbook, or if you find an error, a typographical mistake, or an inaccuracy in NCCER's *Contren®* textbooks, please write us, using this form or a photocopy. Be sure to include the exact module number, page number, a detailed description, and the correction, if applicable. Your input will be brought to the attention of the Technical Review Committee. Thank you for your assistance.

Instructors – If you found that additional materials were necessary in order to teach this module effectively, please let us know so that we may include them in the Equipment/Materials list in the Annotated Instructor's Guide.

Write: Product Development
National Center for Construction Education and Research
P.O. Box 141104, Gainesville, FL 32614-1104

Fax: 352-334-0932

E-mail: curriculum@nccer.org

Craft ______________________ Module Name ______________________

Copyright Date __________ Module Number __________ Page Number(s) __________

Description

____

____

____

(Optional) Correction

____

____

(Optional) Your Name and Address

____

____

Grounding

26203-05

Gaylord Texan Resort

Grapevine, Texas
Exterior Finish Award Winner
Triangle Plastering Systems, Inc.

26203-05

Grounding

Topics to be presented in this module include:

Overview

One of the most comprehensive articles of the *National Electrical Code®* is *Article 250,* also known as the Grounding Article. It is so comprehensive because of the important role that grounding plays in the safe operation of electrical systems.

NEC Article 250 describes two forms of grounding: system grounding and equipment grounding. System grounding is the intentional connection of one of the current-carrying conductors to a grounding electrode, while equipment grounding is the physical, conducting connection of any noncurrent-carrying metal parts to a grounding electrode. The interconnection of all noncurrent-carrying metal parts to each other is called bonding. Once the bonding system is connected to the grounding electrode, the equipment grounding system is effectively grounded.

System grounding is essential in the electrical operation of some grounded systems. In addition, both system grounding and equipment grounding provide a level of personal protection against electrical shock should a fault occur in the electrical system.

Note: *National Electrical Code®* and *NEC®* are registered trademarks of the National Fire Protection Association, Inc., Quincy, MA 02269. All *National Electrical Code®* and *NEC®* references in this module refer to the 2005 edition of the *National Electrical Code®*.

Objectives

When you have completed this module, you will be able to do the following:

1. Explain the purpose of grounding and the scope of *NEC Article 250.*
2. Distinguish between a short circuit and a ground fault.
3. Define the *NEC®* ground-related terms.
4. Distinguish between system grounding and equipment grounding.
5. Use *NEC Table 250.66* to size the grounding electrode conductor for various AC systems.
6. Explain the *NEC®* requirements for the installation and physical protection of grounding electrode conductors.
7. Explain the function of the grounding electrode system and determine which grounding electrodes must be used.
8. Define electrodes and explain the resistance requirements for electrodes using *NEC Section 250.56.*
9. Use *NEC Table 250.122* to size the equipment grounding conductor for raceways and equipment.
10. Explain the function of the main bonding jumper and system bonding jumpers in the grounding system and size the bonding jumpers for various applications.
11. Size the main bonding jumper for a service utilizing multiple service disconnecting means.
12. Explain the *NEC®* requirements for bonding of enclosures and equipment.
13. Explain the *NEC®* requirements for grounding of enclosures and equipment.
14. Explain effectively grounded and its importance in clearing ground faults and short circuits.
15. Explain the purposes of the grounded conductor (neutral) in the operation of overcurrent devices.
16. Explain the *NEC®* requirements for grounding separately derived systems, including transformers and generators.
17. Explain the *NEC®* requirements for grounding at more than one building.
18. Explain the *NEC®* grounding requirements for systems over 600 volts.

Prerequisites

Before you begin this module, it is recommended that you successfully complete *Core Curriculum; Electrical Level One;* and *Electrical Level Two*, Modules 26201-05 and 26202-05.

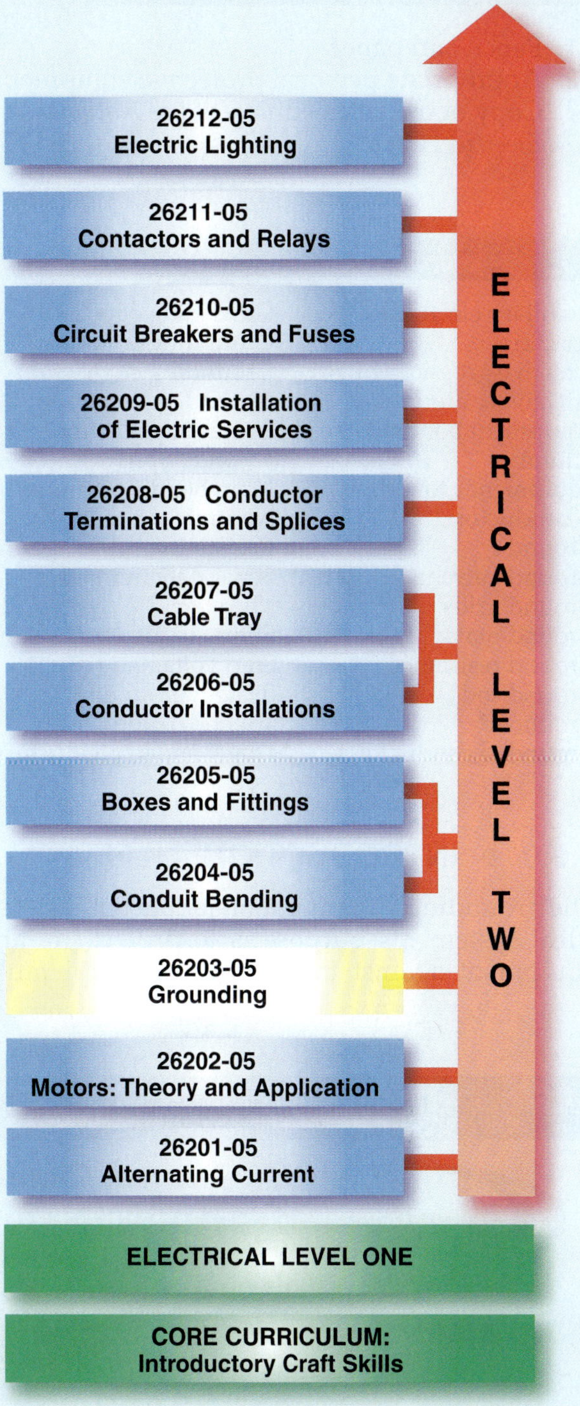

203CMAP.EPS

This course map shows all of the modules in *Electrical Level Two.* The suggested training order begins at the bottom and proceeds up. Skill levels increase as you advance on the course map. The local Training Program Sponsor may adjust the training order.

Required Trainee Materials

1. Pencil and paper
2. Appropriate personal protective equipment
3. Copy of the latest edition of the *National Electrical Code®*

Trade Terms

Auxiliary electrodes
Bonding
Effective grounding path
Effectively grounded
Equipment bonding jumper
Equipment grounding conductor
Ground
Ground current
Ground grids
Ground mats
Ground resistance
Ground rod
Grounded
Grounded conductor
Grounding clip
Grounding conductor
Grounding connections
Grounding electrode
Grounding electrode conductor
Main bonding jumper
Neutral
Resistivity
Separately derived system
Short circuit
Step voltage
System grounding
Touch voltage
Ungrounded conductors

1.0.0 ◆ INTRODUCTION

The grounding system is a major part of the electrical system. Its purpose is to protect life and equipment against the various electrical faults that can occur. It is sometimes possible for higher-than-normal voltages to appear at certain points in an electrical system or in the electrical equipment connected to the system. Proper grounding ensures that the electrical charges that cause these high voltages are channeled to earth or ground before damaging equipment or causing danger to human life. Therefore, a circuit is **grounded** to limit the voltage on the circuit and improve overall operation of the electrical system and continuity of service.

When we refer to **ground,** we are talking about ground potential or earth ground. If a conductor is connected to the earth or to some conducting body that serves in place of the earth, such as a driven **ground rod** (electrode) or cold-water pipe, the conductor is said to be grounded. The **neutral** in a three- or four-wire service, for example, is intentionally grounded and therefore becomes a **grounded conductor.** However, a wire used to connect this neutral conductor to a grounding electrode or electrodes is referred to as a **grounding conductor.** Note the difference in the two meanings; one is grounded, while the other is grounding.

The **equipment grounding conductor** is the conductor used to connect the noncurrent-carrying metal parts of equipment, raceways, and other enclosures to the system grounded conductor, the **grounding electrode conductor,** or both, at the service equipment or at the source of a **separately derived system.** See *Figure 1.*

INSIDE TRACK

Neutral Conductor Size

The grounded neutral conductor is permitted to be smaller than the hot ungrounded phase conductors only when the neutral conductor is properly and adequately sized to carry the maximum load imbalance as computed according to ***NEC Sections 215.2, 220.61, and 230.42.***

It must be pointed out that per ***NEC Section 250.24(C)*** the grounded conductor must not be smaller than the size of the grounding electrode conductor. This is important when you have three-phase services with little or no neutral loads to use as a basis for sizing the grounded conductor.

NEC Section 215.2 (feeders) and ***NEC Section 230.42*** (service-entrance conductors) reference ***NEC Article 220*** for minimum rating and size computation requirements. ***NEC Article 220*** states that the feeder neutral load must be the maximum imbalance of the load, which is the maximum net computed load between the neutral and any one ungrounded conductor.

In summary, both the neutral current, as determined in ***NEC Articles 215, 220, and 230,*** and the minimum grounding electrode size, as determined in ***NEC Article 250,*** must be used to determine the actual neutral wire size.

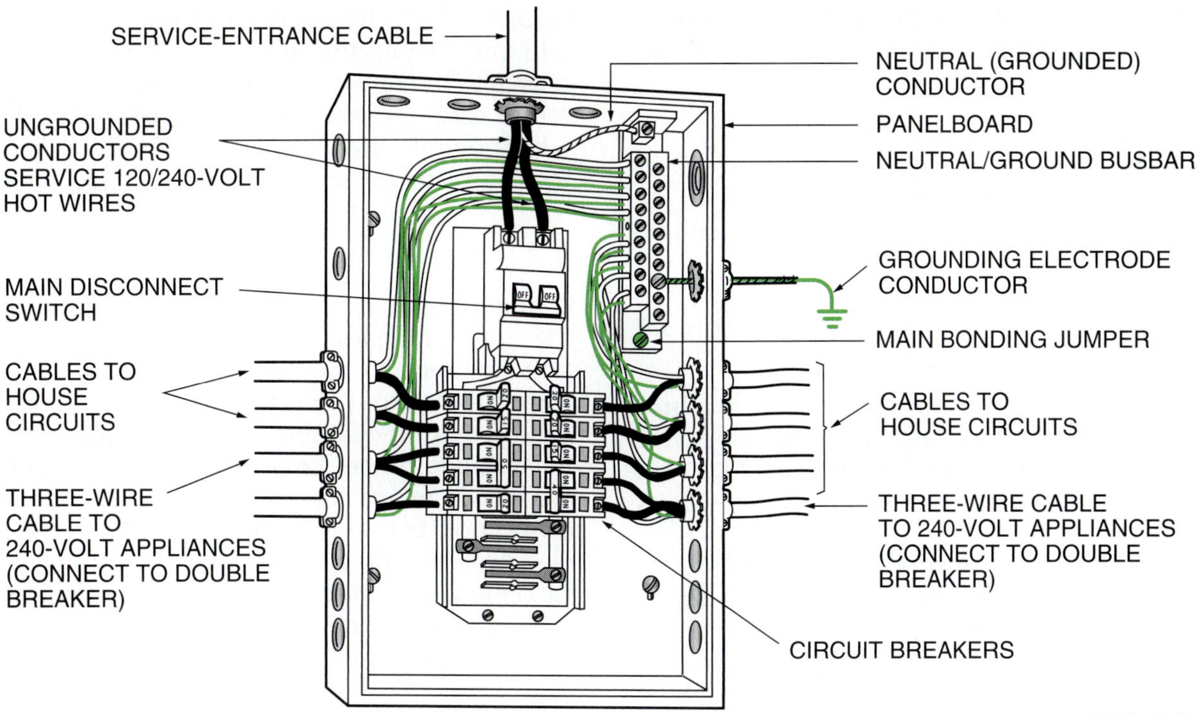

Figure 1 ◆ Panelboard showing grounded and grounding conductors.

This module is designed to explain what a ground is and to teach proper grounding techniques. *NEC®* regulations and the testing of grounding systems will be thoroughly covered. Upon completion of this module, you should have a solid foundation in the principles of grounding, one of the most important aspects of an electrical system for the protection of life and property.

2.0.0 ◆ PURPOSE OF GROUNDING

Systems are solidly grounded to limit the voltage to ground during normal operation and to prevent excessive voltages due to lightning and line surges. They are also grounded to stabilize the voltage to ground during normal operation. Conductive materials enclosing electrical conductors or equipment, or that form a part of the equipment, are grounded. This limits the voltage to ground on these materials. The conductive materials are bonded to facilitate the operation of overcurrent devices under ground fault conditions.

3.0.0 ◆ *NEC®* REQUIREMENTS

NEC Article 250 is the primary governing article for the proper use and installation of grounding and bonding. This article covers general requirements for grounding and **bonding** of electrical installations. It presents the following specific requirements:

- Systems, circuits, and equipment that are required, permitted, or not permitted to be grounded
- Circuit conductor to be grounded on grounded systems
- Location of **grounding connections**
- Types and sizes of grounding and bonding conductors and electrodes
- Methods of grounding and bonding
- Conditions under which guards, isolation, or insulation may be substituted for grounding

What's wrong with this picture?

NEC Section 250.4(A)(5) requires that the path to ground from circuits, equipment, and metal enclosures shall:

- Be permanent and continuous
- Have the capacity to safely conduct any fault current likely to be imposed on it
- Have sufficiently low impedance to facilitate the operation of the protective devices in the circuit
- Be capable of safely carrying the maximum ground fault likely to be imposed

Permanent, reliable, and continuous grounding systems are vital to the safety of electrical systems. Intermittent connections are likely to be unpredictable and may result in hazardous situations.

The minimum size (**grounding electrode**) conductor and equipment grounding conductor is necessary to ensure that the proper capacity to protect the system is in place. Methods to determine the proper size of these conductors will be discussed later in this module.

A low-impedance conductor is necessary because the higher the impedance, the more resistance there is to current flow. The current should flow with the least amount of resistance to ensure that personnel and equipment are protected when a fault does occur.

4.0.0 ◆ SHORT CIRCUIT VERSUS GROUND FAULT

Short circuits and ground faults are commonly misunderstood. Both faults stem from a failure of insulation resistance. It is important to have a common language for better understanding of these conditions.

4.1.0 Short Circuit

A short circuit is a conducting connection, whether intentional or accidental, between any of the conductors of an electrical system, whether it is from line-to-line or from line-to-ground. See *Figure 2*.

The failure might occur from one phase conductor to another phase conductor or from one phase conductor to the grounded conductor or neutral. The maximum value of fault current is dependent on the available capacity the system can deliver to the point of fault. The maximum value of short circuit current from line-to-ground will vary depending upon the distance from the source to the fault. The available short circuit current is further limited by the impedance of the arc where one is established, plus the impedance of the conductors to the point of short circuit.

4.2.0 Ground Fault

A ground fault, as defined in *NEC Section 250.2,* is the unintentional electrically conducting connection between an ungrounded conductor of an electrical circuit and the normally noncurrent-carrying conductors, metallic enclosures, metallic raceways, metallic equipment, or earth. While not specified in the *NEC®*, an unintentional electrically conducting connection between a grounded conductor of an electrical circuit and the normally noncurrent-carrying conductors, metallic enclosures, metallic raceways, metallic equipment, or earth would also be considered a ground fault. While the first instance may result in a large amount of fault current to flow in a properly installed grounded system, the latter may result in parallel paths being formed between the grounded conductor and the grounding system. These parallel paths would result in an unwanted current flow on the grounding system. See *Figure 3*.

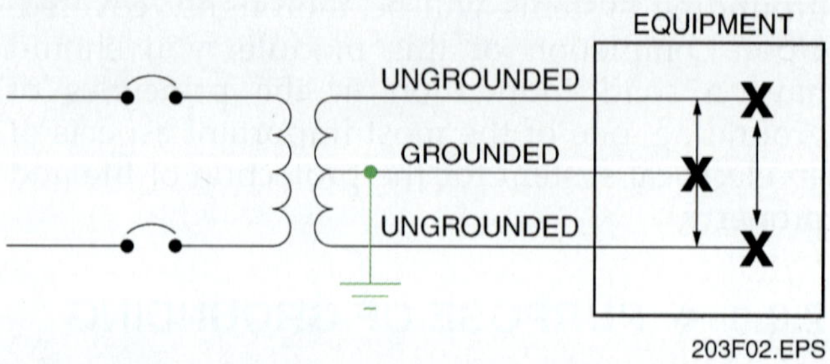

Figure 2 ◆ Short circuit.

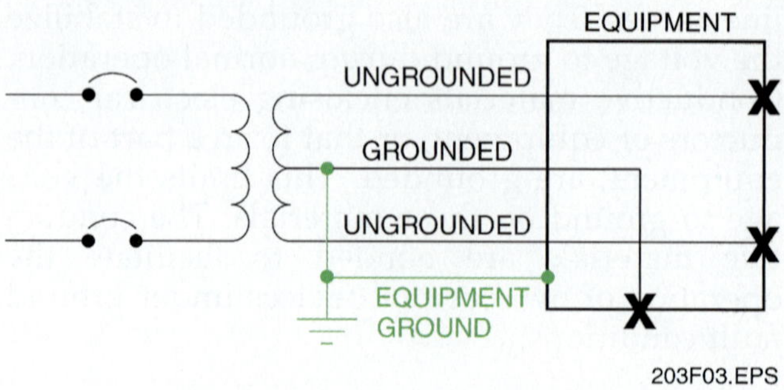

Figure 3 ◆ Ground fault.

5.0.0 ◆ TYPES OF GROUNDING SYSTEMS

The two general classifications of protective grounding are system and equipment grounding.

System grounding and bonding relates to the service-entrance equipment and its interrelated and bonded components. That is, one system conductor is grounded to limit voltages due to lightning, line surges, or unintentional contact with higher voltage lines, as well as to stabilize the voltage to ground during normal operation.

Equipment grounding conductors are used to connect the noncurrent-carrying metal parts of equipment, conduit, outlet boxes, and other enclosures to the system grounded conductor, the grounding electrode conductor, or both, at the service equipment or at the source of a separately derived system. Equipment grounding conductors are bonded to the system grounded conductor to provide a low impedance path for fault current that will facilitate the operation of overcurrent devices under ground fault conditions. *NEC Article 250* covers general requirements for grounding and bonding.

5.1.0 Single-Phase Systems

To better understand a complete grounding system, we will examine a conventional residential or small commercial system beginning at the power company's high-voltage lines and transformer, as shown in *Figure 4*. The pole-mounted transformer is fed with a two-wire 7,200V system that is transformed and stepped down to a three-wire, 120/240V, single-phase electric service suitable for residential use. A wiring diagram of the transformer connections is shown in *Figure 5*. Note that the voltage between Leg A and Leg B is 240V.

However, by connecting a third wire (neutral) on the secondary winding of the transformer between the other two, the 240V splits in half, giving 120V between either Leg A or Leg B and the neutral conductor. Consequently, 240V is available for household appliances such as ranges, hot water heaters, clothes dryers, and the like, while 120V is available for lights, small appliances, televisions, and similar appliances.

Referring again to the diagram in *Figure 5*, conductors A and B are **ungrounded conductors,** while the neutral is a grounded conductor. If only 240V loads were connected, the **neutral or grounded conductor** would carry no current. However, since 120V loads are present, the neutral will carry the unbalanced load and become a current-carrying conductor. For example, if Leg A carries 60A and Leg B carries 50A, the neutral conductor would carry only 10A (60 − 50 = 10). This is why the *NEC*® sometimes allows the neutral conductor in an electric service to be smaller than the ungrounded conductors.

The typical pole-mounted service entrance is normally routed by a grounded (neutral) messenger cable from a point on the pole to a point on the building being served, terminating at the service drop. Service-entrance conductors are routed

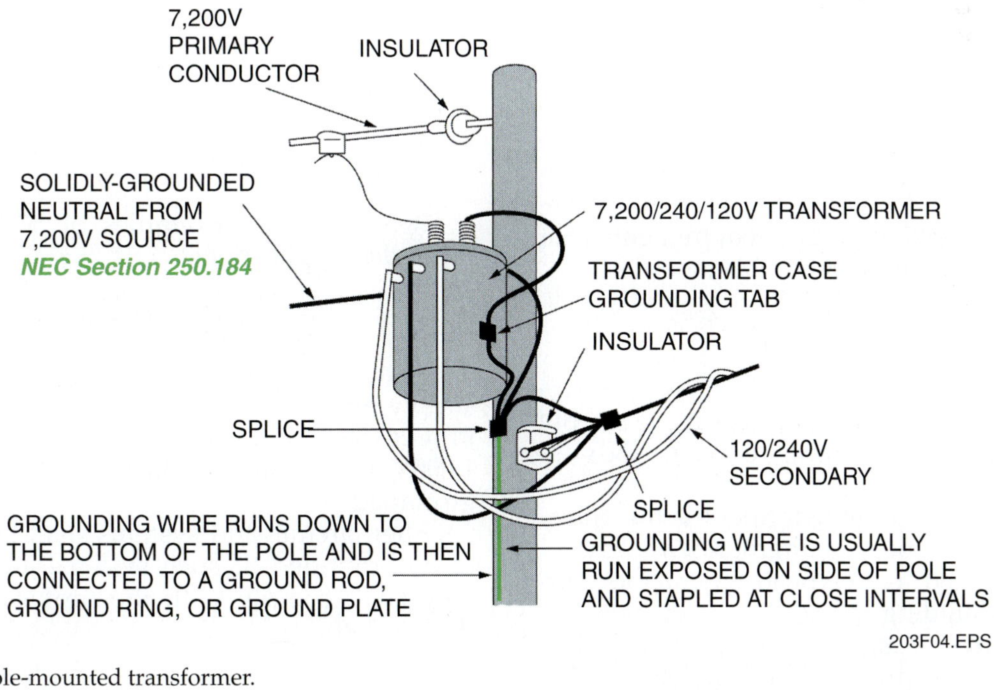

Figure 4 ◆ Pole-mounted transformer.

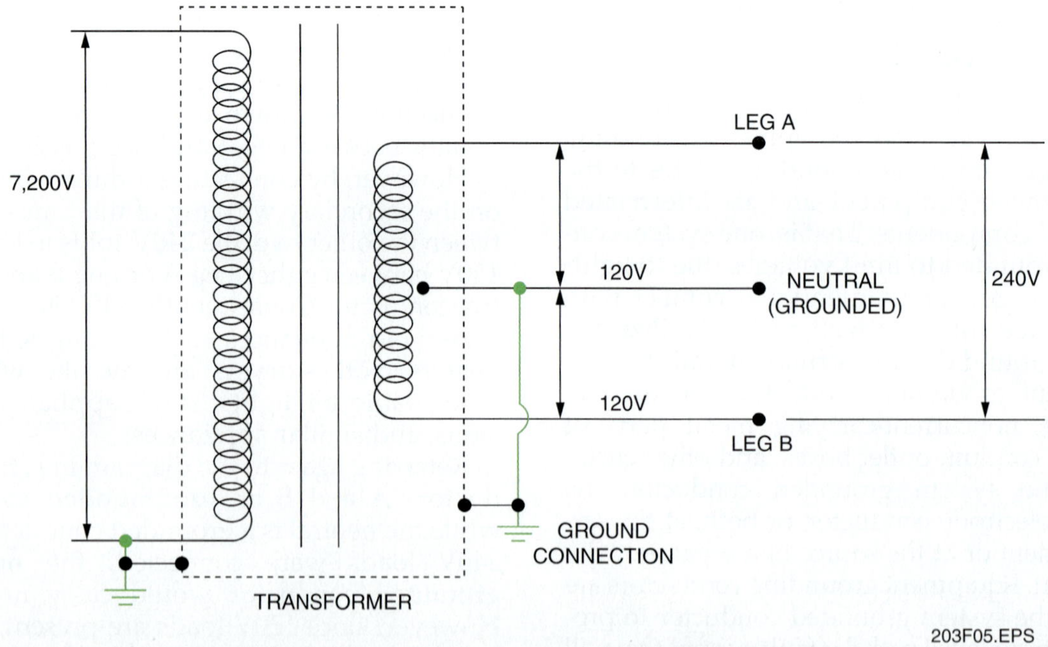

Figure 5 ◆ Wiring diagram of a 7,200V to 120/240V single-phase transformer connection.

between the meter housing and the main service switch or panelboard. This is the point where most systems are grounded—the neutral bus in the main panelboard. See *Figure 6*.

WARNING!

Exercise extreme caution when disconnecting a ground. Never grab a disconnected ground wire with one hand and the grounding electrode with the other. Your body will act as a conductor for any fault current; the results could be fatal.

5.2.0 Grounding Requirements

NEC Sections 250.20 and 250.21 provide the grounding requirements for AC systems. The *NEC*® should always be referenced when working with grounded and ungrounded systems. There are various exceptions to the requirements that will not be covered in detail in this module.

5.2.1 Systems Less Than 50 Volts

Grounding is required:

- Where supplied by transformers if the supply voltage to the transformer exceeds 150V to ground
- Where supplied by transformers if the transformer supply system is ungrounded
- Where installed as overhead conductors outside of the building
- In other circuits and systems provided they comply with the provisions of *NEC Article 250*

Figure 7 shows examples of these requirements.

5.2.2 50-Volt to 1,000-Volt Systems

AC systems of 50V to 1,000V supplying premises wiring and premises wiring systems shall be grounded under any of the following conditions:

- Where the system can be grounded so that the maximum voltage to ground on the ungrounded conductors does not exceed 150 volts
- Where the system is three-phase, four-wire, wye-connected and the neutral is used as a circuit conductor
- Where the system is three-phase, four-wire, delta-connected and the midpoint of one phase winding is used as a circuit conductor
- In other circuits and systems provided they comply with the provisions of *NEC Article 250*

Figure 8 shows examples of 50V to 1,000V grounded applications.

5.2.3 AC Systems 1kV and Over

AC systems over 1,000V must be grounded if they supply mobile or portable equipment as covered in *NEC Section 250.188.* Other AC systems over 1,000V are permitted (but not required) to be grounded.

Other circuits and systems shall be permitted to be grounded. If such systems are grounded, they shall comply with the provisions of *NEC Article 250.*

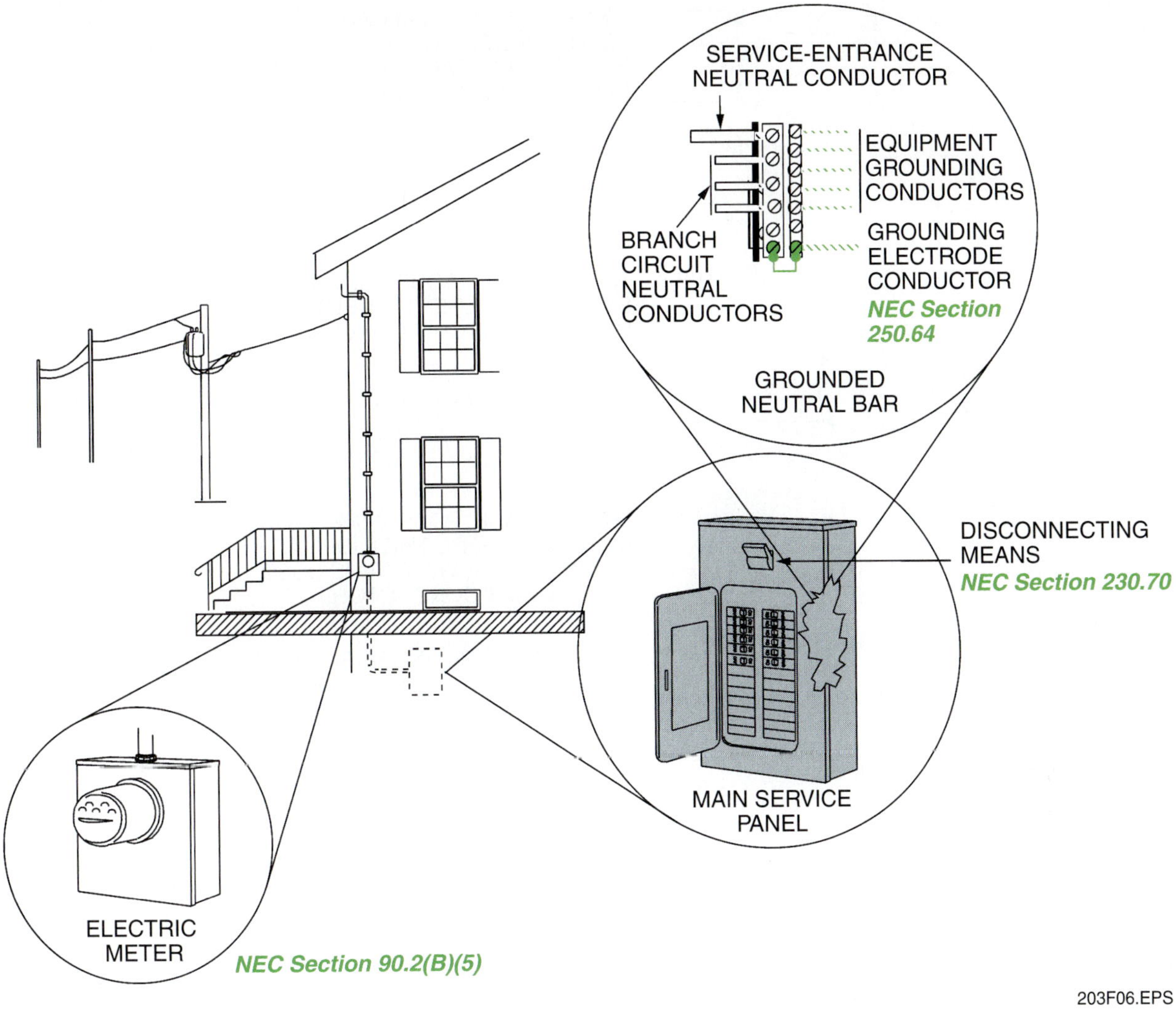

Figure 6 ◆ Typical service entrance and related service equipment.

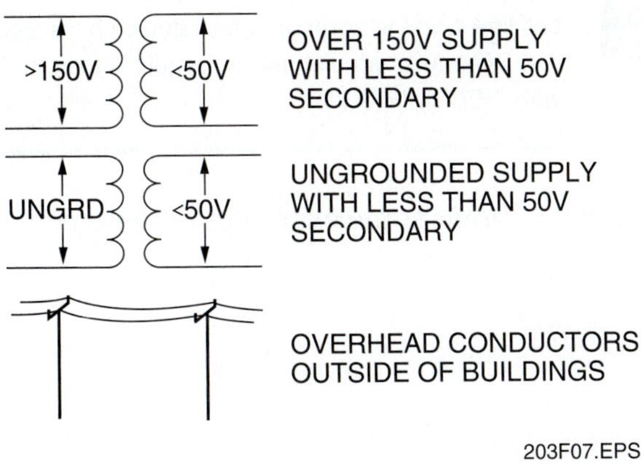

Figure 7 ◆ Systems less than 50V that must be grounded.

5.2.4 Separately Derived Systems

Electrical systems derived from a generator, transformer, or converter windings that have no direct electrical connection to the supply conductors originating in another system must be grounded if the system that is derived complies with the conditions previously described. The separately derived system must then be grounded as specified in NEC Section 250.30.

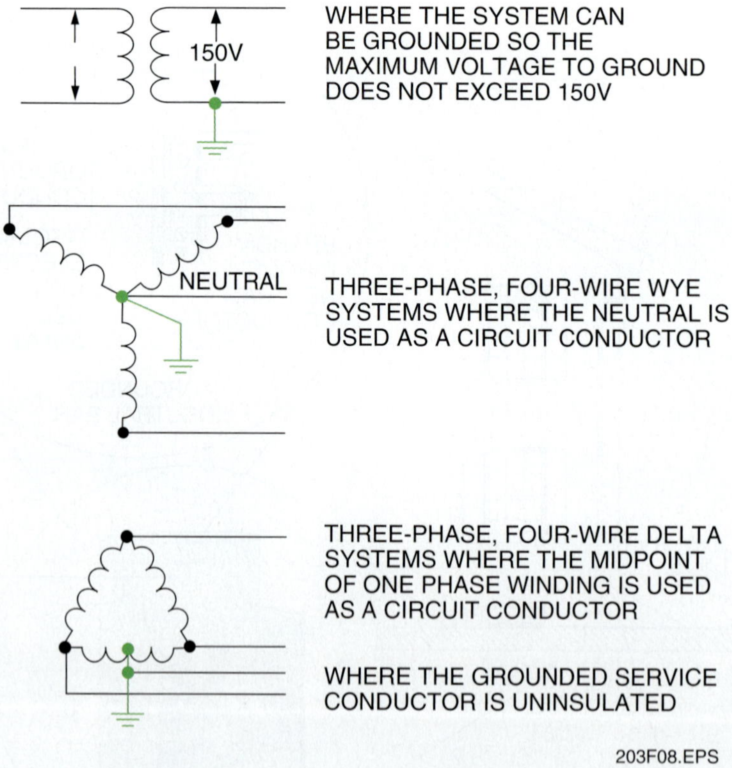

Figure 8 ◆ Systems that must be grounded.

Examples of separately derived systems include:

- Transformers with no direct electrical connection between the primary and secondary
- Generator systems that supply power where the neutral is not connected to the utility system, such as for carnivals, rock crushers, or batch plants
- Generator systems used for emergency, required standby, or optional standby power that have all conductors, including a neutral, isolated from the neutral or grounded conductor of another system (usually via a transfer switch)
- AC or DC systems derived from inverters or rectifiers

5.2.5 Circuits That Cannot Be Grounded

According to *NEC Section 250.22*, only four types of circuits are not permitted to be grounded:

- Circuits for cranes that operate over combustible fibers in Class III locations, as provided in *NEC Section 503.155*
- For healthcare facilities, those isolated power circuits in hazardous (classified) locations as provided in *NEC Section 517.61*
- Circuits for electrolytic cells as provided in *NEC Article 668*
- Secondary circuits on lighting systems as provided in *NEC Article 411.5(A)*

NOTE

Equipment located or used within the electrolytic cell line working zone or associated with the cell line DC power circuits are not required to comply with ***NEC Article 250.***

Figure 9 shows some examples of these circuits.

Inside Track: Systems Permitted to Be Grounded but Not Required to Be Grounded

NEC Section 250.21 states that the following AC systems of 50V to 1,000V shall be permitted to be grounded but shall not be required to be grounded:

- Electric systems used exclusively to supply industrial electric furnaces for melting, refining, tempering, and the like
- Separately derived systems used exclusively for rectifiers that supply only adjustable-speed drives
- Separately derived systems supplied by transformers that have a primary voltage rating of less than 1,000V, provided that all of the conditions listed in *NEC Section 250.21* are met
- Isolated systems as permitted or required in *NEC Articles 517 and 668*
- High-impedance grounded neutral systems as specified in *NEC Section 250.36*

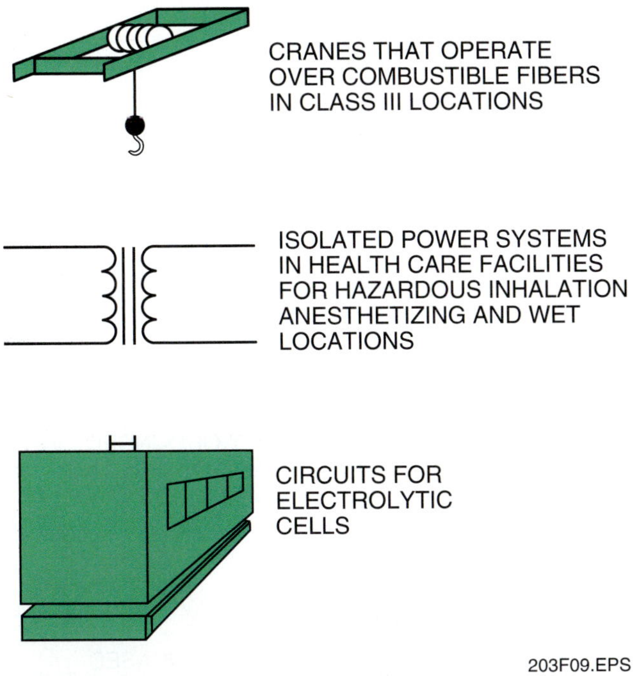

Figure 9 ◆ Circuits that are not permitted to be grounded.

6.0.0 ◆ *NEC®* REQUIREMENTS

The grounding equipment requirements established by Underwriters Laboratories, Inc. (UL) have served as the basis for approval for grounding in the *NEC®*. The *NEC®*, in turn, provides the grounding standard for the Occupational Safety and Health Administration (OSHA).

All electrical systems must be grounded and bonded in a manner prescribed by the *NEC®* to protect personnel and equipment. To be totally effective, a grounding system must limit the voltage on the electrical system and protect it from:

- Lightning
- Voltage surges higher than that for which the circuit is designed
- An increase in maximum potential to ground due to abnormal voltages

6.1.0 Grounding Methods

The requirements for grounding of services are found in *NEC Sections 250.24(A), (B), and (C).* Methods of grounding an electric service are covered in *NEC Article 250, Part III.* In general, all of the following electrodes (if present) must be bonded together to form the grounding electrode system:

- An underground metal water pipe in direct contact with the earth for no less than 10'
- The metal frame of a building where **effectively grounded** per *NEC Section 250.52(A)(2)*
- An electrode encased by at least 2" of concrete, located within and near the bottom of a concrete foundation or footing that is in direct contact with the earth

NOTE

This electrode must be at least 20' long and must be made of electrically conductive coated steel reinforcing bars or rods of not less than ½" in diameter, or consisting of at least 20' of bare copper conductor not smaller than No. 4 AWG wire size.

- A ground ring encircling the building or structure, in direct contact with the earth at a depth not less than 2.5' below grade
- Rod, pipe, or plate electrodes
- Other local metal underground systems or structures

> **NOTE**
> This ring must consist of at least 20' of bare copper conductor not smaller than No. 2 AWG wire size. See *Figure 10.*

Grounding systems used in industrial buildings will frequently use all of the methods shown in *Figure 10*, and the methods used will often surpass the *NEC®*, depending upon the manufacturing process and the calculated requirements made by plant engineers.

Figure 11 shows a floor plan of a typical industrial grounding system.

The building in *Figure 12* has a metal underground water pipe that is in direct contact with the earth for more than 10', so this is one valid grounding source. The building also has a metal underground gas piping system, but this may not be used as a grounding electrode [*NEC Section 250.52(B)*].

NEC Section 250.53(D)(2) further states that the underground water pipe must be supplemented by an additional electrode of a type specified in *NEC Section 250.52(A)(2)*. Since a grounded metal building frame, concrete-encased electrode, or

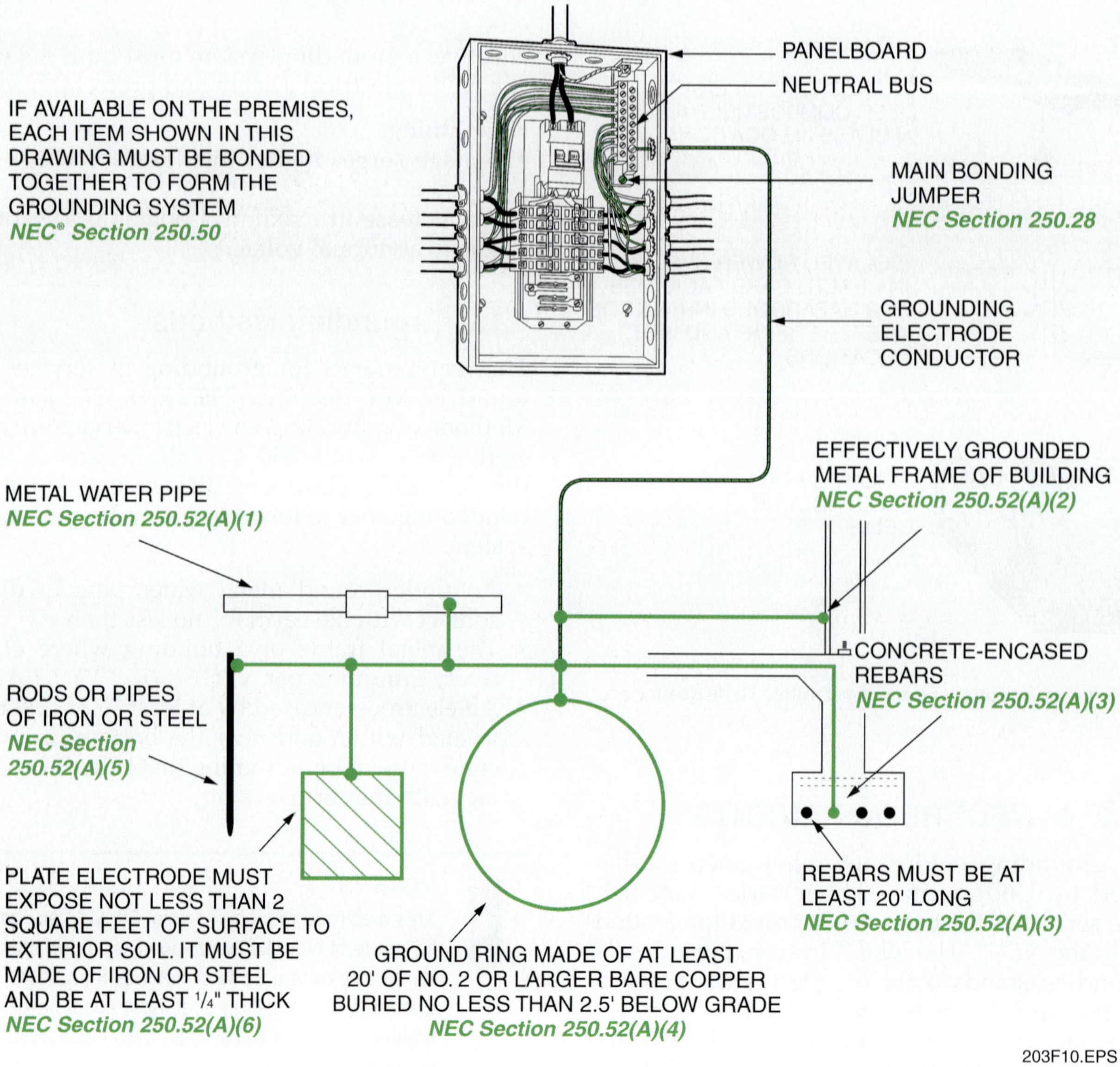

Figure 10 ◆ *NEC®*-approved grounding electrodes.

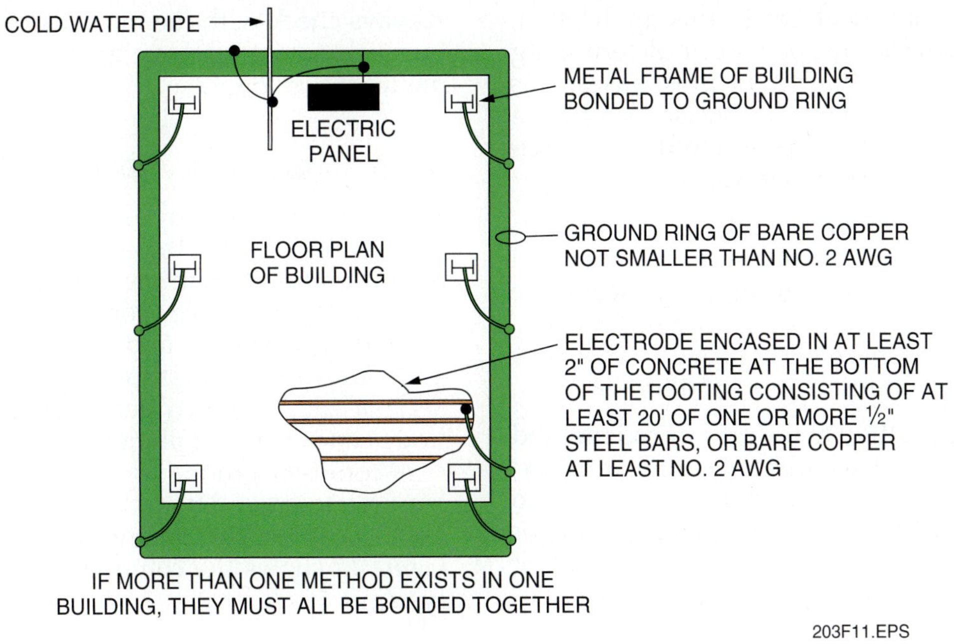

Figure 11 ◆ Floor plan of the grounding system for an industrial building.

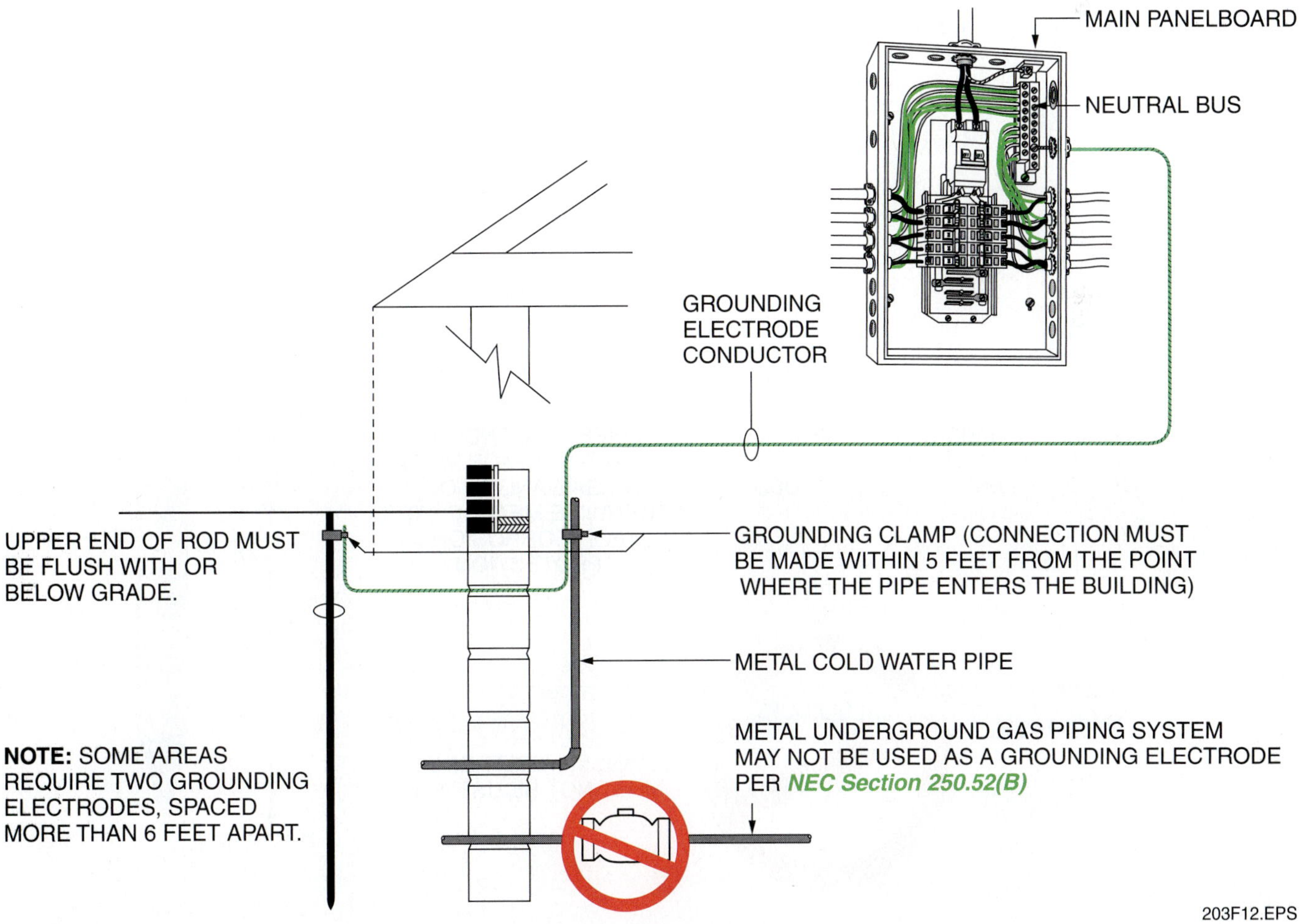

Figure 12 ◆ Grounding requirements for non-industrial buildings.

ground ring is not available in this application, *NEC Section 250.50* must be used in determining the supplemental electrode. In most cases, this supplemental electrode will consist of either a driven rod (*Figure 13*) or pipe electrode, the specifications for which are as follows:

- Withstand and dissipate repeated surge circuits
- Provide corrosion resistance to various soil chemistries to ensure continuous performance for the life of the equipment being protected
- Provide rugged mechanical properties for easy driving with minimum effort and rod damage

An alternate method to the pipe or rod method is a plate electrode. Each plate electrode must expose not less than two square feet of surface to the surrounding earth. Plates made of iron or steel must be at least ¼" thick, while plates of nonferrous metal like copper need only be .06" thick.

Rod, pipe, and plate electrodes must have a resistance to ground of 25 ohms (Ω) or less. If not, they must be augmented by an additional electrode spaced not less than 6' from the first. Many locations require two electrodes regardless of the resistance to ground. This is not an *NEC*® requirement, but is required by some power companies and local ordinances in some cities and counties. Always check with the local inspection authority for such rules that may go beyond the requirements of the *NEC*®.

6.2.0 Grounding Electrode Conductor

The grounding electrode conductor is the sole connection between the grounding electrode and the grounded system conductor for a grounded system, or the sole connection between the grounding electrode and the service equipment enclosure for an ungrounded system. See *NEC Section 250.64,* which describes grounding electrode conductor installation.

A common grounding electrode conductor is required to ground both the circuit grounded conductor and the equipment grounding conductor. *Figure 14* shows the common connection for this system.

6.2.1 Sizing

The grounding electrode conductor must be sized in accordance with *NEC Table 250.66.* The size of this conductor is based on the size of the service-entrance conductor.

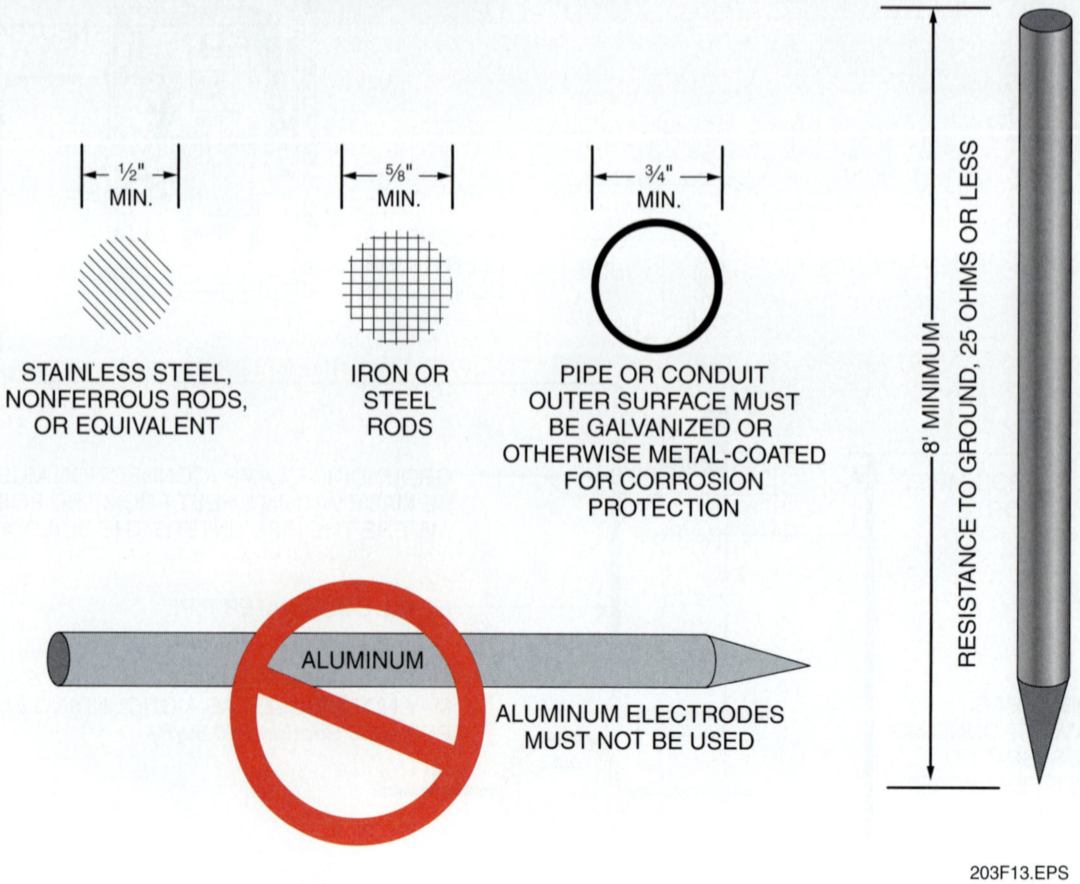

Figure 13 ◆ Requirements for ground rods.

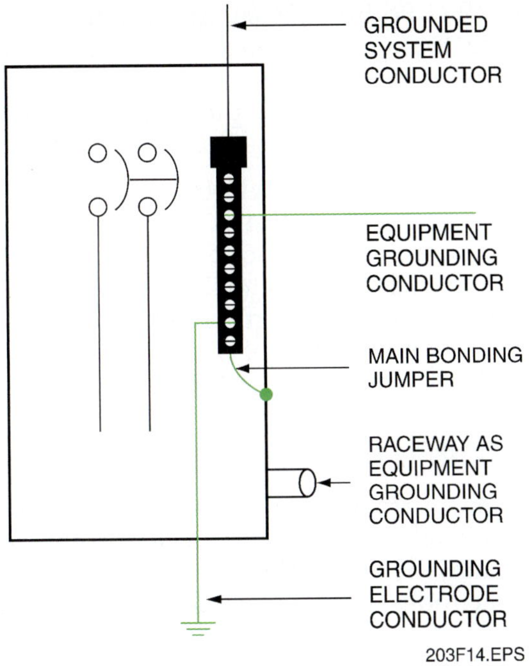

Figure 14 ◆ Grounding electrode connection for grounded system.

The size of the service-entrance conductor is determined from the ampacity of that system. *NEC Table 310.16* is used to determine the size of the service conductor from the ampacity.

The majority of applications use 75°C copper conductors. This information is necessary when using *NEC Table 310.16.* For a 100A service application, the size of the service conductor is No. 3 AWG copper.

Now that the size of the service conductors is known, the size of the grounding electrode conductor can be determined from *NEC Table 250.66.*

Using the No. 3 AWG from the 100A service example, it can be determined from the table that a No. 8 AWG copper grounding electrode conductor is required.

Another example that is fairly common would be to determine the size of the grounding electrode conductor for a 200A, 208V, three-phase system.

Step 1 Determine the size of the service conductor from *NEC Table 310.16.* It is 3/0 AWG.

Step 2 Go down the left column of *NEC Table 250.66* and find 3/0.

Step 3 Go across to find the size of the grounding conductor. It is a No. 4 AWG conductor.

NOTE

Remember, it is the total size of the service conductor that determines the size of the grounding electrode conductor.

6.2.2 Installation and Protection

NEC Section 250.64(B) requires the grounding electrode conductor or its enclosure to be securely fastened to the surface on which it is carried.

No. 4 AWG or larger conductors require protection where exposed to severe physical damage. No. 6 AWG conductors may be run along the surface of the building and be securely fastened. Otherwise, the conductor must be protected by installation in rigid or intermediate metal conduit, rigid nonmetallic conduit, EMT, or cable armor. Smaller conductors shall be protected in conduit or armor.

THINK ABOUT IT

Ground Rods

Suppose you are driving an 8' rod electrode into the ground and you hit a rock after it has been driven to a depth of 6'. Further effort to drive the rod deeper is of no avail. What are your options?

Sizing Electrode Conductors

NEC Table 310.16 shows that the service conductor size (copper wire at 75°C) for a 150A service is 1/0 AWG. Which of the three AWG copper electrode conductor sizes listed below would be used with this service?

No. 4
No. 6
No. 8

6.3.0 Grounding Electrode Conductors

Both the grounding electrode conductor, which connects the grounded conductor and the panelboard neutral bus to ground, and the grounding electrodes must be of either copper, aluminum, or copper-clad aluminum. Furthermore, the material selected must be resistant to any corrosive condition existing at the installation, or it must be suitably protected against corrosion. The conductor may be either solid or stranded, and covered or bare, but it must be in one continuous length without a splice or joint, except for the following conditions:

- Splices in busbars are permitted.
- Where a service consists of more than one single enclosure, it is permissible to connect taps to the grounding electrode conductor.
- Bonding jumper(s) from grounding electrode(s) and GECs shall be permitted to be connected to an aluminum or copper busbar not less than ¼" × 2". See *NEC Section 250.64(C).*

> **WARNING!**
> Exothermic welding is a hazardous process that should only be performed by qualified personnel. Refer to your company's safety procedures and the manufacturer's recommendations before using exothermic welding equipment.

- Per *NEC Section 250.64,* grounding electrode conductors may be spliced at any location by means of irreversible compression-type connectors listed for the purpose or using the exothermic welding process (*Figure 15*).

The size of grounding electrode conductors depends on service-entrance size; that is, the size of the largest service-entrance conductor or equivalent for parallel conductors. *NEC Table 250.66* gives the proper sizes of grounding electrode conductors for various sizes of electric services. *Figure 16* shows the various protection guidelines.

6.4.0 Other Electrodes

When an electrode that meets the requirements of *NEC Section 250.50* is not present, other electrodes may be used. These may be the rod and pipe electrodes discussed earlier, plate electrodes, or other local metal underground systems such as piping and underground tanks. *NEC Sections 250.52(A)(5), 250.52(A)(6), and 250.52(A)(7)* provide information and requirements for these electrodes. *Figure 17* shows other electrodes.

The specific requirements for other electrodes are:

- *Local systems* – Local metallic underground systems such as piping, tanks, etc.
- *Pipe electrodes* – Pipe or conduit electrodes shall be not less than 8' in length nor smaller than ¾"

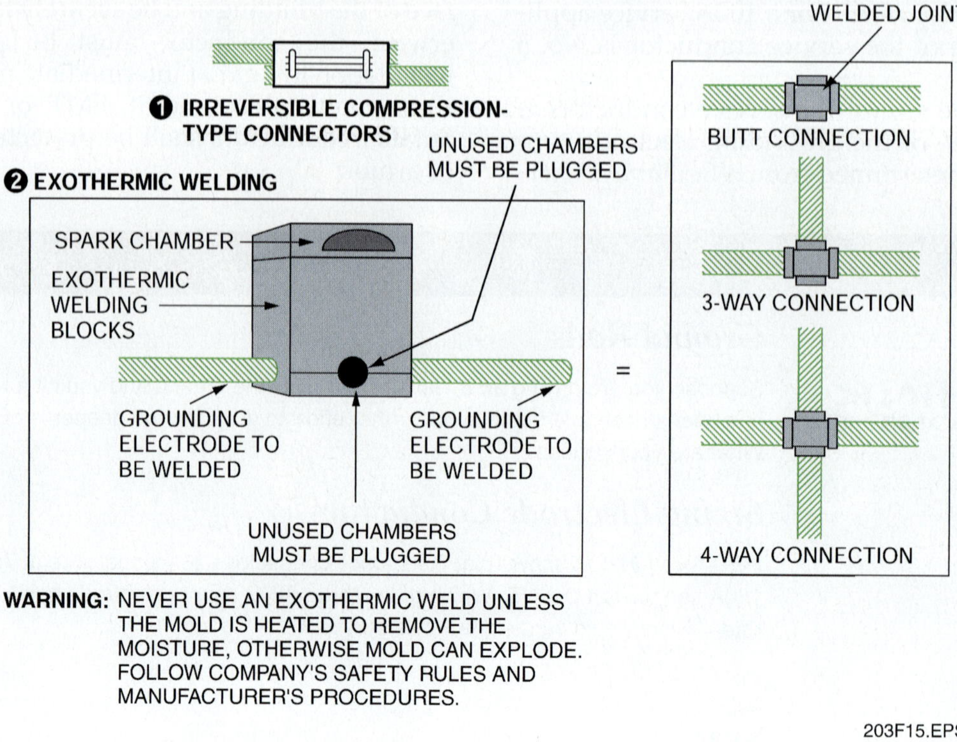

Figure 15 ◆ Methods of splicing grounding conductors.

Protection of Grounding Electrode Conductors

Grounding electrode conductor or enclosure must be securely fastened to the surface.

No. 4 or larger – If exposed to severe physical damage.

No. 6 – Run along surface, securely fastened or protected.

Smaller than No. 6 – Must be protected from damage (must be enclosed in rigid metal conduit).

Bare aluminum – Not allowed in contact with masonry, where subject to corrosive conditions, or within 18 inches of the earth.

203F16.EPS

Figure 16 ◆ Protection of grounding electrode conductors.

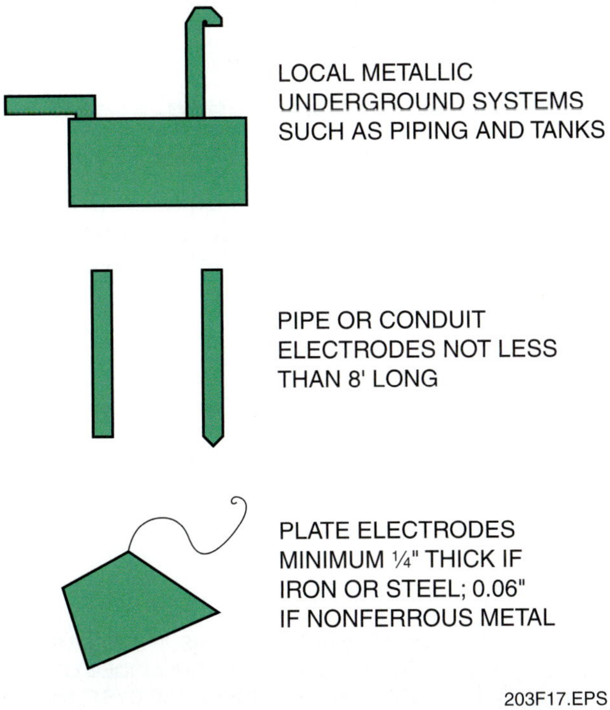

Figure 17 ◆ Other electrodes.

trade size, and if of iron or steel, shall be galvanized or metal-coated for corrosion protection.

- *Rod and pipe electrodes* – Electrodes of steel or iron shall be at least ⅝" in diameter. Rods of nonferrous metal or stainless steel that are less than ⅝" in diameter shall be listed and must be at least ½" in diameter.
- *Plate electrodes* – Electrodes shall have at least two square feet of surface in contact with exterior soil. If of iron or steel, the plate shall be at least ¼" thick. If of nonferrous metal, it shall be at least 0.06" thick.

NOTE

Underground metal gas piping systems are not permitted to be used as grounding electrodes. However, this does not eliminate the requirement that metal gas piping systems be bonded.

INSIDE TRACK

Exothermic Welding

Normally, the connections made by exothermic welding are permanent. They perform better than any crimped or bolted connection because the copper-to-copper or copper-to-steel bond is molecular, eliminating the risk of loosening or corrosion. Exothermic welded connections will carry more current than the conductor, resist repeated fault currents, and will not deteriorate with age.

Plate Electrodes

THINK ABOUT IT

Why does the *NEC*® require that 2' square electrode plates made of iron or steel be at least ¼" (0.25") thick, while 2' square electrode plates made of copper (a nonferrous metal) need only be 0.06" thick?

6.4.1 Installation of Rod, Pipe, and Plate Electrodes

Rod and pipe electrodes must be installed so at least 8' are in contact with the soil. Rods must be driven vertically unless rock obstruction is encountered. If an obstruction is encountered, the rod may be driven at not more than a 45° angle to clear the rock. The other possibility is for the rod to be buried in a trench that is at least 2.5' deep.

The upper end of the rod must be at or below surface level. The rod may be above ground, but at least 8' of rod must be beneath the surface.

7.0.0 ◆ EQUIPMENT GROUNDING

Figure 18 summarizes the equipment grounding rules for most types of equipment required to be grounded. These general *NEC*® regulations apply to all installations except for specific equipment (special applications) as indicated in the code. The *NEC*® also lists specific equipment that is to be grounded regardless of voltage.

In all occupancies, major appliances and many hand-held appliances and tools are required to be grounded. The appliances include refrigerators, freezers, air conditioners, clothes dryers, washing machines, dishwashing machines, sump pumps, and electrical aquarium equipment. Other tools likely to be used outdoors and in wet or damp locations must be grounded or have a system of double insulation.

7.1.0 Equipment Grounding Conductor

The equipment grounding conductor (EGC) is the conductor used to connect the noncurrent-carrying metal parts of equipment, raceways, and

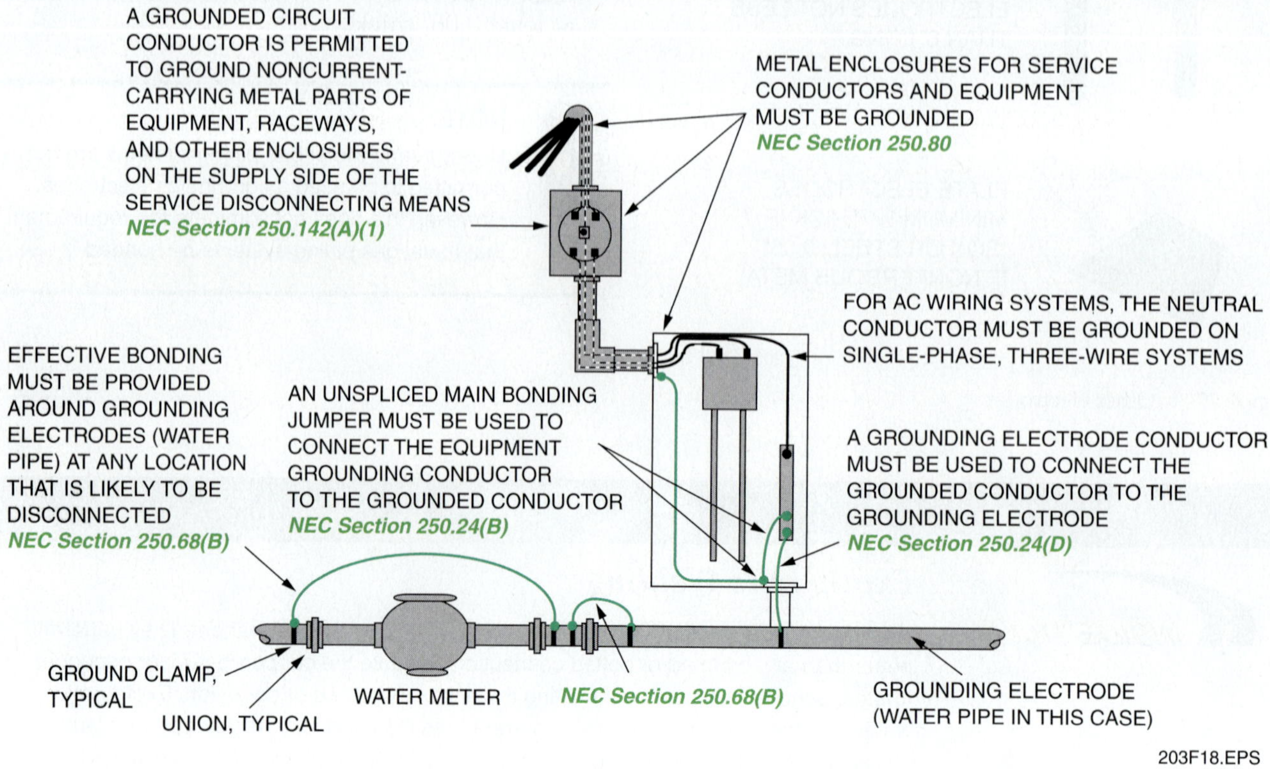

Figure 18 ◆ Equipment grounding summary.

other enclosures to the system grounded conductor and/or the grounding electrode conductor at the service equipment or at the source of a separately derived system.

7.1.1 Grounded and Ungrounded Systems

The EGC or path must extend from the farthest point on the circuit to the service equipment where it is connected to the grounded conductor. *Figure 19* shows an EGC and its path to ground when a ground fault occurs.

7.1.2 Sizing an EGC

EGCs carry fault current from the load to the grounded terminal bar of the service equipment. The size of the EGC is determined from the size of the overcurrent device that is protecting that particular system. *NEC Table 250.122* is used to determine the size of the EGC.

For example, if the ampacity is 400A, the overcurrent device is set at 400A. Use *NEC Table 250.122* and go down the left column until the desired ampacity or the next highest ampacity is found. For the 400A system, the size of the EGC is No. 3 copper wire or No. 1 aluminum or copper-clad wire.

> **NOTE**
> The size of the EGC is determined by the overcurrent protective device.

Determine the size of the EGC for a flexible metal conduit connection to a motor supplied with a 30A circuit.

Step 1 Using *NEC Section 348.60,* verify that an EGC is needed.

Step 2 Since this is an installation that requires flexibility, an EGC is required.

Step 3 The 30A circuit will have a 30A overcurrent protective device feeding the circuit.

Step 4 Find 30A in *NEC Table 250.122.*

Step 5 A No. 10 copper or No. 8 aluminum or copper-clad EGC is required for 30A.

Figure 20 shows the requirements for using flexible metal conduit as an equipment ground.

Another example to analyze would be a liquidtight flexible metal conduit connection to a machine supplied by a 100A circuit.

> **CASE HISTORY**
> **Mechanical Grounds**
>
> A technician was electrocuted recently while servicing a refrigeration unit for a walk-in cooler. The insulation on one of the power conductors inside the flexible metal conduit was damaged, resulting in electrical arcing to a conduit connector on the unit starter box. The conduit connection from the unit to the starter box was loose, thereby effectively disconnecting the mechanical ground from the unit. As the technician was servicing the unit, the thermostat caused the starter to close, energizing the surfaces of the unit and fatally shocking the technician.
>
> **The Bottom Line:** This accident might have been prevented if one or more of the following procedures had been implemented:
>
> - Proper preventive maintenance checks of all wiring and connections
> - Proper lockout/tagout procedures
> - Equipment grounding conductor provided along with the power feed conductors

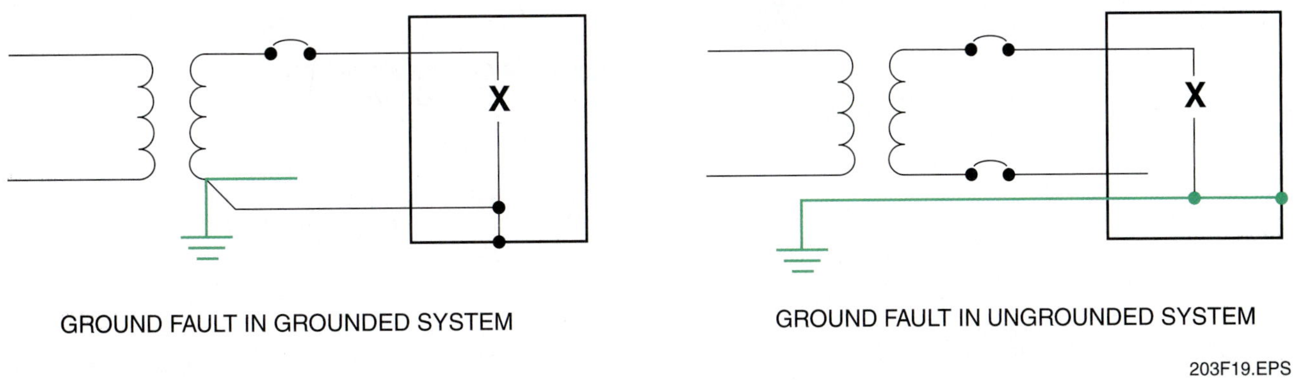

Figure 19 ◆ Ground faults.

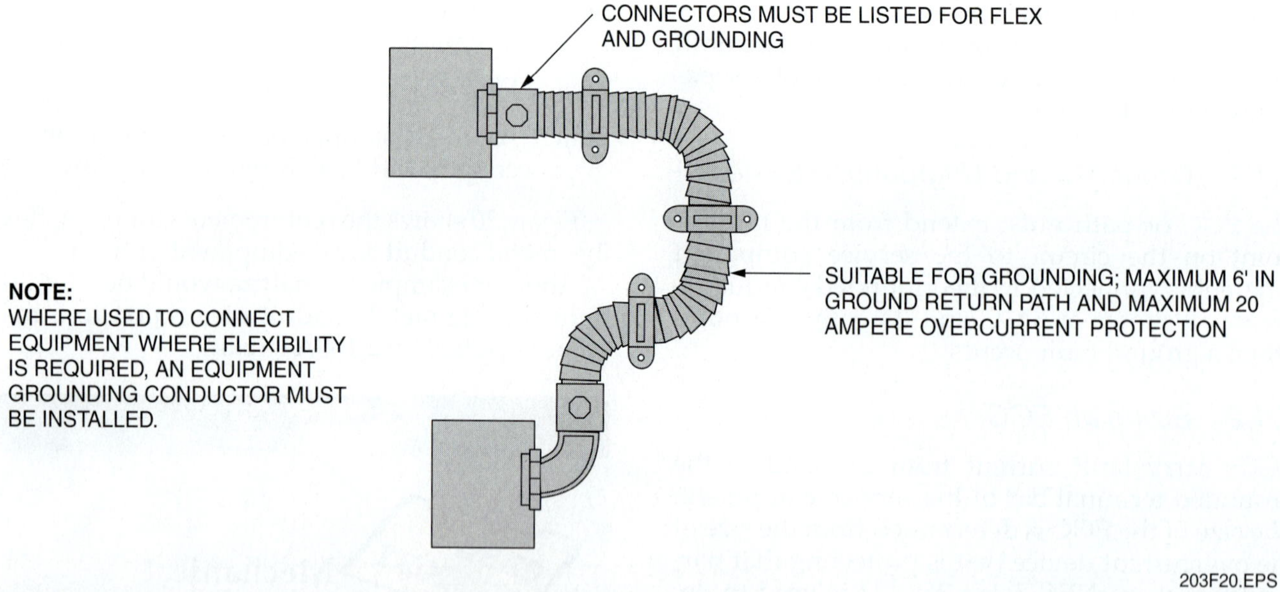

Figure 20 ◆ Installation requirements for flexible metal conduit.

Figure 21 details the requirements for using the conduit as the grounding conductor. After determining the requirements and the ampacity of this application, a No. 8 copper EGC would be required.

7.2.0 Grounding Enclosures

As previously mentioned, equipment grounding covers the metallic noncurrent-carrying parts of an electrical system. Such parts include metallic conduit, outlet boxes, enclosures, and frames on motors, and other electrically operated equipment. These items are bonded together to ensure operation of overcurrent devices in case of ground faults.

A bare or green-insulated grounding wire is usually attached to the metal frame or cabinet of the equipment. When connected to the circuit, this grounding wire is attached to the equipment grounding system that, in turn, was originally bonded to the system neutral busbar at the service-entrance equipment. When properly connected, if a live ungrounded wire should make contact with the frame or cabinet of a motor, appliance, or other metallic object in the system, a

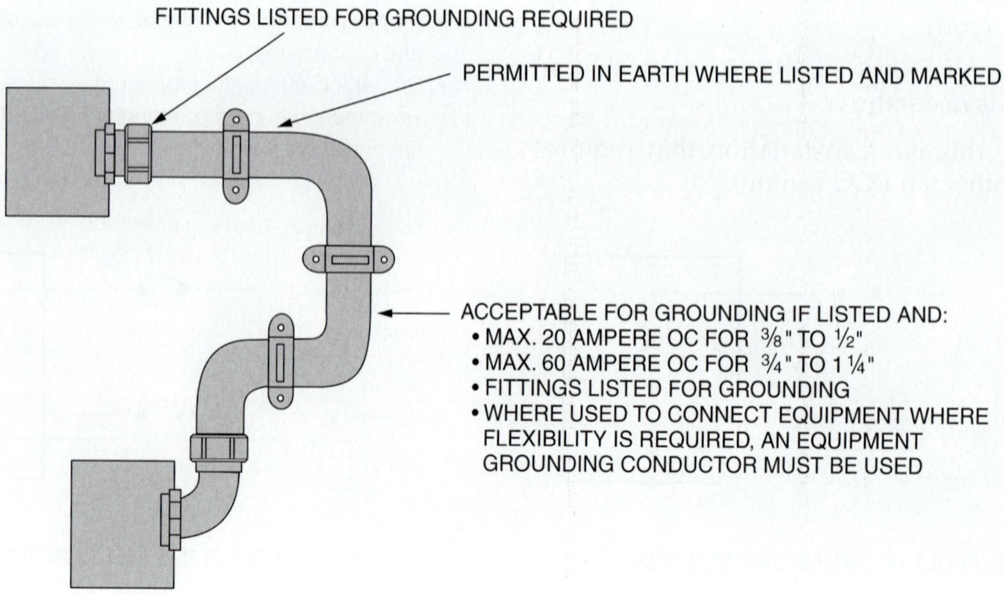

Figure 21 ◆ Installation requirements for liquidtight flexible metal conduit.

ground fault will occur and open the overcurrent protective device that is protecting the circuit. Consequently, equipment grounding is one of the best methods of protecting life and property should a ground fault occur.

7.2.1 Grounding Outlet Boxes and Devices

All receptacles used in residential, commercial, or industrial applications must be of the grounding type, which means that one of the receptacle openings connects the equipment ground to the appliance, tool, or other apparatus that is connected to the receptacles.

Ensuring continuity of the equipment grounding system to each receptacle can be handled in different ways, depending upon the wiring method used. For example, in most residential wiring systems, Type NM cable will be used. If metallic device boxes are used, the bare or green-insulated grounding wire must be attached to the box. This is accomplished by using either a **grounding clip,** as shown in *Figure 22*, or a screw designed for this purpose that is secured to the outlet box. *NEC Section 250.8* prohibits the use of sheet metal screws to connect grounding conductors to enclosures.

Figure 23 shows methods of providing a grounding conductor to duplex receptacles using

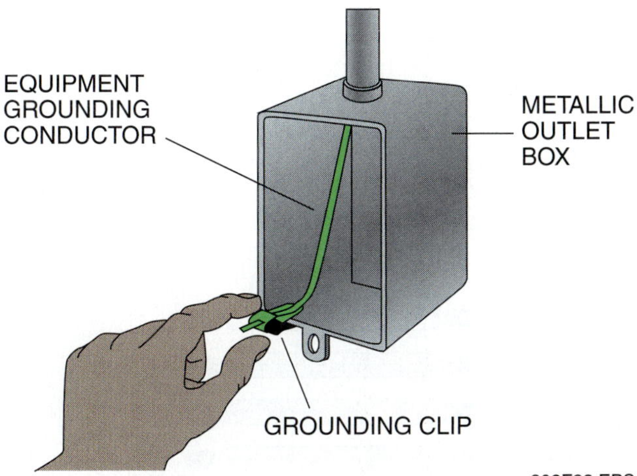

Figure 22 ◆ Grounding clip.

several wiring methods. For example, if nonmetallic boxes are used with Type NM cable, no connection to the box itself is required. However, the equipment grounding conductor must be connected to the grounding terminal on the receptacle. If two or more cables enter the nonmetallic box, each of the equipment grounding wires must be connected with an approved connector, and one wire must then be attached to the grounding terminal on the receptacle. The terminal itself may not contain more than one grounding wire. Therefore, if more than one grounding wire enters the box, they must be spliced independently from the device, and then only one conductor may be attached to the device.

When metallic boxes are used, these equipment grounding wires must be attached to the box, along with the wiring device (receptacle). As mentioned previously, this may be accomplished with either a grounding clip or a grounding screw.

Any of the following are recognized by *NEC Section 250.118* as being adequate for use as an equipment grounding conductor:

- A copper or other corrosion-resistant conductor (solid or stranded; insulated, covered, or bare; and in the form of a wire or a busbar of any shape)
- Rigid metal conduit
- Intermediate metal conduit
- Liquidtight flexible metal conduit
- Electrical metallic tubing
- Flexible tubing
- Flexible metal conduit as permitted by *NEC Sections 250.118(5), 250.118(6), and 250.118(7)*
- Armor of Type AC cable
- The copper sheath of mineral-insulated, metal-sheathed cable
- The metallic sheath or the combined metallic sheath and grounding conductors of Type MC cable
- Cable trays as permitted in *NEC Sections 392.3(C) and 392.7*
- Cable bus framework as permitted in *NEC Section 370.3*
- Other electrically continuous metal raceways listed for grounding

THINK ABOUT IT

Sheet Metal Screws

Why does the *NEC*® prohibit the use of sheet metal screws to connect grounding conductors to enclosures?

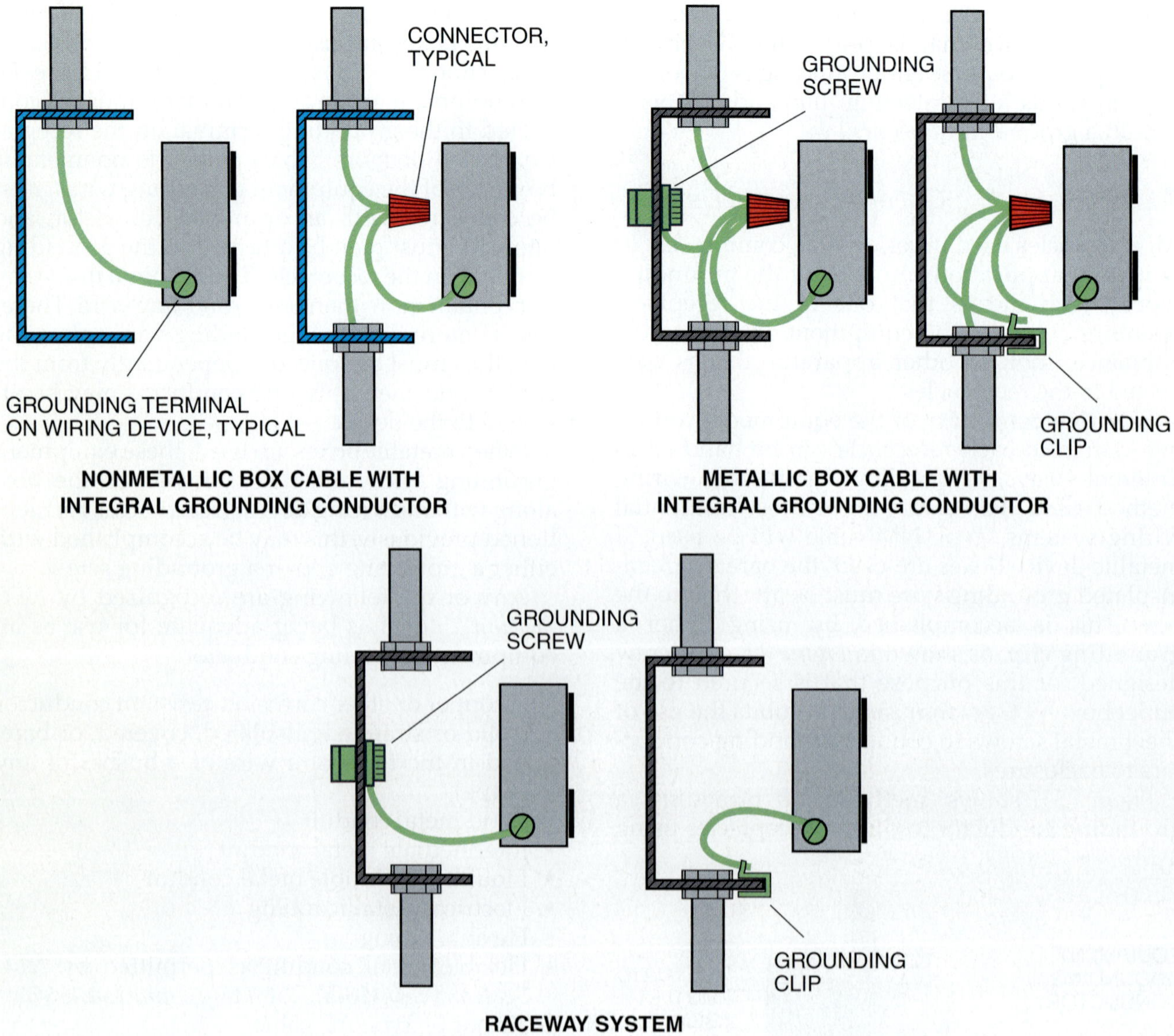

Figure 23 ◆ Grounding receptacles with different wiring methods.

7.3.0 Main Bonding Jumper

When a listed switchboard or panelboard is installed, the **main bonding jumper** that is provided with the equipment is rated for the size of conductors used for the service. The proper size of the main bonding jumper is verified by the listing agency and can be used without calculating the size.

Since the main bonding jumper must carry the full ground fault current of the system back to the grounded service conductor, its size must relate to the rating of the service conductors that supply the service. *NEC Table 250.66* is used to determine the size of the main bonding jumper. *Figure 24* shows how to calculate the size of the main bonding jumper when parallel runs are used.

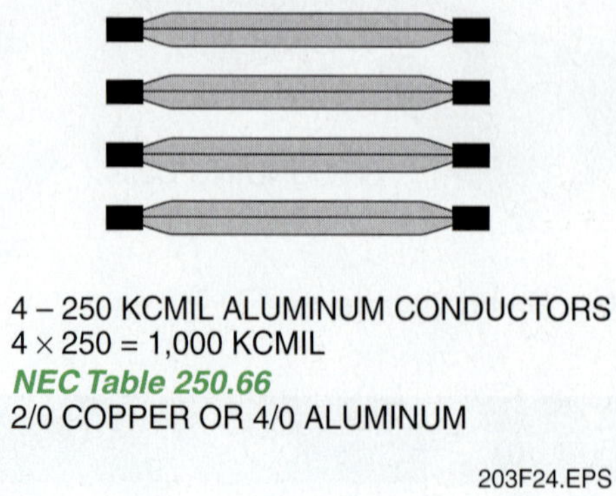

Figure 24 ◆ Main bonding jumper for parallel runs.

THINK ABOUT IT

Grounding and Bonding

When measuring voltage at a grounded duplex receptacle installed in a metal outlet box, you measure 120VAC between the hot and neutral terminals, 120VAC between the hot and ground terminals, and 120VAC between the hot terminal and the outlet box enclosure. You also measure 0VAC between the neutral and ground terminals. Does this represent a short circuit or ground fault?

Where the multiple conductors are larger than the maximum given in *NEC Table 250.66,* the jumper cannot be less than 12.5% of the area at the largest phase conductor [*NEC Section 250.28(D)*].

An example of three 500 kcmil copper conductors gives a total area of 1,500 kcmil. Because this is larger than that of *NEC Table 250.66,* we must take 12.5% of the total area. This is 187.5 kcmil. Since this is not a standard size conductor, we must refer to *NEC Chapter 9, Table 8.*

The table shows that the next larger conductor size in circular mils is 211,600. This gives a conductor of 4/0 AWG.

INSIDE TRACK

Bonding Jumpers

Use a bonding jumper to provide continuity at cable tray joints.

203P0302.EPS

8.0.0 ◆ BONDING SERVICE EQUIPMENT

Electrical continuity is the key to successful ground fault circuits. This continuity between service equipment and enclosures must be maintained by bonding. The items that are required to be bonded together are listed in *NEC Section 250.92(A).* They include:

- The service raceways, cable trays, cable bus framework, or service cable armor or sheath
- All service equipment enclosures containing service conductors, including meter fittings, boxes, etc., interposed in the service raceway or armor
- Any metallic raceway or armor that encloses the grounding electrode conductor

NEC Section 250.92(B) defines the methods permitted to be used to ensure continuity at service equipment. These methods are:

- Bonding equipment to the grounded service conductor in a manner provided by *NEC Section 250.8*
- Connections utilizing threaded couplings or threaded bosses on enclosures when made up wrenchtight
- Threadless couplings and connectors when made up wrenchtight
- Other approved devices, such as bonding-type locknuts

The exception to this requirement is found in *NEC Section 250.84(A).* This refers to an underground service cable that is metallically connected to the underground service conduit. The code states that if a service cable contains a metal armor, and if the service cable also contains an uninsulated grounded service conductor that is in continuous electrical contact with its metallic armor, then the metal covering of the cable is considered to be adequately grounded.

8.1.0 Bonding Multiple Service Disconnecting Means

Multiple disconnecting means can take several forms. *NEC Article 230* discusses the various forms.

Because this bonding jumper is located on the supply side of the service disconnect, the basic rule for sizing the **equipment bonding jumper** for bonding these various configurations is found in *NEC Section 250.102(C).* This section requires the use of *NEC Table 250.66* for selecting the size. However, if the size of the service conductor is larger than 1,100 kcmil copper or 1,750 kcmil aluminum, the bonding jumper must have a cross-sectional area not less than 12.5% of the largest phase conductor. Where the service conductors are paralleled in two or more raceways, the equipment bonding jumper shall be run in parallel. The size of each bonding jumper shall be based on the largest phase conductor in each raceway or cable.

Table 1 shows various sizes of bonding jumpers based on the size of the service-entrance conductor.

Table 1 Bonding of Multiple Service Disconnecting Means

Service-Entrance Conductor	Bonding Jumper (Copper)
500 kcmil in service mast	1/0
1,000 kcmil in wireway	2/0
300 kcmil to 300A service	No. 2
3/0 to 200A service	No. 4
No. 2 to 125A service	No. 8

INSIDE TRACK

Multiple Load Centers

For installations in which the main service panel is located at a distance from areas that have many circuits and/or a heavy load concentration, subpanels may be installed near these loads. Doing so allows the branch circuit wiring runs to be shorter, resulting in lower line voltage losses than would occur if the branch circuits had been run all the way back to the main panel.

8.2.0 Bonding of Enclosures and Equipment

8.2.1 Bonding of Services Over 250 Volts

NEC Section 250.97 defines the methods permitted to be used on circuits over 250V. The methods permitted for these circuits are the same as for services, with the exception of using the service grounded conductors beyond the service.

Figure 25 shows examples of these bonding methods.

8.2.2 Bonding Multiple Raceway Systems

NEC Section 250.102(D) allows the use of a single conductor to bond two or more raceways or cables where the bonding jumper is sized in accordance with *NEC Table 250.122* for the largest overcurrent device supplying circuits therein. *Figure 26* shows bonding of multiple raceways.

It is acceptable to install one equipment bonding jumper individually from each raceway to the equipment grounding terminals of the equipment. This is shown in *Figure 27*. Each bonding conductor is sized per *NEC Table 250.122.*

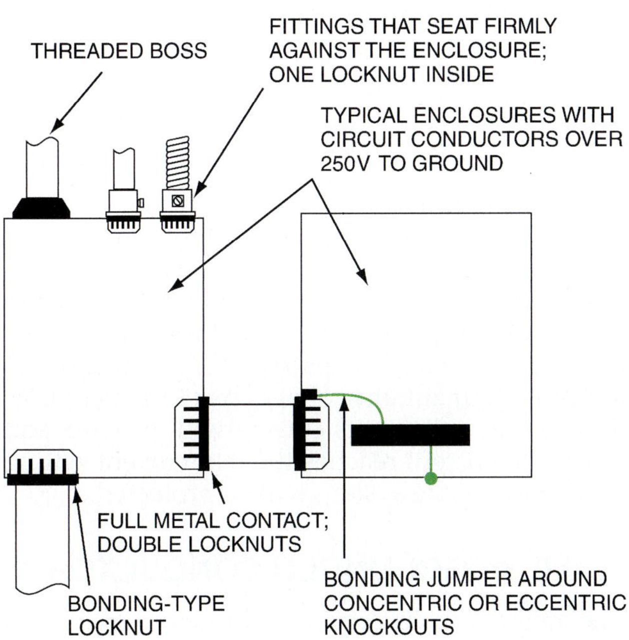

Figure 25 ◆ Bonding for circuits over 250 volts.

8.2.3 Installation of Equipment Bonding Jumper

The equipment bonding jumper is permitted to be installed either inside or outside of a raceway or enclosure. When the jumper is installed outside, the length is limited to not more than 6'. The bonding jumper must also follow the raceway routing. This is vital to keep the impedance of the equipment bonding jumper as low as possible.

Figure 28 shows a bonding jumper installed inside an enclosure.

> **NOTE**
> The equipment bonding jumper must be sized for the overcurrent device.

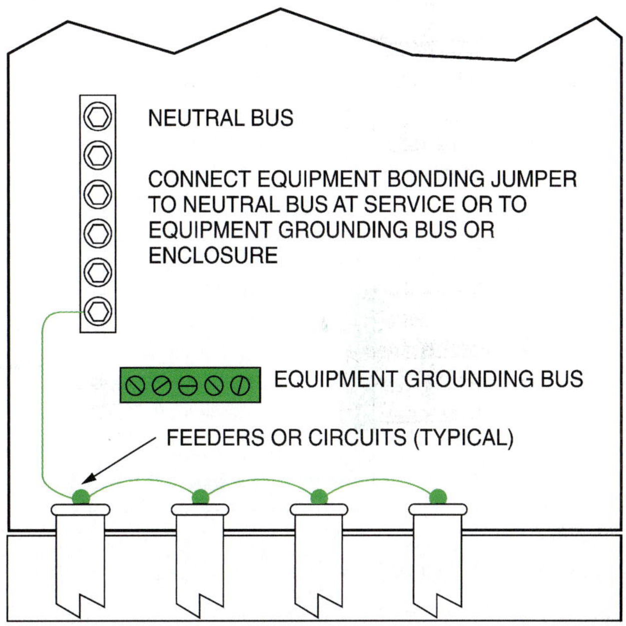

Figure 26 ◆ Bonding multiple raceways.

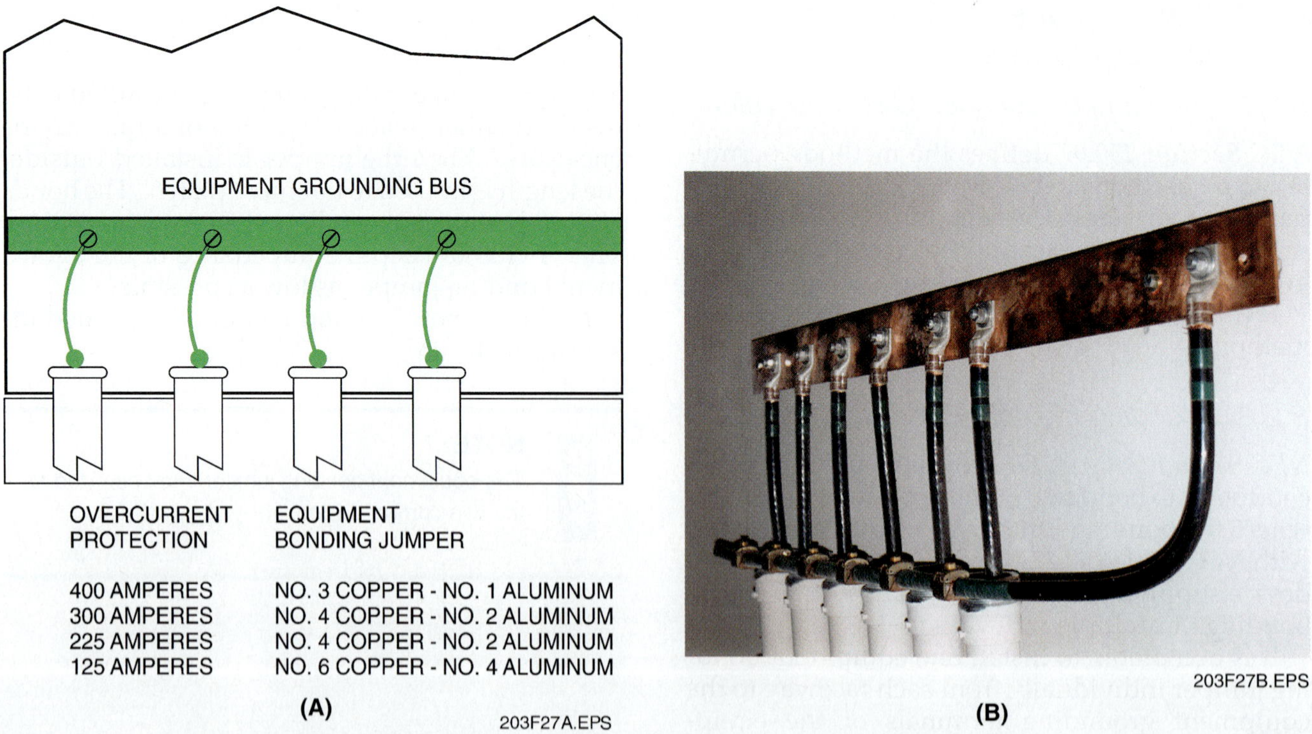

Figure 27 ◆ Individual bonding jumpers.

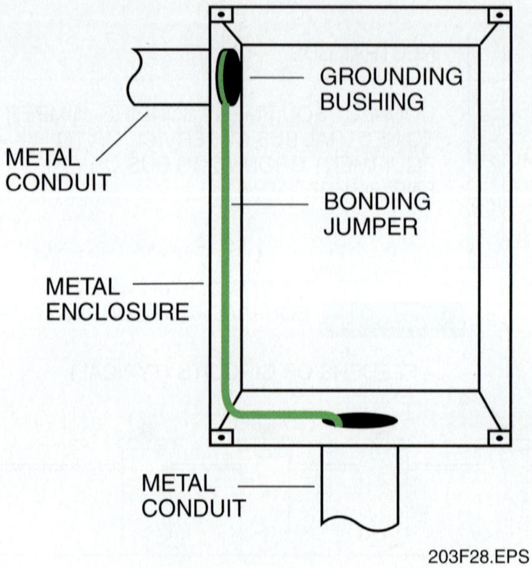

Figure 28 ◆ Bonding jumper.

9.0.0 ◆ EFFECTIVE GROUNDING PATH

As previously stated, to have effective grounding for both a grounded and an ungrounded system, the grounding system shall have an **effective grounding path** to provide a permanent and continuous path, sized to safely conduct any fault current that is likely to occur, and it shall have impedance that is low enough to limit the voltage to ground and to facilitate the operation of the circuit protective devices in the circuit.

Low impedance means that the grounding path and the circuit conductors must always be within the same raceway, or the raceway may provide the grounding path.

9.1.0 Clearing Ground Faults

A ground fault is a different circuit that becomes dependent on the impedance of the ground circuit. The current flow depends on the impedance. The higher the current, the harder it becomes to maintain voltage in the system. It is at this voltage that the short circuit capacity plays its role by attempting to maintain voltage. As the system tries to maintain voltage, its current will increase, and when the current reaches the overcurrent setting of the device, the system will be protected.

10.0.0 ◆ GROUNDED CONDUCTOR

The neutral in a grounded system serves two main purposes:

- It permits utilization of power at line-to-neutral voltage. This will allow the current-carrying conductor to carry any unbalanced current.
- It provides a low-impedance return path for the flow of fault current to the source to facilitate the operation of the overcurrent devices in the circuit.

Separately Derived Systems

INSIDE TRACK

The requirements of *NEC Section 250.30* pertaining to the grounding of separately derived AC systems are most commonly applied to 480Y/277V transformers that transform a 480V supply to a 208Y/120V system for lighting and appliance loads. The requirements provide for a low-impedance path to ground, so that line-to-ground faults on circuits supplied by the transformer will result in sufficient current flow to operate the overcurrent devices.

Figure 29 shows the path to ground for this neutral. If the neutral is not needed for voltage requirements, it still must be run to the service and connected to the grounded electrode conductor and equipment grounding conductor at the service.

11.0.0 ◆ SEPARATELY DERIVED SYSTEMS

Figure 30 shows a transformer-type separately derived system. This system has no direct electrical connection with the premises' source of electrical power. *NEC Section 250.20(D)* discusses whether or not a separately derived system is required to be grounded. Each voltage level that is required to be grounded must meet the requirements of *NEC Section 250.30(A)(1).* The grounding electrode conductor is connected to the grounded conductor within the enclosure. That connection will become a common point for the grounded conductor and the equipment grounding conductor where the common conductor shall be attached and run to the grounding electrode.

In *Figure 30*, the supply to the transformer will be at one voltage level while the secondary is at either a higher or a lower voltage level.

11.1.0 Grounding of Separately Derived Systems

For separately derived systems that are required to be grounded, an equipment grounding conductor must be supplied with the primary circuit. This will provide a low-impedance fault-current path from the transformer case to the

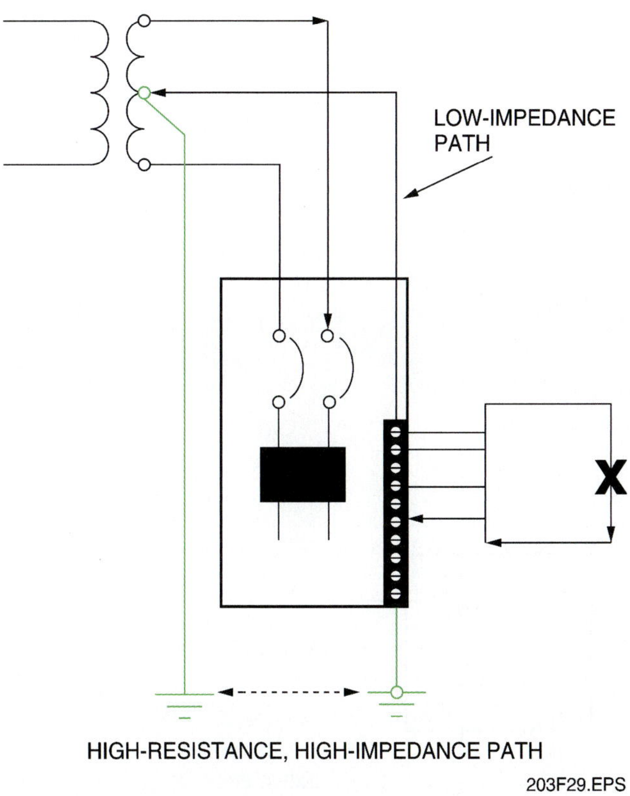

Figure 29 ◆ Grounded service conductor run to service.

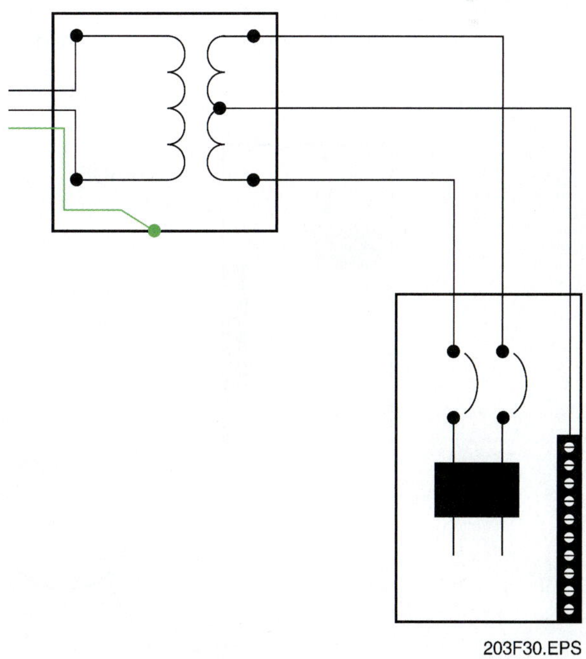

Figure 30 ◆ Transformer-type separately derived system.

main service. Along with the equipment grounding conductor, *NEC Section 250.30(A)(1)* requires a bonding jumper be installed for the derived system required to be grounded. This bonding jumper may be placed in either the transformer or the panel, but not in both. Installing it in both locations would create a parallel path with the grounded conductor, resulting in an unwanted current flow on the grounding conductor. Next, a grounding electrode conductor must be installed. This is illustrated in *Figure 31*.

The grounding electrode for separately derived systems must be:

- As near as practicable to the system
- The nearest available effectively grounded structural metal member or the nearest available effectively grounded water pipe
- Other electrodes where the above are not available (*NEC Sections 250.50 and 250.52*)

11.2.0 Generator-Type Separately Derived System

A simple generator-type separately derived system is shown in *Figure 32*.

The way to determine that a generator is a separately derived system is to examine the transfer switch. If the neutral and all phase conductors are switched, then the system is separately derived. If the neutral is not switched, but solidly connected, then the system is not a separately derived system.

The neutral bonding jumper must be installed either at the generator or any point in between. A grounding electrode conductor must be installed between the neutral and a grounding electrode.

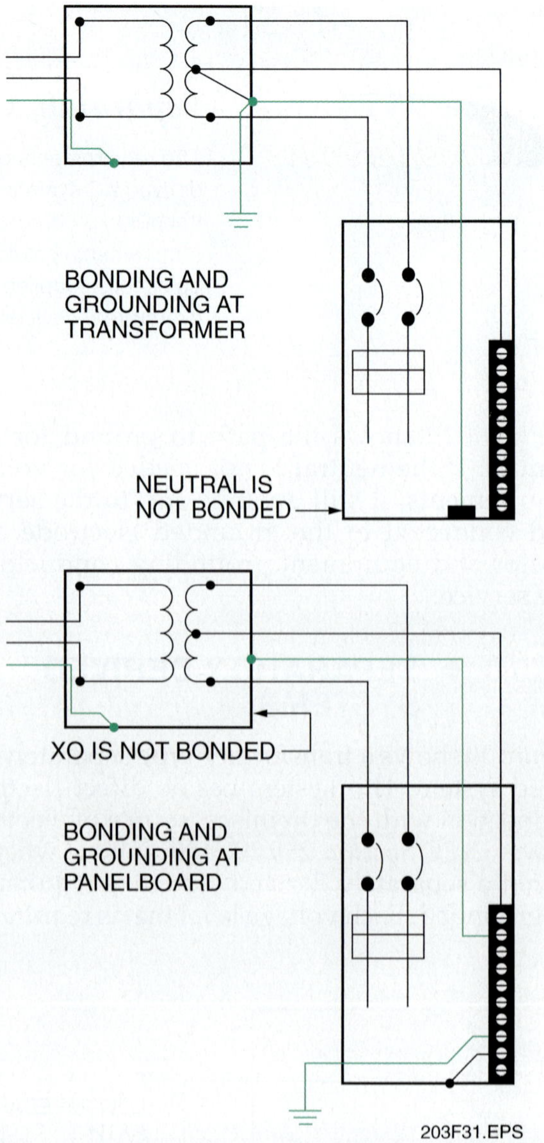

Figure 31 ◆ Separately derived system grounding and bonding locations.

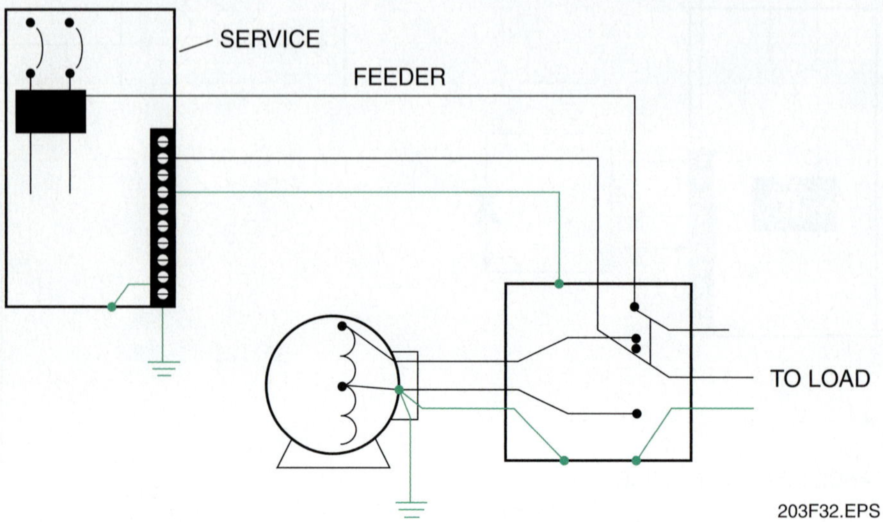

Figure 32 ◆ Generator-type separately derived system.

When the generator is not a separately derived system, the neutral bonding jumper must be removed from the generator and the neutral must not be grounded. Grounding is accomplished through direct connection to the neutral of the premises wiring.

12.0.0 ◆ GROUNDING AT MORE THAN ONE BUILDING

NEC Section 250.32 provides two general methods for grounding electrical systems at additional buildings on the premises.

In the first method, the grounded circuit conductor and the equipment grounding conductor are extended to the second building. The grounded conductor is terminated on an insulated bus isolated from the metal cabinet, while the grounding conductor from the main building is terminated on an equipment grounding terminal bus, directly connected to the cabinet. The equipment grounding conductor would be sized using *NEC Table 250.122* based on the overcurrent device supplying the second structure. All grounding electrodes present at the second building would connect to the grounding bus. The GEC in the second structure is sized using *NEC Table 250.66* based on the size of the feeder conductors. *Table 2* shows the number and type of conductors that must be taken from the first structure to where service is located in the second structure.

When no grounding electrodes are at the additional structures, a grounding electrode must be installed. These electrodes include underground metal pipes, the effectively grounded metal frame of the structure, or concrete-encased electrodes. Rod, pipe, or plate electrodes may also be permitted, in accordance with *NEC Section 250.52.*

In the second method, shown in *Figure 33*, no equipment ground is extended to the second building. The grounded conductor is extended to and terminated on a terminal bus that is bonded to the equipment, similar to the connection made at the grounded conductor at the service equipment. The grounded system will be connected to all grounding electrodes at the second building. In this case, the grounded conductor is sized using *NEC Sections 220.22 and 250.122.* This second method may only be used when an equipment grounding conductor is not run with the supply to the second building, there are no continuous metallic paths bonded to the grounding system in both buildings, and ground fault protection has not been installed on the common electrical service in the first building. These additional provisions prevent the possibility of parallel current paths between grounded and grounding conductors, as well as nuisance tripping of a ground fault protective device.

Table 2 Equipment Grounding Conductor Not Installed

System	Ungrounded	Grounded	Equipment Ground
120V	1	1	1
120/240V	2	1	1
208/120V	3	1	1
480/277V	3	1	1

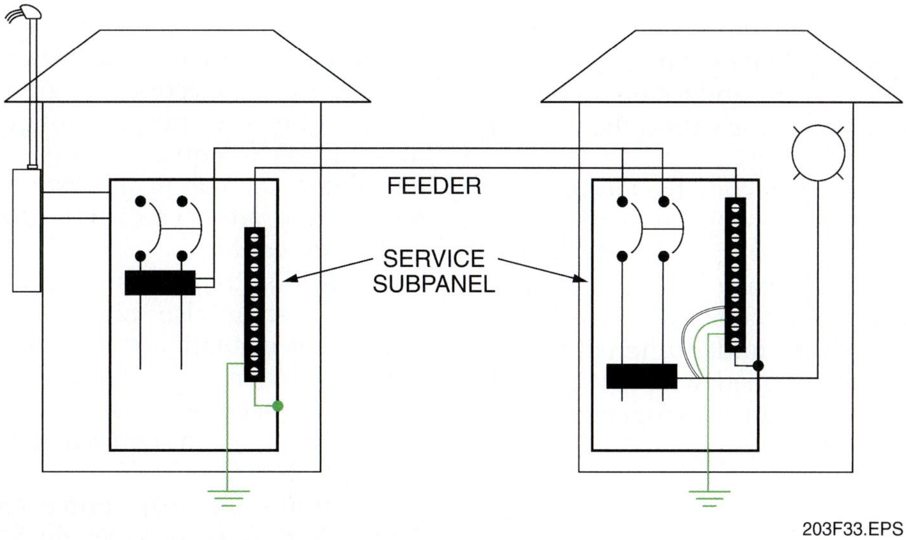

Figure 33 ◆ Grounding neutral at second building.

13.0.0 ◆ SYSTEMS OVER 1,000 VOLTS

NEC Article 250, Part X discusses the grounding requirements for high-voltage systems. It also lists a few additional requirements for derived neutral systems, solidly grounded neutral systems, and impedance-grounded neutral systems.

14.0.0 ◆ TESTING FOR EFFECTIVE GROUNDS

An earth ground resistance tester (*Figure 34*) may be used to make soil **resistivity** measurements or to measure the resistance to earth of the installed grounding electrode system.

NOTE

An ordinary ohmmeter is not used to measure the resistance of a grounding electrode to earth because it does not provide for adequate levels of current and voltage.

WARNING!

Ground testers can be hazardous to both equipment and personnel if improperly used. Always check with your supervisor before using a ground tester.

One use of the ground tester is for testing electrical systems after they are installed and before normal voltage is applied. This test is made after all the conductors, fuses or circuit breakers, panelboards, outlets, etc., are in place and connected. The current used for testing is produced by a small generator within the ground tester that generates DC power, either by turning a crank handle (also a part of the ground tester), or by using a small electric DC motor within the ground tester.

The test is made by connecting the terminals to the two points between which the test is to be made and then rapidly turning the handle on the ground tester. The resistance in ohms can then be read from the meter dial. Satisfactory insulation resistance values will vary under different conditions, and the charts supplied with the ground tester should be consulted for the proper value for a particular installation.

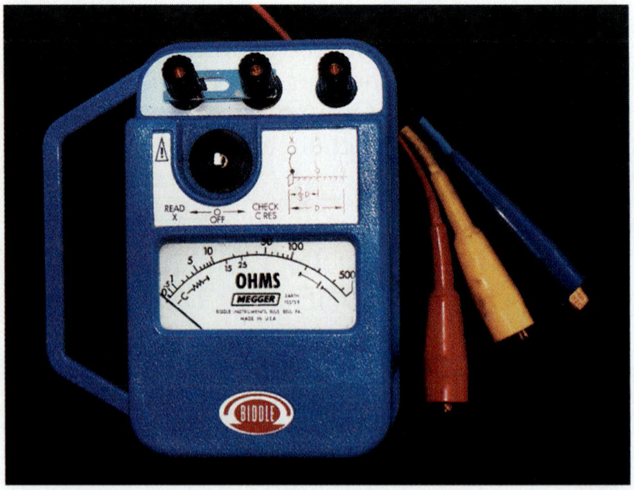

203F34.EPS

Figure 34 ◆ Earth ground resistance tester.

15.0.0 ◆ MEASURING EARTH RESISTANCE

An earth ground is commonly used as an electrical conductor for system returns. Although the resistivity of the earth is high compared to a metal conductor, its overall resistance can be quite low because of the large cross-sectional area of the electrical path.

Connections to the earth are made with grounding electrodes, **ground grids,** and **ground mats.** The resistance of these devices varies proportionately with the earth's resistivity, which in turn depends on the composition, compactness, temperature, and moisture content of the soil.

A good grounding system limits system-to-ground resistance to an acceptably low value. This protects personnel from a dangerously high voltage during a fault in the equipment. Furthermore, equipment damage can be limited by using this **ground current** to operate protective devices.

Ground testers measure the **ground resistance** of a grounding electrode or ground grid system. Some of the major purposes of ground testing are to verify the adequacy of a new grounding system, detect changes in an existing system, and determine the presence of hazardous **step voltage** and **touch voltage.**

In addition to personnel safety considerations, ground testing also provides information for equipment insulation ratings. Equipment can be damaged by an overvoltage that exceeds the rating of the insulation system. *Figure 35* depicts a poorly grounded system where the ground resistance (R_1) is 10Ω.

Assuming a power source resistance (R_2) of 40Ω, a short circuit between the 5,000V power line

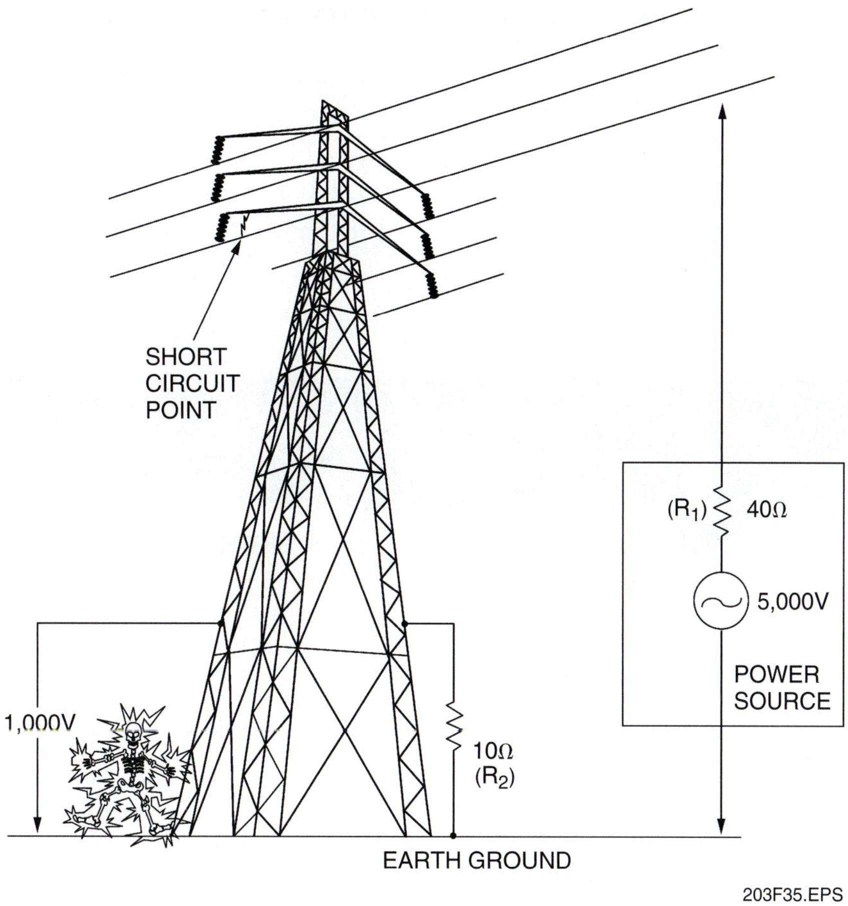

Figure 35 ◆ Poorly grounded system.

and the steel tower would produce 100A of short circuit current (I).

$$I = \frac{E}{R_1 + R_2} = \frac{5{,}000}{40 + 10} = 100\text{A of short circuit current}$$

A person touching the tower would be subjected to the voltage (E') developed across the ground resistance:

$$E' = IR = 100 \times 10 = 1{,}000 \text{ volts between power and ground}$$

Statistics vary widely concerning what may be considered a dangerous voltage. This depends largely on body resistance and other conditions. However, to limit the touch voltage for this situation to 100V, the ground resistance for the tower would have to be less than 1Ω.

15.1.0 How the Ground Tester Works

Several methods are used for measuring the resistance to earth of a grounding electrode; one of the most common is the fall-of-potential method. See *Figure 36*.

In this method, **auxiliary electrodes** 1 and 2 are placed at sufficient distances from grounding electrode X. A current (I) is passed through the earth between the grounding electrode X and auxiliary current electrode 2 and is measured by the ammeter. The voltage drop (E) between the grounding electrode X and the auxiliary potential electrode 1 is indicated on the voltmeter. Resistance (R) can therefore be calculated as follows:

$$R = \frac{E}{I}$$

Certain problems may arise in measuring with the simple system shown in *Figure 36*:

- Natural currents in the soil caused by electrolytic action can cause the voltmeter to read either high or low, depending on polarity.
- Induced currents in the soil, instrument, or electrical leads can cause vibration of the meter pointer, interfering with readability.
- Resistance in the auxiliary electrode and electrical leads can introduce error into the voltmeter reading.

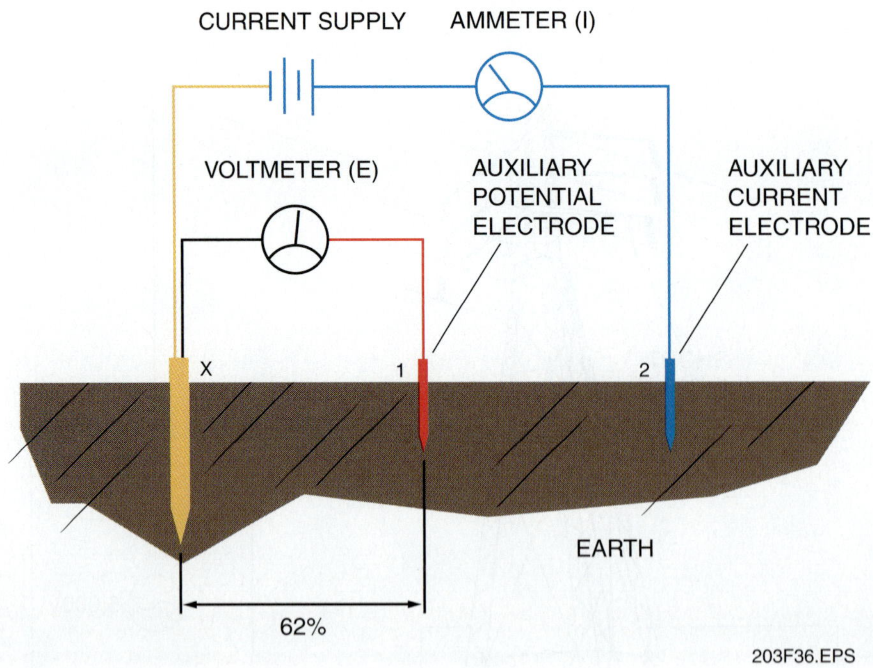

Figure 36 ◆ Fall-of-potential method of testing.

Most ground testers use a null balance metering system. Unlike the separate voltmeter and ammeter method, this instrument provides a readout directly in ohms, thus eliminating calculation. Although the integrated systems of the ground testers are sophisticated, they still perform the basic functions for fall-of-potential testing.

15.2.0 Current Supply

As in the simple circuit, the ground tester also has a current supply circuit (see *Figure 37*). This may be traced from grounding electrode X through terminal X, potentiometer R1, the secondary of power transformer T1, terminal 2, and auxiliary current electrode 2. This produces a current in the earth between electrodes X and 2.

When switch S1 is closed, battery B energizes the coil of vibrator V. Vibrator reed V1 begins oscillating, thereby producing an alternating current in the primary and secondary windings of T1. The negative battery terminal is connected alternately across first one and then the other half of the primary winding.

15.3.0 Voltmeter Circuit

The voltmeter circuit can be traced from grounding electrode X through terminal X, the T2 secondary, switch S2, resistors R2 and R3 (paralleled by the meter and V2 contacts), capacitor C, terminal 1, and auxiliary potential electrode 1. The current in the earth between grounding electrode X and auxiliary current electrode 2 creates a voltage drop due to the earth's resistance. With auxiliary potential electrode 1 placed at any distance between grounding electrode X and auxiliary current electrode 2, the voltage drop causes a current in the voltmeter circuit through balanced resistors R2 and R3. The voltage drop across these resistors causes galvanometer M to deflect from zero center scale.

Vibrator reed V2 operates at the same frequency as V1, thereby functioning as a mechanical rectifier for galvanometer M. The vibrator is tuned to operate at 97.5 hertz (Hz), a frequency unrelated to commercial power line frequencies and their harmonics. Thus, currents induced in the earth by power lines are rejected by most ground testers and have virtually no effect on their accuracy. Stray direct current in the earth is blocked out of the voltmeter circuit by capacitor C.

The current in the primary of T2 can be adjusted with potentiometer R1. Primary current in T2 induces a voltage in the secondary of T2, which is opposite in polarity to the voltage drop caused by current in the voltmeter circuit.

With R1 adjusted so the primary and opposing secondary voltages of T2 are equal, current in the voltmeter circuit is zero, and the galvanometer reads zero. The resistance of grounding electrode X can then be read on the calibrated dial of the potentiometer.

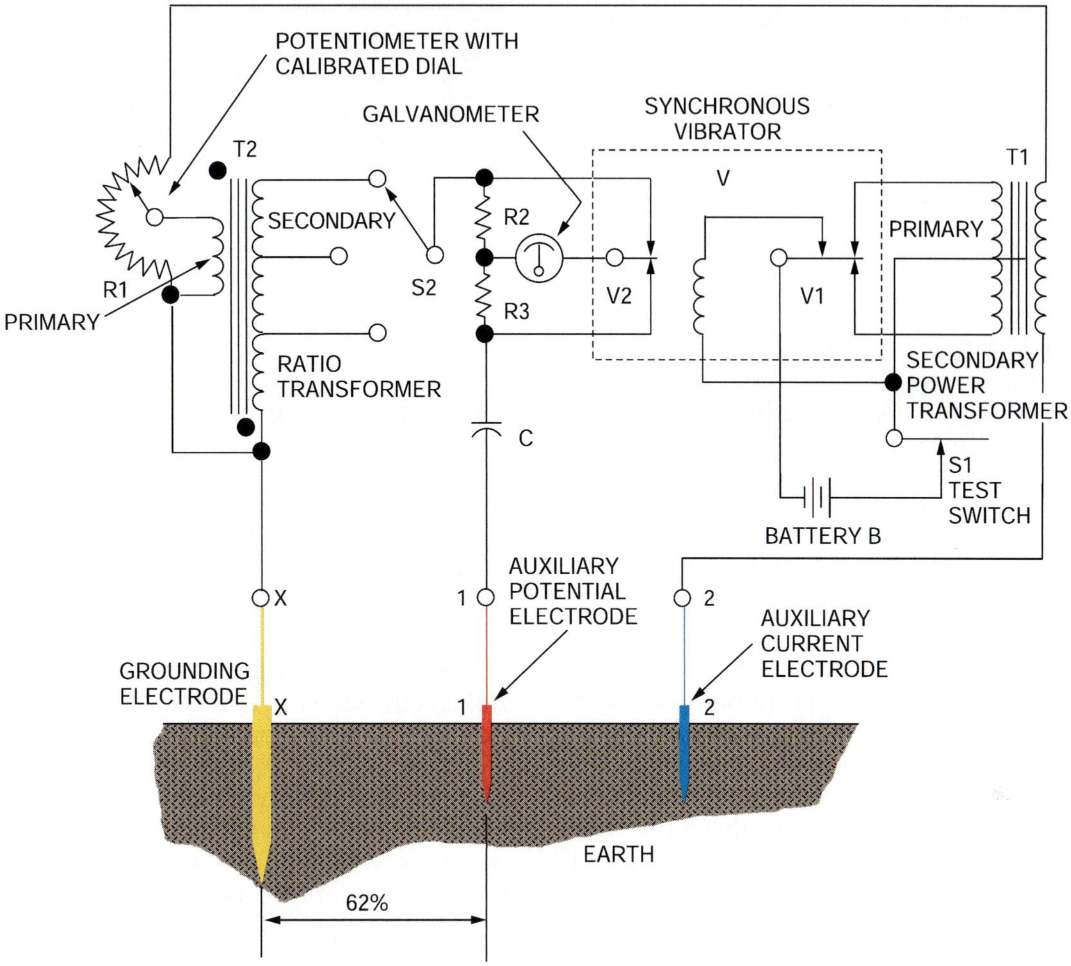

Figure 37 ◆ Three-point testing using a ground tester.

With no current in the voltmeter circuit, the lead resistance of the auxiliary potential electrode 1 has no bearing on accuracy. (With no current, there is no voltage drop in the leads.) Resistance to earth of the current electrode results only in a reduction of current, and consequently, a loss of sensitivity. Therefore, the auxiliary electrodes need only be inserted into the earth 6" to 8" to make sufficient contact. In some locations, where the soil is very dry, it may be necessary to pour water around the current electrode to lower the resistance to a practical value.

15.4.0 Nature of Earth Electrode Resistance

Current in a grounding system is primarily determined by the voltage and impedance of the electrical equipment. However, the resistance of the grounding system is very important in determining the voltage rise between the ground electrode and the earth as well as the voltage gradients that will occur in the vicinity when current is present.

THINK ABOUT IT

Size and Depth of Ground Electrode

What effect do the diameter and driven depth of a ground electrode have on its resistance?

Three components constitute the resistance of a grounding system:

- Resistance of the conductor connecting the ground electrode
- Contact resistance between the ground electrode and the soil
- Resistance of the body of earth immediately surrounding the electrode

Resistance of the connecting conductor can be dealt with separately since this is a function of the conductor cross-sectional area and length. Contact resistance is usually negligibly small if the electrode is free from paint or grease. Therefore, the main resistance is that of the body of earth immediately surrounding the electrode. Current from a grounding electrode flows in all directions via the surrounding earth. It is as though the current flows through a series of concentric spherical shells, all of equal thickness.

The shell immediately surrounding the electrode has the smallest cross-sectional area, and therefore its resistance is highest. As the distance from the electrode is increased, each shell becomes correspondingly larger; thus, the resistance becomes smaller. Finally, a distance from the electrode is reached where additional shells do not add significantly to the total resistance. From a theoretical viewpoint, total resistance is included only when the distance is infinite. For practical purposes, only the volume that contributes the major part of the resistance need be considered. This is known as the effective resistance area and depends on electrode diameter and driven depth.

16.0.0 ◆ THREE-POINT TESTING PROCEDURE

To measure resistance of a grounding electrode with most ground testers, an auxiliary current electrode and an auxiliary potential electrode are required (see *Figures 38* and *39*). The current electrode is placed a suitable distance from the grounding electrode under test, and the potential electrode is then placed at 62% of the current electrode distance.

NOTE

The 62% figure has been arrived at by empirical data gathered by many authorities on ground resistance measurement, and in some cases, has been computed based on analysis of an equivalent hemisphere.

If the current electrode is too near the grounding electrode, their effective resistance areas will overlap, as shown in *Figure 38*. If a series of measurements are made with the potential electrode driven at various distances in a straight line between the current electrode and grounding electrode, the readings will yield a curve as illustrated in *Figure 38*.

In *Figure 39*, a curve is plotted with the current electrode at a sufficient distance from the grounding electrode. Note that the curve is relatively flat between points B and C, which are usually considered to be at ±10' with respect to the 62% point. Usually, a tolerance is established for the maximum allowable deviation for the second and third readings with relation to the initial reading at 62%. This tolerance is a certain ±% of the initial reading, such as ±1%, ±2%, etc.

No definite distance from the current electrode can be forecast since the optimum distance is based on the homogeneity of the earth, depth of the grounding electrode, diameter, etc. However, for a starting point for a single driven grounding electrode, the effective radius of the equivalent hemisphere can be computed. The curve of *Figure 40* can then be used for initial placement of the auxiliary electrodes. Using the practical method of moving the auxiliary potential electrode 10' to either side of the 62% point, a curve can be plotted to determine if the current electrode spacing is adequate, as depicted by the curve flattening out between points B and C in *Figure 39*.

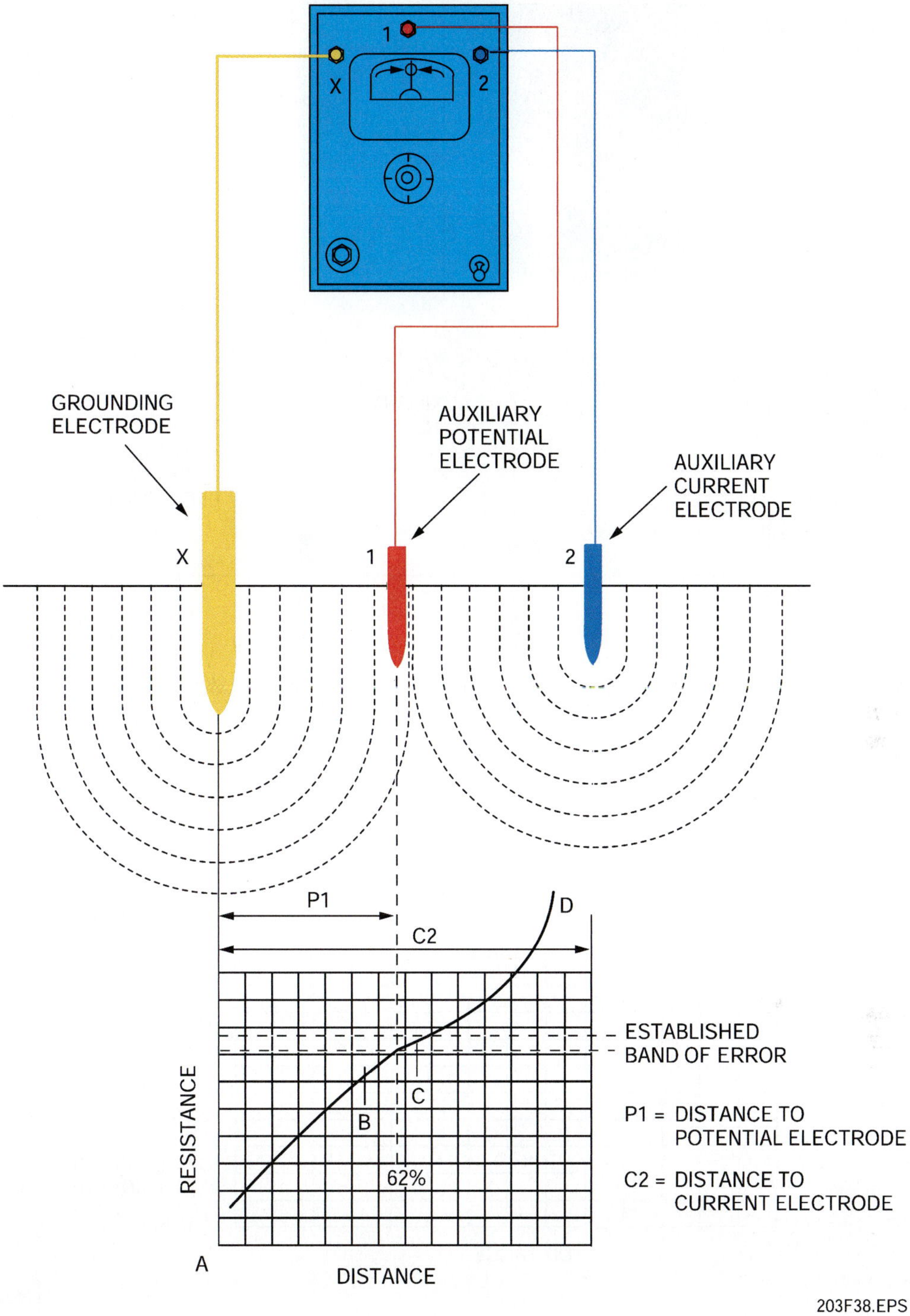

Figure 38 ◆ Plotted curve showing insufficient electrode spacing.

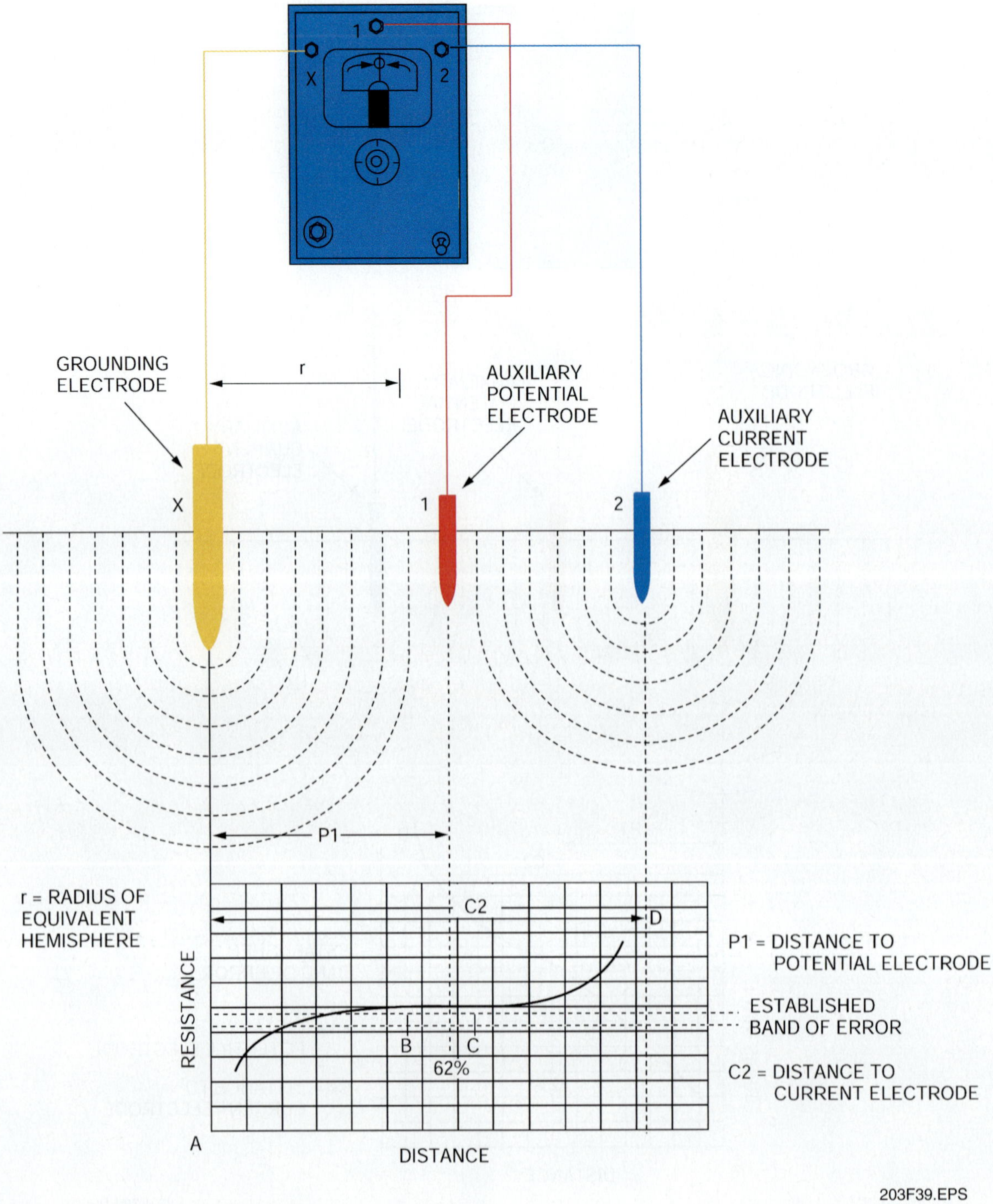

Figure 39 ◆ Plotted curve showing adequate electrode spacing.

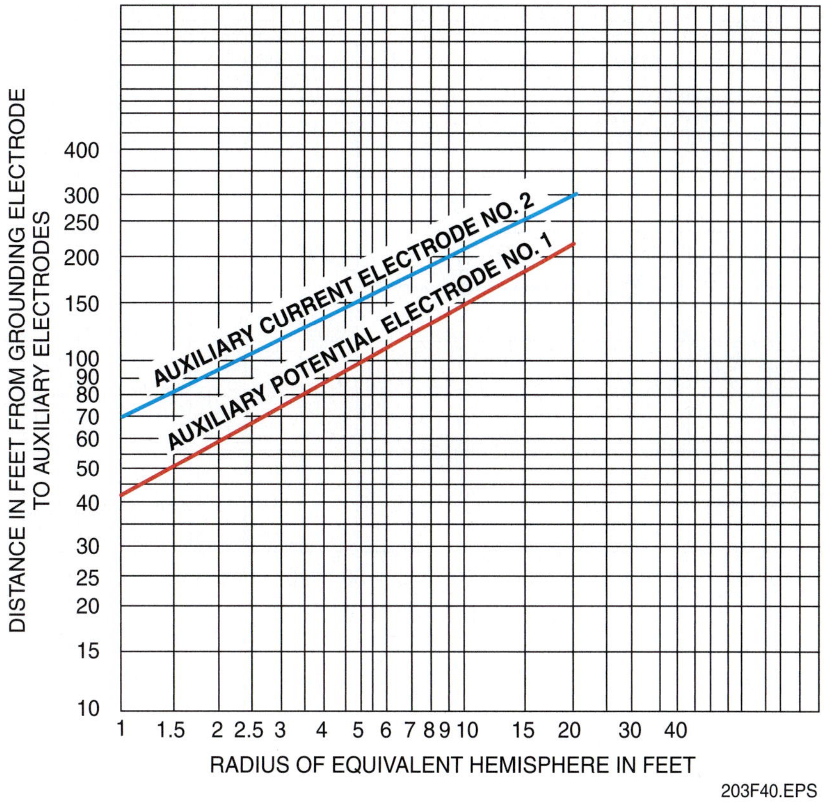

Figure 40 ◆ Auxiliary electrode distance/radii chart.

Testing Station Grounds

INSIDE TRACK

A lethal potential can exist between the station ground and a remote ground if a system fault involving the station ground occurs while earth resistance tests are being made. Since one of the objectives of tests on a station ground is to establish the location of an effectively remote point for both current and potential electrodes, the leads to the electrodes must be treated as though a possible potential could exist between these test leads and any point on the station ground grid.

For example, assume that a 1" (2.54 cm) diameter grounding electrode is buried 10' (3.0 m) deep. The equivalent hemisphere radius is 1.7' (51.8 cm). Using *Figure 40,* the current electrode would be established at approximately 90' (27 m), and the potential electrode at approximately 55' (17 m) for the initial reading. The results obtained when the potential electrode is moved (to 45' [14 m] and then to 65' [20 m]) will determine if the current electrode, and consequently the potential electrode, must be spaced at a greater distance.

It is usually advisable to plot a complete curve for each season of the year. See *Figure 41.* These curves should be retained for comparison purposes. Measurements at established intervals in the future need only be made at the 62% point and, if desired, 10' on each side, providing there is no erratic deviation from the original curve. Serious deviation, other than seasonal, could mean corrosion has eaten away some of the electrode.

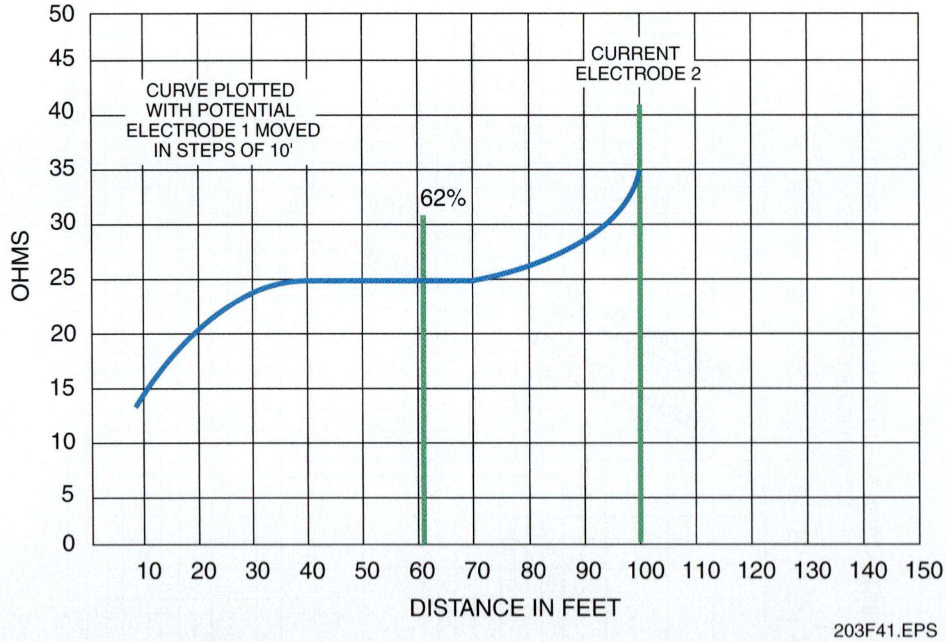

Figure 41 ◆ Typical grounding resistance curve to be recorded and retained.

16.1.0 Operating Procedure

Establish the proper location for the auxiliary current electrode and auxiliary potential electrode as follows:

Step 1 Determine the effective radius of the equivalent hemisphere in feet for the grounding electrode depth in feet, as well as the grounding electrode diameter in inches, using the curves in *Figure 42*.

Step 2 Using this effective radius and the data in Figure 40, determine the distances for the auxiliary current electrode No. 2 and auxiliary potential electrode No. 1 from the ground electrode.

Step 3 Position the ground tester in a suitable location near the grounding system to be tested.

Step 4 Using a test lead of less than 0.05Ω resistance, connect the tester terminal X to the grounding electrode or grounding grid to be measured, terminal 1 to the auxiliary potential electrode, and terminal 2 to the auxiliary current electrode (less than 1Ω).

Step 5 If the general range of resistance to be measured is anticipated, then set the MULTIPLY BY switch to a multiplying factor within this range. If unknown, select the highest multiplier range first.

Step 6 Move the TEST switch to the ADJ position. Rotate the OHMS control knob until the galvanometer indicates balance at center scale.

NOTE

Meter instructions will vary depending on the instrument used. Consult the manufacturer's operating manual.

Step 7 Should the galvanometer indication remain to the right of center scale with the OHMS control at zero, set the MULTIPLY BY switch to the adjacent lower range.

Step 8 Should the galvanometer indication remain to the left of center scale with the OHMS control fully clockwise, set the MULTIPLY BY switch to the adjacent higher range.

Step 9 After balance is achieved with the TEST switch in the ADJ position, move this switch to the READ position and rebalance the galvanometer using the OHMS control.

Step 10 Read the resistance on the calibrated OHMS scale. If this indication is less than 1⁄10 of full range, set the MULTIPLY BY switch to the adjacent lower range and rebalance the galvanometer. (This provides better readability and resolution.)

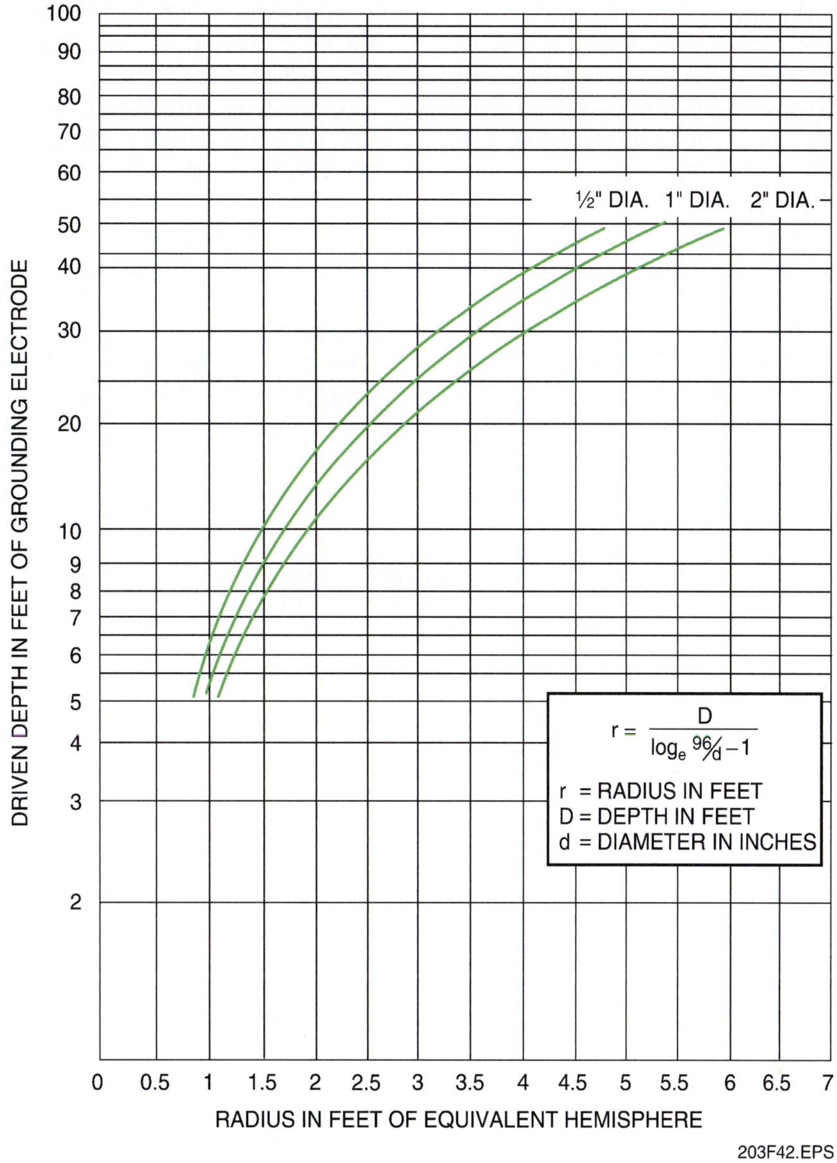

Figure 42 ◆ Ground rod depth versus equivalent hemisphere radii.

Step 11 Multiply the OHMS scale indication by the factor indicated by the MULTIPLY BY switch. This product is the resistance of the grounding system connected to terminal X. Record this resistance value. (If balance cannot be achieved under any conditions, check the test leads. Ensure that the leads are properly connected, as described in Step 4. If the trouble persists, the grounding system may be inadequate due to a high resistance beyond the range of the tester.)

Step 12 Move the auxiliary potential electrode 10' on each side of the 62% point, as shown by points B and C in *Figure 39*, and repeat Steps 5 through 11 for each. Record this resistance value.

Step 13 Compare the resistance values obtained in Steps 11 and 12. They should be within the established tolerance band as explained previously. If they are not within the established tolerance, move the auxiliary current and potential electrodes to a point farther away from the ground under test and repeat Steps 4 through 12. Record the resistance values obtained.

A single driven rod is an economical and simple means of making a grounding electrode. In general, however, a single driven rod does not provide a sufficiently low resistance, and consequently, several rods must be driven and connected in parallel by a grounding conductor or cable. Although these rods are connected as they would be in parallel resistance, their total resistance does not follow the usual law for computing resistances in parallel. To attain the full effect of resistances in parallel, these rods would have to be spaced at such distances that the effective resistance areas immediately surrounding them would not overlap. The extent to which the areas overlap determines how much of their effectiveness is lost. The curves of *Figure 43* show a change in resistance for two, four, and eight parallel rods with various spacing.

When two, three, or four rods are used in parallel, they are usually driven in a straight line and connected together. In cases where more than four are used, they are usually driven in a hollow-square formation and connected in parallel. Distances between the rods are usually made equal, as in the hollow-square formation of eight rods shown in *Figure 44*.

When measurements are to be made on large formations or on substations contained within a fence that is part of the grounding system, the required distances for the auxiliary electrodes should be

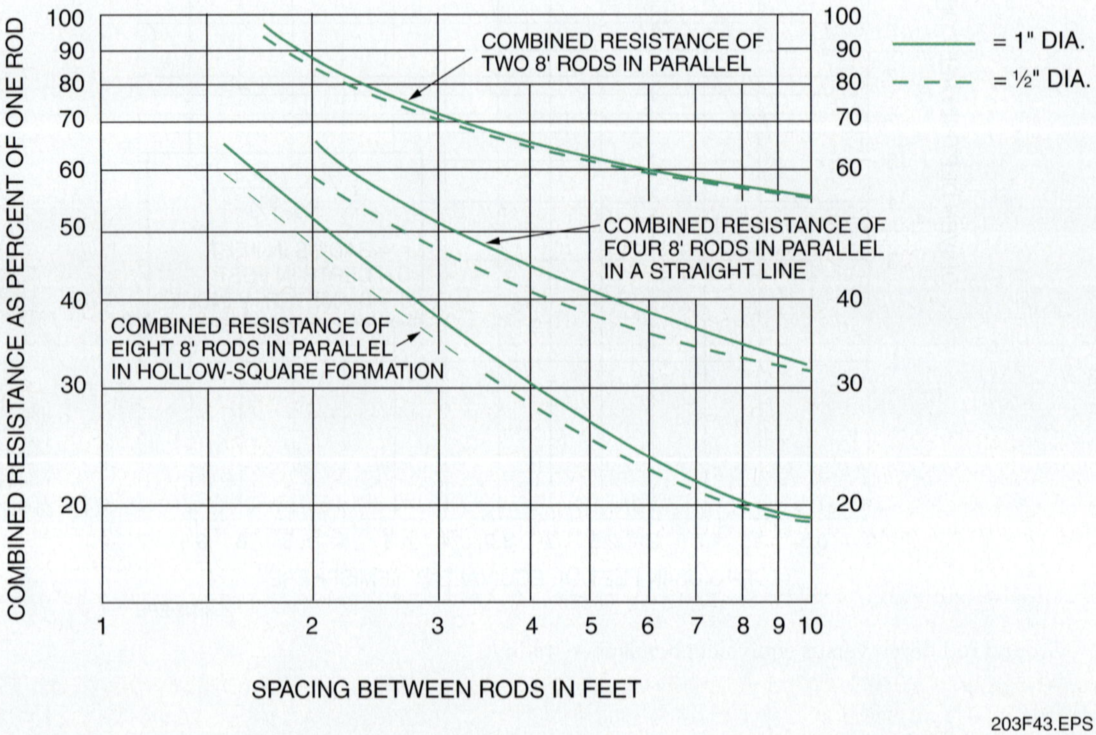

Figure 43 ◆ Combined resistances of rods in parallel.

INSIDE TRACK

Excessive Electrode Resistance During Testing

Excessive electrode resistance or transient noise during testing may indicate an incorrect measurement. If this problem occurs, try the following remedies:

- Check the integrity of all test connections between the leads and the electrodes.
- Make sure that the ground electrodes Y and Z are properly inserted in a quality ground and completely buried, if possible.
- Select a lower test current setting.
- If stray currents are suspected, try moving both electrodes Y and Z in an arc relative to the ground electrode (e.g., move them 90° and test again).

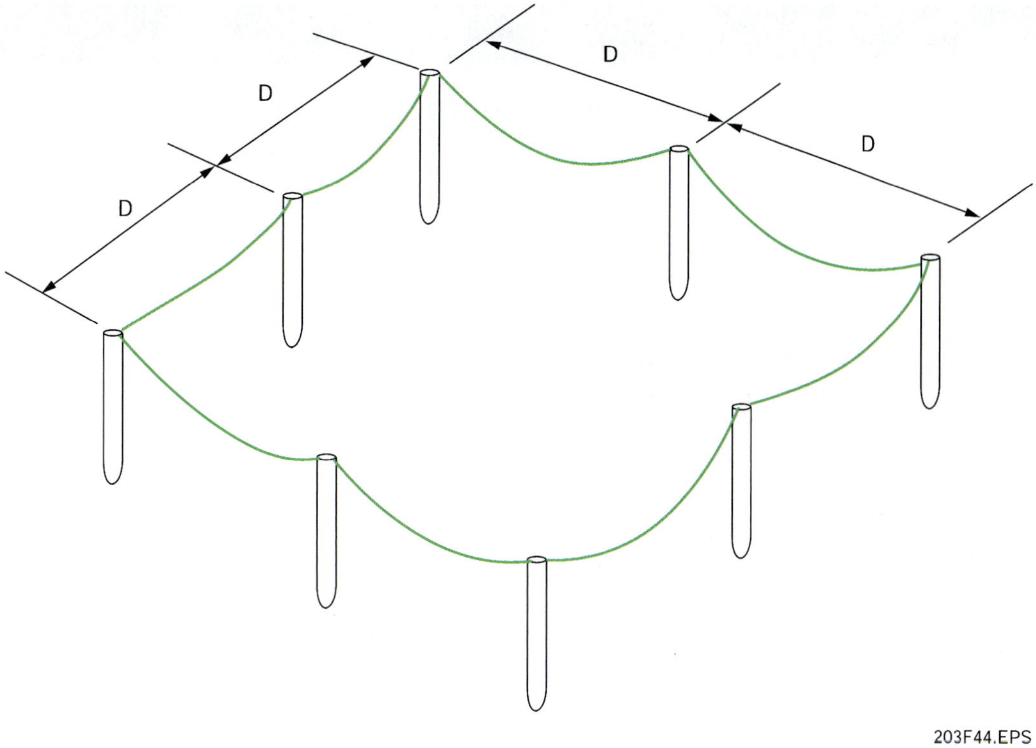

Figure 44 ◆ Hollow-square formation of grounding electrodes.

based on the diagonal (or longest dimension) of the entire grounding system. These distances can be determined from *Figure 45*, which is based on the computed hemisphere for various numbers of rods arranged in hollow-square formation.

For example, a hollow-square formation having 40' (12.2 m) sides would have a diagonal of about 57' (17.4 m). *Figure 45* shows that current and potential electrodes should be spaced at about 370' (113 m) and 230' (70 m), respectively.

Major difficulties are encountered when measuring large grid systems due to the fact that theory and computations are usually based on homogeneous soil. Differences in the soil will affect penetration depth and consequently distort the curve being plotted. In extremely large ground grid systems using several configurations of electrodes, the true center from which to properly space the current and potential electrodes is difficult to determine.

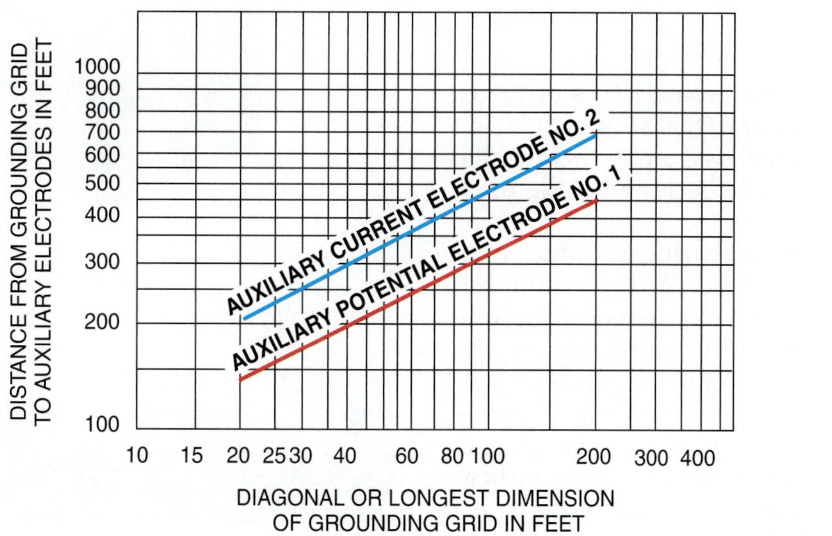

Figure 45 ◆ Auxiliary electrode distances versus longest grid dimension.

THINK ABOUT IT

Poor Soil Conductivity

What can be done to lower the ground electrode (rod) resistance when it is driven into soil with poor conductivity?

When measuring the resistance of any grounding system, single-driven electrodes, or grid systems, it must be realized that other common grounding connections will influence the total measured resistance. For instance, if a ground grid system consisting of multiple-driven rods connected in parallel is in turn connected to a fence surrounding a substation, the fence can be considered connected in parallel with the grid system. The total resistance measured will be that of the entire grounding system. Whether these grounds can be isolated and measured separately is primarily a matter of regulations, the danger involved while they are isolated, and practicality.

16.2.0 Electrode Arrangements

The angle at which the electrodes are placed with respect to a ground grid system is primarily a matter of choice or requirement due to surrounding obstructions. However, there are pros and cons as to whether the electrodes should be placed at 90° with respect to the sides of the grid or diagonally from a corner. Many times, one placement as opposed to the other has yielded different test results. This may very well be due to the homogeneity of the soil of one area versus the other. Curves have also been plotted with the current electrode on one side of the grounding electrode or ground grid system, and the potential electrode on the opposite side (180° apart). This arrangement yields a curve similar to that shown in *Figure 46*. This arrangement may give erroneous results, because there is no way of checking for the presence of errors during the test.

Another method of measuring ground resistance is the two-point method. With this method, a reference ground system of known or negligible resistance is used. Here, the resistance of the grounding electrode and reference ground system are measured in series. (See *Figure 47*.) The resistance of the known reference ground system is then subtracted from the total resistance indication to yield the resistance of the grounding electrode.

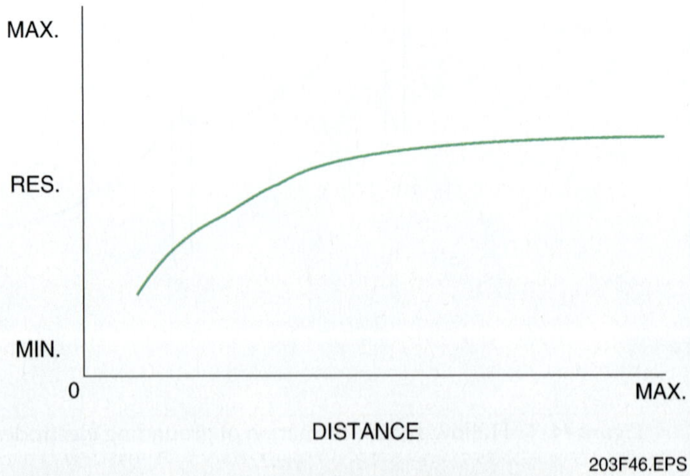

Figure 46 ◆ Plotted curve with current and potential electrodes spaced 180° apart.

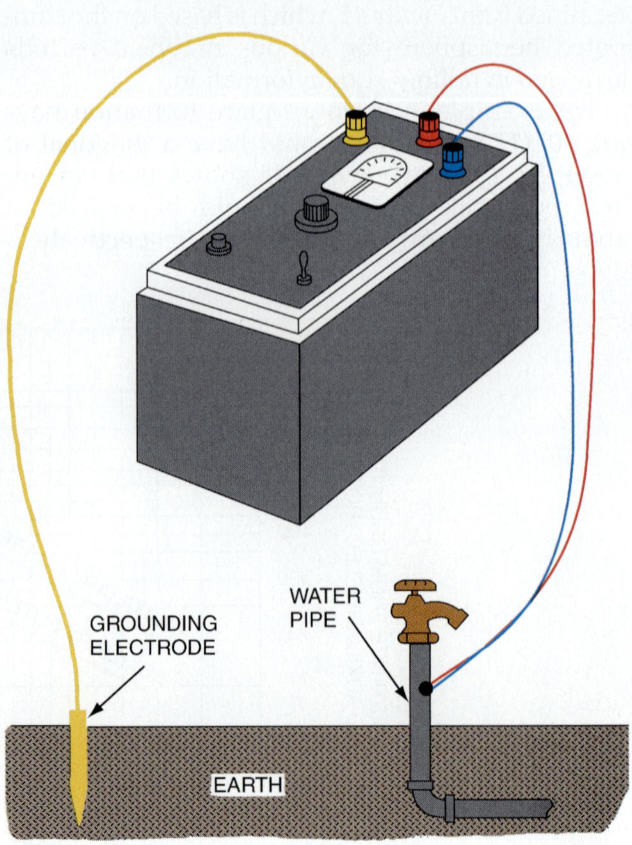

Figure 47 ◆ Two-point method of measurement.

Clamp-On Resistance Testing

INSIDE TRACK

Clamp-on resistance testers offer the ability to measure the resistance without disconnecting the ground. This type of measurement also has the advantage of including the bonding to ground and the overall grounding connection resistance.

There are several objections to this method. One is that the resistance of reference ground systems such as water pipes or other structures is usually unknown. Another objection is that if the reference ground resistance is known, it may be excessively high in comparison, and thus the small fraction end results may be very doubtful. Still another objection is the fact that the reference ground may be extensive and the resistance areas of the reference ground and grounding electrode may overlap considerably.

16.3.0 Equipotential Grounding

Safety grounds should be applied in such a way that a zone of equal potential is formed in the work area. This equipotential zone is formed when fault current is bypassed around the work area by metallic conductors. In this situation, the worker is bypassed by the low-resistance metallic conductors of the safety ground.

Putting It All Together

THINK ABOUT IT

A low-impedance ground is essential to the performance of any electrical protection system. The ground must dissipate electrical transients and surges in order to minimize the chance of damage or injury. Proper grounding, which includes bonding and connections, protects personnel from the danger of shock and protects equipment and buildings from hazardous voltages. Proper grounding also contributes to the reduction of electrical noise and provides a reference for circuit conductors to stabilize their voltage to ground during normal operation. Examine the grounding system of the electrical service at your home or workplace. Is it adequate? If not, how can it be corrected to meet current *NEC*® requirements?

Review Questions

1. The use of _____ as a grounding electrode is in violation of the *NEC®*.
   a. copper cold-water pipe
   b. galvanized pipe
   c. an underground gas line
   d. a grounding ring

2. Which of the following copper AWG sizes should be used for the grounding conductor for a 200A service using 3/0 copper conductors?
   a. No. 2
   b. No. 4
   c. No. 6
   d. No. 8

3. The maximum distance allowed by the *NEC®* to connect a grounding conductor to a water pipe after the pipe enters the building is _____.
   a. 5'
   b. 10'
   c. 15'
   d. 20'

4. A _____ is used to connect a ground rod with a grounding conductor.
   a. grounding clip
   b. butt-and-slide connector
   c. wire nut
   d. grounding clamp

5. The _____ conductor(s) are grounded on a single-phase, three-wire, 120/240V system.
   a. hot
   b. current-carrying
   c. neutral
   d. black

6. Which of the following should you keep in mind when disconnecting a ground?
   a. Exercise extreme caution.
   b. A grounding conductor never carries current.
   c. All grounds are dead (have zero potential) and require no special precautions.
   d. Make sure your body completes a circuit between the two points.

7. The minimum length allowed for driven ground rods under normal conditions is _____.
   a. 4'
   b. 8'
   c. 0'
   d. 16'

8. All of the following are suitable materials for ground rods *except* _____.
   a. ½" solid stainless steel rod
   b. ¾" galvanized pipe
   c. ⅝" copperweld solid rod
   d. ½" solid aluminum rod

9. When a metal water pipe is used as a grounding electrode, _____ must be provided at the water meter.
   a. bonding jumpers
   b. a floor drain
   c. a grounding locknut
   d. a grounding bushing

10. Most residential systems are grounded at the _____.
    a. pole-mounted transformer
    b. service drop
    c. neutral bus in the main panelboard
    d. meter outdoors

Summary

There is no subject in the electrical industry more important than grounding. It is the chief means of protecting life and property from electrical hazards. It also ensures proper operation of the system and helps other protective devices to function properly.

The term grounded means connected to earth by a conductor or to some conducting body that serves in place of the earth. The earth as a whole is properly classed as a conductor. For convenience, its electric potential is assumed to be zero. When a metal object is grounded, it too is thereby forced to take the same zero potential as the earth. Therefore, the main purpose of grounding is to ensure that the grounded object cannot take on a potential differing sufficiently from earth potential to be hazardous.

Notes

Trade Terms Introduced in This Module

Auxiliary electrodes: Metallic electrodes pushed or driven into the earth to provide electrical contact for the purpose of performing measurements on grounding electrodes or ground grid systems.

Bonding: The permanent joining of metallic parts to form an electrically conductive path that will assure electrical continuity and the capacity to safely conduct any current likely to be imposed on it.

Effective grounding path: An intentionally constructed, permanent, low-impedance electrically conductive path designed and intended to carry current under ground-fault conditions from the point of a ground fault on a wiring system to the electrical supply source and that facilitates the operation of the overcurrent protective device or ground-fault detectors on a high-impedance grounded system.

Effectively grounded: Intentionally connected to the earth through a ground connection or connections of sufficiently low impedance and having sufficient current-carrying capacity to prevent the buildup of voltages that may result in undue hazards to personnel or connected equipment.

Equipment bonding jumper: The connection between two or more portions of the equipment grounding conductor.

Equipment grounding conductor: The conductor used to connect the noncurrent-carrying metal parts of equipment, raceways, and other enclosures to the system grounded conductor, the grounding electrode conductor, or both, at the service equipment or at the source of a separately derived system.

Ground: A conducting connection, whether intentional or accidental, between an electrical circuit or equipment and the earth, or to some conducting body that serves in place of the earth.

Ground current: Current in the earth or grounding connection.

Ground grids: System of grounding electrodes interconnected by bare cables buried in the earth to provide lower resistance than a single grounding electrode.

Ground mats: System of bare conductors, on or below the surface of the earth, connected to a ground or ground grid to provide protection from dangerous touch voltage.

Ground resistance: The ohmic resistance between a grounding electrode and a remote or reference grounding electrode that are spaced such that their mutual resistance is essentially zero.

Ground rod: A metal rod or pipe used as a grounding electrode.

Grounded: Connected to earth or to some conducting body that serves in place of the earth.

Grounded conductor: A system or circuit conductor that is intentionally grounded.

Grounding clip: A spring clip used to secure a bonding conductor to an outlet box.

Grounding conductor: A conductor used to connect equipment or the grounded circuit of a wiring system to a grounding electrode or electrodes.

Grounding connections: Connections used to establish a ground; they consist of a grounding conductor, a grounding electrode, and the earth surrounding the electrode.

Grounding electrode: A grounding electrode consists of (1) the nearest available effectively grounded structural metal member of the structure; or (2) the nearest available effectively grounded metal water pipe; or (3) other electrodes as specified in ***NEC Sections 250.50 and 250.52.***

Grounding electrode conductor: The conductor used to connect the grounding electrode to the equipment grounding conductor, the grounded conductor, or both, of the circuit at the service equipment or at the source of a separately derived system.

Main bonding jumper: The connection between the grounded circuit conductor and the equipment grounding conductor at the service.

Neutral: A grounded conductor in an electrical system that does not carry current until the system is unbalanced.

Resistivity: Resistance between opposite faces of a unit cube. Expressed in ohm-centimeters or ohms per cubic centimeter.

Separately derived system: A premises wiring system whose power is derived from a source of electric energy or equipment other than a service. Such systems have no direct electrical connection, including a solidly connected ground circuit conductor, to supply conductors originating in another system.

Short circuit: An often unintended low-resistance path through which current flows around, rather than through, a component or circuit.

Step voltage: The potential difference between two points on the earth's surface separated by a distance of one pace, or about three feet.

System grounding: Intentional connection of one of the circuit conductors of an electrical system to ground potential.

Touch voltage: The potential difference between a grounded metallic structure and a point on the earth's surface equal to the normal maximum horizontal reach—approximately three feet.

Ungrounded conductors: Conductors in an electrical system that are not intentionally grounded.

Additional Resources

This module is intended to present thorough resources for task training. The following reference works are suggested for further study. These are optional materials for continued education rather than for task training.

American Electrician's Handbook, Latest Edition. New York: Croft and Summers, McGraw-Hill.

National Electrical Code® Handbook, Latest Edition. Quincy, MA: National Fire Protection Association.

CONTREN® LEARNING SERIES — USER FEEDBACK

The NCCER makes every effort to keep these textbooks up-to-date and free of technical errors. We appreciate your help in this process. If you have an idea for improving this textbook, or if you find an error, a typographical mistake, or an inaccuracy in NCCER's *Contren®* textbooks, please write us, using this form or a photocopy. Be sure to include the exact module number, page number, a detailed description, and the correction, if applicable. Your input will be brought to the attention of the Technical Review Committee. Thank you for your assistance.

Instructors – If you found that additional materials were necessary in order to teach this module effectively, please let us know so that we may include them in the Equipment/Materials list in the Annotated Instructor's Guide.

Write: Product Development
National Center for Construction Education and Research
P.O. Box 141104, Gainesville, FL 32614-1104

Fax: 352-334-0932

E-mail: curriculum@nccer.org

Craft ______ Module Name ______

Copyright Date ______ Module Number ______ Page Number(s) ______

Description ______

(Optional) Correction ______

(Optional) Your Name and Address ______

Conduit Bending
26204-05

Latus Motors Harley-Davidson, Inc.
Gladstone, Oregon
Pre-Engineered Building $2–10 Million Award Winner
Engineered Structures, Inc.

26204-05

Conduit Bending

Topics to be presented in this module include:

Overview

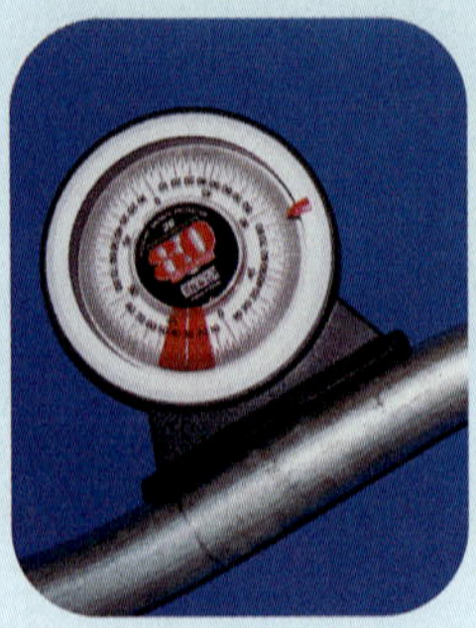

Simple one-shot bends in small-diameter conduit can be made using a hand bender, with little or no calculations required. However, when complete conduit systems must be installed on exposed surfaces or pipe racks, many calculated bends may be required using hand- or power-bending equipment.

The *National Electrical Code*® regulates the number of degrees that can be bent into a single run of conduit. A run of conduit is considered to be any conduit installed between outlet or junction boxes, cabinets, or panels. The *NEC*® also states that all rigid conduit bends must be made so that the conduit is not damaged and the internal diameter of the conduit is not reduced. Additional *NEC*® regulations address the bending radius.

Electricians must know how to create offsets, saddles, kicks, and other bends in conduit without relying on the trial-and-error method, which results in wasted conduit and time. Precision bends can be made the first time by knowing and applying formulas such as those associated with the right triangle.

Note: *National Electrical Code*® and *NEC*® are registered trademarks of the National Fire Protection Association, Inc., Quincy, MA 02269. All *National Electrical Code*® and *NEC*® references in this module refer to the 2005 edition of the *National Electrical Code*®.

Objectives

When you have completed this module, you will be able to do the following:

1. Describe the process of conduit bending using power tools.
2. Identify all parts of popular electric and hydraulic benders.
3. Avoid excessive waste when working with conduit systems.
4. Bend offsets, kicks, saddles, and segmented and parallel bends.
5. Explain the requirements of the *NEC*® for bending conduit.
6. Compute the radius, degrees in bend, developed length, and gain for conduit up to six inches.
7. Explain how to correct damaged conduit and modify existing bends.

Trade Terms

Approximate ram travel
Back-to-back bend
Bending protractor
Bending shot
Concentric bending
Conduit
Degree indicator
Developed length
Elbow
Gain
Inside diameter (ID)
Kicks
Leg length
Ninety-degree bend
Offsets
One-shot shoe
Outside diameter (OD)
Radius
Rise
Segment bend
Segmented bending shoe
Springback
Stub-up
Sweep bend
Take-up (comeback)

Required Trainee Materials

1. Pencil and paper
2. Appropriate personal protective equipment
3. Copy of the latest edition of the *National Electrical Code*®

Prerequisites

Before you begin this module, it is recommended that you successfully complete *Core Curriculum; Electrical Level One; Electrical Level Two,* Modules 26201-05 through 26203-05. You should also read *NEC Articles 342, 344, 352, and 358.*

This course map shows all of the modules in *Electrical Level Two.* The suggested training order begins at the bottom and proceeds up. Skill levels increase as you advance on the course map. The local Training Program Sponsor may adjust the training order.

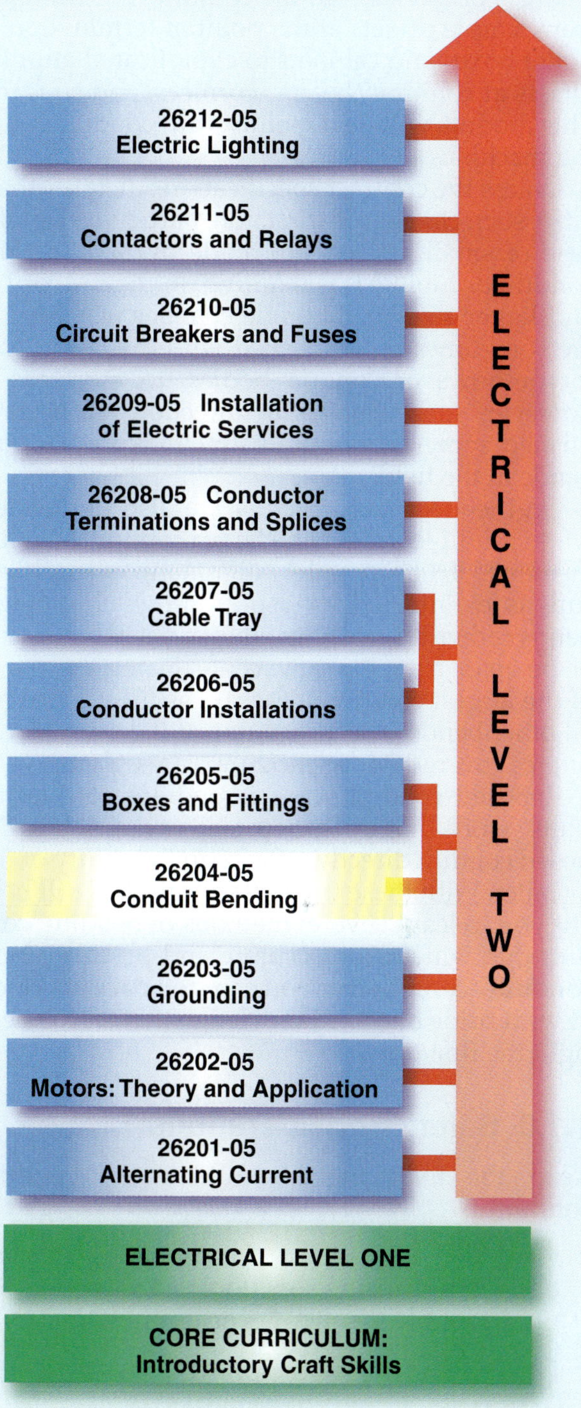

204CMAP.EPS

1.0.0 ◆ INTRODUCTION

The normal installation of intermediate metal **conduit** (IMC), rigid metal conduit (RMC), and electrical metallic tubing (EMT) requires many changes of direction in the conduit runs, ranging from simple **offsets** at the point of termination at outlet boxes and cabinets to complicated angular offsets at columns, beams, cornices, and so forth. This module will deal mainly with conduit bending machines.

Unless the contract specifications dictate otherwise, such changes in direction, particularly in the case of smaller sizes, are made by bending the conduit or tubing as is required. In the case of 1¼" and larger sizes, right angle changes of direction are sometimes accomplished with the use of factory **elbows** or conduit bodies. In most cases, however, such changes in direction are accomplished more economically by making conduit bends in the field.

On-the-job conduit bends are also performed when multiple runs of the larger conduit sizes are installed. Truer parallel alignment of multiple runs is maintained by using on-the-job conduit bends rather than factory elbows. Such bends can all be made from the same center, using the bends of the largest conduit in the run as the pattern for all other bends. This is just one of the useful techniques covered in this module.

Exposed conduit work is one area of electricians' work that puts their skills on display. Exposed conduit directly reflects on the ability of the installer. With these thoughts in mind, it will benefit you to learn several methods of bending conduit that will ensure accurate and precisely-bent conduit—conduit that you can step back and look at with pride and the knowledge that it was bent right the first time.

INSIDE TRACK

Take Pride in Your Work

Even though the conduit in this installation will be hidden by the flooring and the final wall finish, notice how the installing electrician made a special effort to keep the bends perfectly spaced and aligned. Doing a good job even when it might never be seen or admired is the mark of a true professional.

204P0401.EPS

1.1.0 Safety Considerations

Keep the following in mind when bending conduit:

- Read and understand the operating instructions before using any tool. Always use the right tool for the job; never use bending tools for any other purpose.
- Never operate tools with damaged or missing parts.
- Watch for pinch points on power bending tools.
- Use tools in a well-lit, uncluttered area, with enough space for the conduit to move while bending.
- Make sure all hydraulic connections are clean and tight. Replace damaged or worn hoses before using.
- Be prepared for the unexpected.
- Make sure tools are complete and properly assembled before operating.
- Bending shoes and follower bars should be treated as precision instruments. Never toss them into a tool chest or allow them to become damaged or bent out of shape. The quality of the final job will depend on using the proper shoes and follower bars in good condition.
- When working with PVC conduit, make sure adequate ventilation is provided to carry off fumes from joint cement or glue.

2.0.0 ◆ *NEC*® REQUIREMENTS

NEC Section 344.24 requires that all rigid metal conduit bends be made so that the conduit will not be damaged and the internal diameter of the conduit will not be effectively reduced. To accomplish this, the *NEC*® further specifies that the minimum **radius** (*Figure 1*) to the centerline of the conduit shall not be less than that listed in *Table 1*. There is a good reason for this rule. When bends are too tight, pulling becomes extremely difficult and the insulation on the conductors may be damaged.

The *NEC*® requirements for other bends are shown in *Table 2*.

2.1.0 Number of Bends per Run

The *NEC*® specifically states that no more than four quarter bends (360° total) may be made in any one conduit run between outlet boxes, cabinets, panels, or junction boxes; that is, between pull points.

Some electricians believe that offsets, **kicks,** and saddles are not bends, especially in areas where the electrical inspectors are lax. These electricians count only those bends that are actually a quarter circle (90°). The misconception of this is quickly apparent when wires are pulled. Offsets and saddles add just as much resistance to pulling conductors as any 90° elbow. An offset, for example, takes two 45° bends, which equal one 90° bend. A saddle should be counted as two quarter bends or 180°.

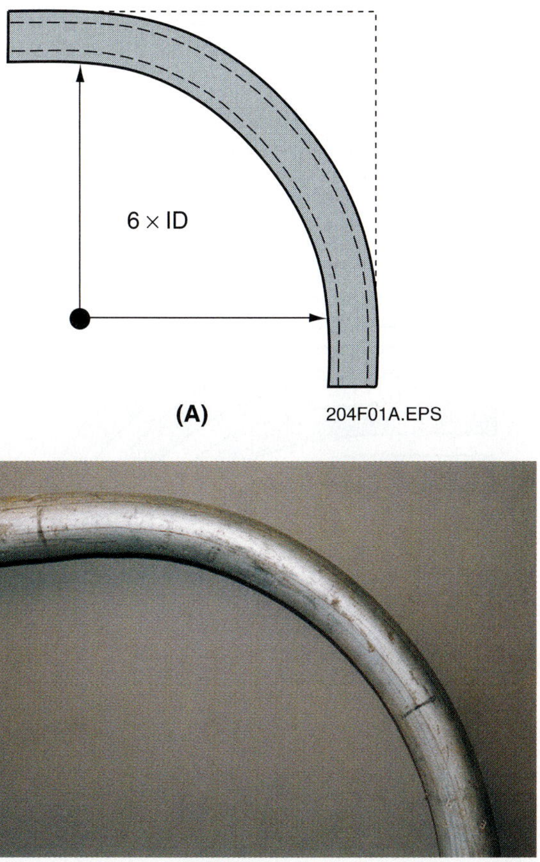

(A) 204F01A.EPS

(B) 204F01B.EPS

Figure 1 ◆ Inside radius requirements.

Table 1 *NEC*® Minimum Requirements for Radius of Conduit Bends – One-Shot and Full-Shoe Benders

Trade Size (Inches)	Radius to Center of Conduit (Inches)
½	4
¾	4½
1	5¾
1¼	7¼
1½	8¼
2	9½
2½	10½
3	13
3½	15
4	16
5	24
6	30

Reprinted with permission from NFPA 70, the *National Electrical Code*®. Copyright © 2004, National Fire Protection Association, Quincy, MA 02269. This reprinted material is not the complete and official position of the National Fire Protection Association on the referenced subject, which is represented only by the standard in its entirety.

Table 2 *NEC*® Minimum Requirements for Radius of Other Conduit Bends

Trade Size (Inches)	Other Bends (Inches)
½	4
¾	5
1	6
1¼	8
1½	10
2	12
2½	15
3	18
3½	21
4	24
5	30
6	36

Reprinted with permission from NFPA 70, the *National Electrical Code*®. Copyright © 2004, National Fire Protection Association, Quincy, MA 02269. This reprinted material is not the complete and official position of the National Fire Protection Association on the referenced subject, which is represented only by the standard in its entirety.

INSIDE TRACK

International Differences

Everywhere except in the United States, steel conduit is measured using the metric system, so it might be helpful to familiarize yourself with common metric conversion calculations, especially if you live near the Canadian or Mexican border. For example, to convert inches to centimeters (cm), multiply inches by 2.54; to convert yards to meters (m), multiply yards by 0.9.

Don't Ignore the Outside Diameter

Although conduit measurements are based on the inside diameter, don't forget that the outside diameter determines the size of the opening that the conduit must penetrate.

A 15° kick in a conduit run may seem insignificant, but after several dozen of these kicks are incorporated into the run, the difficulty of pulling wire becomes apparent. The number of degrees in each kick should be included in the total count, and in no case should the total number (number of bends × number of degrees in each bend) exceed 360°. This is the maximum number of degrees allowed. This means a maximum of twelve 15° kicks and two offsets; twelve kicks and one saddle; etc. Many electricians prefer to install pull boxes at closer intervals to reduce the number of bends, especially when the larger conductor sizes are being pulled. The additional cost of the pull boxes and the labor to install them is often offset by the labor saved in pulling the conductors during the pull.

3.0.0 ◆ TYPES OF BENDS

There are various types of conduit bends: elbows, offset bends, **back-to-back bends,** saddles, etc. A brief review of each follows:

- *Elbow* – An elbow, or ell, is a 90° bend that is used when a conduit must turn at a 90° angle. In single conduit runs when the larger sizes of conduit are being installed, factory elbows are frequently used to save labor on setting up a power bending machine, calculating and marking the conduit for bending, and finally, making the bend. However, in multiple conduit runs, a neater job will result if on-the-job **sweep bends** or concentric bends are properly calculated and installed. See *Figure 2*.

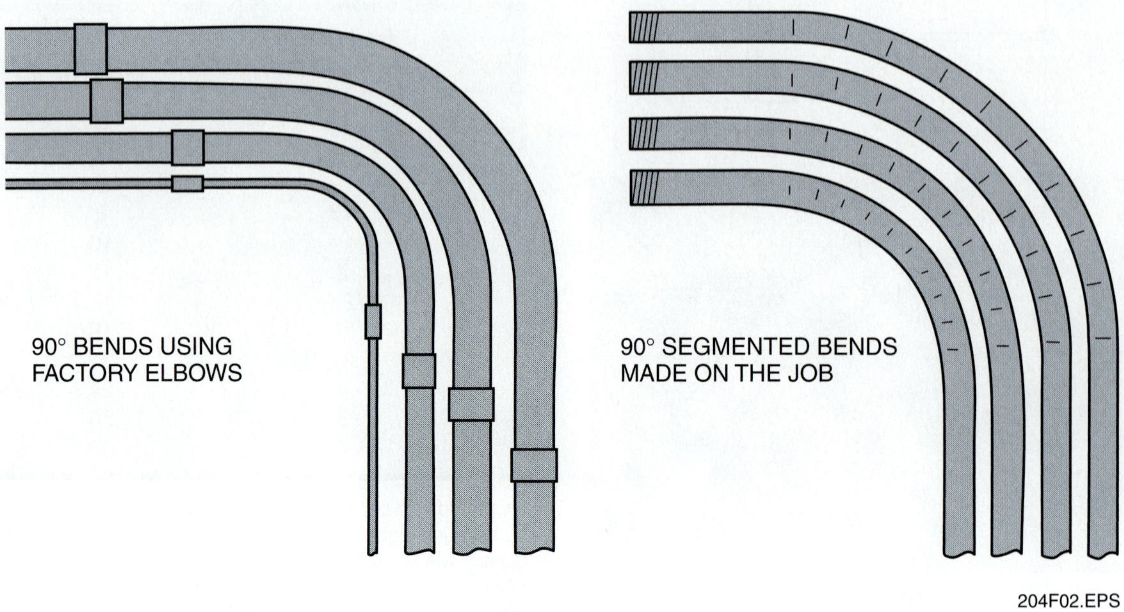

Figure 2 ◆ Typical 90° bends.

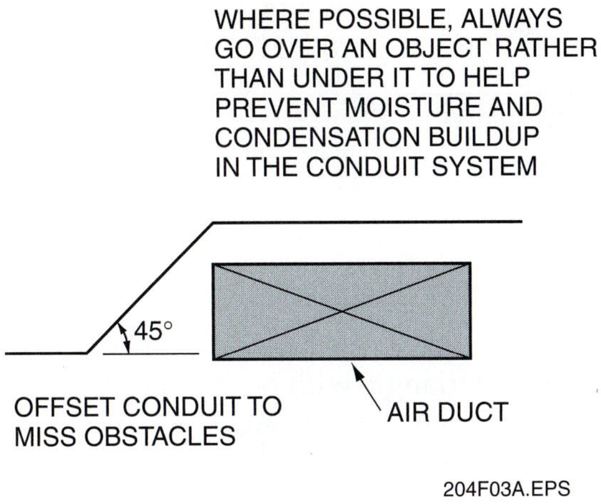

(A)

(B)

Figure 3 ◆ Applications of conduit offsets.

- *Offset* – An offset consists of two equal bends and is used when the conduit run must go over, under, or around an obstacle. An offset is also used at outlet boxes, cabinets, panelboards, and pull boxes. See *Figure 3*.
- *Saddle* – A saddle is used to cross a small obstruction or other runs of conduit. A saddle is made by marking the conduit at a point where the saddle is required and placing a bender a few inches ahead of this point. Bends are made as shown in *Figure 4* in approximately two equal/opposing increments. In most cases, the bends should be as close together as the bender will permit.
- *Kick* – A kick is a change in direction of a conduit run of less than 90°. It is used mostly where the conduit run will be concealed in deck work. The first bend in an offset, for example, is really a kick, as shown in *Figure 5;* another kick in the opposite direction transforms the bend into an offset.

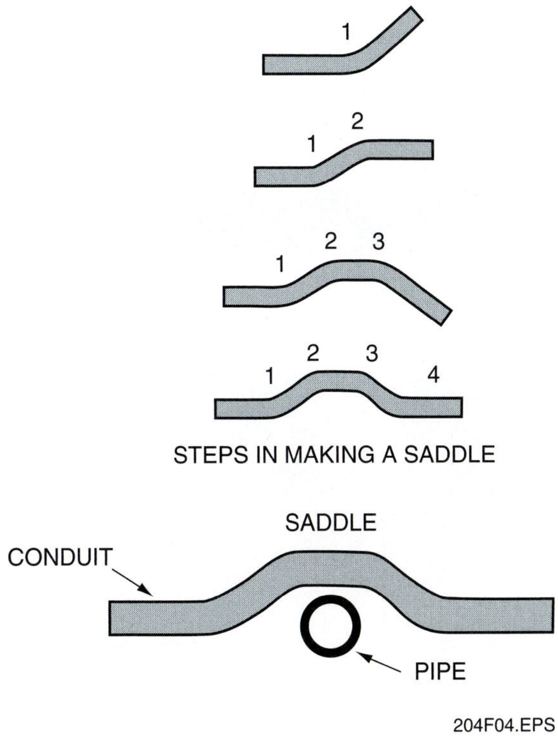

Figure 4 ◆ Practical application of a saddle bend.

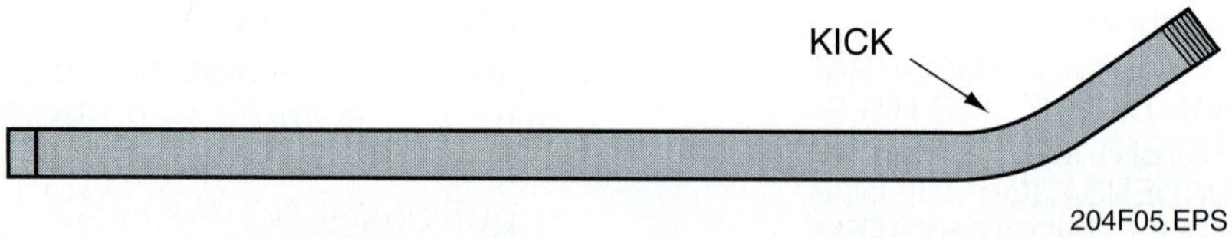

Figure 5 ◆ Kick.

4.0.0 ◆ THE GEOMETRY OF BENDING CONDUIT

Learning to bend conduit involves a thorough knowledge of some basic geometry (the science that deals with the properties of lines, angles, surfaces, and solids). The basic bends discussed previously will handle the majority of the electrician's needs, as these simple bends are merely combined to form more complex bends. When these basic skills are mastered, you will be able to calculate and bend conduit to fit most situations.

The 90° or right angle bend is probably the most basic of all and is used much of the time, regardless of the type of conduit being installed. All other bends are made with angles less than 90°, with the possible exception of short goosenecks used to support a lighting fixture from an outlet box. These goosenecks, however, are seldom seen on modern construction jobs, as there are now a large array of fittings to choose from that obtain the same results in less time and provide for easier conductor installation.

4.1.0 Right Triangle

A right triangle, as shown in *Figure 6(A)*, is defined as any triangle with one 90° angle. The side directly opposite the 90° angle is called the hypotenuse and the side on which the triangle sits is the base. The vertical side is called the height or altitude. For offset bends, right triangle characteristics can also be applied because the offset forms the hypotenuse of a right triangle, as shown in *Figure 6(B)*. There are reference tables available for sizing offset bends based on these characteristics.

4.1.1 Trigonometry Fundamentals

Right triangles are used to develop trigonometric equations (*Figure 7*). These equations can be used to determine the **rise** or **stub-up** distance between bend points, the distance between bend joints, the distance a kick needs to be above the surface, or the amount of additional conduit required. There

INSIDE TRACK

Check It Out

When making complex bends, it is a good idea to test your bends using a piece of wire first, then use the bent wire as a template for your bend. This will give you an idea of whether or not the bend will do the job before wasting expensive conduit.

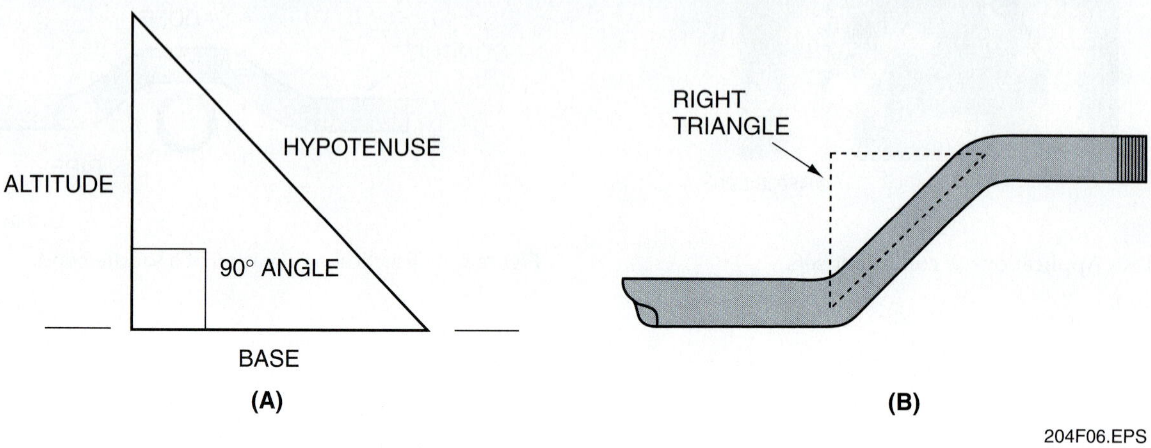

Figure 6 ◆ A right triangle and its relationship to a conduit offset.

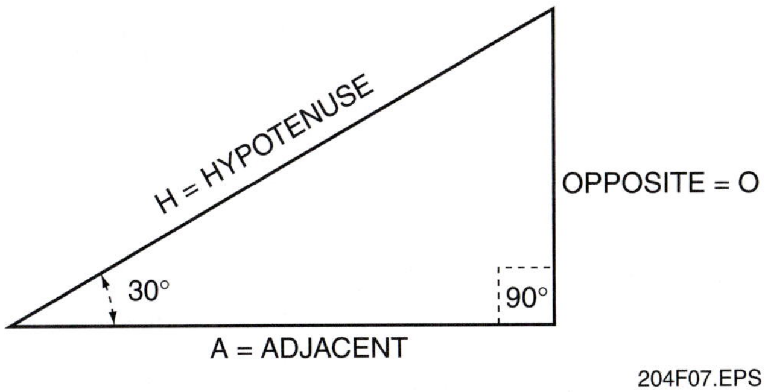

Figure 7 ◆ Trigonometry fundamentals of a right triangle.

are six basic trigonometric functions that will be required:

- Sine
- Cosine
- Tangent
- Cotangent
- Secant
- Cosecant

The sine function can be computed by dividing the hypotenuse by the side opposite the angle being considered. This function can be represented by the following equation:

$$\text{Sine} = \frac{\text{opposite}}{\text{hypotenuse}} = \frac{O}{H}$$

The equations for the other functions are as follows:

$$\text{Cosine} = \frac{\text{adjacent}}{\text{hypotenuse}} = \frac{A}{H}$$

$$\text{Tangent} = \frac{\text{opposite}}{\text{adjacent}} = \frac{O}{A}$$

$$\text{Cotangent} = \frac{\text{adjacent}}{\text{opposite}} = \frac{A}{O}$$

$$\text{Secant} = \frac{\text{hypotenuse}}{\text{adjacent}} = \frac{H}{A}$$

$$\text{Cosecant} = \frac{\text{hypotenuse}}{\text{opposite}} = \frac{H}{O}$$

Where:

H = hypotenuse, side facing the right (90°) angle
O = side opposite the angle you are working with
A = side adjacent to the angle you are working with, but not the hypotenuse

Therefore, in *Figure 6(B)*, the hypotenuse or distance between bends could be computed by using the following cosecant trigonometric function:

$$\text{Cosecant} = \frac{\text{hypotenuse}}{\text{opposite}} = \frac{H}{O}$$

Let's say that the angle of the first bend is 30° and the rise or side opposite is 12". Therefore:

$$\text{Cosecant } 30° = \frac{H}{12"}$$

The cosecant of 30° is 2 (check using your calculator).

$$2 = \frac{H}{12"}$$

Cross multiply:

$$H = 2 \times 12"$$
$$H = 24"$$

Therefore, the hypotenuse or distance between bends would be 24".

Now let us complete another example. In this case, a kick is to be made on the end of a piece of conduit (*Figure 8*). The angle of the kick is to be 30°. The hypotenuse of the kick is 20". Determine how far off the surface the end of the kick needs to be.

When calculating the hypotenuse of a bend, the cosecant of the angle is multiplied by the side opposite. If the angle and the hypotenuse are known, then the inverse can be used to determine the side opposite. Therefore, divide 20" by the cosecant of 30°, or:

$$20" \div 2 = 10"$$

The end of the conduit needs to be brought 10" off the surface to acquire a 30° bend. This process eliminates the need for a protractor level.

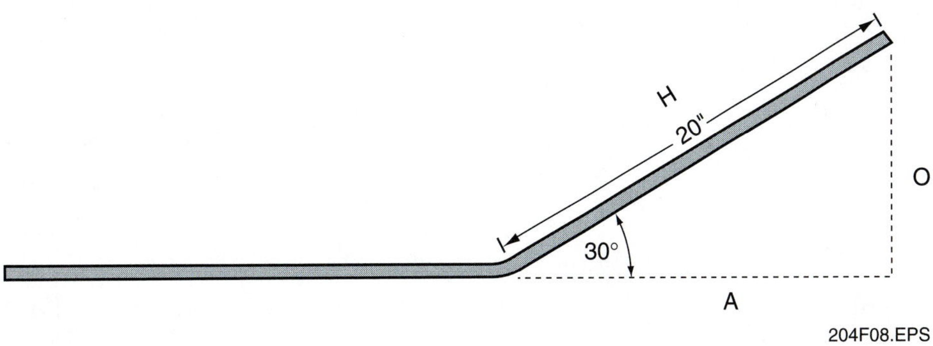

Figure 8 ◆ Kick example.

4.2.0 Circle

A circle is defined as a closed curved line whose points are all the same distance from its center, as shown in *Figure 9(A)*. The distance from the center point to the edge of the circle is called the radius and the length of a straight line from one edge through the center to the other edge is called the diameter. The distance around the circle is called the circumference. A circle can be divided into four equal quadrants, as shown in *Figure 9(B)*. Each quadrant accounts for 90°, making a total of 360°. When making 90° bends, we will be interested in ¼ of a circle, or one quadrant.

Concentric circles, shown in *Figure 9(C)*, are several circles that have a common center but different radii. The concept of concentric circles can be applied to concentric 90° bends in conduit. Such bends have the same center point, but the radius of each is different. *Figure 10* shows how parts of a circle relate to a 90° conduit bend.

For bending conduit, it is necessary to understand the dynamics of the unit circle (*Figure 11*). The unit circle is a circle with a given radius of 1.

In calculating the circumference of a circle, the following formula is used:

$$C = 2\pi R$$

Where:

C = circumference
π = 3.14 (pi)
R = radius

By taking the radius of the unit circle and substituting it into the formula, the result is:

$$C = 2\pi \times 1$$

Any number multiplied by the number 1 is equal to that number, or:

$$1 \times 2 = 2$$

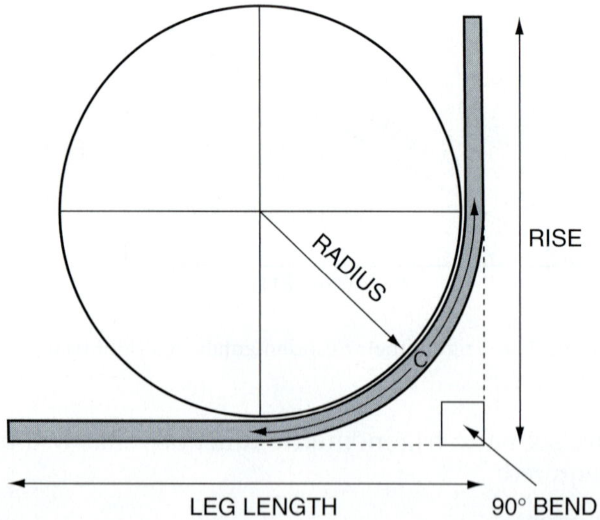

Figure 10 ◆ Parts of a circle related to conduit bending.

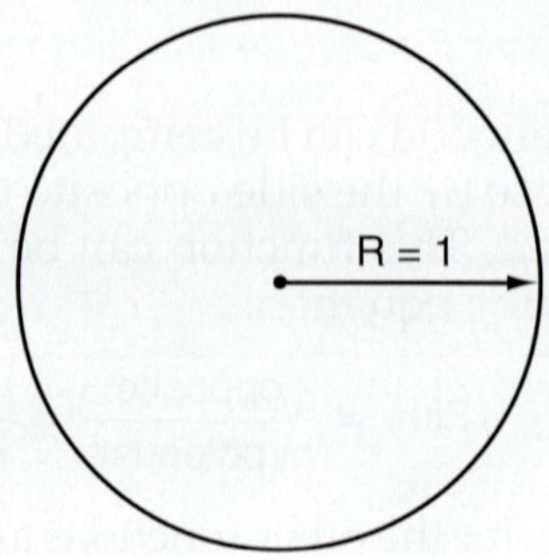

Figure 11 ◆ Unit circle.

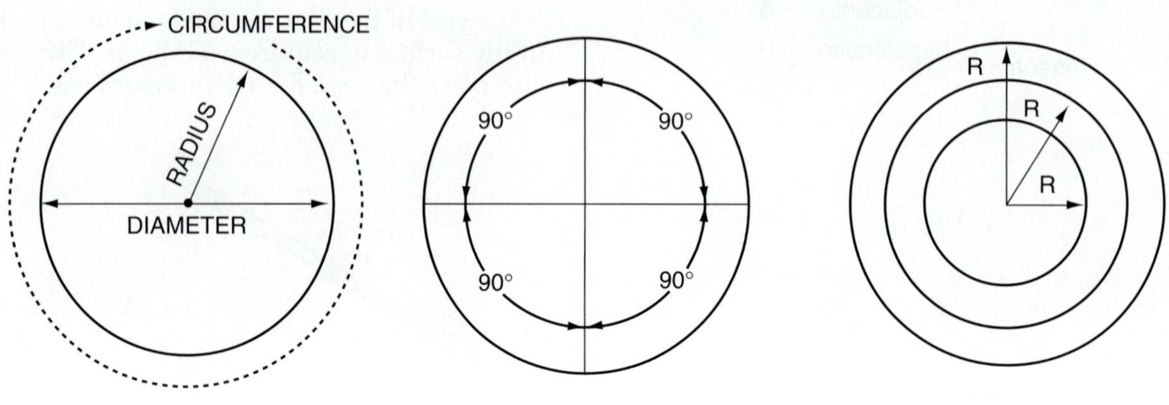

Figure 9 ◆ Characteristics of a circle.

Therefore, $2\pi \times 1 = 2\pi$. This means that in terms of pi, the circumference of the circle is 2π; 360° is 2π and 180° = π. See *Figure 12*.

If we look at 90° in terms of π, 90° is half of 180° or half of π, so 90° is equal to ½ π or $\pi \div 2$. Again, π is a symbol for the numerical value 3.14 (rounded). $\pi \div 2$ is $3.14 \div 2$ or 1.57. Therefore, 90° is represented by the numerical value 1.57. Looking back at the circumference formula tells us that multiplying $2 \times \pi$ or 6.28 times the given radius of the circle will give us the linear distance around the circle. With that in mind, if we multiply 1.57 times the given radius, it will give us the linear distance from 0° to 90° on the circle. This linear distance is known as the **developed length** in regard to the amount of conduit that is required to make a 90° sweep.

4.2.1 Types of Angular Representations

There are three basic ways that angles can be represented or specified. Angles can be stated in degrees, as already mentioned; radians; or gradients.

Angles are measured in degrees from 0° to 360°. They may also be measured in radians from 0 to 6.28 or 0 to 2π. The third way is to specify angles in gradients, from 0 to 400 gradients (grads). These measurements can be equated as follows:

$$90° = \pi \div 2 = 1.57 \text{ radians} = 100 \text{ grads}$$

$$180° = \pi = 3.14 \text{ radians} = 200 \text{ grads}$$

$$270° = 3\pi \div 2 = 4.71 \text{ radians} = 300 \text{ grads}$$

$$360° = 2\pi = 6.28 \text{ radians} = 400 \text{ grads}$$

4.3.0 Equations

To calculate conduit bends accurately, you need to use some basic equations. The equations given here are the most relevant ones for the electrician. Examples are provided for clarification. To calculate the circumference of a circle, use the following equation:

$$C = \pi D$$

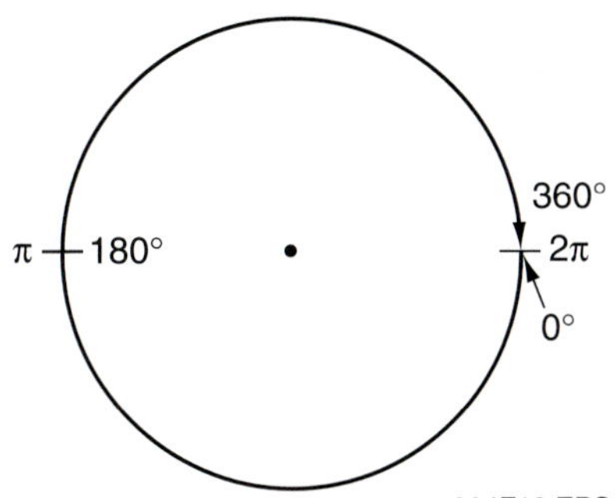

Figure 12 ◆ π and 2π.

Where:

C = circumference

D = diameter

π = 3.14

As discussed earlier, another equation for finding the circumference of a circle is as follows:

$$C = 2\pi R$$

Where:

C = circumference

R = radius (½ the diameter)

π = 3.14

If the radius in a circle (and conduit bend) measures two feet, the circumference of the circle may be found as follows:

$$C = 2\pi R$$

$$C = 2 \times 3.14 \times 2' = 12.56'$$

The arc of a quadrant equals ¼ the circumference of the circle, and its length is found as follows:

$$\text{Length of arc} = (0.25)(2\pi R)$$

$$\text{Length of arc} = (0.25)(2 \times 3.14 \times R)$$

$$\text{Length of arc} = 1.57R$$

Therefore, the length of the arc in the quadrant, if the radius is 2', may be found as follows:

$$\text{Length of arc} = 1.57R$$

$$\text{Length of arc} = 1.57 \times 2' = 3.14'$$

If the radius of the bend and the outside diameter of the conduit are known, then the distance C (the length of the arc) can be calculated. This length, as related to conduit bending, is known as the developed length; it is the actual length of the bend.

Conduit bends with the circumference of the circle (*Figure 13*) and not at right angles. Therefore, the length of the conduit needed for a bend will not equal the right angle distances A and B. **Gain** is the difference between the right angle distances A and B and the shorter distance C, the length of conduit actually needed for the bend.

The gain for a 90° bend is arrived at by multiplying the radius of the bend × 0.43. Therefore, if the radius of the bend in *Figure 13* is 2', the gain may be found as follows:

$$\text{Gain} = 2' \times 0.43 = 0.86' = 10.32''$$

4.3.1 Making a 90° Bend

The 90° stub bend is probably the most basic bend of all. The stub bend is used much of the time, regardless of the type of conduit being installed.

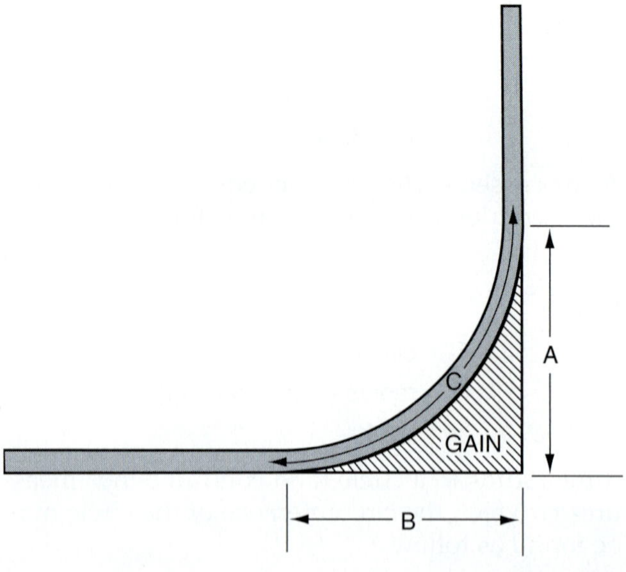

Figure 13 ◆ Gain.

Before beginning to make the bend, you need to know two measurements:

- The desired rise or stub-up
- The **take-up (comeback)** distance of the bender

The desired rise is the height of the stub-up. The take-up is the amount of conduit the bender will use to form the bend. Take-up distances are usually listed in the manufacturer's instruction manual. Once the take-up has been determined, subtract it from the stub-up height. Mark that distance on the conduit (all the way around) at that distance from the end. The mark will indicate the point at which you will begin to bend the conduit. Line up the starting point on the conduit with the starting point on the bender. Most benders have a mark, like an arrow, to indicate the starting point.

NOTE

When bending conduit using the take-up method, the bender is always placed on the conduit and the bend is made facing the end of the conduit from which the measurements were taken.

As discussed above, if the radius of the bend in *Figure 13* is 2', the gain is 10.32".

Table 3 shows the equivalent fractions for many decimals. For 10.32", search the decimal column for 0.32. The closest decimal in the table is 0.3125, equivalent to $\frac{5}{16}$. Therefore, the number in question becomes 10$\frac{5}{16}$".

A decimal may also be mathematically converted to a fraction. A decimal whose denominator is contained in the numerator without a remainder can easily be converted to a fraction by removing the decimal point from the numeral, which then becomes the numerator (the top numeral of the fraction). The denominator is always one plus as many zeros as there are decimal places in the decimal. For example:

$$0.75 = \frac{75}{100} = \frac{3}{4}$$

or:

$$0.375 = \frac{375}{1000} = \frac{3}{8}$$

NOTE

Decimals may also be converted to fractions by using an electronic calculator, provided the calculator has a fraction key. The exact procedure will vary with the different models of calculators.

Gain factors for 0° to 90° bends are shown in *Table 4*. To demonstrate the use of this table, assume that it is desired to find the gain on a 45° conduit bend with a 15" centerline radius. Referring to *Table 4*, look in the left-hand column; glance down the column until the number 40° is found. Since the bend is 45°, read to the right in this row until the column titled 5° is found; note the figure, 0.043. Therefore, the gain factor for a 45° bend is 0.043.

Multiply the gain factor by the centerline radius (15") to obtain the full gain of a 45° bend.

$$0.043 \times 15" = 0.645"$$

To convert this figure to a readable figure on the foot rule, convert the decimal to a common fraction: $\frac{645}{1000}$ = approximately $\frac{5}{8}$. Thus, the full gain of the 45° bend is $\frac{5}{8}$".

For example, the gain for a given bender is 2½". In *Figure 14*, there are two back-to-back 90° bends.

Figure 14 ◆ Back-to-back 90° bends.

Table 3 Decimal Equivalents of Some Common Fractions

Fraction	Decimal	MM	Fraction	Decimal	MM
1/64	0.015625	0.397	33/64	0.515625	13.097
1/32	0.03125	0.794	17/32	0.53125	13.494
3/64	0.046875	1.191	35/64	0.546875	13.891
1/16	0.0625	1.588	9/16	0.5625	14.288
5/64	0.078125	1.984	37/64	0.578125	14.684
3/32	0.09375	2.381	19/32	0.59375	15.081
7/64	0.109375	2.778	39/64	0.609375	15.478
1/8	0.125	3.175	5/8	0.625	15.875
9/64	0.140625	3.572	41/64	0.640625	16.272
5/32	0.15625	3.969	21/32	0.65625	16.669
11/64	0.171875	4.366	43/64	0.671875	17.066
3/16	0.1875	4.763	11/16	0.6875	17.463
13/64	0.203125	5.159	45/64	0.703125	17.859
7/32	0.21875	5.556	23/32	0.71875	18.256
15/64	0.234375	5.953	47/64	0.734375	18.653
1/4	0.25	6.35	3/4	0.75	19.05
17/64	0.265625	6.747	49/64	0.765625	19.447
9/32	0.28125	7.144	25/32	0.78125	19.844
19/64	0.296875	7.54	51/64	0.796875	20.241
5/16	0.3125	7.938	13/16	0.8125	20.638
21/64	0.32812	8.334	53/64	0.828125	21.034
11/32	0.34375	8.731	27/32	0.84375	21.431
23/64	0.359375	9.128	55/64	0.859375	21.828
3/8	0.375	9.525	7/8	0.875	22.225
25/64	0.390625	9.922	57/64	0.890625	22.622
13/32	0.40625	10.319	29/32	0.90625	23.019
27/64	0.421875	10.716	59/64	0.921875	23.416
7/16	0.4375	11.113	15/16	0.9375	23.813
29/64	0.453125	11.509	61/64	0.953125	24.209
15/32	0.46875	11.906	31/32	0.96875	24.606
31/64	0.484375	12.303	63/64	0.984375	25.003
1/2	0.5	12.7	1	1	25.400

Table 4 Gain Factors

	—	1°	2°	3°	4°	5°	6°	7°	8°	9°
0°	0	0	0	0	0	0	0.0001	0.0001	0.0003	0.0003
10°	0.0005	0.0006	0.0008	0.001	0.0013	0.0015	0.0018	0.0022	0.0026	0.0031
20°	0.0036	0.0042	0.0048	0.0055	0.0062	0.0071	0.0079	0.009	0.01	0.0111
30°	0.0126	0.0136	0.015	0.0165	0.0181	0.0197	0.0215	0.0234	0.0254	0.0276
40°	0.0298	0.0322	0.0347	0.0373	0.04	0.043	0.0461	0.0493	0.0527	0.0562
50°	0.06	0.0637	0.0679	0.0721	0.0766	0.0812	0.086	0.0911	0.0963	0.1018
60°	0.1075	0.1134	0.1196	0.126	0.1327	0.1397	0.1469	0.1544	0.1622	0.1703
70°	0.1787	0.1874	0.1964	0.2058	0.2156	0.2257	0.2361	0.247	0.2582	0.2699
80°	0.2819	0.2944	0.3074	0.3208	0.3347	0.3491	0.364	0.3795	0.3955	0.4121
90°	0.4292	—	—	—	—	—	—	—	—	—

The rise for the first 90° bend is 2' and the rise for the second 90° bend is 3'. The back-to-back measurement for these bends is 4'. The task is to determine the amount of straight conduit required to accomplish this task. The conduit is to be cut to length and threaded before bending.

The gain is the amount of conduit that is saved by the radius of the bend. The amount of conduit saved is 2½" per 90° bend. In this situation, there are two 90° bends, or a savings of 5". The total straight lengths add up to 9' of conduit. Therefore:

9' – 5" = 8'-7"

5.0.0 ◆ MECHANICAL BENDERS

Conduit bends are normally made in the smaller sizes of conduit and tubing by hand with the use of hickeys or EMT bending tools. However, on many projects, an advantage can be gained by the use of mechanical bending equipment with suitable adjustable stops and guides. With the use of such equipment, the exact bend can be duplicated in quantity with a minimum of effort. The angle of the bend and the location of the bend in relation to the end of the length of conduit are preset.

A popular mechanical bender is shown in *Figure 15*. This type of bender was originally called the Chicago bender, as it was made by the Chicago Equipment and Manufacturing Company. Today, however, this type of bender is manufactured by several different companies and the correct name is portable mechanical conduit bender. In any event, you may still hear the term Chicago bender on many jobs.

This type of mechanical bender is very popular with many electrical contractors. To use it, a length of conduit is placed in position and secured in place; a long bending handle is then pulled around and the bend completed. This type of bender may be used as a one-shot bender for the smaller sizes of conduit (bypassing the ratchet mechanism). The ratchet mechanism, however, is usually activated when bending the larger sizes of conduit to make the work easier. It is suitable for making bends in conduit sizes up to 2" EMT, 1¼" IMC, and 1½" rigid conduit, provided the proper bending accessories are used (e.g., bending shoes, follower bars, etc.).

Bending shoes and follower bars are designed to form a particular radius bend for a certain type and size of conduit, that is, EMT, IMC, or rigid conduit. These accessories should be treated as precision instruments; any damage will result in inaccurate bends, kinks, and so on. The first consideration is to use only the proper shoe and follower bar for the type and size of conduit being bent. For example, never use an EMT shoe for bending rigid conduit or a 3" shoe for bending 2½" conduit. In general, make certain that the bending shoes and follower bar are compatible with the type of conduit to be bent; to do otherwise may damage the tool and result in inaccurate bends.

The ratchet feature is normally engaged for the larger sizes of conduit, whereas a spring-loaded pawl engages the ratchet for easier bending in segments. For the smaller sizes of conduit, however, the ratchet may be bypassed so that the bend can be made in one shot.

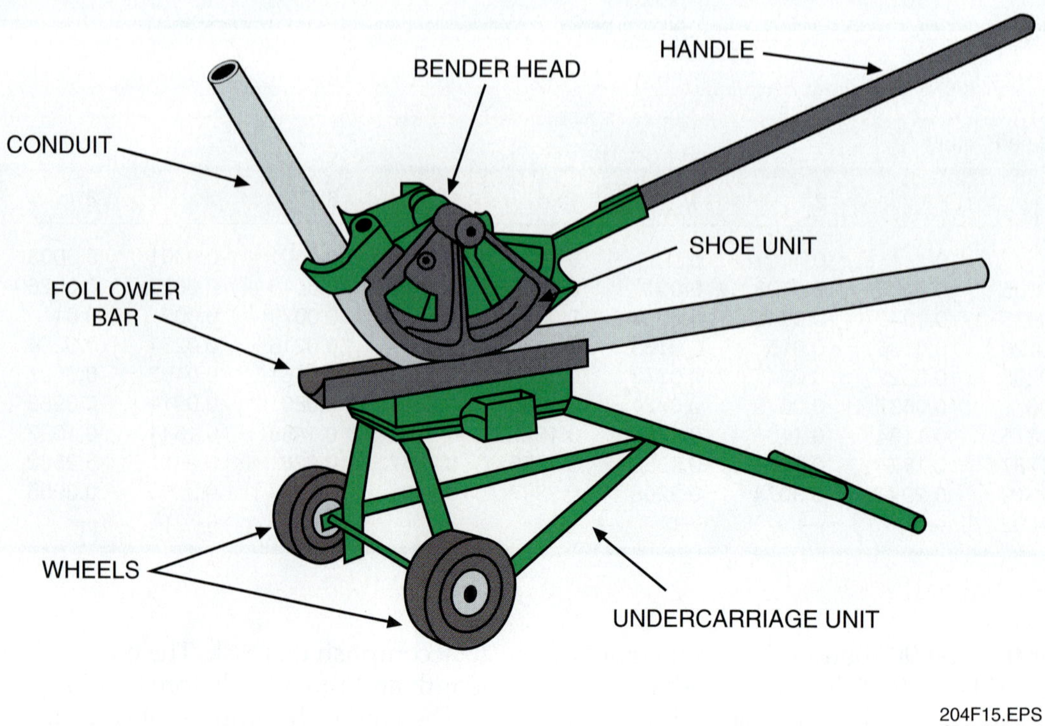

Figure 15 ◆ Typical mechanical bender.

A bending gauge with an adjustable pointer on the bender is helpful when making multiple bends at the same angle. This pointer is set at the desired angle and then the setscrew is tightened. As the bend is being made, the handle is operated until the pointer reaches the index mark. To ensure the correct angle of bend, the first bend should be checked with a **bending protractor** (*Figure 16*) and any necessary adjustments made to the bending gauge pointer before continuing. All successive bends will be exactly the same as the first.

For example, follow this process for charting a bender. This chart will contain minimum size, gain, and centerline distances from the arrow on the bender to the center of the bend for 15°, 30°, 45°, and 60° bends. This is to be accomplished with one scrap piece of conduit.

For this discussion, the scrap piece will be ½" conduit and will be 3' in length (*Figure 17*).

Place a mark on the conduit at a given distance from the end of the conduit. For this discussion, the mark will be 10" in from the end of the conduit (*Figure 18*).

Take a ½" conduit bender and place the arrow, which represents the take-up or minimum rise of the bender, on the mark 10" in from the end of the bender.

Place a protractor level on the conduit in front of the bender. This will be on the piece of conduit marked for 10" back from the end. Bend the 10" portion until the protractor level reads 15°. Remove the bender from the conduit (*Figure 19*).

Now take a straightedge, such as a ruler or torpedo level, and lay it on the inside of the bend so the straightedge lies across the bend

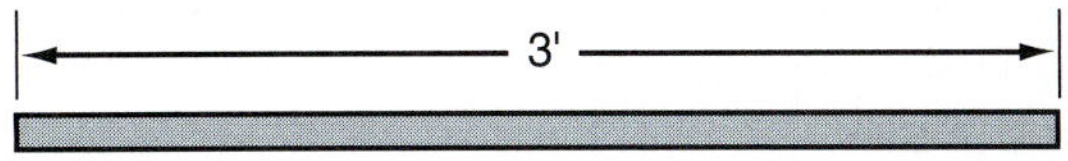

Figure 17 ◆ Conduit.

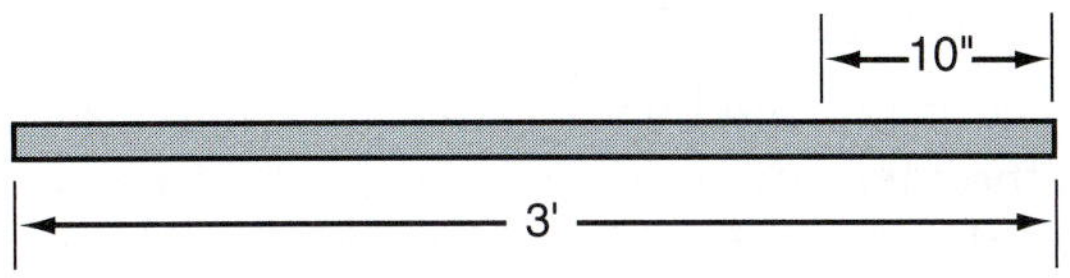

Figure 18 ◆ Conduit with 10" mark.

Figure 19 ◆ Kick of 15°.

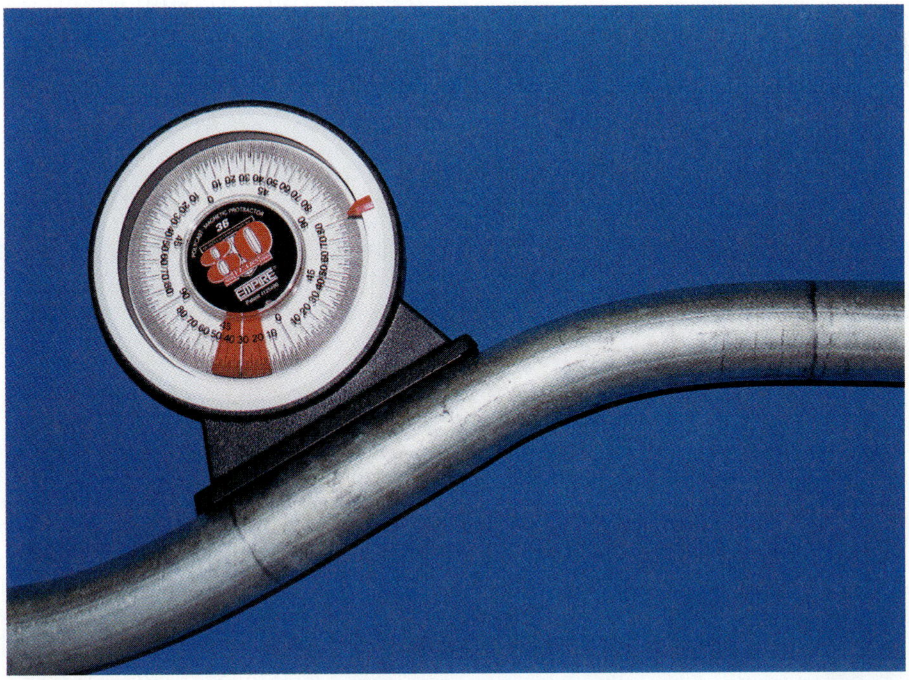

Figure 16 ◆ Bending protractor.

and rests against the straight portion of the conduit (*Figure 20*).

With a sharp pencil, scribe a line across the bend of the conduit.

Now take the straightedge and lay it across the conduit so the straightedge is against the side adjacent to the previous position. It should extend across the bend once again (*Figure 21*).

Now scribe a line across the bend of the conduit. The two pencil lines should cross to form an X on the conduit. This represents the center of the bend (*Figure 22*).

Now take a tape measure and measure from the 10" mark to the centerline of the bend. Record this distance as follows:

15° – 1" (assuming 1" is the measured distance)

Now place the bender back on the conduit so that the take-up arrow is on the 10" mark on the conduit. Place the protractor level back on the conduit. Bend the conduit until the protractor level reads 30°. Take the bender off the conduit and repeat the line crossing process that was discussed for 15°, using the same straightedge. Once again, measure from the 10" mark to the point where the pencil lines cross in the center of the bend. Record this measurement.

15° – 1"

30° – 1½" (assuming this is the measured distance)

Repeat the previous process for 45°. Record this measurement.

15° – 1"

30° – 1½"

45° – 2" (assuming this is the measured distance)

Repeat the previous process for 60°. Record this measurement.

15° – 1"

30° – 1½"

45° – 2"

60° – 2¼" (assuming this is the measured distance)

The centerline distances for the different bends from the take-up mark of the bend have now been recorded. This portion of the chart is now completed.

The next step is to determine the minimum rise of the bender. Place the bender back on the conduit so the take-up mark of the bender is on the 10" mark on the conduit. Place the protractor level back on the conduit and bend the conduit until the protractor level reads 90°. Take the bender off the conduit. Measure from the back of the conduit to the 10" mark. The reading on the tape is 5". This is the minimum 90° stub length (*Figure 23*). Record this measurement.

15° – 1"

30° – 1½"

45° – 2"

60° – 2½"

Minimum rise – 5" (assuming this is the measured distance)

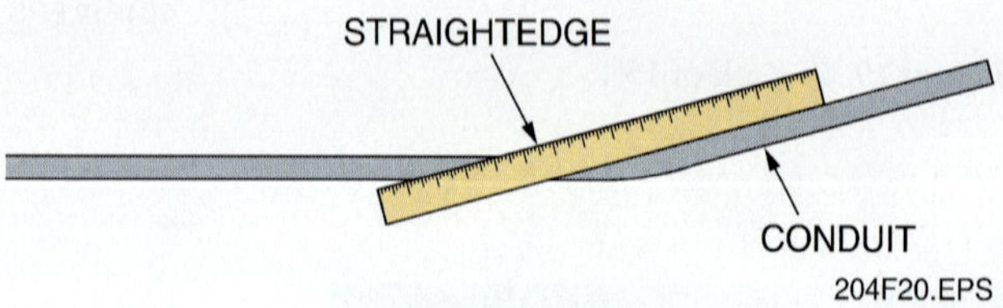

Figure 20 ◆ Conduit and straightedge.

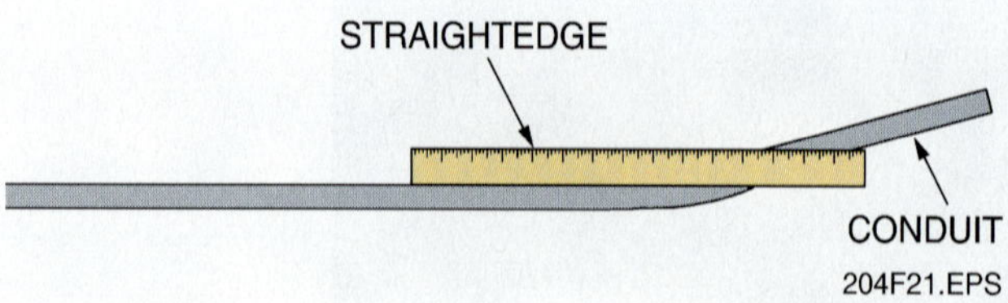

Figure 21 ◆ Conduit and horizontal straightedge.

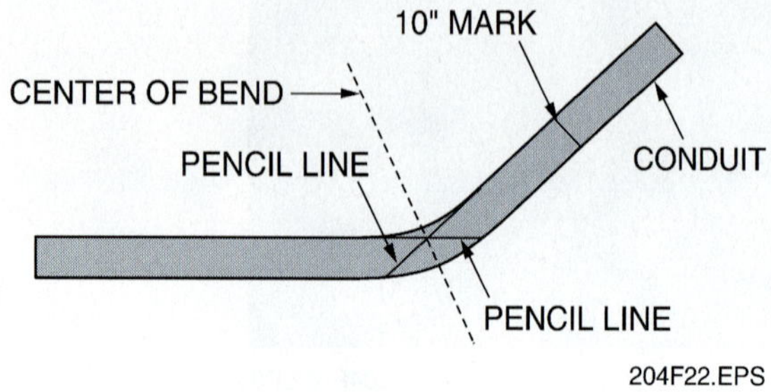

Figure 22 ◆ Center of bend.

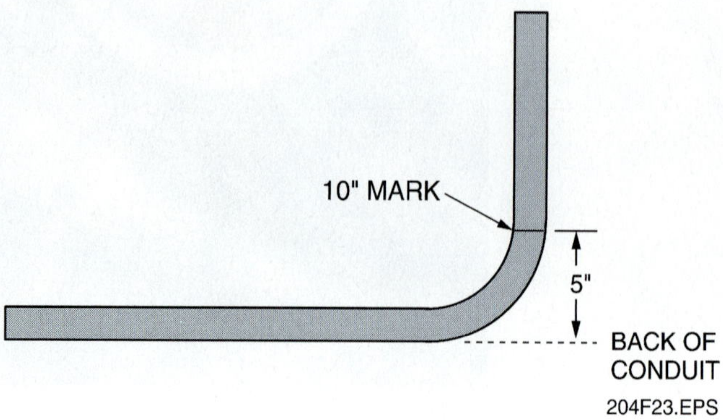

Figure 23 ◆ 90° stub-up.

The next step is to determine the gain of the bender for 90°. Measure the length of both sides of the scrap piece of conduit (*Figure 24*).

Add the two measured stub lengths together.

15" + 1'-11⅝" = 3'-2⅝"

Subtract the original length of the conduit, which was 3', from 3'-2⅝".

3'-2⅝" – 3' = 2⅝"

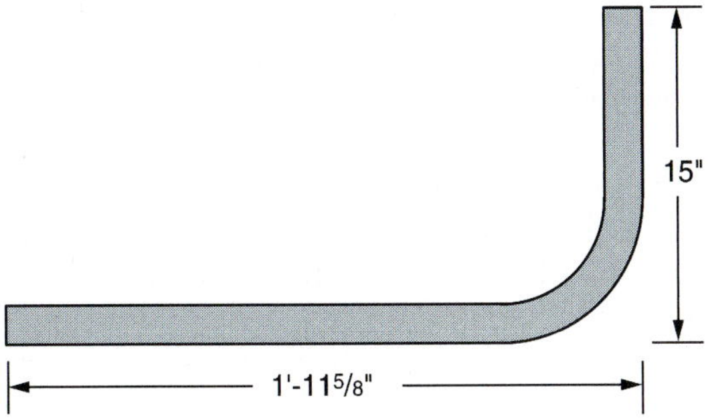

204F24.EPS

Figure 24 ◆ 90° elbow.

This is the gain for this particular ½" conduit bender. Record this information.

15° – 1"

30° – 1½"

45° – 2"

60° – 2½"

Minimum rise – 5"

Gain – 2⅝" (assuming this is the measured distance)

The ½" bender has now been charted. This same process would need to be repeated for a ¾" bender, 1" bender, 1¼" bender, 1½" bender, etc.

This charting process should help you to understand how the manufacturer of a bender comes up with the marks that determine the ability to bend exact 90° bends of any length, which is the minimum rise mark. The star mark seen on many benders is used for back-to-back bends. The star mark is 2⅝" back from the minimum rise mark (take-up). In other words, it is the measured gain distance back from the minimum rise mark. The

INSIDE TRACK

Hickeys and Hand Benders

Hickeys and hand benders are not interchangeable. Hickeys do not support the conduit walls the way hand (shoe) benders do, so they are typically used for making segmented bends rather than one-shot 90° bends. Making too large of a bend at one time with a hickey may cause the conduit to kink.

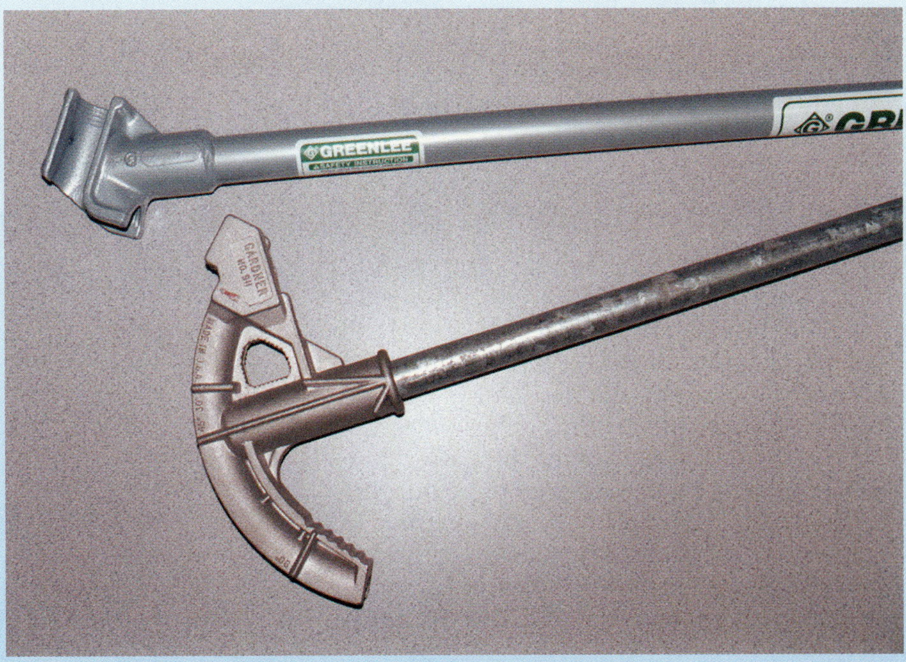

204P0402.EPS

centerline marks are used in lining up the centers of bends for various sizes of conduit. The centerline marks can also be used in lining up saddles on the centerlines of I-beams and process piping. This information is good for charting any type of wrap-around bender.

6.0.0 ◆ MECHANICAL STUB-UPS

Stub-ups are quickly and easily made with mechanical benders. A deduct decal is provided on many benders, but sometimes these decals become damaged, making them difficult to read. Therefore, backup charts should be provided on all jobs. With this deduct chart on hand, the following is an example of making a 90° stub-up to a given height.

Assume that you are working on a deck job and need a number of 1" rigid stub-ups with a rise of 15" each. When you check the deduct chart on the bender for a 15" stub-up using 1" rigid conduit, you note that 11" should be deducted from the total rise of 15". Since 15" − 11" = 4", measure back from the end of the conduit by 4" and make a mark. Encircle the entire conduit at this point so you will not lose the mark once the conduit is placed in the bender. Many electricians like to use a black felt-tip marker for marking conduit.

Load the conduit into the bender with the mark lined up with the front of the bender hook. Engage the ratchet and start pumping the bender handle until the bender pointer reaches the preset index mark for 90°. Move the bender handle forward, then remove the conduit from the bender and check its height. It should be exactly 15". If the height of the bend is slightly off, make the necessary adjustments before continuing. Once the correct height is reached, the remaining bends will also be correct.

7.0.0 ◆ MECHANICAL OFFSETS

The decal chart on the bender that provides deduct information for the stub-ups also contains data for making offsets that require 20°, 30°, and 45° bends. This chart is necessary to make perfect offsets every time using the mechanical bender.

Assume that you are running a raceway system with ½" rigid conduit and an air duct must be bypassed, requiring the conduit run to be offset. After taking measurements on the job, we find that an offset of 12" is needed to clear the air duct.

Measure the distance from the end of the conduit to the start of the first bend; mark the conduit as before. Referring to the chart on the bender with offset information, we decide to make the offset with two 45° bends. The chart indicates that the distance between bends is 16 5/16". Therefore, measure and mark this distance back from the first mark.

Insert the conduit into the bender and line up the first mark with the front of the bender hook. The ratchet may be used, but for ½" conduit, the ratchet override on the front of the bender shoe is normally employed. Make the first 45° bend. Move the bender handle forward to release the conduit. Now slide the conduit forward through the bender hook until the second mark lines up with the front of the bender hook, and then turn the bend over so the end of the conduit is pointing downward toward the deck. Also, make sure that the first bend lines up with the next bend to be made to prevent a crooked bend (commonly referred to as a dog leg) in the conduit. Once everything is aligned, engage the bender handle and make another 45° bend. The height of the offset should be exactly 12" (see *Figure 25*).

The distances between marks for offset bends will vary depending upon the size and type of conduit being bent, so always check the offset information on the bender.

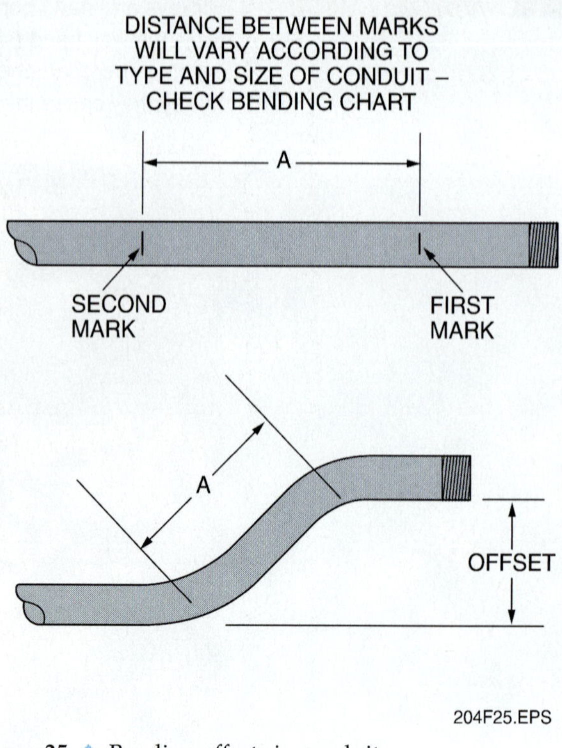

Figure 25 ◆ Bending offsets in conduit.

INSIDE TRACK

Making Accurate Offsets

Be sure to use the same angle for both bends of an offset. To avoid dog legs, make sure your first bend lines up evenly with the rest of the conduit, the handle, and the bender. Take your time, and be sure of proper alignment before making the second bend. A No-Dog® is a simple, pocket-sized device that may be used to prevent crooked bends. It is screwed onto the end of the conduit and has a built-in level to ensure straight bends.

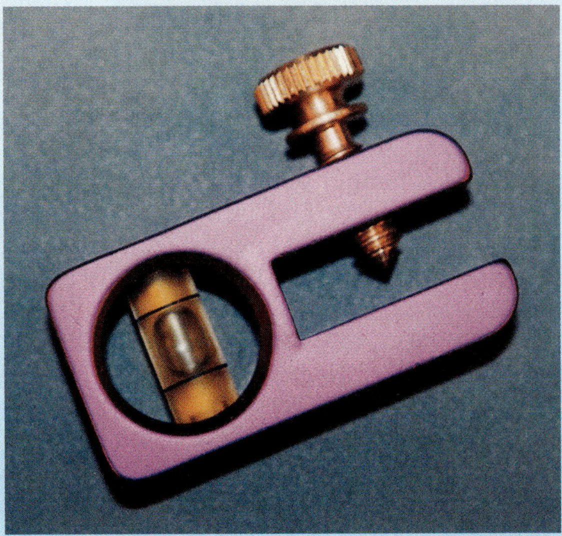

204P0403.EPS

7.1.0 General Tips for Bending with the Chicago Bender

The following are some general tips to keep in mind when bending with the portable mechanical conduit bender:

- An engineer's rule marked in hundredths will simplify formula bending by eliminating the need to convert to fractions.
- When minimum length stubs are being bent, the shoe tends to creep and deforms the end of the conduit and threads. Screwing a coupling onto the pipe stops the shoe from creeping forward and protects the threads.
- When bending offsets, the front of the bender can be temporarily elevated for clearance requirements.
- Most bender shoes are made of cast aluminum and are easily pitted and gouged if foreign material gets between the shoe and the pipe. For longer shoe life, keep the pipe clean and the shoes wiped down.
- When the remaining pipe length is not long enough to reach the roller or pipe support, a larger diameter conduit can be slid over the pipe being bent to complete the bend; or, if the pipe has threads, screw on a coupling and a short piece of scrap pipe.
- Segment and **concentric bending** of smaller sizes of pipe can be performed with this bender. Bend a scrap piece of pipe and measure from the center of the bend to the front of the bending shoe. Use this measurement to adjust the start mark using the segment and concentric bending procedures.

NOTE

For matching bends in sizes ½", ¾", and 1" rigid conduit and IMC, bend all pipes with the 1" shoe.

8.0.0 ◆ ELECTRIC AND HYDRAULIC CONDUIT BENDERS

Bends in larger sizes of conduit (over 2" EMT, 1¼" IMC, or 1½" rigid conduit) are normally made using hydraulic benders. In some instances, bending tables have been developed for use with bending tools or hydraulic benders to simplify making bends to certain dimensions. Various adaptations of benders have been developed to serve certain specific purposes. For example, one type of hydraulic bender is designed for making bends in a section of conduit that has been installed in a raceway system.

8.1.0 Electric Conduit Benders

Electric conduit benders operate on the same basic principle as the mechanical benders described previously except that the bending is accomplished by a gear motor rather than manual power. There are several types of electric benders on the market. A typical electric unit is shown in *Figure 26.* This bender will make stub-ups, offsets, and saddles quickly and easily at any location where a 120V receptacle is available. Furthermore, this bender will make 180° one-shot bends in ½" through 2" EMT, IMC, or rigid conduit.

Bending charts are often included in the form of decals attached to the bender for quick reference on the job. These include deduct and **springback** figures along with information for making offsets.

Many electric benders are sold as a power unit and the bending shoes for various types of conduit are sold as accessories. These shoes quickly snap into place on the power unit, ready for use in seconds.

To describe the operation of the example bender, assume that we need a 16" stub-up in a piece of 1¼" rigid conduit. Measure and mark a piece of conduit 16" from one end. When we check the deduct chart on the bender for this size conduit, the figure is 12¾". Therefore, 12¾" must be deducted from 16", which leaves 3¼". So a mark is made at this distance from the end of the conduit. Encircle the conduit with this mark so it will be plainly visible during the bend.

Insert the conduit in the bender with the mark lined up with the front of the bender hook. For this particular bend, the mark is 3¼" from the end of the conduit.

The three-position operating switch on the example bender is attached to a flexible cord. The center position de-energizes the machine, the up position (forward) is for bending, and the down position (reverse) is for unloading the conduit after the bend has been made. When released, the switch automatically springs to the center or off position. A pointer on the shoe indicates the degree of bend as the bend is being made.

The machine is jogged or inched up until the shoe pointer lines up with the zero mark on the degree scale. The switch is then pressed upward and held in this position to start the bend. As the pointer approaches the 90° mark, refer to the springback chart for the size and type of conduit being bent. In this case, 1¼" rigid conduit, the chart indicates 95°. Therefore, the pointer should pass the 90° mark and stop at 95° to allow for springback when the conduit is removed from the bender; the stub-up is ready for installation.

An offset is made in the electric bender similar to the method described for the mechanical bender; that is, the first mark is located on the conduit and then offset information is obtained from the bender chart. In the case of a 16" offset in a length of 1¼" rigid conduit, the chart indicates a distance of 22⅝" between marks for a 16" offset using 45° bends.

Mark the conduit and insert the conduit into the bending shoe, positioning the mark so that it is even with the front of the bender hook. Start the bend as discussed previously until the pointer reaches the 45° mark on the scale; allow for any springback as indicated in the chart. Reverse the motor until the conduit is loose, then turn the conduit upside down. Position the second mark at the front of the bender hook, making certain the conduit is aligned to prevent a dog leg in the bend. Start the second bend until the pointer reaches the 45° mark, reverse the bender, and unload the conduit. A magnetic torpedo level placed on the side of the conduit will help align bends for offsets. An anti-dog device can also be used for this.

Finally, check the bends for accuracy. Although rebending is possible, it is not a good practice. Rebending puts considerable strain on the conduit, and while it may not break, the coating may crack and cause corrosion.

Figure 26 ◆ Typical electric bender.

8.1.1 Speed Benders

A speed bender operates basically the same as the standard bender except the speed bender utilizes remote digital control with easy-to-use bending charts and instructions to ensure fast, accurate, and consistent bends in conduit sizes from ½" to 2".

This bender can be operated upright or laid on its back for large offsets and saddles. In operation, rather than holding the switch up in the bend position as with the standard bender, the operator sets the bend wanted via digital controls and the bender automatically bends to that degree once the conduit is placed in the bending shoe and the start control is activated. Otherwise, the conduit is marked and bent in the same way as described for the standard bender.

8.2.0 Hydraulic Conduit Benders

The hydraulic bender is indispensable for bending large conduit. Only by the addition of the hydraulic pump and cylinder to a bender frame is the necessary power available for bending rigid conduit up to 6" trade size.

The typical hydraulic bender (*Figure 27*) provides extended bending capability, but also requires additional responsibility. Hydraulic bending equipment represents a large initial investment, and the bender must be properly cared for if it is to last and give the service necessary to justify the cost.

> **WARNING!**
> Dangerous pressures are used in hydraulic benders. Exercise care when working with or around hydraulic benders.

The bender frame, pivot shoes, and bending shoes will require little more than occasional wiping down to remove the dirt and oil film. The pump, cylinder, and hydraulic hoses, however, will demand more attention.

The pump should have its fluid level checked periodically. Low fluid pressure will keep the pump from developing its rated hydraulic pressure. When the level is low and fluid must be added, use only approved hydraulic oil. Ordinary oil will cause damage to the pump assembly.

Hydraulic benders have two types of bending shoes, one-shot and segment.

The **one-shot shoe** has a full 90° radius. The conduit can be formed around the shoe to a full 90° bend without collapsing the walls of the pipe. One-shot benders require no new bending techniques. The take-up is figured to the center of the bend and the bender shoe will have an indicating mark at its center. If take-up values are not listed on the bender frame or bender storage case, scrap pipe can be bent and take-up figures found. All the methods and layout techniques discussed for EMT and rigid metal conduit can be used with a hydraulic bender and one-shot shoes. Offsets will have to be adjusted (less angle of bend) so the spacing between the bend marks is far enough apart to allow the first bend to be rolled 180° and advanced enough to clear the pivot shoe.

Segment shoes are shorter and have a radius that is far less than 90°. A **ninety-degree bend** cannot be made in one operation, as the conduit walls would collapse. Bends, then, must be made in several steps (as few as four and as many as 30) to form a smooth radius. The **segmented bending shoe** allows pipe to be bent to larger size radii.

Segmented shoes are used for concentric bending (bending several conduits with increasing or decreasing radii). One-shot shoes can be used for a **segment bend,** but are not as convenient. Concentric bending is covered in detail later in this module.

Accurate bending of large conduit is possible but requires practice, patience, and ability. With few exceptions, all formulas and bending techniques discussed to this point will apply.

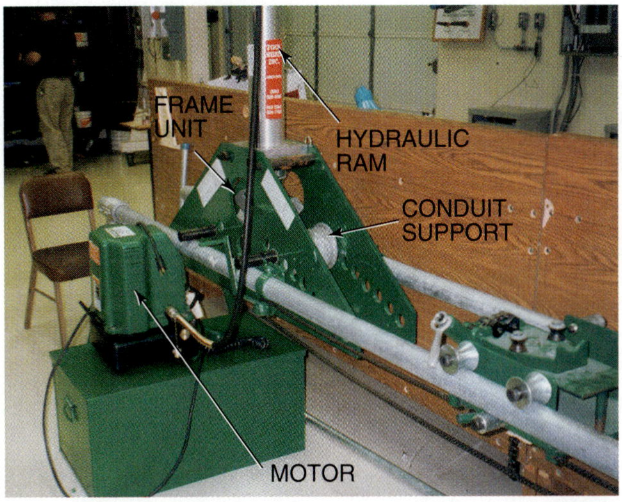

204F27.EPS

Figure 27 ◆ Hydraulic conduit bender.

8.2.1 Bending Tips for Rigid Aluminum

Aluminum conduit is available in all trade sizes from ½" through 6". It is lightweight, corrosion resistant, and has low ground impedance. It is, however, difficult to bend consistently and accurately. Two pipes out of the same bundle will act differently when bent. Even if two pipes are bent using the same layout, they do not always come out the same. Do not be discouraged by this; it is the nature of the metal and it cannot be helped.

Another disadvantage of aluminum is that a one-shot bending shoe will dig in and score the pipe. Also, where the pipe rides on the pivot shoe, it is prone to wrinkling and scoring. Applying petroleum jelly or a lubricant such as WD-40® to the shoe will allow the pipe to slide without the shoe digging in. Petroleum jelly will also make it easier to remove the conduit when the bend is complete.

> INSIDE TRACK
>
> **Hydraulic Bender Safety**
>
> A hydraulic bender generates a tremendous amount of power and operates under dangerous pressures. Do not attempt to use a hydraulic bender unless you have been properly trained and are thoroughly familiar with the unit's operating and safety instructions. Even then, exercise extreme caution and always work in conjunction with a bending partner.

8.2.2 One-Shot Bending

Accurate one-shot stub-ups are easily made on hydraulic benders by applying a little basic geometry in making calculations, and then knowing the operating principles of the bender. *Figure 28* shows the reference points of a common 90° bend. To make one-shot 90° bends, first determine the **leg length** and rise, the gain, the radius of the bend, and the half-gain.

Use the following procedure for laying out accurate stub-ups:

Step 1 Determine lengths A and B.

Step 2 Add lengths A and B. Subtract X for the length of pipe required.

Step 3 Subtract Y from length A or B to get the center of the bend.

Step 4 Calculate the developed length and the length of conduit required.

Step 5 Determine the center of the bend. This can be done by taking the half-gain value from the distance A or B. See *Table 5*.

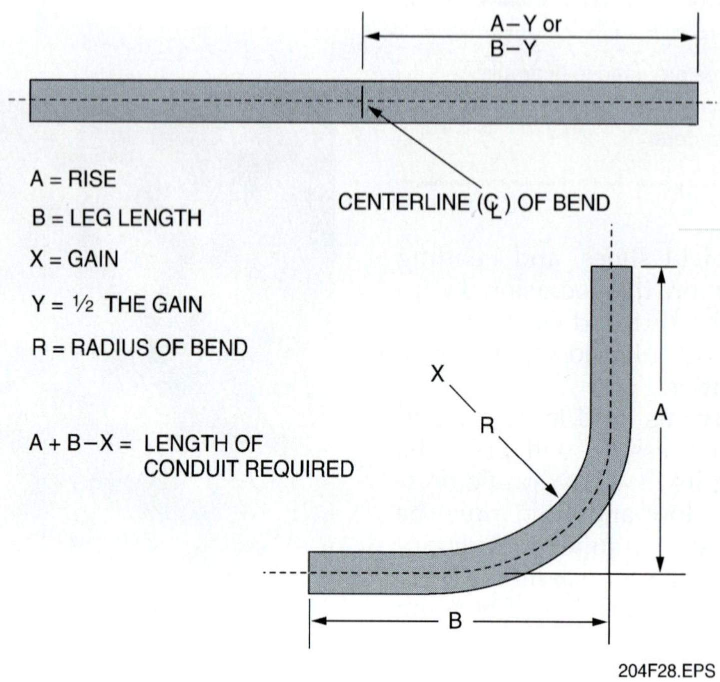

Figure 28 ◆ Laying out stub-ups.

Table 5 Dimensions of Stub-Ups for Various Sizes of Conduit

Pipe and Conduit Size	Radius of Bend R	Minimum Developed Length 90°	Gain X	½ Gain Y
½"	4"	6 5⁄16"	1 11⁄16"	27⁄32"
¾"	4½"	7 1⁄16"	1 15⁄16"	31⁄32"
1"	5¾"	9"	2½"	1¼"
1¼"	7¼"	11⅜"	3⅛"	1 9⁄16"
1½"	8¼"	13"	3½"	1¾"
2"	9½"	14 15⁄16"	4 1⁄16"	2 1⁄32"

8.2.3 90° Segment Bends

When bending conduit in segments with a hydraulic bender, the following factors must be determined:

- The size of conduit to be bent
- The radius of the bend
- The total number of degrees in the bend
- The developed length
- The gain of the bend

To determine the developed length for a 90° bend, multiply the radius by 1.57. The next step is to locate the center of the bend. Most benders have the center mark indicated on the bending shoes. Once the center mark on the conduit is found, it is easy to locate the other bend marks (see *Figure 29*).

You must now determine the number of **bending shots** that will make the bend to suit the requirements, preferably an odd number so that there are an equal number of bends on each side of the center mark. Next, calculate the width of the spaces for each segment bend and make the layout on the conduit. Make an equal number of spaces on each side of the center mark. The gain need only be determined if the bend is being fitted between two existing conduit runs or junction boxes.

To determine the bending data for a 90° bend using 3" conduit with a rise of 48", a leg length of 46", and a centerline radius of 30", proceed as follows (see *Figure 30*).

Step 1 Multiply the radius by 1.57 to determine the developed length. Therefore:

30" × 1.57 = 47.10" (47⅛") developed length

Step 2 Determine the gain for a 90° bend.

Gain = (2 × R) − developed length

Gain = (2 × 30") − 47⅛" = 12⅞"

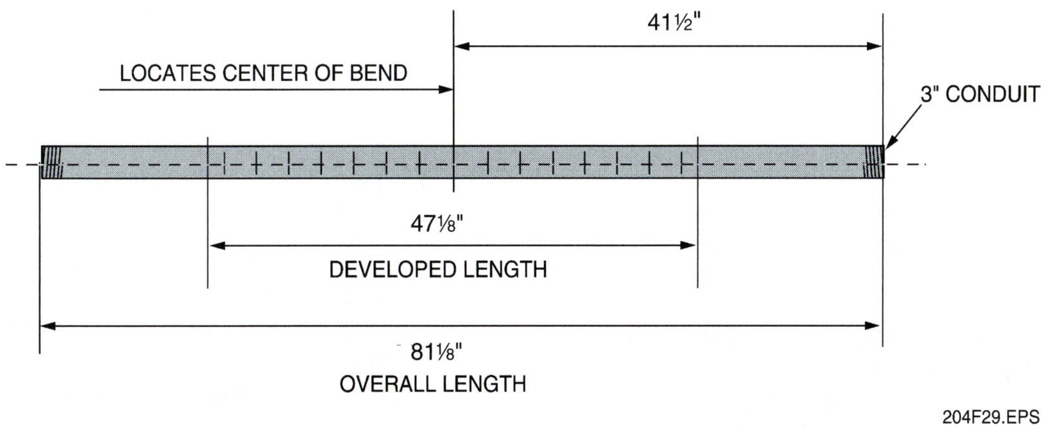

Figure 29 ◆ Laying out segment bends.

THINK ABOUT IT

Developed Length

Why is the radius multiplied by a factor of 1.57 to determine the developed length for a 90° bend?

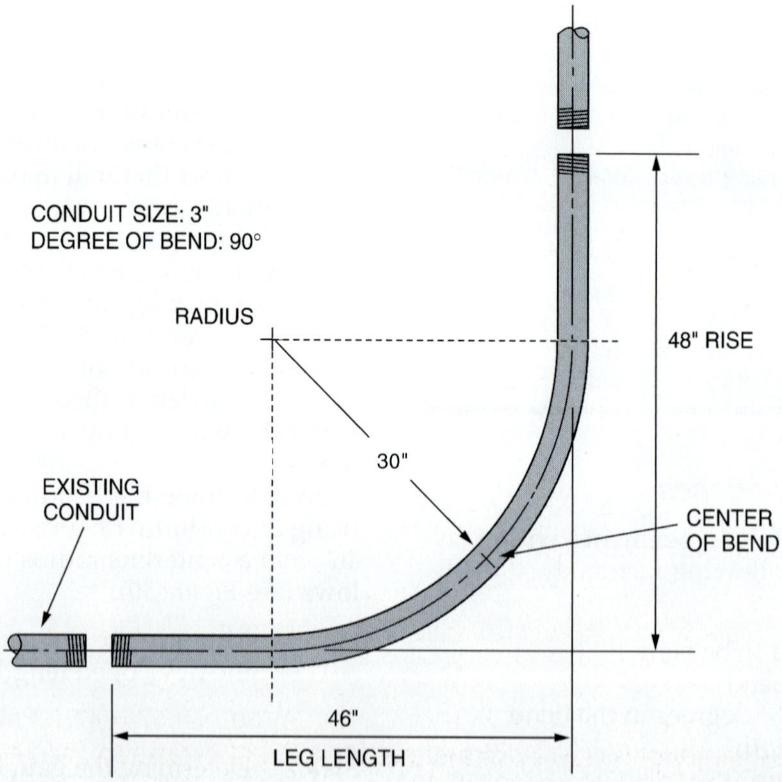

Figure 30 ◆ Specifications for sample bend.

Step 3 To calculate the overall length (OL) of the conduit, add the leg and stub-up lengths and subtract the gain. See *Figure 30.*

OL = leg length + rise − gain

OL = 46" + 48" − 12⅞"

OL = 81⅛"

Step 4 Now locate the center of the required bend. First, determine one-half of the developed length:

½ (47⅛") = 23.56" = 23½"

Use the rise or stub-up dimension of 48". Subtract the radius (30") and add one-half of the developed length:

48" − 30" + 23½" = 41½"

Step 5 As a rule, 6° or less per bend will produce a good bend for a 30" radius. In this case, 6° per bend will be used, making 15 segment bends (90 ÷ 6 = 15). An odd number of segment bends is easy to lay out after finding the center mark because there will be an equal number of spaces on each side of the center mark.

Step 6 To determine the space between the segment marks, divide the developed length by the total number of segments:

47⅛" = 47.125"

47.125" ÷ 15 = 3.14"

3.14" = 3⅛"

Step 7 Position the conduit in the pipe holders, making sure to clamp them securely.

Step 8 Place the center mark 41½" from one end of the conduit. Next, mark seven points on each side of the center point, 3⅛" apart, for a total of 15 marks. These are the centers of the segment bends.

Step 9 It is a good idea to check the distance between the first and last bend marks to be sure the layout is correct before starting the first bend. The distance from the first mark to the last is the developed length minus the length of one bend. (Actually, you are subtracting one-half of a segment bend from each end of the conduit.)

47⅛" − 3⅛" = 44"

Step 10 After positioning the conduit in the bender (*Figure 31*), attach the pipe bending **degree indicator** in a convenient location.

Step 11 Attach the pipe supports with the proper face toward the conduit and insert the pipe support pins. Lock them in position by turning the small lock pin. Now, proceed to make the series of bends.

Step 12 Begin by bending 6° on the first mark. When this is done, the indicator will read 6°. Release the pressure and check the springback; if any is found, overbend by the same amount.

Step 13 When using a bender with a rigid frame, move the pipe support one hole position in (toward the ram) on the side that you have bent the conduit.

Step 14 Continue to bend to 12° on the second mark. Check for springback. When the first bend in the conduit is moved past the one pipe support, the **approximate ram travel** for the remaining bends will be exactly the same.

Step 15 Follow this procedure until you get to the last mark, where you will be bending to 90°. Stop at exactly 90°, release the pressure, and check for springback, correcting if necessary. The result will be a 90° bend without any bows or twists.

For example, suppose the task is to bend a 90° sweep with a given radius of 25". This particular sweep has no definite height. This is to be done on a hydraulic bender using a segmented bending process.

The first step is to determine the linear length or developed length of the conduit required to be used for the bend. This can be done by multiplying the radius of 25" by $\pi \div 2$ or $3.14 \div 2 = 1.57$. The answer is 39.25, as shown below:

$$1.57 \times 25" = 39.25"$$

Therefore, it takes 39.25" of conduit to accomplish a 90° bend with a 25" radius.

The next step is to decide how to lay out the number of segments to be bent to attain the 90° sweep. A rule of thumb is that there will be 20 segments or 21 shots. A shot is the actual bending process. There are 21 shots because the sweep is always laid out from the center of the developed length (*Figure 32*).

Once the center of the developed length is established, half of the developed length is measured out in each direction from the centerline.

To establish the linear distance between segments, divide the developed length by the number of shots (39.25 ÷ 20) or 1.9625" between segments (*Figure 33*).

Once the 20 segments are laid out, the number of degrees per shot (bend) needs to be determined. There are 21 shots, so dividing 90° by 21 will give you the number of degrees per shot:

$$90° \div 21 = 4.29°$$

204F31.EPS

Figure 31 ◆ Conduit placed in hydraulic bender for segment bends.

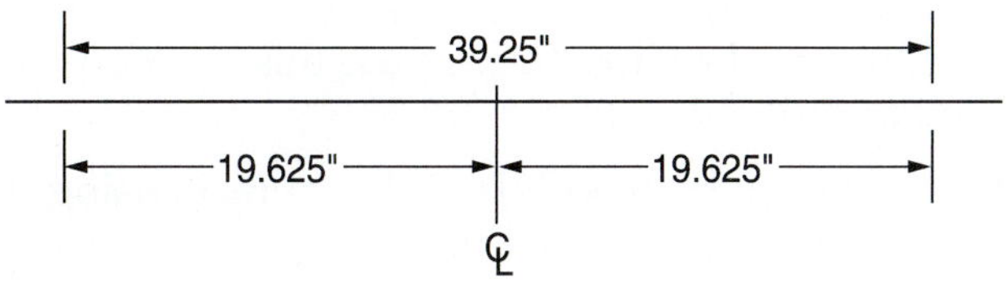

204F32.TIF

Figure 32 ◆ Conduit center.

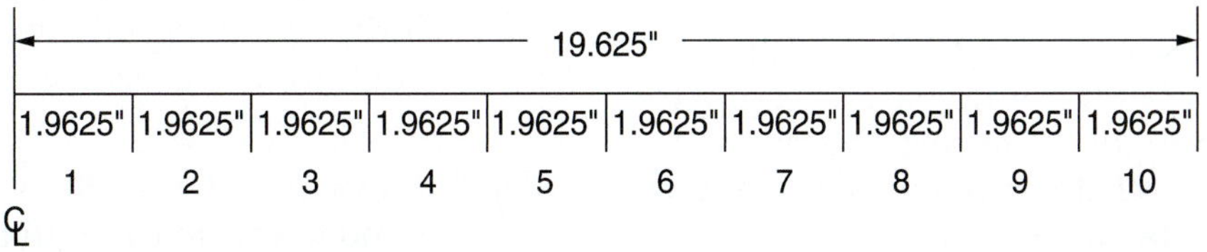

204F33.TIF

Figure 33 ◆ Conduit segments.

You are now ready to do the bending. Starting at one end of the developed length, place the centerline of the bending shoe on the first shot mark and bend to 4.29°. Repeat the process for the remaining 20 bend marks. If the bend is not quite 90°, make the final adjustment in the last bend.

There is another way to lay out the segment marks for the developed length of a bend. Take a piece of white elastic band and lay it out on a table. Make sure that it is not stretched. From one end of the elastic, measure 10 marks spaced evenly apart and mark these points with a fine-tip ink pen. Now place the end of the elastic tape on the centerline of the developed bend length. Stretch out the elastic until the 10th mark is on the end of the developed length. Place a mark on the conduit beside each mark on the elastic tape. Repeat this process on the other half of the developed length. You are now ready to do the segmented bends.

9.0.0 ◆ SEGMENT BENDING TECHNIQUES

Now that the procedure for laying out the conduit with bending marks is understood, the next step is bending. Because segment bending requires several small angle bends to complete a 90° stub, some method to measure the amount of bend will be required. This can be done in four ways:

- Bend degree protractor
- Magnetic angle finder
- Amount of travel method
- Number of pumps method

Bend degree protractor – This is a device that hooks onto the pipe being bent. The circular face is divided into four sections (18, 20, 21, and 30 shots) and is capable of being rotated to whichever scale is to be used. The indicating pointer is weighted and swings free. To use this device, proceed as follows:

Step 1 Level the conduit and secure it using a pipe vise or other means.

Step 2 Rotate the face to the desired scale that corresponds to the number of shots being used.

Step 3 Adjust the scale so the pointer is on 0.

Step 4 Bend the pipe until the pointer reaches the first mark. (Bend a little past to compensate for springback, release the pressure, and then check the pointer. Only a few bends will be needed to find out how much you must bend past the mark to account for springback.)

Step 5 Move the pipe forward in the bender to the second bend mark, and bend until the pointer reaches the second mark on the protractor face. (Again, bend past for springback.)

Step 6 Move the conduit to the third mark and bend the pipe so the pointer is at the third mark (after allowing for springback). Follow this procedure at each bend mark until the 90° stub is achieved, the pipe is level, and the bender is in a vertical position.

NOTE

It is a good idea to check the developing stub length before the last few bends are made. Make spacing corrections as required (e.g., shorten the spacing if the stub is coming up short). If the stub length is reached before the stub is plumb, do not bend at any of the remaining marks. Instead, move the pipe in the bender and bend at the start mark. This will make the stub plumb without adding to the stub length.

Magnetic angle finder – With the magnetic angle finder, the pipe must be kept level and bent vertically (the bender is in vertical position). When a magnetic angle finder is used, care must be taken with each bend as a very small error may become multiplied by 15 to 30 times, becoming a large error. To use the angle finder:

Step 1 Level the conduit and place the angle finder on the stub end. The angle finder will indicate the number of degrees in each bend as determined by the number of shots (e.g., for 20 shots, each bend is 4.5°).

Step 2 Bend the pipe until the angle finder indicates that the bend is just past the desired degree of bend. This allows for springback. Release hydraulic pressure. If the right amount of overbend was made to allow for springback, the angle finder should read the desired angle of bend. If you bent too much or too little, make the adjustment on the next bend. In two or three bends, you will find the right amount to overbend at each mark to allow for springback.

Step 3 Move the pipe to the second bend mark. Bend at this mark until the next setting on the angle finder (allowing for springback) is indicated by the angle finder pointer.

(For example, first bend 4½°, second bend 9°, third bend 13½°, etc.)

Step 4 Bend the pipe at each successive bend mark using the angle finder to indicate the proper degree of bend.

> **NOTE**
> Again, check the developing stub length before bending the last few bends. Adjust the spacing or amount of bend as required.

Amount of travel method – The pipe may be bent in any position. To find the amount of theoretical travel for a 90° bend, proceed as follows:

Step 1 Set up the bender with the pivot shoes in the proper holes for the conduit to be bent.

Step 2 Measure the distance (D) center-to-center between the pivot shoe pins (i.e., from center of pin A to center of pin B). The plunger (also called the ram) will have to travel half this distance to bend a full 90° stub.

Step 3 The travel per shot is one-half the distance from pin to pin divided by the number of shots. For example, the distance from the center of the pins is 24" and the pipe is to be bent in 18 shots. The travel per shot equals one-half the distance from the center of the pins (12") divided by the number of shots (18), which equals 0.666 or approximately ⅔" travel per shot.

Step 4 Bending with this method will follow a slightly different procedure.

Place the center of the bending shoe on the first bending mark (not the start mark) and activate the pump until ⅔" of ram travel is measured. Do not allow for springback with this method.

Move the pipe to the next mark and activate the pump until another ⅔" of ram travel has been measured. Continue to move the pipe and measure ram travel.

As you approach the last few marks (4 or 5), check both the developing stub length and the angle of the bend. The spacing and/or amount of travel can be adjusted on these last marks, as required.

> **NOTE**
> Once you have found the amount of travel for 90°, you can also find the amount of travel for any other angle for offsets, kicks, etc.

Number of pumps method – The pipe may be bent in any position. This method depends on the fact that a given hydraulic pump will produce the same amount of bend for a given number of pumps of the handle; however, it does not take into account that the number of pumps will change with a change in fluid level, the condition of the pump, and the condition of the O-rings.

For example, if it takes 40 pumps to bend a 90° stub, it should only take 20 pumps to achieve 45°, 10 pumps for 22½°, 2 pumps for 4½°, etc. As you can see, this method should work very well, but it will require additional time to determine pump/degree values. However, this can be offset by not having to measure ram travel at each bend as required by the amount of travel method.

Use the same procedure for bending as outlined in the amount of travel method. Check the developing stub length and degree of bend prior to bending the last few shots. Use these remaining shots for final adjustments in stub length and for checking the 90° bend.

> **NOTE**
> Use whichever one of the four methods is the most convenient, but bear in mind that extreme care and attention to detail are required in all methods. A small amount of error at each bend will compound itself, and accuracy in bending will be impossible to achieve.

Using a bending table – To make hydraulic bending easier, a bending table, either a commercial model or one constructed on the job, is a necessity. The table will hold the conduit and the bender, make leveling and plumbing easier, and produce more accurate bends. The table will also eliminate the need to continually wrestle with the conduit and bender.

Hydraulic bending example – For example, suppose that the task is to bend an offset containing two sweeping bends. The radius for the two sweeping bends is 30". This is to be done on a hydraulic bender using a segmented bending process. The angles of the sweeping bends are to be 30° each. The height of the offset is to be 30" (*Figure 34*).

In this problem, we continue to work with the concept of the unit circle. This information is brought together with the cosecant trigonometric function.

In continuing with the concept of the unit circle and the linear distance on the circumference of the circle in terms of pi, it is necessary to understand

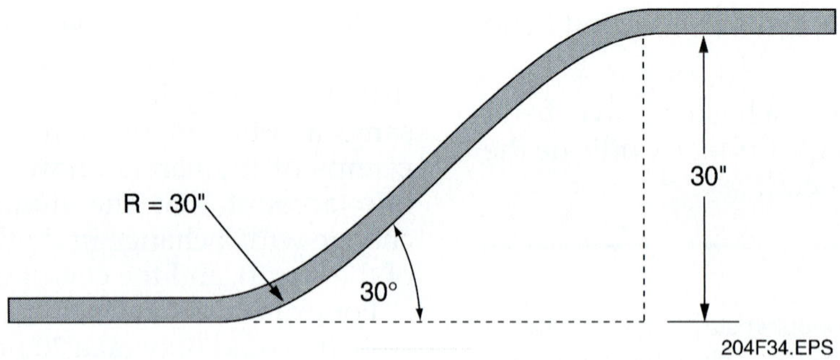

Figure 34 ◆ Two 30° sweeping bends.

the development of the pi relationship from 0° to 90°. All of these relationships from 0° to 90° and other interval angles in the circle are always in reference to 180°. For example: 360° = 2π or 2 × 180°; 180° = π or 1 × 180°; and 90° = π ÷ 2 or ½ × 180°.

The same process holds true for any angle from 0° through 90°. For example: 45° = π ÷ 4 or ¼ × 180°; 60° = ⅓π or ⅓ × 180°; 30° = ⅙π or ⅙ × 180°; and 15° = 1/12π or 1/12 × 180° (*Figure 35*).

To figure the developed length of the conduit to be used by one swing in our 30" radius, we simply multiply π ÷ 6, which represents pi at 30°, times the 30" radius. It is nothing more than a converted circumference formula.

$C = 2\pi R$ (linear distance around a complete circle)

$C = \frac{\pi}{6}R$ (linear distance from 0° to 30°)

$C = \frac{\pi}{6}R$ or $C = 0.523 \times R$ or $C = 0.523 \times 30" = 15.7"$

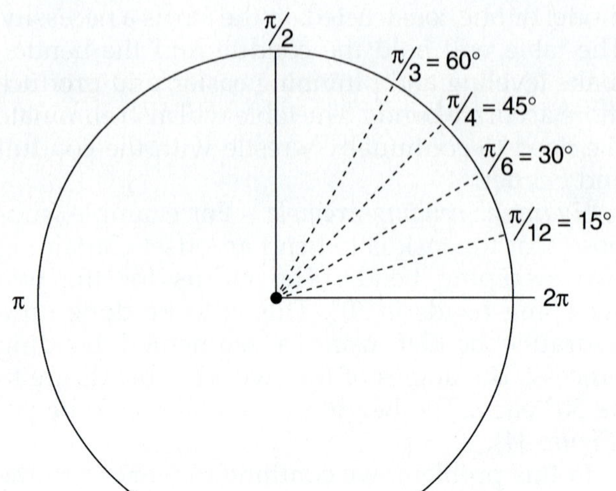

Figure 35 ◆ Radians and degrees.

The developed length for the first radius is 15.7". That means that the developed length for the second sweeping radius is also 15.7". The total amount of conduit used to develop the two sweeps is 31.4".

The next step is to figure out the distance between the centerline of each of these bends on the hypotenuse of our imaginary triangle. To do this, we simply use the cosecant of the angle trigonometric function or Cosecant = H ÷ O. Converting the formula, we get H = O × Cosecant or H = 30" × 2 or 60". The layout of these measurements is shown in *Figure 36*.

Again, laying out the segments for the bending process is done from the centerline of the developed length of each sweeping bend (*Figure 37*). Therefore:

$$15.7" \div 2 = 7.85"$$

When we build a 90° sweep, the rule of thumb is 21 shots or 20 segments. This gives the conduit the appearance of being a natural sweep and not a line of straight segments. (The same odd shot/even segment process will appear again.) If we choose to have seven shots or six segments within the 15.7" of developed length, each segment will be 15.7" ÷ 6 or 2.62" long. Because there are seven shots (bends), each shot will be 30 ÷ 7 or 4.29°. The layout of the bend is shown in *Figure 38*.

Once you have both sweeps laid out as described previously and the centerlines of these sweeps are 60" apart, you are ready to start the bending process. Again, start at one end of the offset and swing the bender segmented shoe down on the conduit. Because you are working an offset, an anti-dog device should be attached to one end of the conduit. This device maintains the centerline for each bend. Begin bending the first sweep, and once this is accomplished, roll the conduit in the bender and re-level the anti-dog device. Bend the second sweep and your offset

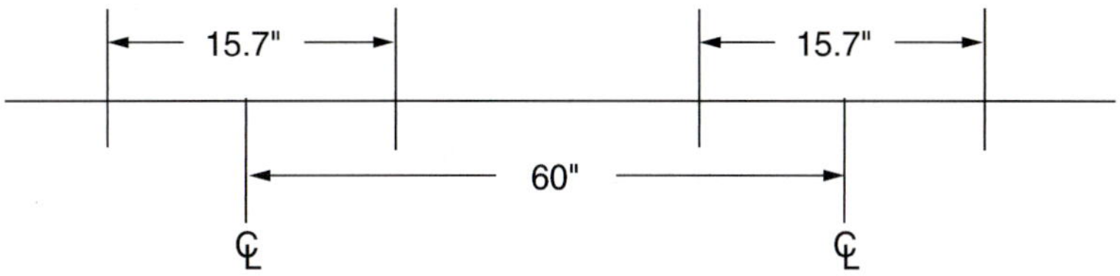

204F36.TIF

Figure 36 ◆ Bend centerline distance.

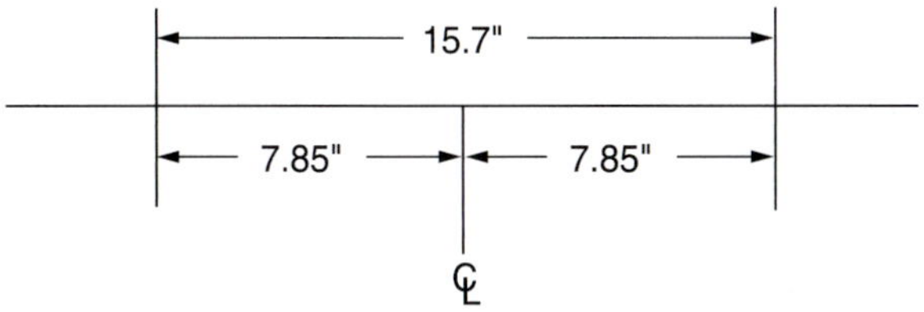

204F37.TIF

Figure 37 ◆ Bend centerline.

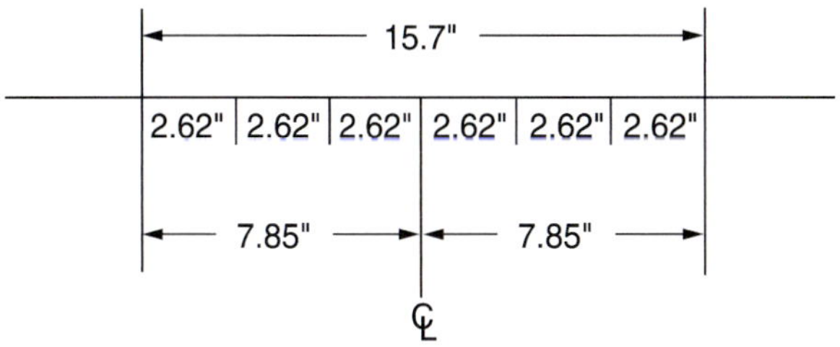

204F38.TIF

Figure 38 ◆ Bend segments.

is built. The final measurements should be as follows:

Developed length of each sweep = 15.7"

Distance between centerline of each sweep = 60"

Distance between shot marks in each sweep = 2.62"

Degrees per shot in each sweep = 4.29"

9.1.0 Concentric Bending

If two or more parallel runs of conduit must be bent in the same direction, as shown in *Figure 39*, the best results will be obtained by using concentric bends. When laying out concentric bends, the bend for the innermost conduit is calculated first. In the example in *Figure 39*, the first bend has a radius of 20". If this dimension is multiplied by 1.57, it yields a value of 31$\frac{13}{32}$". This is the developed length of the shortest radius bend.

The second radius is found by adding the **outside diameter (OD)** size of the first pipe to the radius of the first pipe and the desired spacing between pipes. In this example, the radius of the second bend would be equal to 24⅜" (see *Figure 39*).

Radius of first pipe = 20"

OD size of first pipe = 2⅜"

Spacing between pipes = 2"

Radius of second pipe:
20" + 2⅜" + 2" = 24⅜" or 24.375"

Once the developed length of the second radius is found, the spacing between marks can be determined. The start mark must be the same distance from the end of the second pipe as it was for the first pipe (*Figure 39*). This will be true for each additional pipe as well. The spacing will increase between marks as the radius of the pipes increases. This is due to increasing developed length. The pipes in *Figure 39* are all laid out for 15 shots.

The equation for the developed length (DL) is 1.57 × R (radius). The developed length for the second bend is found by multiplying 1.57 × 24.375" = 38.27".

The radius and developed length of each successive bend are found in a similar manner. After determining the developed length, you must establish the number of segment bends needed to form each 90° bend. In concentric bending, every bend must receive the same number of segment bends to maintain concentricity. As illustrated, 15 segment bends of 6° each will total 90°.

> **NOTE**
>
> The radius change from one bend to another affects the spacing of segment bends. To find the segment bend spacing, divide the developed length of each bend by 15. For the first bend illustrated, the spacing for each segment bend is 2$\frac{3}{32}$"; for the second bend, it is 2$\frac{9}{16}$", and so on, as shown in *Figure 39*.

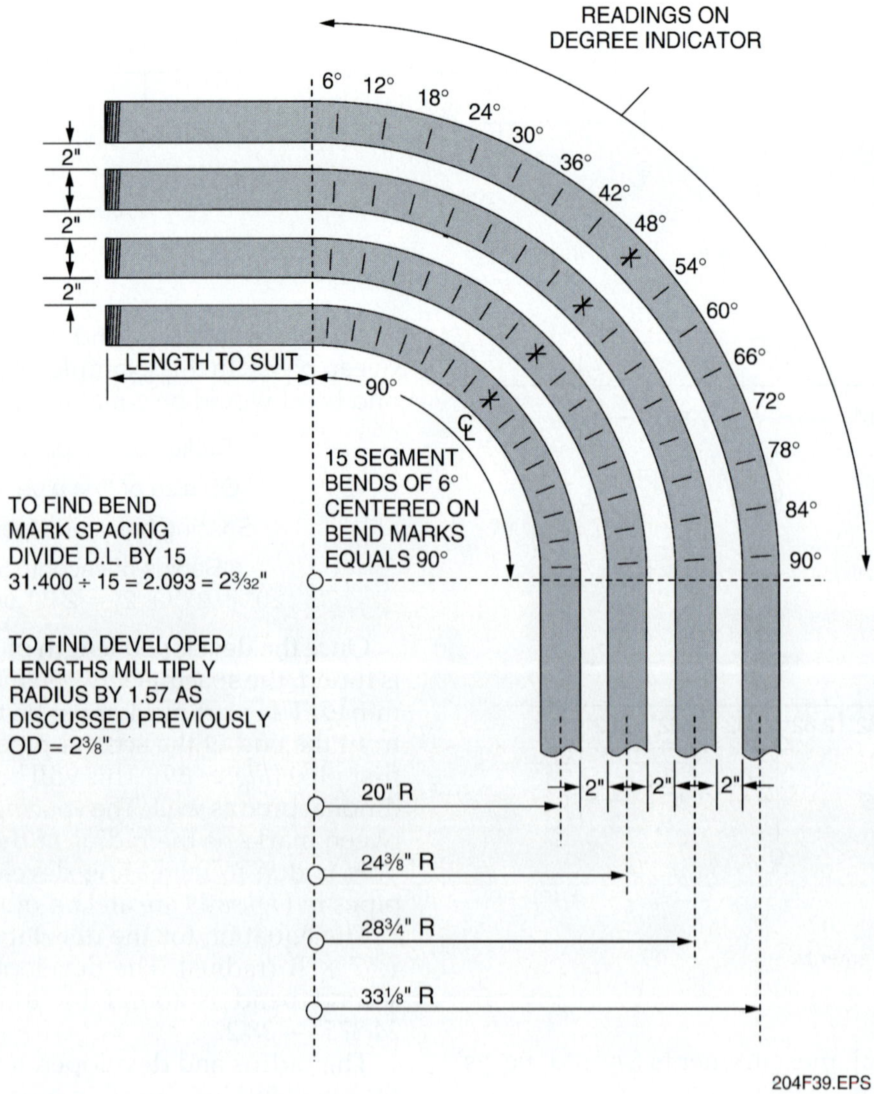

Figure 39 ◆ Principles of concentric bending.

If the legs of the bends have to be a certain length, the gain must be considered just as in any segment bending procedure. When the runs of conduit are not the same size, the radius of each successive bend can be found as follows:

Step 1 Determine the radius of the innermost bend.

Step 2 Calculate one-half the outside diameter of the innermost conduit and of the next adjacent conduit.

Step 3 Note the distance between the two runs of conduit.

Step 4 Add these quantities.

INSIDE TRACK

Segmented Bends

Most electricians find the amount-of-travel method of making segmented bends to be the simplest and most reliable. The number of pumps method can change with the condition of the pump. The bending protractor and mechanical angle methods are accurate but also time consuming.

9.2.0 Offset Bends

Many situations require a conduit to be bent so that it can pass by or over objects such as beams and other conduit or to enter panelboards and junction boxes. The bends used for this purpose are called offsets. To produce an offset, two equal bends of less than 90° are required at a specified distance apart. This distance is determined by the angle of the two bends and can be calculated by using the following procedure and the table in *Figure 40*.

First, determine the offset needed, then find the degree of bend. Next, multiply the offset measurements by the figure directly under the degree of bend (see *Figure 40*). This applies to all sizes of conduit. For example, to form an 18" offset with two 45° bends, first make the following calculation to determine the distance between bends:

$$18" \times 1.414 = 25\tfrac{1}{2}"$$

This is the distance between bends and is labeled side *L* (see *Figure 40*). To connect the two ends of an offset to two pieces of conduit that are already in place, it is necessary to know the total or overall length (OL) of the offset from end to end before bending.

Note the following equation:

$$OL = (A + L + B) - (2 \times \text{gain})$$

Where:

$A = 36"$

$L = 25\tfrac{1}{2}"$

$B = 48"$

$2 \times \text{gain} = 1\tfrac{9}{32}"$

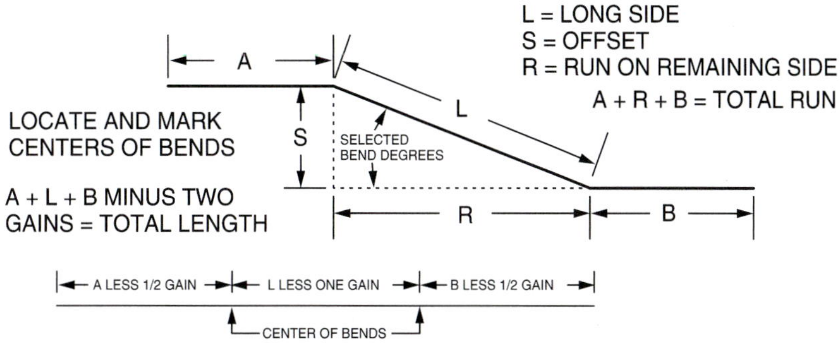

TO FIND UNKNOWN	KNOWN	TIMES CORRESPONDING MULTIPLIER EQUALS — TABLE OF MULTIPLIERS FOR SELECTED DEGREES OF BEND								UNKNOWN
		5-5/8°	11-1/4°	15°	22-1/2°	30°	37-1/2°	45°	60°	
L	S	10.207	5.126	3.864	2.613	2.00	1.643	1.414	1.155	L
S	L	.098	.195	.259	.383	.50	.609	.707	.866	S
R	S	10.158	5.027	3.732	2.414	1.732	1.303	1.00	.577	R
S	R	.098	.199	.268	.414	.577	.767	1.00	1.732	S
L	R	1.005	1.02	1.035	1.082	1.155	1.260	1.414	2.00	L
R	L	.995	.981	.966	.933	.866	.793	.707	.50	R
GAIN PER BEND	RADIUS OF SHOE	.0002	.0006	.0015	.0051	.0124	.0212	.0430	.1076	GAIN PER BEND

204F40.TIF

Figure 40 ◆ Conduit offset bending table.

Leave Room for System Expansion

INSIDE TRACK

It's a good idea to make your first concentric bend large enough to fit another bend or two inside it in case you have to add additional conduit runs later on. Another method is to add any additional conduit runs on the outside of the original bend. Both methods produce a neat and professional installation.

THINK ABOUT IT

Concentric Bends

What does concentric mean? Why can't you just bend all your conduit sections to the same radius and lay them side by side?

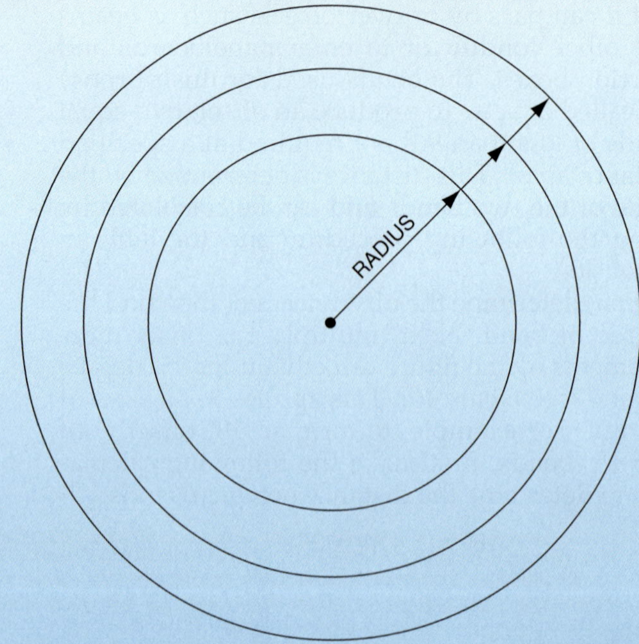

204UA0401.EPS

The gain is calculated by multiplying the shoe radius by the decimal figures shown on the last line under DEGREES OF BEND in *Figure 40*. Let us calculate how an offset might be made in a length of 3" conduit using 45° bends:

$$OL = (A + L + B) - (2 \times gain)$$

$$OL = 109\tfrac{1}{2}" - 1\tfrac{9}{32}" = 108\tfrac{7}{32}"$$

> **NOTE**
>
> This offset uses a 15" radius.

15" × 0.0430 (from the table in *Figure 40*) = 0.645°

For two gains, we have:

$$1.290 = 1\tfrac{9}{32}$$

Since you already know the long dimension of the offset (25"), to find the amount of offset, refer to the table in *Figure 40*, second line down (to find S), then move across the row until you come to the 45° column. Note that the multiplier is 0.707.

Therefore:

$$25" \times 0.707 = 17.675"$$

This is the height of the offset. To make the offset, the conduit is positioned in the hydraulic bender, as shown in *Figure 41*. The bending shoe makes one 45° bend on the first center mark, the conduit is reversed in the bender, and the next 45° bend is made at the second mark. When making bends, always refer to the bending charts that accompany the bender being used for the operation.

For example, an offset is to be built to go over an I-beam that is 24" in height. The distance from the end of the conduit run to the J box on the opposite side of the I-beam is 8' (see *Figure 42*).

The task is to determine the distance between bend points where a 45° bend is being used. Determine the exact length of conduit needed so threading can occur before bending. The conduit will have no clearance over the top of the I-beam.

To find the distance between bend points, multiply the amount of offset required by the cosecant of the bending angle (in this case, 45°):

$$24" \times 1.41 = 33.84"$$

The straight distance from the end of the conduit run to the J box is 8'. To find the amount of conduit to add to this because of the 45° bend, use the cosine formula (C = A ÷ H). The hypotenuse or the distance between bend points is 33.84". The side adjacent is found by converting C = A ÷ H to A = C × H, or:

$$A = 0.707 \times 33.84" = 23.92"$$

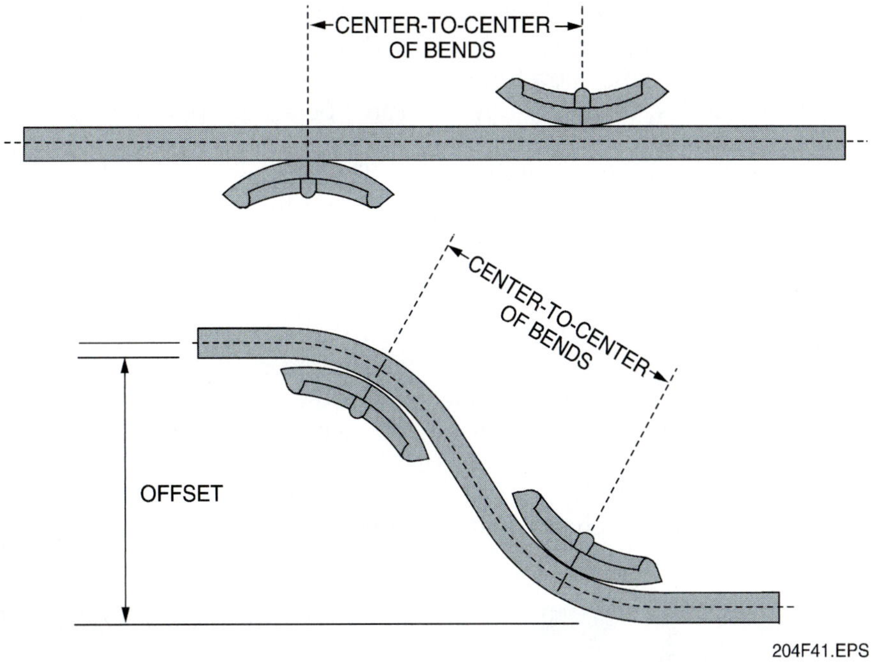

Figure 41 ◆ Position of conduit in bender for making offsets.

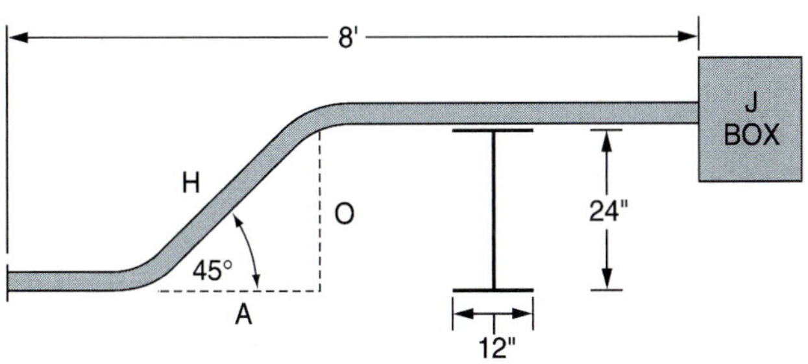

Figure 42 ◆ 24" offset.

Subtract 23.92" from 33.84" and this gives the amount of extra conduit needed when making this 45° bend:

33.84" − 23.92" = 9.92" or 9 92/100"

Add the 9.92" to 8' and the amount of straight conduit required has been determined:

8' + 9.92" = 8'-9.92" of straight conduit

The final measurements should be as follows:

Distance between bend points = 33 21/25" or 33.84"

Length of straight conduit = 8'-9.92" or 8'-9 92/100"

Now let us complete another example. In this case, an offset is to be made without the use of a protractor level. The offset is to be 20" in height (*Figure 43*). The bend angle is 45°. Explain the process required to accomplish this.

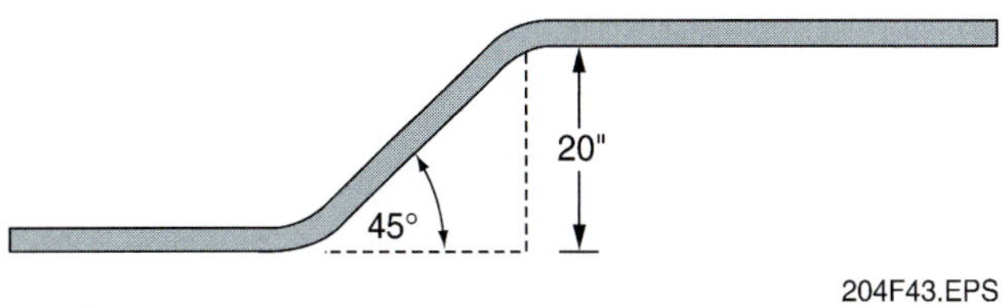

Figure 43 ◆ 20" offset.

The cosecant for 45° is 1.41. The distance between bend points is 1.41 × 20". Therefore:

1.41 × 20" = 28.2"

The first mark on the conduit is placed 14.1" back from the end of the conduit. The end of the conduit needs to be brought up 10" to accomplish a 45° bend. The second mark on the conduit is placed 28.2" down the conduit from the first mark. The bender is placed on the second mark. The conduit is bent until the two surfaces are parallel. At this point, a 20" offset has been created.

9.3.0 Concentric Offsets

To make concentric offsets, use the same procedure as for 90° stubs. Increase the radius of the second pipe by an amount equal to the radius of the first pipe plus the OD size of the first pipe plus the spacing desired between pipes. Although the examples used to explain offset bending had several shots per bend, fewer shots can be used.

Successful offsets can be made with only two or three shots per bend. However, with offsets, 90° bends, or any other segment bends, the extra time spent bending smoother radius bends (more shots) is returned when the conductors are pulled into the completed raceway. Parallel offsets of different pipe sizes can still be bent and aligned perfectly if all pipes are bent with the same one-shot shoe.

9.3.1 Saddle Bends

As mentioned previously, saddle bends are used to cross small obstructions or other runs of conduit. Saddle bends in conduit up to four inches **inside diameter (ID)** may be made in hydraulic conduit benders. Always refer to the charts that accompany the bender. For example, to make a saddle bend on a length of 2" rigid conduit so that it can pass over a 3" water pipe with ¼" clearance, three bending operations are required, as shown in *Figure 44*.

The bend is calculated by referring to appropriate bending charts. The one shown in *Table 6* gives measurements for 2" conduit with ¼" clearance. Therefore, look in the left column under straight-run conduit and go down the column until the 3" row is reached. Moving to the right in this row, it can be seen that the spacing between bends is 15⅝" on center, and bends No. 1 and No. 2 will be 15°, while the third bend is 30°.

Using the information found in the bending table, mark the conduit accordingly. Insert the conduit in the bender and make bends No. 1 and

> **INSIDE TRACK**
>
> ***Offset Bending***
>
> Offsets in small conduit (e.g., ½" or ¾" EMT or ½" rigid or IMC) may be easier to bend with the bender head in the air. However, it may be easier to avoid dog legs if the second bend is made on the floor whenever possible.
>
> ***Straight Offsets***
>
> Use the seams on EMT as a guide for making straight offsets. The best bend angles are:
>
> 10° for 1" depths
>
> 22½° for 2" depths
>
> 30° for 3" and 4" depths
>
> 45° for 5" and greater depths

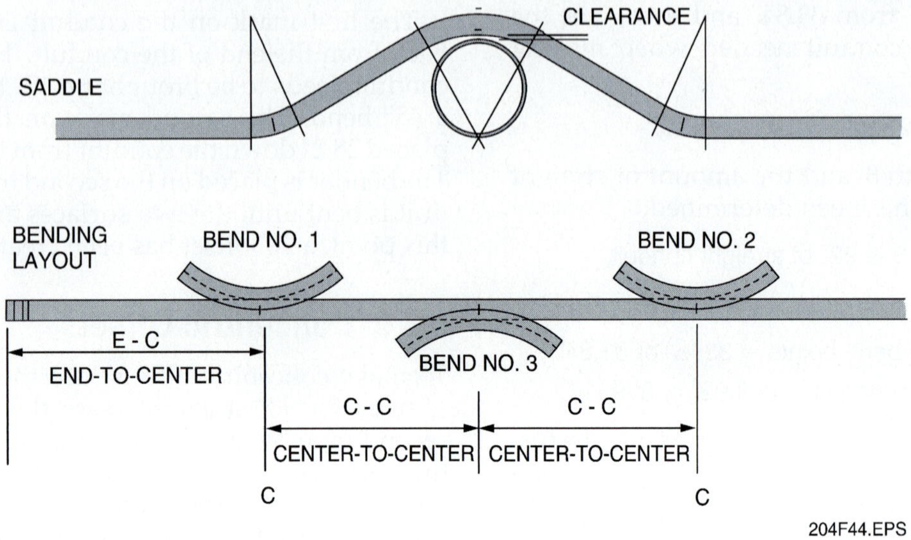

Figure 44 ◆ Principles of saddle bending.

Table 6 Saddle Table

Straight-Run Conduit	Minimum Length	Bend Spacing	Bend Degrees	Bend Degrees	Bend Degrees
—	E-C	C-C	No. 1	No. 2	No. 3
1"	20"	16"	6	6	12
1¼"	20"	16"	7	7	14
1½"	20"	16"	8	8	16
2"	20"	15⅞"	10	10	20
2½"	20"	15¾"	12½	12½	25
3"	20"	15⅝"	15	15	30
3½"	20"	15½"	18	18	36
4"	20"	15½"	20	20	40

No. 2 first, both at 15°. Back off the pump pressure, reverse the conduit in the bender, and then make the third bend at 30°. Release the pump pressure, remove the conduit from the bender, and check the saddle for accuracy.

For another example, suppose a saddle is to be built to go over an I-beam that is 36" in height (*Figure 45*). The distance from the end of the conduit to the center of the I-beam is 5'.

The task is to determine the distance between bend points when a 45° bend is being used. Determine the exact length of conduit needed so threading can be completed before bending. The conduit is to clear the top of the I-beam by 1". Determine the centerline layout of the saddle (*Figure 46*).

The 16" between bend points in this problem is an arbitrary measurement. It is determined by the person doing the bending. Enough conduit should be used to extend past the outside dimensions of the 12" I-beam so the radius of each bend does not run into the I-beam.

The distance between bend points on the hypotenuse of the imaginary triangle is calculated by multiplying the total height of the I-beam (36" + 1" clearance = 37") by the cosecant of the 45° angle of the bend or 1.41 × 37":

$$1.41 \times 37" = 52^{17}/_{100}" \text{ or } 52.17"$$

The centerline layout mark for the saddle would be 5' if the conduit were to be installed on a straight line from the end of the existing conduit run to the center of the I-beam. However, the conduit is not being installed on a straight line. Therefore, the measurement in question is the amount of additional conduit needed because of the 45° angle of deviation (*Figure 47*). The amount of additional conduit is equal to the side adjacent subtracted from the hypotenuse.

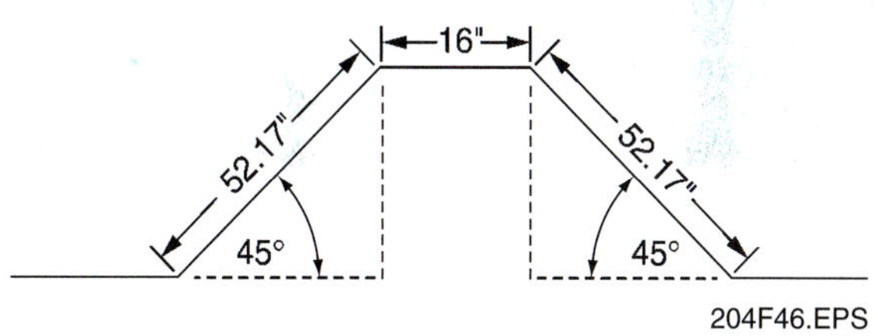

Figure 46 ◆ Saddle.

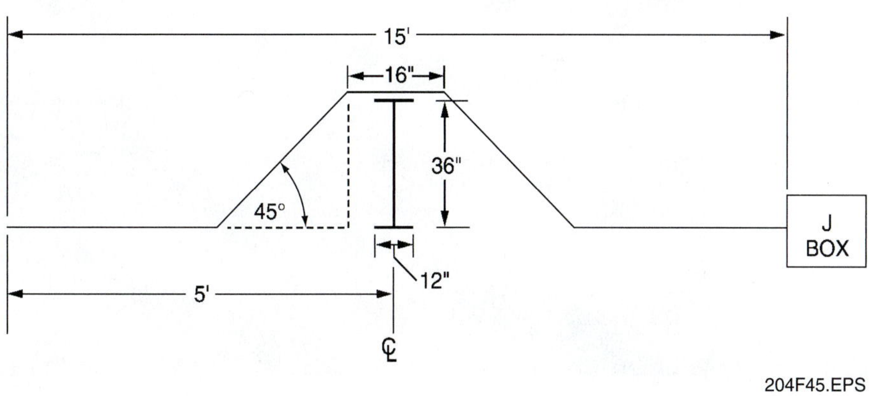

Figure 45 ◆ 36" saddle.

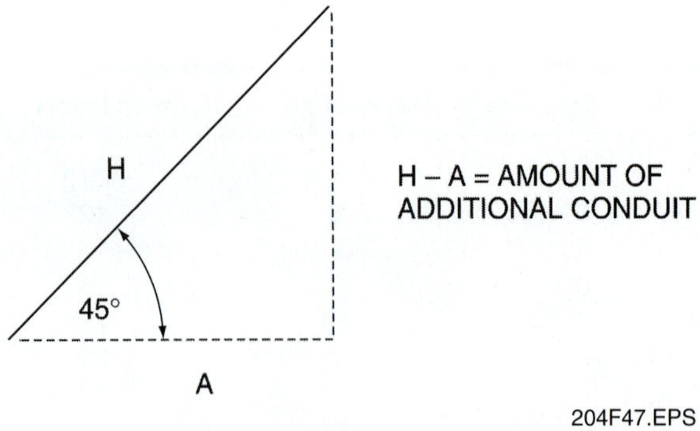

Figure 47 ◆ Cosine function.

The side adjacent (A) first needs to be calculated, then that measurement can be subtracted from the hypotenuse (52.17"). To calculate the side adjacent, use the cosine formula (C = A ÷ H). Because we are looking for A, convert the formula to A = C × H, or:

$$A = 0.707 \times 52.17"$$
$$A = 36.88"$$

Subtract 36.88 from 57.17, or:

$$52.17" - 36.88" = 15.29"$$

That is how much extra conduit is needed to get the center of the saddle to the center of the I-beam. Therefore, the layout mark for the saddle is:

$$5' + 15.29" = 6'\text{-}3.29"$$

The layout of the conduit is shown in *Figure 48*.

To find the total amount of conduit to be used from the end of the existing conduit to the J box, simply take the straight line distance of 15' and add the difference between the hypotenuse and side adjacent to 15' two times, or:

$$15' + 15.29" + 15.29" = 15' + 30.58" = 17'\text{-}6.58"$$

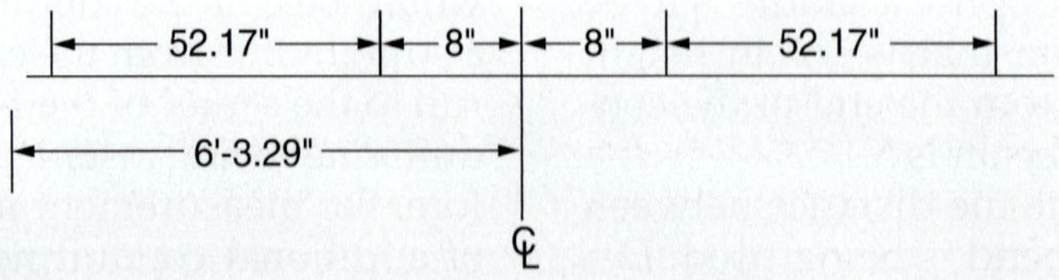

Figure 48 ◆ Conduit layout.

INSIDE TRACK

Saddle Bends

Whenever possible, a three-bend saddle should not be sharper than 30°. If the object you are trying to saddle is more than 6" in diameter, use a four-bend saddle rather than a three-bend saddle. In this installation, two three-bend saddles were required.

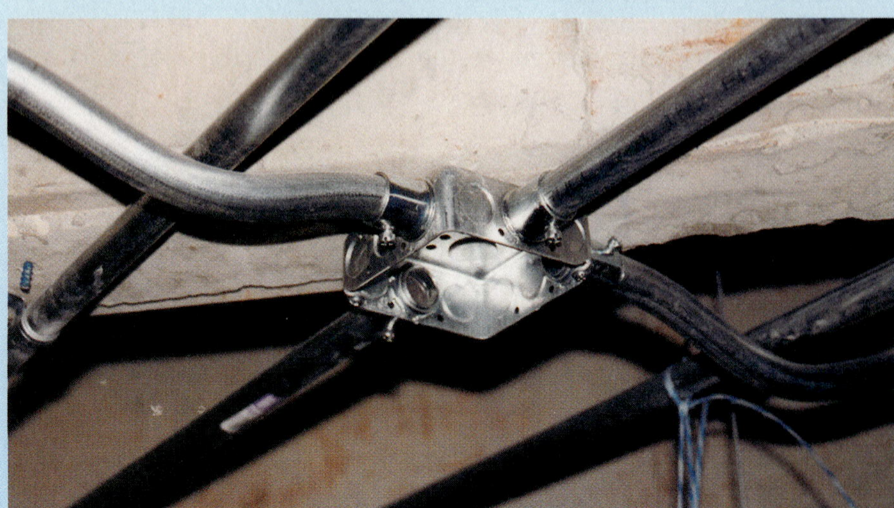

204P0404.EPS

The final measurements should be as follows:

Distance between bend points on the top portion of the saddle = 16"

Distance between bend points that form each hypotenuse = $52\frac{17}{100}$" or 52.17"

Distance from the end of the conduit to the centerline layout mark of the saddle = 6'-3.29" or 6'-$3\frac{29}{100}$"

Overall length of the conduit required to reach the J box = 17'-$6\frac{58}{100}$" or 17'-6.58"

10.0.0 ◆ TRICKS OF THE TRADE

There are many tricks of the trade that you will learn during your career as an electrician. Some of these will be handed down to you by experienced workers; others will be learned through experience. Most professionals are constantly seeking new methods to improve their efficiency and to make the work go smoother and faster without sacrificing workmanship.

One of the handiest personal tools for bending conduit is the small, magnetic torpedo level with a 35° and 45° bubble. This tool should be in the tool pouch of every electrician. In many cases, it can make the difference between a good job and a poor job, that is, obtaining level and plumb conduit runs or not.

Workers have also designed their own custom tools to help in conduit installations. One tip is to take a pair of vice grips and weld a small, flat piece of iron on the top of the jaws for use with the magnetic level. During multiple bends, the vice grips are positioned at the desired point on the end of a piece of conduit, locked in place, and then the magnetic level is placed on the flat plate so the bubble may be watched during the bend. There are also some new commercial tools that are designed for leveling conduit bends; keep up with these developments by reading trade journals.

10.1.0 Eliminating Dog Legs

One of the problems that exists when bending an offset or saddle is the creation of a dog leg. A dog leg occurs when the centerline that runs the length of the conduit is not maintained for both bends of an offset and for the four bends of a saddle. Dog legs can be eliminated using an anti-dog device, as discussed earlier, or by following the procedure described here.

See *Figure 49*. When the conduit is snugged up by the bender, take a piece of Unistrut® and clamp it to the end of the conduit farthest from the bending shoe. Before tightening the conduit strap, level the Unistrut®. Once this is done, make your first bend. Release the conduit from the bending shoe and move the conduit forward to the next bending mark. Roll the conduit 180° and re-level

INSIDE TRACK

Magnetic Level

A magnetic torpedo level can be attached directly to the conduit. It will remain where you put it, even if the conduit is turned over.

204P0405.EPS

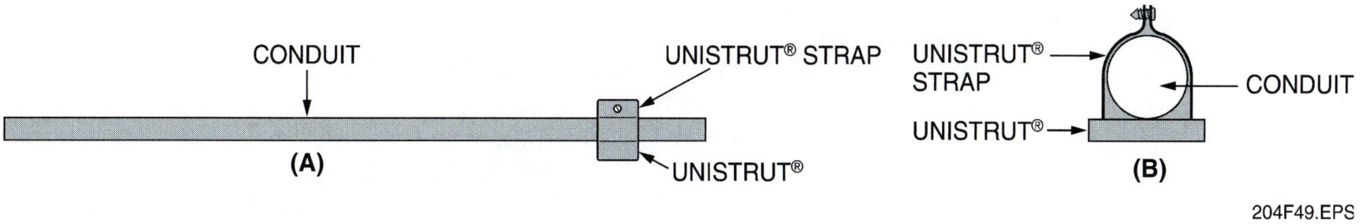

204F49.EPS

Figure 49 ◆ Dog leg elimination.

the Unistrut®. Tighten the conduit in the bender and again make sure that the Unistrut® is level. Bend the second bend and you will have an offset with no dog leg.

10.2.0 Using a Table Corner or Plywood Sheet for Calculating Added Length

The task is to use a table corner or the corner of a sheet of plywood and demonstrate how to come up with the amount of additional conduit needed to go from one point to another once a bend has been made. The distance between points is 8'. The amount of offset to be made is 25" (*Figure 50*). The degree of bend is 30°.

See *Figure 51*. Measure 25" up one side of the sheet of plywood or the corner of a table. The hypotenuse of the triangle will be 2 × 25" or 50" (2 is the cosecant for 30°). Measure from the 25" mark to the surface that is tangent to the side that is 25" long, until the 50" mark on the tape measure crosses the tangent side. Mark this 50" point. Then, place the end of the ruler on the 50" mark and let the 50" point on the ruler extend beyond the end of the plywood or table. The amount of tape that extends beyond the plywood or table edge is the amount of extra conduit that is needed to accomplish this bend. This is 6⅝" more conduit than if going on a straight line for 8' or 8'-6⅝" of conduit required to accomplish the task.

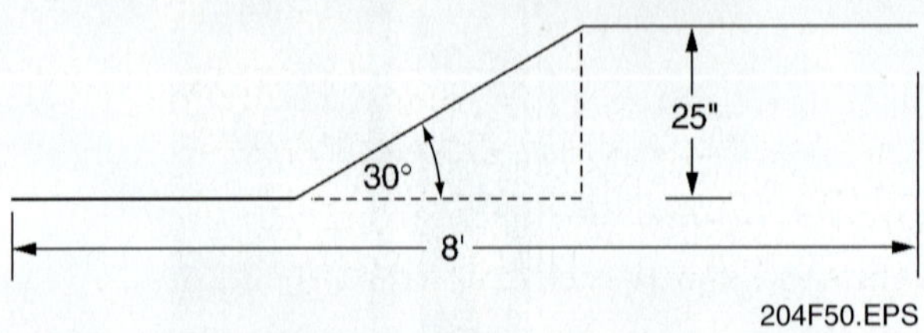

Figure 50 ◆ 25" offset.

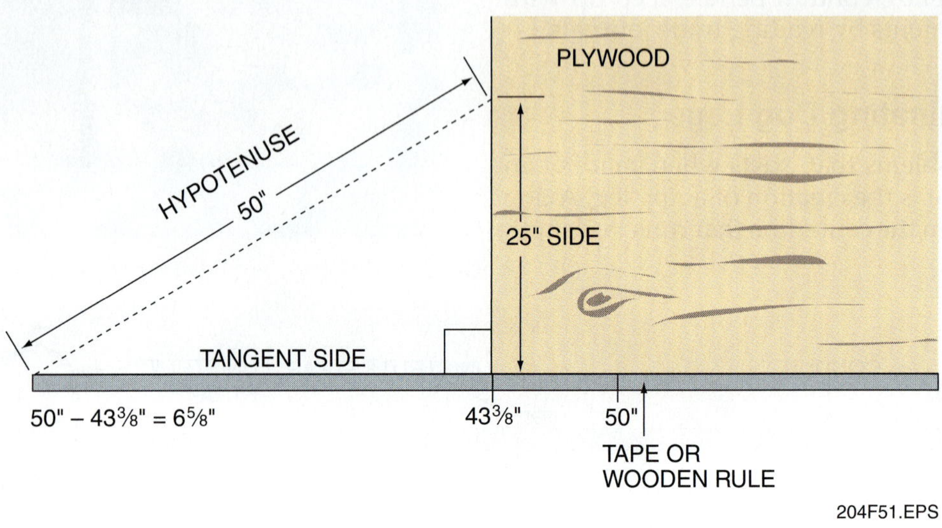

Figure 51 ◆ Extra conduit.

11.0.0 ◆ PVC CONDUIT INSTALLATIONS

Rigid nonmetallic conduit and fittings (PVC electric conduit) may be used where the potential is 600V or less in direct earth burial; in walls, floors, and ceilings of buildings; in cinder fill; and in damp and dry locations. Its use is prohibited by the *NEC®* in certain hazardous locations, for support of fixtures or other equipment, where subject to physical damage, and under certain other conditions. See *NEC Article 352.*

PVC conduit can be cut easily at the job site without special tools, although PVC cutters help in cutting square ends. Sizes from ½" through 1½" can be cut with a fine-tooth saw. For sizes 2" through 6", a miter box or similar saw guide should be used to keep the conduit steady and ensure a square cut. To ensure satisfactory joining, care should be taken not to distort the end of the conduit when cutting.

After cutting, deburr the pipe ends and wipe clean of dust, dirt, and plastic shavings. Deburring is accomplished easily using a pocket knife or file.

One of the important advantages of PVC conduit, in comparison with other rigid conduit materials, is the ease and speed with which solvent-cemented joints can be made. The following steps are required for a proper joint:

Step 1 The conduit should be wiped clean and dry.

WARNING!
The cement used with PVC can be hazardous. Provide adequate ventilation, avoid skin contact, and always follow the manufacturer's usage and safety instructions.

Step 2 Apply a full, even coat of PVC cement to the end of the conduit and the fitting. The cement should cover the area that will be inserted in the socket.

Step 3 Push the conduit and fitting firmly together with a slight twisting action until it bottoms and then rotate the conduit in the fitting (about a half turn) to distribute the cement evenly. Avoid cement buildup inside the conduit. The cementing and joining operation should not exceed more than 20 seconds.

Step 4 When the proper amount of cement has been applied, a bead of cement will form at the joint. Wipe the joint with a brush to remove any excess cement. The joint should not be disturbed for 10 minutes.

12.0.0 ◆ BENDING PVC CONDUIT

Most manufacturers of PVC conduit offer various radius bends in a number of segments. Where special bends are required, PVC conduit is easy to bend on the job. Stub-ups, saddles, concentric bends, offsets, and kicks are all possible with PVC conduit, just as with metallic conduit.

PVC conduit is bent with the aid of a heating unit. The PVC must be heated evenly over the entire length of the curve. Heating units are available from various sources that are designed specifically for the purpose in sizes to accommodate all conduit diameters (*Figure 52*). While some heaters use gas for the heat source, most employ infrared heat energy, which is more quickly absorbed in the conduit. Small sizes are ready to bend after a few seconds in the hotbox. Larger diameters require two or three minutes, depending on the conditions. Other methods of heating PVC conduit for bending include heating blankets and hot air blowers. The use of torches or other flame-type devices is not recommended. PVC conduit exposed to excessively high temperatures may take on a brownish color. Sections showing evidence of such scorching should be discarded.

WARNING!
Always wear gloves when working with heat.

If a number of identical bends are required, a jig can be helpful (*Figure 53*). A simple jig can be made by sawing a sheet of plywood to match the

204F52.EPS

Figure 52 ◆ PVC heating units.

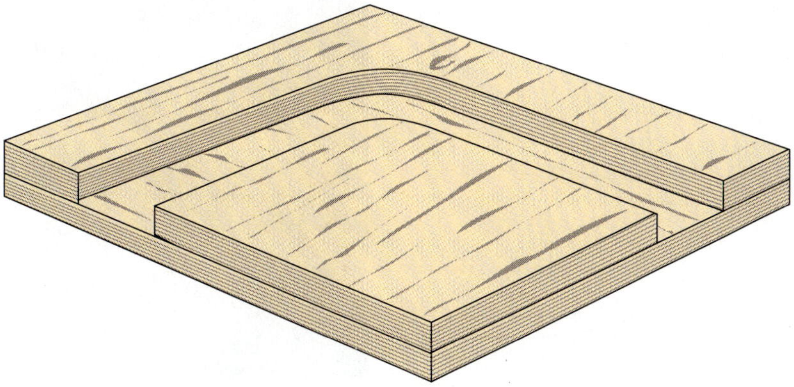

204F53.EPS

Figure 53 ◆ Plywood jig.

desired bend. Nail to a second sheet of plywood. The heated conduit section is placed in the jig, sponged with water to cool, and is then ready to install. Care should be taken to fully maintain the ID of the conduit when bending.

If only a few bends are needed, scribe a chalk line on the floor or workbench. Then, match the heated conduit to the chalk line and cool. The conduit must be held in the desired position until it is relatively cool since the PVC material will tend to revert to its original shape. Templates are also available for many bends.

Another method is to take the heated conduit section to the point of installation and form it to fit the actual installation (*Figure 54*). This method is especially effective for making blind bends or compound bends using smaller sizes of PVC. After bending, wipe a wet rag over the bend (*Figure 55*) to cool it.

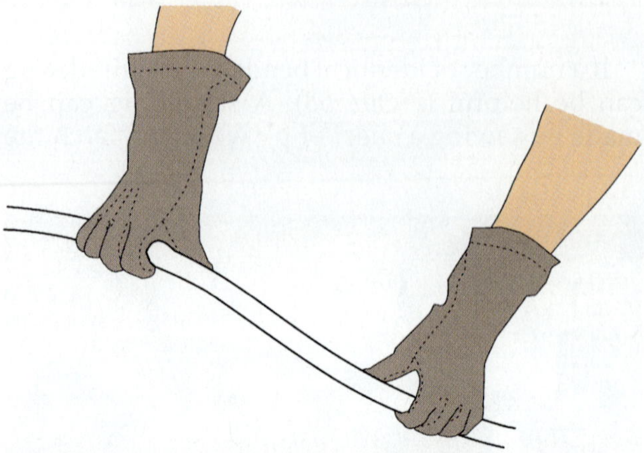

204F54.EPS

Figure 54 ◆ Some PVC bends may be formed by hand.

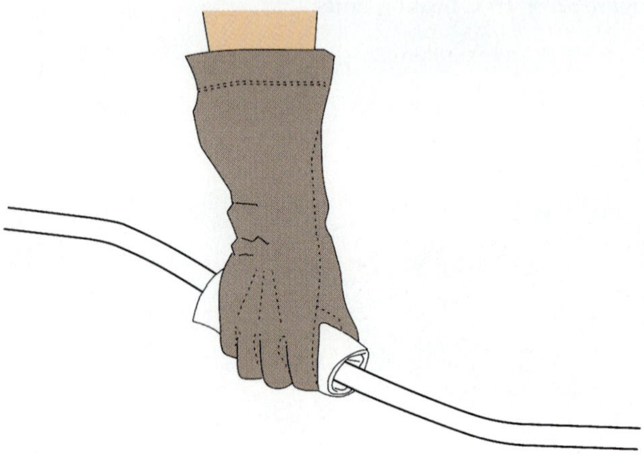

204F55.EPS

Figure 55 ◆ After the bend is formed, wipe a wet rag over the bend to cool it.

Bends in small-diameter PVC conduit (½" to 1½") require no filling for code-approved radii. When bending PVC of 2" or larger diameter, there is a risk of wrinkling or flattening the bend. To help eliminate this problem, a plug set is used. A plug is inserted in each end of the piece of PVC being bent, and then a hand pump is used to pressurize the conduit before bending it. The pressure is about three to five pounds per square inch.

Place airtight plugs (*Figure 56*) in each end of the conduit section before heating. The retained air will expand during the heating process and hold the conduit open during bending. Do not remove the plugs until the conduit has cooled.

In applications where the conduit installation is subject to constantly changing temperatures and the runs are long, precautions should be taken to allow for expansion and contraction of PVC conduit. When expansion and contraction are factors, an O-ring expansion coupling should be installed near the fixed end of the run, or fixture, to take up any expansion or contraction that may occur. Confirm the expansion and contraction lengths available in these fittings as they may vary by manufacturer. Charts are available that indicate what expansion can be expected at various temperature levels. The coefficient of linear expansion of PVC conduit is 0.0034" per 10' per °F. *Table 7* lists the expansion rates for various temperatures.

Expansion couplings are seldom required in underground or slab applications. Expansion and contraction may generally be controlled by bowing the conduit slightly or burying the conduit immediately. After the conduit is buried, expansion and contraction cease to be factors. Care should be taken, however, in constructing a buried installation. If the conduit should be left exposed for an

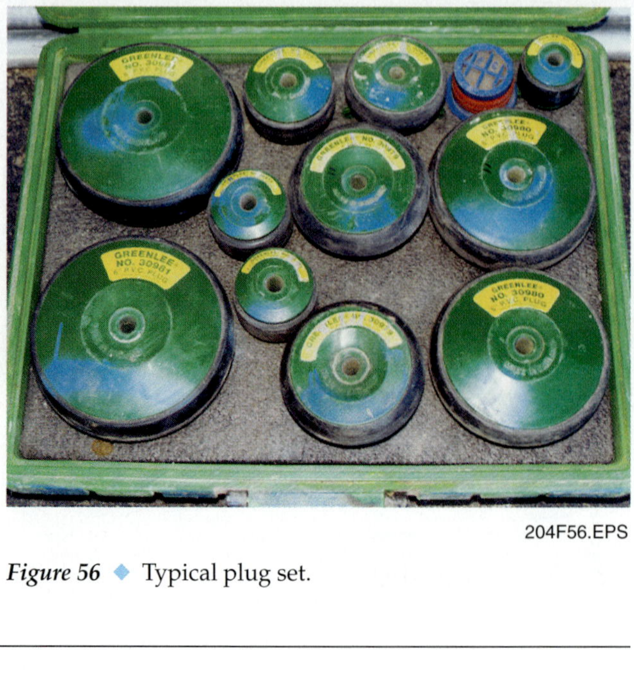

204F56.EPS

Figure 56 ◆ Typical plug set.

Table 7 PVC Expansion Rates

Temperature Change in °F	Length Change in Inches per 100 Ft of PVC Conduit	Temperature Change in °F	Length Change in Inches per 100 Ft of PVC Conduit
5	0.2	105	4.2
10	0.4	110	4.5
15	0.6	115	4.7
20	0.8	120	4.9
25	1.0	125	5.1
30	1.2	130	5.3
35	1.4	135	5.5
40	1.6	140	5.7
45	1.8	145	5.9
50	2.0	150	6.1
55	2.2	155	6.3
60	2.4	160	6.5
65	2.6	165	6.7
70	2.8	170	6.9
75	3.0	175	7.1
80	3.2	180	7.3
85	3.4	185	7.5
90	3.6	190	7.7
95	3.8	195	7.9
100	4.1	200	8.1

extended period of time during widely-variable temperature conditions, an allowance should be made for expansion and contraction.

In above-ground installations, care should be taken to provide proper support of PVC conduit because of its semi-rigidity. This is particularly important at high temperatures. The distance between supports should be based on temperatures encountered at the specific installation. Charts are available that clearly outline at what intervals support is required for PVC conduit at various temperature levels.

INSIDE TRACK

PVC Conduit Bending

PVC conduit has a memory. As soon as you have finished bending the PVC, cool it down with cold, wet towels or hose it down with cold water. Otherwise, it will begin to return to its original position.

THINK ABOUT IT

Putting It All Together

Examine the visible conduit bends in your home or workplace. Are they neat and professionally made? If not, what might you have done differently to improve the installation?

Review Questions

1. The minimum radius of a 90° bend in rigid conduit is _____ times the internal diameter of the conduit.
   a. four
   b. five
   c. six
   d. seven

2. The minimum bending radius for 1" rigid conduit without a lead sheath is _____.
   a. 4"
   b. 6"
   c. 8"
   d. 10"

3. The maximum number of 90° bends allowed between pull points in a conduit system is _____.
   a. one
   b. two
   c. three
   d. four

4. A saddle bend is counted as two ¼ bends or _____.
   a. 90°
   b. 160°
   c. 270°
   d. 180°

5. When referring to the two bends required to make an offset in a length of conduit, which of the following is always true?
   a. The degree of bend for each must be exactly 45°.
   b. The degree of bend for each must be exactly 30°.
   c. The degree of bend for each must be equal.
   d. The degree of bend for each must be unequal.

6. The reason for making a saddle bend in a run of conduit is to _____.
   a. change direction in the height of a conduit run
   b. make a 90° bend
   c. cross a small obstacle
   d. make a conduit termination in an outlet box

7. The equation used to calculate the circumference of a circle is _____.
   a. $C = \pi D$
   b. $C = 1\pi R$
   c. $C = \pi R^2$
   d. $C = \pi^2$

8. The fractional equivalent of the decimal 0.015625 is _____.
   a. 1/64
   b. 1/32
   c. 3/64
   d. 1/16

9. The rise in a conduit bend is best described as the _____.
   a. horizontal distance that runs from the point of bend and parallel with the deck
   b. leg length
   c. radius
   d. height of the stub-up

10. The best type of bending shoe for making concentric bends is the _____ shoe.
    a. concentric
    b. one-shot
    c. segmented
    d. long

11. Aluminum conduit is available in standard sizes from _____.
    a. ½" to 4"
    b. ½" to 6"
    c. ¾" to 6"
    d. ¾" to 4"

12. Two or more parallel bends that are bent in the same direction are known as _____.
    a. stubs
    b. perpendicular bends
    c. concentric bends
    d. axial bends

13. Which of the following formulas should be used to find the side adjacent?
    a. $S = O \times H$
    b. $S = O \div H$
    c. $C = A \div H$
    d. $C = A \times H$

Review Questions

14. An important safety precaution to use when working with PVC conduit is to _____.
    a. provide proper ventilation to carry off fumes from the joint cement
    b. use lengths of PVC of not more than 8'
    c. use inside diameters of 4" or less
    d. never wear gloves that may adhere to the joint cement

15. When bending larger diameter PVC pipe, _____ is/are used to prevent flattening.
    a. plugs and air pressure
    b. sand fill
    c. water fill
    d. higher temperature

Summary

Many conduit installations are visible; that is, they run exposed. Consequently, electricians must take special care to ensure that all exposed conduit runs are parallel, level, and plumb. Nothing else will do. This is one phase of the electrical construction industry where electricians have an opportunity to show off. In fact, an expert installation of a conduit system is similar to a work of art.

Learn the basics of conduit bending and installations, put your knowledge to practical use, and take pride in your abilities to perform a professional conduit installation. Make your work second to none. Of course, contractors and clients want speed, but if you have a good basic knowledge of conduit bending and then put this knowledge to use, your craftsmanship will let you bend conduit smarter and faster, giving a good-looking installation along with speed to satisfy your employer, the building owners, and all concerned.

Notes

Trade Terms Introduced in This Module

Approximate ram travel: The distance the ram of a hydraulic bender travels to make a bend. To simplify and speed bending operations, many benders are equipped with a scale that shows ram travel. Using a simple table (supplied with many benders), the degree of bend can easily be converted to inches of ram travel. This measurement, however, can only be approximated because of the variation in springback of the conduit being bent.

Back-to-back bend: Any bend formed by two 90° bends with a straight section of conduit between the bends.

Bending protractor: Made for use with benders mounted on a bending table and used to measure degrees; also has a scale for 18, 20, 21, and 22 shots when using it to make a large sweep bend.

Bending shot: The number of shots needed to produce a specific bend.

Concentric bending: The process of making 90° bends in parallel runs of conduit. This requires increasing the radius of each conduit from the inside of the bend toward the outside.

Conduit: Piping designed especially for pulling electrical conductors. Types include rigid, IMC, EMT, PVC, aluminum, and other materials.

Degree indicator: An instrument designed to indicate the exact degree of bend while it is being made.

Developed length: The amount of straight pipe needed to bend a given radius. Also, the actual length of the conduit that will be bent.

Elbow: A 90° bend.

Gain: The amount of pipe saved by bending on a radius as opposed to right angles. Because conduit bends in a radius and not at right angles, the length of conduit needed for a bend will not equal the total determined length. Gain is the difference between the right angle distances A and B and the shorter distance C—the length of conduit actually needed for the bend.

Inside diameter (ID): The inside diameter of conduit. All electrical conduit is measured in this manner. The outside dimensions, however, will vary with the type of conduit used.

Kicks: Bends in a piece of conduit, usually less than 90°, made to change the direction of the conduit.

Leg length: The distance from the end of the straight section of conduit to the bend, measured to the centerline or to the inside or outside of the bend or rise.

Ninety-degree bend: A bend in a piece of conduit that changes its direction by 90°.

Offsets: Two equal bends made to avoid an obstruction blocking the run of the conduit.

One-shot shoe: A large bending shoe that is designed to make 90° bends in conduit.

Outside diameter (OD): The size of any piece of conduit measured on the outside diameter.

Radius: The relative size of the bent portion of a pipe.

Rise: The length of the bent section of conduit measured from the bottom, centerline, or top of the straight section to the end of the bent section.

Segment bend: Any bend formed by a series of bends of a few degrees each, rather than a single one-shot bend.

Segmented bending shoe: A smaller type of shoe designed for bending segmented bends only (always less than 15°).

Springback: The amount, measured in degrees, that a bent conduit tends to straighten after pressure is released on the bender ram. For example, a 90° bend, after pressure is released, will pull back about 2° to 88°.

Stub-up: Another name for the rise in a section of conduit.

Sweep bend: A 90° bend with a radius larger than that produced by a standard one-shot shoe.

Take-up (comeback): The amount that must be subtracted from the desired stub length to make the bend come out correctly using a point of reference on the bender or bending shoe.

Additional Resources

This module is intended to present thorough resources for task training. The following reference works are suggested for further study. These are optional materials for continued education rather than for task training.

Benfield Conduit Bending Manual, 2nd Edition. KS: EC&M Books.

National Electrical Code® Handbook, Latest Edition. Quincy, MA: National Fire Protection Association.

Tom Henry's Conduit Bending Package (includes video, book, and bending chart). Winter Park, FL: Code Electrical Classes, Inc.

CONTREN® LEARNING SERIES — USER FEEDBACK

The NCCER makes every effort to keep these textbooks up-to-date and free of technical errors. We appreciate your help in this process. If you have an idea for improving this textbook, or if you find an error, a typographical mistake, or an inaccuracy in NCCER's *Contren®* textbooks, please write us, using this form or a photocopy. Be sure to include the exact module number, page number, a detailed description, and the correction, if applicable. Your input will be brought to the attention of the Technical Review Committee. Thank you for your assistance.

Instructors – If you found that additional materials were necessary in order to teach this module effectively, please let us know so that we may include them in the Equipment/Materials list in the Annotated Instructor's Guide.

Write: Product Development
National Center for Construction Education and Research
P.O. Box 141104, Gainesville, FL 32614-1104

Fax: 352-334-0932

E-mail: curriculum@nccer.org

Craft ______ Module Name ______

Copyright Date ______ Module Number ______ Page Number(s) ______

Description ______

______

______

______

(Optional) Correction ______

______

______

(Optional) Your Name and Address ______

______

______

Boxes and Fittings

26205-05

ExxonMobil URC Training Center
Houston, Texas
Commercial $10–25 Million Award Winner
D. E. Harvey Builders

26205-05
Boxes and Fittings

Topics to be presented in this module include:

Overview

The outlet and pull boxes used in an electrical system are selected according to their volume capacity. This volume capacity, called box fill, is measured in cubic inches or centimeters and is regulated by the *National Electrical Code*®.

There are many types of outlet and pull boxes to choose from. It is essential to select the right box for the job in order to make conductor installation easier, and to comply with *NEC*® regulations that govern the types and sizes of boxes. Some of the factors affecting box selection include environmental conditions, mounting surfaces or materials, and raceway design, as well as the size, type, and number of conductors.

Conduit bodies, or condulets as they are often called, also play an important role in conductor installation. The *NEC*® regulates the type of conduit and conduit bodies that may be installed in hazardous locations.

Note: *National Electrical Code*® and *NEC*® are registered trademarks of the National Fire Protection Association, Inc., Quincy, MA 02269. All *National Electrical Code*® and *NEC*® references in this module refer to the 2005 edition of the *National Electrical Code*®.

Objectives

When you have completed this module, you will be able to do the following:

1. Describe the different types of nonmetallic and metallic boxes.
2. Understand the *NEC*® requirements for box fill.
3. Calculate the required box size for any number and size of conductors.
4. Explain the *NEC*® regulations for volume required per conductor in outlet boxes.
5. Properly locate, install, and support boxes of all types.
6. Describe the *NEC*® regulations governing pull and junction boxes.
7. Explain the radius rule when installing conductors in pull boxes.
8. Understand the *NEC*® requirements for boxes supporting lighting fixtures.
9. Describe the purpose of conduit bodies and Type FS boxes.
10. Install the different types of fittings used in conjunction with boxes.
11. Describe the installation rules for installing boxes and fittings in hazardous areas.
12. Explain how boxes and fittings are selected and installed.
13. Describe the various types of box supports.

Required Trainee Materials

1. Pencil and paper
2. Appropriate personal protective equipment
3. Copy of the latest edition of the *National Electrical Code*®

Prerequisites

Before you begin this module, it is recommended that you successfully complete *Core Curriculum; Electrical Level One;* and *Electrical Level Two,* Modules 26201-05 through 26204-05.

This course map shows all of the modules in *Electrical Level Two.* The suggested training order begins at the bottom and proceeds up. Skill levels increase as you advance on the course map. The local Training Program Sponsor may adjust the training order.

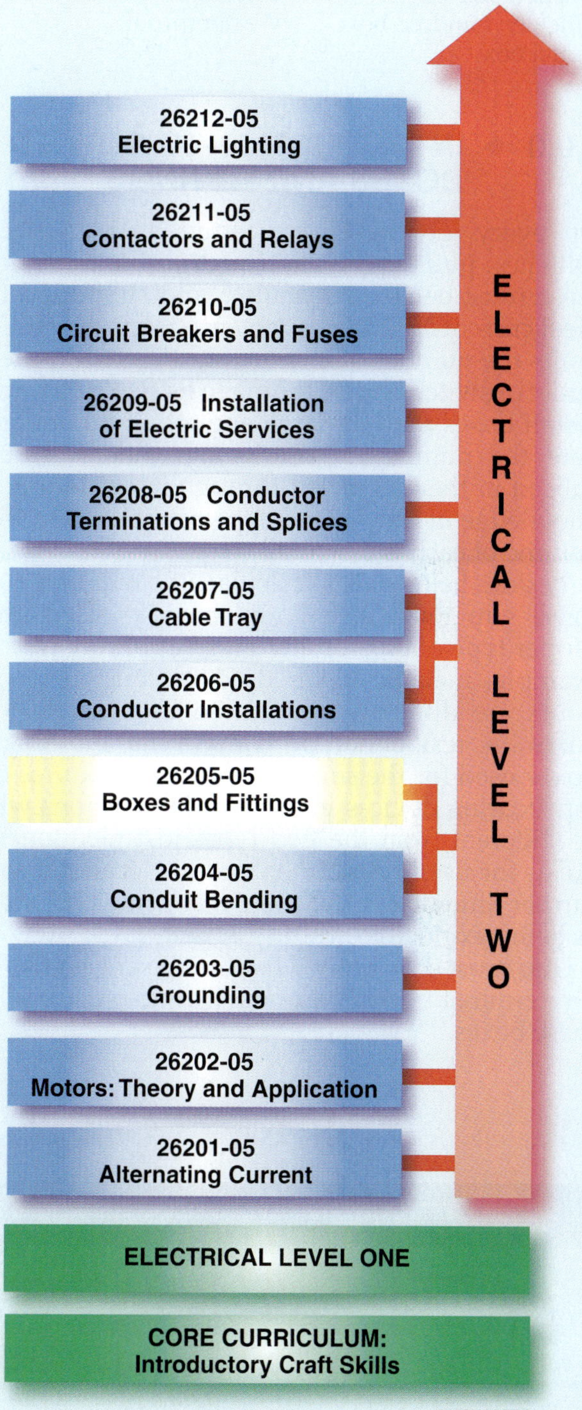

205CMAP.EPS

Trade Terms

Conduit body	Pull box
Connector	Raintight
Explosion-proof	Sealing compound
Handy box	Waterproof
Junction box	Watertight
Nineteen hundred box	Weatherproof
Outlet box	

1.0.0 ◆ INTRODUCTION TO OUTLET BOXES AND FITTINGS

On every job, boxes are required for outlets, switches, **pull boxes,** and **junction boxes.** All of these must be sized, installed, and supported to meet current *NEC*® requirements. Since the *NEC*® limits the number of conductors allowed in each outlet or switch box according to its size, you must install boxes that are large enough to accommodate the number of conductors that must be spliced in the box or fed through it. Therefore, a knowledge of the various types of boxes and the volume of each is essential.

Besides being able to calculate the required box sizes, you must also know how to select the proper type of box for any given application. For example, metallic boxes used in concrete deck pours are different from those used as device boxes in residential or commercial buildings. Boxes used for the support of lighting fixtures or for securing devices in outdoor installations will be different from the two types just mentioned. Boxes for use in certain hazardous locations will further differ in construction; many must be rated as being **explosion-proof.**

You must also know what fittings are available for terminating the various wiring methods in these boxes.

Electrical drawings rarely indicate the exact types of **outlet boxes** to be used in a given area, with the possible exception of boxes used in hazardous locations. Electricians who lack practical on-the-job experience may not always choose the best box for a given application. The use of improper boxes and other ill-adapted materials will cause excessive time to be taken on the job. For example, outlet boxes for use with Type AC cable should contain built-in clamps; many times boxes will be ordered with knockouts only. This latter case requires the use of additional **connectors,** which may only take a few additional seconds per connection, but when these are added up over the period of a large project, much additional time is wasted. Because the cost of labor is an expensive item, any excess labor required will more than offset any savings gained from the use of inadequate materials.

Excess labor is often consumed due to obstructions that enter conduit installations during the general construction work. Most of these obstructions can be avoided by plugging all conduit terminals with capped bushings or some similar means of protection (*Figure 1*). Care should also be taken to thoroughly tighten all fittings, couplings, and so on. In the case of outlet boxes installed in poured concrete structures, painting the inside of the boxes with grease and using capped bushings will greatly aid in keeping concrete out of the raceway system. If any concrete should enter the box, the grease will prevent it from adhering to the metal box.

This module is designed to cover boxes and fittings in depth, based on the applicable *NEC*® regulations. A knowledge of this material should enable you to approach your work in an efficient manner.

INSIDE TRACK

Special Considerations

Before you can effectively plan and route conductors to their termination points and then select and install the correct types and sizes of boxes, you will need to have the following information:

- Length and number of conductors
- Conductor ampacity
- Allowances for voltage drops
- Type(s) of hazard(s) in the environment

205F01.EPS

Figure 1 ◆ Conduit caps.

2.0.0 ◆ TYPES OF BOXES

Outlet boxes normally fall into three categories:

- Pressed steel boxes with knockouts of various sizes for raceway or cable entrances
- Cast iron, aluminum, or brass boxes with threaded hubs of various sizes and locations for raceway entrances
- Nonmetallic boxes

Pressed steel boxes also fall into two categories:

- Boxes with conduit, electric metallic tubing, and cable
- Boxes designed for use with specific types of surface metal moldings

Outlet boxes vary in size and shape depending upon their use, the size of the raceway, the number of conductors entering the box, the type of building construction, the atmospheric conditions of the area, and special requirements.

Outlet box covers are usually required to adapt the box to the particular use it is to serve. For example, a 4" square box is adapted to one-gang or two-gang switches or receptacles by the use of either one-gang or two-gang flush device covers. A one-gang cast hub box can be adapted to provide a vapor-proof switch or a vapor-proof receptacle cover.

Special outlet box hangers are available to facilitate their installation, particularly in frame building construction.

The types of enclosures used as outlet and device boxes for the support of fixtures or for securing devices, such as switches, receptacles, or other equipment, on the same yoke or strap are available in various sizes and shapes. These enclosures may be used in the one-gang, two-gang, three-gang, or four-gang types. Ceiling outlet boxes are available in various shapes. Where a square metal box is used, it may be used for the support of a fixture or device, depending on the type of raised cover used. Device boxes are those that are usually installed to enclose receptacles and switches.

Boxes installed for the support of a lighting fixture are required to be specifically designed for the purpose. Most device boxes are typically not designed or listed for use to support lighting fixtures. The use of a device box to support fixtures is addressed in *NEC Section 314.27(A), Exception*.

A floor box that is listed specifically for installation in a floor is required where receptacles or junction boxes are installed in a floor. Listed floor boxes are provided with covers and gaskets to exclude surface water and cleaning compounds.

A box used at fan outlets is not permitted to be used as the sole support for ceiling (paddle) fans, unless it is specifically listed for the application as the sole means of support. Where a listed ceiling fan does not exceed 35 pounds in weight, with or without its accessories, it is permitted to be supported by outlet boxes identified for such use. The support of this box must be made so that it is rigidly supported from a structural member of the building. A paddle fan box and its related accessories are shown in *Figure 2*.

INSIDE TRACK

Identifying Boxes

After installing all the boxes in a complex installation, you can save yourself much time and trouble later on if you label each box location according to type. Using a can of spray paint, paint small dots on the floor below each box. Use one dot for a plug receptacle, two dots for a switch box, and three dots for a wall light. After the drywall installers are finished, you'll be able to easily identify any wall box.

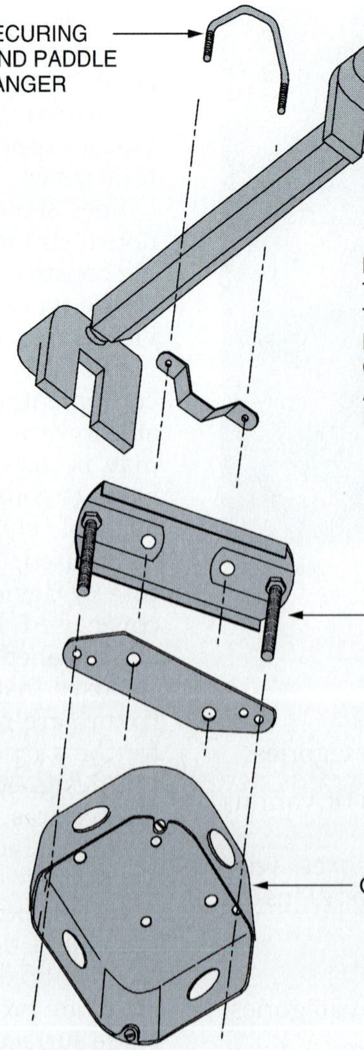

Figure 2 ◆ Typical paddle fan box for installation after the ceiling is installed.

2.1.0 Octagon and Round Boxes

Metallic octagon boxes are available with knockouts for use with either conduit (using locknuts and bushings) or cable box connectors. They are also available with both Type AC and NM cable clamps. The standard width of octagon boxes is four inches, with depths available in 1¼", 1½", or 2⅛". Extension rings are also available for increasing the depth.

Round boxes are available in the same dimensions, but *NEC Section 314.2* prevents using such boxes where conduit or connectors—requiring locknuts and bushings—are connected to the side of the box. The conduit or box connector must terminate in the top of such boxes. Round boxes with cable clamps, however, are permitted for use with Type AC and NM cables that may terminate in either the side or top of the box.

Nonmetallic round boxes are permitted only with open wiring on insulators, concealed knob-and-tube wiring, Type NM cable, and nonmetallic raceways.

Figure 3 shows typical metallic octagon boxes and an octagon extension ring. *Figure 3(A)* shows a box with concentric knockouts for conduit or box connectors, while *Figure 3(B)* shows a box that utilizes cable clamps. *Figure 3(C)* shows the extension ring. *Figure 4* shows a nonmetallic round box with a bar hanger for mounting between studs or joists.

Octagon and round boxes are used mostly for ceiling and wall-mounted incandescent lighting fixtures. However, covers are available for octagon boxes that will support receptacles and switches. Blank covers are also available when the box is used as a junction box.

Locating Boxes on Walls

The Universal Building Code (UBC) requires boxes on opposite sides of walls between occupancies to be separated by at least 24" in order to maintain the fire rating. (See *NEC Section 300.21.*)

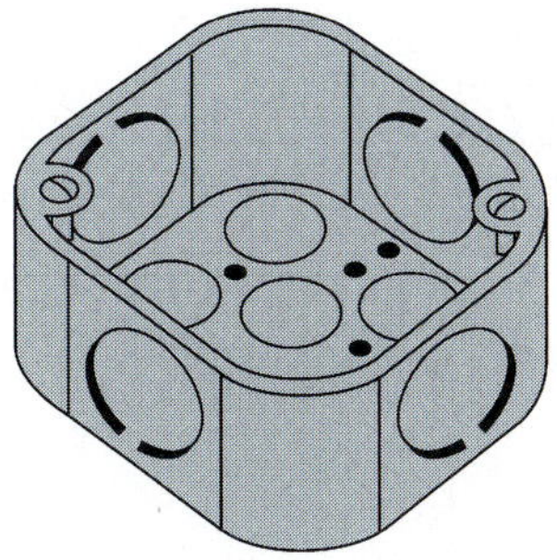

(A)

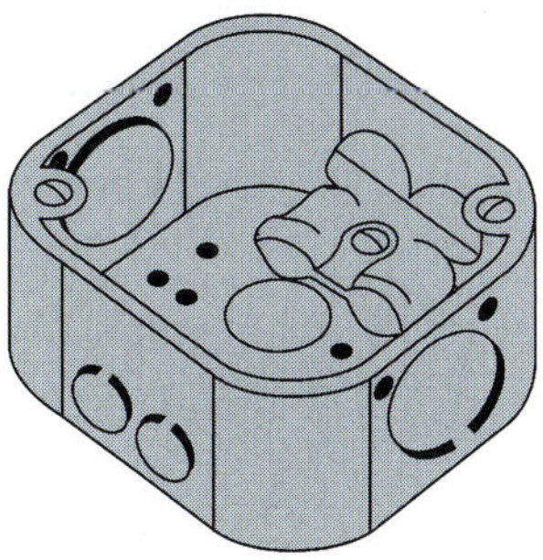

(B)

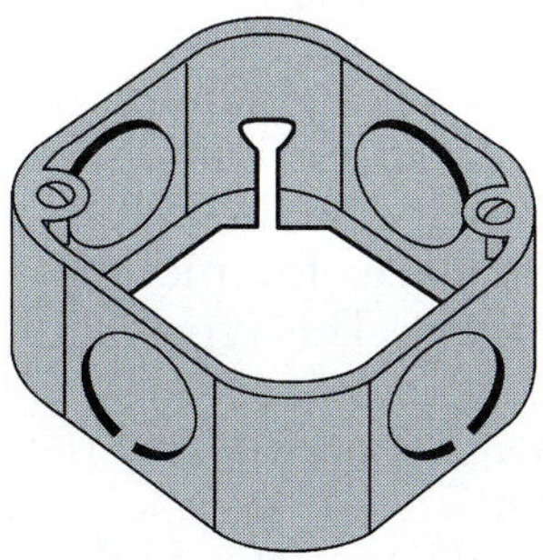

(C)

205F03.EPS

Figure 3 ◆ Typical octagon boxes and octagon extension ring.

Figure 5 shows a cross section of a 3½" round shallow box used for supporting lighting fixtures that have integral wire termination space. *NEC Section 314.24* requires that this and all boxes have a minimum depth of ½". Boxes intended to enclose flush devices must not have a depth of less than 15⁄16".

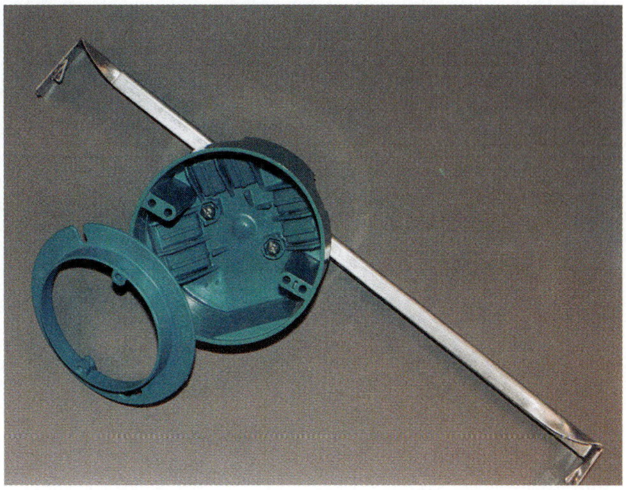

205F04.EPS

Figure 4 ◆ Nonmetallic round box with mounting bracket.

2.2.0 Square Boxes

Square boxes are available in 4" and 411⁄16" square sizes. Both types are available in depths of 1¼", 1½", and 2⅛". Extension rings are also available to further increase the depth. These boxes are available with or without mounting brackets (for nailing to wooden structural members). Boxes designed for use with cable have either Type AC or NM clamps for securing the cable at the box entrance points. Either type of square box may be used with a single or two-gang device ring (e.g., plaster ring or tile ring) for mounting receptacles or switches. A ring with a round opening is also available for mounting lighting fixtures. Blank covers are available when the boxes are used as junction boxes.

Square boxes have been dubbed **nineteen hundred boxes** by some electricians because one of the early manufacturers gave its square boxes a model number of 1900. *Figure 6* shows a square box with a rectangular extension ring that is used to bring the box to the masonry surface.

Square boxes are used for mounting two wiring devices such as junction boxes, and when the number of wires warrants more capacity than is available in other types of boxes.

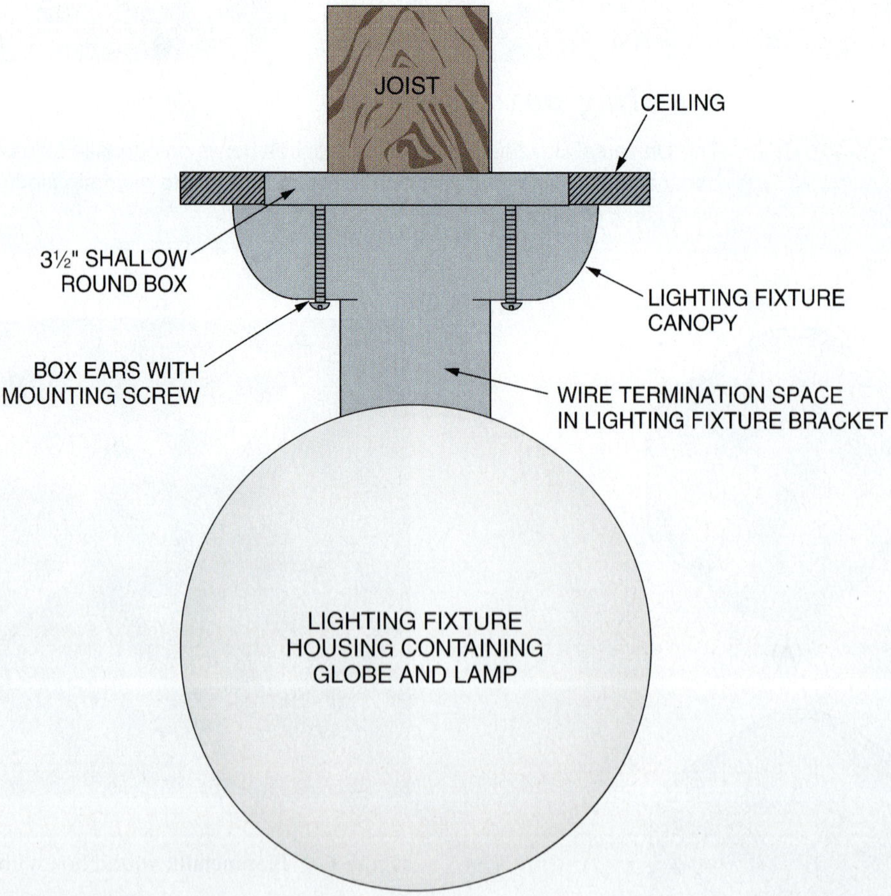

Figure 5 ◆ Shallow round box used for mounting a lighting fixture.

Figure 6 ◆ Square box with extension ring.

2.3.0 Device Boxes

Device boxes are designed for flush mounting mainly in residential and some commercial applications. They are available with or without cable clamps and brackets for mounting to wooden structural members. This type of box is also available with plaster ears for installation in finished wall partitions. Several types of device boxes are shown in *Figure 7*.

A special single-gang box with rounded edges is also used and is called a **handy box** (also known as a utility box). Such boxes are available in depths of 1½", 1⅞", and 2⅛". Care must be exercised when using these boxes as their limited volume restricts the number of conductors permitted in the box.

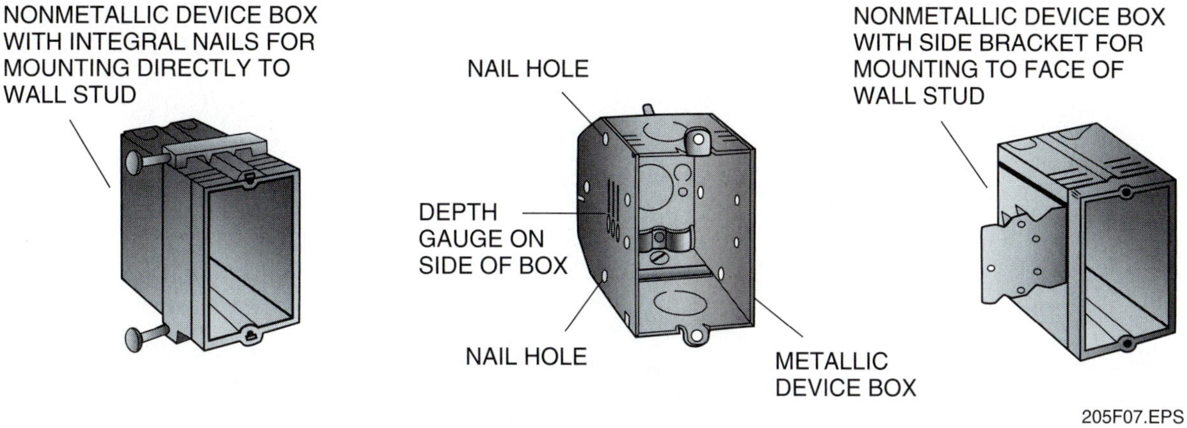

Figure 7 ◆ Typical device boxes.

2.4.0 Concrete Boxes

Special boxes designed for use in concrete pours are used in flat-slab construction jobs. These boxes consist of a sleeve with external ears and a plate which is attached after the sleeve is nailed to the deck.

Concrete boxes are manufactured in different heights and care should be taken to use boxes of sufficient height to allow the knockouts to come well above the reinforcing rods. This eliminates the need for offsets in the conduit where it enters a box. *Figure 8* shows a practical application of a concrete box.

2.5.0 Boxes for Damp and Wet Locations

In damp or wet locations, boxes and fittings must be placed or equipped to prevent moisture or water from entering and accumulating within the box or fitting. It is recommended that approved boxes of nonconductive material be used with nonmetallic sheathed cable or approved nonmetallic conduit when the cable or conduit is used in locations where there is likely to be occasional moisture present. Boxes installed in wet locations must be approved for the purpose per *NEC Section 314.15(A)*.

A wet location is any location subject to saturation with water or other liquids, such as locations exposed to weather or water, washrooms, garages, and interiors that might be hosed down. Underground installations or those in concrete slabs or masonry in direct contact with the earth must be considered to be wet locations. **Raintight, waterproof,** or **watertight** equipment (including fittings) may satisfy the requirements for **weatherproof** equipment. Boxes with threaded conduit hubs and gasketed covers will normally prevent water from entering the box except for condensation within the box.

A damp location is a location subject to some degree of moisture. Such locations include partially protected outdoor locations—such as under canopies, marquees, and roofed open porches. It also includes interior locations subject to moderate degrees of moisture—such as some basements, some barns, and cold storage warehouses.

Weatherproof covers for outdoor receptacles must be chosen with care. If the receptacle feeds a permanently connected load (such as a lighting fixture), the entire enclosure must be weatherproof with the plug inserted. If the receptacle is

INSIDE TRACK

Use of Round Boxes Is Limited

Round boxes are not to be used in cases where conduit must enter from the side of the box because it is difficult to make a good connection with a locknut or bushing on a rounded surface.

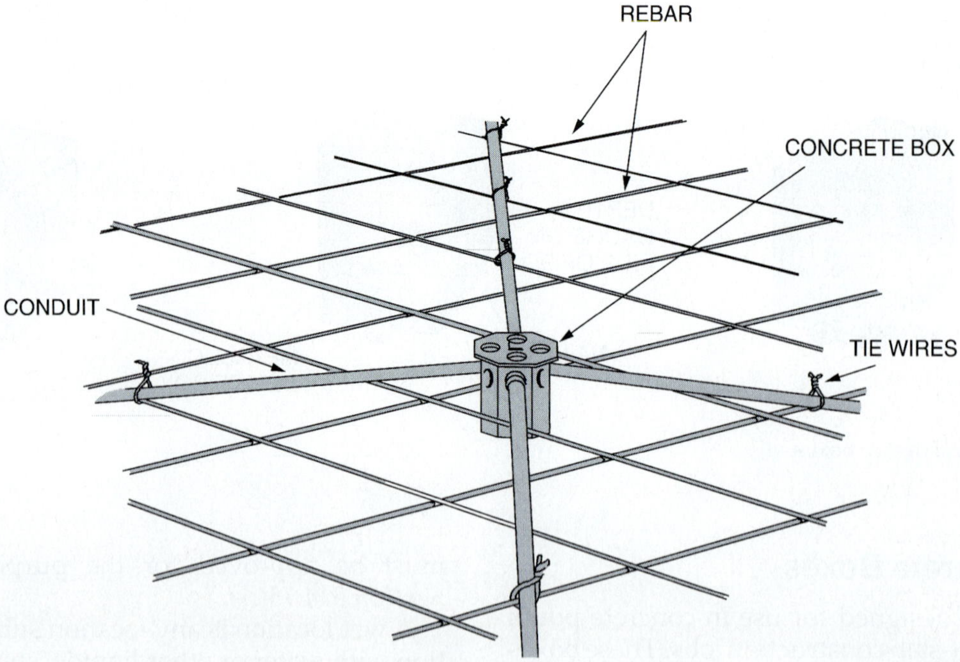

Figure 8 ◆ Practical application of concrete boxes.

Knockouts

INSIDE TRACK

When completing building renovations, it may be difficult to remove a knockout from a previously installed box without dislodging the box. One way to do it is to drill into the knockout and partially insert a self-tapping screw. Then use diagonal or side-cut pliers to pull the knockout from the box.

used only with portable tools or other portable equipment, the enclosure must be weatherproof with the cover closed and the cover must be self-closing. *NEC Section 406.8* covers installation of receptacles in damp or wet locations.

3.0.0 ◆ SIZING OUTLET BOXES

In general, the maximum number of conductors permitted in standard outlet boxes is listed in *NEC Table 314.16(A).* These figures apply where no fittings or devices such as fixture studs, cable clamps, switches, or receptacles are contained in the box and where no grounding conductors are part of the wiring within the box. Obviously, in all modern residential wiring systems there will be one or more of these items contained in the outlet box. Therefore, where one or more of the above-mentioned items are present, the number of conductors is reduced by one less than that shown in the table for each type of fitting and by two for each device strap. For example, a deduction of two conductors must be made for each strap containing a device such as a switch or duplex receptacle; a further deduction of one conductor shall be made for one or more grounding conductors entering the box. For example, a 3" × 2" × 3½" box is listed in the table as containing a maximum number of eight No. 12 wires. If the box contains cable clamps and a duplex receptacle, three wires will have to be deducted from the total of eight—providing for only five No. 12 wires. If a ground wire is used, only four No. 12 wires may be used, which might be the case when a three-wire cable with ground is used to feed a three-way wall switch.

A pictorial definition of stipulated conditions as they apply to *NEC Section 314.16* is shown in the following illustrations. *Figure 9* illustrates an assortment of raised covers and outlet box extensions. These components, when combined with the appropriate outlet boxes, serve to increase the

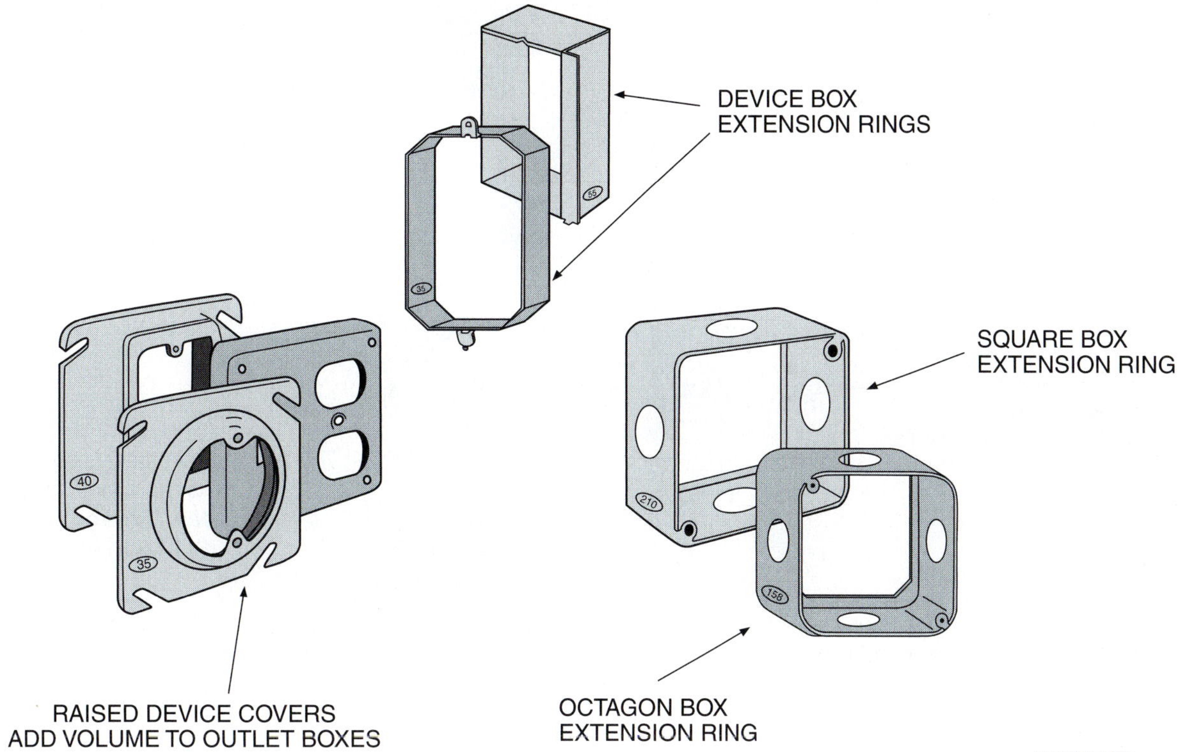

Figure 9 ◆ Devices or components that add to outlet box capacity.

usable work space. Each type is marked with its cubic inch capacity, which may be added to the figures in *NEC Table 314.16(A)* to calculate the increased number of conductors allowed.

Figure 10 shows typical wiring configurations, which must be counted as conductors when calculating the total capacity of outlet boxes. A wire passing through the box without a splice or tap is counted as one conductor. Therefore, a cable containing two wires that passes in and out of an outlet box with a splice or tap is counted as two conductors. However, wires that enter a box and are either spliced or connected to a terminal, and then exit again, are counted as two conductors. In the case of two cables that each have two wires, the total conductors charged will be four. Wires that enter and terminate in the same box are charged as individual conductors and in this case, the total charge would be two conductors. Remember, when one or more grounding wires enter the box and are joined, a deduction on only one is required, regardless of their number.

Further components that require deduction adjustments from those specified in *NEC Table 314.16(A)* include fixture studs, hickeys, and fixture stud extensions *[NEC Section 314.16(B)(3)]*. One conductor must be deducted from the total for each type of fitting used. Two conductors must be deducted for each strap-mounted device, such as duplex receptacles and wall switches; a deduction of one conductor is made when one or more internally mounted cable clamps are used *[NEC Section 314.16(B)(2) and (4)]*.

INSIDE TRACK

Outdoor Boxes

Outdoor wiring must be able to resist the entry of water. Outdoor boxes are either drip tight, which means sealed against falling water from above, or watertight, which means sealed against water from any direction. Drip-tight boxes simply have lids that deflect rain; they are not waterproof. Watertight boxes are sealed with gaskets to prevent the entry of water from any angle.

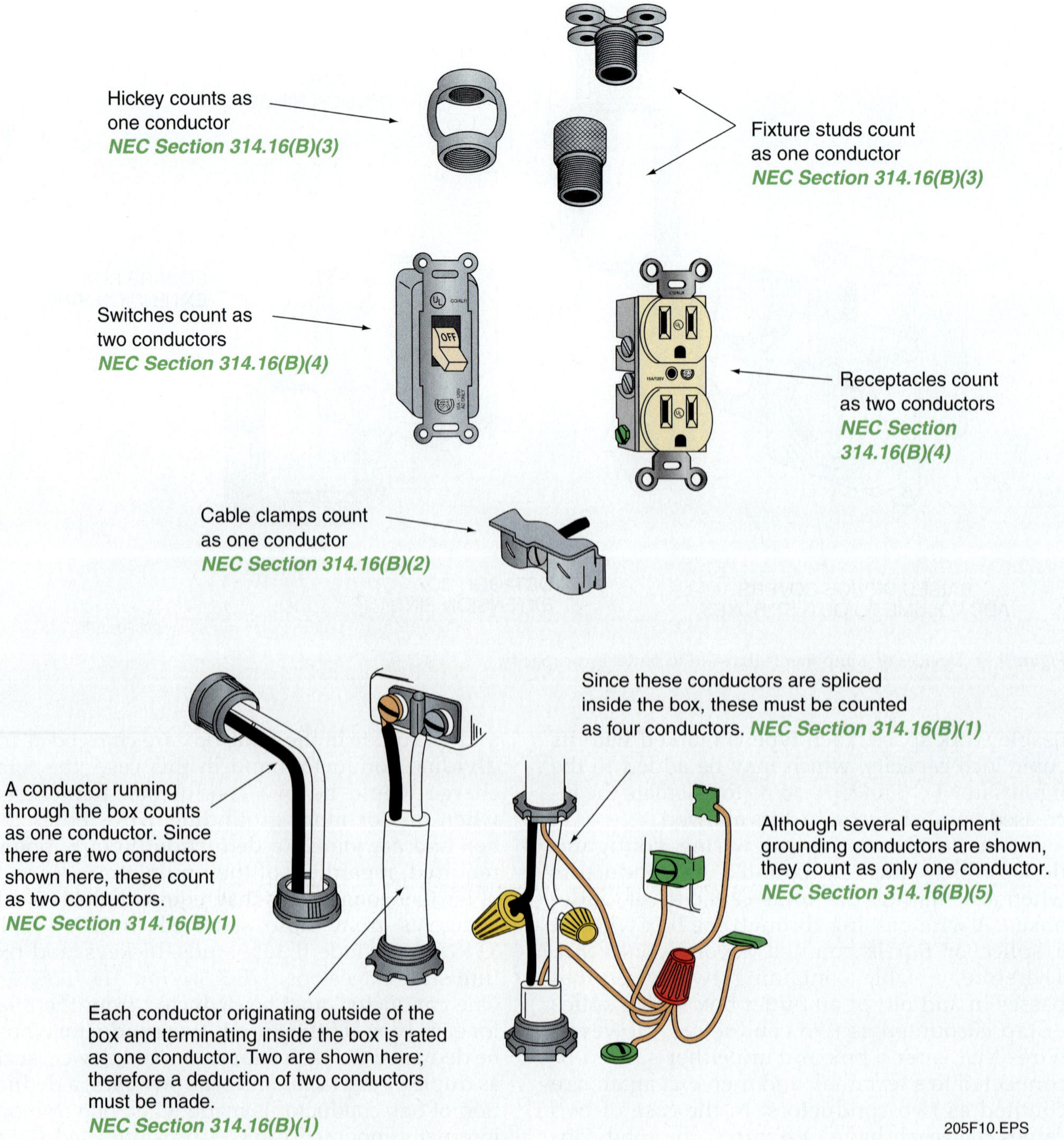

Figure 10 ◆ Devices and components that require deductions in outlet box capacity.

Figure 11 shows components that may be used in outlet boxes without affecting the total number of conductors. Such items include grounding clips and screws, wire nuts, and cable connectors when the latter are inserted through knockout holes in the outlet box and secured with locknuts. Prewired fixture wires are not counted against the total number of allowable conductors in an outlet box; neither are conductors originating and ending in the box.

To better understand how outlet boxes are sized, we will take two No. 12 AWG conductors installed in ½" EMT and terminating into a metallic outlet box containing one duplex receptacle. What size outlet box will meet *NEC®* requirements?

The first step is to count the total number of conductors and equivalents that will be used in the box *(NEC Section 314.16)*.

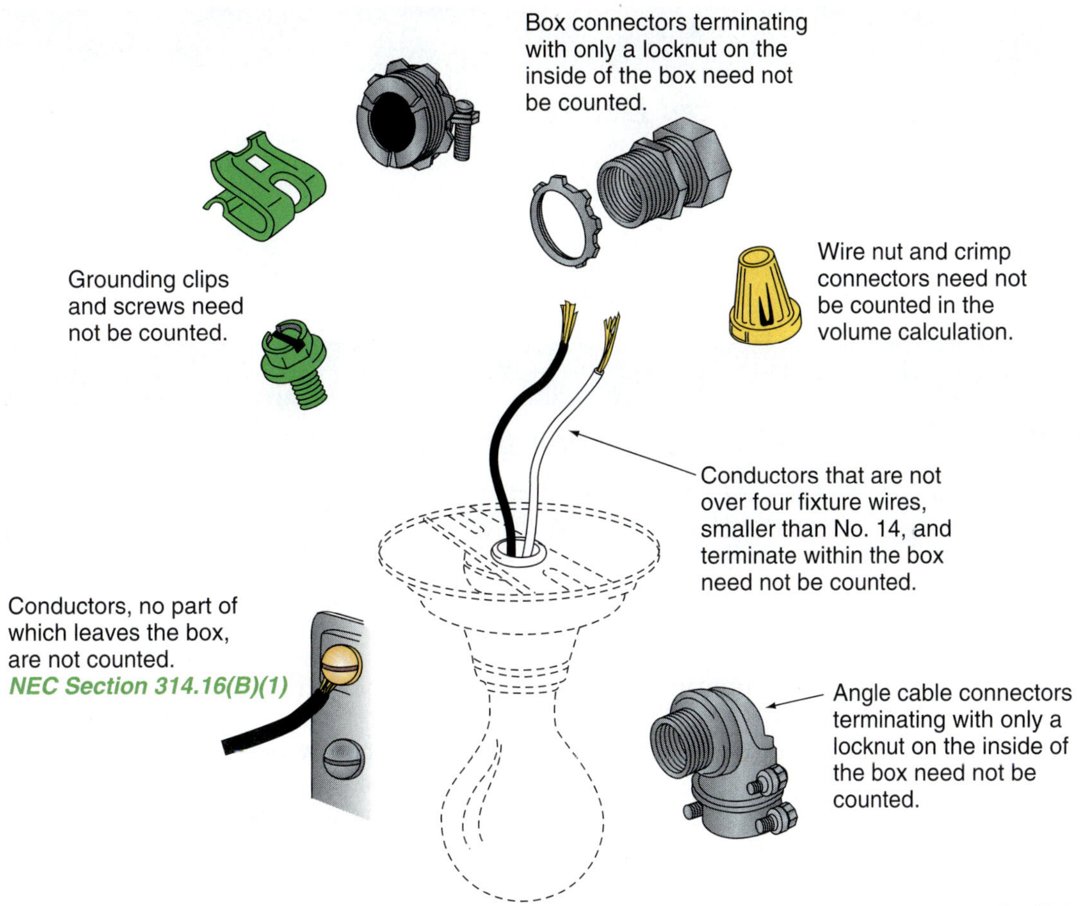

Figure 11 ◆ Items that may be disregarded when calculating outlet box capacity.

INSIDE TRACK

Allocating Space in a Box

Here is a summary of the *NEC®* rules for determining the capacity of a box. Add the number of conductor equivalents indicated for each device:

Each conductor	1
Each strap containing a device	2
Each fixture, stud, or hickey	1
One or more grounding conductors	1
One or more cable clamps	1

Also add one extra allowance for an additional grounding conductor when it is an isolating ground conductor.

Step 1 Calculate the total number of conductors and their equivalents:

One receptacle = 2
Two #12 conductors = 2
Total #12 conductors = 4

Step 2 Determine the amount of space required for each conductor. *NEC Table 314.16(B)* gives the box volume required for each conductor:

No. 12 AWG = 2.25 cubic inches

Think About It

Oversized Devices

Some devices, such as GFCI receptacles and dimmer switches, are larger than the standard boxes. How does the *NEC®* deal with this issue?

Inside Track

Metal Boxes Must Be Grounded

Metal boxes are good conductors. Therefore, when metal boxes are used, they must be grounded to the circuit grounding system.

Step 3 Calculate the outlet box space required by multiplying the number of cubic inches required for each conductor by the number of conductors found in Step 1 above.

$$4 \times 2.25 = 9.00 \text{ cubic inches}$$

Step 4 Once you have determined the required box capacity, again refer to *NEC Table 314.16(A)* and note that a 3" × 2" × 2" box comes closest to our requirements. This box size is rated for 10.0 cubic inches.

For another example, if four No. 12 conductors enter the box, two additional No. 12 conductors must be added to our previous count for a total of six conductors.

$$6 \times 2.25 = 13.5 \text{ cubic inches}$$

Again, refer to *NEC Table 314.16(A)* and note that a 3" × 2" × 2¾" device box with a rated capacity of 14.0 cubic inches is the closest device box that meets *NEC®* requirements. Of course, any box with a larger capacity is permitted.

4.0.0 ◆ PULL AND JUNCTION BOXES

Pull and junction boxes are provided in an electrical installation to facilitate the installation of conductors, or to provide a junction point for the connection of conductors, or both. In some instances, the location and size of pull boxes are designated on the drawings. In most cases, however, the electricians on the job will have to determine the proper number, location, and sizes of pull or junction boxes to facilitate conductor installation.

Pull boxes should be as large as possible. Workers need space within the box for both hands and in the case of the larger wire sizes, workers will need room for their arms to feed the wire. *NEC Section 314.28* specifies that pull and junction boxes must provide adequate space and dimensions for the installation of conductors. For raceways containing conductors of No. 4 or larger, and for cables containing conductors of No. 4 or larger, the minimum dimensions of pull or junction boxes installed in a raceway or cable run shall comply with the following:

- In straight pulls, the length of the box shall not be less than eight times the trade diameter of the largest raceway.
- Where angle or U pulls are made, the distance between each raceway entry inside the box and the opposite wall of the box shall not be less than six times the trade diameter of the largest raceway in a row. This distance shall be increased for additional entries by the amount of the sum of the diameter of all other raceway entries in the same row on the same wall of the box. Each row shall be calculated individually, and the single row that provides the maximum distance shall be used.
- The distance between raceway entries enclosing the same conductor shall not be less than six times the trade diameter of the larger raceway.
- When transposing cable size into raceway size, the minimum trade size raceway required for the number and size of conductors in the cable shall be used.

Long runs of conductors should not be made in one pull. Pull boxes, installed at convenient intervals, will relieve much of the strain on the conductors. The length of the pull, in many cases, is left to the judgment of the workers or their supervisor, and the condition under which the work is installed.

The installation of pull boxes may seem to cause a great deal of extra work and trouble, but they save a considerable amount of time and hard work when pulling conductors. Properly placed, they eliminate many bends and elbows and do away with the necessity of fishing from both ends of a conduit run.

If possible, pull boxes should be installed in a location that allows electricians to work easily and conveniently. For example, in an installation where the conduit comes up a corner of a wall and changes direction at the ceiling, a pull box that is installed too high will force the electrician to stand on a ladder when feeding conductors, and will allow no room for supporting the weight of the wire loop or for the cable-pulling tools.

Unless the contract drawings or project engineer state otherwise, it is just as easy for the pull boxes to be placed at a convenient height that allows workers to stand on the floor with sufficient room for both wire loop and tools.

In some electrical installations, a number of junction boxes must be installed to route the conduit in the shortest, most economical way. The *NEC®* requires all junction boxes to be readily accessible. This means that a person must be able to get to the conductors inside the box without removing plaster, wall covering, or any other part of the building.

Junction boxes detract from the decorative scheme of a building. Therefore, where such boxes will be used in areas open to the public or in other areas where the boxes will be unattractive, they should be installed above suspended ceilings, in closets, or at least in corners of the room or area.

Junction boxes or pull boxes must be securely fastened in place on walls or ceilings or adequately suspended.

While certain sizes of factory-constructed boxes are available with concentric knockouts, in many instances it will be necessary to have them custom built to meet the job requirements. When it is not possible to accurately anticipate the raceway entrance requirements, it will be necessary to cut the required knockouts on the job using hydraulic knockout cutters to keep labor to a minimum.

In the case of large pull boxes and troughs, shop drawings should be prepared prior to the construction of these items with all required knockouts accurately indicated in relation to the conduit run requirements.

4.1.0 Sizing Pull and Junction Boxes

Figure 12 shows a junction box with several runs of conduit. Since this is a straight pull, and 4" conduit is the largest size in the group, the minimum length required for the box can be determined by the following calculation:

Trade size of conduit × 8 [per ***NEC Section 314.28(A)(1)***] = minimum length of box

or:

$$4" \times 8 = 32"$$

Therefore, this particular pull box must be at least 32" in length. The width of the box, however, need only be of sufficient size to enable locknuts and bushings to be installed on all the conduits or connectors entering the enclosure.

Junction or pull boxes in which the conductors are pulled at an angle (*Figure 13*) must have a distance of not less than six times the trade diameter of the largest conduit *[NEC Section 314.28(A)(2)]*. The distance must be increased for additional conduit entries by the amount of the sum of the diameter of all other conduits entering the box on the same side. The distance between raceway entries enclosing the same conductors must not be less than six times the trade diameter of the largest conduit.

Since the 4" conduit is the largest size in this case:

$$L_1 = 6 \times 4" + (3 + 2) = 29"$$

Since the same conduit runs are located on the adjacent wall of the box, L_2 is calculated in the same way; therefore, $L_2 = 29"$.

INSIDE TRACK

Using Pull Boxes

Pull boxes make it easier to install conductors. They can also be installed to avoid having more than 360° worth of bends in a single run. (Remember, if a pull box is used, it is considered the end of the run for the purposes of the *NEC®* 360° rule.)

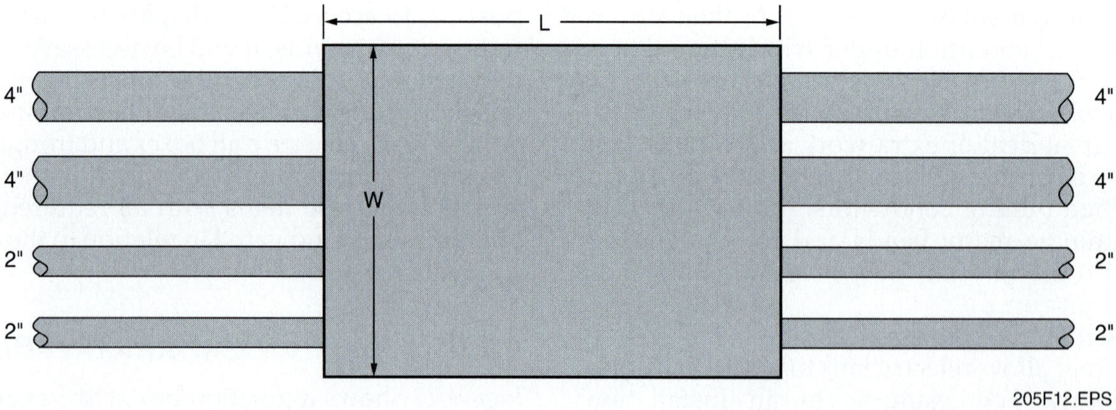

Figure 12 ◆ Pull box with two 4" and two 2" conduit runs.

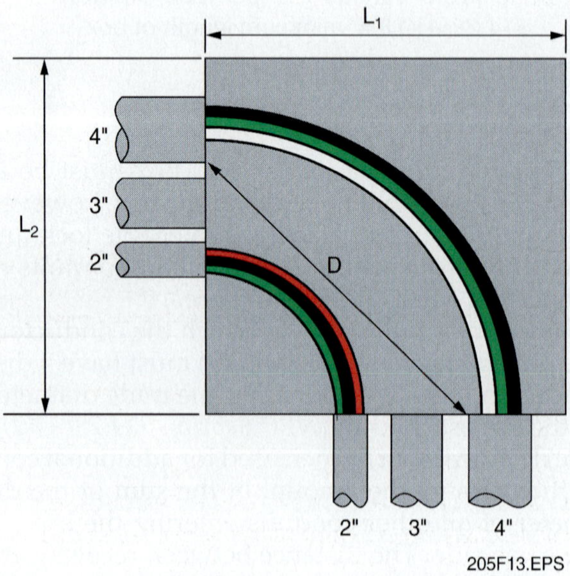

Figure 13 ◆ Pull box with conduit runs entering at right angles.

The distance (D) = 6 × 4" or 24". This is the minimum distance permitted between conduit entries enclosing the same conductor.

The depth of the box need only be of sufficient size to permit locknuts and bushings to be properly installed. In this case, a 6" deep box would suffice.

5.0.0 ◆ CONDUIT BODIES

Conduit bodies, also called condulets, are defined in *NEC Article 100* as a separate portion of a conduit or tubing system that provides access through a removable cover to the interior of the system at a junction of two or more sections of the system or at a terminal point of the system. Conduit bodies are usually used with RMC and IMC. The cost of conduit bodies, because they are cast, is significantly higher than the stamped steel boxes. Splicing in conduit bodies is typically not recommended; however, it is permitted under certain conditions as specified in *NEC Section 314.16(C)(2)*.

As an electrical trainee, you will hear such terms as LL, LR, as well as other letters to distinguish between the various types of conduit bodies. To identify certain conduit bodies, an old trick of the trade is to hold the conduit body like a pistol (*Figure 14*). When doing so, if the oval-shaped opening of the conduit body is to your left, it is called an LL—the first L stands for elbow and the second L stands for left. If the opening is on your right, it is called an LR—the R stands for right. If the opening is facing upward, this type of conduit body is called an LB—the B stands for back. If there is an opening on both sides, it is called an LRL—for both left and right. The other popular

INSIDE TRACK

Label Junction Boxes

It's a good idea to label every junction box plate with the circuit number, the panel it came from, and its destination. The next person to service the installation will be grateful for this extra help.

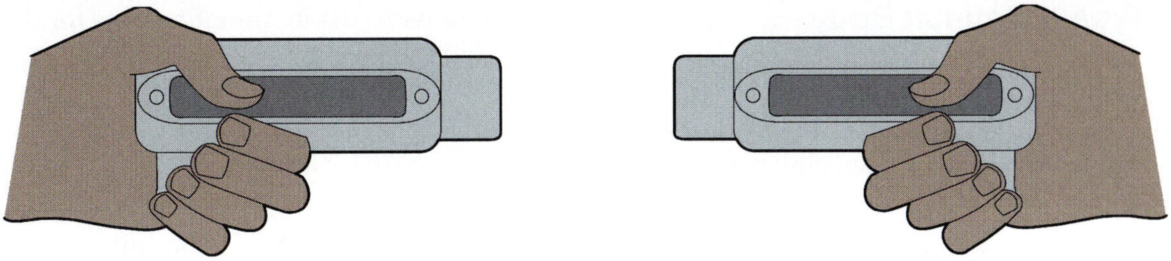

Figure 14 ◆ Identifying conduit bodies.

shapes are named for their letter look-alikes, that is, T and X. The only exception is the C conduit body. Let us take a closer look at each of these.

5.1.0 Type C Conduit Body

A Type C conduit body (*Figure 15*) may be used to provide a pull point in a long conduit run or a conduit run that has bends totaling more than 360° (see *NEC Sections 342.26, 344.26, and 358.26*). In this application, the Type C conduit body is used as a pull point.

5.2.0 Type L Conduit Bodies

A Type L conduit body (*Figure 16*) is used as a pulling point for conduit that requires a 90° change in direction. (Again, the letter L is short for elbow.) To use a Type L conduit body, the cover is removed, the wire is pulled out and coiled on the ground (or floor), and then it is reinserted into the other opening and pulled. Type L conduit bodies are available with the cover on the back (Type LB), on the sides (Type LL and Type LR), or on both sides (Type LRL). The cover and gasket for conduit body fittings must be ordered separately; do not assume that these parts come with the conduit body when it is ordered.

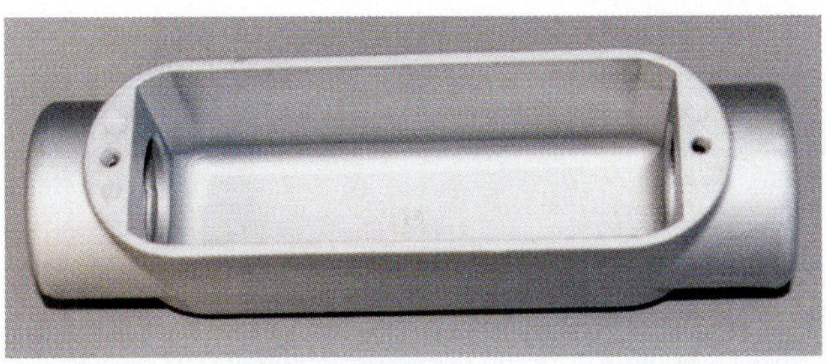

Figure 15 ◆ Type C conduit body.

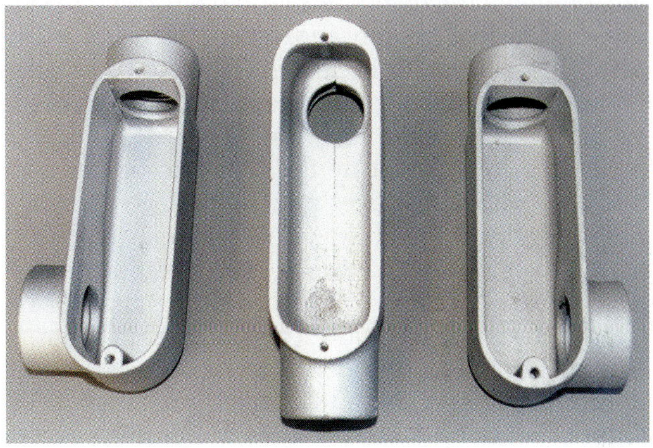

Figure 16 ◆ Type L conduit bodies.

5.3.0 Type T Conduit Body

A Type T conduit body, also known as a tee, is used to provide a junction point for three intersecting conduits (*Figure 17*). Tees are used extensively in rigid conduit systems. The cost of a tee conduit body is more than twice that of a standard 4" square box with a cover. Therefore, the use of Type T conduit bodies with EMT is limited. According to *NEC Section 314.16(C)(2),* conductor splicing is permitted in a Type T conduit body if the necessary conditions are satisfied.

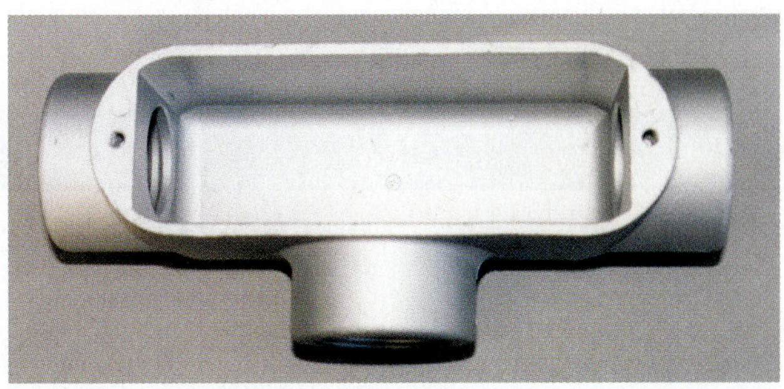

Figure 17 ◆ Type T conduit body.

5.4.0 Type X Conduit Body

A Type X conduit body is used to provide a junction point for four intersecting conduits. The removable cover provides access to the interior of the X so that wire pulling and splicing may be performed (see *Figure 18*).

5.5.0 FS and FD Boxes

FS boxes are cast boxes available in single-gang, two-gang, and three-gang configurations. They are sized to permit the installation of switches and receptacles. Covers for switches and receptacles are available for FS boxes that have formed openings much like switch and receptacle plates. FD boxes are similar to FS boxes. The letter D in the FD box indicates it is a deeper box (2½" deep versus 1⅝" deep for an FS box). Neither FS nor FD boxes are considered by the *NEC*® to be conduit bodies. FS and FD boxes may be used in environments defined by NEMA 1 (dry, clean environments); NEMA 3R (outdoor, wet environments); and NEMA 12 (dusty, oily environments). NEMA stands for the National Electrical Manufacturers Association.

Engineers specify and electricians install FS and FD boxes for a reason. Never alter these boxes by drilling mounting holes in them. Most are provided with cast-in mounting eyes for this purpose. Mount these boxes only as recommended by the manufacturer.

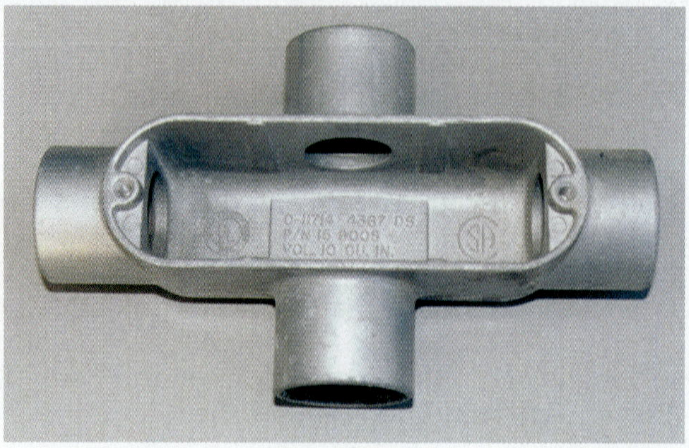

205F18.EPS

Figure 18 ◆ Type X conduit body.

5.6.0 Pulling Elbows

Pulling elbows are used exclusively for pulling wire at a corner point of a conduit run. The volume of a pulling elbow is too low to permit splicing wire. See *Figure 19(A)*.

5.7.0 Entrance Ell (SLB)

An entrance ell, or SLB, is built with an offset so that it may be attached directly to the surface that is to have a conduit penetration. A cover on the back of the SLB permits wire to be pulled out and reinserted into the conduit that penetrates the support surface. See *Figure 19(B)*.

6.0.0 ◆ OUTLET BOXES IN HAZARDOUS LOCATIONS

Any area in which the atmosphere or a material in the area is such that the arcing of operating electrical contacts, components, and equipment may cause an explosion or fire is considered a hazardous location. In all such cases, explosion-proof equipment, raceways, and fittings are used to provide an explosion-proof wiring system, including the outlet boxes.

Hazardous locations have been classified in the *NEC*® into certain class locations. Various atmospheric groups have been established (on the basis of the explosive character of the atmosphere) for the testing and approval of equipment for use in the various groups.

6.1.0 Class I Locations

Those locations in which flammable gases or vapors may be present in the air in quantities sufficient to produce explosive or ignitable mixtures are classified as Class I locations. Examples of such locations are interiors of spray paint booths

INSIDE TRACK

FS and FD Boxes

FS and FD boxes are precision-molded boxes with a cover and gasket designed to provide a tight seal. Because they cost significantly more than standard boxes, they are typically used only in industrial environments where the boxes are subjected to moisture, dirt, dust, and corrosion.

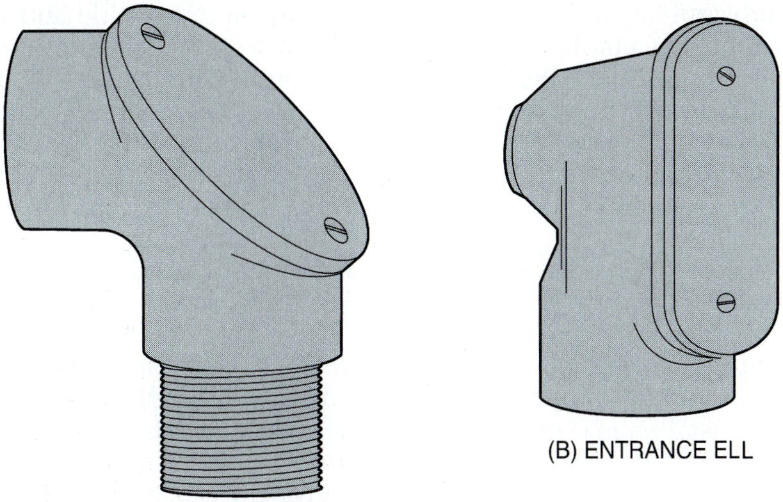

(B) ENTRANCE ELL

(A) PULLING ELBOW

205F19.EPS

Figure 19 ◆ Elbows.

INSIDE TRACK

Making Splices in an Entrance Ell (SLB)

Never make splices in an entrance ell (SLB). Splices can be made in a conduit body that has the cubic inches marked, but never in an SLB.

where volatile, flammable solvents are used, inadequately ventilated pump rooms where flammable gas is pumped, and drying rooms for the evaporation of flammable solvents.

6.2.0 Class II Locations

Class II locations are those that are hazardous because of the presence of combustible dust. Class II, Division 1 locations are areas where combustible dust under normal operating conditions may be present in the air in quantities sufficient to produce explosive or ignitable mixtures; examples are working areas of grain handling and storage plants and rooms containing grinders or pulverizers. Class II, Division 2 locations are areas where dangerous concentrations of suspended dust are not likely, but where dust accumulations might form.

6.3.0 Class III Locations

These locations are those areas that are hazardous because of the presence of easily ignitable fibers or flyings, but such fibers and flyings are not likely to be suspended in the air in these locations in quantities sufficient to produce ignitable mixtures. Such locations usually include some parts of rayon, cotton, and textile mills; clothing manufacturing plants; and woodworking plants.

The wide assortment of explosion-proof equipment now available makes it possible to provide adequate electrical installations under any of these hazardous conditions; however, you must be thoroughly familiar with all *NEC®* requirements. You must also know what fittings are available, how to install them properly, and where and when to use them.

The usual construction documents (drawings and specifications) for a hazardous area are drawn the same as the layout of an electrical system for a non-hazardous area—the only distinction is a note on the drawings stating that the wiring in this particular area must conform to the *NEC®* requirements for hazardous locations. The drafter or designer will sometimes add the letters EXP or XP next to all the symbols of the outlets that are to be explosion-proof. A few large-scale detail drawings may also be present. However, few engineers or drafters detail their drawings for hazardous areas sufficiently for the electricians to proceed with the installation without additional study and layout work on the job site. Therefore, you must be familiar with the layout and installation procedures before attempting such an installation. For other than very simple installations, it may be advisable to make rough, detailed wiring layouts of the proposed installation, even if they are merely sketches on the original working drawings.

When an electrical system in a hazardous location is designed or installed, the type of building structure and finish must be considered. If the building is under construction, this information may be obtained from the architectural drawings and specifications. If the installation is made in an existing building, a preliminary job site investigation is often necessary. The location of the explosion-proof outlets, whether concealed or exposed, and the class of hazardous locations should appear in the electrical drawings and specifications. If such information is not provided, the contractor or electrician will have to determine this information from the architect, owner, or local inspection authority.

In general, rigid metallic conduit is required for all hazardous locations, except for special flexible terminations and as otherwise permitted in the *NEC®*. The conduit should be threaded with a standard conduit butting die that provides ¾" taper per foot. The conduit should be made up wrench-tight in order to minimize sparking in the event that fault current flows through the conduit system [*NEC Section 500.8(D)*]. Where it is impractical to make a threaded joint tight, a bonding jumper should be used. All boxes, fittings, and joints must be threaded for connection to the conduit system and must be of an approved, explosion-proof type. Threaded joints must be made up with at least five threads fully engaged. Where it becomes necessary to employ flexible connectors at motor or fixture terminals, flexible fittings approved for the particular class location must be used.

6.4.0 Seal-Off Fittings

Seal-off fittings are required in conduit systems to minimize the passage of gases and vapors and prevent the passage of flames from one portion of the electrical installation to another through the conduit. For Class I, Division 1 locations, *NEC Section 501.15(A)(1)* states that in each conduit run entering an explosion-proof enclosure containing switches, circuit breakers, fuses, relays, resistors, or other apparatus that may produce arcs, sparks, or high temperatures, seals shall be placed as close as practicable and in no case more than 18" from such enclosures. There shall be no junction box or similar enclosure in the conduit run between the sealing fitting and the apparatus enclosure.

Conduit seals are also required in each conduit run of 2" or larger entering the enclosure or fitting housing terminals, splices, or taps, and within 18" of such enclosures or fittings.

In each conduit run leaving the Class I, Division 1 hazardous area, the sealing fitting may be located on either side of the boundary of the hazardous area within 10' of the boundary, but shall be so designed and installed that any gases or vapors that may enter the conduit system, within the Division 1 hazardous area, will not enter or be communicated to the conduit beyond the seal. There shall be no union, coupling, box, or fitting, except approved explosion-proof reducers at the sealing fitting, in the conduit between the sealing fitting and the point at which the conduit leaves the area.

Sealing compound must be approved for the purpose, must not be affected by the surrounding atmosphere or liquids, and must not have a melting point of less than 200°F (93°C). Most sealing compound kits contain a powder in a polyethylene bag within an outer container. To mix, remove the bag of powder, fill the outside container with water up to the marked line on the container, pour in the powder, and mix.

To pack the seal-off, remove the threaded plug or plugs from the fitting and insert the fiber supplied with the packing kit. Tamp the fiber between the wires and the hub before pouring the sealing compound into the fitting. Then, pour in the sealing cement and reset the threaded plug tightly. The fiber packing prevents the sealing compound (in the liquid state) from entering the conduit lines.

The seal-off (sealing) fittings shown in *Figure 20* are typical of those used. *Figure 20(A)* is for vertical mounting and is provided with a threaded, plugged opening into which the sealing cement is poured. The seal-off in *Figure 20(B)* is for either horizontal or vertical runs, and has an additional plugged opening in the lower hub to facilitate packing fiber around the conductors in order to form a dam for the sealing cement.

Choosing the correct type of seal-off compound is very important. Always use a type that is compatible with the type of conductor insulation and is recommended by the manufacturer of the seal-off itself. There are certain types on the market that are designed for certain applications; make sure you use the right one.

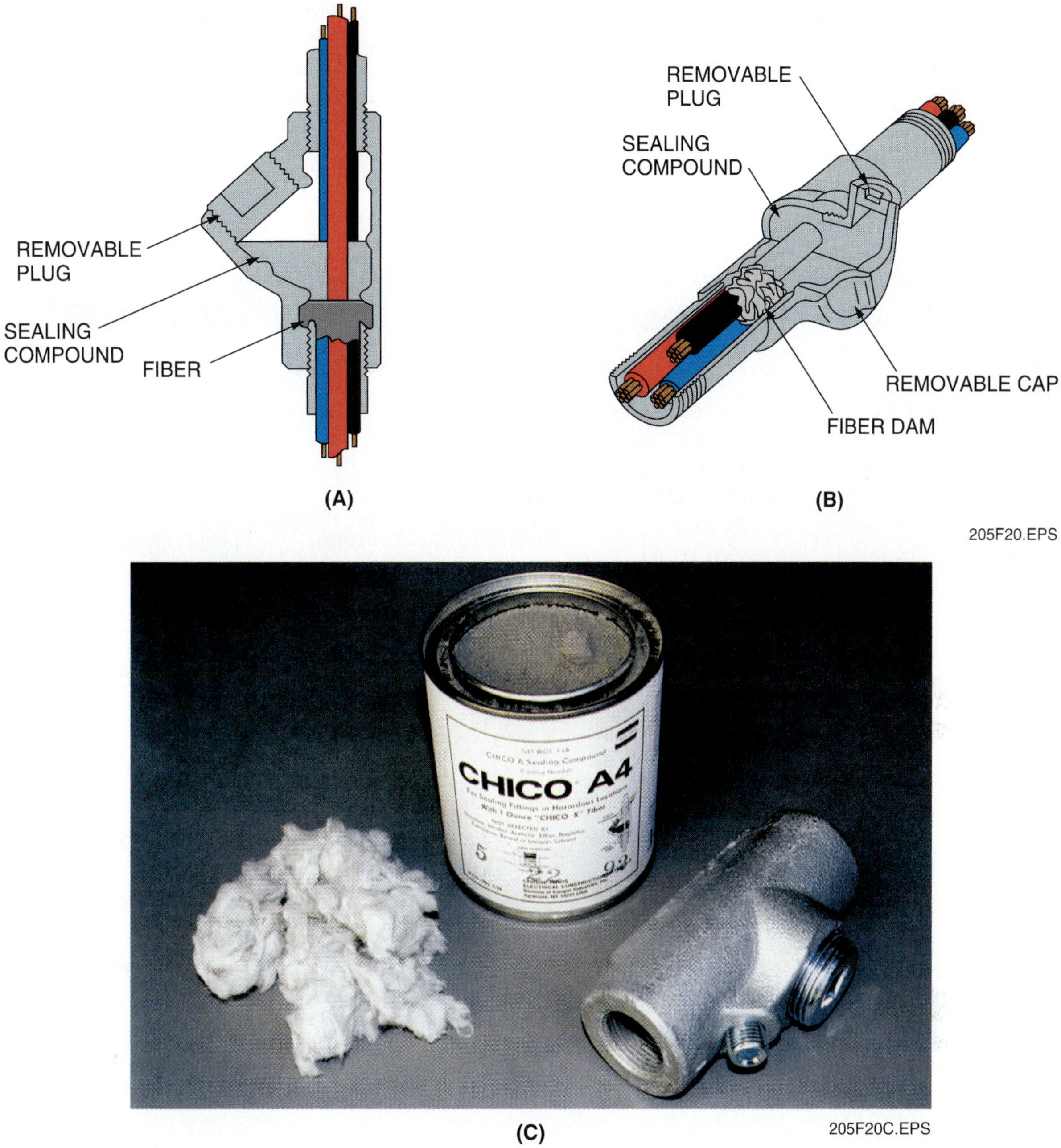

Figure 20 ◆ Typical seal-off fittings.

Most other explosion-proof fittings are provided with threaded hubs for securing the conduit. Typical fittings include switch and junction boxes, conduit bodies, union end connectors, flexible couplings, explosion-proof lighting fixtures, receptacles, and panelboard and motor starter enclosures. A practical representation of these and other fittings appears in *Figure 21*.

In certain hazardous locations, flexible couplings are used to connect vibrating machines, such as motors. Such a connection is shown in *Figure 22*.

INSIDE TRACK

Hazardous Locations Summarized

This table provides an overview of how hazards are organized by class and division. Refer to ***NEC Article 500*** for more detail.

Class	Division	Hazard	Example
I	1	Flammable gas or vapors exist under normal conditions	Spray paint facilities where flammable solvents are used; rooms where flammable gas is pumped
II	2	Flammable gas or vapors exist but are normally contained	Areas adjacent to Division 1 facilities; areas where flammable gases or vapors are handled in a closed system; similar areas where proper ventilation is provided
III	1	High concentrations of airborne combustible dust	Grain handling or storage plants
	2	Combustible dust accumulation on surfaces	Same as Division 1
	1	Easily ignitable fibers or airborne particles released through a manufacturing process	Textile mills; woodworking plants
	2	Ignitable fiber storage	Same as Division 1

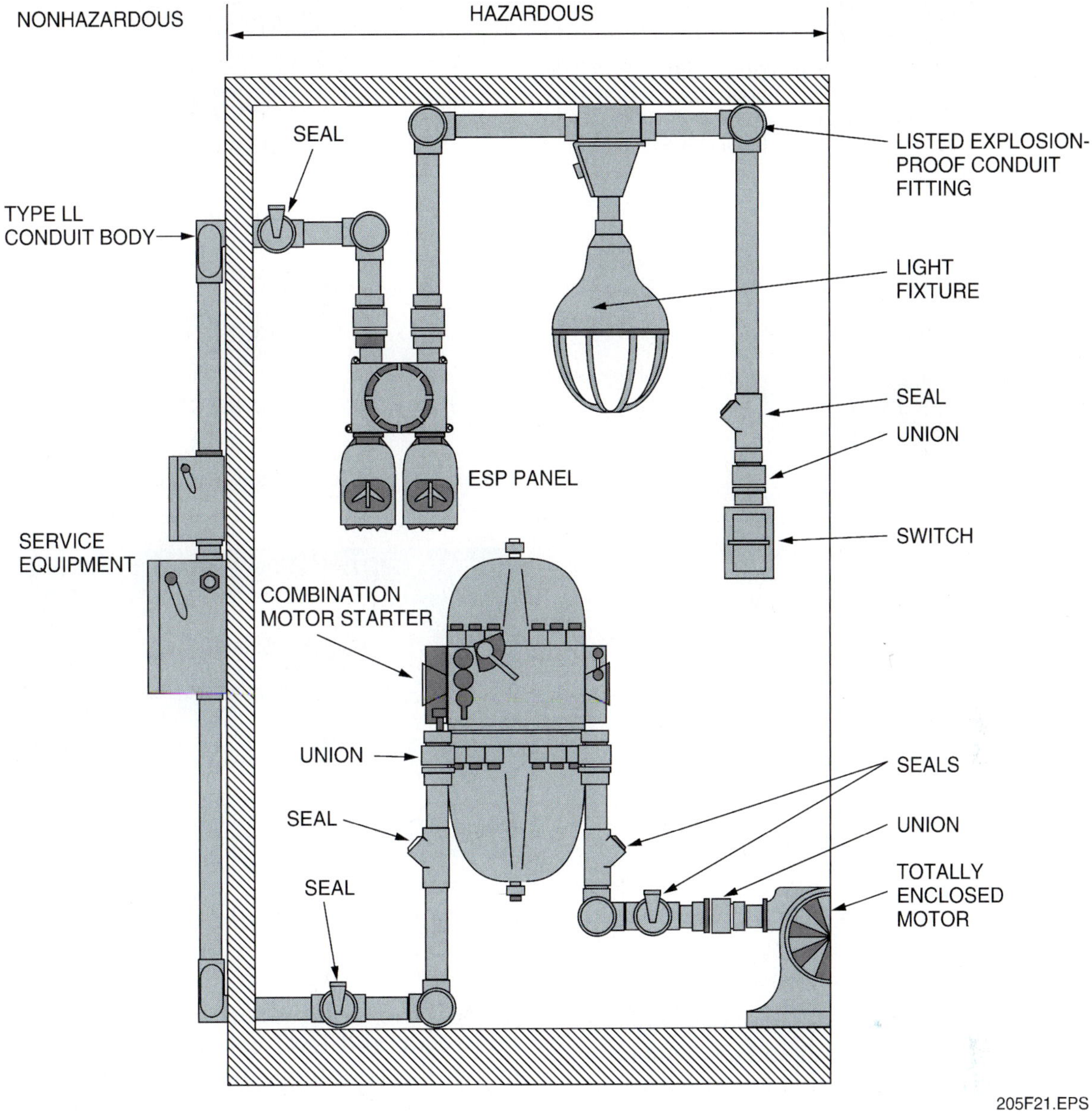

Figure 21 ◆ Practical representation of explosion-proof fittings in Class I, Division 1.

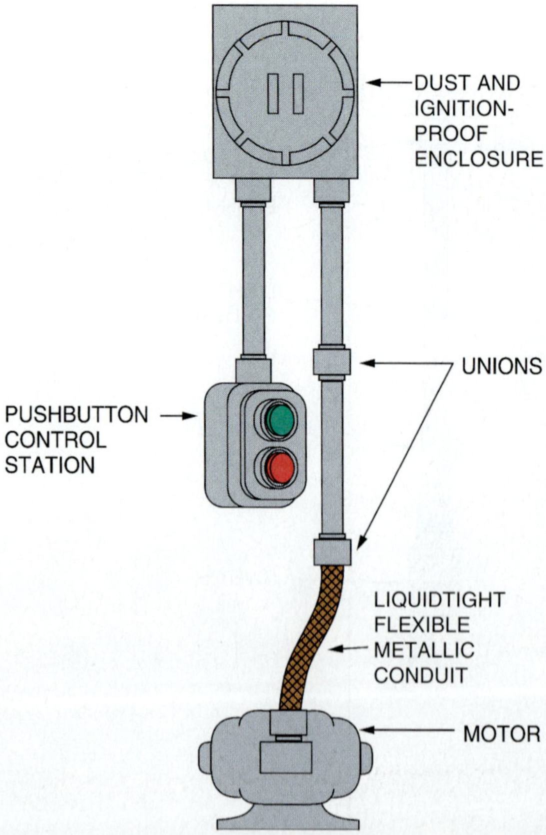

Figure 22 ◆ Power diagram for Class II, Division 2 installation with flexible conduit to motor.

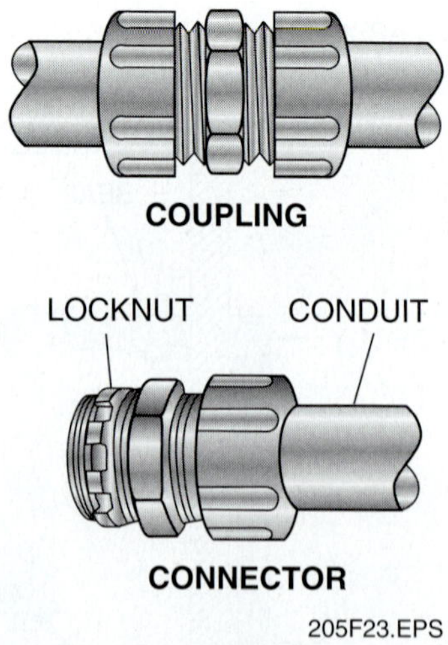

Figure 23 ◆ EMT compression fittings.

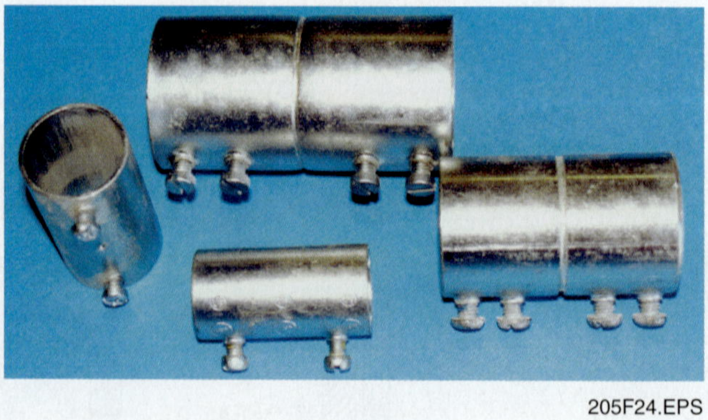

Figure 24 ◆ Setscrew fittings.

7.0.0 ◆ FITTINGS

Certain fittings are required in every raceway system for joining runs of conduit and also when the raceway terminates in an outlet box or other enclosure. Most metallic raceways qualify as an equipment grounding conductor provided they are tightly connected at each joint and termination point to provide a continuous grounding path.

7.1.0 EMT Fittings

Because EMT or thinwall is too thin for threads, special fittings must be used. For wet or damp locations, compression fittings such as those shown in *Figure 23* are used. This type of fitting contains compression rings made of plastic or other soft material that forms a watertight seal.

EMT fittings for dry locations can be either the setscrew type or the compression type. To use the setscrew type, the reamed ends of the EMT are inserted into the sleeve and the setscrews are tightened with a screwdriver to secure them and the conduit in place. Various types of setscrew couplings are shown in *Figure 24*.

EMT also requires connectors at each termination point, and EMT connectors are available to match the couplings described previously, that is, compression and setscrew types. They are similar to the couplings except that one end of the connector is threaded to accept a locknut and bushing.

7.2.0 Rigid, Aluminum, and IMC Fittings

Rigid metal conduit, aluminum conduit, and intermediate metal conduit all have sufficient wall thicknesses to permit threading. Consequently, all three types may be joined with threaded couplings (*Figure 25*) and when any of these types terminate into an outlet box or other enclosure, double locknuts are used to secure the conduit to the box opening. Running threads are not permitted for connection at couplings.

Sometimes rigid conduit or tubing must be connected to flexible metal conduit for connection to electric motors and other machinery that may

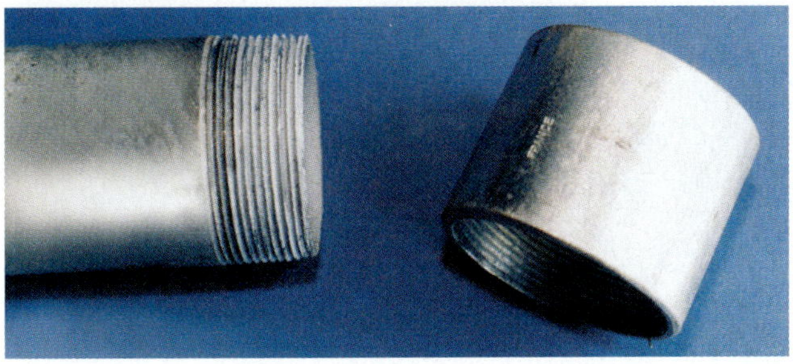

205F25.EPS

Figure 25 ◆ Rigid metal conduit with coupling.

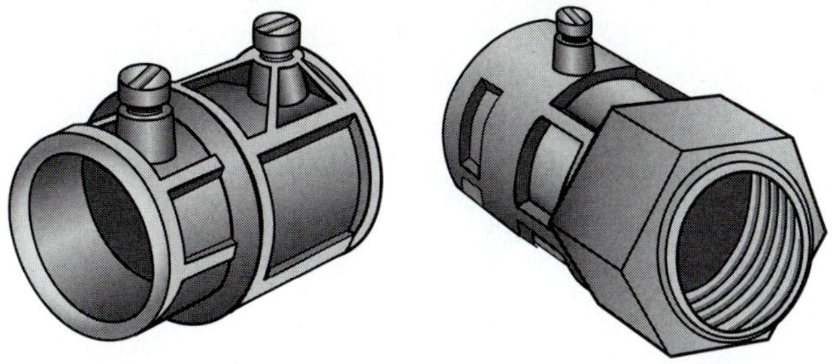

205F26.EPS

Figure 26 ◆ Combination couplings.

vibrate during operation. Combination couplings (*Figure 26*) are used to make the transition. When using combination couplings, be sure the flexible conduit is pushed as far as possible into the coupling. This covers the sharp edges of the conduit to protect the conductors from damage.

Threadless couplings and connectors may also be used under certain conditions with rigid, IMC, and aluminum conduit. When used, they must be made up wrench-tight and where buried in masonry or concrete, they must be concrete-tight. Where installed in wet locations, they must be rainproof. This type of coupling is not permitted in most hazardous locations.

Other types of fittings used with raceway systems are shown in *Figure 27*.

THINK ABOUT IT

Components Used in Hazardous Locations

Photo (A) shows a special motor starter designed for use in a hazardous location. Photo (B) shows the same starter with the cover removed. Photo (C) shows an explosion-proof seal. Examine these components. How do they differ from standard components?

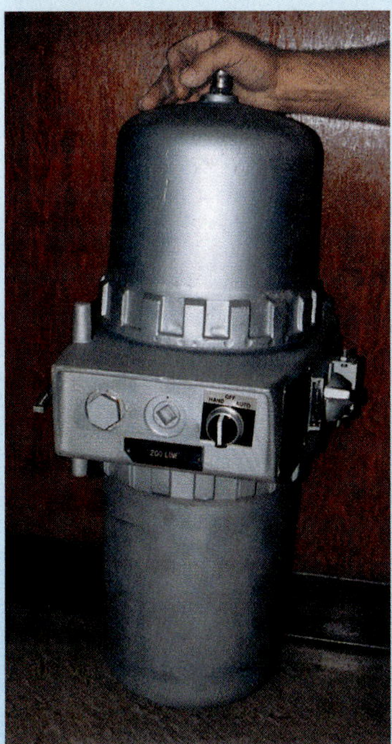

205P0501.EPS

(A)

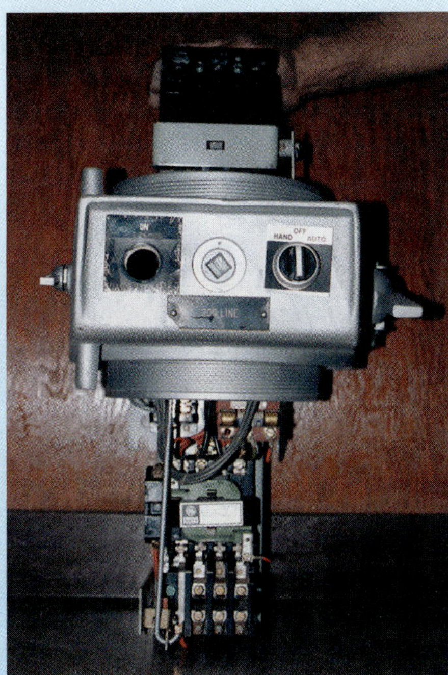

205P0502.EPS

(B)

205P0503.EPS

(C)

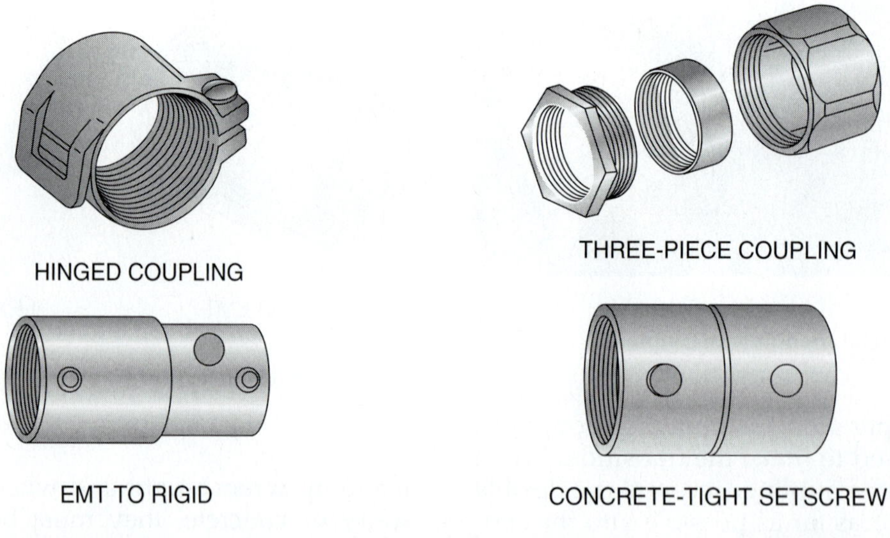

Figure 27 ◆ Metal conduit couplings.

7.3.0 Locknuts and Bushings

In general, locknuts are used on the inside and outside walls of outlet boxes or other enclosures to which threaded conduit is connected. When conduit connectors are used, such as EMT connectors, only one locknut is required on the threads that protrude inside the box. A grounding locknut may be needed if bonding jumpers are used inside the box or enclosure. Special sealing locknuts are also available for use in wet locations. Locknuts are shown in *Figure 28*.

Bushings protect the wires from the sharp edges of the conduit or connector. Bushings are usually made of plastic, fiber, or metal. Some metal bushings have a grounding screw to permit an equipment or bonding jumper wire to be installed. Several types of bushings are shown in *Figure 29*.

INSIDE TRACK

Explosion-proof Flex

This photo shows a special type of metal conduit for use in hazardous locations. Note that all conduit of this type comes with factory-made fittings for optimum protection under extreme conditions.

205P0504.EPS

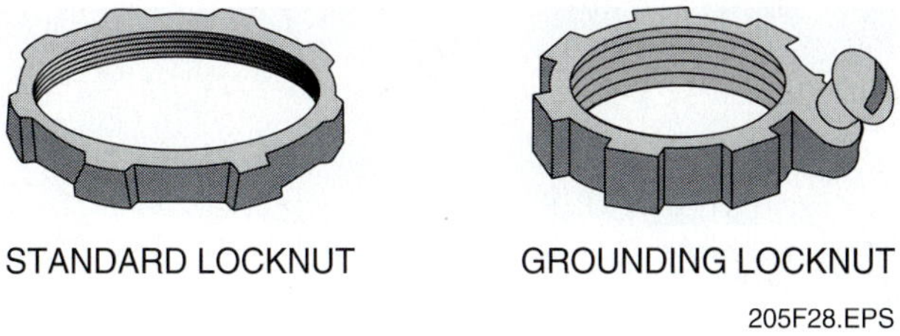

Figure 28 ◆ Common types of locknuts.

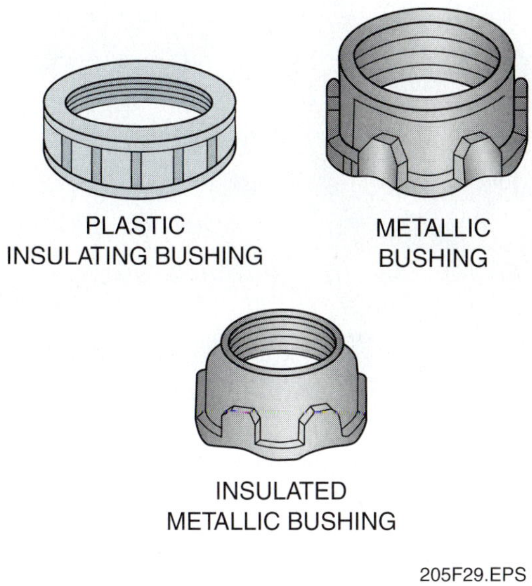

Figure 29 ◆ Typical bushings used at termination points.

An insulating bushing is installed on the threaded end of conduit that enters a sheet metal enclosure. The purpose of the bushing is to protect the conductors from being damaged by the sharp edges of the threaded conduit end. Any ungrounded conductor, No. 4 AWG or larger, that enters a raceway, box, or enclosure must be protected with an insulating bushing, as required in ***NEC Sections 300.4(F), 312.6(C), and 314.17(D).***

A grounding insulated bushing has provisions for protecting conductors and also has provisions for the connection of a ground wire. The ground wire, once connected to the grounding bushing, may be connected to the box to which the conduit is connected. See *Figure 30*.

An opening must be provided in the outlet box or enclosure for the entrance of conduit and connectors when raceway systems terminate. Most boxes and enclosures are provided with an adequate number of concentric knockouts. However, some may not have precut knockouts or the ones that are available may not be in the required location. In these cases, a knockout punch must be used to make a hole for the conduit connection. A hand-operated knockout punch is shown in *Figure 31*.

To use the knockout punch, the center of the hole is located in the box or enclosure and marked with a center punch. A pilot hole is then drilled to accept the threaded drive bolt of the knockout punch. The punch is separated from the drive screw, the screw is then placed through the pilot hole with the die on one side of the box wall, and the punch is screwed onto the drive screw on the opposite side of the wall. The punch is then aligned and screwed onto the drive screw hand-tight, in which case the punch should lightly bite into the wall of the enclosure. A wrench is then used to tighten the drive nut until the punch is drawn through the enclosure wall, making a neat circular opening.

Where many such openings must be made, or when knockouts for the larger sizes of conduit must be cut, contractors normally furnish workers with power knockout tools to facilitate the operation.

THINK ABOUT IT

Setscrew and Compression Fittings

This picture shows both setscrew and compression fittings. Which type provides a better connection? Why?

205P0505.EPS

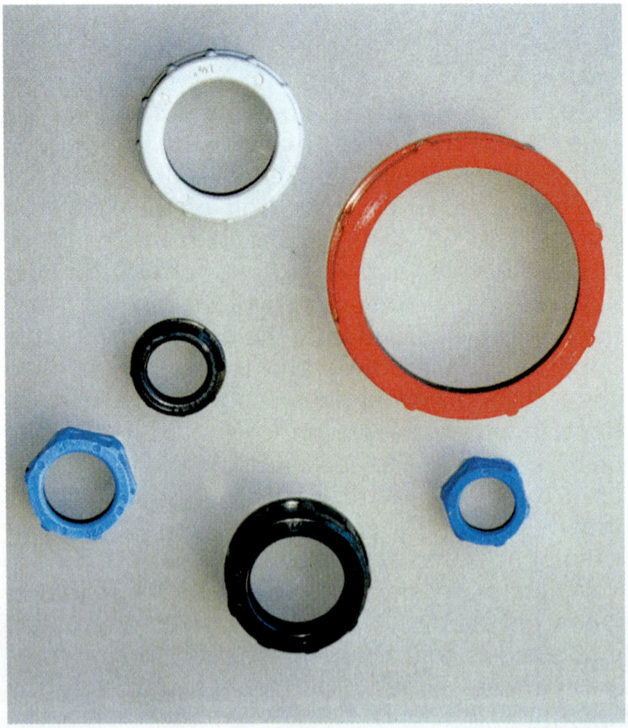

205F30A.EPS

(A)

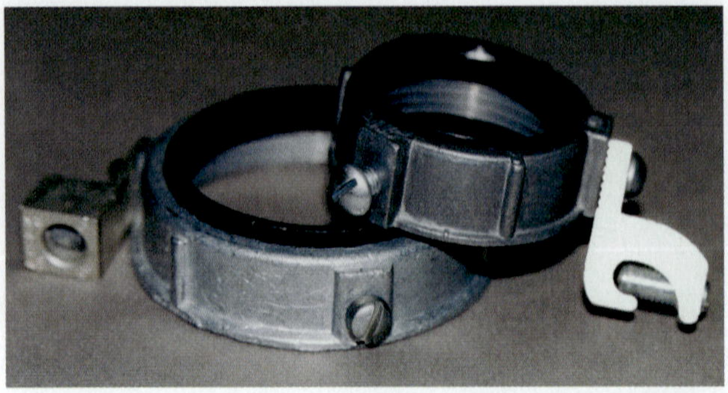

205F30B.EPS

(B)

Figure 30 ◆ (A) Regular insulating bushings and (B) grounding insulating bushings.

205F31.EPS

Figure 31 ◆ Knockout punch kit.

THINK ABOUT IT

Putting It All Together

Turn off the power in one area of your home and then remove some of the switch and receptacle plates. Examine the wiring inside each box. Is the box adequately sized for the number of wires and devices?

Review Questions

1. The maximum weight allowed by the *NEC*® when ceiling fans are mounted directly to an approved outlet box is _____ pounds.
   a. 25
   b. 35
   c. 45
   d. 55

2. Square outlet boxes are available in 4" and _____ sizes.
   a. 4¹¹⁄₁₆"
   b. 5"
   c. 5¼"
   d. 6"

3. Concrete boxes are normally used _____.
   a. in residential wiring to house wiring devices
   b. to feed lay-in troffer lighting fixtures
   c. in flat-slab construction projects
   d. in airport runway lighting

4. Deduct _____ conductor(s) for each strap-mounted device in a device box.
   a. one
   b. two
   c. three
   d. four

5. Using _____ adds to the capacity of an outlet box.
   a. fixture studs
   b. wire nuts
   c. strap-mounted devices
   d. raised device covers

6. When calculating the pull box size for straight pulls, the length of the box must not be less than _____ times the trade diameter of the largest raceway.
   a. two
   b. four
   c. six
   d. eight

7. If the largest trade diameter of a raceway entering a pull box is 3", and it is a straight pull, the minimum size box allowed is _____.
   a. 20"
   b. 24"
   c. 30"
   d. 36"

8. A _____ conduit body has openings on four different sides plus an access opening.
   a. Type X
   b. Type C
   c. Type T
   d. Type LL

9. Which of the following best describes how Type FS boxes should be installed?
   a. Holes should be drilled in back of the box for mounting screws.
   b. Holes should be drilled on the sides of the box only for mounting screws.
   c. No holes should be drilled in the box for mounting.
   d. Holes may be drilled only in existing installations.

10. The purpose of using seals or seal-off fittings in hazardous areas is to prevent _____.
    a. other conductors from being pulled into the raceway
    b. the passage of gases or vapors
    c. water from entering the raceway system
    d. the conductors from being removed from the raceway system

Summary

Electricians work with boxes and fittings almost every day on every project. Consequently, you must have a thorough knowledge of the types available and their applications. Portions of *NEC Chapters 3 and 4* cover the installation of boxes and fittings, so refer to appropriate sections in these chapters whenever a question arises.

One of the best ways to learn about boxes and fittings is to study manufacturers' catalogs. You will find a wealth of information in each of these, including detailed instructions (in many of them) on installation techniques. Some of these catalogs also provide simplified *NEC*® explanations of the use of the manufacturer's products.

Notes

Trade Terms Introduced in This Module

Conduit body: A separate portion of a conduit or tubing system that provides access through a removable cover (or covers) to the interior of the system at a junction of two or more sections of the system or at a terminal point of the system. Boxes such as FS and FD or larger cast or sheet metal boxes are not classified as conduit bodies.

Connector: Device used to physically connect conduit or cable to an outlet box, cabinet, or other enclosure.

Explosion-proof: Designed and constructed to withstand an internal explosion without creating an external explosion or fire.

Handy box: Single-gang outlet box used for surface mounting to enclose receptacles or wall switches on concrete or concrete block construction of industrial and commercial buildings; nongangable; also made for recessed mounting; also known as a utility box.

Junction box: An enclosure where one or more raceways or cables enter, and in which electrical conductors can be, or are, spliced.

Nineteen hundred box: A term commonly used to refer to any two-gang, four-inch-square outlet box.

Outlet box: A metallic or nonmetallic box installed in an electrical wiring system from which current is taken to supply some apparatus or device.

Pull box: A sheet metal box-like enclosure used in conduit runs to facilitate the pulling of cables from point to point in long runs, or to provide for the installation of conduit support bushings needed to support the weight of long riser cables, or to provide for turns in multiple-conduit runs.

Raintight: Constructed or protected so that exposure to a beating rain will not result in the entrance of water under specified test conditions.

Sealing compound: The material poured into an electrical fitting (seal-off) to seal and minimize the passage of vapors.

Waterproof: Constructed so that moisture will not interfere with successful operation.

Watertight: Constructed so that water will not enter the enclosure under specified test conditions.

Weatherproof: Constructed or protected so that exposure to the weather will not interfere with successful operation.

Additional Resources

This module is intended to present thorough resources for task training. The following reference works are suggested for further study. These are optional materials for continued education rather than for task training.

American Electrician's Handbook, Latest Edition. New York: Croft and Summers, McGraw-Hill.

National Electrical Code® Handbook, Latest Edition. Quincy, MA: National Fire Protection Association.

CONTREN® LEARNING SERIES — USER FEEDBACK

The NCCER makes every effort to keep these textbooks up-to-date and free of technical errors. We appreciate your help in this process. If you have an idea for improving this textbook, or if you find an error, a typographical mistake, or an inaccuracy in NCCER's *Contren®* textbooks, please write us, using this form or a photocopy. Be sure to include the exact module number, page number, a detailed description, and the correction, if applicable. Your input will be brought to the attention of the Technical Review Committee. Thank you for your assistance.

Instructors – If you found that additional materials were necessary in order to teach this module effectively, please let us know so that we may include them in the Equipment/Materials list in the Annotated Instructor's Guide.

Write: Product Development
National Center for Construction Education and Research
P.O. Box 141104, Gainesville, FL 32614-1104

Fax: 352-334-0932

E-mail: curriculum@nccer.org

Craft ____________ Module Name ____________

Copyright Date ____________ Module Number ____________ Page Number(s) ____________

Description

____

____

____

(Optional) Correction

____

____

(Optional) Your Name and Address

____

____

Conductor Installations

26206-05

Inn at the Ballpark
Houston, Texas
Renovation $10–99 Million Award Winner
HOAR Construction, LLC

26206-05

Conductor Installations

Topics to be presented in this module include:

Overview

The first step in any conductor installation is planning. Most small conductors can be pulled into a conduit using manual techniques such as the two-person method in which one person pulls the conductors and another feeds the conductors at the supply end. A common manual pulling tool is the fish tape, which may be made of spring steel or fiberglass. The stripped ends of the conductors are fastened to the end of the fish tape and secured in place with electrical tape. As the person on the pulling end pulls, the person on the feeding end feeds the conductors into the conduit in unison with the pull, protecting the conductors from damage.

Wire-pulling ropes are often used in place of fish tapes because of their flexibility and non-conductive properties. To initially install a pulling rope in a conduit system, either a vacuum or blower system is used to install a lightweight string into the system. The pulling rope is then attached to the string and pulled through the conduit system. The conductors to be installed are attached to the pulling rope and either manually pulled through the raceway system or installed using a power pulling device.

There are several types of power pulling devices available. Most of them operate on the same principle as a power winch. Only properly trained personnel may install conductors using a power winch system, as high torques can be developed that can cause personal injury, damage to conductors, or both.

Note: *National Electrical Code*® and *NEC*® are registered trademarks of the National Fire Protection Association, Inc., Quincy, MA 02269. All *National Electrical Code*® and *NEC*® references in this module refer to the 2005 edition of the *National Electrical Code*®.

Objectives

When you have completed this module, you will be able to do the following:

1. Describe the various methods of installing conductors in conduit.
2. Plan and set up for a cable pull.
3. Understand the importance of selecting the proper location for cable pulls.
4. Describe how cable reels are transported to the pulling site.
5. Set up reel stands and spindles for a wire-pulling installation.
6. Explain how mandrels, swabs, and brushes are used to prepare conduit for conductors.
7. Properly install a pull line for a cable-pulling operation.
8. Explain the operation of power fish tape systems.
9. Prepare the ends of conductors for pulling.
10. Describe the types of cable pullers.
11. Describe the process of high-force cable pulling.
12. Explain how to support conductors in vertical conduit runs.
13. Describe the installation of cables in cable trays.
14. Explain the importance of communication during a cable-pulling operation.
15. Calculate the probable stress or tension in cable pulls.

Required Trainee Materials

1. Paper and pencil
2. Appropriate personal protective equipment
3. Copy of the latest edition of the *National Electrical Code*®

Prerequisites

Before you begin this module, it is recommended that you successfully complete *Core Curriculum; Electrical Level One;* and *Electrical Level Two,* Modules 26201-05 through 26205-05.

This course map shows all of the modules in *Electrical Level Two.* The suggested training order begins at the bottom and proceeds up. Skill levels increase as you advance on the course map. The local Training Program Sponsor may adjust the training order.

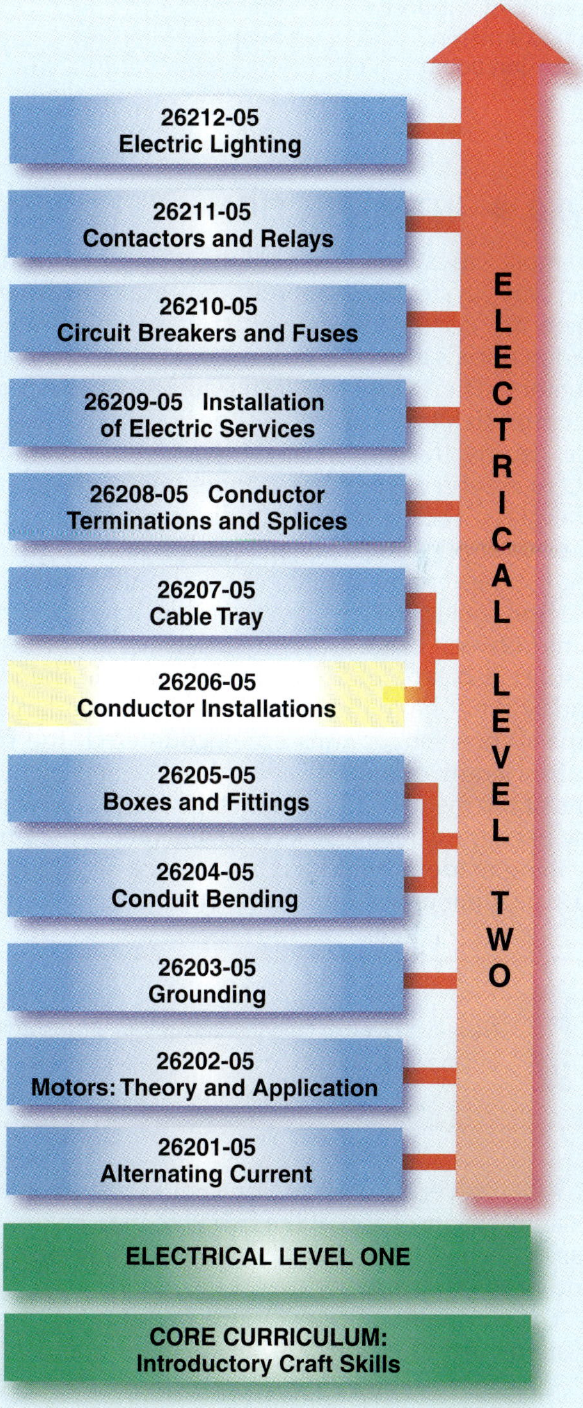

206CMAP.EPS

Trade Terms

- Basket grip
- Cable grip
- Capstan
- Clevis
- Conductor support
- Conduit piston
- Extension boom
- Extension bushing
- Fish line
- Fish tape
- Setscrew grip
- Sheave
- Soap

1.0.0 ◆ INTRODUCTION

In most cases, the installation of conductors in raceway systems is merely routine. However, there are certain practices that can reduce labor and materials and help prevent damage to the conductors. The use of modern equipment, such as vacuum **fish tape** systems, is one way to reduce labor during this phase of the wiring installation. There are three types of fish tape: steel, nylon, and fiberglass. They also come in different weights for various applications. The proper size and length of the fish tape, as well as the type, should be one of the first considerations. For example, if most of the runs between branch circuit outlets are 20' or less, a short fish tape of 25' will easily handle the job and will not have the weight and bulk of a larger tape. When longer runs are encountered, the required length of the fish tape should be enclosed in one of the metal or plastic fish tape reels. This way, the fish tape can be rewound on the reel as the pull is being made to avoid having an excessive length of tape lying around on the floor or deck.

WARNING!

Never fish a steel tape through or into enclosures or raceways containing an energized conductor.

When several bends are present in the raceway system, the insertion of the fish tape may be made easier by using flexible fish tape leaders on the end of the fish tape.

The combination blower and vacuum fish tape systems are ideal for use on long runs and can save much time. Basically, the system consists of a tank and air pump with accessories. An electrician can vacuum or blow a line or tape in any size conduit from ½" through 4", or even up to 6" conduit with optional accessories.

After the fish tape is inserted in the raceway system, the conductors must be firmly attached by some approved means. On short runs, where only a few conductors are involved, all that is necessary is to strip the insulation from the ends of the wires, bend these ends around the hook in the fish tape, and securely tape them in place. Where several wires are to be pulled together, the wires should be staggered and the fish tape securely taped at the point of attachment so that the overall diameter is not increased any more than is absolutely necessary.

Basket grips (*Figure 1*) are available in many sizes for almost any size and combination of conductors. They are designed to hold the conductors firmly to the fish tape and can save much time and trouble that would be required when taping wires.

In all but very short runs, the wires should be lubricated with wire lubricant prior to attempting the pull, as well as during the pull. Some of this lubricant should also be applied to the inside of the conduit.

Wire dispensers are great aids in keeping the conductors straight and facilitating the pull. Many different types of wire dispensers are now marketed that handle virtually any size spool of wire or cable. Some of the smaller dispensers can handle up to ten spools of wire from No. 22 to No. 10 AWG; the larger ones can handle a lot more. These dispensers are sometimes called wire caddies (see *Figure 2*).

206F01.EPS

Figure 1 ◆ Basket grip.

INSIDE TRACK

Nonconductive Fish Tape

When fishing conductors in a conduit or raceway that already contains existing energized (live) conductors, the safest method is to turn off and lock out/tag the power sources for all the live conductors. However, in some rare instances, fishing a conductor through a conduit or raceway containing other live conductors may be unavoidable. In this case, always request your supervisor's approval before proceeding and always use a nonconductive fish tape made of nylon or fiberglass.

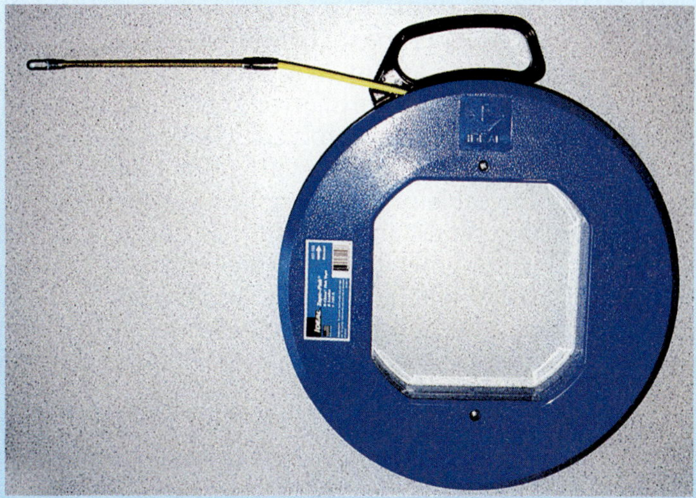

206P0601.EPS

206F02.EPS

Figure 2 ◆ Typical wire caddy.

2.0.0 ◆ PLANNING THE INSTALLATION

The importance of planning any wire-pulling installation cannot be over-stressed. Proper planning will make the work go easier and much labor will be saved.

Large sizes of conductors are usually shipped on reels, involving considerable weight and bulk. Consequently, setting up these reels for the pull, measuring cable run lengths, and similar preliminary steps will often involve a relatively large amount of the total cable installation time. Therefore, consideration must be given to reel setup space, proper equipment, and moving the cable reels into place.

Whenever possible, the conductors should be pulled directly from the shipping reels without pre-handling them. This can usually be done through proper coordination of the ordering of the conductors with the job requirements. While doing so requires extremely close checking of the drawings and on-the-job measurements (allowing for adequate lengths of conductors in pull boxes, elbows, troughs, connections, and splices), the extra effort is well worth the time to all involved.

When the lengths of cable have been established, the length of cable per reel can be ordered so that the total length per reel will be equal to the total of a given number of raceway lengths, and the reel so identified.

In most cases, the individual cables of the proper length for a given number of runs are reeled separately onto two or more reels at the factory, depending on the number of conductors in the runs.

When individual conductors are shipped on separate reels, it is necessary to set up for the same number of reels as the number of conductors to be pulled into a given run, as shown in *Figure 3*.

As an extra precaution against error in calculating the lengths of conductors involved, it is a good idea to actually measure all runs with a fish tape before starting the cable pull, adding for makeup to reach the terminations, and accounting for discarding the cable underneath the pulling sleeve or pulling wrap, as it will have been excessively stressed during the pull. Check these totals against the totals indicated on the reels. Under normal cable delivery schedules, when the feeder raceways have been installed at a relatively early stage of the overall building construction, it may not delay the final completion of the electrical installation to delay ordering the cables until the raceways can actually be measured.

When pulling conductors directly from the reels, care must be taken that each given run be cut off from the reel so that there is a minimum amount of waste. In other words, preclude the possibility of the final run of cable taken from the reel being too short for that run.

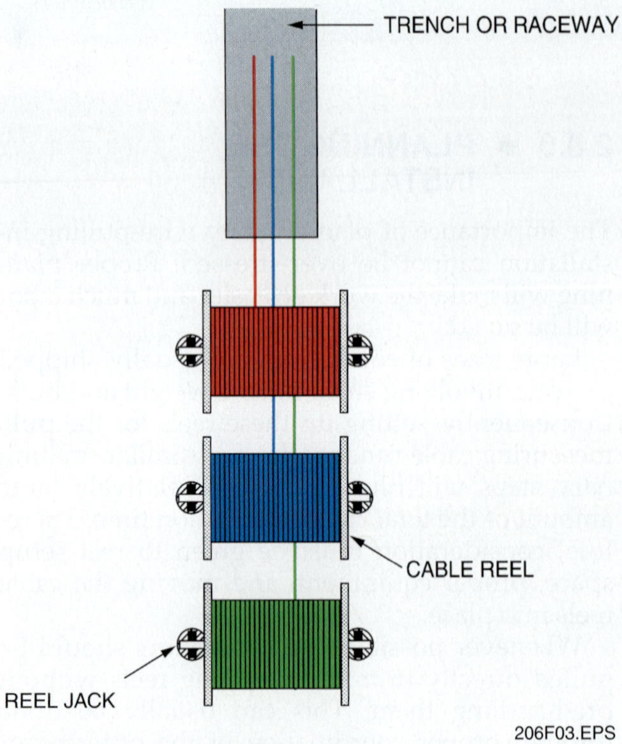

Figure 3 ◆ Multi-cable pull.

2.1.0 Pulling Location

Each job will have to be judged separately as to the best location of pulling setups; the number of setups should be reduced to a minimum in line with the best direction of the pull. The pulling location for a particular job is determined by the weight of the cable, height of the pull, practicality of moving the equipment to the pulling location, number of setups required, as well as the location of any bends. Also, a separate setup might have to be made at the top of each rise, whereas a single setup might be made at a ground floor pull box location from which several feeders are served with the same size and number of conductors.

The location of the pulling equipment determines the number of workers required for the job. A piece of equipment that can be moved in and set up on the first floor by four workers in an hour's time may require six workers working two hours when set up in basements or parking levels of buildings.

Every employer expects a day's work for a day's pay, but no employer should expect workers to do the work of two or three. Therefore, when planning a cable pull, make certain enough workers are on hand to adequately handle the installation.

It is a simple operation for a few workers to roll cable reels from a loading platform to a first floor setup, whereas moving them to upper floors involves much handling and usually requires a crane or other hoists. In addition, the reel jacks have to be moved to the setup point when a downward pull is made, and after the pulling operation is completed, they must be taken back to the first floor.

THINK ABOUT IT

Multiple Reels

Why are multiple reel pulling setups used?

Measuring Tape

INSIDE TRACK

A waterproof polyester tape with permanent markings every foot from 0' to 3,000' is available on reels for use in measuring conduit/raceway runs. It can be fished through the run manually or using a power fishing system.

2.2.0 Cable Pull Operations

These operations are performed to a lesser or greater degree in almost all cable pulls with larger sizes of conductors:

Step 1 Measure or re-check runs and establish communications between both ends of the pull.

Step 2 Provide pulling equipment.

Step 3 Receive and unload pulling equipment.

Step 4 Move pulling equipment to the pulling location.

Step 5 Set up and anchor pulling equipment.

Step 6 Remove to another location or move to a loading platform and load on a truck or forklift.

Step 7 Receive and store cable; may be moved directly to setup location if job conditions permit.

Step 8 Move to setup point.

Step 9 Move reel jacks and mandrel to setup point.

Step 10 Jack up reels.

Step 11 Prepare cable ends.

Step 12 Install fish tape.

Step 13 Install pulling line or cable.

Step 14 Connect pulling line.

Step 15 Lubricate with proper lubricants.

Step 16 Pull cable.

Step 17 Disconnect pulling line.

Step 18 Remove reels.

Step 19 Identify conductor legs.

Step 20 Rack cables in pull boxes and troughs.

Step 21 Splice or connect cables.

Step 22 Check and test.

In some instances, the following additional operations are involved, depending upon the exact details of the project. Other items of importance will be discussed later in this module.

Step 23 Remove lagging or other protective covering from reels.

Step 24 Unreel cable and cut it to length.

Step 25 Re-reel cable for pulling.

Step 26 Replace lagging on reels.

Step 27 Operate such auxiliary equipment as guide-through cabinets or pull boxes and signal systems between reel and pulling setups.

Each of these pulling operations is discussed in detail in this module (also see *Figure 4*).

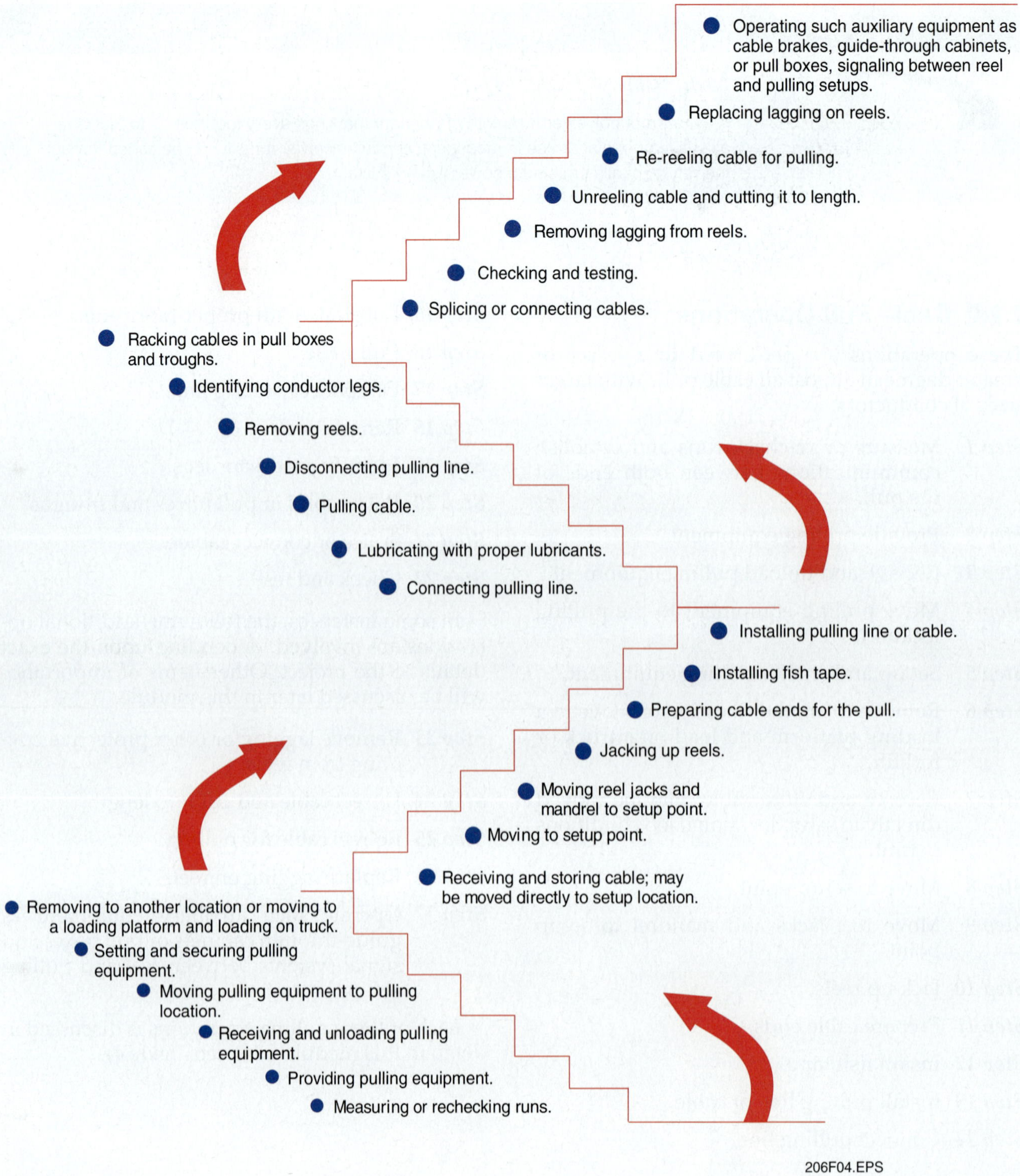

Figure 4 ◆ Basic steps of a wire-pulling operation.

3.0.0 ◆ SETTING UP FOR WIRE PULLING

As mentioned previously, much planning is required for pulling the larger sizes of conductors in raceways. There are several preliminary steps required before the actual pull begins.

The proper use of appropriate equipment is crucial to a successful cable installation. The equipment needed for most installations is shown in the checklist in *Figure 5*. Some projects may require all of these items, while others may require only some of them. Each cable-pulling project must be taken on an individual basis and analyzed accordingly. Seldom will two pulls require identical procedures.

> **NOTE**
> Think of everything that can go wrong and take every precaution.

3.1.0 Setting Up the Cable Reels

When reels of cable arrive at the job site, it is best to move them directly to the setup location if at all possible. This prevents having to handle the reels more than necessary. However, if this is not practical, arrangements must be made for storage until the cable is needed.

The exact method of handling reels of cable depends on their size and the available tools and equipment. In many cases, the reels may be rolled to the pulling location by one or more workers. For reels up to 24" wide × 40" in diameter, a cable reel transporter can be used to transport the cable reel; it also acts as a dispenser during the pulling operation. When available, a forklift is ideal for lifting and transporting cable reels. See *Figure 6*.

However, for very large reels (48" or more in diameter), a crane or similar hoisting apparatus is usually necessary for lifting the reels onto reel jacks supported by jack stands to acquire the necessary height. *Figure 7* shows a summary of proper and improper ways to transport reels of wire or cable on the job site.

Figure 8 shows several types of reel jacks, including the spindle. For a complete setup, two stands and a spindle are required for each reel. The reel jacks or stands are available in various sizes from 13" to 54" high to accommodate reel diameters up to 96". Extension stands used in conjunction with reel stands can accommodate 8", 10", 12", and 14" reels.

Reel-stand spindles are commonly available in diameters from $2\frac{3}{8}$" to $3\frac{1}{2}$" and from 59" to 100" in length for carrying reel loads up to 7,500 pounds. However, some heavy-duty spindles are rated for loads up to 15,000 pounds.

EQUIPMENT CHECKLIST

- ❑ PORTABLE ELECTRIC GENERATOR
- ❑ EXTENSION CORDS AND GFCI
- ❑ PUMP, DIAPHRAGM
- ❑ MAKEUP BLOWER AND HOSE
- ❑ MANHOLE COVER HOOKS
- ❑ WARNING FLAGS, SIGNS
- ❑ ELECTROSTATIC kV TESTER
- ❑ ELECTRIC SAFETY BLANKETS AND CLAMPS
- ❑ RADIOS OR TELEPHONES
- ❑ GLOVES
- ❑ FLOOD LAMPS
- ❑ FISH TAPE OR STRING BLOWER/VACUUM
- ❑ HAND LINE
- ❑ DUCT-CLEANING MANDRELS
- ❑ DUCT-TESTING MANDRELS
- ❑ CAPSTAN-TYPE PULLER
- ❑ SNATCH BLOCKS
- ❑ SHORT ROPES FOR TEMPORARY TIE-OFFS
- ❑ GUIDE-IN FLEXIBLE TUBING (ELEPHANT TRUNKS)
- ❑ SEVERAL WIRE ROPE SLINGS OF VARIOUS LENGTHS
- ❑ SHACKLES/ROPE CLEVIS
- ❑ GANG ROLLERS WITH AT LEAST 4' EFFECTIVE RADIUS
- ❑ HAND WINCHES
- ❑ MANHOLE EDGE SHEAVE
- ❑ PULLING ROPE
- ❑ SWIVELS
- ❑ BASKET GRIP PULLERS
- ❑ 0-1/5/10 KIP DYNAMOMETER
- ❑ REEL ANCHOR
- ❑ REEL JACKS
- ❑ CABLE CUTTERS
- ❑ LINT-FREE RAGS
- ❑ CABLE-PULLING LUBRICANTS
- ❑ PRELUBING DEVICES
- ❑ PLYWOOD SHEETS
- ❑ DIAMETER TAPE
- ❑ 50' MEASURING TAPE
- ❑ SILICONE CAULKING (TO SEAL CABLE ENDS)

206F05.EPS

Figure 5 ◆ Wire-pulling equipment checklist.

Figure 6 ◆ Two methods of transporting cable reels.

3.2.0 Preparing Raceways for Conductors

Another preliminary step prior to pulling conductors in raceway systems is to inspect the raceway itself. Few things are more frustrating than to pull four 1,000 kcmil conductors through a conduit and find out when the pull is almost done that the conduit is blocked or damaged. Such a situation usually requires pulling the conductors back out, repairing the fault, and starting all over again.

These problems are not too serious with exposed banks of conduit, as the conduit can usually be separated at the fault, the fault corrected, and another piece of conduit installed, using unions if necessary. However, in underground conduit runs—especially those encased in concrete—the situation can be both time-consuming and costly.

Figure 9 shows several devices used to inspect raceway systems, as well as to prepare raceways for easier and safer conductor pulls. Go and no-go steel and aluminum mandrels are available for pulling through runs of conduit before the cable installation. Mandrels should be approximately 80% of the conduit size (twice the 40% fill factor).

A test pull will detect any hidden obstructions in the conduit prior to the pull (*Figure 10*). If any are found, they can be corrected before wasting time on an installation that might result in conductor damage and the possibility of having to repull the conductors.

The conduit swab in *Figure 9* may be used ahead of the conductors during a pull. Its main purpose is to swab out water and debris from the raceway and spread a uniform film of pulling compound inside the conduit for easier pulling.

The conduit brush in *Figure 9* will help clean and polish the interior of conduit before pulling the cable. Such brushes remove sand and other light obstructions. Note that this brush has a pulling eye on one end and a threaded rod on the opposite end, enabling it to be pushed or pulled through the conduit.

One final step before starting the pull is to measure the length of the raceway, including all turns in junction boxes and the like. A fish tape may be pushed through the raceway system and a piece of tape used to mark the end. When it is pulled back out, a tape measure may be used to measure the exact length. An easier way, however, is to use a power fishing system to push or pull a measuring tape through the conduit run. Details of this operation are explained in the next section.

When measuring the conductor length, be sure to allow sufficient room where measurements are made through a pull box. Conductors should enter and leave pull boxes in such a manner as to allow the greatest possible sweep for the conductors. Large conductors are especially difficult to bend but with proper planning, you can simplify the feeding of these conductors from one conduit to another.

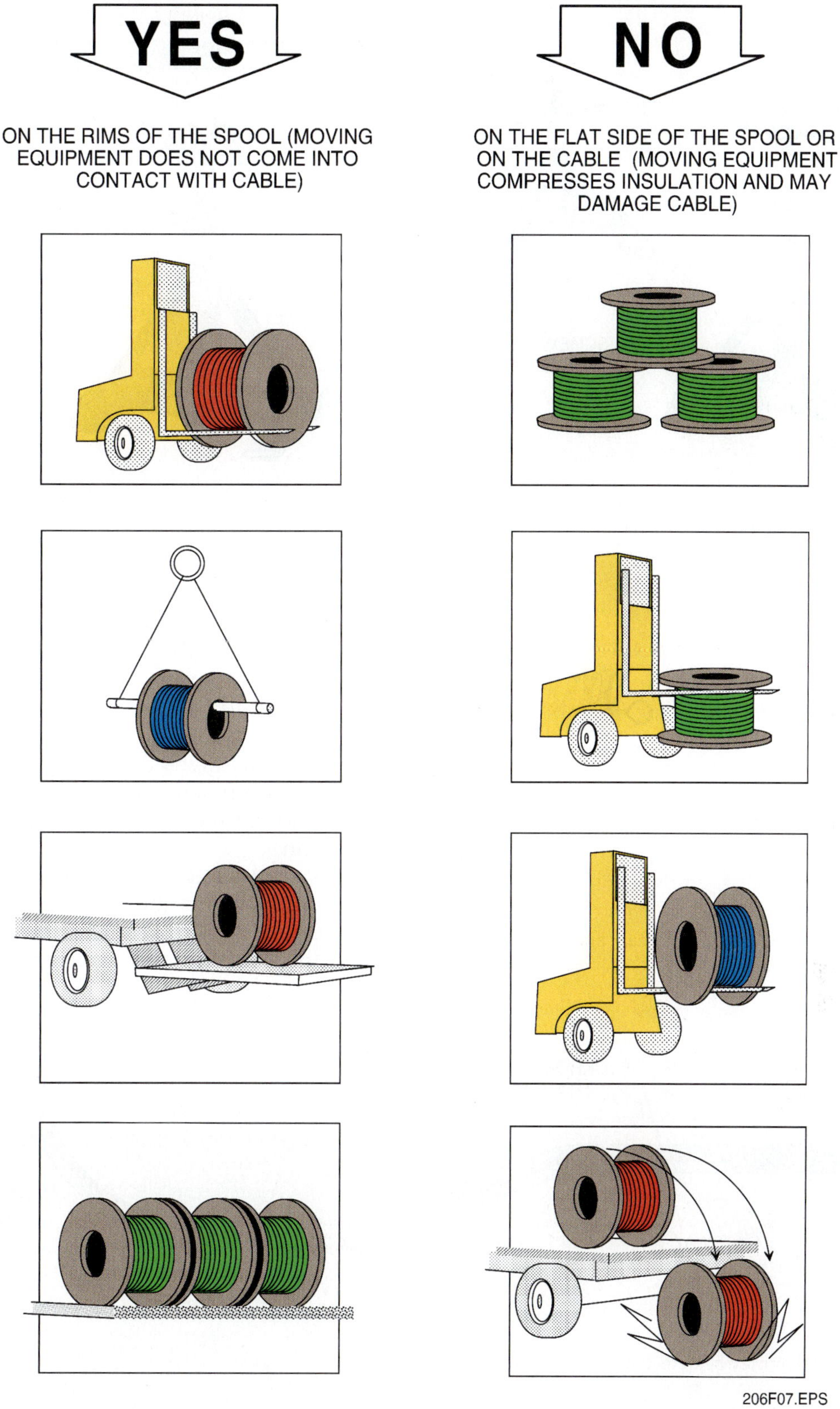

Figure 7 ◆ Proper and improper ways of transporting cable reels.

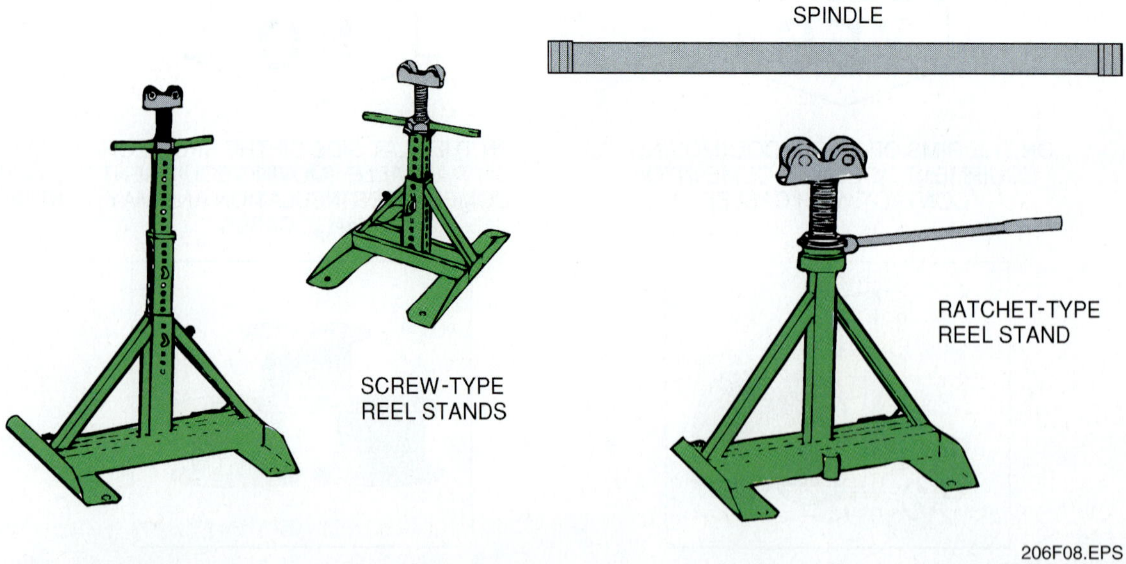

Figure 8 ◆ Typical reel stands.

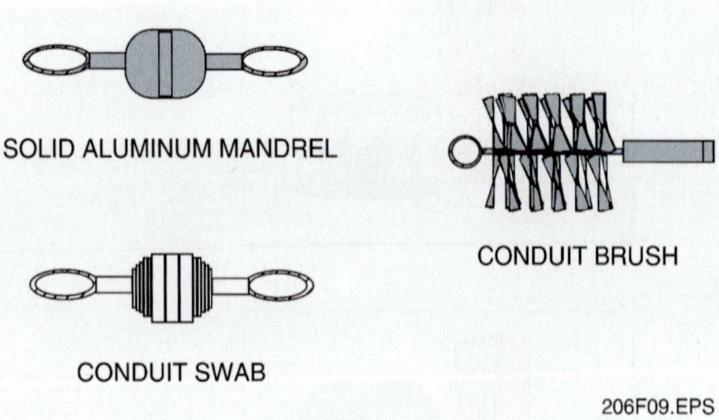

Figure 9 ◆ Devices used to inspect, clean, and lubricate raceway systems.

For example, if a conduit run makes a right-angle bend through a pull box, the conduit for a given feeder should come into the box at the lower left-hand corner and leave diagonally opposite at the upper right-hand corner, as shown in *Figure 11*. This gives the conductors the greatest possible sweep with the box, eliminating sharp bends and consequent damage to the conductor insulation.

Runs should also be calculated to allow for splices and terminations in junction boxes, panelboards, motor-control centers, the cable discarded under the sleeve or pulling wrap, etc.

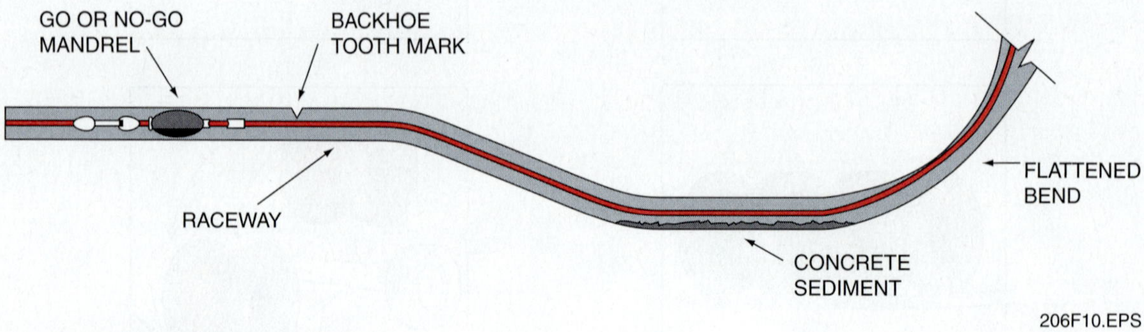

Figure 10 ◆ Faults that may be detected with a conduit mandrel.

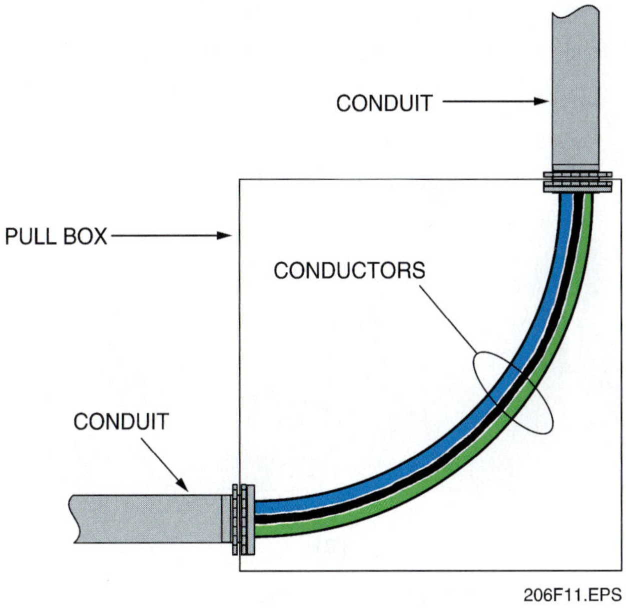

Figure 11 ◆ Obtaining the greatest possible conductor sweep in a pull box.

3.3.0 Installing the Pull Line

At one time, pull-in wires were frequently placed in conduit runs as the raceways were installed. However, in recent times, with modern cable-fishing equipment, this practice is seldom used.

Pull lines are sometimes manually fished in with a steel fish tape, but much time can be saved by using a blower/vacuum fish tape system. In general, a **conduit piston**—sometimes referred to as a mouse or missile—is blown with air pressure or vacuumed through the run. The foam piston is sized to the conduit and has a loop on both ends. In most cases, **fish line** or measuring tape is attached to the piston as it is blown or vacuumed through the conduit run. The measuring tape serves two purposes: it provides an accurate measurement of the conduit run, and the tape is used to pull the cable-pulling rope into the conduit run. In some cases, if the run is suitable, the pulling rope is attached directly to the piston and vacuumed into the run. *Figure 12* shows a blower/vacuum fish tape system being used to vacuum a pull line in a conduit while *Figure 13* shows the same apparatus blowing the piston through. Most of these units provide enough pressure to clean dirt or water from conduit during the fishing operation. *Figure 14* shows two types of pistons used with this system. The one on the right utilizes air-guide vanes to prevent the piston from tumbling inside the larger sizes of conduit.

3.4.0 Preparing Cable Ends for Pulling

The pulling-in line or cable must be attached to the cable or conductors in such a manner that it cannot part from the cable during the pull. Two common methods include direct connection with the cable conductors themselves and connection by means of pulling grips or baskets placed over the cable or group of conductors. The use of the proper type of grip or basket will facilitate the pull, but in many cases—especially on long pulls—workers prefer to use three- or four-hole **cable grips** with setscrews that secure each conductor to the pulling block. *Figure 15* shows several types of pulling grips.

Most pulling blocks have a rope **clevis** as an integral part of the block. However, when using pulling grips or baskets in 2" or larger conduit, a rope clevis is normally used to facilitate connecting the pulling rope to the wire grip. Two types are currently used: the straight clevis and the swivel clevis.

3.4.1 Stripping the Cable Ends

When the type of grip being used requires that the ends of the cable insulation be stripped from the conductors, conventional methods are used—the same as for terminating conductors for splices or connections to terminal lugs in panelboards, switchgear, etc.

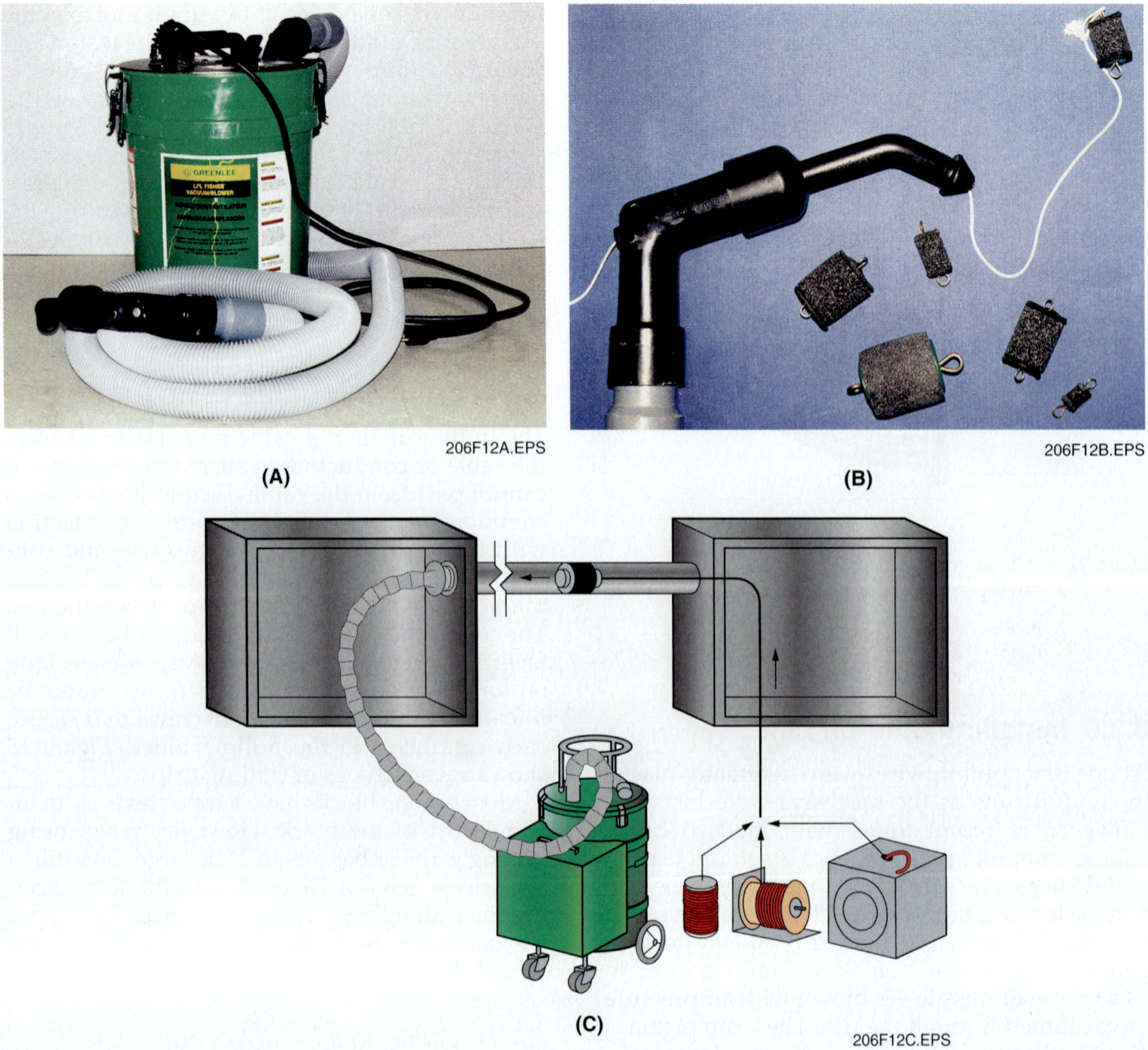

Figure 12 ◆ Power fishing system.

Cable Blowing

INSIDE TRACK

Lightweight cables, especially fiber optic cable, can be floated through conduit using a special high-pressure blower unit.

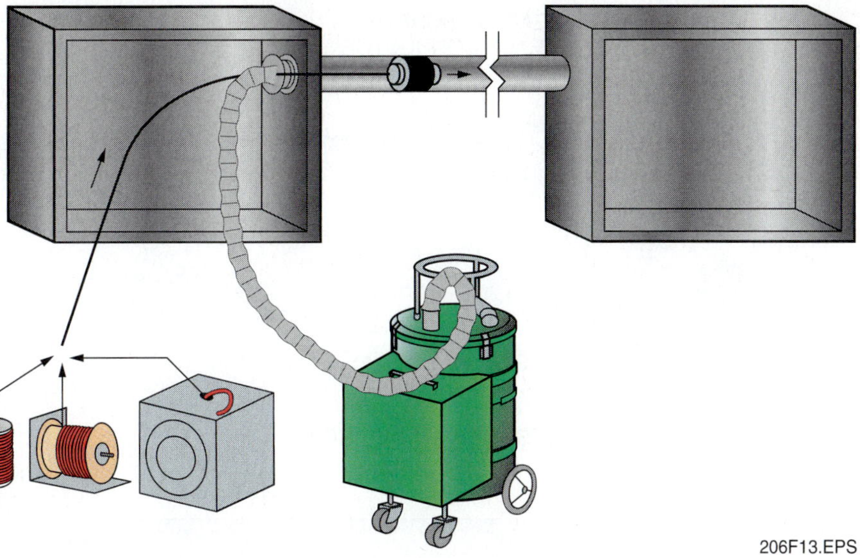

206F13.EPS

Figure 13 ◆ Blower/vacuum fish tape system used to blow a pull line in conduit.

(A)

206F14A.EPS

(B)

206F14B.EPS

Figure 14 ◆ Types of pistons in common use.

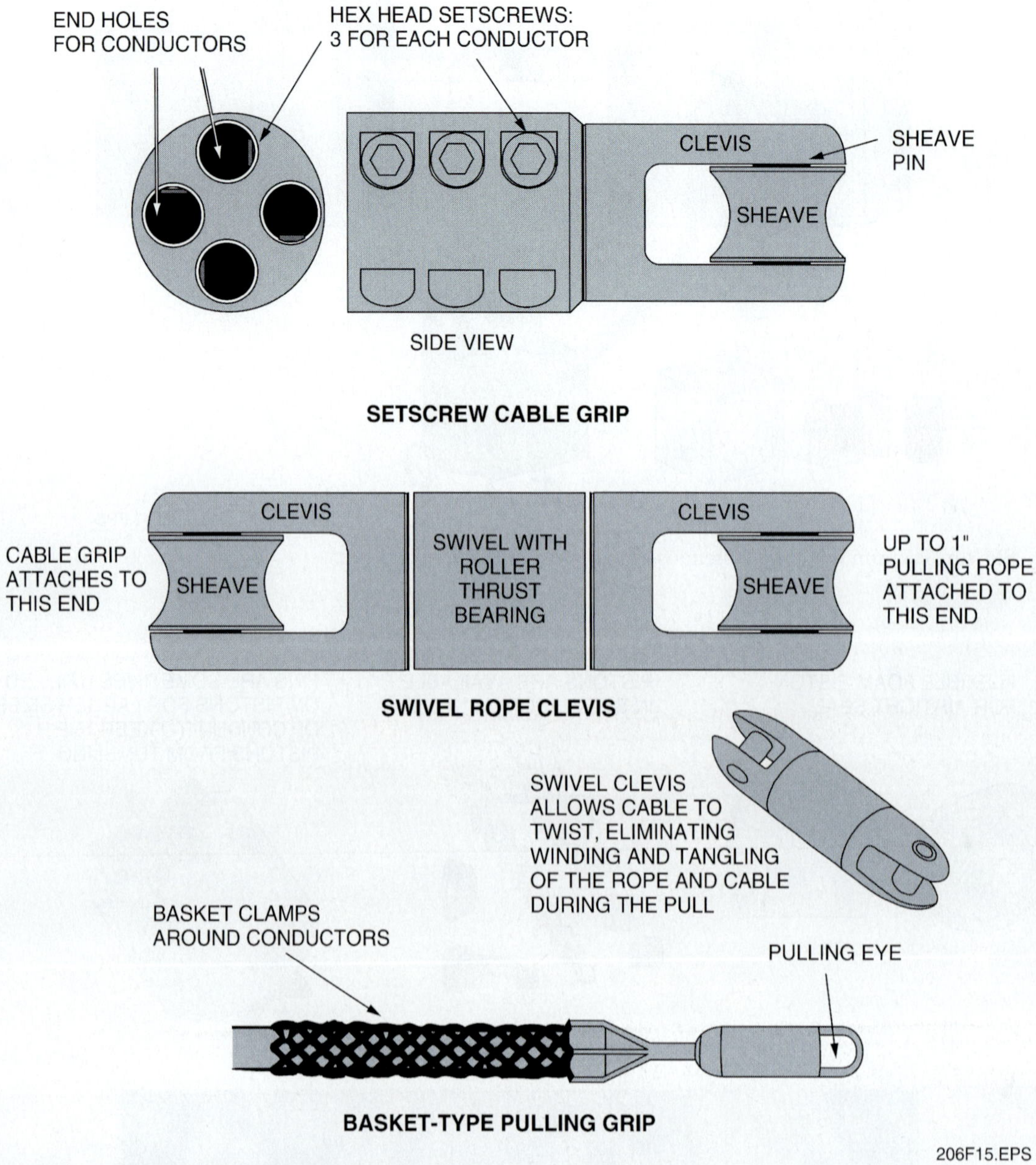

Figure 15 ◆ Various types of pulling grips used during conductor installation.

In general, the ends of conductors should first be trimmed. Cable cutters capable of cutting conductors through 1,000 kcmil save workers much time over using a hacksaw. There are also cable strippers, adjustable from 1/0 AWG through 1,000 kcmil, that handle midspan and termination stripping of THHN, THWN, XHHW, and similar insulation. These tools are excellent for stripping conductors for use in setscrew clamp-type pulling grips.

To use a stripping tool, first mark the required distance from the ends of the conductors, using the pulling grip as a gauge. Close the jaws of the stripping tool on the cable and twist. These self-feeding devices ensure positive progression down the cable to any position desired. To stop stripping, apply back pressure to the stripper until a full circle has been completed.

Once the conductor ends have been stripped, insert one conductor at a time into the **setscrew grip.** Make sure the end of the bare conductor is firmly in place, and then tighten the setscrews with a hex wrench. Continue on to the next conductor, and so on, until all conductors are secured in the pulling grip.

INSIDE TRACK

Pulling Eyes, Basket Grips, and Swivels

Conductors can be attached to a pulling line using various methods, including pulling eyes or basket grips.

Smaller conductors may require only a pulling eye attached to the fish tape or pulling line. To attach the pulling eye, the conductors are first stripped, exposing a length of bare wire. These bare wires are inserted through the eye and twisted back onto themselves. The exposed twisted wires of the conductors can then be wrapped with smaller copper wire to prevent them from untwisting. They are then completely taped with three layers of electrical tape, starting from the conductor insulation, to prevent snagging as they are drawn through the conduit run.

If a basket grip is used to pull the conductors, make sure to securely tape several inches of the grip to several inches of the conductor insulation beyond the grip using multiple layers of tape. This will prevent the grip from coming off the conductors in the middle of the run if the tension on the pulling line is released for one reason or another.

No matter what method is used, always insert a swivel of some sort between the pulling line or fish tape and the conductors to alleviate twisting of the conductors.

206P0602.EPS

3.4.2 Safety with Cutters

When using a stripping tool for the first time, make sure that you read and understand all instructions and warnings before using the tool. The following should also be observed when working with cable terminators:

- Wear eye protection.
- Inspect tools before using, and replace damaged, missing, or worn parts.
- Be prepared for the unexpected. Make sure your footing and body position are such that you will not lose your balance.
- Use only the type and size material in the stated capacity.
- Do not use the tool on or near live circuits.

3.5.0 Types of Pulling Lines

Wire-pulling ropes have come a long way from the hemp ropes used by electricians a couple of decades ago. The most common wire-pulling ropes on the market include the following:

- Nylon
- Polypropylene
- Multiplex polyester
- Double-braided polyester composite

The type of rope selected will depend mainly on the pulling load; that is, the weight of the cable, the length of the pull, and the total resistance to the pull. For example, Greenlee's Multiplex cable-pulling rope is designed for low-force cable pullers. It has a low stretch characteristic that makes it suitable for pulls up to 2,000 pounds. Lengths are available from 100 to 1,200 feet. Greenlee's double-braided composite rope for high-force cable pullers is designed for pulls up to 6,500 pounds.

CAUTION

Any equipment associated with the pull must have a working load rating in excess of the force applied during the pull. All equipment must be used and mounted in strict accordance with the manufacturer's instructions.

Care must be used in selecting the proper rope for the pull, and then every precaution must be taken to make sure that the cable-pulling force does not exceed the rope capacity. There are several reasons for this, but the main one is safety. For example, think of a 100' length of nylon rope with a 10,000-pound breaking strength. Such ropes can stretch 40' before breaking, releasing 200,000 foot-pounds of energy in the process. Think of the damage this amount of energy could do to a raceway system and to nearby workers if the rope broke under this amount of force.

In some cases, a power blower/vacuum fish tape system is used to vacuum the piston through the raceway with the pulling rope attached. In other cases, a line is first blown or vacuumed through the conduit and then the pulling rope is attached to the line for pulling it through the raceway system. In either case, there are certain precautions that should be taken when using a power fish tape system:

- Read and understand all instructions and warnings before using the tool.
- Never fish in runs that might contain live power.
- Be prepared for the unexpected. Make sure your footing and body position are such that you will not lose your balance in any unexpected event.
- Use blower/vacuum systems only for specified light fishing and exploring the raceway system.
- Never use pliers or other devices that are not designed to pull a fish tape. They can kink or nick the tape, creating a weak spot.

4.0.0 ◆ CABLE-PULLING EQUIPMENT

Except for short cable pulls, hand-operated or power-operated cable pullers or winches are used to furnish the pulling power. In general, cable reels are set in place at one end of the raceway system, and the cable puller is set up at the opposite end. One end of the previously-installed pulling rope is attached to a clevis, basket, or other cable grip to which the conductors are attached. The other end of the rope (at the cable puller) is wrapped around the rotating drum on the cable puller (*Figure 16*).

INSIDE TRACK

Be Sure to Check the Pulling Rope Rating

Many pulling ropes available today are made of synthetic materials designed for pulling by hand or with a winch-type puller. However, if you are using certain synthetic pulling ropes on friction-type capstan power pullers, make sure the rope is rated by the manufacturer for this use so that it will be able to withstand the heat generated by any extended slippage of the rope on the capstan. During high-force pulling operations, melting damage and possible failure of unrated synthetic rope can occur quickly during periods of capstan slippage.

> **WARNING!**
> Always use a wire-pulling lubricant that is compatible with the type of cable being pulled. Check with the cable manufacturer for their recommendations, and always contact the lubricant manufacturer about the compatibility of their products with specific cables. Also check the product's MSDS for any applicable safety requirements.

Wire-pulling lubricant—sometimes referred to as **soap**—is inserted into the empty conduit as well as wiped thoroughly onto the front of the cable. One or more operators are on hand to feed the cable, while one worker is usually all that is required on the pulling end.

The number of wraps on the puller drum decides the amount of force applied to the pull. For example, the operator needs to apply only 10 pounds of force to the pulling rope in all cases. With this amount of force applied by the operator, and with one wrap around the rotating drum, 21 pounds of pulling force will be applied to the pulling rope; 2 wraps, 48 pounds; 3 wraps, 106 pounds, etc. This principle is known as the **capstan** theory and is the same principle applied to block-and-tackle hoists or the lone cowboy who is able to rope and hold a 2,000-pound bronco by wrapping his lariat around the center post in a corral. *Table 1* gives the amount of pulling force with various numbers of wraps when the operator applies only 10 pounds of tailing force.

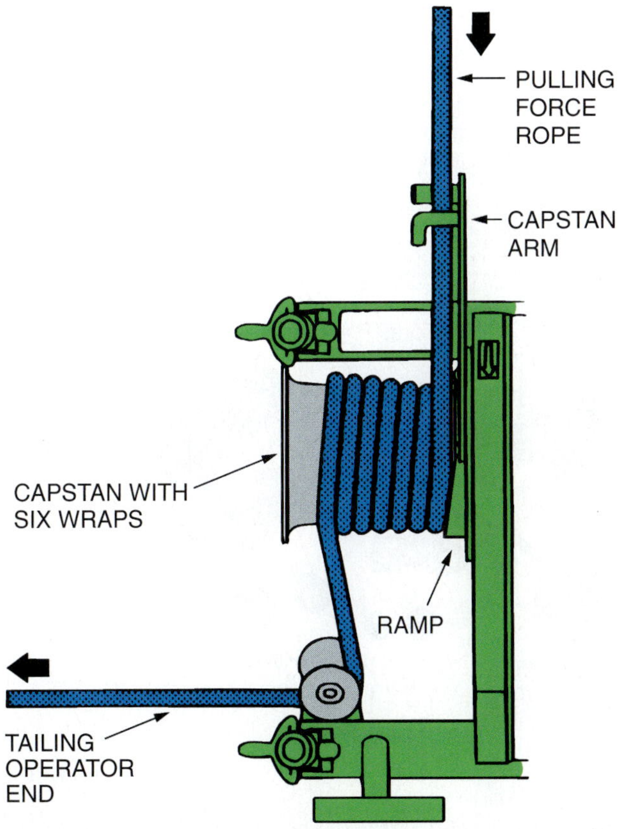

Figure 16 ◆ Basic parts of a power cable puller.

4.1.0 Pulling Safety

Adhere to the following precautions when using power cable-pulling equipment:

- Read and understand all instructions and warnings before using the tool.
- Use compatible equipment; that is, use the properly-rated cable puller for the job, along with the proper rope and accessories.
- Always be prepared for the unexpected. Make sure your footing and body position are such that you will not lose your balance. Keep out of the direct line of force.
- Make sure all cable-pulling systems, accessories, and rope have the proper rating for the pull.
- Inspect tools, rope, and accessories before using; replace damaged, missing, or worn parts.
- Personally inspect the cable-pulling setup, rope, and accessories before beginning the cable pull. Make sure that all equipment is properly and securely rigged.
- Make sure all electrical connections are properly grounded and adequate for the load.
- Use cable-pulling equipment only in uncluttered areas.

> **NOTE**
> The strain placed on the wrapped cable during the pull may weaken this part of the cable. Be sure to discard the wrapped cable after making the pull.

Table 1 Pulling Forces for Various Wraps

Number of Wraps	Operator Force (Lbs)	Pulling Force (Lbs)
1	10	21
2	10	48
3	10	106
4	10	233
5	10	512
6	10	1,127
7	10	2,478

> **CAUTION**
> Make absolutely certain that all communications equipment is in working order prior to the pull. Place personnel at strategic points with operable communications equipment to stop and start the pull as conditions warrant. Anyone involved with the pull has the authority to stop the pull at the first sign of danger to personnel or equipment.

4.2.0 Types of Cable Pullers

There are several types of cable pullers on the market. Most, however, operate on the same principle. The self-contained hand-crank wire puller in *Figure 17(A)* is designed to pull up to 1,500 pounds with only 30 pounds of handle force. It is used mostly on projects where only a few cable runs need to be pulled (i.e., those projects that do not warrant the setup time required for a power cable puller).

An electric cable puller is lightweight and easy to set up. It is also designed to pull up to 1,500 pounds. An electric cable puller is shown in *Figure 17(B)*.

> **WARNING!**
> Make absolutely certain that all cable-pulling equipment is anchored properly. Follow the manufacturer's recommendations for the type of puller being used.

4.2.1 Setting Up Cable Pullers

Figure 18 shows the basic setup for an up cable pull where the conduit is concealed. The first step is to remove the locknut from the end of the conduit and attach the pipe adapter with the proper **extension bushing** to the end of the conduit. Then, attach the pipe adapter **sheave** to the flexible pipe adapter. Add the mobile T-boom and cable puller for a flush-mounted pull, resting the **extension boom** on the floor.

Figure 19 shows a setup for a down pull through concealed conduit. The setup is similar to the one described for the up pull except the flexible pipe adapter and pipe adapter sheave are attached to the top conduit rather than the bottom conduit.

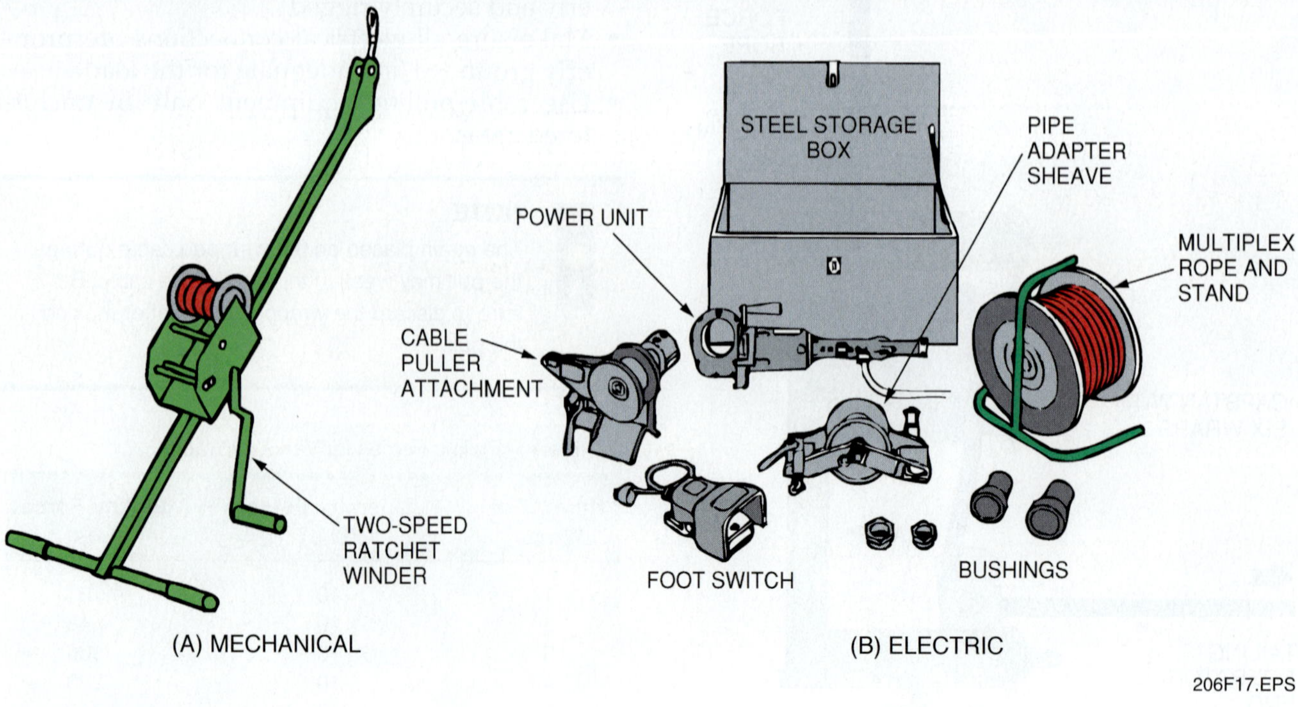

Figure 17 ◆ Cable pullers.

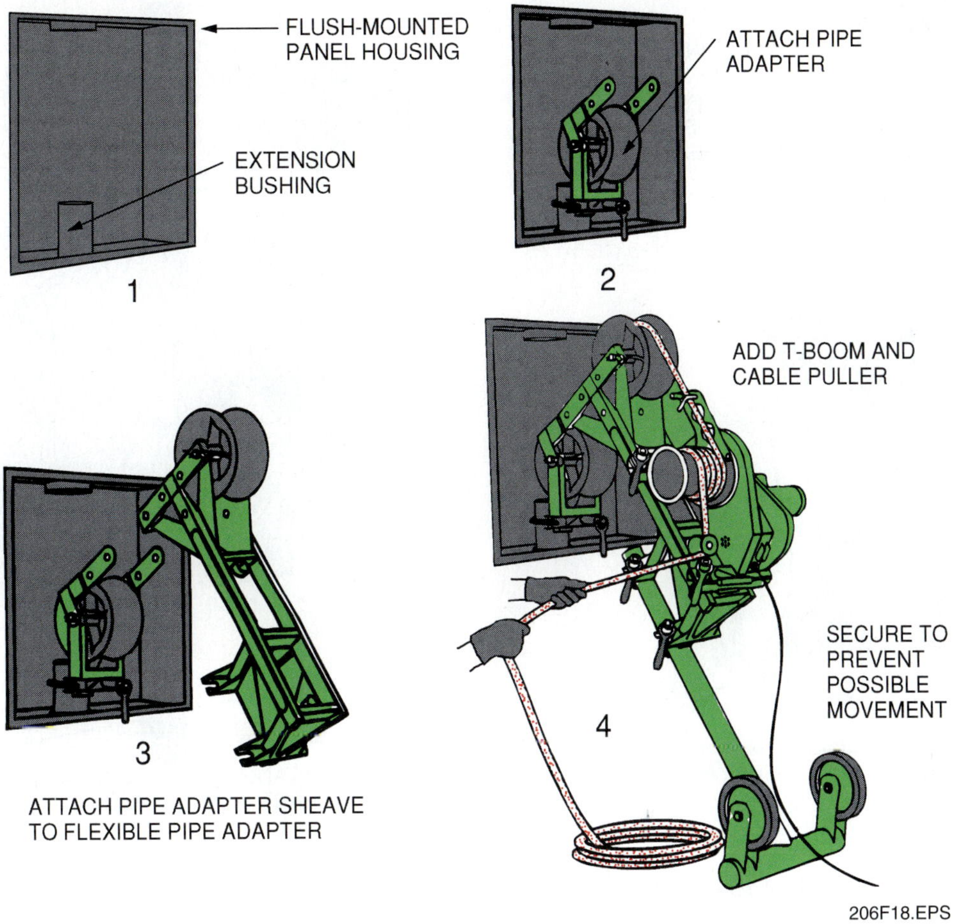

Figure 18 ◆ Steps in setting up cable puller for concealed conduit pull.

INSIDE TRACK

Guiding and Lubricating Conductors

When guiding conductors into a conduit/raceway during a pull, the conductors may tend to twist, overlap, or become crossed during the pulling operation, especially if fed from boxes instead of reels. Excessive twisting, overlaps, crossovers, etc. can cause binding of the conductors in conduit/raceway turns, create bunching obstructions in the conduit/raceway, and can contribute to insulation burns. Operators at the feeding end of the pull must attempt to keep the conductors as straight as possible during the pulling operation and lubricants should be applied liberally during the pull to allow the conductors to slide against each other and the sides of the run.

Monitored Cable Pullers

Besides basic power cable pullers, relatively expensive pullers with a dynamometer and an automatic over-tension shutoff device are also available. These units prevent you from accidentally exceeding the maximum recommended pulling tension for the conductor(s) being pulled.

Figure 19 ◆ Puller setup for down pull in concealed conduit system.

Figure 20 ◆ Puller setup for down pull in exposed conduit system.

Attaching the cable puller to exposed conduit is done in a somewhat different way. The exposed conduit itself is used for support of the pulling apparatus (*Figure 20*), and no floor support is necessary.

5.0.0 ◆ HIGH-FORCE CABLE PULLING

High-force cable pulling is not done every day. It is actually a small part of any electrical raceway installation. Workers may take days, weeks, or even months to install the complete raceway system; the cable pulling operation may only take an hour or two for each run.

There are several items to consider during high-force cable pulling:

- The design of the raceway system must be studied by consulting the working drawings and by examining the installed system to ensure that it is properly installed for the type of conductors that will be pulled through the system. Items to consider include conduit sizes, number and size of conductors, number of bends, sufficient pull boxes, and adequate supports.
- The conductors need to be matched to the proper pulling equipment.
- The proper pulling rope must be selected. Choose a rope that has at least four times the strength of the required pulling force; that is, if the estimated pulling force is 1,200 pounds, the rope should be rated for no less than 4,800 pounds. Also check the rope carefully for wear or damage prior to the pull. Remember that a rope is only as strong as its weakest point.
- A decision must be made as to the best end of the raceway for pulling. The reel setup must also be carefully placed. In general, conductors that are to be installed downward should be fed off the top of the reel and where conductors are to be fed upward, the best method is to feed from the bottom of the reel. This eliminates sharp kinks or bends in the conductors.
- The appropriate pulling equipment must be used. The equipment must be of the proper capacity for the job. Space consideration for the equipment is also important, as is the particular type of mounting.

- Having enough workers is critical. Never be caught short. This is where experienced workers earn their pay. In high-force wire pulling, experience is the best teacher.
- Safety must be foremost in everyone's mind.

When planning the cable pull, make sure that enough cable is on hand for the run. Cable may be verified while still on the reel by using a cable-length meter. Of course, the conduit run length should have already been checked, as discussed previously. Do not proceed further until everyone is assured that enough cable is on hand for all bends, sweeps in junction boxes, etc. If the cable falls short, take steps to correct the situation before continuing.

Equipment with feeding sheaves will help provide a smooth guide for the cable. Also, have sufficient lubricant on hand and use it both before and during the pull.

Select and install proper cable grips on the cable ends. Gripping must be adequate to handle the imposed force.

When pulling cables in a horizontal run, the worker simply has to reduce the amount of operator pull in order to slow down the cable pull. Releasing all operator pull stops the pulling force entirely. However, when pulling vertical runs, the rope must be tied off after stopping the pull to keep the cable from reversing in the raceway.

The entire operation should be supervised from start to finish. Therefore, communications equipment must be utilized on both ends of the pull at all times.

During the pull, make sure that the pulling rope remains free and is not wrapped around any part of the body.

A typical cable-pulling setup is shown in *Figure 21*, including cable reels, reel stands, a power cable puller, conduit system, and other accessories necessary for the pull.

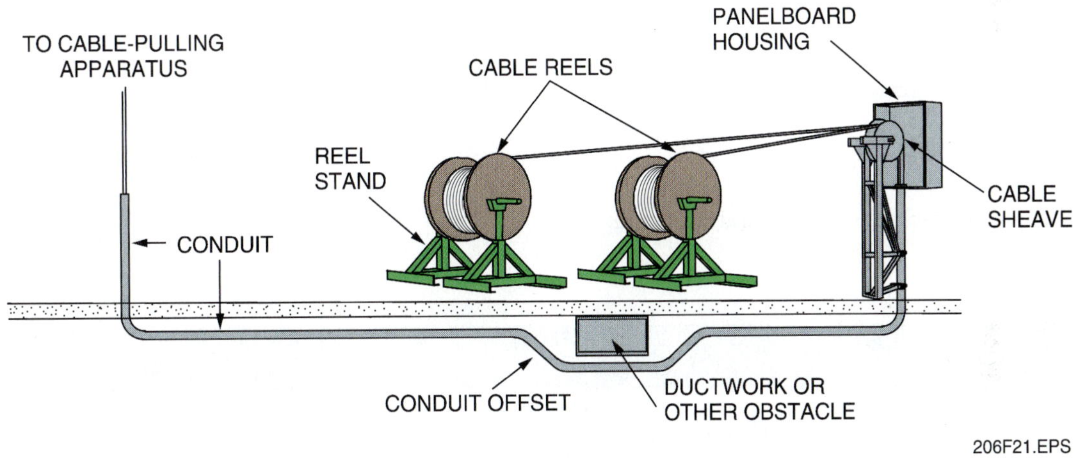

Figure 21 ◆ Typical cable-pulling setup.

THINK ABOUT IT

Communications

What type of communications equipment would most likely be used during cable-pulling operations?

5.1.0 The Feeding End—Sheaves and Rollers

The feed-in setup should unreel the cable along its natural curvature, as shown in *Figure 22(A)*, as opposed to a reverse S curvature, as shown in *Figure 22(B)*. Feed-in setups are shown in *Figure 23* for manhole, underfloor duct, and overhead cable tray. Note the use of auxiliary equipment in some of these drawings, that is, cable reels, guide-in tubes, sheaves, and rollers.

Single sheaves may be used only for guiding cables. Multiple blocks should be arranged to hold the cable-bending radius wherever the cable is deflected, as shown in *Figure 24(A)*. For pulling around bends, use conveyor sheave assemblies of the appropriate radius series, as shown in *Figure 24(B)*.

Sheaves and pulleys must be positioned to ensure the effective curvature is smooth and deflected evenly at each pulley. Never allow a polygon curvature to occur, as shown in *Figure 25*.

CAUTION

Use the radius of the surface over which the cable is bent, not the outside flange diameter of the pulley. For example, a 10" cable sheave typically has an inside (bending) radius of 3".

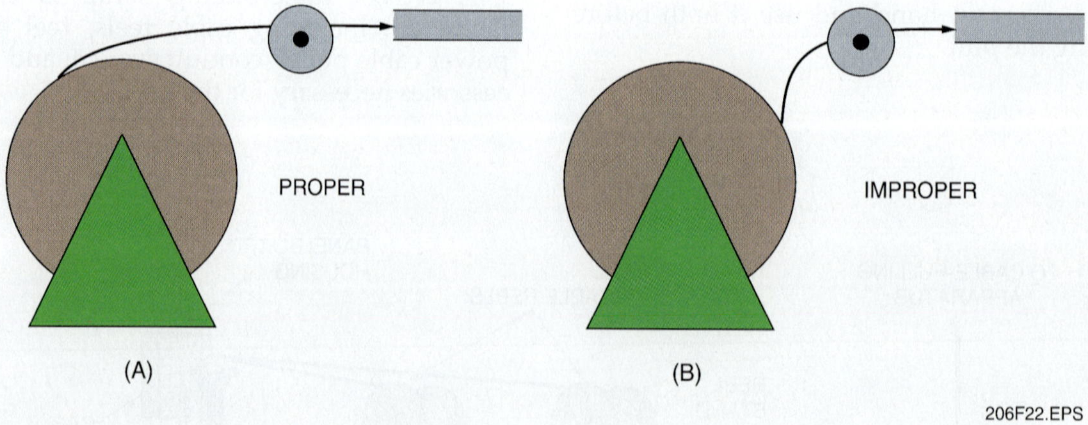

Figure 22 ◆ Unreel the cable along its natural curvature.

Feed-In Setups

THINK ABOUT IT

Why is an S-curve setup not desirable?

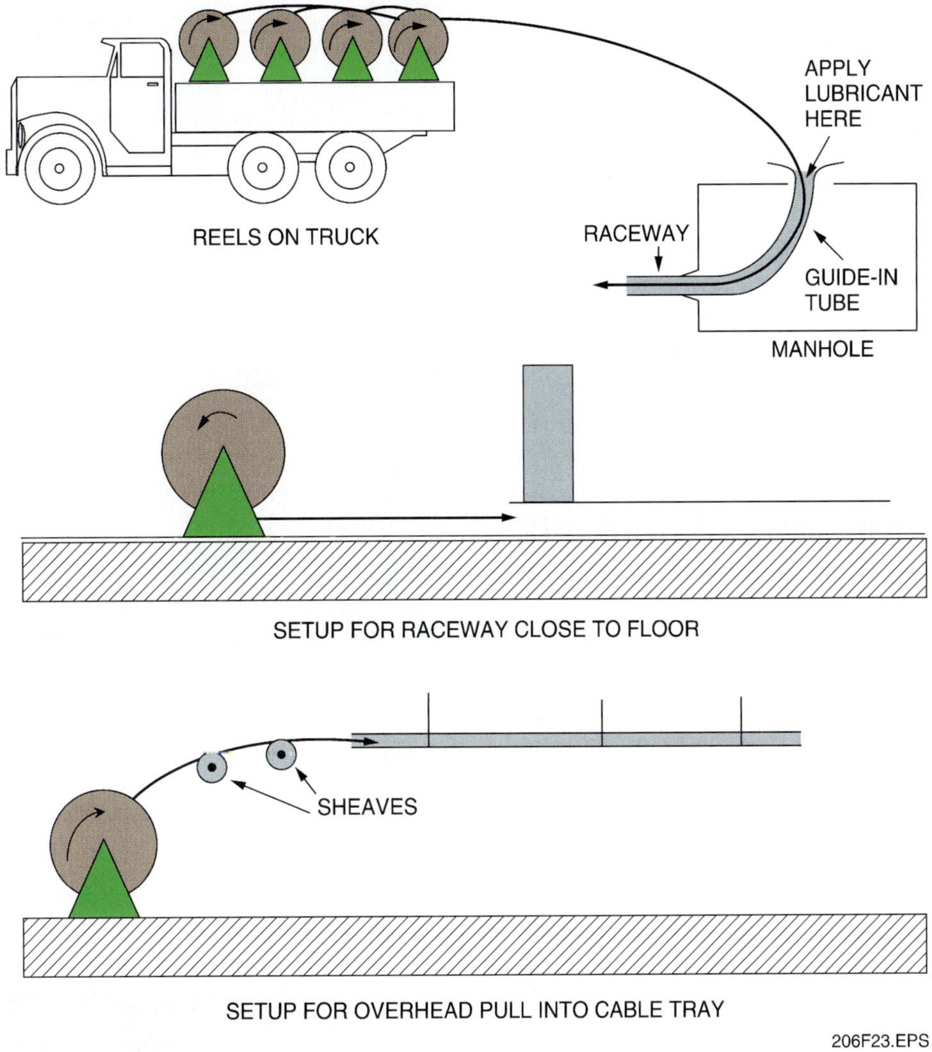

Figure 23 ◆ Cable feed-in setups.

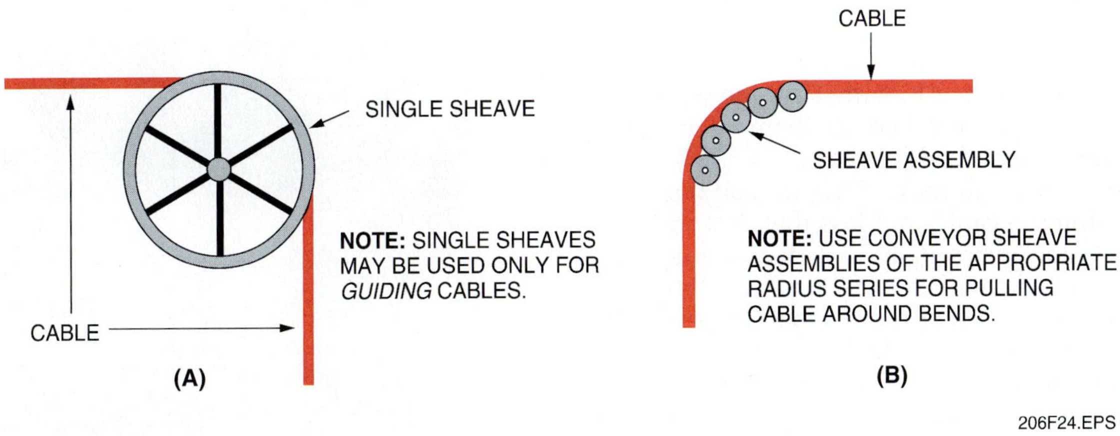

Figure 24 ◆ Arrangement of cable pulleys and sheaves.

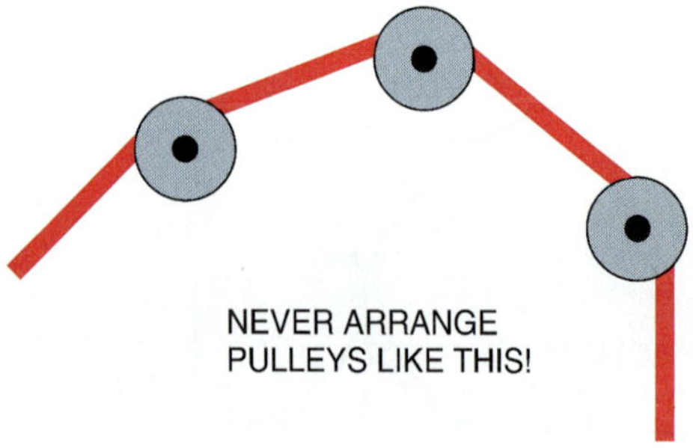

Figure 25 ◆ Never allow a polygon curvature to occur in a cable-pulling operation.

6.0.0 ◆ SUPPORTING CONDUCTORS

Conductors in vertical raceways must be supported in accordance with *NEC Section 300.19* if the vertical rise exceeds the values in *Table 2*. In general, one **conductor support** must be provided at the top of the vertical raceway or as close to the top as practical. Intermediate supports must also be provided as necessary to limit supported conductor lengths to not greater than those values specified in *NEC Table 300.19(A).*

The *NEC*® allows several different methods of supporting conductors in vertical raceways; the following are typical:

- Conductors may be supported by clamping devices constructed of or employing insulating wedges inserted in the ends of the conduit, as shown in *Figure 26*. Where clamping of insulation does not adequately support the cable, the conductor itself must also be clamped.
- Conductors may be supported by inserting boxes at the required intervals in which insulating supports are installed and secured in a satisfactory manner to withstand the weight of the conductors attached to them. The boxes must be provided with covers.
- Cables may be supported in junction boxes by deflecting the cables not less than 90° and carrying them horizontally to a distance not less than twice the diameter of the cable. The cables are carried on two or more insulating supports and additionally secured to these supports by tie wires, if desired. Where this method is used, cables must be supported at intervals not greater than 20% of those mentioned in *NEC Table 300.19(A).*
- Cables may be supported by a method of equal effectiveness.

Table 2 Spacings for Conductor Supports in Vertical Raceways

Size of Wire	Support (Aluminum or Copper-Clad Aluminum Conductor)	Support (Copper Conductor)
18 AWG through 8 AWG	≤ 100'	≤ 100'
6 AWG through 1/0 AWG	≤ 200'	≤ 100'
2/0 AWG through 4/0 AWG	≤ 180'	≤ 80'
Over 4/0 through 350 kcmil	≤ 135'	≤ 60'
Over 350 kcmil through 500 kcmil	≤ 120'	≤ 50'
Over 500 kcmil through 750 kcmil	≤ 95'	≤ 40'
Over 750 kcmil	≤ 85'	≤ 35'

CAUTION

When installing conductors in vertical raceways, install proper supporting devices to hold the conductors before removing the pulling equipment or cutting the conductors.

206F26.EPS

Figure 26 ◆ Conductors supported with wedges.

A variety of supports are manufactured specifically for supporting cable in vertical conduit runs, and many ideas can be obtained from the manufacturers' catalogs. Therefore, make an effort to obtain such catalogs from electrical suppliers or manufacturer's representatives. In fact, manufacturers' catalogs of electrical tools and equipment are excellent study guides for any phase of the electrical industry. Manufacturers of electrical equipment want their equipment used, and they have found that one of the best ways to accomplish this is to provide easy-to-understand instructions and examples of practical applications. Most manufacturers go to great expense to provide this information, but it is usually free of charge for those working in the industry.

There are several precautionary measures that must be taken when pulling cables in vertical runs. The worst danger is runaways; that is, the weight of the cable combined with gravity exceeds the speed of the pull and falls at a rapid rate down the raceway run. Such a situation can cause injury to workers on both ends of the pull. Consequently, braking systems should be used on all long vertical cable pulls. To do otherwise is asking for trouble. Get specific instructions from your supervisor before beginning a vertical pull—either up or down.

WARNING!

Guard against runaways on all vertical pulls. Make certain that proper braking equipment is used to stop a conductor fall should a runaway occur.

7.0.0 ◆ PULLING CABLE IN CABLE TRAYS

When long lengths of cable are to be installed in raceways or cable trays, problems are frequently encountered, particularly when the cable has to be pulled directly into the tray and changes in direction of the tray sections are involved. An entire cable-pulling system has to be planned and set up so that the cable may be pulled into the trays without scuffing or cutting the sheathing and insulation, and also to avoid damaging the cable trays or the tray hangers. To accomplish a successful cable tray pull, a complete line of installation tools is available for pulling lengths of cable up to 1,000 feet or longer. These tools consist mainly of conveyor sheaves and cable rollers. *Figure 27* shows a partial cable tray system with sheaves and rollers in place.

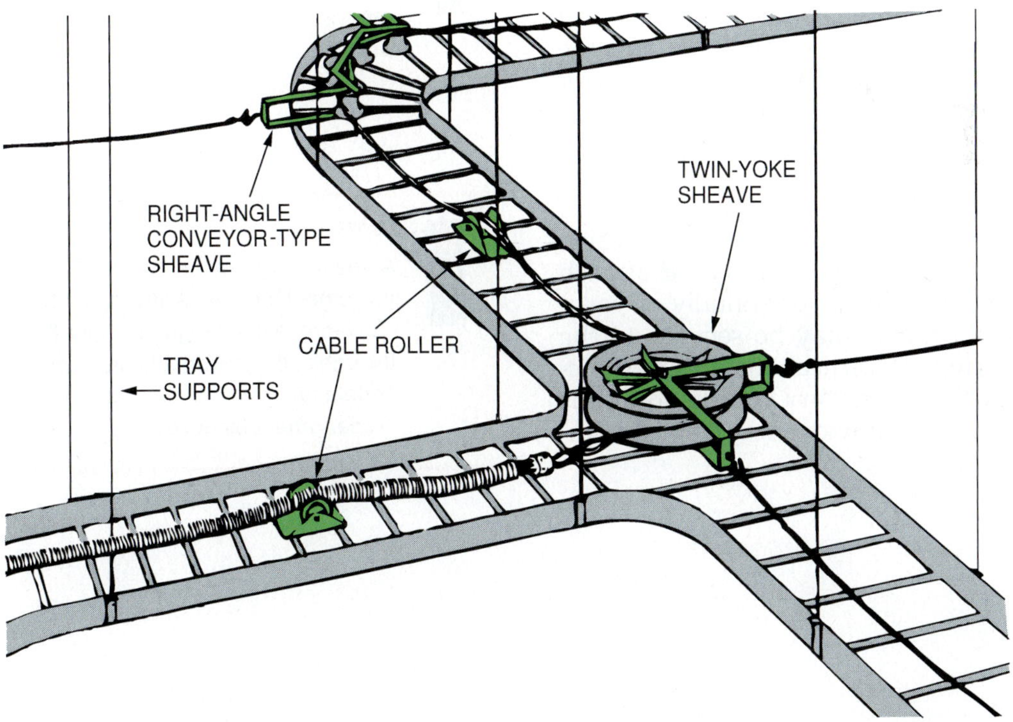

Figure 27 ◆ Typical cable tray cable-pulling arrangement.

Short lengths of cable can be laid in place without tools or pulled with a basket grip. Long lengths of small cable (2" or less in diameter) can also be pulled with a basket grip. Larger cables, however, should be pulled by the conductor and the braid, sheath, or armor. This is done with a pulling eye applied at the cable factory or by tying the conductor to the eye of a basket grip and taping the tail end of the grip to the outside of the cable.

In general, the pull exerted on the cables pulled with a basket grip that is not attached to the conductor should not exceed 1,000 pounds. For heavier pulls, care should be taken not to stretch the insulation, jacket, or armor beyond the end of the conductor nor bend the ladder, trough, or channel out of shape.

The bending radius of the cable should not be less than the values recommended by the cable manufacturer, which range from four times the diameter for a rubber-insulated cable with a 1" maximum outside diameter without lead, shield, or armor, to eight times the diameter for interlocked armor cable. Cables or special construction such as wire armor and high-voltage cables require a larger radius bend.

Best results are obtained in installing long lengths of cable up to 1,000 feet with as many as a dozen bends by pulling the cable in one continuous operation at a speed of 20 to 25 feet per minute. It may be necessary to brake the reel to reduce sagging of the cables between rollers and sheaves.

The pulling line diameter and length will, of course, depend on the pull to be made and the tools and equipment available. Winch and power units must be of adequate size for the job and capable of developing the high pulling speeds required for the best and most economical results.

In general, single or multi-cable rollers are placed in the bottom of trays to protect the cable as it is pulled along. Sheaves are placed at each change of direction—either horizontally or vertically. The bottom rollers may be secured to the tray bottoms except at vertical changes in direction. Extra support is necessary at these locations to prevent damage or movement of the tray system (see *Figure 28*).

NOTE

If single cables are to be installed, always place them on the outside of a bend to allow room on the inside of the bend for pulling other cables.

Sheaves must be supported in the opposite direction of the pull. For example, all right-angle conveyor sheaves should be supported at two locations, as shown in *Figure 29,* to compensate for the pull of the cable.

NOTE

Power cable pulls should not be stopped unless absolutely necessary. However, anyone associated with the pull—upon evidence of danger to either the cable or the workers—may stop a cable-pulling operation. Communication is the most important factor in these cases.

CAUTION

Workers feeding a cable pull must carefully inspect the cable as it is paid off the cable reel. Any visible defects in the cable at the feeding end warrants stopping the pull.

WARNING!

At the first sign of any type of malfunctioning equipment, broken sheaves, or other events that could present a danger to either the workers or the cable, the pull should be stopped. Make certain that all communications equipment is in proper order before starting a pull.

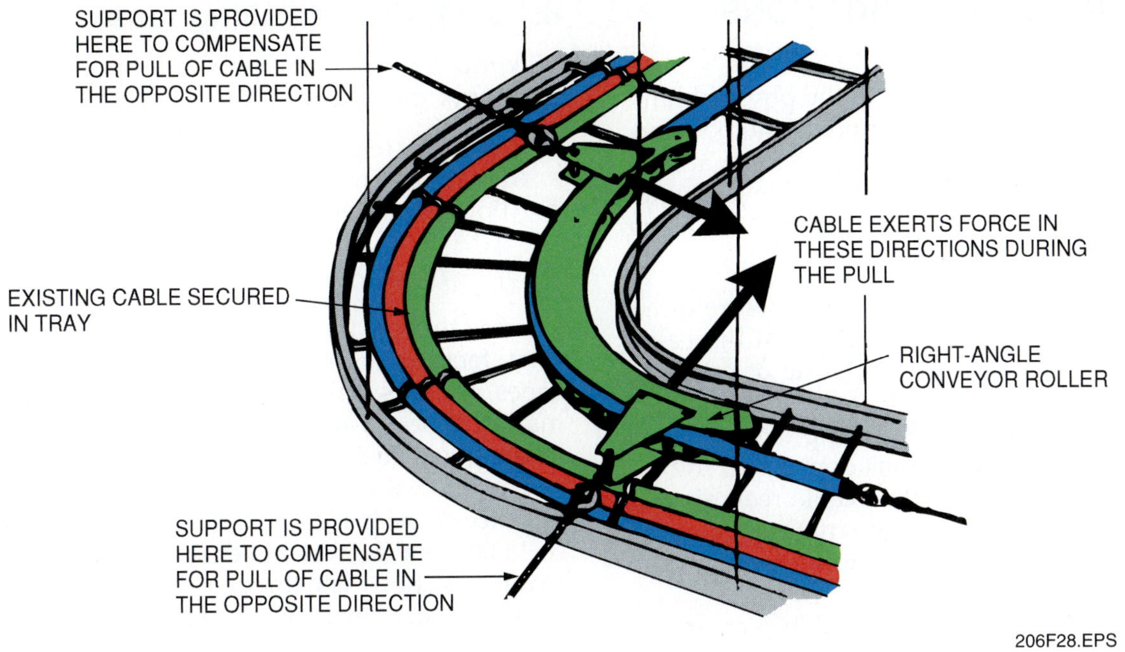

Figure 28 ◆ Support points for right-angle turns.

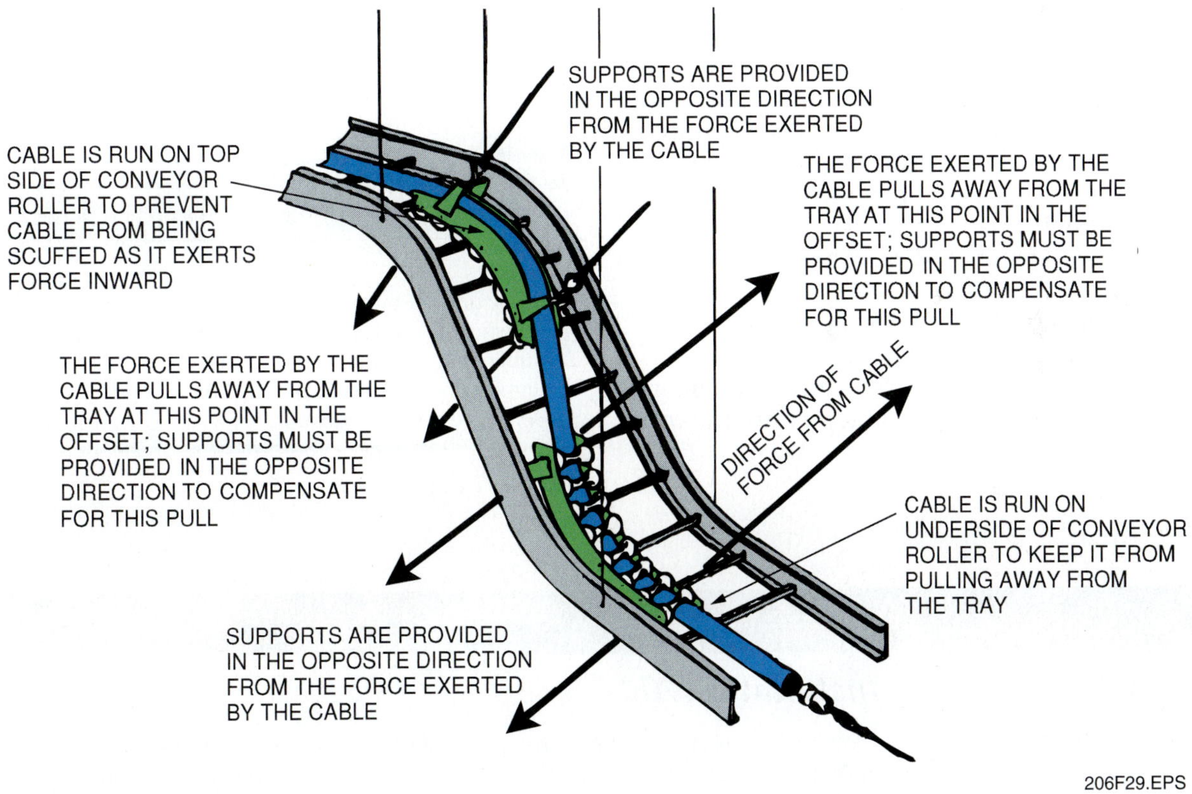

Figure 29 ◆ Typical support points for a cable tray offset.

8.0.0 ◆ PHYSICAL LIMITATIONS OF CABLE

Consideration must be given to the physical limitations of a cable as it is being pulled into position. Pulling subjects cable to extreme stress and, if done improperly, can displace a cable's components. Thus, it is important that the following guidelines be observed:

- While reels are in storage—either before or after a pull—the conductor ends must be sealed to prevent moisture from entering or creeping into the cable ends.
- The minimum ambient working temperatures for cable-pulling operations depend on the cable jackets. In cold weather, cable reels should be stored in a warm area overnight so that the cable jackets will be at the proper temperature for pulling the next day.
- Calculate and stay within the cable's maximum pulling tension, maximum sidewall load, and minimum bending radii.
- Ensure that the raceway joints are aligned and that the wiring space is sufficient.
- Train the cable to avoid dragging on the edge of the raceway; also avoid laying or dragging cable on the ground.
- If using a basket grip, secure it to the cable with steel strapping and cut well behind the areas it covers once the cable is in place. The portion of the cable under the grip should be discarded.
- Ensure that the elongation of the pull rope minimizes jerking.
- Pull with a capstan and no faster than 40 feet per minute.
- Do not stop a pull unless absolutely necessary.
- Never pull the middle of the cable.
- Seal the ends with appropriate putty or silicone caulking and overwrap with tape until the conductors are terminated.

8.1.0 Maximum Pulling Tension

Maximum pulling tension should not exceed the smaller of the following values:

- Allowable tension on pulling device
- Allowable tension on conductor
- Allowable sidewall load

8.1.1 Allowable Tension on Pulling Device

Do not exceed the working load stated by the manufacturer of the pulling devices (pulling eyes, ropes, anchors, basket grips, etc.). If catalog information is not available, work at 10% of the rated braking tensile strength.

The allowable tension with a basket grip must not exceed the lbs/cmil value (as shown in *Table 3*) or 1,000 pounds, whichever is smaller. Exceptions to this rule, however, do occur, but seldom will this figure rise to over 1,250 pounds.

8.1.2 Allowable Tension on Conductors

The metallic phase conductors are the tensile members of the cables and should bear all of the pulling force. Never use shielding drain wires or braids for pulling. Table 3 provides the allowable pulling tensions of various types of conductors. The listed values should never be exceeded.

Table 3 Physical Limitations of Cable

Material	Cable Type	Temper	Lbs/Cmil
Copper	All	Soft	0.008
Aluminum	Power	Hard	0.008
Aluminum	Power	¾ Hard	0.006
Aluminum	Power	AWM	0.005
Aluminum	URD (solid)	Soft (½ hard)	0.003
All	Thermocouple	—	0.008

INSIDE TRACK

Installation Aids

Other tools available to simplify the pulling of cables into cable tray systems include triple pulleys, bull wheels, and both wide and narrow rollers. Be sure to position these pulleys and rollers at the proper locations to prevent damage to the cables during the installation and also to help the installation proceed as quickly as possible.

CAUTION

When smaller conductors are pulled with large conductors, the smaller conductors may be damaged.

Reduce the maximum pulling tension by 20% to 40% if several conductors are being pulled simultaneously since the tension is not always evenly distributed among the conductors.

8.2.0 Calculating Pulling Tension

Normally, the maximum tension for a specific type of cable can be found using data from the cable manufacturer.

In general, the maximum tension for a single-conductor cable should not exceed 6,000 pounds. The maximum tension for two or more conductors should not exceed 10,000 pounds.

The maximum stress for leaded cables must not exceed 1,500 pounds per square inch of lead sheath area when pulled with a basket grip.

The maximum tension must not exceed 1,000 pounds per square inch of insulation area for non-leaded cables when pulled with a basket grip.

The maximum tension at a bend must not exceed 500 times the radius of curvature of the conduit or duct expressed in feet.

8.2.1 Tension in Horizontal Pulls

The pulling tension in a given horizontal raceway section may be calculated as follows:

For a straight section, the pulling tension is equal to the length of the duct run multiplied by the weight per foot of the cable and the coefficient of friction, which will vary depending on the type and amount of lubrication used. Therefore, the equation is as follows:

$$T = L \times w \times f$$

Where:

T = total pulling tension
L = length of raceway run in feet
w = weight of cable in pounds per foot
f = coefficient of friction

For ducts having curved sections, the following equation applies:

$$T_{OUT} = T_{in}\, e^{fa}$$

Where:

T_{OUT} = tension of bend
T_{in} = tension into bend
f = coefficient of friction
e = Napierian logarithm base 2.718
a = angle of bends in radians

To aid in solving the above equation, values of e^{fa} for specific angles of bend and coefficients of friction are listed in *Table 4*. For more precise values, tables are available from cable manufacturers.

8.3.0 Sidewall Loading

Before actually getting into a practical application of calculating cable-pulling tension, we will discuss one more factor involved during the process of wire pulling—sidewall loading or sidewall bearing pressure.

The sidewall load is the radial force exerted on a cable being pulled around a conduit bend or sheave. Excessive sidewall loading can crush a cable and is, therefore, one of the most restrictive factors in installations having bends or high tensions. *Figure 30* shows a section taken across a conduit run in a 90° bend. Note that pulling tension is exerted parallel with the walls of the

Table 4 Angle of Bend vs. Coefficients of Friction

Angle of Bend (degrees/radians)	Values of e^{fa} for Coefficients of Friction		
	f = 0.75	f = 0.50	f = 0.35
15/0.2618	1.22	1.14	1.10
30/0.5236	1.48	1.30	1.20
45/0.7854	1.80	1.48	1.32
60/1.0472	2.19	1.68	1.44
90/1.5708	3.25	2.20	1.73

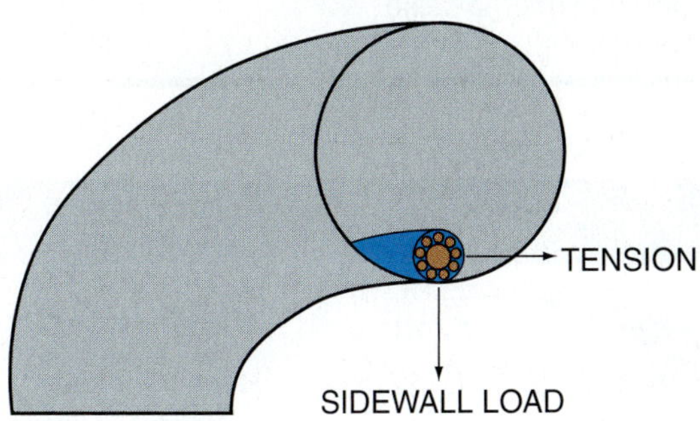

Figure 30 ◆ Sidewall loading.

conduit. However, due to the 90° bend, pressure is also exerted downward against the wall of the conduit. Once again, this is known as the sidewall load. Sidewall loading is reduced by increasing the radius of bends.

In general, the sidewall load on any raceway run should not exceed 500 pounds/foot of bend radius. This pressure, however, must be reduced on some types of cables. For example, *Table 5* shows one manufacturer's recommendations for the maximum sidewall pressures permitted for various types of cables. Always refer to the cable manufacturer's instructions for the type of cable being used.

8.4.0 Practical Applications

Figure 31 shows a typical raceway system containing three 500 kcmil lead sheath copper conductors. Note the straight 300' run from A to B, a 45° kick, and then another 100' straight run from C to D. We will find the calculated pulling tension from D to A.

Referring to cable data in the manufacturer's catalogs, we find that this cable weighs 8 pounds per foot and has a 0.141" lead sheath. The outside diameter of the three-conductor cable assembly is 3". We will use 0.5 as the coefficient of friction and proceed with our calculation of the pulling tension from A to B *(Figure 31)* as follows:

Table 5 Maximum Sidewall Pressures for Various Types of Cable

Cable Type	Sidewall Pressure in Pounds/Foot of Bend Radius
600V nonshielded control	300
600V and a kV nonshielded EP power	500
5kV and 15kV EP power	500
25kV and 35kV power	300
Interlocked armored cable (all voltages)	300

Step 1 Find the tension between points A and B.

$$\text{Tension at B} = T_1 = L_1 \times w \times f$$

L_1 is the length between points A and B while w is the weight of the cable per foot, and f is the coefficient of friction, which we have determined to be 0.5. Substituting the known values in the equation, we have:

$$T_1 = 300 \times 8 \times 0.5 = 1{,}200 \text{ pounds}$$

Step 2 Find the tension between points B and C.

$$\text{Tension at C} = T_1\, e^{fa}$$

Since the distance between B and C involves an angle, refer to *Table 4* of this module. Looking in the left column of the table, we see that 45° equals 0.7854 radians; this figure multiplied by the coefficient of friction (0.5) equals 0.3927. Radians of angles may also be found with electronic pocket calculators if they have scientific functions. Follow the instructions in the manual accompanying the calculator. The exact key strokes will vary with the brand of calculator, but most require pressing the degree key, entering the numeral for degrees, then pressing the convert key, and finally pressing the radian or RAD key. The radians of the entered angle will be displayed.

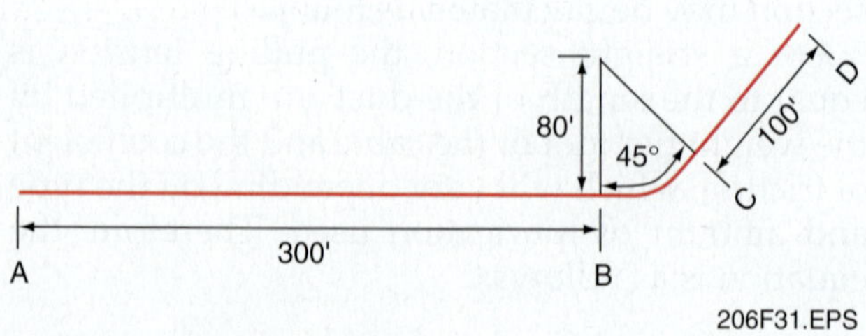

Figure 31 ◆ Sample conduit run.

INSIDE TRACK

Software Programs

Software programs are now available to take the math out of calculating pulling tensions. These programs typically calculate pulling tension and sidewall pressures based on conductor size using their own data on friction coefficients. These programs also determine conduit fill, conductor configuration, jam ratios, and the amount of lubricant required for the pull.

Once again, refer to *Table 4*. Find the 45° angle of bend in the left column; read across the row to the column under 0.50—our coefficient of friction figure. The number is 1.48, the value of e^{fa}. Substituting these values in the equation, we have the following:

$$A = 45° = 0.7854 \text{ radians}$$
$$fa = 0.7854 \times 0.5 = 0.3927$$
$$e^{fa} = 1.48$$
$$1{,}200 \times 1.48 = 1{,}776 \text{ pounds}$$

Therefore, the tension at C is 1,776 pounds.

Step 3 Find the tension from points C to D.

$$\text{Tension from C to D} = T_2 = L_2 \times w \times f$$
$$T_2 = 100 \times 8 \times 0.5 = 400 \text{ pounds}$$

Step 4 Find the total pulling tension by adding the figures obtained previously.

$$T = T_2 + T_1e^{fa} = 400 + 1{,}776$$
$$= 2{,}176 \text{ pounds}$$

The maximum pulling force using a basket grip for this size cable should not exceed 1,900 pounds. Therefore, if the pull is made from point A to point D, a pulling eye will have to be used since the total pulling tension exceeds 1,900 pounds. However, if the pull is reversed—pulling from point D to point A—the total pulling tension will be reduced since the distance from point D to the 45° angle (point C) is ⅓ the distance from point A to the 45° angle (point B).

$$\text{Tension at C} = 400 \text{ pounds}$$
$$\text{Tension at B} = 400 \times 1.48 = 592 \text{ pounds}$$
$$\text{Total tension at A} = 1{,}200 + 592$$
$$= 1{,}792 \text{ pounds}$$

Therefore, if the cable is pulled from point D to point A, either a pulling eye or basket grip may be used.

NOTE

A lower tension is obtained by feeding the pull from the end nearest the bend.

9.0.0 ◆ CABLE PULLING INSTRUMENTS

There are several instruments used in conjunction with cable pulling operations. Since details of operation vary with the manufacturer, these instruments will only be briefly discussed in this module. However, your instructor will cover exact operating details of one or more of these instruments during your training sessions. Operating manuals are also packaged with each instrument before shipping. Try to obtain manuals for the instruments used by your employer and study these to obtain a better understanding of the instruments you will be using on the job.

- *Cable length meters* – Cable length meters are available for direct reading in lengths from 2,000' to 20,000'. Most are calibrated for different wire sizes whereas the sizes are selected with a selector switch on the instrument. Controls may also be set for either copper or aluminum conductors. These instruments are ideal for determining the exact length of conductors on reels prior to making a pull.
- *Wire sorter* – This instrument is for use on unenergized circuits for tracing conductors. In use, one lead is attached to ground while the numbered leads are attached to the wires being traced. The wires are then marked to correspond with the instrument leads. The instrument is then taken to the opposite end of the circuit. Attach the identifier to ground and then touch each wire end with the probe to read the wire number. This end of the run is then marked accordingly.
- *Dynamometer* – This type of meter is designed to read dynamic friction during a cable pull. Many are designed for a specific cable-pulling tool and are shipped as an integral or optional part of the cable puller. For example, one common cable puller electronically displays both actual speed and actual force while running. Furthermore, this puller automatically shuts down when maximum preset force is reached to prevent damage to the cable being pulled. Others are portable units for use with any type of cable-pulling equipment. Such instruments are invaluable for use during high-force cable pulls to avoid damage to conductors during the pull. When the instrument indicates that the maximum pulling force has been reached, the pull can be stopped before damage occurs to the cable or conductors.

CASE HISTORY

Cable Pulling

At a large commercial job site this past summer, workers began a complex pull shortly before the end of the day. At 5 P.M., they left for the day, having completed only a portion of the pulling operation. The next day, they resumed work promptly at 8 A.M., but the pulling lubricant had already dried in the conduit. As a result of the excess friction on the pull, the rope broke, destroying thousands of feet of expensive cable.

The Bottom Line: Don't stop a pull in the middle of a job.

INSIDE TRACK

Complex Installations

Complex cable-pulling installations are best handled by highly experienced contractors. This is one small portion of an award-winning installation that took over 70,000 hours to complete.

206P0603.EPS

Review Questions

1. Who is allowed to stop the pull at the first sign of danger to the cable, raceway system, or personnel?
   a. Only the project supervisor
   b. Only workers on the feeding end of the run
   c. Only workers on the pulling end of the run
   d. Anyone involved with the pull

2. After some length of conductor has been used from a cable reel, what should be done before placing the reel in storage?
   a. A stress test should be made to see if the cable has been damaged.
   b. An insulation test should be made using a megger.
   c. The ends of the conductors should be sealed.
   d. The conductor jacket should be warmed before storing.

3. Which of the following precautions should be taken when making a cable pull in extremely cold weather?
   a. The cable should be stored in a warm area overnight.
   b. The cable should be stored outside overnight.
   c. Antifreeze should be mixed with the pulling lubricant.
   d. Electric warming blankets should be used during the pull.

4. Which of the following pulling methods should be used when the pulling tension exceeds 1,000 pounds?
   a. A basket grip by itself
   b. A pulling eye
   c. Conductors bent around a hook and taped
   d. Conductors bent around a snake hook and left untaped

5. Which of the following methods is *not* true with regard to transporting or storing cable reels?
   a. Reels should be stored in an upright position.
   b. Reels should be laid flat on their sides.
   c. Reels should be lifted by a rigged crane with a spindle through the reel.
   d. Reels should be carried with a forklift so that the forks do not touch the cable.

6. Where is the best place to seek information about a wire-pulling lubricant that is compatible with a particular type of cable insulation?
   a. The lubricant manufacturer
   b. The cable manufacturer
   c. Both the cable and lubricant manufacturers
   d. The *NEC*®

7. How should the feed-in setup be handled during a cable installation?
   a. The cable should unreel with a reverse S curvature.
   b. It makes no difference.
   c. Only the pulling end is important when unreeling the cable.
   d. The cable should unreel along its natural curvature.

8. Which of the following is *not* used to guide cable during a pull?
   a. A conveyor sheave
   b. A polygon setup
   c. A guide-in tube
   d. A cable reel

9. Which of the following is the most important consideration when pulling cables in vertical pulls?
   a. Calculate the cable weight before installation.
   b. Install the cable from the bottom up.
   c. Guard against runaways.
   d. Use less wire lubricant than in horizontal applications.

10. Which of the following best describes the location of sheave supports for pulling cable in cable trays?
    a. They should be supported in the opposite direction of the pull.
    b. They should be supported in the same direction as the pull.
    c. They should be supported only by the tray assembly itself.
    d. No support is necessary when sheaves are used in cable trays.

Summary

A cable-pulling operation involves careful planning. Furthermore, the proper use of appropriate equipment is crucial to a successful cable installation. Think of everything that can go wrong, and then plan accordingly.

Communications equipment is another crucial part of a cable installation. Workers on both ends must be in constant contact with each other in case something goes wrong. The workers feeding the cable must carefully inspect the cable as it is paid off the reels and stop the pull immediately if the cable appears damaged. In fact, anyone involved with the pull should be able to stop the pull whenever the safety of the cable or personnel is threatened. Again, good communication is crucial.

Safety precautions must be followed exactly during any cable installation. Use the correct size and type of equipment for the job. Never exceed the maximum force of the weakest component. Do not position yourself in a direct line with a cable pull.

Notes

Trade Terms Introduced in This Module

Basket grip: A flexible steel mesh grip that is used on the ends of cable and conductors for attaching the pulling rope. The more force exerted on the pull, the tighter the grip wraps around the cable.

Cable grip: A device used to secure ends of cables to a pulling rope during cable pulls.

Capstan: The turning drum of a cable puller on which the rope is wrapped and pulled. An increase in the number of wraps increases the pulling force of the cable puller.

Clevis: A device used in cable pulls to facilitate connecting the pulling rope to the cable grip.

Conductor support: The act of providing support in vertical conduit runs to support the cables or conductors. The *NEC*® gives several methods in which cables may be supported, including wedges in the tops of conduits, supports to change the direction of cable in pull boxes, etc.

Conduit piston: A cylinder of foam rubber that fits inside the conduit and is then propelled by compressed air or vacuumed through the conduit run to pull a line, rope, or measuring tape. Also called a *mouse.*

Extension boom: Part of a cable-pulling apparatus that is usually used when pulling conductors through concealed conduit runs.

Extension bushing: Used to extend the end of the conduit in panels and other enclosures to provide a grip for the flexible adapter sheave.

Fish line: Light cord used in conjunction with vacuum/blower power fishing systems that attaches to the conduit piston to be pushed or pulled through the conduit. Once through, a pulling rope is attached to one end and pulled back through the conduit for use in pulling conductors.

Fish tape: A flat iron wire or fiber cord used to pull conductors or a pulling rope through conduit.

Setscrew grip: A cable grip, usually with built-in clevis, in which the cable ends are inserted in holes and secured with one or more setscrews.

Sheave: A pulley-like device used in cable pulls in both conduit and cable tray systems.

Soap: Slang for wire-pulling lubricant.

Additional Resources

This module is intended to present thorough resources for task training. The following reference works are suggested for further study. These are optional materials for continued education rather than for task training.

American Electrician's Handbook, Latest Edition. New York: Croft and Summers, McGraw-Hill.

Cable Installation Manual, Latest Edition. New York: Cablec Corp.

National Electrical Code® Handbook, Latest Edition. Quincy, MA: National Fire Protection Association.

CONTREN® LEARNING SERIES — USER FEEDBACK

The NCCER makes every effort to keep these textbooks up-to-date and free of technical errors. We appreciate your help in this process. If you have an idea for improving this textbook, or if you find an error, a typographical mistake, or an inaccuracy in NCCER's *Contren®* textbooks, please write us, using this form or a photocopy. Be sure to include the exact module number, page number, a detailed description, and the correction, if applicable. Your input will be brought to the attention of the Technical Review Committee. Thank you for your assistance.

Instructors – If you found that additional materials were necessary in order to teach this module effectively, please let us know so that we may include them in the Equipment/Materials list in the Annotated Instructor's Guide.

Write: Product Development
National Center for Construction Education and Research
P.O. Box 141104, Gainesville, FL 32614-1104

Fax: 352-334-0932

E-mail: curriculum@nccer.org

Craft ______________ Module Name ______________

Copyright Date ______________ Module Number ______________ Page Number(s) ______________

Description

____

____

____

(Optional) Correction

____

____

(Optional) Your Name and Address

____

____

Cable Tray
26207-05

City of Phoenix Fire Station No. 50
Phoenix, Arizona
Commercial $2–5 Million Award Winner
FCI Constructors, Inc.

26207-05
Cable Tray

Topics to be presented in this module include:

Overview

Electricians who regularly work on commercial and industrial installations projects may not only be called upon to install conduit systems, but may also find themselves installing elevated or suspended cable trays to support cables or conductors, especially in industrial locations.

Cable tray systems can be accessorized with fittings that either snap or bolt into place to provide tees, wyes, crosses, reducers, barriers, covers, and even box connectors. Cable tray installations, like any raceway system, are regulated by the *National Electrical Code®*.

The support of cable tray is an important consideration and is determined by the number and size of conductors or cables to be installed in the tray. If the support is not adequate for the load at various stress points, overloading can cause rung and/or rail failure. To prevent these failures, load stress points must always be identified before making any splices in straight cable tray sections. Methods used to hang and support cable tray assembles include trapeze, direct rod, and wall mounting.

Note: *National Electrical Code®* and *NEC®* are registered trademarks of the National Fire Protection Association, Inc., Quincy, MA 02269. All *National Electrical Code®* and *NEC®* references in this module refer to the 2005 edition of the *National Electrical Code®*.

Objectives

When you have completed this module, you will be able to do the following:

1. Describe the components that make up a cable tray assembly.
2. Explain the methods used to hang and secure cable tray.
3. Describe how cable enters and exits cable tray.
4. Select the proper cable tray fitting for the situation.
5. Explain the NEMA standards for cable tray installations.
6. Explain the *NEC*® requirements for cable tray installations.
7. Select the required fittings to ensure equipment grounding continuity in cable tray systems.
8. Interpret electrical working drawings showing cable tray fittings.
9. Size cable tray for the number and type of conductors contained in the system.
10. Select rollers and sheaves for pulling cable in specific cable tray situations.
11. Designate the required locations of rollers and sheaves for a specific cable pull.

Required Trainee Materials

1. Paper and pencil
2. Appropriate personal protective equipment
3. Copy of the latest edition of the *National Electrical Code*®

Prerequisites

Before you begin this module, it is recommended that you successfully complete *Core Curriculum; Electrical Level One;* and *Electrical Level Two,* Modules 26201-05 through 26205-05.

This course map shows all of the modules in *Electrical Level Two.* The suggested training order begins at the bottom and proceeds up. Skill levels increase as you advance on the course map. The local Training Program Sponsor may adjust the training order.

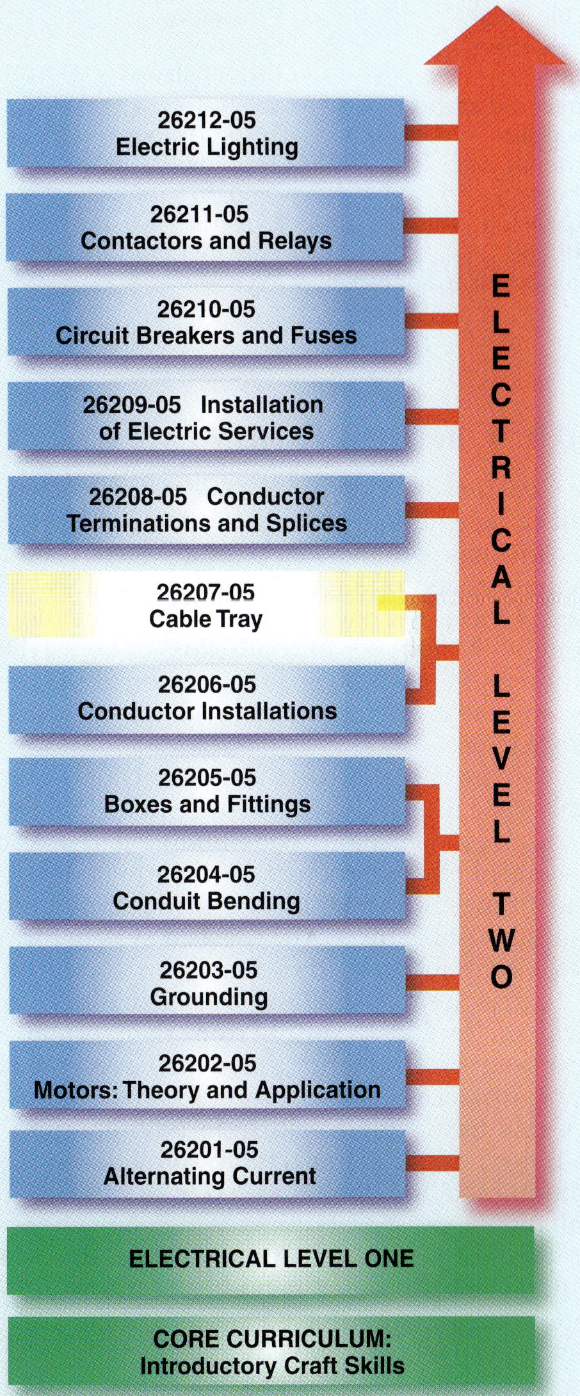

207CMAP.EPS

Trade Terms

Barrier strip	Ladder tray
Cable pulley	Pipe racks
Cable roller	Sheave
Cross	Swivel plates
Direct rod suspension	Tee
Dropout	Trapeze mounting
Dropout plate	Tray cover
Elbow	Trough
Expansion joints	Unistrut®
Fittings	Wall mounting
Interlocked armor cable	Wye

1.0.0 ◆ INTRODUCTION

NEC Section 392.2 defines a cable tray system as a unit or assembly of units or sections and associated **fittings** forming a structural system used to securely fasten or support cables and raceways. *NEC Article 392* covers cable tray installations, along with the types of conductors to be used in various systems. Whenever a question arises concerning cable tray installations, this is the *NEC®* article to use.

Cable trays are the usual means of supporting cable systems in industrial applications. The trays themselves are usually made up into a system of assembled, interconnected sections and associated fittings, all of which are made of metal or noncombustible units. The finished system forms into a continuous rigid assembly for supporting and carrying single, multiconductor, or other electrical cables and raceways from their origin to their point of termination, frequently over considerable distances.

Cable tray is fabricated from both aluminum and steel. Some manufacturers provide an aluminum cable tray that is coated with PVC for installation in caustic environments. Relatively new all-nonmetallic trays are also available; this type of tray is ideally suited for use in corrosive areas and in areas requiring voltage isolation. Cable tray is available in various forms, including ladder, **trough,** center rail, and solid bottom, and can be supported by either side mounts or center mounts.

Ladder tray, as the name implies, consists of two parallel channels connected by rungs, similar in appearance to a conventional straight or extension ladder. Trough types consist of two parallel channels (side rails) having a corrugated, ventilated bottom. The solid bottom cable tray is similar to the trough. All of these types are shown in *Figure 1*. Ladder, trough, and solid bottom trays are completely interchangeable; that is, all three types can be used in the same run when needed.

Cable tray is manufactured in 12' and 24' lengths. Common widths range from 6" to 36". All sizes are provided in either 3", 4", 5", or 6" depths.

Cable tray sections are interconnected with various types of fittings. Fittings are also used to provide a means of changing the direction or dimension of the cable tray system. Some of the more common fittings include:

- Horizontal and vertical **tees**
- **Wyes**
- Horizontal and vertical bends
- Horizontal **crosses**
- Reducers
- **Barrier strips**
- Covers
- Splice plates
- Box connectors

The area of a cable tray cross section that is usable for cables is defined by width (W) × depth (D), as shown in *Figure 2*. The overall dimensions of a cable tray, however, are greater than W times D because of the side flanges and seams. Therefore, overall dimensions vary according to the tray design. Cables rest on the bottom of the tray and are held within the tray area by two longitudinal side rails, as shown in *Figure 3*.

A channel is used to carry one or more cables from the main tray system to the vicinity of the cable termination (see *Figure 4*). Conduit is then used to finish the run from the channel to the actual termination.

Certain *NEC®* regulations and National Electrical Manufacturers Association (NEMA) standards should be followed when designing or installing cable tray. Consequently, practically all projects of any great size will have detailed drawings and specifications for the workers to follow. Shop drawings may also be provided.

INSIDE TRACK

Cable Tray Installation

Cable trays are widely used in many commercial and industrial installations.

207P0701.EPS

Special-Application Cable Tray

This lightweight, flexible nonmetallic cable tray is used to route wiring to the large motors in a manufacturing facility.

207P0702.EPS

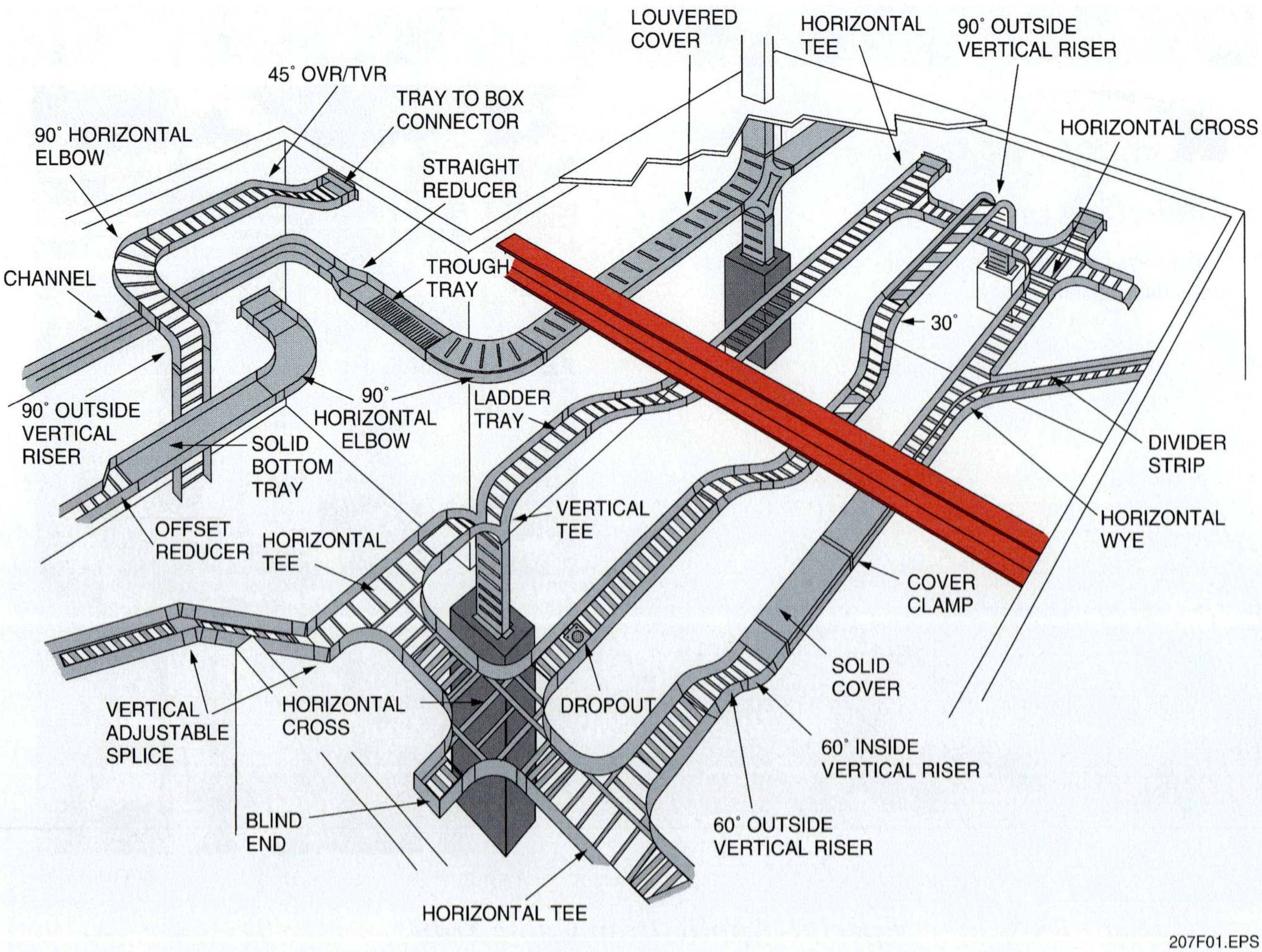

Figure 1 ◆ Typical cable tray system.

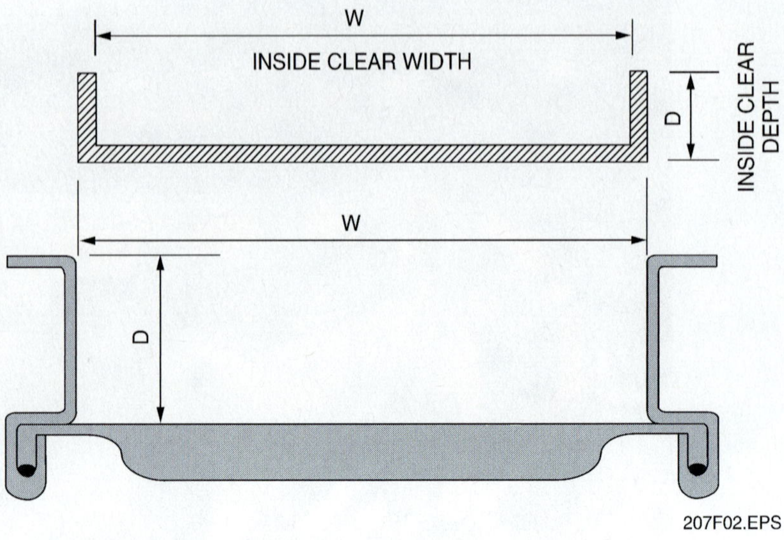

Figure 2 ◆ Cross section of cable tray comparing usable dimensions to overall dimensions.

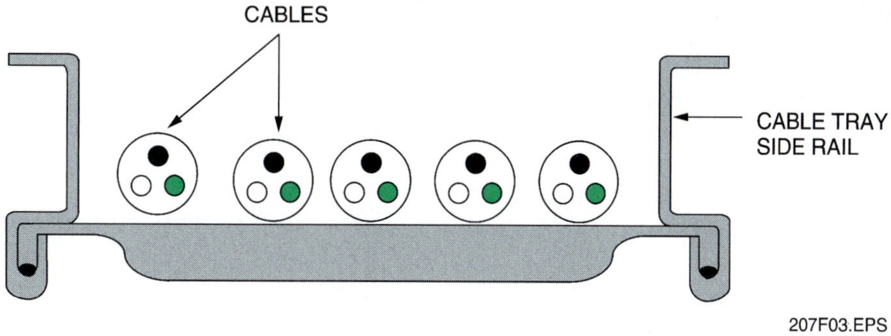

Figure 3 ◆ Cables rest on the bottom of the tray and are held in place by the longitudinal side rails.

INSIDE TRACK

Ladder Tray

Ladder tray is used for about three-fourths of today's installations. It offers numerous advantages:

- Ladder tray has greater air circulation than other trays, which helps to dissipate heat.
- The ladder rungs are convenient for tying down conductors.
- Ladder tray does not collect moisture.
- Conductors can be dropped through the bottom of the tray at any point.

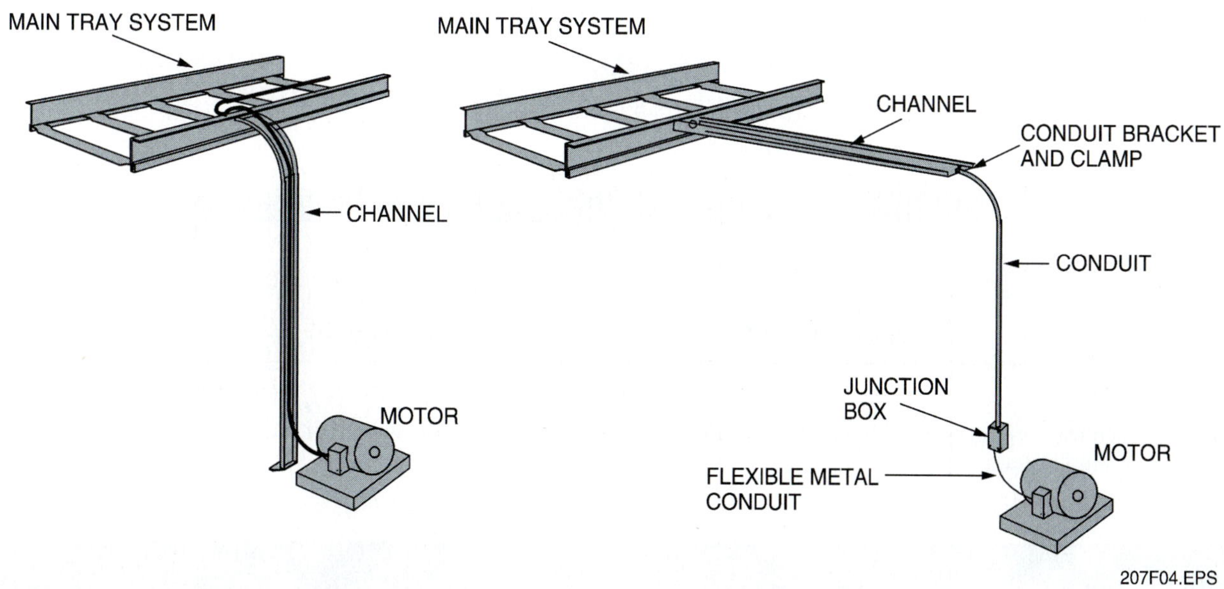

Figure 4 ◆ Two applications of cable tray channel.

2.0.0 ◆ NEMA STANDARDS

NEMA is a nonprofit organization supported by the manufacturers of electrical equipment and supplies. NEMA develops standards that are used when purchasing and installing electrical equipment. These standards help to ensure that the product will match the application.

2.1.0 Load Factors

The load capacity of cable tray varies with each manufacturer and depends on the shape and thickness of the side rails, the shape and thickness of the bottom members, spacing of rungs (if any), material used, safety factor used, method used to determine the allowable load, method of supporting the tray, and volume capacity. Consequently, each manufacturer publishes load data.

2.2.0 Determining Fill

The density of fill can only be determined by personnel laying out the system. In doing so, however, be aware that cables packed closely together can impair each other's efficiency (see *Figure 5*).

2.3.0 Determining the Load on Supports

Each support should be capable of safely supporting approximately 1.25 times the full weight of the cable and tray on a typical span, as shown in *Figure 6*.

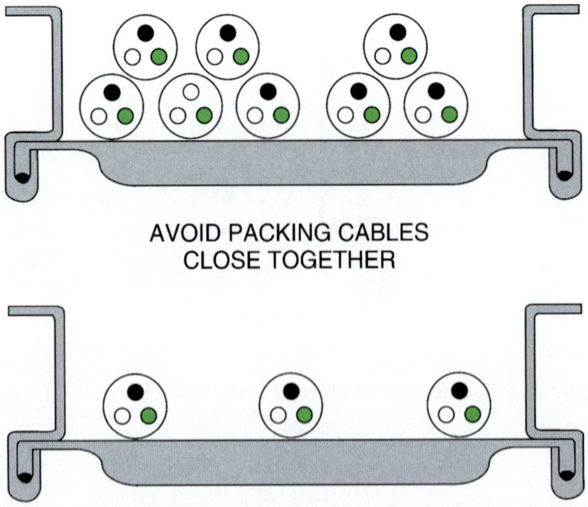

Figure 5 ◆ Cables packed closely together can impair efficiency.

2.4.0 Deflection Under Load

Consider the case of a cable tray spanning only two supports (simple span). As the tray is loaded, the side rails take the deflected form. Simultaneously, in cross section, the side rails rotate inward or outward and the tray bottom deflects. The amount of inward or outward movement of the side rails is a critical factor in the ability of the tray to carry a load.

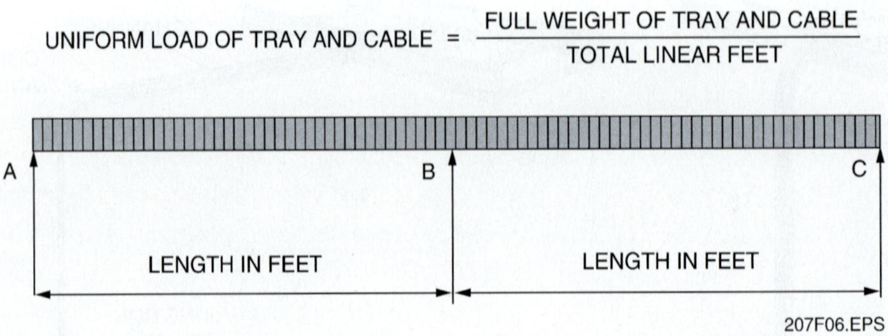

Figure 6 ◆ Each tray support should be capable of safely supporting 1.25 times the weight of the entire tray and cable assembly span.

THINK ABOUT IT

Cable Fill

In what way can cables installed too closely together impair each other's efficiency?

2.5.0 Failure Under Load

There are two types of failures that can occur with a loaded cable tray. These are longitudinal (side rail) failures and transverse (rung) failures. Transverse (rung) failures occur when the load applied causes the fibers on the tension side (bottom edge) of the rung to stretch and permanently deform. Simultaneously, fibers on the compression side (top edge) are permanently crushed. Longitudinal (side rail) failures occur either as bending failures when the tray is supported on a larger span, or, on longer spans, buckling failures may occur because the side rails of the tray have little resistance to inward or outward movement. As the tray deflects, the side rails rotate and the top (compression) flanges of the tray buckle. Bending failures occur on short spans because the side rails of the tray have greater resistance to rotation and remain reasonably upright. The tray does not fail until the load is such that it causes the fibers on the tension side (bottom edge) of the side rail to stretch and permanently deform. Simultaneously, fibers on the compression side (top edge) are permanently crushed. For any tray on an intermediate span, it is difficult to anticipate whether a bending or buckling failure would occur (see *Figure 7*).

2.6.0 Splicing Straight Sections

In *Figure 8*, the load on the cable tray creates bending moments along the spans. The stress in the

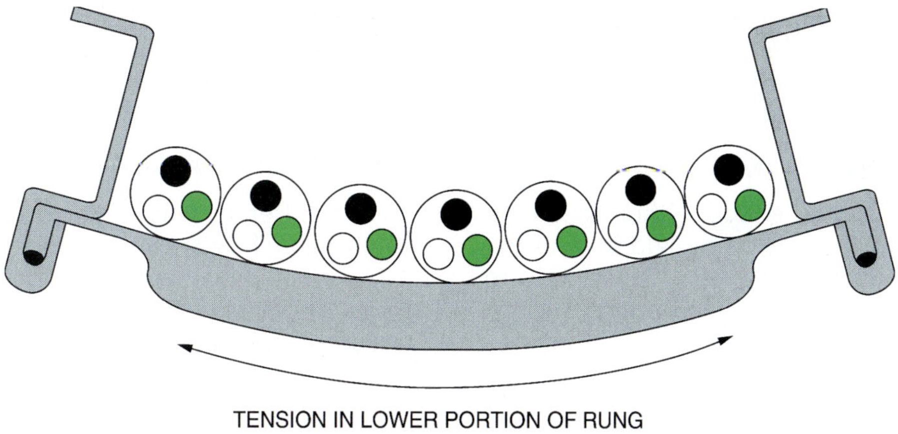

Figure 7 ◆ Bending of loaded tray.

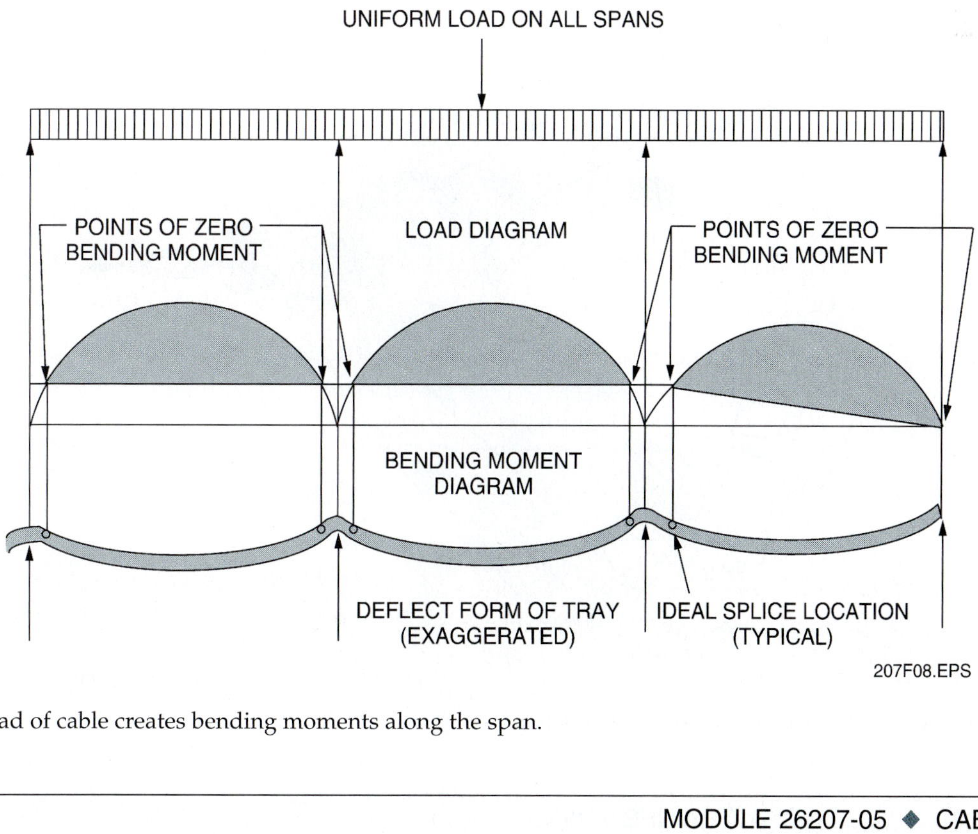

Figure 8 ◆ Load of cable creates bending moments along the span.

side rails of the tray is directly related to the bending moments at all points along the tray. The magnitude of the bending moment at any point is determined by measuring the vertical height of the shaded portion. In any cable tray system, a splice is a point of weakness. Consequently, splices should be located at the points of least stress. Ideally, splices would be located at the points of zero bending moment, and the strength of the tray system would be at a maximum. In actual practice, if the splice is located within one-fourth of the span's distance from the support, the result will be close to ideal.

Locating the splice within the one-fourth points of the span requires extra labor on the part of the installer. When a splice occurs within the central length of a span between the one-fourth points, the tray will support the load for which it was designed, but the safety factor will be greatly reduced.

2.7.0 Cable Placement in Tray

Cables are placed in the tray either by being pulled along the tray or by being laid in over the side. *Figure 9* shows a **cable pulley** being used to facilitate a cable pull in a tray supported by trapeze hangers.

2.8.0 Cable Tray Cover

A cable **tray cover** is used primarily for two reasons:

- To protect the insulation of the cables against damage that might be caused if an object were to fall into the tray. Prime hazards are tools, discarded cigarettes, and weld splatter.
- To protect certain types of cable insulation against the damaging effects of direct sunlight.

2.8.1 Cover Selection

When maximum protection is desired, solid covers should be used. However, if accumulation of heat from the cables is expected, caution should be used. Ventilated covers should be used if some protection of the cable is desired and provisions must be made to allow the escape of heat developed by the cables.

2.9.0 Cable Exit from Tray

Several different ways in which cables may exit from a cable tray are shown in *Figure 10*. While all of these methods are *NEC*®-approved and endorsed by most cable tray manufacturers, the engineering specifications on some projects may

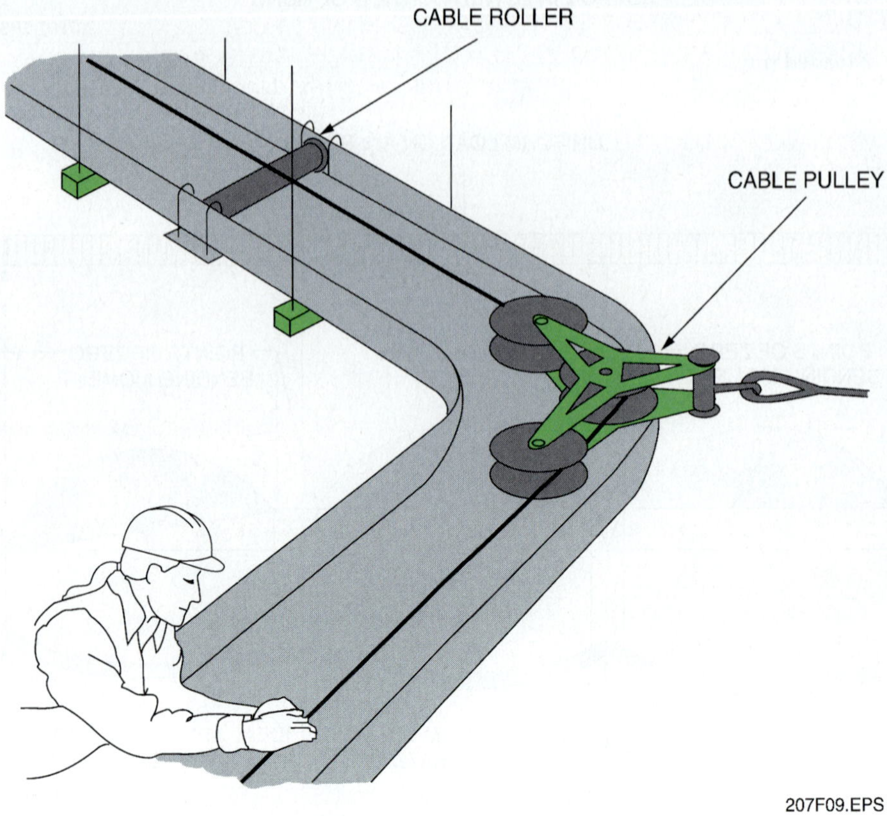

Figure 9 ◆ Cable pulley used to facilitate a cable pull in a tray.

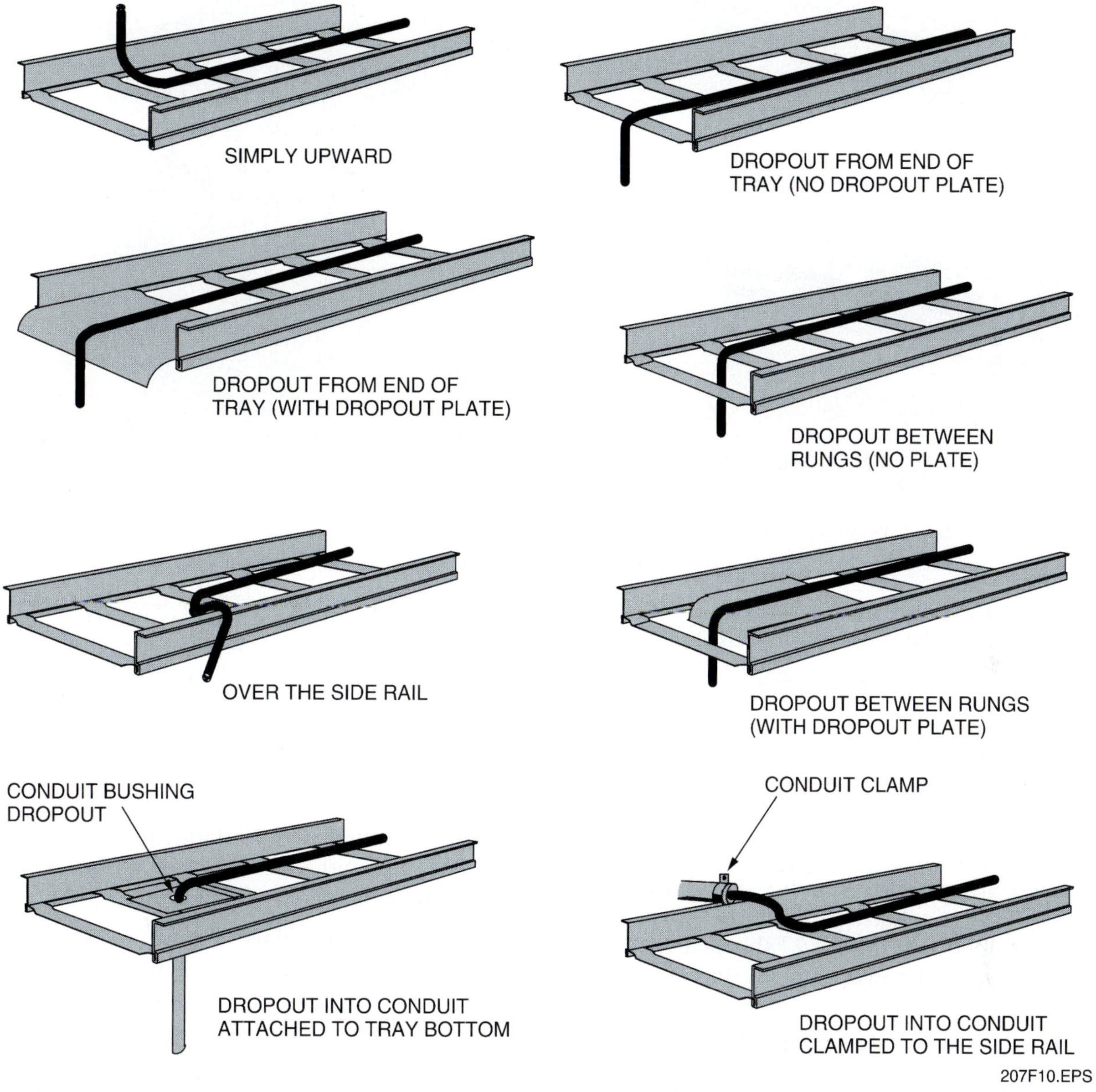

Figure 10 ◆ Several ways in which cables may exit from a cable tray.

prohibit the use of some of these methods. Most notable are the **dropout** between runs method and the dropout from the end of the tray method. The cable radius may be too short with either of these methods. Also, since no **dropout plates** are used, the cable or conductors are not protected. Although *NEC Section 392.5(B)* specifically requires that cable trays have smooth edges to ensure that cable will not be damaged, accidents do occur. For example, a tool might be dropped on a cable tray rung during the installation, which may cause a burr or other sharp edge on the rung. Then, after the cable is installed and the system is in use, vibrating machinery may cause this burr to cut into the cable insulation, resulting in a ground fault and possible power outage. Always review the project specifications carefully and/or check with the project supervisor before using either of these methods.

2.9.1 Dropout Plates

A dropout plate provides a curved surface for the cable to follow as it passes from the tray, as shown in *Figure 11(A)*. Without a dropout plate, cables can be bent sharply, causing damage to the insulation. See *Figure 11(B)*.

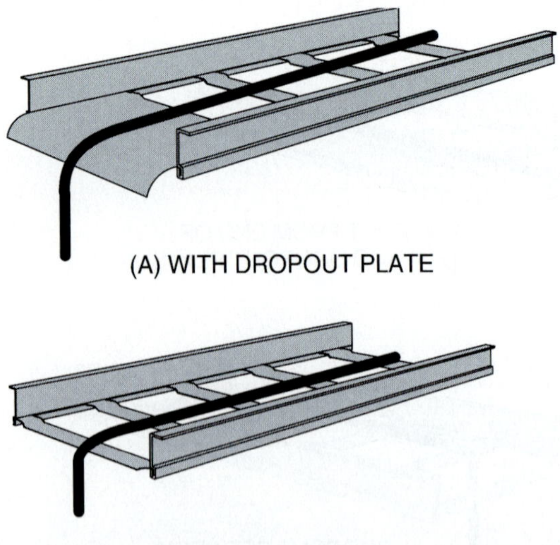

Figure 11 ◆ A dropout plate provides a curved surface for the cable to follow as it leaves the tray.

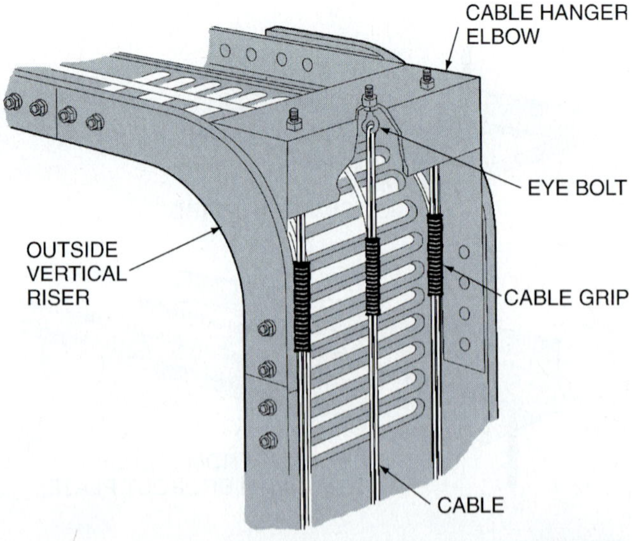

Figure 12 ◆ Typical application of supports in a vertical run.

2.10.0 Cable Supports in Vertical Trays

A cable hanger **elbow** is used to suspend cables in long vertical runs. Care should be taken to ensure that the weight of the suspended length of cable does not exceed the cable manufacturer's recommendation for the maximum allowable tension in the cable.

In short vertical runs, the weight of cables can be supported either by the outside vertical riser elbow or by the vertical straight section when the cables are clamped to it. *Figure 12* shows a typical application of supports in a vertical run.

2.11.0 Cable Edge Protection

The bottom of the solid bottom tray might be convex, concave, or flat. When two pieces of tray are butted together, the bottoms may be out of alignment. An alignment strip, also known as an H bar, is placed between the tray bottoms.

2.12.0 Splice Plates

There are several types of splice plates available, including vertical, horizontal, and expansion plates.

2.12.1 Vertical Adjustment Splice Plates

Vertical adjustment splice plates are used to change the elevation in a run of cable tray. They should not be used when it is important to maintain a cable bending radius. Vertical adjustment splice plates are useful when the change in elevation is so slight or the angle so unusual that it would not be possible to install a standard outside vertical riser elbow and an inside vertical riser elbow in the space available.

In general, four **swivel plates** are used to build an offset in a cable tray system. Once the proper angles have been calculated, proceed as follows:

Step 1 Bolt four swivel plates together at the proper angles, using the inner holes as the center or pivot hole (see *Figure 13*).

Step 2 Using a flat surface such as a bench or concrete deck, space two swivel plates at the proper center-to-center distance apart (refer to *Figure 13*).

Step 3 Measure and cut the amount of tray needed to complete the offset.

2.12.2 Horizontal Adjustment Splice Plates

These plates are sometimes used in place of horizontal elbows to change the direction in a run of cable tray. They are used primarily where there is insufficient space or an unusual angle that prevents the use of a standard elbow.

2.12.3 Expansion Splice Plates

These plates are used at intervals along a straight run of cable tray to allow space for thermal expansion or contraction of the tray to occur, or where offsets or **expansion joints** occur in the supporting structure.

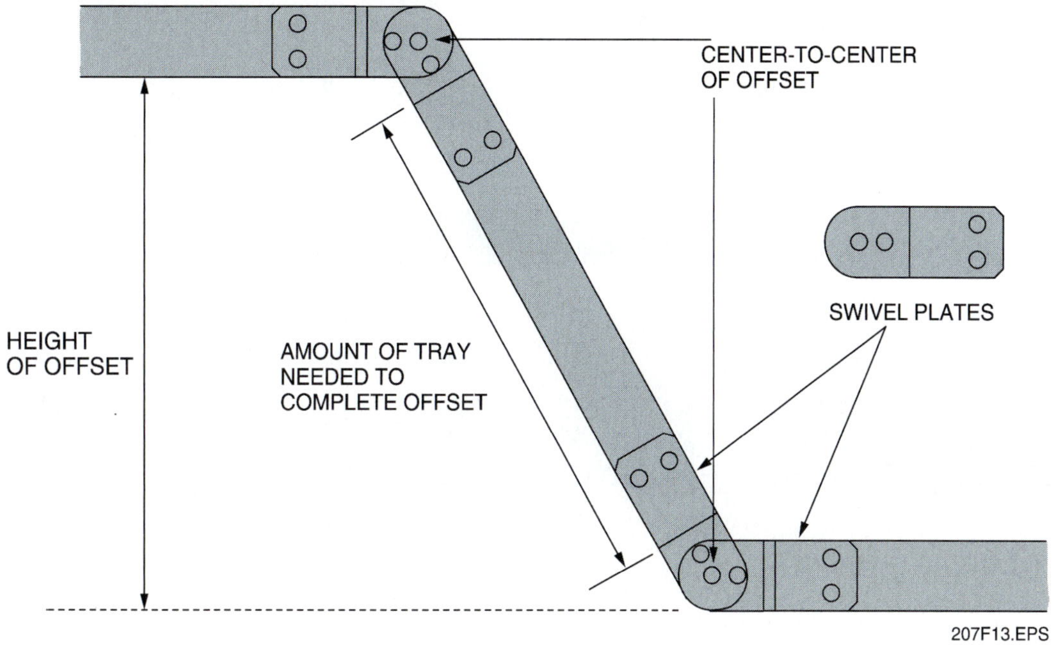

Figure 13 ◆ Fabricating a cable tray offset with swivel plates.

To enable the expansion joint to function properly, the cable tray must be allowed to slide freely on its supports. Any cable tray hold-down device used in an installation subject to expansion or contraction must give clearance to the tray. An expansion joint and splice plates are shown in *Figure 14*.

2.13.0 Barrier Strips

Barrier (divider) strips are used to separate certain types of cable as a result of the nature of the installation, types of circuits used, type of equipment used, local codes, or the *NEC*®. Some reasons for using divider strips are to:

- Separate or isolate electrical circuits
- Separate or isolate cables of different voltages
- Separate cable or wire runs from each other to prevent fire or ground fault damage from spreading to other cables or wires in the same tray
- Aid neatness in the arrangement of the cables
- Warn electricians of the difference between cables on each side of the divider strip

2.13.1 Barrier Strip Cable Protectors

Barrier strip cable protectors are used to bind any raw metallic edge over which a cable is to pass. Their purpose is to protect the cable insulation against damage.

INSIDE TRACK

Contraction and Expansion of Cable Tray

Structural expansion joints do not normally provide a solution to the thermal contraction and expansion of cable trays, because the materials of the structures and trays are different. For example, for a 100°F variance in temperature, a steel tray will require an expansion joint every 128' and an aluminum tray will require a joint every 65'. You must determine the gap setting for the joint according to the temperature at the time of installation. Tables for setting up the gap distances are available from cable tray manufacturers.

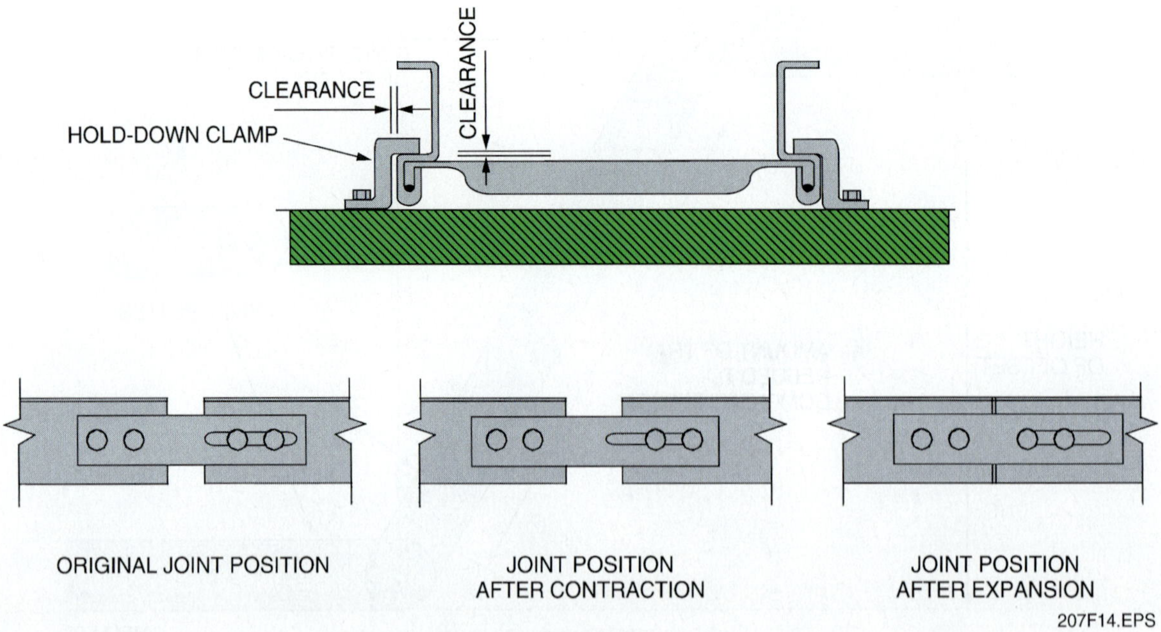

Figure 14 ◆ Expansion joint and splice plates.

3.0.0 ◆ CABLE TRAY SUPPORT

Proper supports for cable tray installations are very important in obtaining a good overall layout. Cable is usually supported in one or more of the following ways:

- **Trapeze mounting**
- **Direct rod suspension**
- **Wall mounting**
- Center hung support
- Pipe rack mounting

3.1.0 Trapeze Mounting

When trapeze mounting is used, a structural member—usually a steel channel or **Unistrut®**—is connected to the vertical supports to provide an appearance similar to a swing or trapeze. The cable tray is mounted to the structural member using anchor clips or J-clamps. Often, the underside of the channel or Unistrut® is used to support conduit (see *Figure 15*).

3.2.0 Direct Rod Suspension

The direct rod suspension method of supporting cable tray uses threaded rods and hanger clamps. The threaded rod is connected to the ceiling or other overhead structure and is connected to the hanger clamps that are attached to the cable tray side rails, as shown in *Figure 16(A)*.

3.3.0 Wall Mounting

Wall mounting is accomplished by supporting the tray with structural members attached to the wall, as shown in *Figure 16(B)*. This method of support is often used in tunnels (mining operations) and other underground or sheltered installations where large numbers of conductors interconnect equipment that is separated by long distances. When using this or any other method of supporting cable tray, always examine the structure to which the hangers are attached, and make absolutely certain that the structure is of adequate strength to support the tray system.

3.4.0 Trapeze Mounting and Center Hung Support

Trapeze mounting of cable tray is similar to direct rod suspension mounting. The difference is in the method of attaching the cable tray to the threaded rods. A structural member, usually a steel channel or strut, is connected to the vertical supports to provide an appearance similar to a swing or tra-

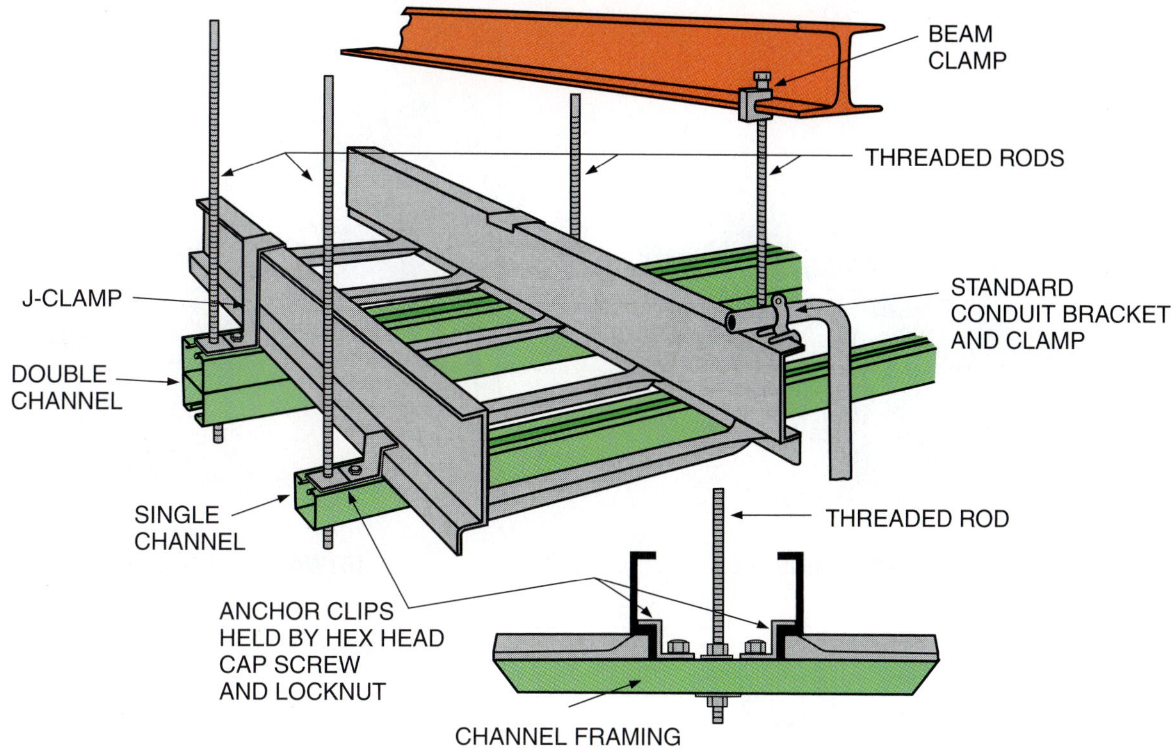

Figure 15 ◆ Channel support details.

INSIDE TRACK

Center Hung Trays

With center hung trays, care must be taken to balance the weight of the cable on either side of the tray to prevent tipping.

peze. The cable tray is mounted to the structural member. Often, the underside of the channel or strut is used to support conduit. A trapeze mounting assembly is shown in *Figure 16(C)*.

A method that is similar to trapeze mounting is a center hung tray support, as shown in *Figure 16(D)*. In this case, only one rod is used and it is centered between the cable tray side rails. A bushing sleeve, such as a short piece of small-diameter PVC, may be used over the center rod to protect the conductors.

3.5.0 Pipe Rack Mounting

Pipe racks are structural frames used to support the piping that interconnects equipment in outdoor industrial facilities. Usually, some space on the rack is reserved for conduit and cable tray. Pipe rack mounting of cable tray is often seen in petrochemical plants where power distribution and instrumentation wiring is routed over a large area and for long distances.

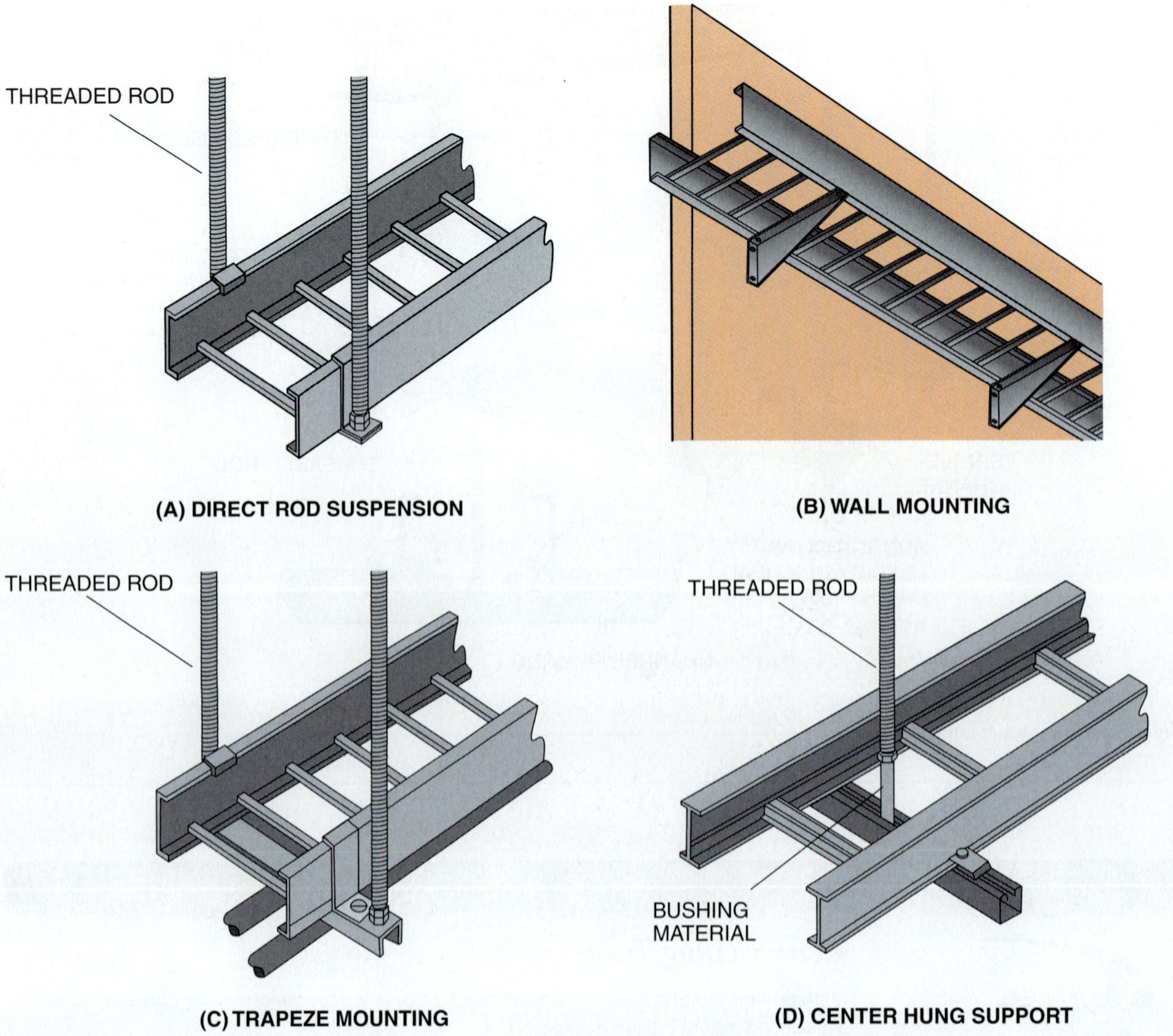

Figure 16 ◆ Alternate ways to hang cable tray.

4.0.0 ◆ CENTER RAIL CABLE TRAY SYSTEMS

Center rail or monorail cable tray systems (*Figure 17*) are light, easy to install, and provide open sides for ready access when changing or adding cables. Center rail cable tray is used in light-duty applications such as sound, telephone, and other communications systems. In addition to its weight limitations, it cannot be used with dividers so its use is restricted to systems where dividers are not necessary.

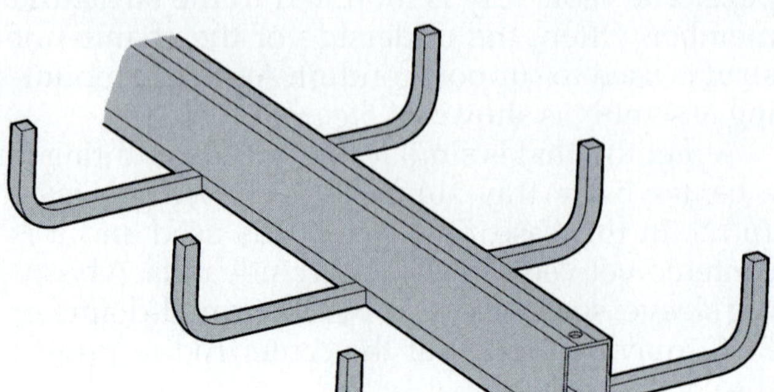

Figure 17 ◆ Center rail cable tray.

5.0.0 ◆ NEC® REQUIREMENTS

NEC Article 392 deals with cable tray systems along with related wire, cable, and raceway installations.

Figures 18, 19, and *20* summarize the *NEC*® requirements for cable tray installations. For in-depth coverage, always refer to the *NEC*®. Cable tray manufacturers also have some excellent reference material available that details the installation of their products. Consult your local dealer for more information.

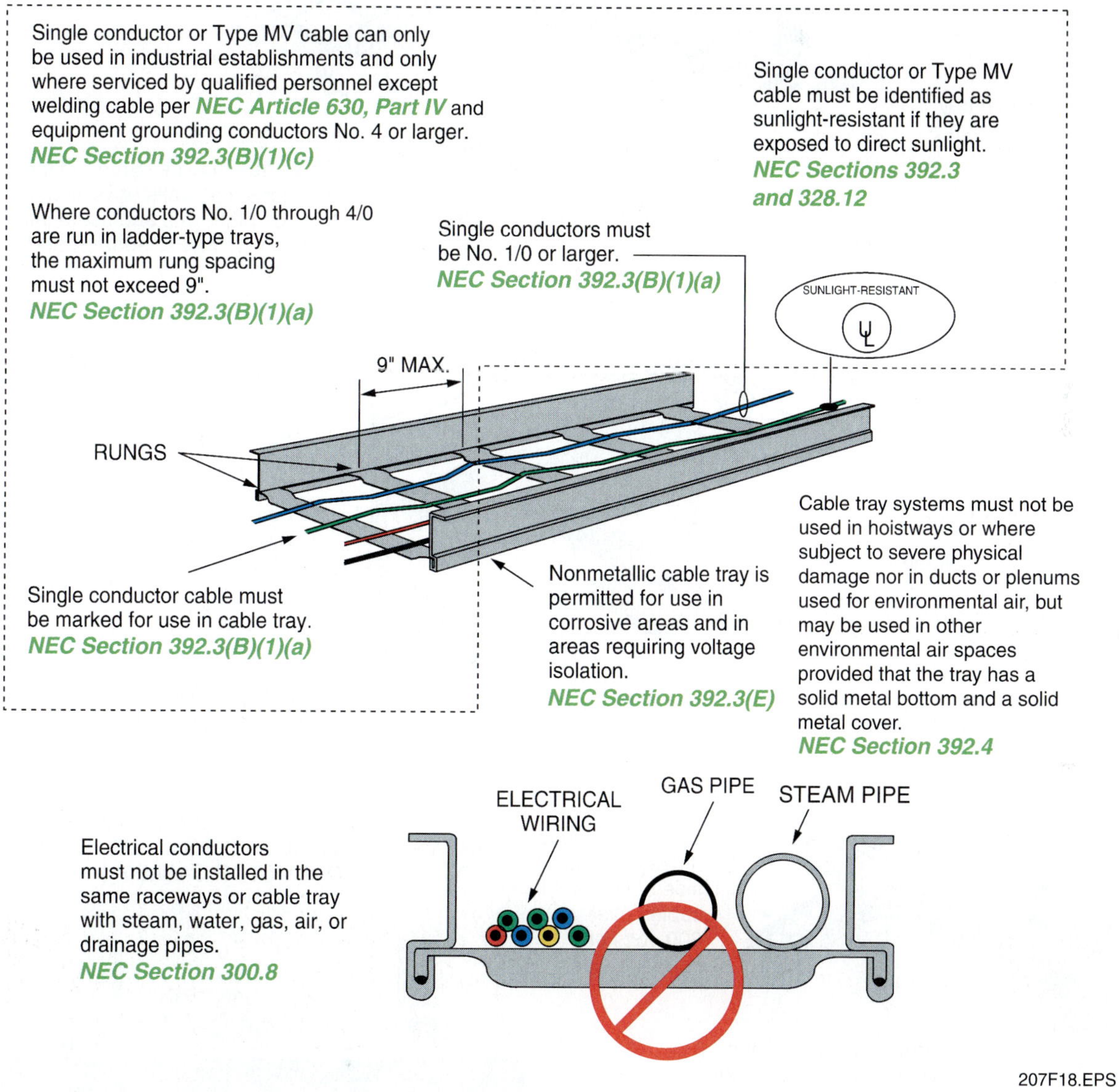

Figure 18 ◆ *NEC*® regulations governing the use of cable tray.

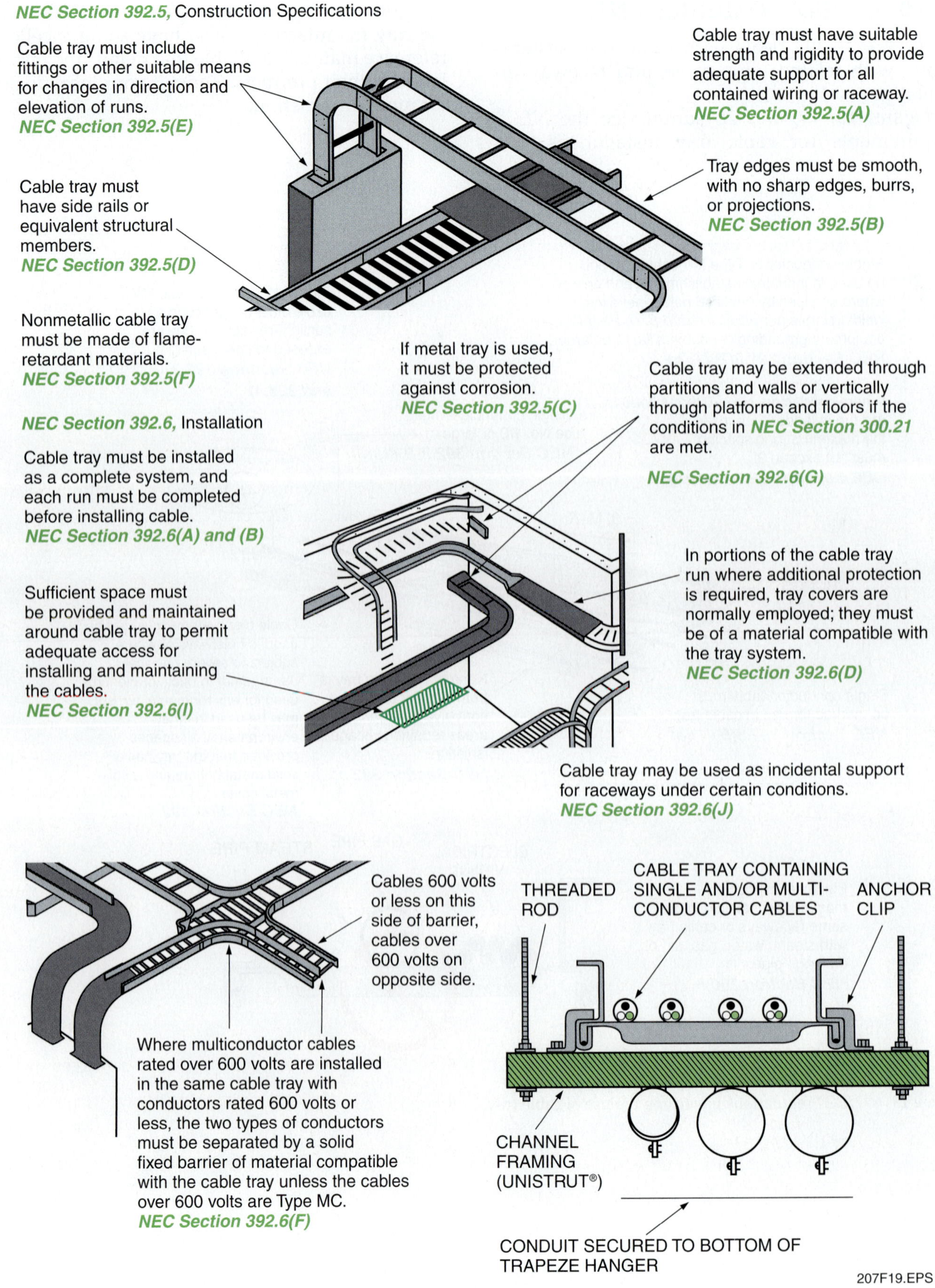

Figure 19 ◆ *NEC*® regulations governing cable tray construction and installation.

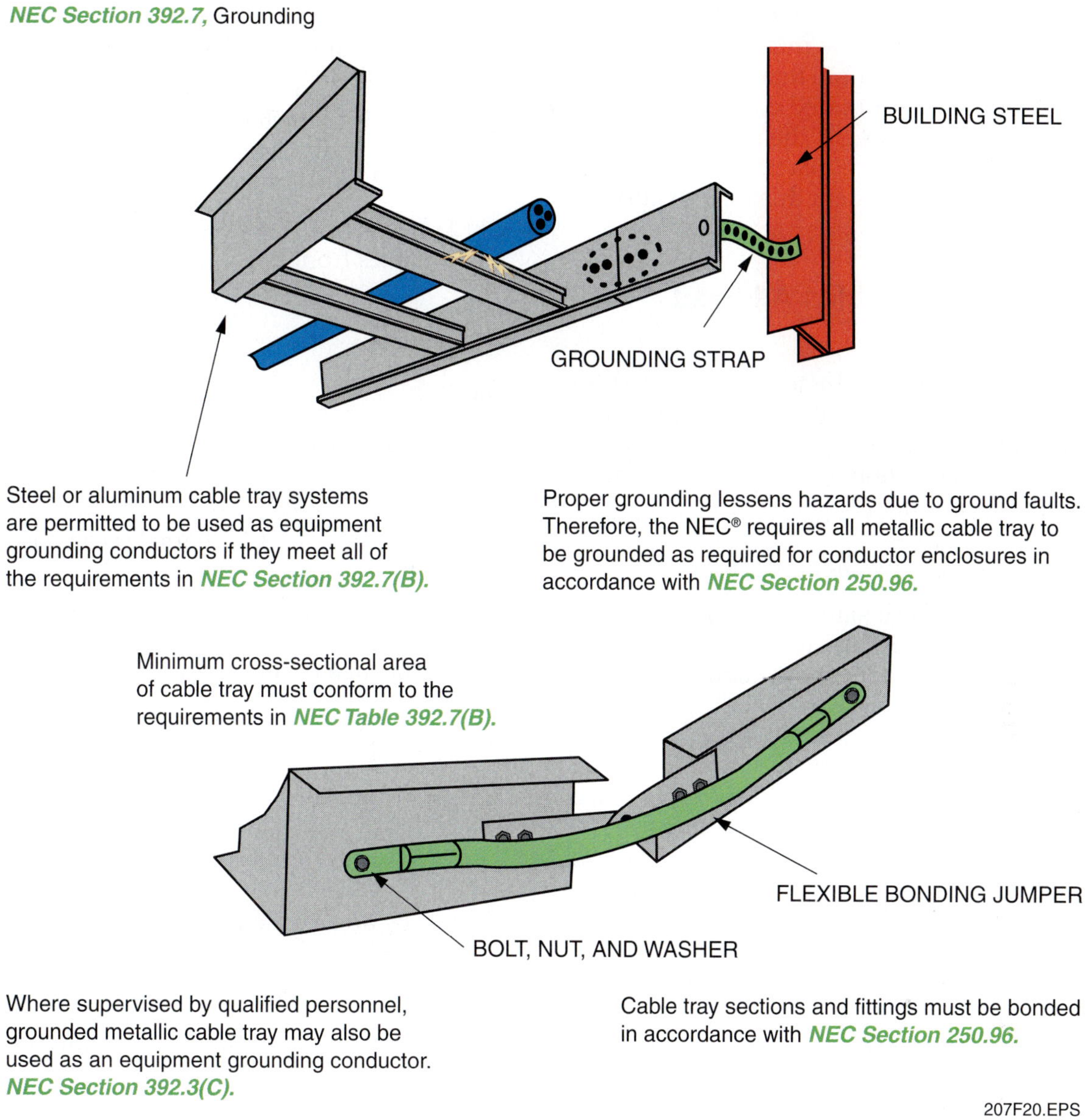

Figure 20 ◆ *NEC*® regulations governing cable tray grounding.

6.0.0 ◆ CABLE INSTALLATION

NEC Section 392.8 covers the general installation requirements for all conductors used in cable tray systems; that is, splicing, securing, and running conductors in parallel. For example, cable splices are permitted in cable trays provided they are made and insulated by *NEC*®-approved methods. Furthermore, any splices must be readily accessible and must not project above the side rails of the tray.

In most horizontal runs, the cables may be laid in the tray without securing them in place. However, on vertical runs or any runs other than horizontal, the cables must be secured to transverse members of the cable tray.

Cables may enter and leave a cable tray system in a number of different ways, as discussed previously. In general, no junction box is required where such cables are installed in bushed conduit or tubing. Where conduit or tubing is used, it must be secured to the tray with the proper fittings. Further precautions must be taken to ensure that the cable is not bent sharply as it enters or leaves the conduit or tubing.

6.1.0 Conductors Connected in Parallel

Where single-conductor cables comprising each phase, neutral, or grounded conductor of a circuit are connected in parallel as permitted in *NEC Section 310.4,* the conductors must be installed in groups consisting of not more than one conductor per phase, neutral, or grounded conductor to prevent a current imbalance in the paralleled conductors due to inductive reactance. This also prevents excessive movement due to fault current magnetic forces.

6.2.0 Number of Cables Allowed in Cable Tray

The number of multiconductor cables rated at 2,000V or less that are permitted in a single cable tray must not exceed the requirements of *NEC Section 392.9.* This section applies to both copper and aluminum conductors.

6.2.1 All Conductors Size 4/0 or Larger

Where all of the cables installed in the tray are 4/0 or larger, the sum of the diameters of all cables shall not exceed the cable tray width, and the cables must be installed in a single layer. For example, if a cable tray installation is to contain three 4/0 multiconductor cables (1.5" in diameter), two 250 kcmil multiconductor cables (1.85"), and two 350 kcmil multiconductor cables (2.5"), the minimum width of the cable tray is determined as follows:

$$3(1.5) + 2(1.85) + 2(2.5) = 13.2"$$

The closest standard cable tray size that meets or exceeds 13.2" is 18". Therefore, this is the size to use.

6.2.2 All Conductors Smaller Than 4/0

Where all of the cables are smaller than 4/0, the sum of the cross-sectional area of all cables smaller than 4/0 must not exceed the maximum allowable cable fill area as specified in Column 1 of *NEC Table 392.9;* this gives the appropriate cable tray width. To use this table, however, you must have the manufacturer's data for the cables being used. This will give the cross-sectional area of the cables.

The steps involved in determining the size of cable tray for multiconductors smaller than 4/0 AWG are as follows:

Step 1 Calculate the total cross-sectional area of all cables used in the tray. Obtain the area of each from the manufacturer's data.

Step 2 Look in Column 1 of *NEC Table 392.9* and find the smallest number that is at least as large as the calculated number.

Step 3 Look at the number to the left of the row selected in Step 2. This is the minimum width of cable tray that may be used.

For example, determine the minimum cable tray width required for the following multiple conductor cables—all less than 4/0 AWG:

- Four at 1.5" diameter
- Five at 1.75" diameter
- Three at 2.15" diameter

Step 1 Determine the cross-sectional area of the cables from the equation:

$$A = \frac{\pi \times D^2}{4}$$

Where:

A = area

D = diameter

The area of a 1.5" diameter cable is:

$$\frac{(3.14159)(1.5^2)}{4} = 1.7671 \text{ square inches}$$

The area of the four 1.5" cables is:

$$4 \times 1.7671 \text{ square inches} = 7.0684 \text{ square inches}$$

The area of 1.75" diameter cable is:

$$\frac{(3.14159)(1.75^2)}{4} = 2.4053 \text{ square inches}$$

The area of the five 1.75" cables is:

$$5 \times 2.4053 \text{ square inches} = 12.0265 \text{ square inches}$$

The area of the three 2.15" cables is:

$$\frac{(3.14159)(2.15^2)}{4} = 3.6305 \text{ square inches}$$

Therefore, the total area of the three cables is:

$$3 \times 3.6305 \text{ square inches} = 10.8915 \text{ square inches}$$

The total cross-sectional area is found by adding the above three totals to obtain:

$$7.0684 + 12.0265 + 10.8915 = 29.9864 \text{ square inches}$$

Step 2 Look in Column 1 of *NEC Table 392.9* and find the smallest number that is at least as large as 29.9864 square inches. The number is 35.

Step 3 Look to the left of 35 and you will see the inside tray width of 30". Therefore, the minimum tray width that can be used for the given group of conductors is 30".

INSIDE TRACK

Low-Voltage Cable

Increasingly, cable trays are being used to carry many low-voltage conductors in communication centers. In a large commercial building, thousands of telecommunications cables are distributed in bundles from the equipment room to the telecommunications closets on each floor. Cable trays offer a convenient means of running these cables through the building as well as a much easier method of allowing for system expansion. Instead of having to access conduit buried within the building walls, a new communication cable can simply be added to the cable tray system, which is normally accessible through a dropped ceiling.

207P0703.EPS

207P0704.EPS

6.2.3 Combination Cables

Where 4/0 or larger cables are installed in the same cable tray with cables smaller than 4/0, the sum of the cross-sectional area of all cables smaller than 4/0 must not exceed the maximum allowable fill area from Column 2 of *NEC Table 392.9* for the appropriate cable tray width. The 4/0 and larger cables must be installed in a single layer, and no other cables can be placed on them.

To determine the tray size for a combination of cables as discussed in the above paragraph, proceed as follows:

Step 1 Repeat the steps from the procedure used previously to determine the minimum tray width required for the multiconductor cables having conductors sized 4/0 and larger.

Step 2 Repeat the steps from the procedure used previously to determine the cross-sectional area of all multiconductor cables having conductors smaller than 4/0 AWG.

Step 3 Multiply the result of Step 1 by the constant 1.2 and add this product to the result of Step 2. Call this sum A. Search Column 2 of *NEC Table 392.9* for the smallest number that is at least as large as A. Look to the left in that row to determine the minimum size cable tray required.

To illustrate these steps, we will assume that we need to find the minimum cable tray width of two multiconductor cables, each with a diameter of 2.54" (conductors size 4/0 or larger); three cables with a diameter of 3.30" (conductors size 4/0 or larger), plus eight cables with a diameter of 1.92" (conductors less than 4/0).

Step 1 The sum of all the diameters of cable having conductors 4/0 or larger is:

$$2(2.54) + 3(3.30) = 14.98"$$

Step 2 The sum of the cross-sectional areas of all cables having conductors smaller than 4/0 is:

$$\frac{(8)(3.14159)(1.92^2)}{4} = 23.1623 \text{ square inches}$$

Step 3 Multiply the result of Step 1 by 1.2 and add this product to the result of Step 2:

$$(1.2 \times 14.98) + 23.1623$$
$$= 41.1383 \text{ square inches}$$

The smallest number in Column 2 of *NEC Table 392.9* that is not larger than 41.1383 is 42. The tray width that corresponds to 42 is 36". Therefore, select a cable tray width of 36".

6.2.4 Solid Bottom Tray

Where solid bottom cable trays contain multiconductor power or lighting cables, or any mixture of multiconductor power, lighting, control, and signal cables, the maximum number of cables must conform to the following:

- Where all of the cables are 4/0 or larger, the sum of the diameters of all cables must not exceed 90% of the cable tray width, and the cables must be installed in a single layer.
- Where all of the cables are smaller than 4/0, the sum of the cross-sectional areas of all cables must not exceed the maximum allowable cable fill area in Column 3 of *NEC Table 392.9* for the appropriate cable tray width.
- Where 4/0 or larger cables are installed in the same cable tray with cables smaller than 4/0, the sum of the cross-sectional areas of all of the smaller cables must not exceed the maximum allowable fill area resulting from the computation in Column 4, *NEC Table 392.9* for the appropriate cable tray width. The 4/0 and larger cables must be installed in a single layer, and no other cables can be placed on them.

Where a solid bottom cable tray with a usable inside depth of 6" or less contains multiconductor control and/or signal cables only, the sum of the cross-sectional areas of all cables at any cross section must not exceed 40% of the interior cross-sectional area of the cable tray. A depth of 6" must be used to compute the allowable interior cross-sectional area of any cable tray that has a usable inside depth of more than 6".

In a previous example, we determined that the minimum tray size for multiconductor cables with all conductors size 4/0 or larger was 18". The sum of all cable diameters for this example was 13.2". To see if an 18" solid bottom tray can be used, multiply the tray width by 0.90 (90%):

$$18 \times 0.9 = 16.2"$$

Therefore, 16.2" is the minimum width allowed for solid bottom tray. Since we are using 18" tray, this meets the requirements of *NEC Section 392.9(C)(1).*

When dealing with solid bottom trays and using *NEC Table 392.9,* use Columns 3 and 4 instead of Columns 1 and 2, as used for ladder and trough-type cable trays.

6.2.5 Single-Conductor Cables

Calculating cable tray widths for single-conductor cables (2,000V or under) is similar to the calculations used for multiconductor cables, with the following exceptions:

- Conductors that are 1,000 kcmil and larger are treated the same as multiconductor cables having conductors size 4/0 or larger.
- Conductors that are smaller than 1,000 kcmil are treated the same as multiconductor cables having conductors smaller than size 4/0.

NEC Section 392.10 covers the details of installing single-conductor cables in cable tray systems.

6.3.0 Ampacity of Cable Tray Conductors

NEC Section 392.11 gives the requirements for cables used in tray systems with rated voltages of 2,000 volts or less. *NEC Section 392.13* covers cables with voltages of 2,001 and over.

THINK ABOUT IT

Ampacity in Cable Trays

Examine ***NEC Sections 392.11 and 392.13.*** What are the factors that affect the permitted ampacity?

7.0.0 ◆ CABLE TRAY DRAWINGS

For an economical and satisfactory installation, working out the details of supports and hangers for a cable tray system is usually done beforehand by the engineering department or project engineer and is seldom left to the judgment of a field force that is not acquainted with the loads and forces to be encountered. As a result, drawings and specifications will usually be furnished to the work crew to provide details about the cable tray system. All workers involved with the installation should know how to interpret these drawings.

The exact method of showing cable tray systems on working drawings will vary, so always consult the symbol list or legend before beginning the installation. Also, study any shop or detail drawings that might accompany the construction documents.

If space permits, many engineers prefer to draw the cable tray system as close to scale as possible, using various symbols to show the different types of cable tray to be installed.

Look at the floor plan drawing in *Figure 21*. The cable tray system in this project originates at several power panels and motor control centers to feed and control motors in other parts of the building. The trays run from the motor control centers, are offset to miss beams and other runs of cable tray, and then branch off to various parts of the building.

Although experienced workers in the industrial electrical trade will have little trouble reading the information in this drawing, new electricians may have some difficulty in visualizing the system. However, if a supplemental drawing were provided with the floor plan drawing in *Figure 21*, it would provide a clearer picture of the system

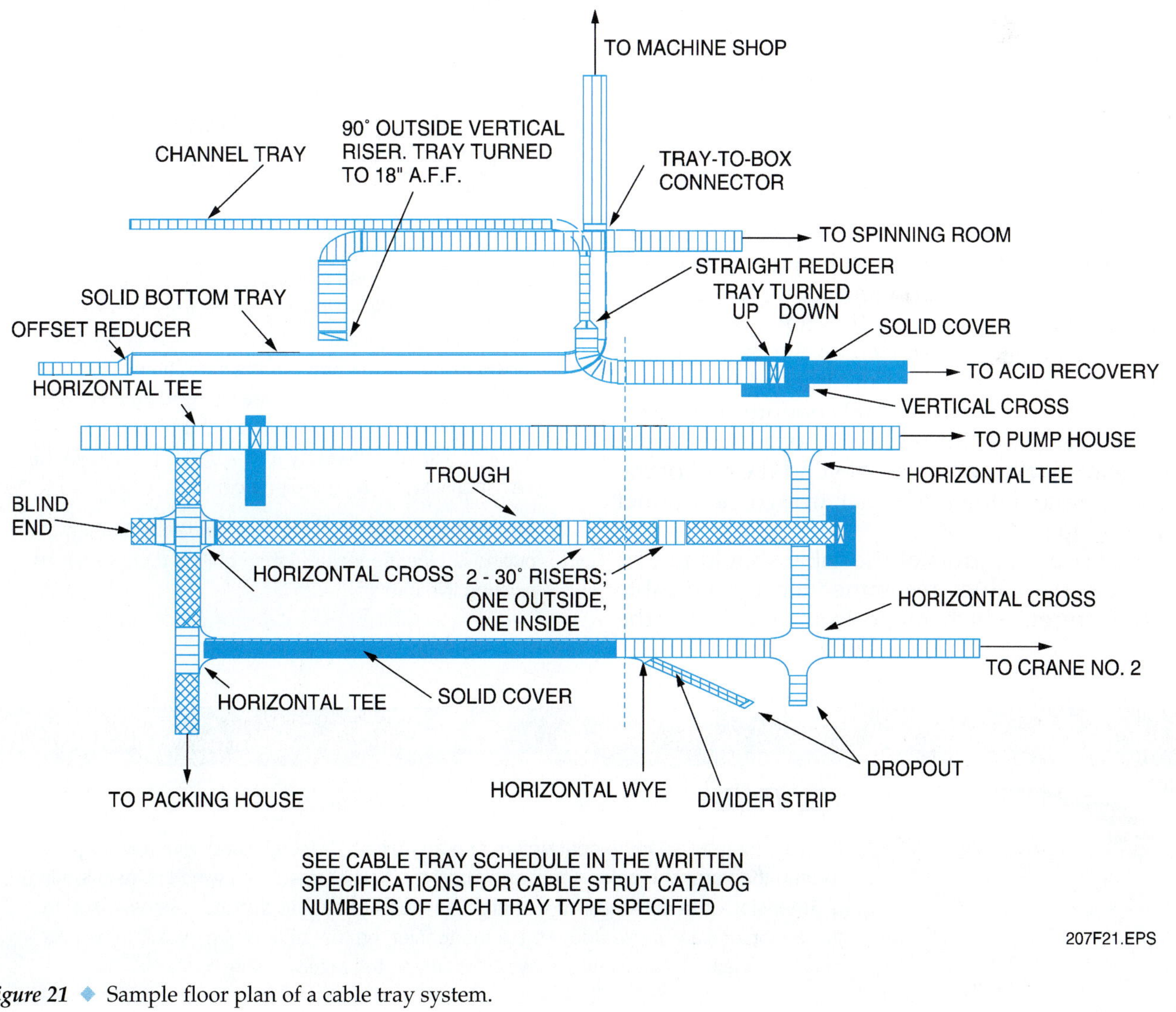

Figure 21 ◆ Sample floor plan of a cable tray system.

and leave little doubt as to how the cable tray system is to be installed. Even new workers in the trade would be able to see how the system should be installed, and this would take some of the load off experienced workers and give them more time to accomplish other tasks.

In actual practice, however, consulting engineering firms seldom furnish the isometric drawing; they merely show the layout in plan view. Consequently, plan views involving the construction details of cable tray systems must be studied carefully during the planning stage—before the work is begun.

8.0.0 ◆ PULLING CABLE IN TRAY SYSTEMS

When installing cables in tray systems, proper precautions must be taken to avoid damaging the cables. A complete line of installation tools is available for pulling long lengths of cable up to 1,000' or longer. These tools save considerable installation time.

Short lengths of cable can be laid in place without power pulling tools, or the cable can be pulled manually using a basket grip and pulling rope. Long lengths of small cable, 2" or less in diameter, can also be pulled with a basket grip and pulling rope. Larger cables, however, should be pulled by the conductor and the braid, sheath, or armor. This is usually done with a pulling eye applied to the cable at the factory, or by tying the conductor to the eye of a basket grip and taping the tail end of the grip to the outside of the cable.

In general, the pull exerted on the cables pulled with a basket grip that is not attached to the conductor should not exceed 1,000 pounds. For heavier pulls, care should be taken not to stretch the insulation, jacket, or armor beyond the end of the conductor nor bend the ladder, trough, or channel out of shape.

The bending radius of the cable should not be less than the values recommended by the cable manufacturer, which range from four times the diameter for a rubber-insulated cable that has a 1" maximum outside diameter without lead, shield, or armor, to eight times the diameter for **interlocked armor cable.** Cables of special construction such as wire armor and high-voltage cables require a larger radius bend.

When installing long lengths of cable up to 1,000' with as many as a dozen bends, best results are obtained by pulling the cable in one continuous operation at a speed of 20' to 25' per minute. The pulling line diameter and length will depend on the pull to be made and construction equipment available. The winch and power unit must be of an adequate size for the job and capable of developing the high pulling speed required for the best and most economical results.

A complete description of a **cable roller, sheave,** and other pulling equipment, along with their setup, was covered in *Conductor Installations.* You will also find information on parallel tuggers

CASE HISTORY: Pulling Cable in a Tray

At a clothing manufacturing plant in New York, an electrician attempted to pull cable in a tray already partly filled with energized cables. He used plenty of cable-pulling lubricant, but did not use sheaves and rollers to isolate the new cables from the existing ones. During the pull, the new cables wedged against the live ones, yanking one hot wire apart at a splice. The wire grounded itself on the tray and arced for 30' down the tray, fatally shocking the electrician.

The Bottom Line: Use sheaves and rollers when pulling cables, especially when pulling them between existing live cables.

INSIDE TRACK: *Big Pulls*

For very long pulls, some contractors use an auxiliary winch. Power winches may exceed their preset tension limits and abort the pull, so a second winch is installed in a high-tension area of the pull in a straight section of tray. The second winch is used to pull a loop of slack cable and reduce the tension on the main winch, which now pulls only the length of slack cable between itself and the auxiliary winch.

and dynamometers (tension meters). Please refer to this module for additional information concerning cable pulls.

9.0.0 ◆ SAFETY

Installing cable tray means working at heights above the floor. Consequently, workers must take the necessary precautions. In general, workers installing cable tray will use ladders, scaffolds, lifts, work from the tray assembly itself, or a combination of all these.

Keep the tray assembly uncluttered during installation. Tools, tray fittings, and the like are ideal obstacles for tripping workers. They may also fall off the tray and injure workers below. To help prevent the latter, set up work barriers beneath the section of tray assembly being worked on. If you are working on the ground, never penetrate or move these barriers until the work above is complete.

Learn to secure your life line properly, and also make sure your full body harness is a proper fit. A harness that is too large may slip, resulting in a serious injury or even death.

THINK ABOUT IT

Putting It All Together

If possible, examine the conduit system at your workplace or school. Did the installers make use of cable tray? If not, is the system large enough to warrant the use of it?

Review Questions

1. A barrier strip is a _____.
   a. division strip installed in a raceway to separate certain types of cables
   b. strip used to provide the tray with added support
   c. glass strip used as a cable tray insulator
   d. metal strip used solely for grounding purposes

2. Which of the following best describes how closely packed cables will be affected when energized?
   a. Their efficiency will be increased.
   b. Their efficiency will be decreased.
   c. No change will be encountered.
   d. The operating temperature of each cable will be lower.

3. When cable exits downward from a tray without a dropout plate, _____ could occur.
   a. current surges
   b. reduced amperage
   c. voltage surges
   d. insulation damage

4. The main purpose of vertical adjustment splice plates is to _____.
   a. increase the strength of a cable tray section
   b. support the cable tray system
   c. change the elevation in a run of cable tray
   d. secure the cable within the tray

5. Which of the following is a violation of *NEC Section 300.8?*
   a. Placing an instrumentation air line in the same tray with electrical conductors
   b. Spacing tray rungs 9" or less apart
   c. Running single conductors in trays that are larger than size 1/0
   d. Using nonmetallic tray systems in corrosive areas

6. When a cable tray section branches off from a main section in two 90° turns, the section is called a(n) _____.
   a. wye
   b. tee
   c. divider strip
   d. ell

7. A section of cable tray that makes a horizontal 90° turn is known as a _____.
   a. vertical tee
   b. vertical ell
   c. horizontal elbow
   d. straight reducer

8. When a cable tray system must be protected from dropping objects that may damage the cable, a _____ is used.
   a. solid bottom tray
   b. trough-type tray
   c. tray cover
   d. blind end

9. Which of the following best describes the type of fitting that will be used with a solid cover cable tray?
   a. An Ericson coupling
   b. Vertical splices
   c. A horizontal cross
   d. A cover clamp

10. A(n) _____ branches in four different directions.
    a. horizontal wye
    b. horizontal cross
    c. inside vertical riser
    d. dropout

Summary

A cable tray system includes the assembly of units or sections and associated fittings that form a rigid structural system used to support cables and raceways. Cable tray systems are commonly used in industrial applications and are normally constructed of either aluminum or steel, although PVC-coated and all-nonmetallic trays are also available.

Cable tray is available in various forms, including ladder, trough, center rail, and solid bottom, and can be supported by either side mounts or center mounts. *NEC Article 392* covers cable tray installations, along with the types of conductors to be used in various cable tray systems.

Notes

Trade Terms Introduced in This Module

Barrier strip: A metal strip constructed to divide a section of cable tray so that certain kinds of cable may be separated from each other.

Cable pulley: A device used to facilitate pulling conductor in cable tray where the tray changes direction. Several types are available (single, triple, etc.) to accommodate almost all pulling situations.

Cable roller: A device used to facilitate cable pulls (lower the pulling resistance) in cable trays. Several types are available, including straight rollers, radius rollers, etc.

Cross: A four-way section of cable tray used when the tray assembly must branch off in four different directions.

Direct rod suspension: A method used to support cable tray by means of threaded rods and hanger clamps. One end of the threaded rod is secured to an overhead structure, while the other end is connected to hanger clamps that are attached to the cable tray side rails.

Dropout: Cable leaving the tray assembly and travelling directly downward; that is, the cable is not routed into a conduit or channel.

Dropout plate: A metal plate used at the end of a cable tray section to ensure a greater cable bending radius as the cable leaves the tray assembly.

Elbow: A section of cable tray used to change the direction of the tray assembly a full quarter turn (90°). Both vertical and horizontal elbows are common.

Expansion joints: Plates used at intervals along a straight run of cable tray to allow space for thermal expansion or contraction of the tray.

Fittings: Devices used to assemble and/or change the direction of cable tray systems.

Interlocked armor cable: Mechanically protected cable; usually a helical winding of metal tape formed so that each convolution locks mechanically upon the previous one (armor interlock).

Ladder tray: A type of cable tray that consists of two parallel channels connected by rungs, similar in appearance to the common straight ladder.

Pipe racks: Structural frames used to support the piping that interconnects equipment in outdoor industrial facilities.

Sheave: A pulley used to facilitate cable pulls in cable tray.

Swivel plates: Devices used to make vertical offsets in cable tray.

Tee: A section of cable tray that branches off the main section in two other directions.

Trapeze mounting: A method of supporting cable tray using metal channel, such as Unistrut®, Kindorf, etc., supported by two threaded rods, and giving the appearance of a swing or trapeze.

Tray cover: A flat piece of metal, fiberglass, or plastic designed to provide a solid covering that is needed in some locations where conductors in the tray system may be damaged.

Trough: A type of cable tray consisting of two parallel channels (side rails) having a corrugated, ventilated bottom or a corrugated, solid bottom.

Unistrut®: A brand of metal channel used as the bottom bracket for hanging cable trays. Double Unistrut® adds strength and stability to the trays and also provides a means of securing future runs of conduit.

Wall mounting: A method of supporting cable tray systems using supports secured directly to the wall.

Wye: A section of cable tray that branches off the main section in one direction.

Additional Resources

This module is intended to present thorough resources for task training. The following reference works are suggested for further study. These are optional materials for continued education rather than for task training.

American Electrician's Handbook, Latest Edition. New York: Croft and Summers, McGraw-Hill.

National Electrical Code® Handbook, Latest Edition. Quincy, MA: National Fire Protection Association.

CONTREN® LEARNING SERIES — USER FEEDBACK

The NCCER makes every effort to keep these textbooks up-to-date and free of technical errors. We appreciate your help in this process. If you have an idea for improving this textbook, or if you find an error, a typographical mistake, or an inaccuracy in NCCER's *Contren®* textbooks, please write us, using this form or a photocopy. Be sure to include the exact module number, page number, a detailed description, and the correction, if applicable. Your input will be brought to the attention of the Technical Review Committee. Thank you for your assistance.

Instructors – If you found that additional materials were necessary in order to teach this module effectively, please let us know so that we may include them in the Equipment/Materials list in the Annotated Instructor's Guide.

Write: Product Development
National Center for Construction Education and Research
P.O. Box 141104, Gainesville, FL 32614-1104

Fax: 352-334-0932

E-mail: curriculum@nccer.org

Craft __________ Module Name __________

Copyright Date __________ Module Number __________ Page Number(s) __________

Description

(Optional) Correction

(Optional) Your Name and Address

Conductor Terminations and Splices

26208-05

Gaylord Texan Resort
Grapevine, Texas
Exterior Finish Award Winner
Triangle Plastering Systems, Inc.

26208-05

Conductor Terminations and Splices

Topics to be presented in this module include:

Overview

The job is not over once the raceway system and conductors are installed. The final assignment is splicing and terminating the conductors. Certain procedures must be followed when preparing the conductors for splicing or termination. A specific length of conductor insulation must be carefully removed to expose the conductor's end. After this, specific cleaning methods intended to remove any remaining insulation residue or contamination must be employed. A properly prepared conductor end provides the maximum conduction and least resistance between the conductor and termination point or splice.

There are many methods that can be used to prepare the conductor for termination or splicing. The tools available for preparing conductor ends range from simple hand-held knives to specialized cable stripping tools used for large power cables and wires. Regardless of the tool or method used to prepare the wire or cable, the conductor must be left undamaged, with a cleanly exposed conductor end. The type of termination or splice used depends on the application. Common connectors include wire nuts, mechanical connectors, and compression (crimp) connectors.

Note: *National Electrical Code*® and *NEC*® are registered trademarks of the National Fire Protection Association, Inc., Quincy, MA 02269. All *National Electrical Code*® and *NEC*® references in this module refer to the 2005 edition of the *National Electrical Code*®.

Objectives

When you have completed this module, you will be able to do the following:

1. Describe how to make a good conductor termination.
2. Prepare cable ends for terminations and splices.
3. Install lugs and connectors onto conductors.
4. Train cable at termination points.
5. Explain the role of the *NEC*® in making cable terminations and splices.
6. Explain why mechanical stress should be avoided at cable termination points.
7. Describe the importance of using proper bolt torque when bolting lugs onto busbars.
8. Describe crimping techniques.
9. Select the proper lug or connector for the job.
10. Describe splicing techniques.
11. Explain how to use hand and power crimping tools.

Trade Terms

AL-CU
Connection
Connector
Insulating tape
Lug
Pressure connector
Reducing connector
Splice
Strand
Tap
Terminal
Termination
Wire nut

Required Trainee Materials

1. Paper and pencil
2. Appropriate personal protective equipment
3. Copy of the latest edition of the *National Electrical Code*®

Prerequisites

Before you begin this module, it is recommended that you successfully complete *Core Curriculum; Electrical Level One;* and *Electrical Level Two,* Modules 26201-05 through 26207-05. You should also read ***NEC Article 312.***

This course map shows all of the modules in *Electrical Level Two.* The suggested training order begins at the bottom and proceeds up. Skill levels increase as you advance on the course map. The local Training Program Sponsor may adjust the training order.

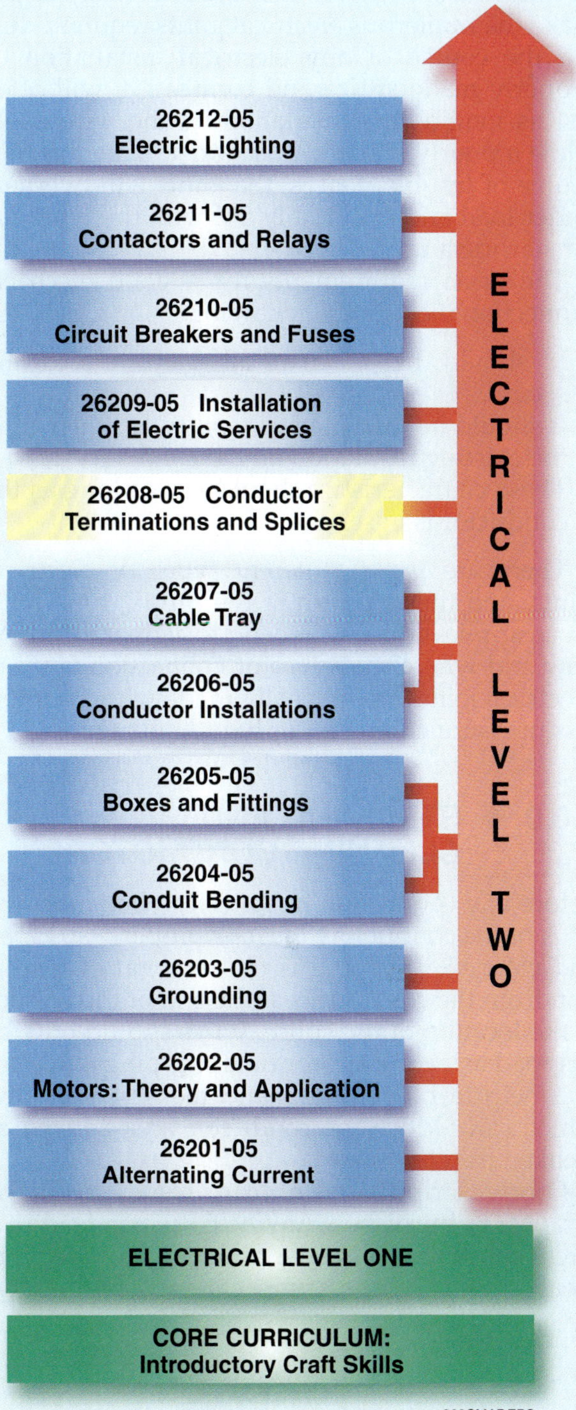

208CMAP.EPS

1.0.0 ◆ INTRODUCTION

Anyone involved with electrical systems of any type should have a good knowledge of wire **connectors** and splicing, as they are both necessary to make the numerous electrical joints required during the course of any electrical installation. A properly made **splice** and **connection** will often last as long as the insulation on the wire itself, while a poorly made connection will always be a source of trouble; that is, the joints will overheat under load and eventually fail with the potential for starting a fire.

The basic requirements for a good electrical connection include the following:

- It should be mechanically and electrically secure.
- It should be insulated as well as or better than the existing insulation on the conductors.
- These characteristics should last as long as the conductor is in service.

There are many different types of electrical joints, and the selection of the proper type for a given application will depend to a great extent on how and where the splice or connection is used. Electrical joints are normally made with a solderless **pressure connector** or **lug** to save time.

2.0.0 ◆ STRIPPING AND CLEANING CONDUCTORS

Before any connection or splice can be made, the ends of the conductors must be properly stripped and cleaned. Stripping is the removal of insulation from the conductors at the end of the wire or at the location of the splice. Wires should only be stripped using the appropriate stripping tool. This will help to prevent cuts and nicks in the wire, which can reduce the conductor area as well as weaken the conductor.

Poorly stripped wire can result in nicks, scrapes, or burnishes. Any of these can lead to a stress concentration at the damaged cross section. Heat, rapid temperature change, mechanical vibration, and oscillatory motion can aggravate the damage, causing faults in the circuitry or even total failure.

Lost **strands** are a problem in splice or crimp-type **terminals,** while exposed strands might be a safety hazard.

Slight burnishes on conductors, as long as they had no sharp edges, were acceptable at one time. Now, however, most experts believe that under certain conditions, removing as little as 40 microinches of conductor plating from some wires can cause a failure.

Faulty stripping can pierce, scuff, or split the insulation. This can cause changes in dielectric strength and lower the wire's resistance to moisture and abrasion. Insulation particles often get trapped in solder and crimp joints. These form the basis for a defective **termination.** A variety of factors determine how precisely a wire can be stripped, including: wire size, insulation concentricity, adherence, and others.

It is a common mistake to believe that a certain gauge of stranded conductor has the same diameter as a solid conductor. This is a very important consideration in selecting the proper blades for strippers. *Table 1* shows the nominal sizes referenced for the different wire gauges.

To eliminate nicking, cutting, and fraying, wires should only be stripped using the appropriate stripping tool. The specific tool used depends on the size and type of wire being stripped.

2.1.0 Stripping Small Wires

There are many kinds of wire strippers available. *Figure 1* shows two common types of wire strippers for small wires. Note that the one on the left is ergonomically designed with a curved soft foam handle. To use these tools, insert the wire into the proper size knife groove, then squeeze the tool handles. The tool cuts the wire insulation, allowing the wire to be easily removed without crushing its stripped end. The length of the strip is regulated by the amount of wire extending beyond the blades when the wire is inserted in the knife groove.

INSIDE TRACK

Conductor Terminations and Splices

Poor electrical connections are responsible for a large percentage of equipment burnouts and fires. Many of these failures are a direct result of improper terminations, poor workmanship, and the use of improper splicing devices.

Table 1 Dimensions of Common Wire Sizes			
		Overall Diameter in Inches	
Size (AWG/kcmil)	Area (Circular Mils)	Solid	Stranded
18	1,620	0.040	0.046
16	2,580	0.051	0.058
14	4,130	0.064	0.073
12	6,530	0.081	0.092
10	10,380	0.102	0.116
8	16,510	0.128	0.146
6	26,240	—	0.184
4	41,740	—	0.232
3	52,620	—	0.260
2	66,360	—	0.292
1	83,690	—	0.332
1/0	105,600	—	0.373
2/0	133,100	—	0.419
3/0	167,800	—	0.470
4/0	211,600	—	0.528
250	—	—	0.575
300	—	—	0.630
350	—	—	0.681
400	—	—	0.728
500	—	—	0.813
600	—	—	0.893
700	—	—	0.964
750	—	—	0.998
800	—	—	1.03
900	—	—	1.09
1,000	—	—	1.15
1,250	—	—	1.29
1,500	—	—	1.41
1,750	—	—	1.52
2,000	—	—	1.63

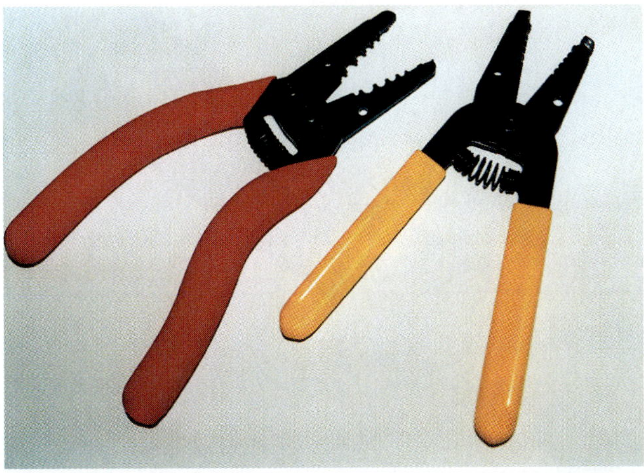
208F01.EPS

Figure 1 ◆ Wire strippers.

2.2.0 Stripping Power Cables and Large Wires

Figure 2 shows a heavy-duty stripper used to strip power cables with outside diameters ranging from ½" to 1¾". It can be used to strip insulation starting from the end of the cable, strip insulation from some point to the end of the cable, or to make a window cut. All stripping tools should be operated according to the manufacturer's instructions. The procedures for using the tool to strip insulation from the end of the cable and to make a window cut are described here.

To strip insulation from the end of the cable (*Figure 3*), proceed as follows:

Step 1 Loosen the locking knob to open the tool to the maximum position. Place the cable in the V-groove and close the tool firmly around the cable. Tighten the locking knob.

Step 2 Turn the cap assembly until the blade reaches the required depth.

CAUTION

Do not allow the blade to contact the conductor because damage to the conductor and/or the blade can result.

Step 3 Rotate the tool around the cable, advancing to the required strip length.

Step 4 Rotate the tool in the reverse direction to produce a square end cut (*Figure 3*).

Step 5 Loosen the locking knob to release the tool and remove it from the cable.

Step 6 Peel off the insulation.

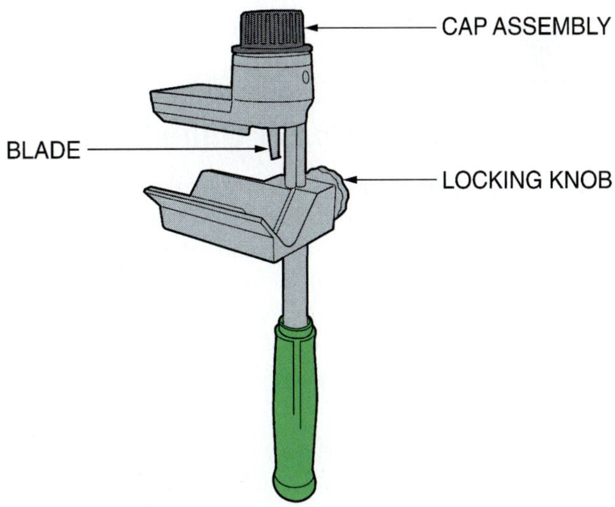

208F02.EPS

Figure 2 ◆ Cable stripper.

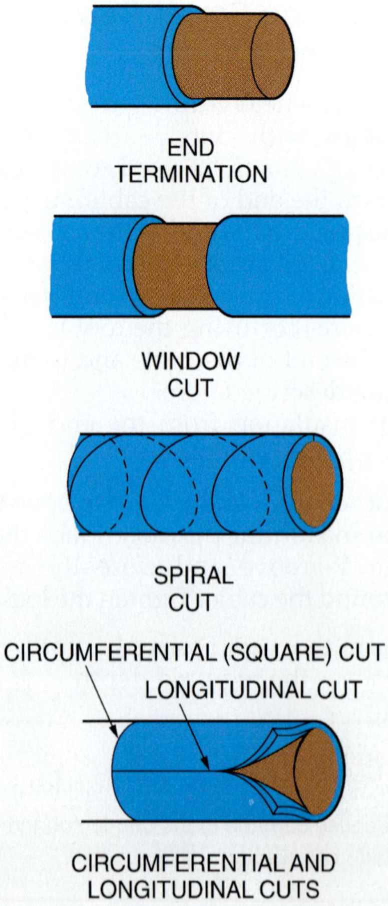

Figure 3 ◆ Types of cable stripping.

To make a window cut (*Figure 3*), proceed as follows:

Step 1 With the tool opened to the maximum position, place the cable in the V-groove and close the tool firmly around the cable. Tighten the locking knob.

Step 2 Turn the cap assembly until the blade reaches the required depth.

Step 3 Rotate the tool to produce the first square cut.

Step 4 Rotate the tool in the reverse direction to cut the required window strip length.

Step 5 Rotate the tool in the original direction to produce the second square cut.

Step 6 Loosen the locking knob assembly to release the tool and remove it from the cable.

Step 7 Peel off the insulation.

Figure 4 shows a round cable slitting and ringing tool that can be used to strip single- or multiple-conductor cables. The tool can be used to cut around the cable (square cut) or slit the length of the cable jacket (longitudinal cut) for easy removal. The tool blade is adjustable to accommodate different jacket thicknesses.

INSIDE TRACK

Cutting Large Power Cable

When cutting large power cables prior to stripping, you can make the job easier by using a ratchet-type or hydraulic-type cable cutter to cut the cable.

208P0801.EPS

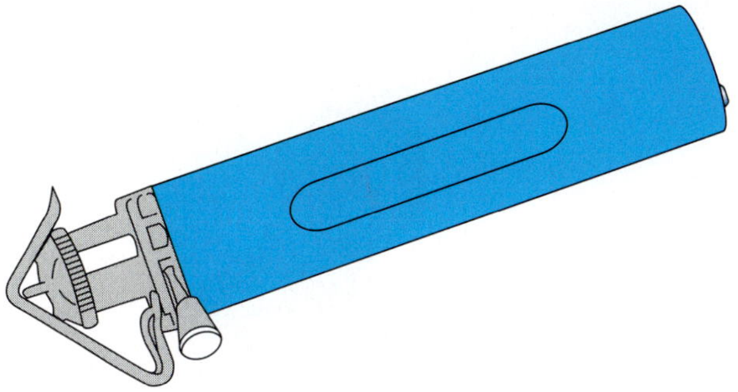

208F04.EPS

Figure 4 ◆ Round cable slitting and ringing tool.

3.0.0 ◆ WIRE CONNECTIONS UNDER 600 VOLTS

NEC Section 110.14 governs electrical connections, including terminations and splices. Wire connections are used to connect a wire or cable to such electrically operated devices as fan coil units, duct heaters, oil burners, motors, pumps, and control circuits of all types.

A variety of wire connectors for stranded wire are shown in *Figure 5*. These connectors are available in various sizes to accommodate wire sizes No. 22 AWG and larger. They can be installed with crimping tools having a single indenter or double indenter. The range is normally stamped on the tongue of each terminal.

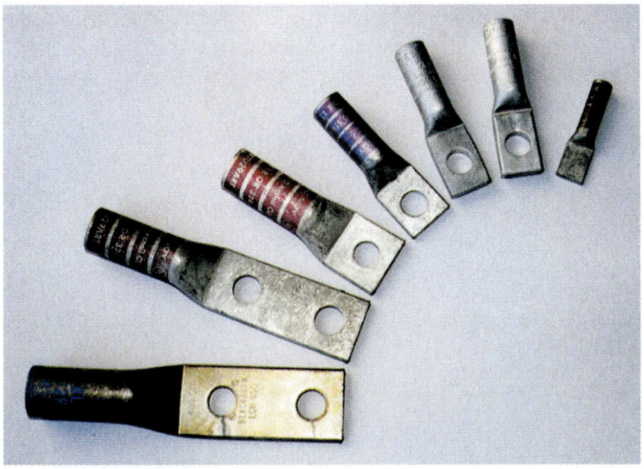

208F05.EPS

Figure 5 ◆ Crimp-on wire lugs.

Mechanical compression-type terminators are also available to accommodate wires from No. 8 AWG through 1,000 kcmil. One-hole lugs, two-hole lugs, split-bolt connectors, and other types are shown in *Figure 6*.

Crimp-type connectors made to accommodate wires smaller than No. 8 are normally made to accept at least two wire sizes and are often color coded. For example, one manufacturer's color code used for such connectors is red for No. 18 or No. 20 wire, blue for No. 16 or No. 14 wire, and yellow for No. 12 or No. 10 wire. Crimp-type connectors used for wire sizes No. 8 and larger, commonly called lugs, are made to accept one specific conductor size. Crimp-type **reducing connectors** are used to connect two different size wires. Mechanical compression-type connectors and lugs are made so that they accommodate a range of different wire sizes.

The parallel-tap connector with an insulated cover shown in *Figure 6* is one example of a pre-insulated, molded mechanical compression connector. There are several kinds available. They come in setscrew/pressure plate and insulation-piercing configurations made for use in a variety of feeder **tap** and splice applications. Because they are equipped with an insulating cover, the requirement for taping the joint is eliminated.

3.1.0 Aluminum Connections

Aluminum has certain properties that are different from copper and must be understood if reliable connections are to be made. These properties are: cold flow, coefficient of thermal expansion, susceptibility to galvanic corrosion, and the formation of oxide film on the surface.

Because of the thermal expansion and cold flow of aluminum, standard copper connectors cannot be safely used on aluminum wire. Most

INSIDE TRACK

Stripping Power and Large Cables

When using any bladed-type cable stripper, always be careful to keep your fingers away from the blade. Also, make sure to replace the blade whenever it becomes dull. Remember, a dull blade is more dangerous than a sharp one because you are more likely to apply undue pressure when using a dull blade, causing it to slip.

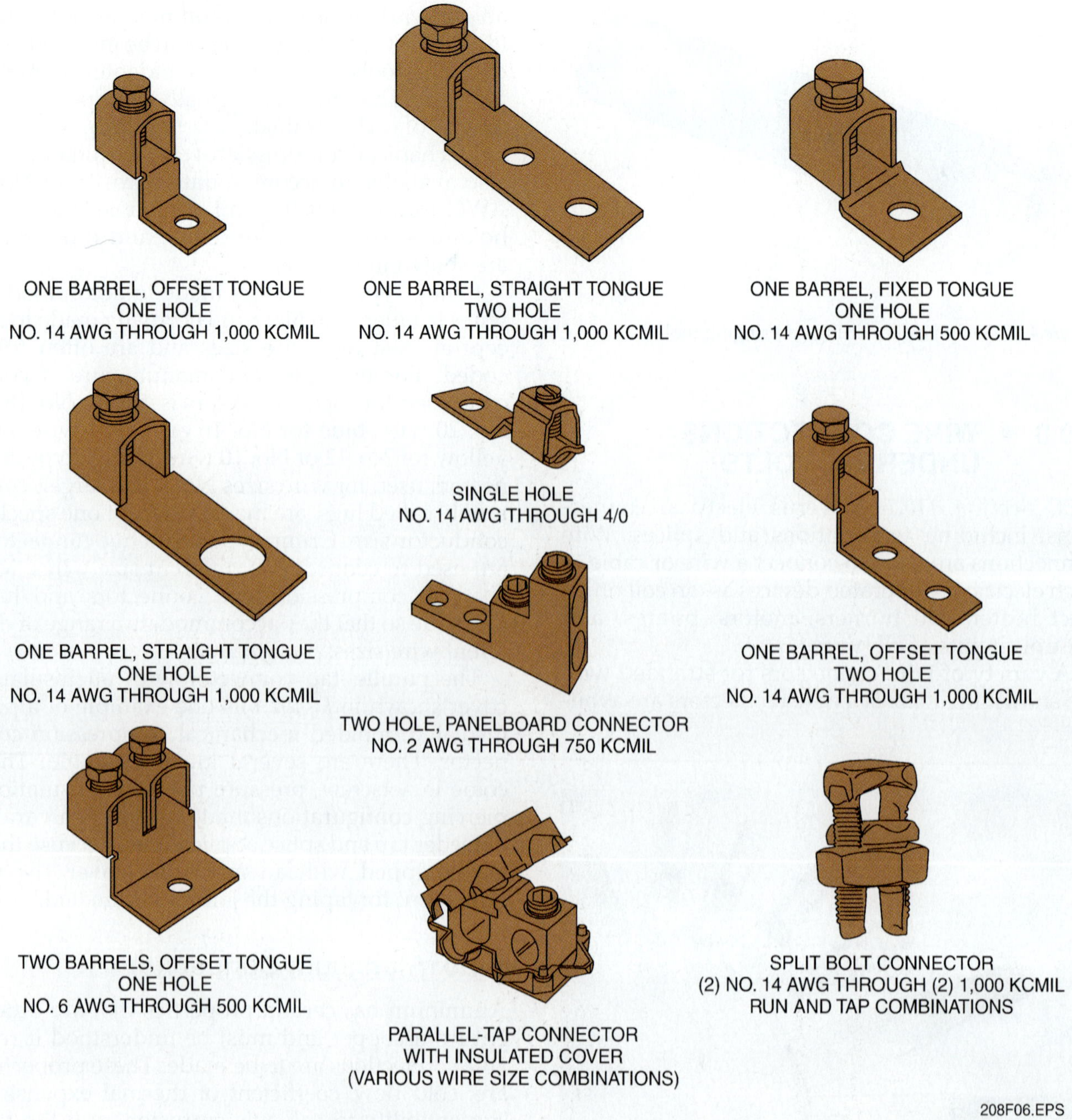

Figure 6 ◆ Various mechanical compression connectors.

THINK ABOUT IT

Mechanical Compression Connectors

What might be a use for a compression connector with a two-hole tongue?

manufacturers design their aluminum connectors with greater contact area to counteract these properties of aluminum. Tongues and barrels of all aluminum connectors are larger or deeper than comparable copper connectors.

The electrolytic action between aluminum and copper can be controlled by plating the aluminum with a neutral metal (usually tin). The plating prevents electrolysis from taking place, and the joint remains tight. As an additional precaution, a joint-sealing compound should be used. Connectors should also be tin-plated and prefilled with an oxide-inhibiting compound.

The insulating aluminum oxide film must be removed or penetrated before a reliable aluminum joint can be made. Aluminum connectors are designed to bite through this film as they are applied to conductors. It is further recommended that the conductor be wire brushed and preferably coated with a joint compound to guarantee a reliable joint.

CAUTION

Never use connectors designed strictly for use on copper conductors on aluminum conductors. Connectors listed for use on both metals will normally be marked AL-CU. All connectors must be applied and installed in the manner for which they are listed and labeled.

NEC Section 110.14 prohibits conductors made of dissimilar metals (copper and aluminum, copper and copper-clad aluminum, or aluminum and copper-clad aluminum) from being intermixed in a terminal or splicing connector unless the device is identified for the purpose and for the conditions under which it may be used. As a general rule:

- Connectors marked with only the wire size should only be used with copper conductors.
- Connectors marked with AL and the wire size should only be used with aluminum wire.
- Connectors marked with **AL-CU** and the wire size may be safely used with either copper or aluminum.

3.2.0 Heat-Shrink Insulators

Heat-shrink insulators for small connectors provide skintight insulation protection and are fast and easy to use. They are designed to slip over wires, taper pins, connectors, terminals, and splices. When heat is applied, the insulation becomes semi-rigid and will provide positive strain relief at the flex point of the conductor. A vapor-proof band will seal and protect the conductor from abrasion, chemicals, dust, gasoline, oil, and moisture. Extreme temperatures, both hot and cold, will not affect the performance of these insulators. The source of heat can be any number of types, but most manufacturers of these insulators

INSIDE TRACK

Tightening Compression Connector Screws and Bolts

Mechanical compression connectors must be tightened to a specified torque using a torque screwdriver or torque wrench. Overtightening can cut the wires or break the fitting, while undertightening may lead to loose connections, resulting in overheating and failure.

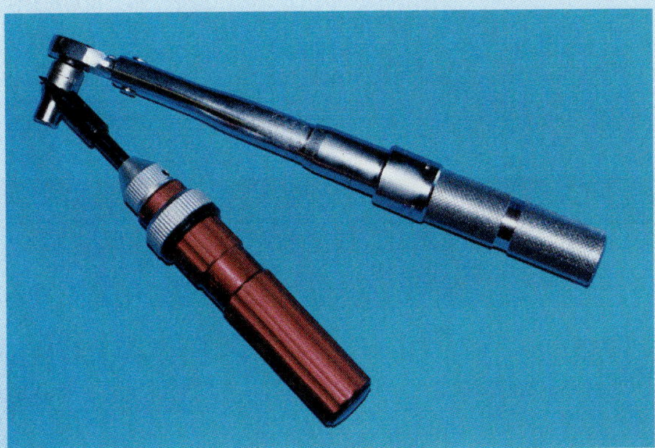

208P0802.EPS

also produce a heat gun especially designed for use on heat-shrink insulators. It closely resembles and operates the same as a conventional hair dryer, as shown in *Figure 7*.

In general, a heat-shrink insulator may be thought of as tubing with a memory. After it is initially manufactured, it is heated and expanded to a predetermined diameter and then cooled. Upon application of heat through various methods, the tubing compound "remembers" its original size and shrinks to that smaller diameter. It is available in a range of sizes and is designed to shrink easily over any wire or device when heat is applied. This property enables it to conform to the contours of any object. The following describes some of the types currently available. A typical tubing selector guide appears in *Table 2*.

- *PVC* – This type is a general-purpose, economical tubing that is widely used in the electronics industry. The PVC compound is irradiated by being bombarded with high-velocity electrons. This results in a denser, cross-linked material with superior electrical and mechanical properties. It also ensures that the tubing will resist cracking and splitting.
- *Polyolefin* – Polyolefin tubing has a wide range of uses for wire bundling, harnessing, strain relief, and other applications where cables and components require additional insulation. It is irradiated, flame-retardant, flexible, and comes in a wide variety of colors.
- *Double wall* – This type is available and designed for outstanding protective characteristics. It is a semi-rigid tubing with an inner wall that melts and an outer wall that shrinks to conform to the melted area.
- *Teflon®* – This type is considered by many users to be the best overall heat-shrink tubing—physically, electrically, and chemically. Its high-temperature rating of 250°C resists brittleness and loss of translucency from extended exposure to high heat and will not support combustion.
- *Neoprene* – Components that warrant extra protection from abrasion require a highly durable yet flexible tubing. Irradiated neoprene tubing offers this optimal coverage.
- *Kynar®* – Irradiated Kynar® is a thin-wall, semirigid tubing with outstanding resistance to abrasion. This transparent tubing enables easy inspection of components that are covered and retains its properties at its rated temperature.

Most tubing is available in a wide variety of colors and configurations. The manufacturer's tubing selector guide can help in the selection of the best tubing for any given application.

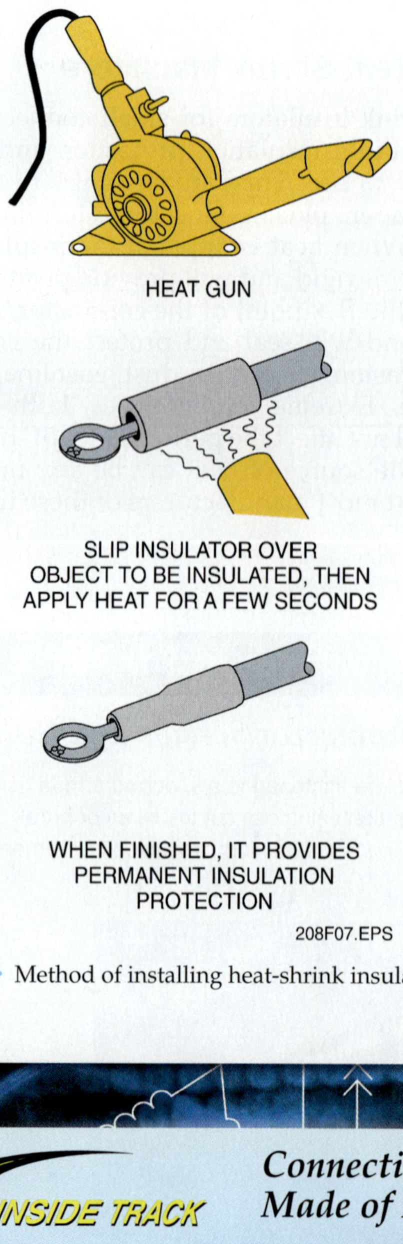

Figure 7 ◆ Method of installing heat-shrink insulators.

INSIDE TRACK

Connection of Conductors Made of Dissimilar Materials

Unless specifically stated on the shipping carton or on the connector itself, wires made of copper, aluminum, or copper-clad aluminum may not be spliced together in the same connector.

Table 2 Tubing Selector Guide

Type	Material	Temp. Range (°C)	Shrink Ratio	Max. Long. Shrinkage (%)	Tensile Strength (psi)	Colors	Dielectric Strength (V/mil)
Nonshrinkable	PVC	+105	—	—	2,700	White, red, clear, black	800
Shrinkable	PVC	−35 to +105	2:1	10	2,700	Clear, black	750
Nonshrinkable	Teflon®	−65 to +260	—	—	2,700	Clear	1,400
Shrinkable	Flexible polyolefin	−55 to +135	2:1	5	2,500	Black, white, red, yellow, blue, clear	1,300
Nonshrinkable	Teflon®	−65 to +260	—	—	7,500	Clear	1,400
Shrinkable	Polyolefin double wall	−55 to +110	6:1	5	2,500	Black	1,100
Shrinkable	Kynar®	−55 to +175	2:1	10	8,000	Clear	1,500
Shrinkable	Teflon®	+250	1.2:2	10	6,000	Clear	1,500
Shrinkable	Teflon®	+250	1½:1	10	6,000	Clear	1,500
Shrinkable	Neoprene	+120	2:1	10	1,500	Black	300

3.3.0 Wire Nuts

Ever since its invention in 1927, the **wire nut,** also known as the *Wirenut*® and the solderless connector (*Figure 8*), has been a favorite wire connector for use on residential and commercial branch circuit applications. Several varieties of wire nuts are available, but the following are the ones used most often:

- Those for use on wiring systems 300V and under
- Those for use on wiring systems 600V and under (1,000V in lighting fixtures and signs)

Most brands are UL listed for aluminum to copper in dry locations only; aluminum to aluminum only; and copper to copper only. The maximum temperature rating is 105°C (221°F).

Wire nuts are frequently used for all types of splices in residential and commercial applications and are considered to be the fastest connectors on the market for this type of work.

To use a wire nut, trim the bare conductors using the appropriate tool, and then screw on the wire nut. The wire nut draws the conductors and insulation into the shirt of the connector, which increases resistance to flashover. The internal spring is designed to tightly thread the conductors into the wire nut and then hold them with a positive grip. Some types of wire nuts have thin wings on each side of the connector to facilitate their installation. See *Figure 9*. Wire nuts are normally made in sizes to accommodate conductors as small as No. 22 AWG up to as large as No. 10 AWG, with practically any combination of those sizes in between.

Specially designed wire nuts are also made specifically for use in wet locations and/or direct burial applications. These wire nuts have a water repellent, non-hardening sealant inside the body that completely seals out moisture to protect the conductors against moisture, fungus, and

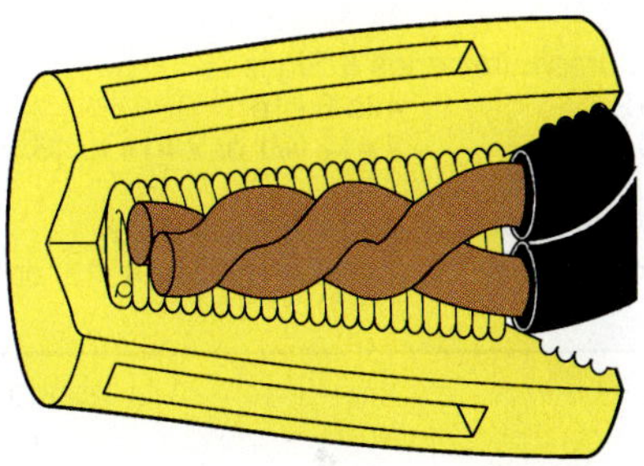

208F08.EPS

Figure 8 ◆ Typical wire nut showing interior arrangement.

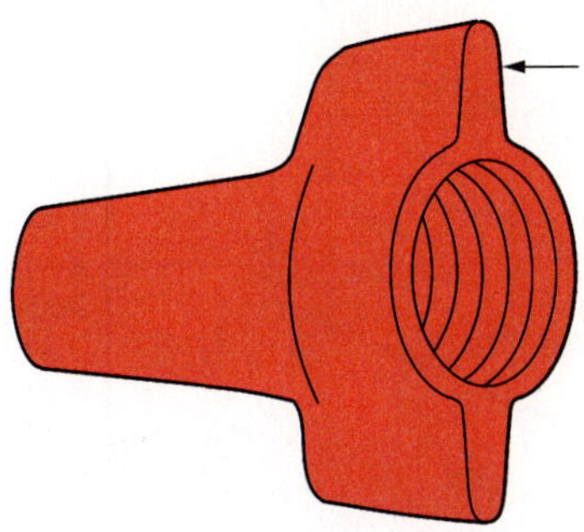

208F09.EPS

Figure 9 ◆ Some wire nuts have thin wings on each side to facilitate installation.

INSIDE TRACK

Insulated Spring Connectors

Insulated spring connectors, commonly called wire nuts or Wirenuts®, are solderless connectors made in various color-coded sizes that allow for splicing the hundreds of different solid or stranded wire combinations typically encountered in branch-circuit and fixture splicing applications.

208P0803.EPS

corrosion. The sealant remains in a gel state and will not melt or run out of the wire nut body throughout the life of the connection. Unlike other types of wire nuts, this type can be used one time only. The wire nut can be backed off, eliminating the need to cut the wires for future or retrofit applications, but once removed, it must be discarded.

The general procedure for splicing wires with wire nuts is as follows:

Step 1 Select the proper size wire nut to accommodate the wires being spliced. Wire nut packages contain charts that list the allowable combinations of wires by size. Refer to the label on the wire nut box or container for this information (*Figure 10*).

Step 2 Strip the insulation from the ends of the wires to be spliced. The length of insulation stripped off is typically about ½"; however, it depends on the wire sizes and the wire nut being used. Follow the manufacturer's directions given on the wire nut package.

Step 3 Stick the ends of the wires into the wire nut and turn clockwise until tight. Note that some manufacturers of wire nuts require that the wires be pre-twisted before screwing on the nut. Also, some manufacturers recommend using a nut driver to tighten the wire nut. Always follow the manufacturer's instructions.

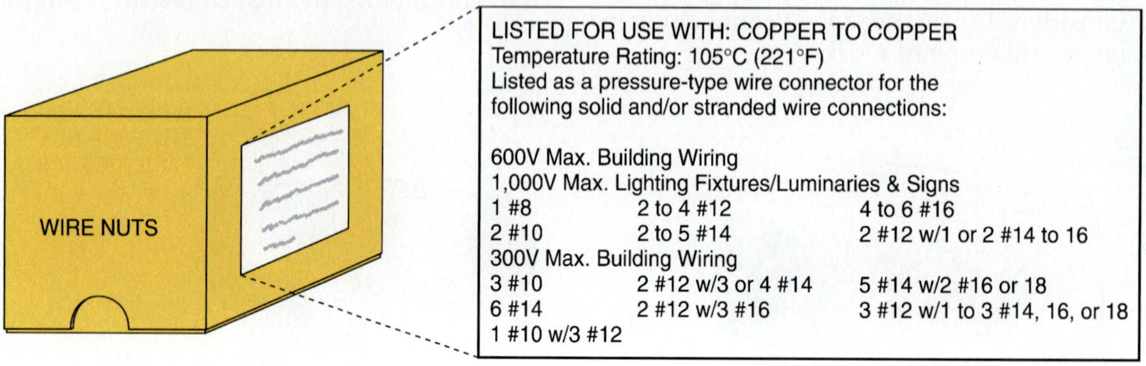

208F10.EPS

Figure 10 ◆ Read package label to find allowable wire combinations.

Making Box Connections

INSIDE TRACK

After making the connections within a box, tuck the wires neatly into the back of the box, as shown here. That way, when the painters and plasterers come along to finish the walls, the wires will not be covered in drywall compound or paint.

208P0804.EPS

4.0.0 ◆ GUIDELINES FOR INSTALLING CONNECTORS

Guidelines for installing the different types of connectors covered in this module are given in the following sections.

4.1.0 Installing Crimp-Type Terminals and Connectors

The task of fastening a compression-type connector to a wire requires the use of the proper connector, crimping tool, and installation procedure.

4.1.1 Crimping Tools

Compression-type connectors and lugs must be attached to wires using the appropriate crimping tool. Since connectors, crimping tools, and crimping tool dies are designed as a unit for specific wire sizes, only the recommended tool and dies should be used. There are a wide variety of crimping tools available, from simple to complex. Most crimping tools are designed for use with either insulated connectors or non-insulated connectors. Quality tools used to crimp larger connectors and wires are typically made with hand-operated ratcheting mechanisms or hydraulic mechanisms that do not release until adequate crimping pressure has been applied. Some crimping tools require the use of related die sets. Some types are dieless. Many manufacturers color code both their connectors and matching crimping tool dies as an aid in selecting the proper die. They also mark their dies with die index numbers to identify the use of the proper connector/die combination.

Figure 11 shows one manufacturer's hand-operated tool. It is designed to crimp connectors to wires ranging in size from No. 8 AWG through

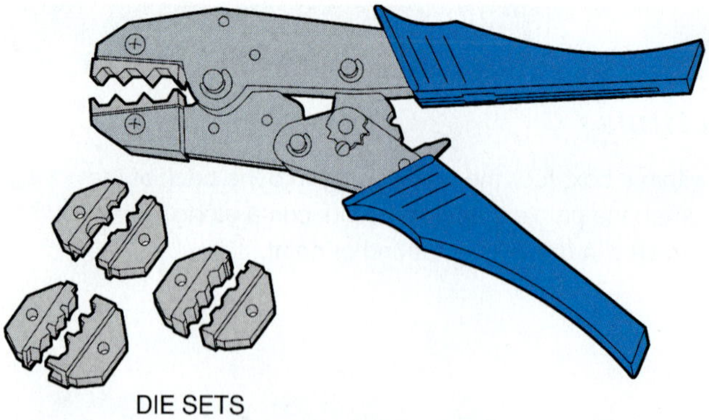

Figure 11 ◆ Crimping tool used to crimp small connectors.

No. 22 AWG stranded. This tool has interchangeable die sets to accommodate different connector sizes. It also has a full-cycle ratchet mechanism that provides for a complete, positive crimp. Once started, the ratchet mechanism does not allow the tool handles to be opened until after the full ratcheting cycle is completed, unless the user actuates the emergency release lever. This ensures a completed crimp every time.

Figure 12 shows hand-operated and hydraulic crimping (compressor) tools typical of those used to crimp connectors for stranded wires ranging from No. 8 AWG to 750 kcmil. Both tools normally develop about 12 tons of compression force at 10,000 pounds per square inch (psi). Guidelines for the use of these tools are described here. Many other tools operate in a similar manner.

To use a hand-operated crimping tool, proceed as follows:

Step 1 Select the proper dies for use with the connector to be crimped. Do not operate the tool without the dies.

Step 2 Push the die release button on the C-head and slide one of the die halves into position until the retainer snaps. Insert the other die half in the piston body by pushing the die release button and sliding the die in until the retainer snaps.

Step 3 Place the tool C-head in position over the connector to be crimped. Pump the handle until compression is complete, as indicated by the dies touching at their flat surfaces nearest the throat of the C-head.

Step 4 Retract the ram and remove the connector after completion of the crimp. This is done by raising the pump handle slightly, rotating it clockwise until it stops, then pushing the handle down in a pumping motion until the pressure release snaps.

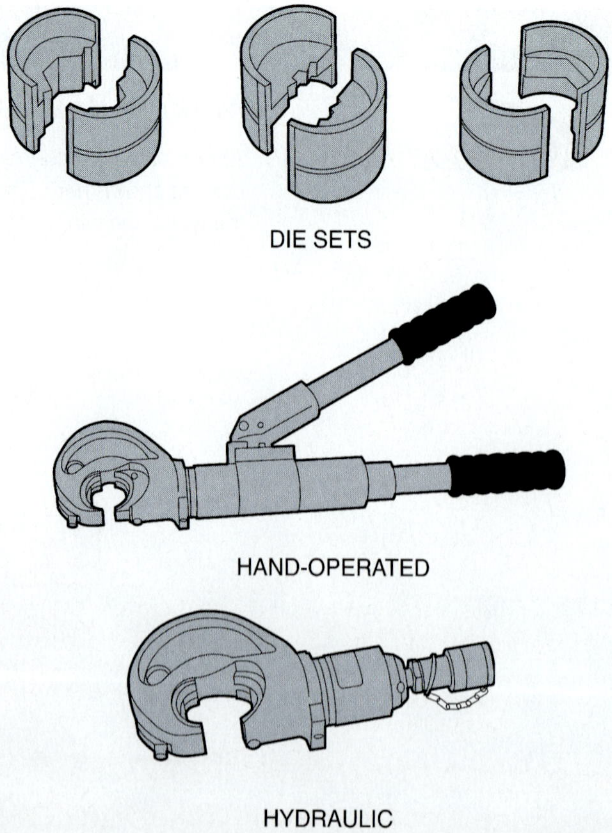

Figure 12 ◆ Crimping (compression) tools used to crimp large connectors.

WARNING!
Always read and follow the manufacturer's instructions when using power tools.

To operate a hydraulic crimping tool, proceed as follows:

Step 1 Using a suitable hydraulic hose, connect the hydraulic pump to the crimping tool.

Step 2 Select the proper dies for use with the connector to be crimped. Do not operate the tool without the dies.

Step 3 Push the die release button on the C-head and slide one of the die halves into position until the retainer pin snaps. In a similar manner, install the other die half in the piston body.

Step 4 Place the tool C-head in position over the connector to be crimped. Operate the remote pump until compression is complete, as indicated by the dies touching on the frame side.

Step 5 Release the pressure at the hydraulic pump to retract the lower die half, then remove the connector from the tool.

4.1.2 General Compression (Crimp) Connector Installation Procedure

Step 1 Select a crimp-type connector of the proper size and of appropriate material for the wire size you are using. Copper connectors should be used with copper wires and aluminum connectors with aluminum wires. Dual-rated connectors may be used with both copper and aluminum wires.

Step 2 Using a suitable wire stripper, remove the insulation from the end of the wire, being careful not to nick the wire. Strip the insulation back far enough so that the bare conductor will go fully into the connector. Make sure not to strip off too much insulation; it should fit close to the connector when the wire is fully inserted into the connector.

Step 3 Clean the stripped portion of the wire. Use a wire brush for large wire sizes. Also clean the related unplated terminal pad and the surface to which the connector will be attached.

Step 4 Obtain the crimping tool and dies made for the type and size of connector to be crimped.

Step 5 Insert the stripped end of the wire completely into the connector. Position the crimping tool in place over the connector, then operate the tool to fully crimp the connector, as directed in the tool manufacturer's instructions. Make sure that the crimping tool jaws are fully closed, indicating that a full compression crimp has been made. Failure to make a secure crimp will create a weak joint.

Step 6 Using a bolt or screw and washers (if required), secure the crimped connector and attached wire to the correct terminal in the equipment. Tighten the terminal bolt and torque to the level specified by the equipment manufacturer. Too little or too much torque can adversely affect the performance of the connection.

Table 3 lists some torque values typical of those used for tightening common sizes of steel and aluminum terminal bolts.

Table 3 Recommended Tightening Torques for Various Bolt Sizes

Steel Hardware		Aluminum Hardware	
Bolt Size	**Recommended Torque (Inch-Pounds)**	**Bolt Size**	**Recommended Torque (Inch-Pounds)**
¼–20	80	½–13	300
5/16–18	180	⅝–11	480
⅜–16	240	¾–10	650
½–13	480	—	—
⅝–11	660	—	—
¾–10	1,900	—	—

4.2.0 Installing Mechanical-Type Terminals and Connectors

The procedure for installing mechanical-type connectors is basically the same as that described above for compression-type connectors, with the following exceptions. Before installing the mechanical connector on the wire, an oxide-inhibiting joint compound should be applied liberally to the conductor to prevent the formation of surface oxides once the connection is made (also apply the compound to any terminal pad). Following this, the connector is installed on the wire, then the connector's bolt/screw is tightened to the torque level specified by the connector manufacturer. Proper torque is important. Too much torque may sever the wires or break the connector; too little torque can cause overheating and failure.

4.3.0 Installing Specialized Cable Connectors

There are a wide variety of cables that require specially designed connectors, commonly called terminators, to secure them to equipment enclosures. Normally, all such specialized connectors are supplied with complete instructions for their installation. This section will introduce you to one type of specialized connector designed for use with metal-clad (Type MC) cable. Its construction and installation are typical of many specialized connectors.

Type MC cable is a factory assembly of one or more insulated circuit conductors with or without optical fiber members. It is enclosed in a metallic sheath of interlocking tape or a smooth or corrugated tube. Throughout the industry, there is increasing use of Type MC cable installed in trays and on racks instead of non-armored cable in conduit. At the appropriate locations, the cables are

Inside Track

Tightening Torques

Many terminations and types of equipment are marked with tightening torque information. For items of equipment, torque information is often marked on the equipment and/or is given in the manufacturer's installation instructions. When specific torque requirements are given, always follow the manufacturer's recommendations.

In cases where no tightening requirements are given, guidelines for tightening screw and bolt-type mechanical compression connectors can be found in the *National Electrical Code® Handbook* comments pertaining to **NEC Section 110.14.** Other good sources of tightening information are manufacturers' catalogs and product literature.

routed from the trays or racks, then along the structure to the various items of equipment. *NEC Article 330* governs the installation of Type MC cable. There are several types of connectors that can be used with Type MC cable. The specific connector used is determined by the size and type of cable and the application.

For the purpose of an example, the procedure for installing one manufacturer's weatherproof connector (*Figure 13*) designed for use with Type MC cable is given here:

Step 1 Select the correct connector size. This is normally done by comparing the physical dimensions of the cable to a cross-reference table given in the manufacturer's product literature and/or installation instructions.

Step 2 Strip back the jacket and armor of the cable as needed to meet equipment requirements. Expose the cable armoring further by stripping the cable jacket for a specified distance (L), as shown in *Figure 13*. This distance will be found in the manufacturer's installation instructions.

Step 3 Make sure that the jacket seal and retaining spring are in their uncompressed state. If necessary, loosen the connector body and the compression nut. Note that it is not necessary to separate the connector parts.

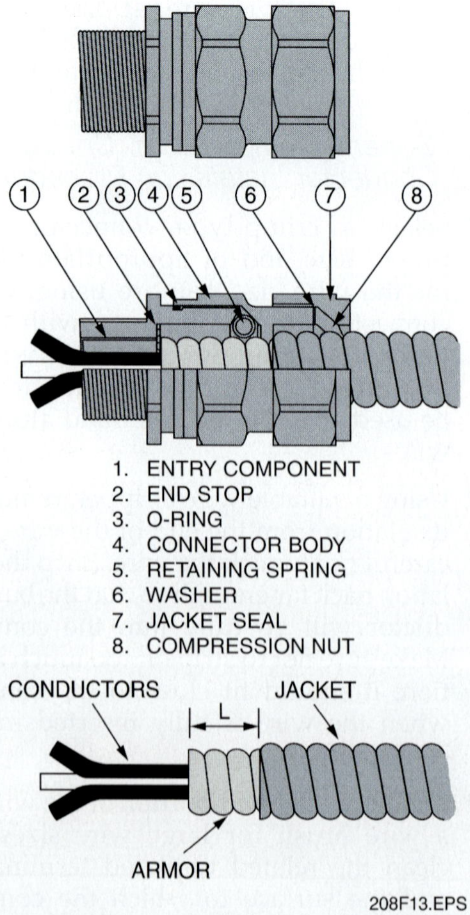

Figure 13 ◆ Weatherproof connector used with Type MC cable.

Step 4 Screw the connector body into the equipment if it has a threaded entry, or secure it with a locknut if it has an unthreaded entry.

Step 5 Pass the cable through the connector until the armor makes contact with the end stop. If it is not possible for the insulated wires to pass through the end stop, then the end stop should be removed so that the wires can move past it and the armor can make contact with the integral end stop within the entry component.

Step 6 Tighten the connector body to compress the retaining spring and secure the armor. Normally, this is hand-tight plus one and a half full turns.

Step 7 Tighten the outer compression nut to form a seal on the cable jacket. Normally, this is hand-tight plus one full turn.

Step 8 If appropriate, terminate the individual wires contained in the cable using compression or mechanical-type connectors, as described above.

5.0.0 ◆ BENDING CABLE AND TRAINING CONDUCTORS

Training is the positioning of cable so that it is not under tension. Bending is the positioning of cable that is under tension. When installing cable or any large conductors, the object is to limit the tension so that the cable's physical and electrical characteristics are maintained for the expected service life. Training conductors, rather than bending them, also reduces the tension on lugs and connectors, extending their service life considerably.

All bends made in cable must comply with the *NEC*®. Minimum bending radii are determined by the cable diameter and, in some instances, by the construction of the cable. For example, bends in Type MC cable must be made so that the cable is not damaged, and the radius of the curve of the inner edge of any bend shall not be less than the following:

- *Smooth sheath* – Ten times the external diameter of the metallic sheath for cable not more than ¾" in external diameter; twelve times the external diameter of the metallic sheath for cable more than ¾" but not more than 1½" in external diameter; and fifteen times the external diameter of the metallic sheath for cable more than 1½" in external diameter.
- *Interlocked-type armor or corrugated sheath* – Seven times the external diameter of the metal sheath.
- *Shielded conductors* – Twelve times the overall diameter of the individual conductors or seven times the overall diameter of the multiconductor cable, whichever is greater.

The two types of cable bending tools in common use are the ratchet bender and the hydraulic bender (*Figure 14*). The ratchet cable bender in *Figure 14(A)* bends 600V copper or aluminum conductors up to 500 kcmil, while the hydraulic bender in *Figure 14(B)* is designed for cables from 350 kcmil through 1,000 kcmil. In addition, the hydraulic bender is capable of one-shot bends up to 90° and automatically unloading the cable when the bend is finished. Either type simplifies and speeds cable installation.

Conductors at terminals or conductors entering or leaving cabinets or cutout boxes and the like must comply with certain *NEC*® requirements, many of which are covered in *NEC Article 312.* The bending radii for various sizes of conductors that do not enter or leave an enclosure through the wall opposite its terminal are shown in *Table 4*. When using this table, the bending space at terminals must be measured in a straight line from the end of the lug or wire connector (in the direction

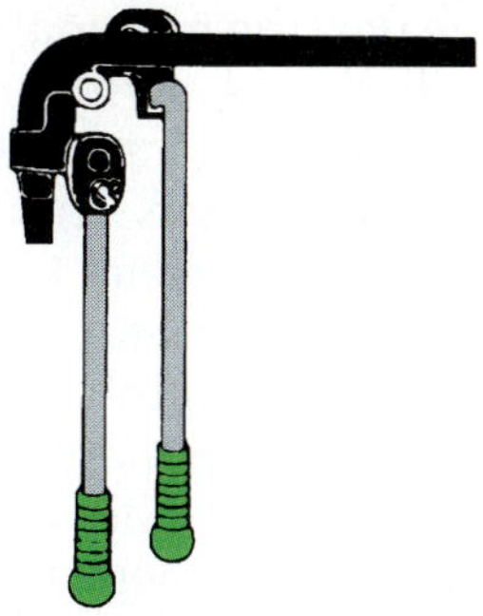

(A) RATCHET CABLE BENDER

(B) HYDRAULIC CABLE BENDER

208F14.EPS

Figure 14 ◆ Types of cable bending tools.

Table 4 Minimum Wire Bending Space and Gutter Width

AWG or Circular-Mil Size of Wire	Wires per Terminal 1	2	3	4	5
14–10	Not Specified	—	—	—	—
8–6	1½	—	—	—	—
4–3	2	—	—	—	—
2	2½	—	—	—	—
1	3	—	—	—	—
1/0–2/0	3½	5	7	—	—
3/0–4/0	4	6	8	—	—
250 kcmil	4½	6	8	10	—
300–350 kcmil	5	8	10	12	—
400–500 kcmil	6	8	10	12	14
600–700 kcmil	8	10	12	14	16
750–900 kcmil	8	12	14	16	18
1,000–1,250 kcmil	10	—	—	—	—
1,500–2,000 kcmil	12	—	—	—	—

Reprinted with permission from NFPA 70, the *National Electrical Code*®. Copyright © 2004, National Fire Protection Association, Quincy, MA 02269. This reprinted material is not the complete and official position of the National Fire Protection Association on the referenced subject, which is represented only by the standard in its entirety.

that the wire leaves the terminal) to the wall, barrier, or obstruction, as shown in *Figure 15*.

An unshielded cable can tolerate a sharper bend than a shielded cable. This is especially true of cables having helical metal tapes, which, when bent too sharply, can separate or buckle and cut into the insulation. The problem is compounded by the fact that most tapes are under jackets that conceal such damage. The shielding bedding tapes or extruded polymers may initially have sufficient conductivity and coverage to pass acceptance testing, but they often fail prematurely at the shield/insulation interface.

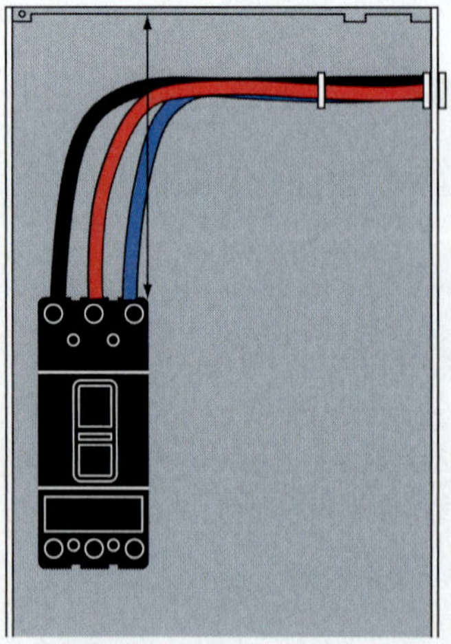

Figure 15 ◆ Bending space at terminals is measured in a straight line.

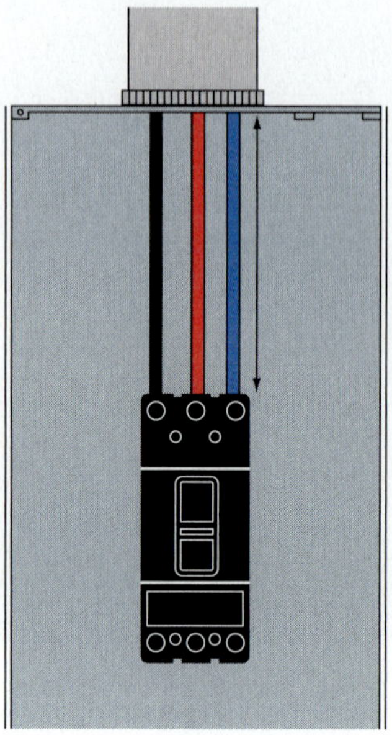

Figure 16 ◆ Conductors entering an enclosure opposite the conductor terminals.

NOTE

Remember that cable offsets are bends.

When conductors enter or leave an enclosure through the wall opposite its terminals (*Figure 16*), *NEC Table 312.6(B)* applies. See *Table 5*. In using this table, the bending space at terminals must be measured in a straight line from the end of the lug or wire connector in a direction perpendicular to the enclosure wall. For removable and lay-in wire terminals intended for only one wire, the bending space in the table may be reduced by the number of inches shown in parentheses.

Table 5 Minimum Wire Bending Space at Terminals

AWG or Circular-Mil Size of Wire	Wires per Terminal			
	1	2	3	4 or More
14–10	Not Specified	—	—	—
8	1½	—	—	—
6	2	—	—	—
4	3	—	—	—
3	3	—	—	—
2	3½	—	—	—
1	4½	—	—	—
1/0	5½	5½	7	—
2/0	6	6	7½	—
3/0	6½ (½)	6½ (½)	8	—
4/0	7 (1)	7½ (1½)	8½ (½)	—
250	8½ (2)	8½ (2)	9 (1)	10
300	10 (3)	10 (2)	11 (1)	12
350	12 (3)	12 (3)	13 (3)	14 (2)
400	13 (3)	13 (3)	14 (3)	15 (3)
500	14 (3)	14 (3)	15 (3)	16 (3)
600	15 (3)	16 (3)	18 (3)	19 (3)
700	16 (3)	18 (3)	20 (3)	22 (3)
750	17 (3)	19 (3)	22 (3)	24 (3)
800	18	20	22	24
900	19	22	24	24
1,000	20	—	—	—
1,250	22	—	—	—
1,500–2,000	24	—	—	—

Reprinted with permission from NFPA 70, the *National Electrical Code*®. Copyright © 2004, National Fire Protection Association, Quincy, MA 02269. This reprinted material is not the complete and official position of the National Fire Protection Association on the referenced subject, which is represented only by the standard in its entirety.

6.0.0 ◆ *NEC®* TERMINATION REQUIREMENTS

There are several *NEC®* requirements governing the termination of conductors as well as the installation of enclosures containing conductors. *NEC Sections 110.14 and 312.6* cover most installations and terminations. However, other sections, such as *NEC Sections 300.4(F) and 430.10*, will apply for specific applications.

6.1.0 Incoming Line Connections

In general, all ungrounded conductors in a motor control center (MCC) installation require some form of overcurrent protection to comply with *NEC Section 240.20.* Such overcurrent protection for the incoming lines to the MCC is usually in the form of fuses or a circuit breaker located at the transformer secondary that supplies the MCC. The conductors from the transformer secondary constitute the feeder to the MCC, and the rules of *NEC Section 240.21* apply. These rules allow the disconnect means and overcurrent protection to be located in the MCC, provided that the feeder taps from the transformer are sufficiently short and other requirements are met.

6.1.1 Main Disconnect

A circuit breaker or a circuit interrupter combined with fuses controlling the power to the entire MCC may provide the required overcurrent protection as described above or there may be a supplementary disconnect (isolation) means. *Figure 17* shows a main disconnect with stab load connectors.

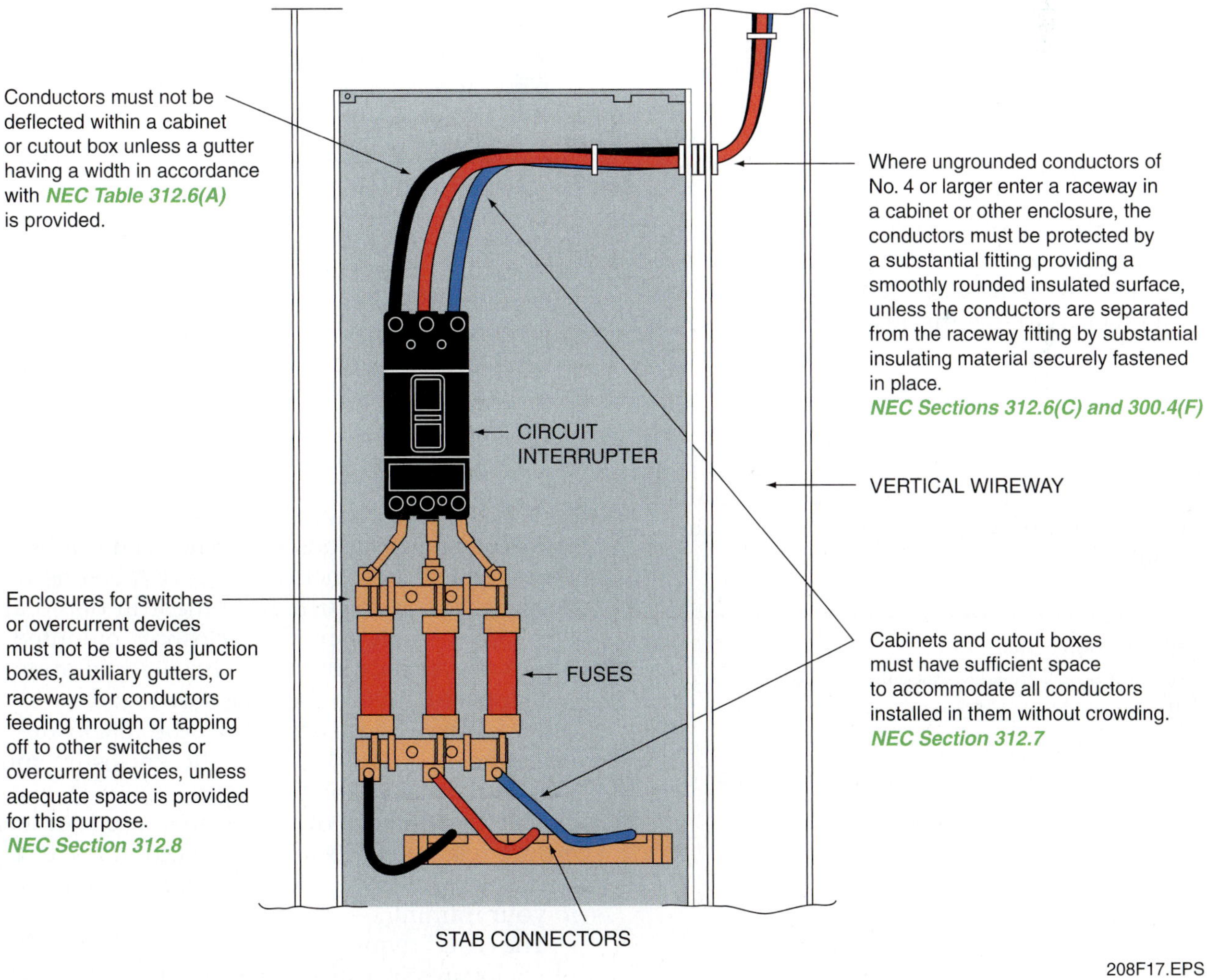

Figure 17 ◆ Main disconnect with stab load connectors.

When the MCC has a main disconnect, the incoming lines (feeders) are brought to the line terminals of the circuit breaker or circuit interrupter. The load side of the circuit breaker or the load side of the fuses associated with the circuit interrupter is usually connected to the MCC busbar distribution system. In cases where the main disconnect is rated at 400A or less, the load connection may be made with stab connections to vertical busbars that connect to the horizontal bus distribution system.

6.1.2 Short Circuit Bracing

All incoming lines to either incoming line lugs or main disconnects must be braced to withstand the mechanical force created by a high fault current. If the cables are not anchored sufficiently or the lugs are not tightened correctly, the connections become the weakest part of a panelboard or motor control center when a fault develops. In most cases, each incoming line compartment is equipped with a two-piece spreader bar located at a certain distance from the conduit entry. This spreader bar should be used along with appropriate lacing material to tie cables together where they can be bundled and to hold them apart where they must be separated. In other words, the incoming line cables should first be positioned and then anchored in place.

Manufacturers of electrical panelboards and motor control centers normally furnish detailed information on recommended methods of short circuit bracing; follow this information exactly.

6.1.3 Making Connections

Before beginning work on incoming line connections, refer to all drawings and specifications dealing with the project at hand. Details of terminations are usually furnished on larger installations.

> **WARNING!**
> All incoming line compartments present an obvious hazard when the door is opened or covers are removed with power on. When working in this area, the incoming feeder should be de-energized.

6.1.4 Making Grounding Conductor Connections

NEC Section 250.8 requires that grounding conductors be connected to boxes and enclosures using exothermic welding, listed pressure conductors, clamps, or other listed means. When more than one equipment grounding conductor enters a box, all the conductors must be spliced or joined either within the box or to the box using devices suitable for this purpose. The arrangement of the grounding connections must be made so that the disconnection or removal of a receptacle, fixture, etc., fed from the box will not interfere with or interrupt the grounding continuity.

When connecting grounding conductors in metal outlet boxes or enclosures, *NEC Section 250.148* requires that a connection be made between the grounding conductors and the metal box by means of a grounding screw that shall not be used for any other purpose, or a listed grounding device.

> **NOTE**
> Sheet metal screws are not permitted to be used as a means of connecting grounding conductors to boxes or enclosures.

Another method approved by the *NEC®* for connecting grounding conductors to a metal box is to use a listed grounding device such as a grounding clip. In nonmetallic boxes, the grounding conductors must be connected so that a connection can be made to any fitting or device in the box that requires grounding.

6.2.0 Mounting Enclosures

NEC Article 312 covers the installation and construction specifications of cabinets, cutout boxes, and meter socket enclosures. Part A begins with the installation of such equipment in damp and wet locations. In general, enclosures mounted in damp or wet locations must have a minimum of ¼" air space between the enclosure and the wall or other supporting surface. If the enclosure is mounted in a wet location, the enclosure must also be weatherproof. See *Figure 18*.

The installation of enclosures used in hazardous (classified) locations must conform to *NEC Articles 500 through 517* and will be covered later in your training.

The *NEC®* requirements governing the position of enclosures in walls is covered in *NEC Section 312.3;* a summary of these requirements appears in *Figure 19*.

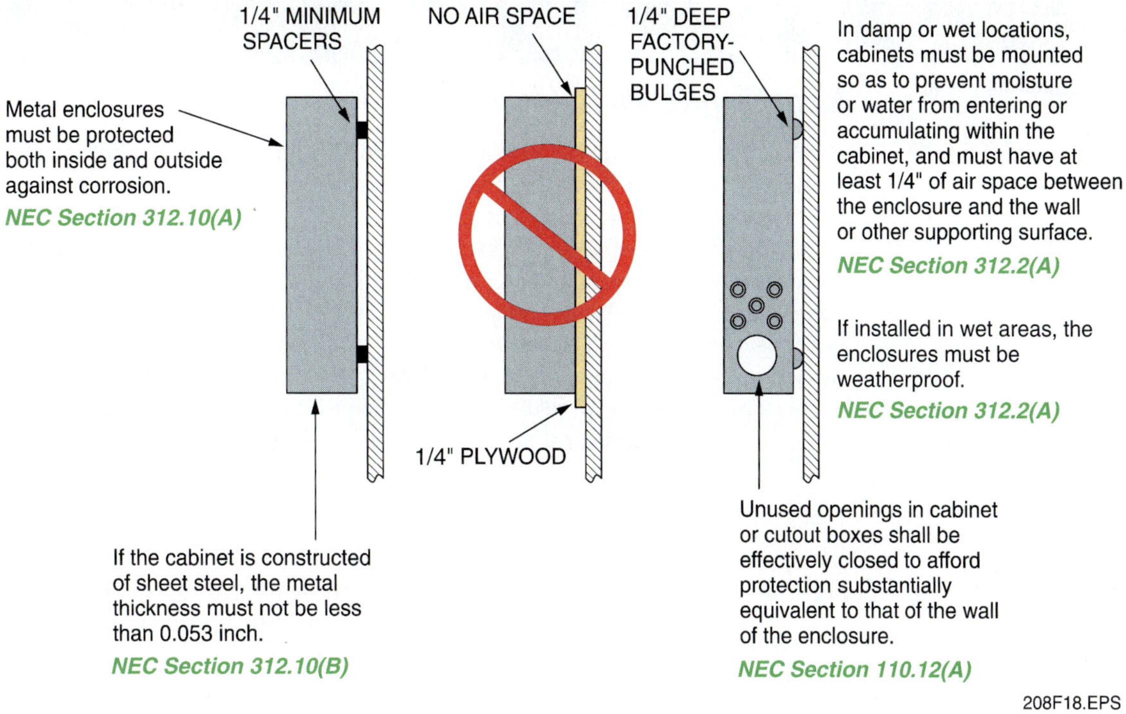

Figure 18 ◆ *NEC*® requirements governing the mounting of enclosures in damp or wet locations.

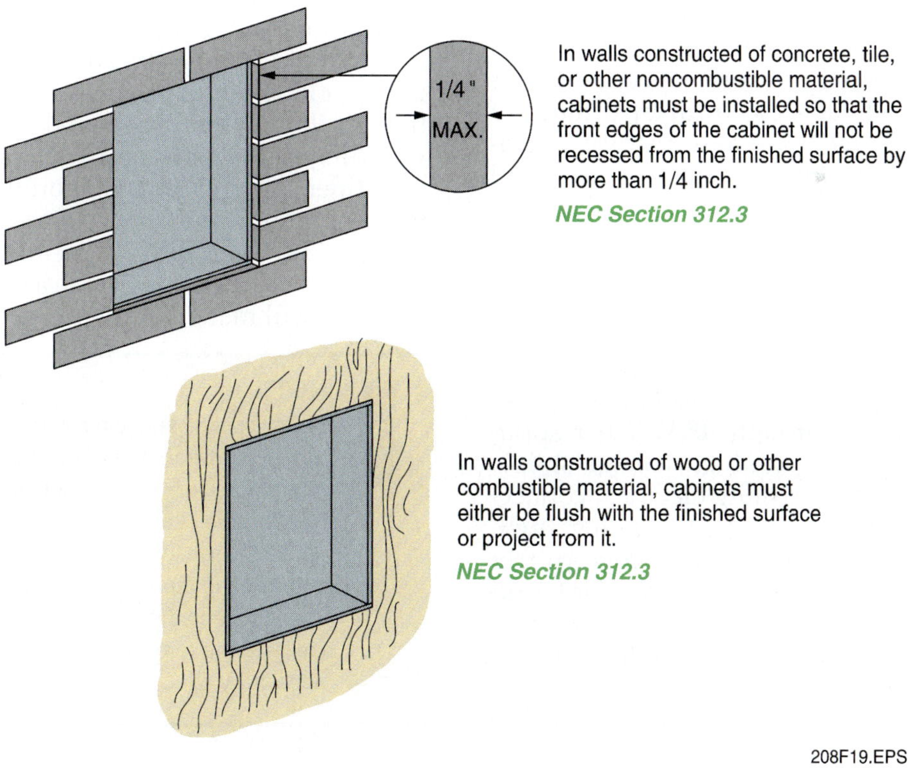

Figure 19 ◆ *NEC*® requirements governing the position of enclosures in walls.

Exothermic Welded Connections

INSIDE TRACK

Per *NEC Section 250.64(C)*, grounding electrode conductors may be spliced at any location by means of irreversible compression-type connectors listed for that purpose or by using the exothermic welding process.

In grounding systems, connections are often the weakest link, especially if they are subjected to high currents and corrosion. For this reason, exothermic welded connections are typically used in commercial and industrial structures to splice and/or connect the grounding electrode conductor system in a building. An exothermic welded connection produces a joint or connection that is better than compression-type connectors because of its welded molecular bond. It normally will not loosen, corrode, or increase in resistance over the lifetime of the installation.

7.0.0 ◆ TAPING ELECTRICAL JOINTS

When it is not practical to protect a spliced joint by some other means, electrical tape may be used to insulate the joint. When tape is used, the joint should be taped carefully to provide the same quality of insulation over the splice as over the rest of the wires.

There is a wide variety of electrical **insulating tapes** made from nonconductive materials for use in specific applications. Some common types of electrical tape include vinyl plastic tape, linerless rubber tape, high-temperature silicone rubber tape, and glass cloth tape. Electrical tapes made of vinyl plastic are widely used as a primary insulation on joints made with thermoplastic-insulated wires. They are used for splices up to 600V and for fixture and wire splices up to 1,000V. Depending on the product, they are made for indoor use, outdoor use, or both. Linerless rubber splicing tape provides for a tight, void-free, moisture-resistant insulation without loss of electrical characteristics. It is typically used as a primary insulation with all solid dielectric cables through 70kV. Other applications include jacketing on high-voltage splices and terminals, moisture-sealing electrical connections, busbar insulations, and end sealing high-voltage cables. High-temperature silicone rubber tapes are used as a protective overwrap for terminating high-voltage cables. Glass cloth electrical tapes provide a heat-stable insulation for hot-spot applications such as furnace and oven controls, motor leads, and switches. They are also used to reinforce insulation where heavy loads cause high heat and breakdown of insulation, such as in motor control exciter feeds, etc. All-metal braid tapes are also available. These are used to continue electrostatic shielding across a splice. When taping a splice, begin by selecting the correct tape for the job. Always follow the tape manufacturer's recommendations.

A general procedure describing one method of taping a splice or joint, such as encountered when connecting motor lugs, is shown in *Figure 20*. A method for taping a split-bolt connector is shown in *Figure 21*. Prior to taping, you should make sure that the joined lugs are securely fastened together with the appropriate hardware. Pieces of a suitable filler tape (or putty) should be wrapped or molded around the lugs and attaching hardware so as to fill the voids and eliminate any sharp edges. It also helps to provide a smooth, even surface that will make taping easier.

NOTE

For all splices and joints where it is likely that the tape will have to be removed at some future date to perform work on the joint, an upside down (that is, adhesive side up) wrap of tape should be applied to the joint before applying the final layers of insulating tape to the joint in the usual manner. This will keep the area free from tape residue and facilitate the removal of the tape later on, if necessary.

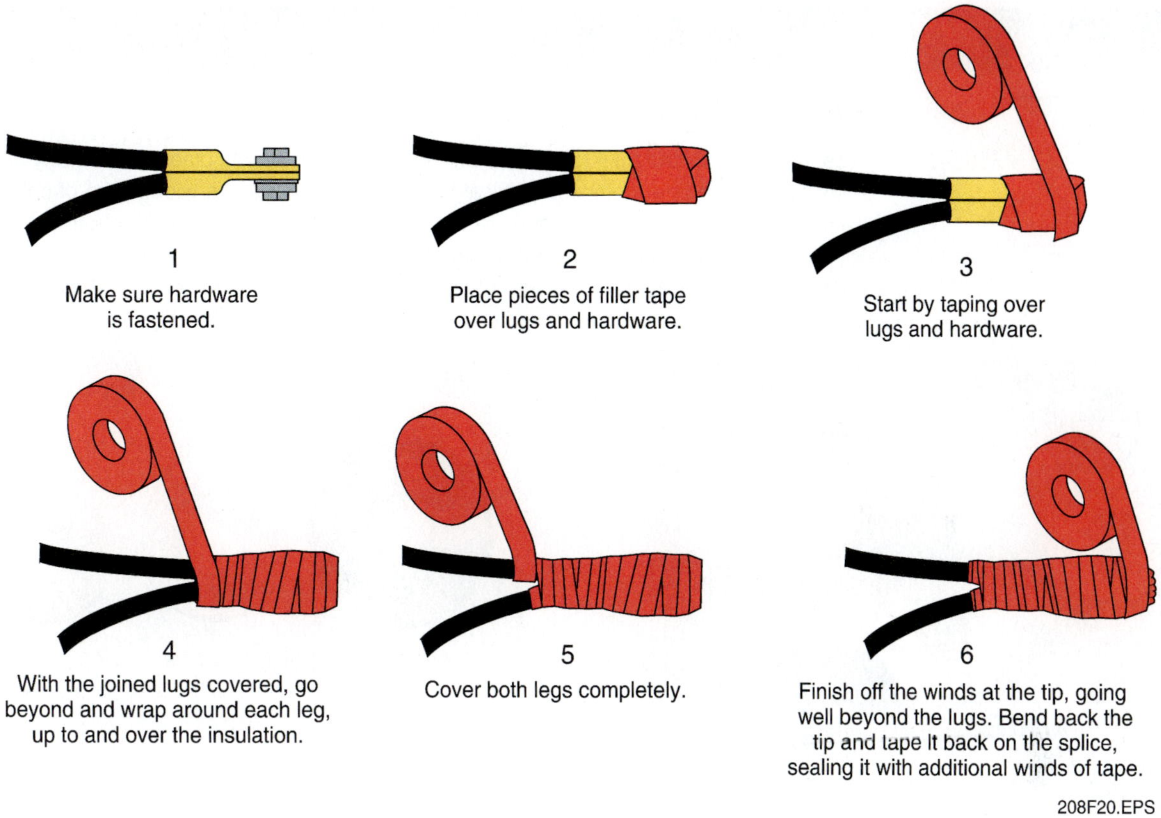

Figure 20 ◆ Typical method for taping motor lug connections.

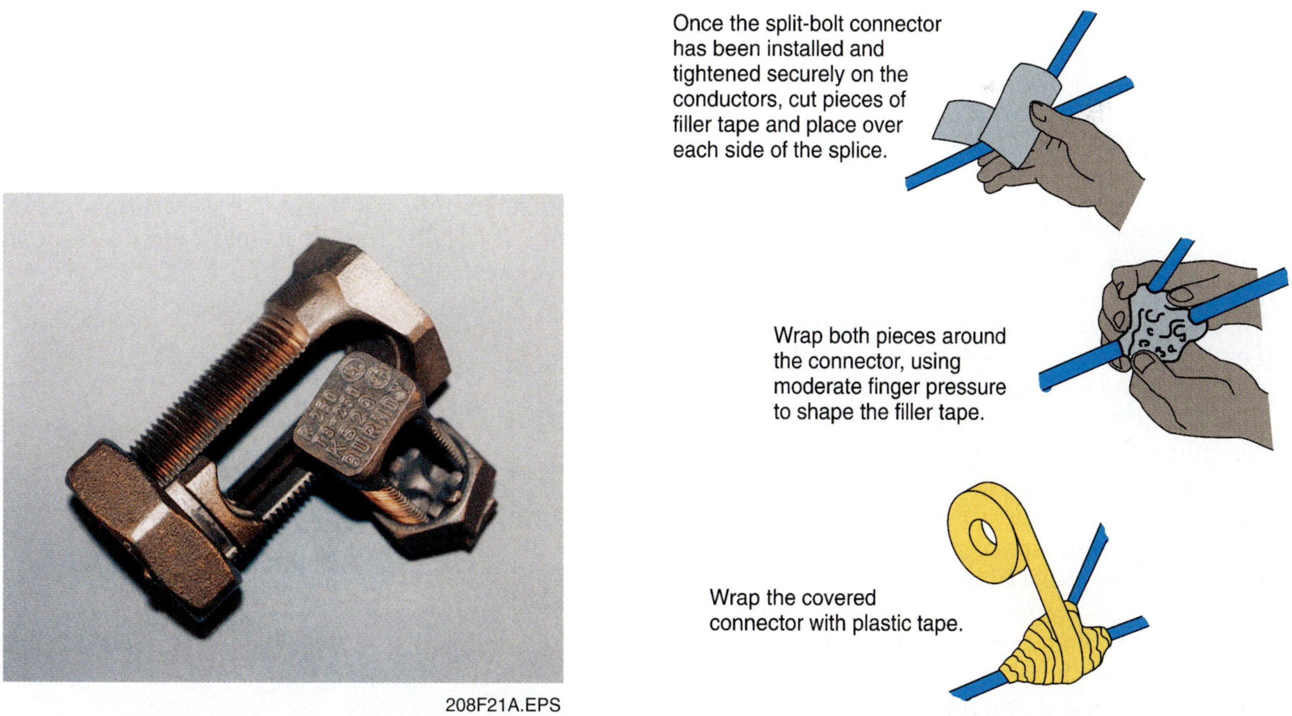

(A) SPLIT-BOLT CONNECTOR (B) TAPING PROCEDURE 208F21B.EPS

Figure 21 ◆ Typical method of taping a split-bolt connector.

Electrical Tape

INSIDE TRACK

Most non-electricians think of electrical tape as only the simple black vinyl variety found in nearly every home toolbox. Electrical tape actually comes in a wide range of colors to be used for labeling various conductors when making terminations. See photo (A). High-voltage electrical tape, as shown in photo (B), can be used to make connections on wires up to 1,000V, depending on the listing of the product being used. Always check the product label to be sure it matches the intended application.

208P0805.EPS

(A)

208P0806.EPS

(B)

8.0.0 ◆ MOTOR CONNECTION KITS

Motor connection kits are available to insulate bolted splice connections, such as those in motor terminal boxes. These kits eliminate the need for taping and the use of filler tape or putty. To aid in joint reentry during rework, the insulator strips off easily, leaving a clean bolt area and thus eliminating the need to remove old tape and putty. Motor connection kits are available for use with stub (butt splice) connections (*Figure 22*) where there is insufficient room to make in-line connections.

They are also made to insulate in-line splice connections where space permits. These insulating kits incorporate a high-voltage mastic, which seals the splice against moisture, dirt, and other contamination. One type of motor connection kit insulator is heat-shrinkable. It installs easily using heat from a propane torch to shrink the insulator in a manner similar to that of heat-shrink tubing (*Figure 23*). Another type of kit used for insulating stub connections comes in the form of an elastomeric insulating cap that is cold-applied by rolling it over the stub splice (*Figure 23*). Always follow the kit manufacturer's recommendations when selecting a kit to use for a particular application. Typically, the kit is selected based on the size of the motor feeder cable.

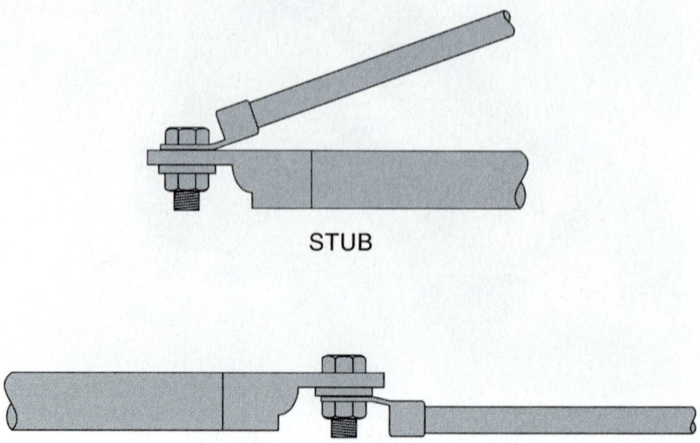

208F22.EPS

Figure 22 ◆ Stub and in-line splice connections.

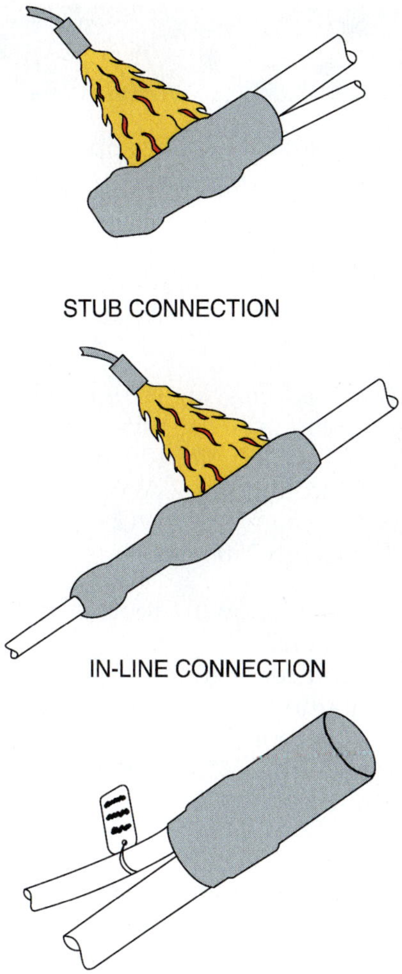

Figure 23 ◆ Motor connection kits installed on splices.

THINK ABOUT IT

Making Connections to Equipment Terminals

Are there any requirements as to how many conductors can be connected to the terminals provided in electric equipment?

Putting It All Together

Examine the wiring connections in your home or workplace. Are they properly made? Can you determine if any connections have the potential for failure?

Review Questions

1. How long should a good conductor splice last?
   a. 60 days
   b. 6 years
   c. As long as current flows
   d. As long as the insulation on the wire itself

2. Which of the following best describes the term stripping as it applies to conductor splicing?
   a. Removal of the insulation from conductors
   b. Removal of the packing material from the carton
   c. Removal of any excess strands of wire from a splice
   d. Removal of the pulling ring from lead sheath conductors

3. Which of the following is *not* a type of stripping?
   a. End termination
   b. Window cut
   c. Spiral cut
   d. Indent cut

4. What is the diameter, in inches, of 2,000 kcmil wire?
   a. 0.893
   b. 0.998
   c. 1.29
   d. 1.63

5. Which of the following best describes the purpose of a reducing connector?
   a. To temporarily connect a circuit
   b. To join two different size conductors
   c. To join conductors of the same size
   d. To make a 90° bend in parallel conductors

6. How long should heat be applied to a heat-shrink insulator in order for it to take the required shape?
   a. A few hours
   b. A few minutes
   c. A few seconds
   d. A few days

7. Which of the following sizes of wire would you normally use wire nuts on?
   a. No. 22 through No. 10 AWG
   b. No. 8 through No. 2 AWG
   c. No. 1 through 1/0 AWG
   d. 4/0 through 250 kcmil

8. Which of the following tools is best for positioning cable?
   a. Split-bolt connectors
   b. A heat gun
   c. A power connector indenter
   d. Wire-bending tools (either ratchet or hydraulic)

9. Which of the following types of electrical tape is best for taping most joints for voltages up to 600V?
   a. High-temperature silicone rubber
   b. Rubber
   c. Glass cloth
   d. Vinyl plastic

10. Crimping tools include all of the following *except* _____.
    a. hand-operated crimping tools
    b. hydraulic compression tools
    c. die sets
    d. pliers

Summary

Solderless connectors (wire nuts) are devices used to join wires without the need for solder. Such connectors are convenient and save much time on the job. Wire nuts are used to splice smaller conductors (No. 10 AWG and smaller) on residential and some commercial installations.

Crimp connectors or terminals are used for larger wire sizes. They are convenient for terminating conductors at terminal boards, control wiring terminals, and the like.

Lugs are provided for the larger wire sizes on panelboards and motor control centers. It is very important to tighten these lugs properly to provide a sound electrical connection as well as to provide short circuit bracing.

There are a variety of insulating tapes available to seal spliced joints. Always match the tape with the application.

Notes

Trade Terms Introduced in This Module

AL-CU: An abbreviation for aluminum and copper, commonly marked on terminals, lugs, and other electrical connectors to indicate that the device is suitable for use with either aluminum conductors or copper conductors.

Connection: That part of a circuit that has negligible impedance and joins components or devices.

Connector: A device used to physically and electrically connect two or more conductors.

Insulating tape: Adhesive tape that has been manufactured from a nonconductive material and is used for covering wire joints and exposed parts.

Lug: A device for terminating a conductor to facilitate the mechanical connection.

Pressure connector: A connector applied using pressure to form a cold weld between the conductor and the connector.

Reducing connector: A connector used to join two different sized conductors.

Splice: The electrical and mechanical connection between two pieces of cable.

Strand: A group of wires, usually stranded or braided.

Tap: A splice connection of a wire to another wire. Also called a tap-off.

Terminal: A device used for connecting cables.

Termination: The connection of a cable.

Wire nut: A form of wire connector that is tightened or loosened by screwing the device clockwise or counterclockwise, respectively. Some wire nuts are provided with two thin, flat wings extending from opposite sides to facilitate tightening the connectors.

Additional Resources

This module is intended to present thorough resources for task training. The following reference works are suggested for further study. These are optional materials for continued education rather than for task training.

American Electrician's Handbook, Latest Edition. New York: Croft and Summers, McGraw-Hill.

National Electrical Code® Handbook, Latest Edition. Quincy, MA: National Fire Protection Association.

The NCCER makes every effort to keep these textbooks up-to-date and free of technical errors. We appreciate your help in this process. If you have an idea for improving this textbook, or if you find an error, a typographical mistake, or an inaccuracy in NCCER's *Contren®* textbooks, please write us, using this form or a photocopy. Be sure to include the exact module number, page number, a detailed description, and the correction, if applicable. Your input will be brought to the attention of the Technical Review Committee. Thank you for your assistance.

Instructors – If you found that additional materials were necessary in order to teach this module effectively, please let us know so that we may include them in the Equipment/Materials list in the Annotated Instructor's Guide.

Write: Product Development
National Center for Construction Education and Research
P.O. Box 141104, Gainesville, FL 32614-1104

Fax: 352-334-0932

E-mail: curriculum@nccer.org

Craft ____________ Module Name ____________

Copyright Date ____________ Module Number ____________ Page Number(s) ____________

Description ____________

(Optional) Correction ____________

(Optional) Your Name and Address ____________

Installation of Electric Services
26209-05

Steven F. Udvar-Hazy Center
National Air and Space Museum
Chantilly, Virginia
Mega-Projects Over $100 Million Award Winner
Hensel Phelps Construction Co.

26209-05

Installation of Electric Services

Topics to be presented in this module include:

Overview

Commercial power becomes the responsibility of the customer at the electric service. Electricians install the electric service but do not generally make the connections at the point where the commercial power connects. This is usually the job of the electric utility provider.

If the commercial power connects to the electric service overhead, the incoming service supply is referred to as a service drop. On the other hand, if the lines are supplied underground, the service supply is said to be a service lateral. The *National Electrical Code*® regulates the installation of both service drops and service laterals, including minimum requirements related to overhead conductor heights, burial depths, raceways, and other installation conditions.

Building an electric service at the consumer's location requires knowledge of materials and service installation skills. If the service supply is a service drop, the service conductors enter the service raceway through a service head or weatherhead, which is designed to keep rain out of the service equipment. Separate service-entrance conductors are usually installed in the service raceway and these connect to the service drop at the point of attachment to the service raceway. The service equipment includes the electrical meter enclosure, service disconnecting means, main circuit breaker panelboard, and grounding system.

Note: *National Electrical Code*® and *NEC*® are registered trademarks of the National Fire Protection Association, Inc., Quincy, MA 02269. All *National Electrical Code*® and *NEC*® references in this module refer to the 2005 edition of the *National Electrical Code*®.

Objectives

When you have completed this module, you will be able to do the following:

1. Describe various types of electric services for commercial and industrial installations.
2. Read electrical blueprints and diagrams describing service installations.
3. Calculate and select service-entrance equipment.
4. Explain the role of the *NEC*® in service installations.
5. Install main disconnect switches, panelboards, and overcurrent protection devices.
6. Identify the circuit loads, number of circuits required, and installation requirements for distribution panels.
7. Explain the types and purposes of service grounding.
8. Explain the purpose and required location(s) of ground fault circuit interrupters.
9. Describe single-phase service connections.
10. Describe both wye- and delta-connected three-phase services.

Trade Terms

Delta-connected	Service equipment
Load center	Service lateral
Service	Service raceway
Service conductors	Substation
Service drop	Wye-connected
Service entrance	

Required Trainee Materials

1. Pencil and paper
2. Appropriate personal protective equipment
3. Copy of the latest edition of the *National Electrical Code*®

Prerequisites

Before you begin this module, it is recommended that you successfully complete *Core Curriculum; Electrical Level One; Electrical Level Two,* Modules 26201-05 through 26208-05. You should also read ***NEC Article 230.***

This course map shows all of the modules in *Electrical Level Two.* The suggested training order begins at the bottom and proceeds up. Skill levels increase as you advance on the course map. The local Training Program Sponsor may adjust the training order.

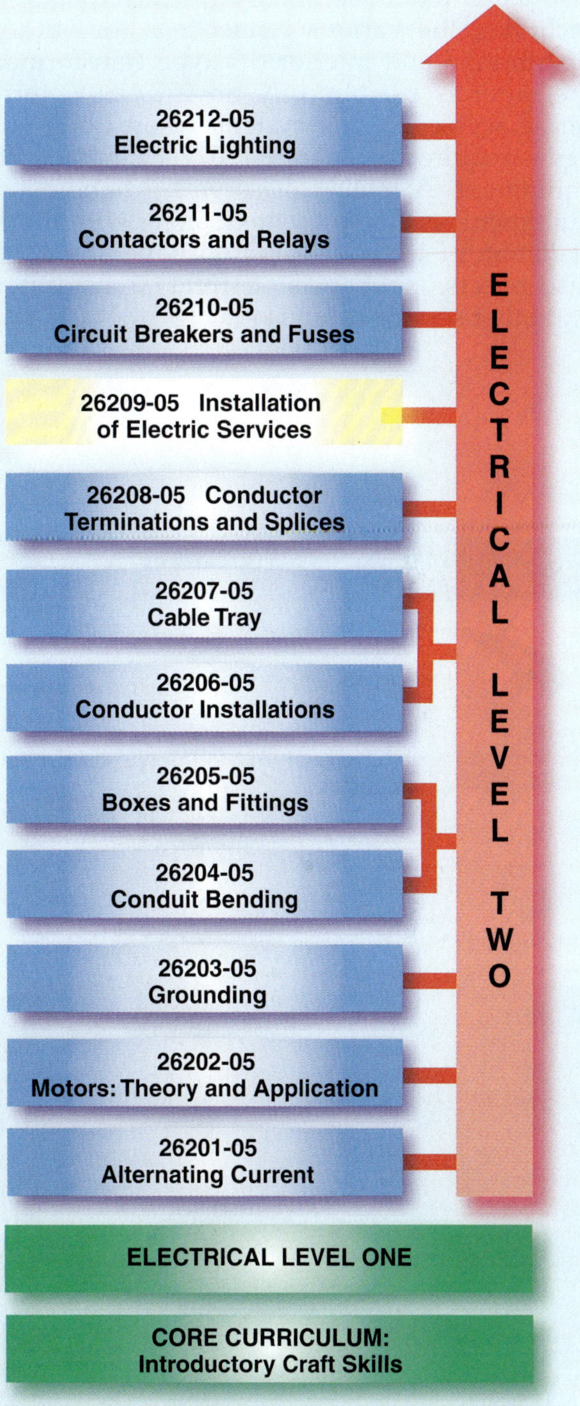

209CMAP.EPS

1.0.0 ◆ INTRODUCTION

This module is designed to cover most electric **service** applications that will be encountered by electricians working on commercial and industrial projects. Detailed installation techniques are also presented for secondary systems up to 600V, including the various connections for outdoor distribution and interior dry-type transformers; the latter is used mostly on 277/480V, three-phase, **wye-connected** systems where lower voltage is required to operate 120/208V outlets and equipment. A review of service grounding requirements is also presented, along with the installation of main distribution panels, multiple disconnects, subpanels, current transformers, and other service equipment.

Electric services can range in size from a small 120V, single-phase, 15A service, which is the minimum allowed by *NEC Section 230.79(A)*, for a roadside vegetable stand, to huge industrial installations involving **substations** dealing with thousands of volts and amperes. Regardless of the size, all electric services are provided for the same purpose: to deliver electrical energy from the supply system to the wiring system on the premises served. Consequently, all establishments containing equipment that utilizes electricity require an electric service.

INSIDE TRACK

Service Entrances

The regulations governing the method of bringing the electric power into a building are established by the local utility company. These regulations can vary considerably between utility companies and regional areas.

209P0901.EPS

Figure 1 shows the basic components of a typical commercial electric service. In this illustration, note that the high-voltage lines terminate on a power pole near the building being served. A bank of transformers is mounted on the pole to reduce the transmission voltage to a usable level (in this case, 120/208V, three-phase, wye-connected).

The remaining sections are:

- *Service drop* – The overhead conductors through which electrical service is supplied between the last power company pole and the point of their connection to the service-entrance conductors located at the building or other support used for the purpose.
- *Service entrance* – All components between the point of termination of the overhead service drop or underground **service lateral** and the building's main disconnecting device, except for metering equipment.
- *Service conductors* – The conductors between the point of termination of the overhead service drop or underground service lateral and the main disconnecting device in the building or on the premises.
- *Service equipment* – The necessary equipment, usually consisting of a circuit breaker or switch and fuses and their accessories, located near the point of entrance of supply conductors to a building and intended to constitute the main control and cutoff means for the electric supply to the building.

When the service conductors to the building are routed underground, as shown in *Figure 2*, these conductors are known as the service lateral, which are the underground conductors through which service is supplied between the power company's distribution facilities and the first point of their connection to the building or area service facilities located at the building or other support used for the purpose.

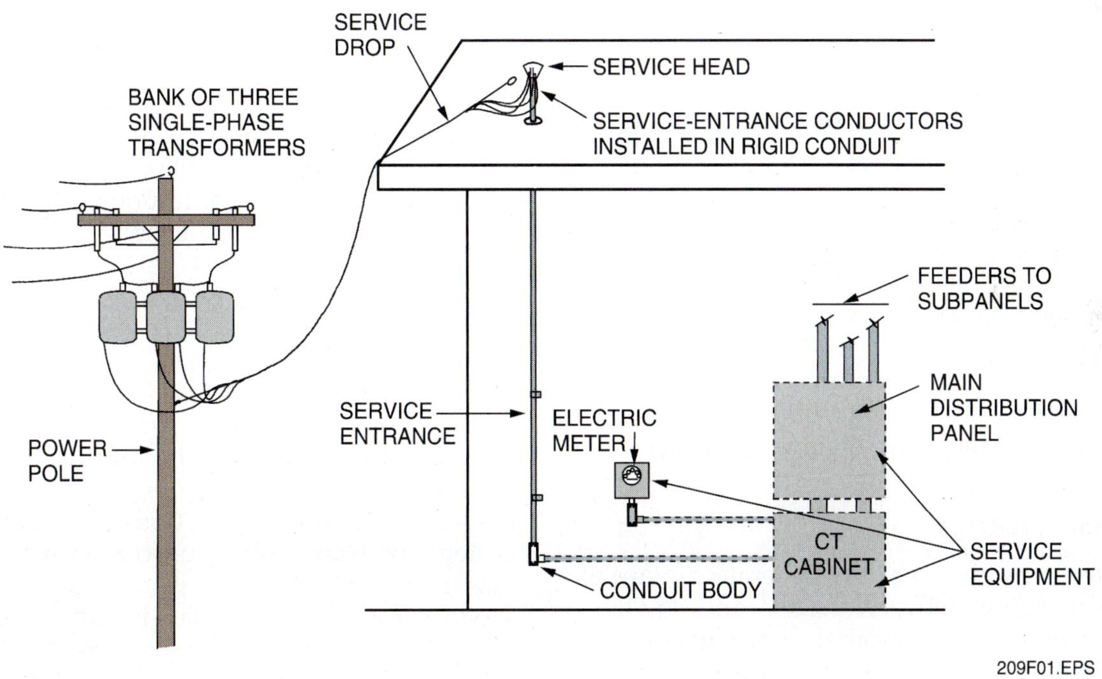

Figure 1 ◆ Typical three-phase overhead service.

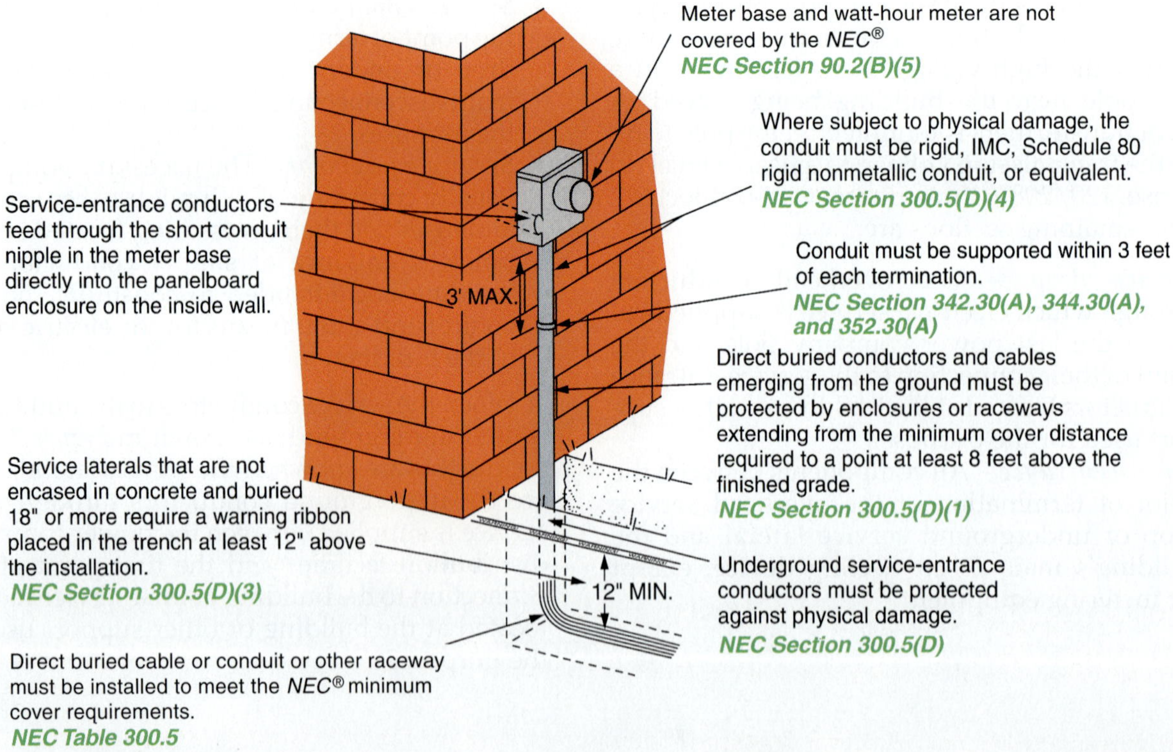

Figure 2 ◆ Single-phase underground service lateral for small commercial building.

2.0.0 ◆ ELECTRICAL GENERATION AND DISTRIBUTION

A review of electrical distribution systems is the best foundation for understanding alternating current and the purpose of electric services.

The essential elements of an AC electrical system capable of producing useful power include generating stations, transformers, substations, transmission lines, and distribution lines. *Figure 3* shows these elements and their relationships.

2.1.0 Generation

Electricity is produced at the generating plant at voltages ranging between 2,400V and 13,200V. Transformers are also located at the generating plant to step up the voltage to hundreds of thousands of volts for transmission.

Electricity is transported from one part of the system to another by cables made up of many strands of wire. The continuous system of conductors through which electricity flows is called the distribution circuit.

2.2.0 Transmission

The system for moving high-voltage electricity is called the transmission system. Transmission lines are interconnected to form a network of lines. Should one line fail, another will take over the load. Such interconnections provide a reliable system for transporting power from generating plants to communities.

Most transmission lines installed by power companies utilize three-phase current—three separate streams of electricity traveling on separate

THINK ABOUT IT

Electrical Utilities

Is the equipment installed and operated by an electrical utility required to comply with the *NEC*®?

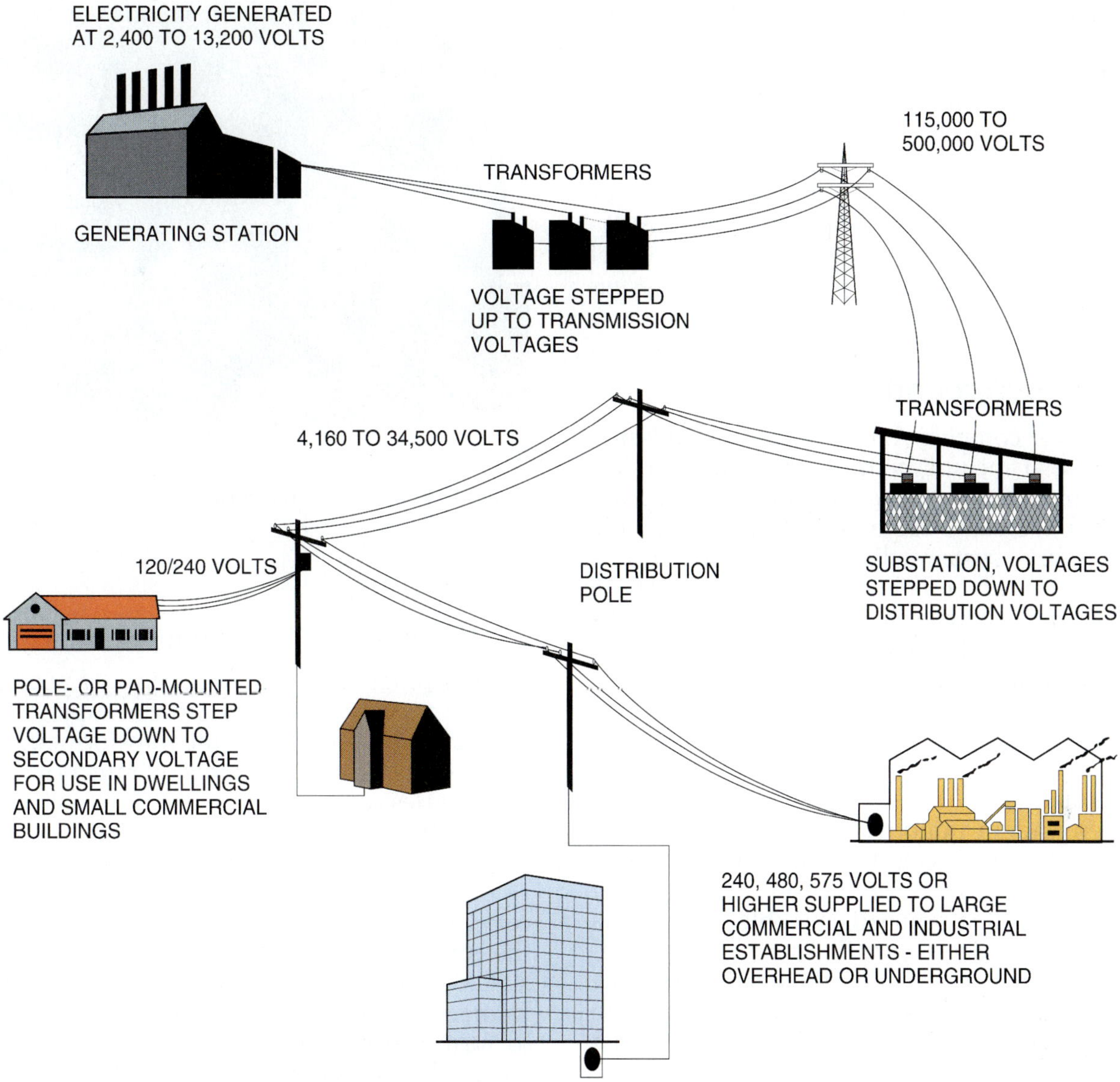

Figure 3 ◆ Parts of a typical electrical distribution system.

conductors. This is an efficient way to transport large quantities of electricity. At various points along the way, transformers step down the transmission voltage at facilities known as substations.

2.3.0 Substations

Substations can be small buildings or fenced-in yards containing switches, transformers, and other electrical equipment and structures. Substations are convenient places to monitor the system and adjust circuits. Devices called regulators, which maintain system voltage as the demand for electricity changes, are also installed in substations. Another device, which momentarily stores energy, is called a capacitor, and is sometimes installed in substations; this device reduces energy losses and improves voltage regulation. Within the substation, rigid tubular or rectangular bars, called busbars or buses, are used as conductors.

At the substation, the transmission voltage is stepped down to voltages below 35,000V, which feed into the distribution system.

The distribution system delivers electrical energy to the user's energy-consuming equipment (i.e., lighting, motors, machines, and appliances).

Conductors called feeders radiate in all directions from the substation, carrying the power from the substation to various distribution centers. At key locations in the distribution system, the voltage is stepped down by transformers to the level needed by the customer. Distribution conductors on the high-voltage side of a transformer are called primary conductors or primaries; those on the low-voltage side are called secondary conductors or secondaries.

Transformers are actually smaller versions of substation distribution transformers that are installed on poles, on concrete pads, or in transformer vaults throughout the distribution system.

Distribution lines carry either three-phase or single-phase current. Single-phase power is normally used for residential and small commercial occupancies, while three-phase power serves most of the other users.

2.4.0 Underground

Most power companies now utilize transmission systems that include both overhead and underground installations. In general, the terms and devices are the same for both. In the case of the underground system, distribution transformers are installed at or below ground level. Those mounted on concrete slabs are called padmounts (*Figure 4*), while those installed in underground vaults are called submersibles.

Buried conductors (cables) are insulated to protect them from soil chemicals and moisture. Many overhead conductors do not require such protective insulation.

When underground transmission or distribution cables terminate and connect with overhead conductors at buses or on the tops of poles, special devices called potheads or cable terminators are employed. See *Figure 5*. These devices prevent moisture from entering the insulation of the cable and also serve to separate the conductors to prevent arcing between them. The cable installation along the length of the pole is known as the cable riser.

209F04.EPS

Figure 4 ◆ Padmount transformer.

2.5.0 Secondary Systems

From a practical standpoint, most of your work will be involved with the power supply on the secondary (usage) side of the transformer.

Two general arrangements of transformers and secondaries are in common use. The first arrangement is the sectional form, in which a unit of load, such as one city street or city block, is served by secondary conductors, with the transformer located in the middle. The second arrangement is the continuous form, in which the primary is installed in one long continuous run, with transformers spaced along it at the most suitable points to form the secondaries. As the load grows or shifts, the transformers spaced along it can be moved or rearranged, if desired. In a sectional

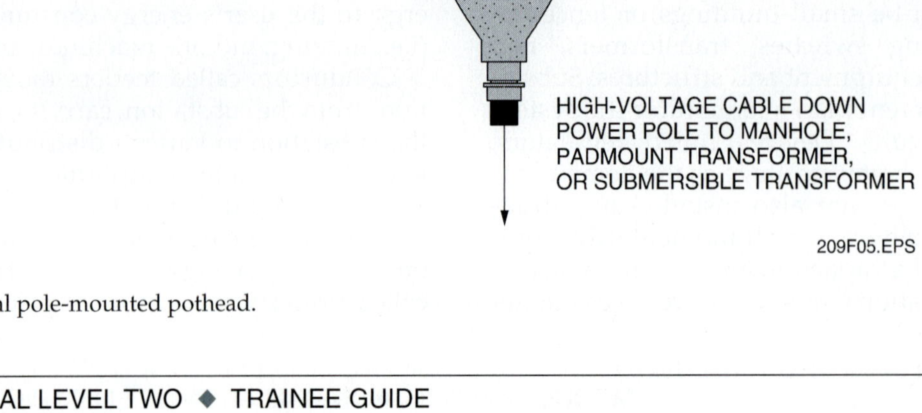

209F05.EPS

Figure 5 ◆ Typical pole-mounted pothead.

arrangement, such a load can be increased only by changing to a larger size of transformer or installing an additional unit in the same section.

One of the greatest advantages of the secondary bank is that the starting currents of motors are divided among transformers, reducing voltage drop and diminishing the resulting lamp flicker at the various outlets.

Power companies all over the United States and Canada are now trying to incorporate networks into their secondary power systems, especially in areas where a high degree of service reliability is necessary. Around cities and industrial applications, most secondary circuits are three-phase, either 120/208V or 277/480V, and are wye-connected. Usually, two to four primary feeders are run into the area and the transformers are alternately connected to them. The feeders are interconnected in a grid or network so that if any feeder goes out of service, the load is still carried by the remaining feeders.

The primary feeders supplying networks are run from substations at the usual primary voltage for the system, such as 4,160V, 4,800V, 6,900V, or 13,200V. Higher voltages are practical if the loads are large enough to warrant them.

2.6.0 Common Power Supplies

The most common power supply used for residential and small commercial applications is 120/240V, single-phase, three-wire service; it is used primarily for light and power, including single-phase motors up to about 7½ horsepower (hp). A diagram of this service is shown in *Figure 6*.

Four-wire, **delta-connected** secondaries (*Figure 7*) and four-wire, wye-connected secondaries (*Figure 8*) are common in industrial and commercial applications.

The characteristics of the electric service and the equipment connected to the service must match. Also, the characteristics of an electric service will often dictate those for the electrical equipment or vice versa.

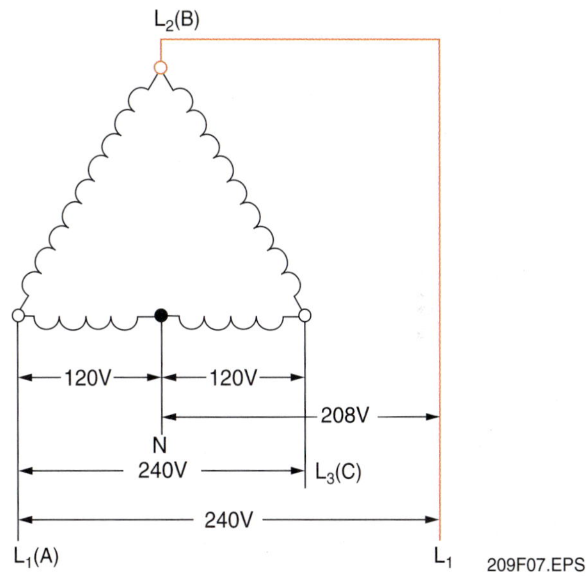

Figure 7 ◆ Three-phase, four-wire, delta-connected secondary.

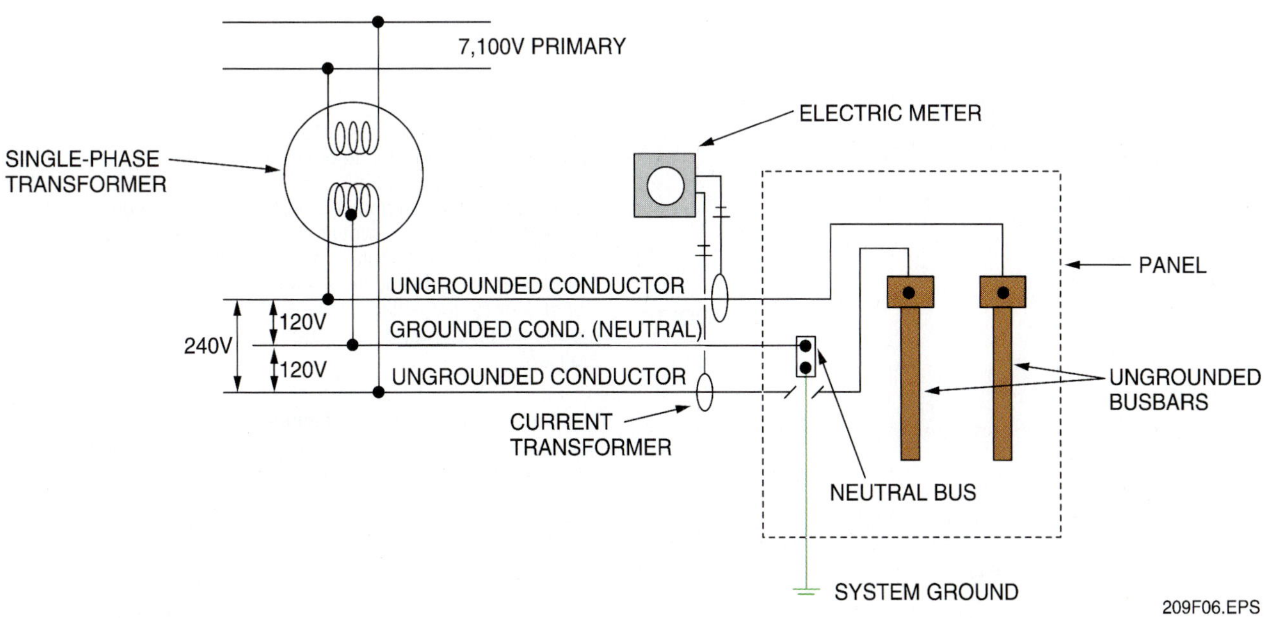

Figure 6 ◆ Single-phase, three-wire, 120/240V electric service.

Referring again to the three-phase, wye-connected service in *Figure 8*, note that the voltage between any one of the three conductors (legs L_1, L_2, and L_3) and the grounded (neutral or N) conductor is 120V. Consequently, you might assume that the voltage between any two of the phase conductors would be 240V. However, this is not the case. When dealing with any three-phase wye system, a factor—the square root of 3 ($\sqrt{3}$ or 1.73)—must be applied to determine the phase-to-phase voltage. Therefore, to find the voltage between any two phase conductors, multiply the voltage of one phase conductor to ground (120V) by the square root of 3:

$$120 \times \sqrt{3} = 207.84\text{V (approx. 208V)}$$

Therefore, feeder and branch circuits connected to 120/208V, three-phase, four-wire systems can supply the following loads:

- 120V, single-phase, two-wire
- 208V, single-phase, two-wire
- 208V, three-phase, three-wire
- 120/208V, three-phase, four-wire

The 120/208V, three-phase, four-wire system yields an electrical supply for loads rated at 120/208V, such as HVAC equipment, cooking units, washers, and dryers.

Another popular wye-connected system is the three-phase, four-wire, 277/480V system.

Feeder and branch circuits connected to 277/480V, three-phase, four-wire systems can supply the following loads:

- 277V, single-phase, two-wire
- 480V, single-phase, two-wire
- 480V, three-phase, three-wire
- 277/480V, three-phase, four-wire

The delta-connected system in *Figure 9* operates a little differently. While the wye-connected system is formed by connecting one terminal from three equal voltage transformer windings together to make a common terminal, the delta-connected system has its windings connected to

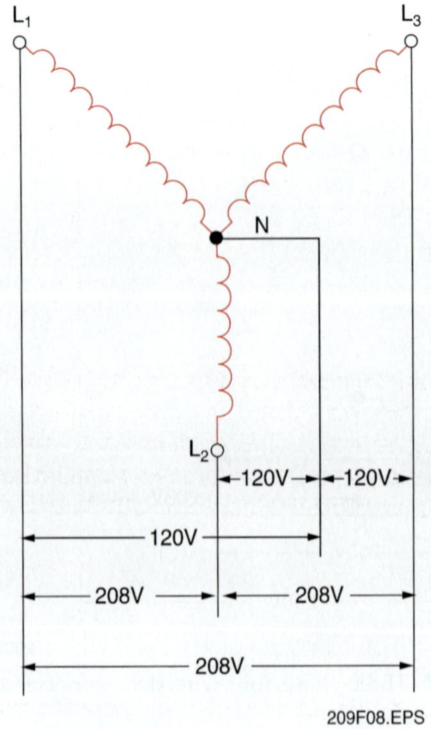

Figure 8 ◆ Four-wire, wye-connected secondary.

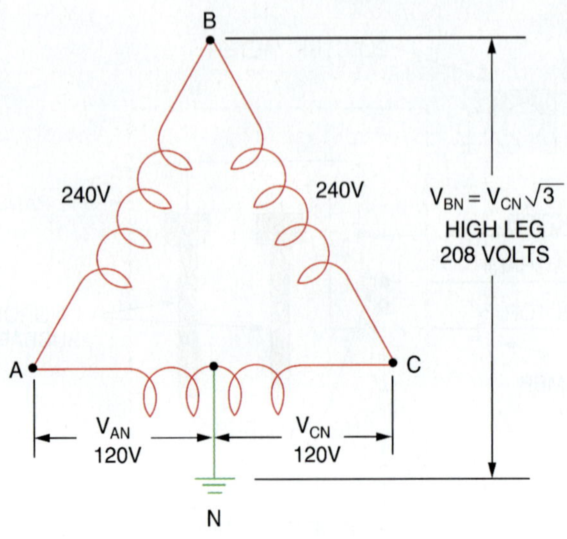

On a three-phase, four-wire 120/240V delta-connected system, the midpoint of one phase winding is grounded to provide 120V between phase A and ground; also between phase C and ground. Between phase B and ground, however, the voltage is higher and may be calculated by multiplying the voltage between C and ground (120V) by the square root of 3 or 1.73. Consequently, the voltage between phase B and ground is approximately 208V (thus, the name *high leg*).

The *NEC*® requires that conductors connected to the high leg of a four-wire delta system be color coded with orange insulation or tape.

209F09.EPS

Figure 9 ◆ Characteristics of a center-tap, delta-connected system.

Four-Wire, Delta-Connected Secondary

THINK ABOUT IT

You are connecting 120V lighting branch circuits in a panelboard equipped with a neutral bus. One of the conductors feeding the panel has an orange finish. Can a bus connected to this feeder be used in conjunction with the neutral to supply a 120V branch lighting circuit?

form a triangle or the Greek delta symbol Δ. Note in *Figure 9* that a center-tap terminal is used on one winding to ground the system. On a 120/240V system, 120V is available between the center-tap terminal and each ungrounded terminal on either side, while 240V is available across the full winding of each phase.

Refer again to *Figure 9*. Point B is known as the high leg or wild leg. This high leg has a higher voltage to ground than the other two phases. The voltage of the high leg can be determined by multiplying the voltage to ground of either of the other two legs by the square root of 3. Therefore, if the voltage between phase A and ground is 120V, the voltage between phase B and ground may be determined as follows:

$$120 \times \sqrt{3} = 207.84\text{V (approx. 208V)}$$

From this, it should be obvious that no single-pole breakers should be connected to the high leg of a center-tapped, four-wire, delta-connected system. In fact, *NEC Sections 230.56 and 408.3(E)* state that the phase busbar or conductor having the higher voltage to ground must be permanently marked by an outer finish that is orange in color. This prevents future workers from connecting 120V single-phase loads to this high leg, which would probably result in damaging any equipment connected to the circuit. Remember the color orange: No 120V loads are to be connected to this phase.

3.0.0 ◆ SERVICE COMPONENTS

Electricians who work mostly on commercial projects will be involved with the service installation from the power company's point of attachment to the building's service equipment, including all wiring and components in between, with the possible exception of the electric meter.

To understand the function of each part of an electric service, we will examine an actual installation—a commercial retail store. Assume that you are in charge of the project and this is your first day on the job. The contractor's superintendent stops by the project site and hands you a set of working drawings and written specifications. It is up to you to determine how the service is to be installed. Furthermore, you will be required to compile a materials list and order all the necessary items to complete the service installation.

This particular project consists of a rectangular building that is a part of a shopping center complex. The concrete block walls have been erected and the building is under roof. The concrete floor will not be poured until all electrical and plumbing work has been installed. However, the permanent electric service is to be installed immediately to provide temporary power for the workers. The remaining wiring in the building will be installed later.

3.1.0 Consulting the Construction Documents

The first order of business is to consult the working drawings and then read the appropriate sections in the written specifications. The floor plan of the building appears as shown in *Figure 10*. Note that the standard panel symbol is used (a solid rectangle) to indicate the location of Panel *A*, the only power panel used on this project. The panel symbol indicates that the panel is to be surface-mounted on the inside rear wall of the building in the storage area. The electric meter, as well as a time clock for controlling night and outside lighting, are also shown on this floor plan. The meter is installed on the outside rear wall, while the time clock is installed on the inside rear wall, next to Panel *A*.

Notes and callout arrows on this floor plan refer to a power-riser diagram and also a panelboard schedule on the same drawing sheet; these appear in *Figures 11* and *12*, respectively. This drawing sheet showing the floor plan, power-riser diagram, and panelboard schedule provides most of the required information so that the service can be installed to meet the project specifications.

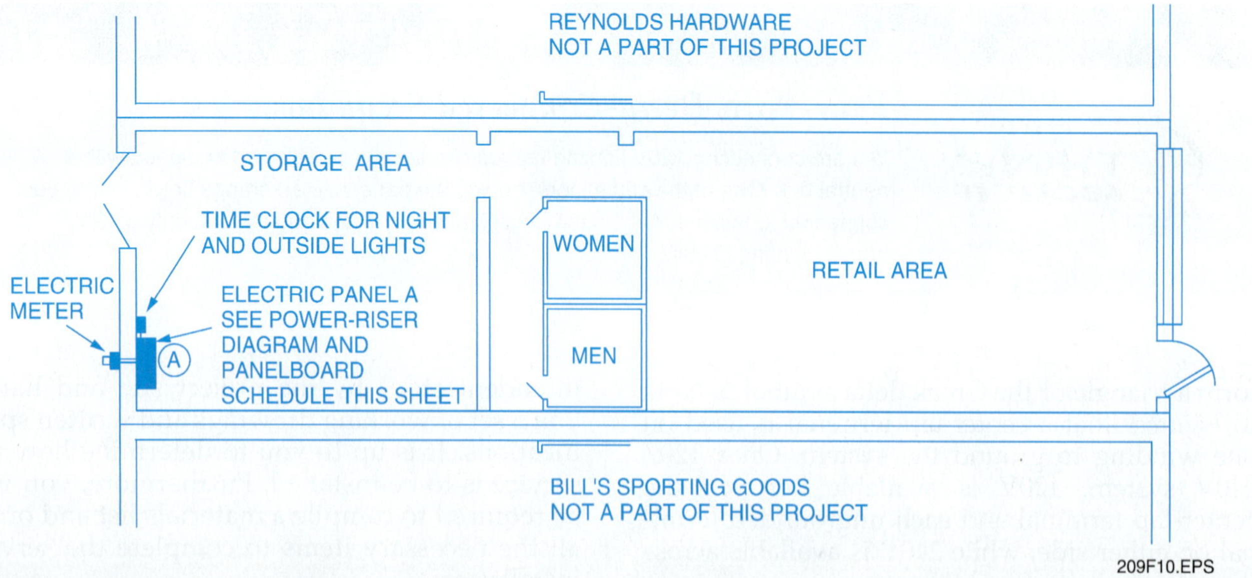

Figure 10 ◆ Floor plan of an example commercial facility.

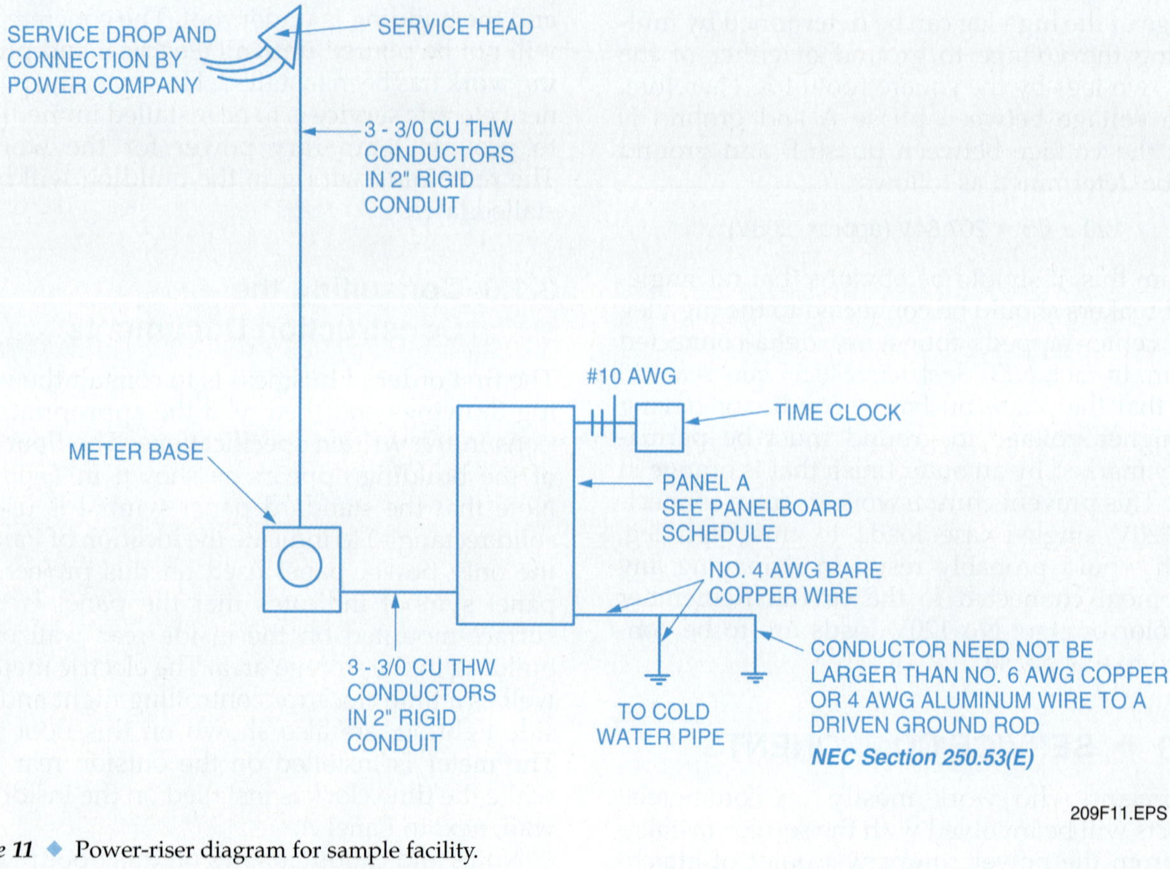

Figure 11 ◆ Power-riser diagram for sample facility.

In most cases, electrical workers are not required to design electrical systems; rather, they are required to interpret the engineer's designs. Consequently, panelboard schedules will vary with each designer. However, once you have a feel for interpreting electrical working drawings, you should have little difficulty in reading schedules.

The written specifications should also be read to make certain that no conflicts exist, and to further verify the information found on the working drawings. A sample specification appears in *Figure 13*.

PANELBOARD SCHEDULE										
PANEL NO.	CABINET TYPE	PANEL MAINS								ITEMS FED OR REMARKS
		AMPS	VOLTS	PHASE	1P	2P	3P	PROT.	FRAME	
A	SURFACE	200A	120/240	1φ, 3 - W	12	-	-	20A	70A	LTS., RECEPTS, W.C.
					-	1	-	60A	100A	CONDENSING UNIT
					-	1	-	30A	70A	WATER HEATER
					-	1	-	20A	70A	AIR-HANDLING UNIT
					-	2	-	20A	70A	TOILET HEATERS
					8	-	-	-	70A	PROVISIONS ONLY

209F12.EPS

Figure 12 ◆ Panelboard schedule for sample facility.

From the information obtained from the drawings and specifications, we know that the service for this project is single-phase, three-wire, 120/240V, 200A. The main panel (Panel A) is a surface-mounted Square D Type NQO (or equivalent) with a 200A main circuit breaker. Furthermore, we can determine the number of spaces required in the panel by totaling the number of circuit breakers listed in the panelboard schedule as follows:

12 single-pole, 20A breakers = 12 spaces
1 two-pole, 60A breaker = 2 spaces
1 two-pole, 30A breaker = 2 spaces
3 two-pole, 20A breakers = 6 spaces
8 provisions only = 8 spaces
Total = 30 spaces

Therefore, a surface-mounted panel (Square D and type NQO with 200A main circuit breaker and provisions for 30 spaces) can be ordered. The required circuit breakers should also be ordered and installed at the same time. This will meet with the project specifications.

3.2.0 Service Head

Referring again to *Figure 11,* start at the top of the service riser. The first item shown is the service head, sometimes called the weatherhead. *NEC Section 230.54(A)* requires the **service raceway** in our example to be equipped with a raintight service head at the point of connection to the service drop conductors.

A service head (*Figure 14*) is a fitting that prevents water from entering the service raceway. This is accomplished by bending the service conductors contained in the raceway downward as they exit from the service head so that any water or moisture will drip from the outside conductors before entering the service head. These conductors are also protected by a plastic or fiber strain insulator or bushing placed at the entrance of the service head to separate the service conductors as required by *NEC Section 230.54(E).* Two types of service heads are in common use: one type has internal threads that enable the service head to be screwed directly onto the conduit; the other type utilizes a clamp with retaining screws. In the latter type, the service head is placed on top of the service raceway and the clamp is tightened with the retaining screws.

Further protection from water and moisture is provided by drip loops, as specified in *NEC Section 230.54(F).* Service heads are required to be located above the service drop attachment. Drip loops are then formed where the service drop conductors are connected to the service conductors, and these drip loops must be located below the service head. *Figure 14* shows how drip loops prevent water from entering the service raceway; that is, water will not flow uphill into the service head, so the water drips from the conductors at the lowest point of the drip loop.

The service-entrance conductors must have a minimum length of 3½' after they leave the service head. This is to ensure a good drip loop and to provide adequate length for splicing onto the service drop.

NOTE

Some states allow a shorter minimum length after leaving the service head. Always check and follow local codes.

PANELBOARDS-CIRCUIT BREAKER

A. GENERAL:

Furnish and install circuit breaker panelboards as indicated in the panelboard schedule and where shown on the drawings. The panelboard shall be a dead-front safety-type equipped with molded case circuit breakers and shall be the type as listed in the panelboard schedule. Service-entrance panelboards shall include a full capacity box bonding strap and be approved for service entrance. The acceptable manufacturers of the panelboards are ITE, General Electric, Cutler-Hammer, and Square D, provided that they are fully equal to the type listed on the drawings. The panelboard shall be listed by Underwriters' Laboratories and bear the UL Label.

B. CIRCUIT BREAKERS:

Provide Type NQO circuit breakers of frame, trip rating, and interrupting capacity as shown on the schedule. Also, provide the number of spaces for future circuit breakers as shown in the schedule. The circuit breakers shall be quick-make, quick-break, thermal-magnetic, trip indicating, and have a common trip on all multipole breakers with internal tie mechanisms.

C. WIRING TERMINALS:

Terminals for feeder conductors to the panelboard mains and neutral shall be suitable for the type of conductor specified. Terminals for branch circuit wiring, both breaker and neutral, shall be suitable for the type of conductor specified.

D. CABINETS AND FRONTS:

The panelboard bus assembly shall be enclosed in a steel cabinet. The size of the wiring gutters and gauge of steel shall be in accordance with NEMA Standards. The box shall be fabricated from galvanized steel or equivalent rust-resistant steel. Fronts shall include door and have flush, brushed stainless steel, spring-loaded door pulls. The flush lock shall not protrude beyond the front of the door. All panelboard locks shall be keyed alike. Fronts shall not be removable with the door in the locked position.

E. DIRECTORY:

On the inside of the door of each cabinet, provide a typewritten directory which will indicate the location of the equipment or outlets supplied by each circuit. The directory shall be mounted in a metal frame with a nonbreakable transparent cover. The panelboard designation shall be typed on the directory card and panel designation stenciled in 1-1/2" high letters on the inside of the door.

F. PANELBOARD INSTALLATION

(1) Before installing panelboards check all of the architectural drawings for possible conflict of space and adjust the location of the panelboard to prevent such conflict with other items.

(2) When the panelboard is recessed into a wall serving an area with accessible ceiling space, provide and install an empty conduit system for future wiring. All 1-1/4" conduit shall be stubbed into the ceiling space above the panelboard and under the panelboard if such accessible ceiling space exists.

(3) The panelboards shall be mounted in accordance with ***NEC Article 312.*** The Electrical Contractor shall furnish all material for mounting the panelboards.

209F13.EPS

Figure 13 ◆ Sample panelboard specifications.

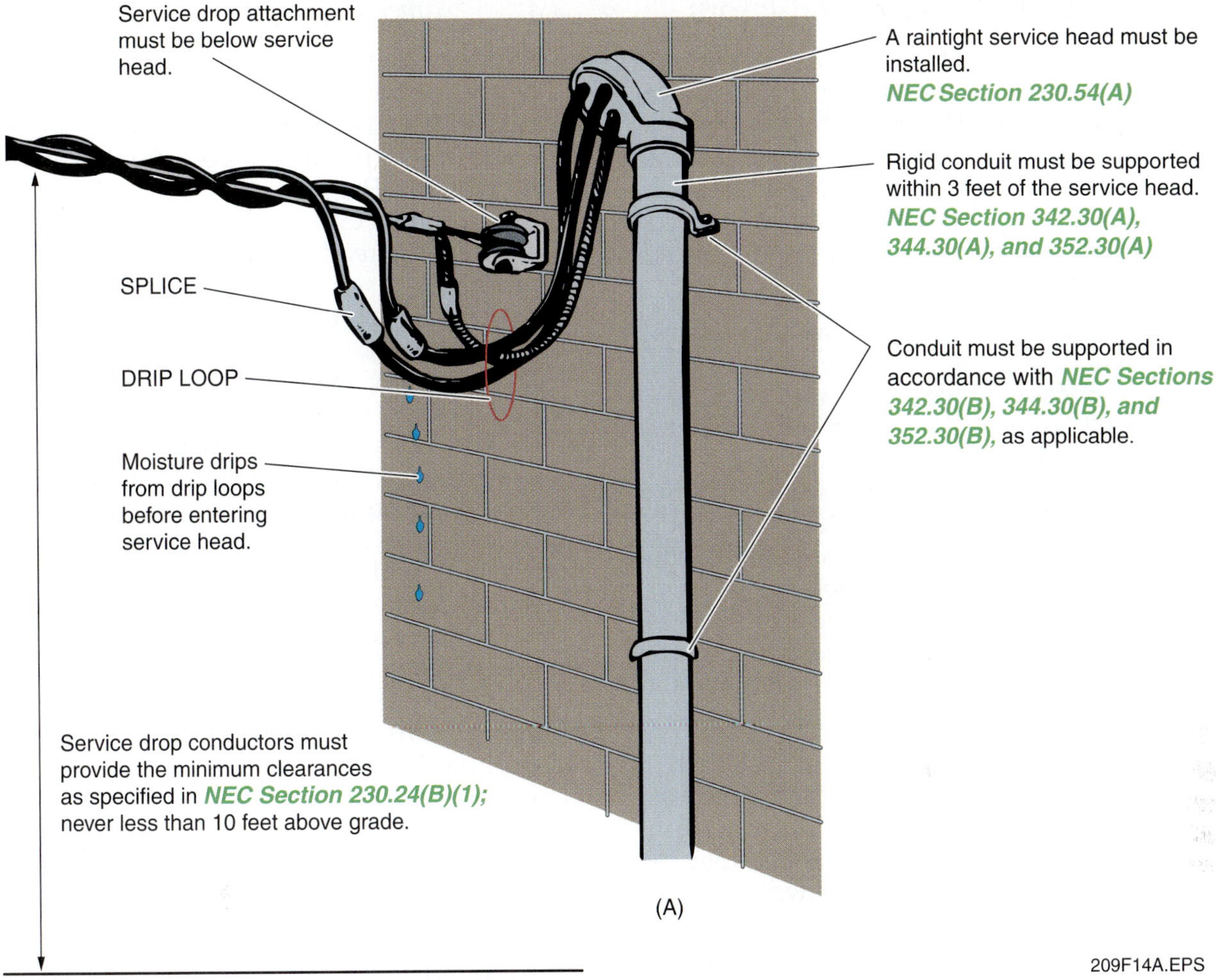

(A)

209F14A.EPS

209F14B.EPS

(B)

Figure 14 ◆ Service head and related components.

3.3.0 Service-Entrance Conductors

The 3/0 AWG, Type THW conductors shown in the power-riser diagram in *Figure 11* are service-entrance conductors. These conductors are run from the main disconnect breaker through the meter to the service head and terminate with splices onto the service drop. The conductors must not be spliced at any place between these points except for the following:

- Clamped or bolted connections in metering equipment
- Where service-entrance conductors are tapped to supply two to six disconnecting means that are grouped at a common location
- Where service conductors are extended from a service drop to an outside meter location and returned to connect to the service-entrance conductors of an existing installation
- Where the service-entrance conductors consist of a busway, connections are permitted as required to assemble the various sections and fittings

Service conductors are normally installed in two different ways: in a raceway system or in a cable assembly. In our sample, the conductors are installed in 2" rigid conduit and extend from the service head down to the threaded weatherproof hub on top of the meter base. The *NEC®* permits the conductors to be spliced at this point, that is, connected to the bolted terminals on the meter base. However, no splices are permitted from the service head to the meter base.

3.4.0 Service Equipment

Equipment and components falling under the heading of service equipment include the main disconnect switch or breaker, circuit breakers, fuses, and other items necessary to control and disconnect the power supply.

3.4.1 Metering Equipment

A watt-hour meter is used by the power company to determine the amount of electrical energy consumed by the customer.

Energy is the product of power (kilowatts) and time (hours). The type of meter connected to most residential and small commercial occupancies provides a reading in kilowatt-hours (kWh). For example, if the meter reads a usage of 500W for a period of six hours, it would register 3kWh ($0.5 \times 6 = 3$).

There are several different types of metering devices in use. The type used in our sample building is known as the feed-through type. This type of meter is used mostly for services up to 200A, although feed-through meters up to 400A are not uncommon in many locations. Services rated above 400A will almost always use separate current transformers enclosed in a current transformer cabinet (CT cabinet). Current transformers are discussed in greater detail later in this module.

A typical watt-hour meter consists of a combination of coils, conductors, and gears—all encased in a housing, as shown in *Figure 15*. The coils are constructed on the same principle as a split-phase induction motor, in that the stationary current coil and the voltage coil are placed so that they produce a rotating magnetic field. A disc near the center of the meter is exposed to the rotating magnetic field. The torque applied to the disc is proportional to the power in the circuit, and the braking action of the eddy currents in the disc makes the speed of the rotation proportional to the rate at which the power is consumed. The disc, through a train of gears, moves the pointers on the register dials to record the amount of power used directly in kilowatt-hours.

Most watt-hour meters use five dials (see *Figure 15*). The dial farthest to the right on the meter counts the kilowatt hours singly. The second dial

INSIDE TRACK

Service Mast Installations

NEC Sections 342.30(A) and 344.30(A) do not require above-the-roof terminations of metal conduit service masts to be secured within 3' of the service head.

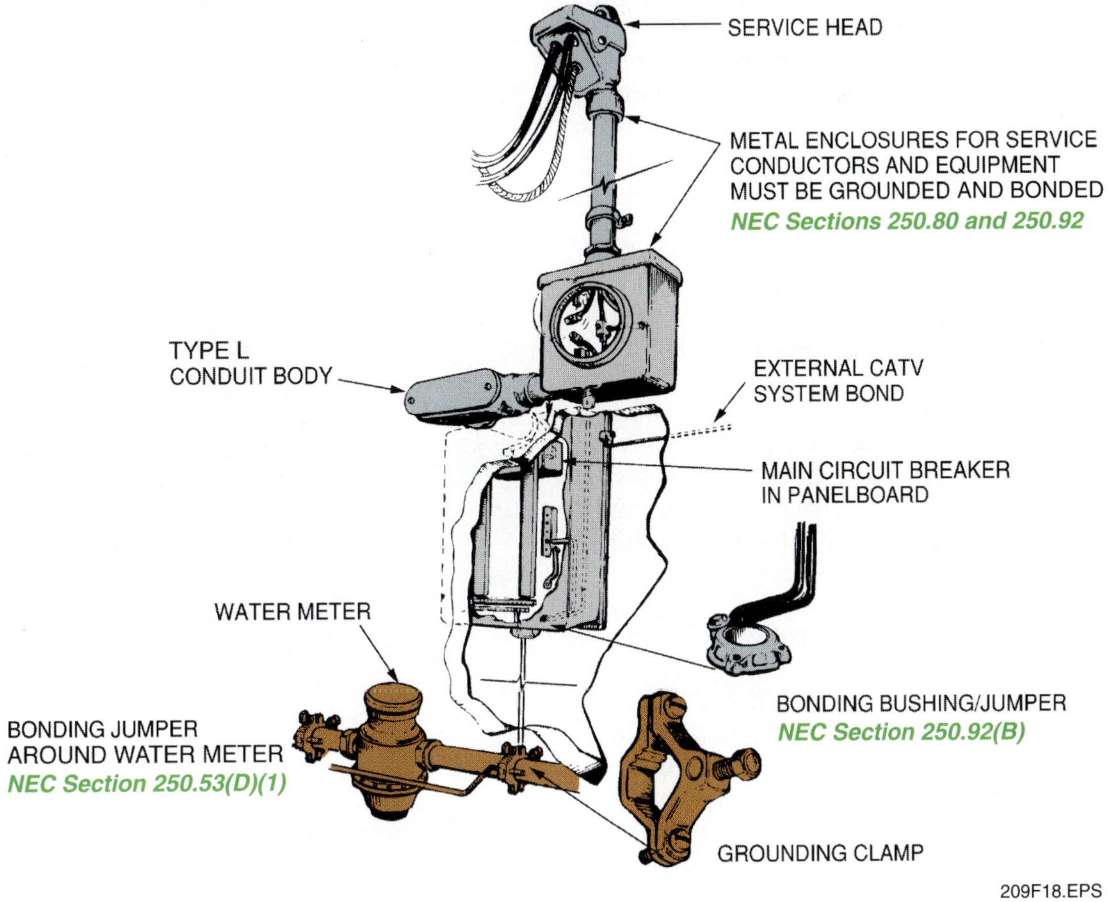

Figure 18 ◆ Summary of *NEC*® service grounding regulations.

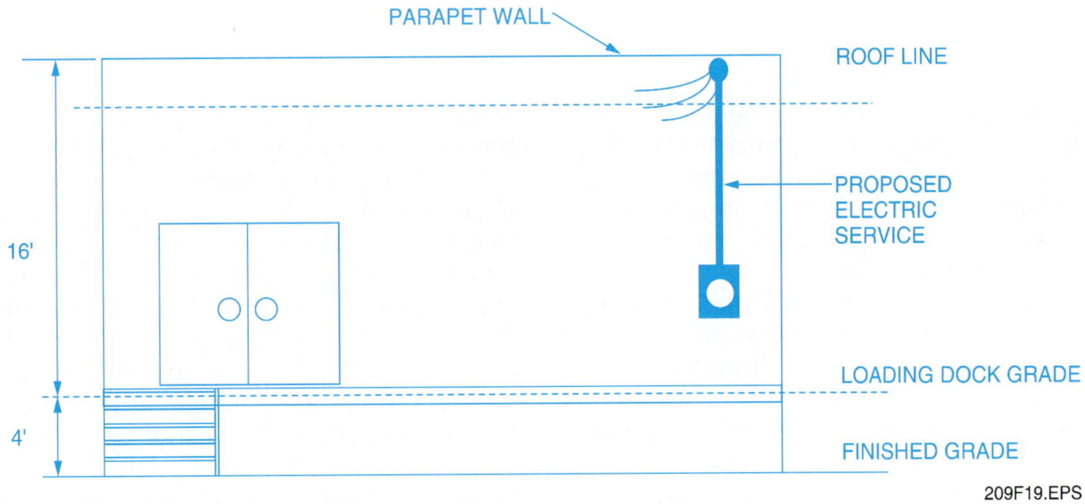

Figure 19 ◆ Rear elevation of sample building.

Consequently, one 10' length of 2" conduit will put the service raceway and service head at the very top of the wall, leaving plenty of room for the power company to connect the service drop below the service head to comply with *NEC Section 230.54(C)*. Furthermore, the service drop will be more than 10' above the deck—complying with *NEC Sections 230.24 and 230.26.*

Once the service configuration has been laid out, it is best to verify the layout with the local power company. This is also a good time to request the meter base that will have to be installed by the electrician on the job.

This is also a good time to start thinking about the tool requirements. Since the service raceway extends 16' above the loading dock, workers will need a 24' extension ladder. A rotary hammer will be needed to drill the holes for the lead anchors and to penetrate the concrete block to insert the 2" nipple between the meter base and the panelboard on the inside wall. Cable cutters will speed up the work when cutting the 3/0 service conductors and a bucket of wire-pulling lubricant will facilitate the installation of the 3/0 conductors into the 2" conduit. For this short length of pull, no special cable-pulling apparatus should be required. The conductors are merely cut to length, and then two workers feed these conductors from the top of the service raceway down to the meter base for connection to the meter base terminals. The same is true of the short run between the meter base and the panelboard on the opposite wall.

Much time can be saved on any project if careful planning is exercised before beginning any work. Even if the material is already on the job site, as taken from the electrical estimator's list, the electricians on the job must know how to utilize this material to its best advantage. Traditionally, more time has been wasted on electrical jobs because of material or tool shortages than any other causes. Therefore, before a project is started, make sure that all the necessary materials and tools are at hand. If not, order them immediately. Any time saved on any project, in most cases, means a savings to someone.

Before continuing with this module, complete the takeoff list in *Table 2*. The purpose of this exercise is to prepare a complete list of materials for the service described for our sample commercial building. Since grounding details have not been covered, the items required to ground the system have been listed. In general, two 1" grounding clamps and a bonding jumper are required around the water meter.

Another 1" grounding clamp is required to connect the grounding electrode conductor to the 1" water pipe, and a ⅝" grounding clamp is required to connect the grounding electrode conductor to a driven ground rod. This requires 20' of No. 4 AWG bare copper wire. Also refer to *Figure 18* to better visualize exactly what is required for the grounding of this service.

Table 2 Material Takeoff Exercise

Quantity	Item
1	2" service head
—	2" rigid conduit
—	2" rigid conduit strap(s)
—	200A meter base (obtain from power company)
—	Raintight hub(s) for meter base
—	2" Type L conduit body
—	2" conduit nipple(s) (10" long)
—	4" chase nipple(s) (from meter hub to Type L conduit body)
—	2" locknut(s)
—	Grounding bushing(s)
—	Square D NQO panelboard(s)
—	One-pole, 20A circuit breaker(s)
—	Two-pole, 60A circuit breaker(s)
—	Two-pole, 30A circuit breaker(s)
—	Two-pole, 20A circuit breaker(s)
—	¼-20 lead anchor(s) and ¼-20 machine screw(s) 1½" long
—	No. 3/0 THW copper wire
3	Grounding clamp(s) for 1" pipe
1	Grounding clamp(s) for ⅝" ground rod
20'	No. 4 AWG bare copper wire
1	⅝" copperweld grounding rod(s), 8' in length

5.0.0 ◆ THREE-PHASE SERVICES

Most services encountered on commercial and industrial projects will consist of three-phase systems with voltage ratings as low as 120/208V to as high as perhaps 4,160V or more.

In single-phase current, only one voltage curve is generated, while three-phase current indicates that there are three voltage curves present on the system simultaneously. There is very little difference between installing a single-phase service (just described) and a three-phase service. The main difference is that there is an extra service conductor for a three-phase, four-wire system, requiring that the service head have four openings instead of three. Of course, the panelboard will also have to be arranged for a three-phase system—requiring an extra ungrounded busbar. Other than these changes, the installation process is essentially the same as that described for the single-phase system.

To illustrate, *Figure 20* shows the power-riser diagram for a 1,200A, three-phase, wye-connected

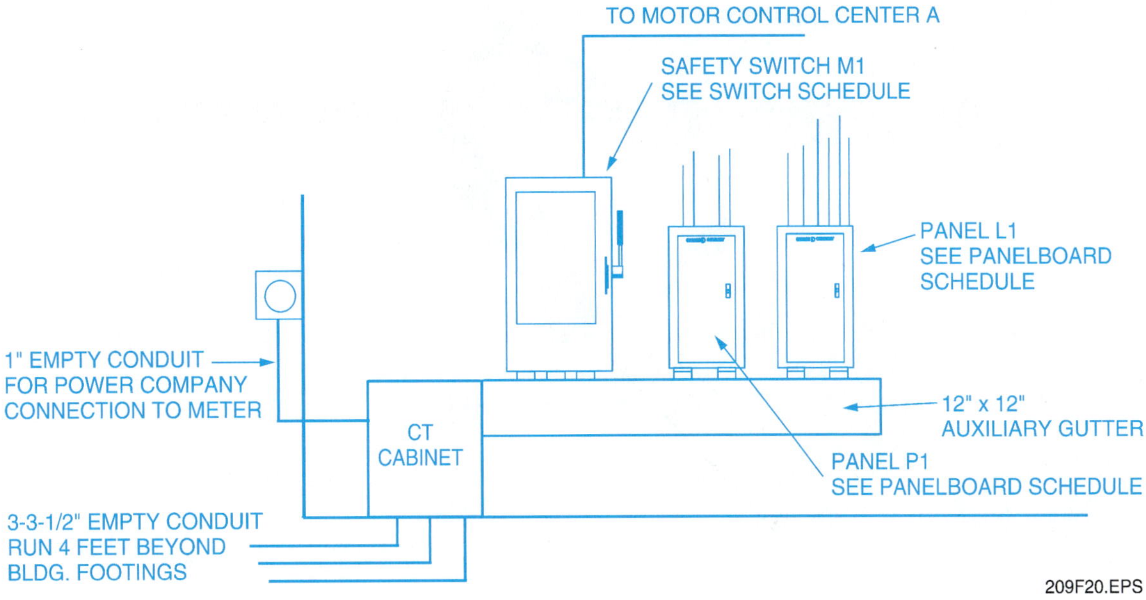

Figure 20 ◆ Power-riser diagram for a three-phase, four-wire, 1,200A service.

service. Note that the service lateral consists of three 3½" empty conduit runs to be installed from the CT (current transformer) cabinet to 4' beyond the building's footings. These are provided for the power company's service lateral.

The service conductors continue through the CT cabinet and terminate in an auxiliary gutter. Three service equipment taps are made in this gutter to feed the following:

- One 800A fusible safety switch, which feeds a motor control center
- One 200A power panel with a main circuit breaker
- One 200A lighting panel with a main circuit breaker

We will take a look at the individual components for this service.

5.1.0 Current Transformers

Meters used by power companies to record the amount of current used by customers usually respond to a current that varies from 0A to 5A. To respond to the actual current of the service, each meter is provided with current transformers. If the peak demand of the service is 100A, a 100:5 current transformer is used. If the peak current demand is expected to be 200A, a 200:5 current transformer is used.

Services above 400A usually utilize a group of current transformers, one for each ungrounded conductor or set of conductors. There are two basic types of current transformers: the busbar type and the doughnut type. The doughnut-type current transformer encircles the ungrounded conductors in the system to read the current flow, much the same as a clamp-on ammeter. See *Figure 21*. The busbar-type current transformer has each transformer connected in series with a busbar and does not encircle the conductor.

Current transformers are normally enclosed in a CT cabinet. *Figure 22* shows a typical CT cabinet with current transformers and their related wiring. In some cases, the current transformers may be mounted exposed on overhead conductors, but this is more the exception than the rule.

Power companies have different requirements for sizes of CT cabinets, but the dimensions shown in *Table 3* are typical for several service sizes.

Power companies also have different specifications for the location and wiring of CTs and CT cabinets, depending on the locale. The following are the requirements of one power company. However, always check the local requirements.

- The meter base and meter may be located on either the side or top of the current transformer cabinet, or they may be located at a distance away, if approved by the power company and as long as the conduit containing the instrument wiring is run exposed.
- In no instance shall more than one set of conductors terminate in the instrument transformer cabinet. Subfeeders and branch circuits are to terminate at the customer's distribution panel. The instrument transformer cabinet shall not be used as a junction box.

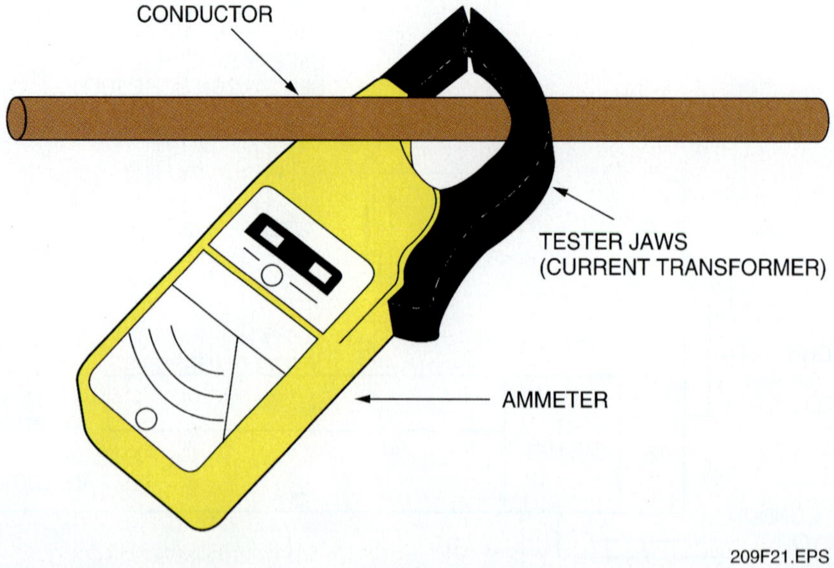

Figure 21 ◆ Clamp-on ammeters operate on the same principle as doughnut-type current transformers.

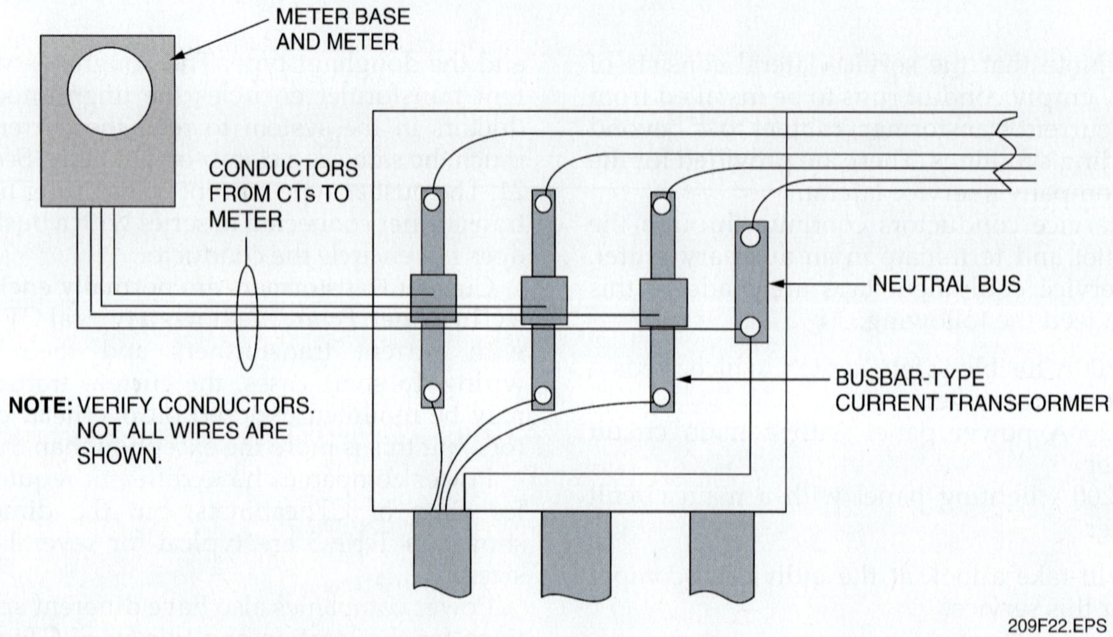

Figure 22 ◆ Typical CT cabinet arrangement.

Table 3 Typical CT Cabinet Sizes

Phase	Service Characteristics	Cabinet Size
Single-phase	120/240V, three-wire	10" × 24" × 32"
Three-phase	120/240V, four-wire	10" × 36" × 42"
Three-phase	120/208V, four-wire	10" × 36" × 42"
Three-phase	480V, three-wire	10" × 36" × 42"
Three-phase	277V/480, four-wire	10" × 36" × 48"

- When service-entrance conductors enter or leave through the back of the cabinet, the size of the CT cabinet must be increased to provide additional working space.
- For services at higher voltages, additional space must be provided in the transformer cabinet for mounting potential transformers. Consult the local power company for dimensions.
- If recording demand instruments are required, increase the height of the meter mounting from 36" to 48".

5.2.0 Gutters

Auxiliary gutters are used to route the service conductors and to provide an enclosure for tapping these conductors for safety switches and panels. In this type of arrangement, appropriate connectors are normally used to make the taps. However, many electrical contractors have found that a bussed gutter saves labor and provides for a neater installation.

A bussed gutter is an assembly of busbars in an enclosure. The enclosure may be rated for outdoor (weatherproof) or indoor installations. Busbars installed in the gutter may be made of aluminum or copper and must have an ampacity rating for the application; that is, if the service conductors are rated for 1,200A (as in our sample building), the busbars must be rated for at least the same ampacity. Furthermore, they must be UL listed.

From an installation or a maintenance/modification viewpoint, bussed gutters are one of the most common types of wiring methods for use with multiswitch services. One advantage of a bussed gutter is the ease of installation and modification. Adding disconnect switches or changing switches is relatively easy. No connectors have to be untaped and reconnected, as with systems using wire connectors on the conductors for taps.

In our example, with a service conductor ampacity of 1,200A, the rating of the busbars in bussed gutters must also be 1,200A if bussed gutters are to be used. See *Figure 23*. Furthermore, the bussed gutter must be rated for the available fault current and must have sufficient wire bending space per *NEC Sections 110.3(A)(3) and 312.(A) through (C).*

5.2.1 Bus Bracing

One characteristic of fault currents is an induced torque in conductors carrying the fault. Because of this torque, the busbars in a bussed gutter must be attached to the enclosure in such a manner as to prevent their being dislodged and/or making contact with the gutter frame during the fault. When busbars are attached in such a manner as to withstand the torque created by the available amount of fault current, they are said to be *braced* for that amount of current. For instance, busbars may be braced for 20,000A, 30,000A, or whatever level of fault current is required, up to 200,000A. The bussed gutter must be labeled by the manufacturer for the amount of fault current the buses are braced to withstand.

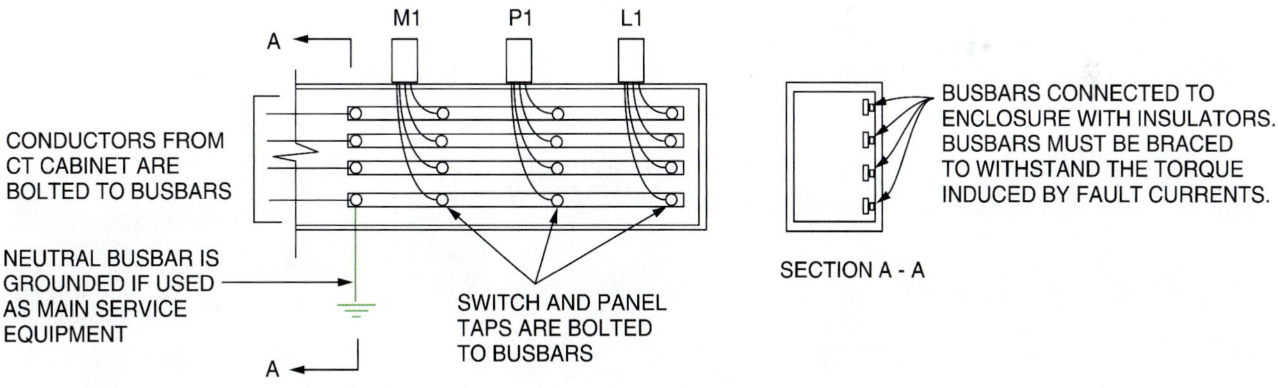

Figure 23 ◆ Three-phase bussed gutter.

Three-Phase Busduct

Three-phase busduct is a modular type of power distribution system that provides power directly through a set of system busbars rather than using power cable. It allows for easy system expansion because whenever a new piece of equipment is added, a new switch can simply be plugged into place and the equipment supplied without the need to run additional power cable through the plant.

209P0906.EPS

209P0907.EPS

209P0908.EPS

6.0.0 ◆ 277/480V SERVICES

Wye-connected, 277/480V services are common in medium to large commercial projects. This type of service is also frequently used in industrial applications to power motors for driving machinery and other apparatus. Such installations frequently use switchgear enclosures such as the one shown in *Figure 24*. Service equipment of this type is made up of vertical sections that connect to a common bus system within the enclosure. These sections contain fusible switches or circuit breakers, metering equipment, or other devices related to the electric service.

NEC Section 230.71(A) allows a maximum of six service disconnecting means per service grouped in any one location. If there are more than six switches or circuit breakers in the switchboard, then a main switch or circuit breaker must be provided to disconnect all service conductors in the building or structure from the power supply.

209F24.EPS

Figure 24 ◆ Typical switchgear.

Typically, the metering equipment (CTs, potential transformers, etc.) is installed in the same enclosure as the main disconnecting means; additional space is usually provided in the switchgear for this equipment. Furthermore, taps are normally provided in the main bus in a barriered section for connection to emergency switches, such as for fire alarm systems, emergency lighting, etc., as allowed by *NEC Section 230.82(5).*

6.1.0 Practical Applications

Figure 25 shows a plot plan of a shopping center facility that utilizes a 277/480V, wye-connected, three-phase service to supply numerous tenant areas. In general, a padmount transformer installed on the property perimeter reduces the distribution voltages to 277/480V. An underground service is installed from this padmount transformer to a switchgear room in one section of the shopping center. A single-line diagram of the electrical system for this project is shown in *Figure 26*. Note that there are only six fusible switches in the main switchgear, so no main disconnecting means is necessary. Each of these six feeders supplies a bussed gutter system (discussed previously), each of which contains six meter bases and a fused safety switch that feeds a 277/480V panel for lighting and HVAC equipment. This latter panel also feeds a 480V to 120/208V dry transformer, which in turn furnishes power to a 120/208V panel for feeding tenant receptacles and display lighting.

In general, this system records the amount of power used by each tenant so that tenants may be billed accordingly. A 277V fluorescent lighting system provides general illumination, which is fed from the 277/480V panel in each tenant space. However, some 120V display lighting is employed. It is fed from the 120/208V panel.

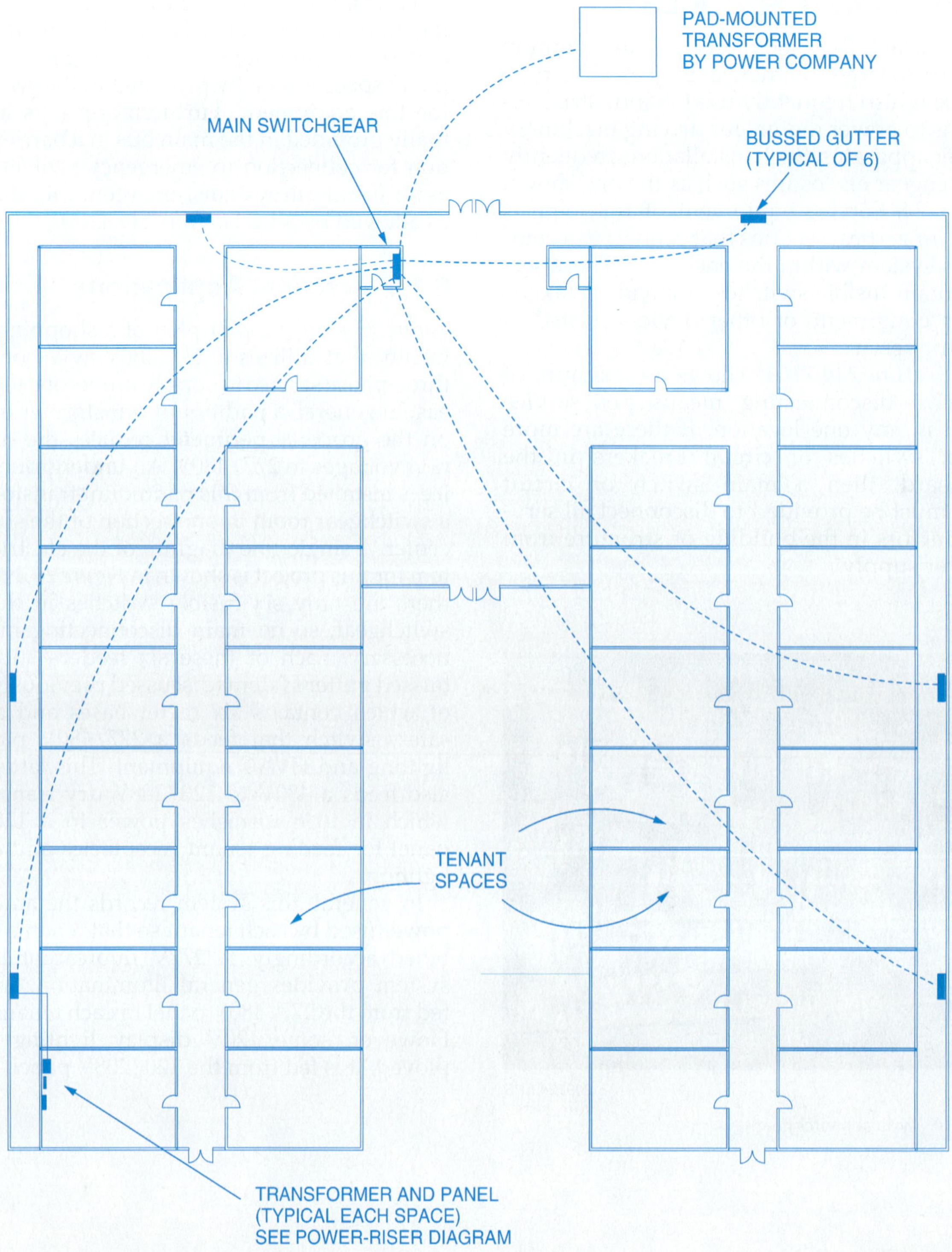

Figure 25 ◆ Plot plan of shopping center installation.

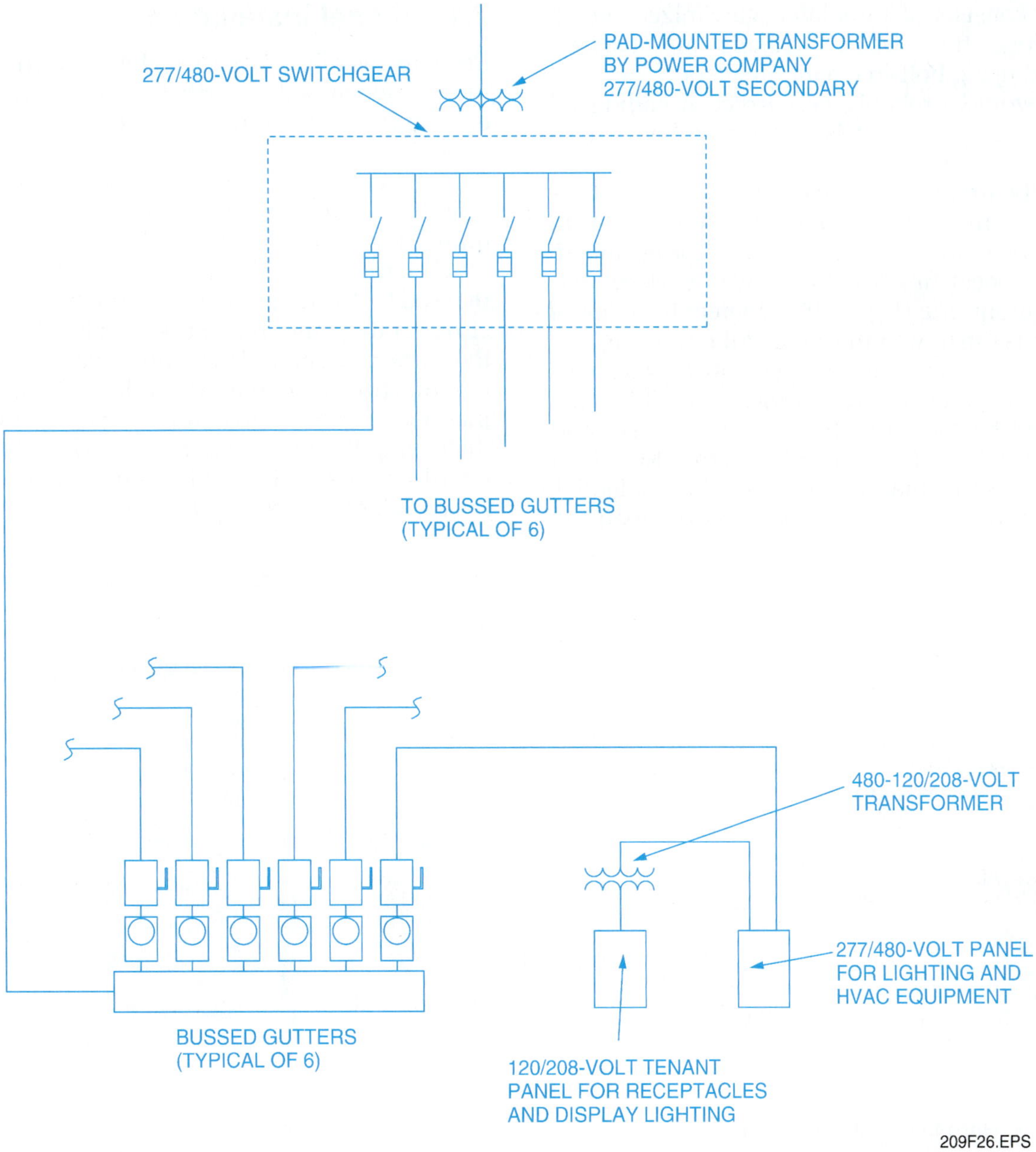

Figure 26 ◆ Power-riser diagram for shopping center installation.

7.0.0 ◆ SWITCHES, PANELBOARDS, AND LOAD CENTERS

Panelboards consist of assemblies of overcurrent protection devices in a metal cabinet, with or without disconnecting devices. The cabinet includes a cover or trim with one or two doors to allow access to the overcurrent and disconnecting devices and, in some types, access to the wiring space in the panelboard.

There is some confusion concerning the differences between load centers and panelboards. Typically, **load centers** are fuse or circuit breaker cabinets used on residential or small commercial projects. They are preassembled units with the interior buses installed at the factory. Upon installation, the required number of plug-in circuit breakers or fuse holders are installed, the circuit conductors terminated, and the front cover installed.

Many electricians classify panelboards as enclosures for overcurrent protection devices that are used on larger commercial and industrial installations. Furthermore, the can or housing usually consists of unpainted galvanized metal. Frequently, the circuit breakers are factory-installed using bolt-in circuit breakers.

You would probably be correct in calling all load centers panelboards, but remember, not all panelboards are load centers.

Panelboards fall into two mounting classifications: flush mounting (*Figure 27*), wherein the trim extends beyond the outside edges of the cabinet to provide a neat finish with the wall surface; and surface mounting (*Figure 28*), wherein the edge of the trim is flush with the edge of the cabinet.

Panelboards fall into two general classifications with regard to overcurrent devices: circuit breaker types and fusible types. Small circuit breaker and fusible panelboards, commonly referred to as load centers, are manufactured for use in residential, small commercial, and small industrial occupancies.

7.1.0 Panel Installation

Prior to installing a panel, the selected location must be examined to verify that proper clearances exist and that the environment is proper for the panel installation.

In general, all panelboards must have a rating of not less than the minimum feeder capacity required for the load computed in accordance with *NEC Article 220.* Panelboards must be durably marked by the manufacturer with the voltage and current rating and the number of phases for which they are designed. They must also include the manufacturer's name or trademark in such a manner as to be visible after installation without disturbing the interior parts or wiring. All panelboard circuits and circuit modifications shall be legibly identified as to purpose or use on a circuit

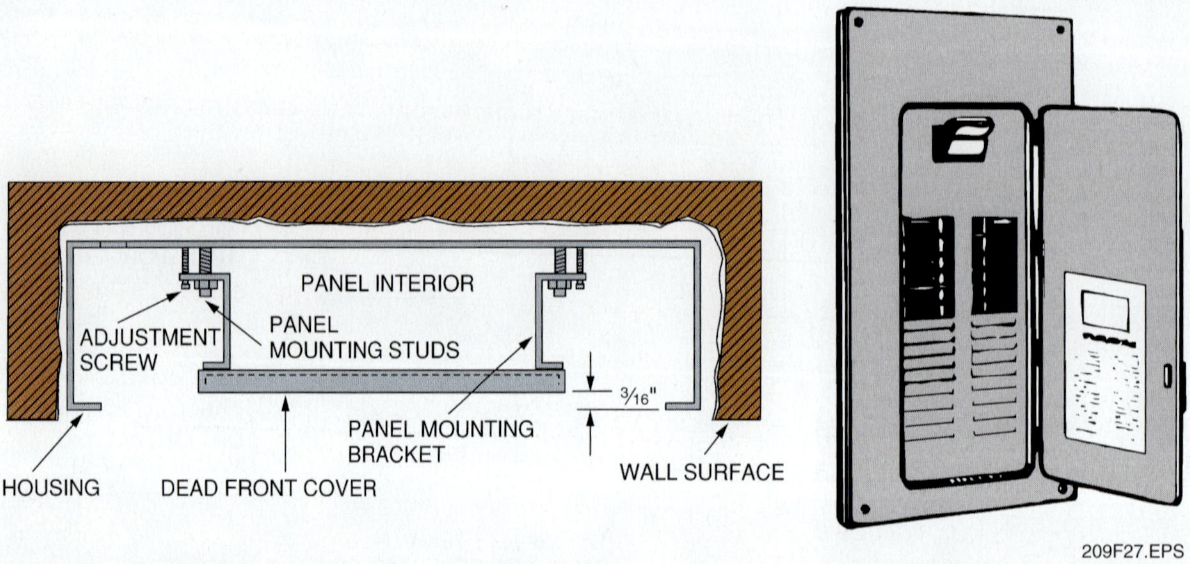

Figure 27 ◆ Flush-mounted panelboard.

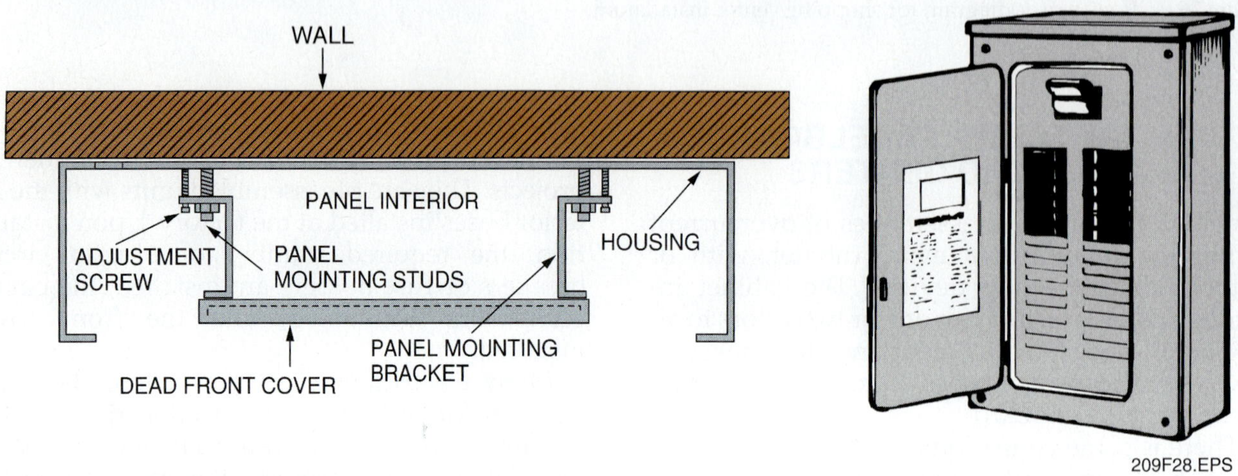

Figure 28 ◆ Surface-mounted panelboard.

directory located on the face or on the inside of the panel doors (*NEC Section 408.4*). The working height near panelboards is 6.5'. See *Figure 29*.

Once a proper location has been determined, the panel is removed from its packing boxes, assembled, and installed. When removing the panel from its packing, verify that all necessary components have been delivered and make sure that any stray packing material has been removed from the panel. Check to make sure that the right panel is to be installed. A checklist might include the following items:

- Is the panel to be top fed or bottom fed? This information should be obtained from the drawings or from the project supervisor.
- Check to verify that the voltage rating of the panel is as specified in the drawings.
- Check to verify that the ampacity of the panel is as specified in the drawings.
- Check to verify that the phase and number of conductors is as specified in the drawings.
- Verify that the panel was not damaged during shipping.
- Verify that the lugs supplied with the panel will fit the conductors being installed.

7.1.1 Installing Flush-Mounted Enclosures

Flush-mounted enclosures installed in noncombustible material must be mounted so that the front edge of the enclosure is not set back farther than ¼" from the finished surface. If installed in other than noncombustible walls, the panel edge must be flush with the finished wall (*NEC Section 312.3*).

7.1.2 Installing Surface-Mounted Enclosures

Surface-mounted enclosures must be securely fastened in place. If the wall structure offers little structural support, as in the case of ¼" wood paneling or ½" gypsum board, the enclosure must be located so that it may be attached to framing members inside the wall covering. In some cases, a framing structure will have to be built to support the panel.

7.1.3 Installing the Panel Interior

Prior to installing the panel interior, check to verify that the enclosure is securely fastened in place and is free of all foreign material. Obtain and study the specifications and instructions that are included with the panel. If no instructions are available, the following is a general installation procedure that may be used:

Step 1 Mount the interior to the enclosure using the four mounting studs installed on the enclosure back.

Step 2 Adjust the depth of the interior with the adjustment screws. The dead front cover should be no farther than $\frac{3}{16}$" from the wall surface for a flush panel, or the same distance from the enclosure face for a surface-mounted panel.

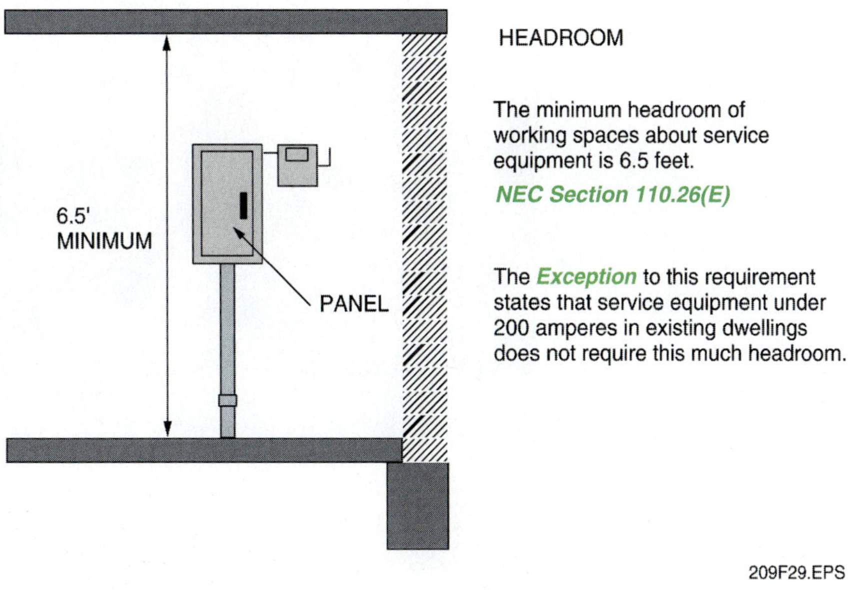

Figure 29 ◆ *NEC*® headroom requirements.

INSIDE TRACK

Panelboards

The selection of the components used in a panelboard is based on the following criteria:

- Feeder voltage and amperage
- Phase (single or three-phase) and number of conductors
- Type of busbar material (copper or aluminum)
- Number of required spaces or blanks
- Whether the panelboard is configured as a main breaker type of panel or main-lugs only (MLO) type of panel

NOTE

MLO panels do not have a MAIN circuit breaker.

After the panelboard is assembled, the appropriate number and type of circuit breakers must be ordered separately and installed in the panel.

209P0909.EPS

7.1.4 Knockouts

A series of concentric or eccentric circular partial openings are usually cut in the top, bottom, and sides of both load center and panelboard housings; some may also be cut in the back. These openings are cut in such a manner that they may be removed by tapping (knocking) them out, usually with a screwdriver blade or punch and hammer, just as you would in a device or fixture box (*Figure 30*).

The direction from which the knockouts can be removed alternates from inside the enclosure to outside the enclosure; that is, ½" knockouts are knocked outward from inside the enclosure; ¾" knockouts are knocked inward from outside the enclosure; 1" knockouts are knocked outward from inside the enclosure, etc. See *Figure 31*.

In most cases, raceways connected to panelboards using the concentric knockout openings have poor equipment grounding connections. Consequently, *NEC Section 250.92(B)* requires bonding jumpers to be used around concentric or eccentric knockouts that are punched or otherwise formed so as to impair the electrical connection to ground. This is accomplished by using a grounding locknut or bushing and then connecting a bonding jumper either to another grounding locknut or bushing, or else to the equipment grounding terminal inside the panelboard.

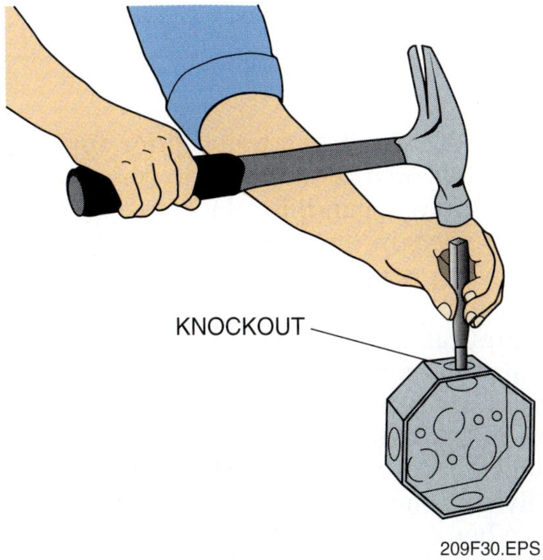

Figure 30 ◆ Removing knockouts.

7.2.0 Panel Connections

Electrical connections in a panelboard fall under two categories:

- Line connections, including termination and routing of service and feeder conductors
- Load connections, including termination and routing of branch circuit and feeder conductors

> **WARNING!**
> Even though the main breaker handle is in the OFF position, the line side of the switch is still energized.

Prior to terminating aluminum conductors, verify that the lugs are stamped CU-AL or a label inside the panel states that the connection of aluminum conductors is permitted.

NEC Section 408.3(E) requires the phase arrangement on three-phase buses to be A, B, and C from front to back, top to bottom, or left to right, as viewed from the front of the panel. The B phase must be the phase with the highest voltage to ground on a three-phase, four-wire, delta-connected system. See *Figure 32*.

7.3.0 Enclosures

The majority of overcurrent devices (fuses and circuit breakers) are used in some type of enclosure (i.e., panelboards, switchboards, motor control centers, individual enclosures, etc.).

NEMA has established enclosure designations because individually enclosed overcurrent protection devices are used in so many different types of locations, weather and water conditions, dust

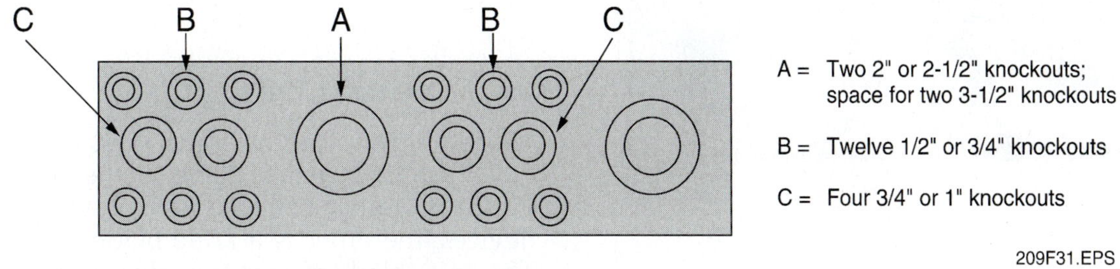

Figure 31 ◆ Typical knockouts in top of panelboard.

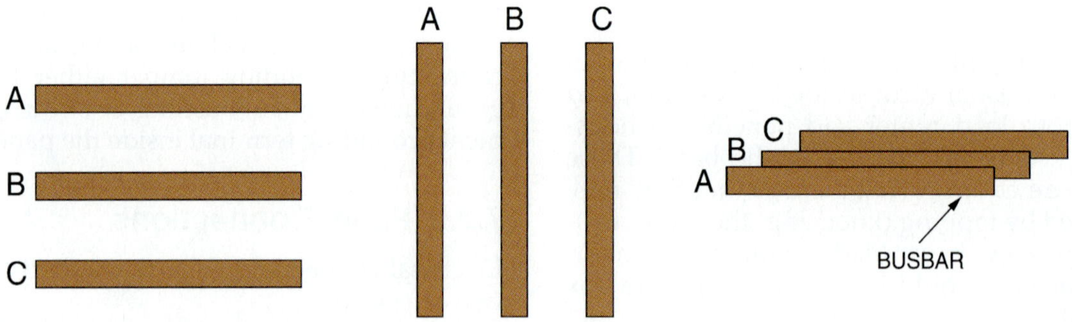

209F32.EPS

Figure 32 ◆ *NEC*®-approved phase arrangements.

and other contaminating conditions, etc. The NEMA designations were recently revised to obtain a clearer and more precise definition of the enclosure needed to meet various standard requirements.

For example, a NEMA Type 12 enclosure can now be substituted in many installations in place of a NEMA Type 5 enclosure. The advantage of this substitution is that the NEMA Type 12 enclosure is much less expensive than the NEMA Type 5 enclosure. *Figure 33* shows two types of NEMA enclosures and *Table 4* provides a brief explanation of NEMA enclosure specifications.

7.4.0 Safety Switches

Most manufacturers of safety switches have at least two complete lines to meet industrial, commercial, and residential requirements. Both types usually have visible blades and safety handles. With visible blades, the contact blades are in full view so you can clearly see their position. Safety handles are always in complete control of the switch blades, so whether the cover is open or closed, when the handle is in the OFF position, the switch is always OFF, that is, on the load side of the switch. See *Figure 34*.

WARNING!

Even though a safety switch handle is in the OFF position, the line side of the switch is still energized.

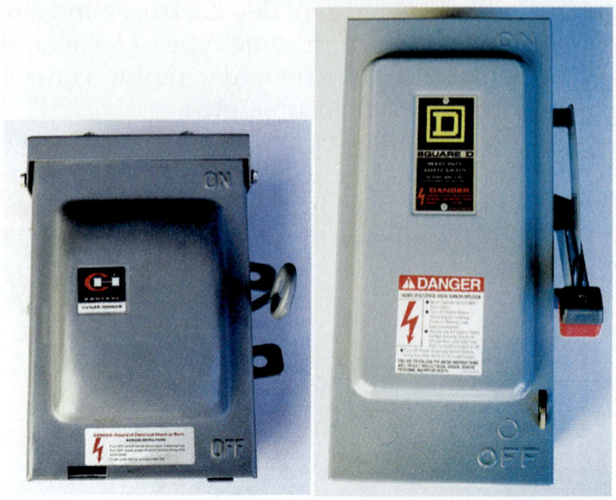

209F33A.EPS 209F33B.EPS

(A) (B)

Figure 33 ◆ NEMA enclosures. (A) General-duty safety switch. (B) Heavy-duty safety switch.

Heavy-duty switches are used for applications where the switch is subjected to frequent operation and rough handling. Heavy-duty switches are also used in atmospheres where a general-duty switch would be unsuitable. Heavy-duty switches are widely used by most heavy industrial applications; motors and HVAC equipment will also be controlled by such switches. Most heavy-duty switches are rated at 30A through 1,200A, 240V to 600V (AC-DC). The switches with horsepower ratings are able to interrupt approximately six times the full-load, motor current ratings. When equipped with Class J or Class R fuses, many heavy-duty safety switches are UL listed for use on systems with up to 200,000A available fault current. Heavy-duty switches are available with NEMA Types 1, 3R, 4, 5, 7, 9, and 12 enclosures.

7.4.1 Switch Contacts

There are two types of switch contacts used in today's safety switches. One is the butt contact, which is similar to those used in circuit breaker devices; the other is a knife blade and jaw type. The knife blade type is considered to be superior to other types on the market.

Table 4 NEMA Enclosure Specifications

Enclosure	Explanation
NEMA Type 1 General Purpose	To prevent accidental contact with enclosed apparatus. Suitable for use indoors where not exposed to unusual service conditions.
NEMA Type 3 Weatherproof (Weather Resistant)	Protects against specified weather hazards. Suitable for use outdoors.
NEMA Type 3R Raintight	Protects against entrance of water from rain. Suitable for general outdoor applications not requiring protection against sleet.
NEMA Type 4 Watertight	Designed to exclude water applied in the form of a hose stream. Protects against stream of water during cleaning operations.
NEMA Type 5 Dusttight	Constructed so that dust will not enter the enclosed area. Being replaced in some equipment by NEMA Type 12.
NEMA Type 7 Hazardous Locations A, B, C, or D Class I–air break letter(s) following type number indicate particular groups of hazardous locations per the *NEC®*.	Designed to meet application requirements of the *NEC®* for Class I, hazardous locations (explosive atmospheres). Circuit interruption occurs in the air.
NEMA Type 9 Hazardous Locations E, F, or G Class II–letter(s) following type number indicate particular groups of hazardous locations per the *NEC®*.	Designed to meet application requirements of the *NEC®* for Class II hazardous locations (combustible dusts, etc.).
NEMA Type 12 Industrial Use	For use in those industries where it is necessary to exclude dust, fibers, and filings, or oil or coolant seepage.

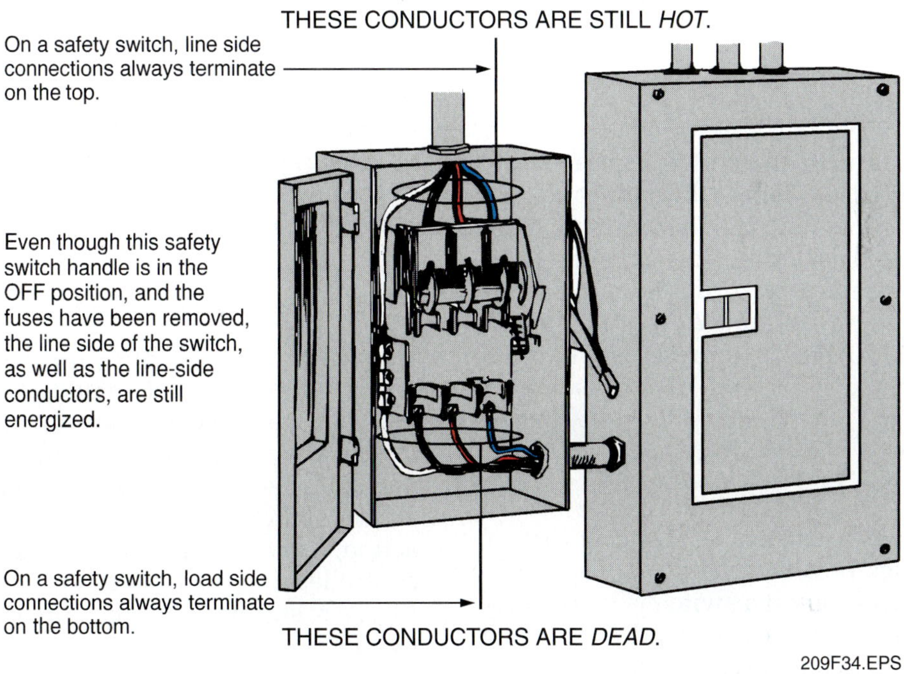

Figure 34 ◆ Using a safety switch as a panelboard disconnect.

All current-carrying parts of safety switches are usually plated with tin, cadmium, or nickel to reduce heating by keeping metal oxidation at a minimum. The switch blade and jaws are made of copper for high conductivity. With knife blade construction, the jaws distribute a uniform clamping pressure over the entire blade-to-jaw contact surface. In the event of a high current fault, the electromagnetic forces that develop tend to squeeze the jaws tightly against the blade. In the butt-type contact, these forces tend to force the contacts apart, causing them to burn severely.

INSIDE TRACK

Safety Switches

When selecting a safety switch for a particular application, always check the manufacturer's product literature to determine if any switch accessories are required for use with the switch. Safety switches often must be equipped with accessories for certain applications. For example, when used with service equipment, they must be labeled as *SUITABLE FOR USE AS SERVICE EQUIPMENT* and a main bonding jumper must be installed. Similarly, when installing NEMA Type 3R, 4, 5, and/or 12 enclosures, the use of bolt-on rainproof or watertight hubs is often required.

Fuse clips are also plated to control corrosion and keep heating to a minimum. All heavy-duty fuse clips have steel reinforcing springs to increase their mechanical strength and provide a firmer contact pressure. As a result, fuses will not work loose due to vibration or rough handling.

7.4.2 Insulating Materials

As the voltage rating of switches is increased, arc suppression becomes more difficult and the choice of insulation material becomes a more critical problem. The arc suppressors used by many manufacturers consist of a housing made of insulation material and one or more magnetic suppressor plates. All arc suppressors are tested to ensure proper control and extinguishing of arcing.

7.4.3 Operating Mechanism

Heavy-duty safety switches have spring-driven, quick-make, quick-break mechanisms. The quick-break action is necessary if a switch is to be safely switched to OFF under a heavy load. The spring action, in addition to making the operation quick-make, quick-break, firmly holds the switch blades in an ON or OFF position. The operating handle is an integral part of the switching mechanism, so if the springs should fail, the switch can still be operated. When the handle is in the OFF position, the switch is always OFF.

A one-piece crossbar is usually employed to offer direct control over all blades simultaneously. The one-piece crossbar provides stability and strength, plus proper alignment for uniform blade operation.

Dual cover interlocks are also standard on all heavy-duty switches. The dual interlock prevents the enclosure door from being opened when the switch is ON and also keeps the switch from being turned ON while the door is open.

7.4.4 General-Duty Safety Switches

General-duty switches are for residential and light commercial applications where the price of the device is a limiting factor. General-duty switches are meant to be used where operation and handling are moderate and where the available fault current is less than 10,000A. Some examples of general-duty switch applications include: residential HVAC equipment, light-duty fan coil circuit disconnects for commercial projects, and the like.

General-duty switches are rated up to 600A at 240V (AC only) in general-purpose (NEMA Type 1) and rainproof (NEMA Type 3R) enclosures. These switches are horsepower-rated and capable of opening a circuit with approximately six times a motor's full-load current rating.

All current-carrying parts of general-duty switches are plated with either tin or cadmium to reduce heating. Switch jaws and blades are made of plated copper for high conductivity. A steel reinforcing spring increases the mechanical strength of the jaws and ensures firm contact pressure between the blade and jaw.

7.4.5 Double-Throw Safety Switches

Double-throw switches are used as transfer switches and are not intended as motor circuit switches; therefore, they are not horsepower-rated.

Safety switches are manually operated and are available as either fused or unfused devices. These switches have quick-make, quick-break action, plated current-carrying parts, a key-controlled interlock mechanism, and screw-type lugs. Arc suppressors are supplied on all switches rated above 250V.

7.5.0 Main Lugs

Main lugs load centers provide distribution of electrical power where a main disconnect with overcurrent protection is provided separately from the load center. All terminals are suitable for aluminum or copper conductors.

Many main-lugs only (MLO) load centers rated at 125A and up have interiors that are reversible for either top or bottom feed. The cover does not need to be reversed when the interior is reversed. Load centers of 125A and above are commonly available in 14" wide boxes for more wire bending space. Some single-phase, three-wire load centers are also approved for three-phase grounded B-type systems at 240VAC. A main lugs load center can also be converted to a main breaker load center provided that it has approved breakers. Follow the manufacturer's instructions.

7.6.0 Main Breaker Load Centers

Main breaker load centers typically have the following characteristics:

- The main breakers are factory-installed, cutting installation costs. There are no lugs to remove; no screws, nuts, or washers to misplace; and no expensive main breakers to lose. Factory-assembled main disconnects also ensure a proper and safe electrical connection.
- Boxes that are 14" wide are available in 100A to 225A main breaker load centers. This offers more side gutter space for wiring and permits flush load center installation between 16" centered studs without an extra mounting support to hold the box in place.
- The line side terminals of the main breakers are suitable for use with copper or aluminum conductors.

7.7.0 Panelboards

Switchgear manufacturers offer complete lines of lighting and distribution panelboards, most of which are available either unassembled from distributor stock or factory-assembled. All types should be UL listed and meet *Federal Specification WP-115a.*

NQO panelboards are rated for use on the following AC services: 120/240V, single-phase, three-wire; 240V, three-phase, three-wire delta; 240V, three-phase delta with grounded B phase; and 120/208V, three-phase, four-wire wye. They carry no DC rating. NQO panelboards are available either factory-assembled or unassembled.

This type of panelboard is suitable for use in industrial buildings, schools, offices, and commercial buildings and institutions in which the largest branch breaker does not exceed 150A and the system voltage is not greater than 240VAC.

NQO panelboards have maximum ratings of 400A (main breaker or main lugs). Branch circuit breakers may be catalog prefix QO, QO-H, QH, Q1, or Q1-H, and one-, two-, or three-pole—having a maximum rating of 150A and featuring plug-in bus connections. QO and Q1 circuit breakers are standard with 10,000 amperage interrupting capacity (AIC) rating and QH breakers with 65,000 AIC rating. Other ratings for specific applications are also available.

Branch circuit breakers with ground fault circuit interruption may also be supplied in Type NQO panelboards. Rated 10,000 AIC symmetrical, these devices provide GFCI ground fault protection as well as overload and short circuit protection for branch circuit wiring.

NQO unassembled panelboards are available as follows:

- Branch breakers
- Interior with solid neutral
- Box (14" wide × 4" deep, 14" wide × 5¾" deep, or 20" wide × 5¾" deep)
- Mono-flat front with door and flush lock
- Accessories

NQO factory-assembled panelboards are identical in construction to the unassembled type. Main ratings and branch circuits are the same. Unlike unassembled panelboards, however, the branches are factory-installed.

Assembled and unassembled NQO boxes are constructed of galvanized steel. Several types of knockouts are provided in each end wall. Interiors having a maximum main lugs rating of 225A are available in 14" × 4", 14" × 5¾", or 20" × 5¾" boxes. 14" × 5¾" or 20" × 5¾" boxes are required for panelboards having a main circuit breaker. Boxes for interiors having 400A mains (breakers or lugs) are 20" × 5¾".

Interiors for standard-width panelboards having a maximum rating of 225A are of the single-bus construction. In this construction, one-, two-, and three-pole catalog prefix Q1 breakers extend the full width of the panelboard and cannot be mounted opposite each other. QO, QO-H, and QH circuit breakers twin-mount on the bus assembly. In other words, a three-pole QO requires three-pole spaces, but a three-pole Q1 requires six QO spaces.

Interiors of 400A panelboards utilize a double-row bus construction. This type of construction consists of two sets of busbars mounted on a single pan. The respective phase buses of each set are paralleled with each other by means of insulated, solid connectors. QO and Q1 breakers mount on a one-for-one basis (i.e., a three-pole QO requires the same spaces as a three-pole Q1).

All current-carrying parts are plated for maximum corrosion resistance and minimum heating at contact surfaces. Main lugs are UL listed for use with either copper or aluminum cable. Main lugs may be replaced by the appropriate crimp lug, when required. Lug catalog numbers and crimping tools are called out on the panelboard wiring diagram. Box-type lugs for circuits on both branch breakers and the solid neutral permit maximum convenience and speed in wiring and are also UL listed for use with either copper or aluminum cable.

8.0.0 ◆ MAINTENANCE

The first requirement in a satisfactory maintenance program for electrical panelboards and switches is safety, including the use of the proper lockout/tagout procedures.

The second is good equipment that has been properly installed. No one can do a good maintenance job on equipment that is not appropriate for the job or that has been installed haphazardly with no eye to future maintenance requirements. If such conditions exist, they should be brought to the attention of the proper party and corrected, rather than trying to establish a maintenance program for them.

The third requirement for a good maintenance program is proper maintenance personnel. Persons who must maintain equipment should have a thorough knowledge of the equipment's operation and have the ability to make thorough inspections and minor repairs of that equipment.

The fourth requirement of a good maintenance program is the establishment of preventive maintenance. This is an all-inclusive phrase for the continuing inspection of equipment, the report and recording of the condition of the equipment, and the repair of the equipment.

The term preventive maintenance has come to mean a system of routine inspections of equipment that is properly recorded for future reference on some type of inspection records. More specifically, the term stands for the heading-off of future equipment problems by making minor repairs in advance of major operating difficulties. In electrical panelboards and switches, a simple tightening of a lug early on can prevent a serious short circuit or a heated terminal later. Due to the aspect of recording all such inspections, some electricians believe that preventive maintenance is merely a system of records. Actually, the records supplement the inspection and are designed to take the place of the maintenance technician's memory. In such extreme cases where only one maintenance worker services a particularly small plant or building, records might be entirely disregarded, although this is not recommended. However, where a number of maintenance personnel are servicing large systems, records are vital to the proper operation of the maintenance inspection routine.

Electrical equipment is more prone to damage by operating conditions than almost any other piece of equipment. Water, dust, heat, cold, humidity, vibration, and countless other conditions can affect the proper operation of electrical equipment. Because of this, there are three cardinal rules to follow in maintaining electrical apparatus. These are:

- Keep it clean.
- Keep it dry.
- Keep it tight.

One of the greatest problems existing in panelboards, switchgear, load centers, and the like are loose connections at both the main lugs and at circuit breakers and/or fuse blocks. Loose connections cause overheating and eventual failure of the terminals and/or conductors. This condition is especially prevalent where aluminum conductors are used. Therefore, periodic checks should be made to ensure that all connections are tight.

CASE HISTORY

Improper Maintenance Procedures Can Be Dangerous

A group of electricians were performing maintenance on AC switchgear. The equipment was placed in the maintenance mode by opening the primary 20kV interrupter switch. This switch was housed in the upper compartment of a two-compartment cabinet. Each compartment had its own door. One of the electricians was assigned to clean the lower compartment, but he began to clean the upper compartment as well. He sprayed cleaning fluid from an aerosol can onto the circuitry. When the aerosol spray contacted the line side of the primary switch, it provided a conductive plasma for the current. The current passed through the spray, spray can, and the electrician's body, which had apparently made contact with one of the switch blades, causing him to be electrocuted and badly burned.

The Bottom Line: This accident could have been prevented if proper safety and maintenance procedures had been followed. No electrician should attempt to work on any equipment, regardless of voltage, without first getting permission from the supervisor and only after having received proper training for the task at hand. This includes training in the use of special tools and safety equipment required for the task.

THINK ABOUT IT

Putting It All Together

Examine the service at your home or workplace. Is the service entrance properly installed and protected? Is the service-entrance cable cracked or frayed? What about the electrical panel? Are the breakers properly labeled? Is there a main service disconnect switch?

Review Questions

1. Which of the following best describes the components between the service drop and the building's main disconnecting device?
   a. Wye connection
   b. Service entrance
   c. Service union
   d. Service lateral

2. The most common electric service for residential and small commercial applications is _____.
   a. single-phase, 120V, two-wire service
   b. single-phase, 120/240V, three-wire service
   c. three-phase, 208/120V, four-wire service
   d. three-phase, 277V/480, four-wire service

3. The approximate voltage between ground and the high leg of a 120/240V delta-connected service with a center tap is _____.
   a. 120V
   b. 163V
   c. 208V
   d. 240V

4. What is used in conjunction with a watt-hour meter to measure the power used in services over 400A?
   a. An autotransformer
   b. A current transformer
   c. An ohmmeter
   d. An ammeter

5. Which of the following is used on most working drawings to give details of the panelboards and overcurrent protective devices used on the project?
   a. A panelboard schedule
   b. An appliance schedule
   c. A cross-sectional detail drawing
   d. A site plan

6. How must overhead service-entrance conductors be arranged from the point where they leave the service head to the point where they are connected to the service drop?
   a. They must be absolutely straight with no sag.
   b. They must terminate in a pothead.
   c. A drip loop must be provided.
   d. They must be color-coded yellow, green, and orange.

7. A(n) _____ is used to determine the amount of electricity consumed by the customer.
   a. ammeter
   b. wattmeter
   c. watt-hour meter
   d. ohmmeter

8. The minimum size copper grounding electrode conductor allowed when used with copper service-entrance cables No. 2 AWG or smaller is _____.
   a. No. 2
   b. No. 4
   c. No. 6
   d. No. 8

9. When a metal water pipe is used as a grounding electrode, a _____ must be provided at the water meter.
   a. bonding jumper
   b. floor drain
   c. grounding locknut
   d. grounding bushing

10. Which of the following is considered to be a part of the metering equipment in a 277/480V, wye-connected service?
    a. Subpanels
    b. Service-entrance conductors
    c. Potential transformers
    d. Autotransformers

Summary

Regardless of the size or complexity, all electric services serve the same purpose: to deliver electrical energy from the supply system to the wiring system on the premises served. The basic components of a commercial electric service include a transformer to convert the transmission voltage to a usable level; a service drop, which includes the conductors that are used to connect the service from the power pole to the service facilities at the building; a service entrance, which includes all components between the termination of the overhead service drop and the building's main disconnecting device, except for the metering equipment; service conductors, which are the conductors used in the service entrance; and service equipment, which provides overcurrent protection, a means of disconnect, and a point of attachment for the metering equipment. *NEC Article 230* covers the requirements for electric services.

Notes

Trade Terms Introduced in This Module

Delta-connected: A three-phase transformer connection in which the terminals are connected in a triangular shape like the Greek letter delta (Δ).

Load center: A specific type of panelboard designed for light-duty, single- or three-phase applications, such as residential or light commercial installations.

Service: The electric power delivered to the premises.

Service conductors: The conductors between the point of termination of the overhead service drop or underground service lateral and the main disconnecting device in the building or on the premises.

Service drop: The overhead conductors through which electrical service is supplied between the last power company pole and the point of their connection to the service-entrance conductors located at the building or other support used for the purpose.

Service entrance: All components between the point of termination of the overhead service drop or underground service lateral and the building's main disconnecting device, except for metering equipment.

Service equipment: The necessary equipment, usually consisting of a circuit breaker or switch and fuses and their accessories, located near the point of entrance of supply conductors to a building and intended to constitute the main control and cutoff means for the electric supply to the building.

Service lateral: The underground service conductors between the street main, including any risers at a pole or other structure or from transformers, and the first point of connection to the service-entrance conductors in a terminal box, meter, or other enclosure with adequate space, inside or outside the building wall.

Service raceway: The rigid conduit, IMC, or other raceway that encloses the service-entrance conductors.

Substation: An installation or area where an assembly of devices and apparatus are used to monitor, control, transform, or modify electrical power. Substations are normally installed at a point on the system where transmission voltages are reduced to distribution voltages.

Wye-connected: A three-phase transformer connection in which all three phases are connected at a central point, forming a Y configuration.

Additional Resources

This module is intended to present thorough resources for task training. The following reference works are suggested for further study. These are optional materials for continued education rather than for task training.

American Electrician's Handbook, Latest Edition. New York: Croft and Summers, McGraw-Hill.

National Electrical Code® Handbook, Latest Edition. Quincy, MA: National Fire Protection Association.

CONTREN® LEARNING SERIES — USER FEEDBACK

The NCCER makes every effort to keep these textbooks up-to-date and free of technical errors. We appreciate your help in this process. If you have an idea for improving this textbook, or if you find an error, a typographical mistake, or an inaccuracy in NCCER's *Contren®* textbooks, please write us, using this form or a photocopy. Be sure to include the exact module number, page number, a detailed description, and the correction, if applicable. Your input will be brought to the attention of the Technical Review Committee. Thank you for your assistance.

Instructors – If you found that additional materials were necessary in order to teach this module effectively, please let us know so that we may include them in the Equipment/Materials list in the Annotated Instructor's Guide.

Write: Product Development
National Center for Construction Education and Research
P.O. Box 141104, Gainesville, FL 32614-1104

Fax: 352-334-0932

E-mail: curriculum@nccer.org

Craft ______________________ Module Name ______________________

Copyright Date __________ Module Number __________ Page Number(s) __________

Description

____

____

____

____

(Optional) Correction

____

____

____

(Optional) Your Name and Address

____

____

____

Circuit Breakers and Fuses

26210-05

St. Vincent's North Tower

Birmingham, Alabama

Health Care $25–99 Million Award Winner

Brasfield & Gorrie, LLC

26210-05

Circuit Breakers and Fuses

Topics to be presented in this module include:

Overview

The primary function of fuses and circuit breakers is to protect people and equipment from excessive current by an unintentional load increase or fault condition. A fault condition can be defined as current finding an unintended path back to its point of origin. A fault may occur from one current-carrying conductor to another (short circuit) or from a current-carrying conductor to ground (ground fault).

There are several ratings associated with circuit breakers and fuses. These protective devices have a maximum voltage rating, current trip rating, interrupting capacity rating, and some have a current-limiting rating. Current-limiting devices react much faster to clear the short circuit or fault current. This added characteristic greatly reduces the amount of short circuit or fault current that is allowed to pass through before the device opens, thus reducing the levels of available fault current at the next device downstream.

Ground fault circuit interrupters (GFCI) do not provide overcurrent protection. They are devices that recognize a ground fault condition and open the circuit in which they are connected.

Note: *National Electrical Code*® and *NEC*® are registered trademarks of the National Fire Protection Association, Inc., Quincy, MA 02269. All *National Electrical Code*® and *NEC*® references in this module refer to the 2005 edition of the *National Electrical Code*®.

Objectives

When you have completed this module, you will be able to do the following:

1. Explain the necessity of overcurrent protection devices in electrical circuits.
2. Define the terms associated with fuses and circuit breakers.
3. Describe the operation of a circuit breaker.
4. Select the most suitable overcurrent device for the application.
5. Explain the role of the *NEC*® in specifying overcurrent devices.
6. Describe the operation of single-element and time-delay fuses.
7. Explain how ground fault circuit interrupters (GFCIs) can save lives.
8. Calculate short circuit currents.
9. Describe troubleshooting and maintenance techniques for overcurrent devices.

Trade Terms

Cartridge fuse
Circuit breaker
Dual-element fuse
Edison-base
Fault current
Frame size
Fuse
Fuse link
Molded-case circuit breaker
Nonrenewable fuse
Overcurrent protection
Plug fuse
Pole
Short circuit

Required Trainee Materials

1. Pencil and paper
2. Appropriate personal protective equipment
3. Copy of the latest edition of the *National Electrical Code*®

Prerequisites

Before you begin this module, it is recommended that you successfully complete the following: *Core Curriculum; Electrical Level One; Electrical Level Two,* Modules 26201-05 through 26209-05.

This course map shows all of the modules in *Electrical Level Two.* The suggested training order begins at the bottom and proceeds up. Skill levels increase as you advance on the course map. The local Training Program Sponsor may adjust the training order.

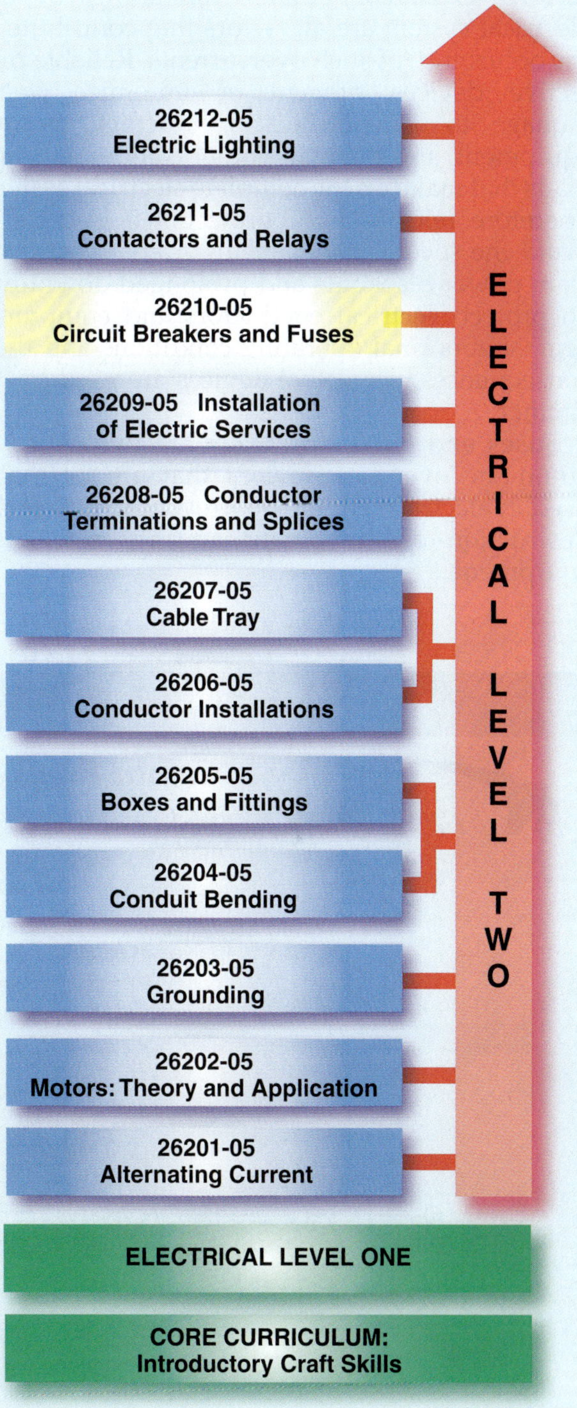

210CMAP.EPS

1.0.0 ◆ INTRODUCTION

All electrical circuits and their related components are subject to destructive overcurrents. Harsh environments, general deterioration, accidental damage, damage from natural causes, excessive expansion, and overloading of the electrical system are all factors that contribute to the occurrence of such overcurrents. Reliable protective devices prevent or minimize costly damage to transformers, conductors, motors, equipment, and the many other components and loads that make up the complete electrical system. Therefore, reliable circuit protection is essential to avoid the severe monetary losses that can result from power blackouts and prolonged downtime. To protect electrical conductors and equipment against abnormal operating conditions and their consequences, protective devices are used in the circuits.

Fuses and **circuit breakers** are two types of automatic overload devices that are normally used in electrical circuits to prevent fires and the destruction of the circuit and its associated equipment.

Basically, a circuit breaker is a device used for closing and interrupting a circuit between separable contacts under both normal and abnormal conditions. This is done manually by switching the handle to the ON or OFF positions. However, the circuit breaker is also designed to open a circuit automatically on a predetermined overload or **fault current** without damage to itself or its associated equipment. As long as a circuit breaker is applied within its rating, it will automatically interrupt any fault and, therefore, must be classified as an inherently safe overcurrent protective device.

The internal arrangement of a circuit breaker is shown in *Figure 1*, while its external operating characteristics are shown in *Figure 2*. Note that the handle on a circuit breaker resembles an ordinary toggle switch. On an overload, the circuit breaker opens itself, or trips. Once tripped, the handle jumps to the middle position (*Figure 2*). To reset, turn the handle to the OFF position and then turn it as far as it will go beyond this position (RESET); finally, turn it to the ON position.

INSIDE TRACK

Circuit Breakers and Fuses

The function of circuit breakers and fuses used in electrical circuits can be compared to that of a pressure relief valve used with a steam boiler. If dangerously high pressures develop within the steam boiler, the pressure relief valve opens to relieve the pressure, thus preventing damage to the boiler. Similarly, circuit breakers and fuses act as safety valves in electrical circuits by opening the circuit should a dangerous overcurrent condition exist as a result of an overload or short circuit, thus preventing damage to the circuit components.

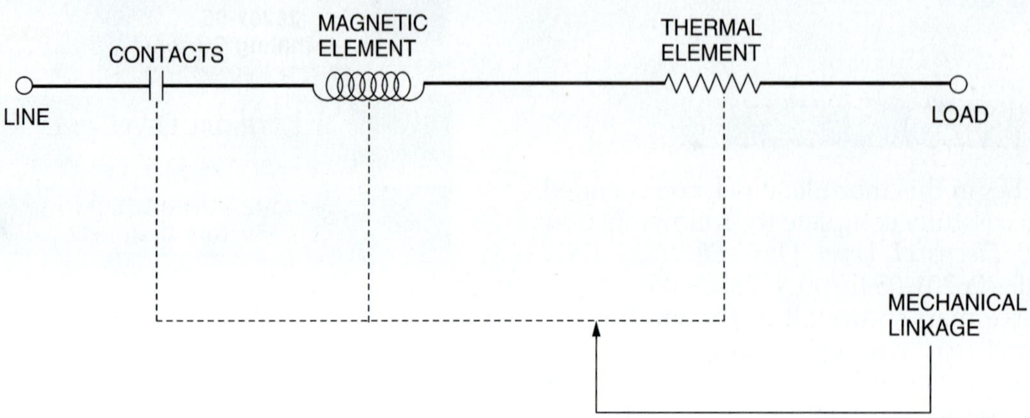

210F01.EPS

Figure 1 ◆ Internal arrangement of a circuit breaker.

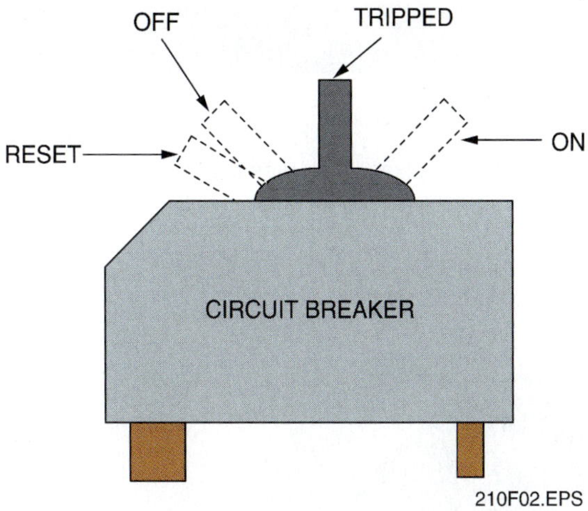

Figure 2 ◆ Operating characteristics of a circuit breaker.

A standard **molded-case circuit breaker** usually contains:

- A set of contacts
- A magnetic trip element
- A thermal trip element
- Line and load terminals
- Bussing used to connect these individual parts
- An enclosing housing of insulating material

The circuit breaker handle manually opens and closes the contacts and resets the automatic trip units after an interruption. Some circuit breakers also contain a manually operated push-to-trip testing mechanism.

Circuit breakers are classified according to given current ranges. Each group is classified by the largest ampere rating of its range. These groups are:

- 15A–100A
- 125A–225A
- 250A–400A
- 500A–1,000A
- 1,200A–2,000A

Therefore, they are classified as 100A, 225A, 400A, 1,000A, and 2,000A frames. These numbers are commonly referred to as frame classifications or **frame sizes** and are terms applied to groups of molded-case circuit breakers that are physically interchangeable with each other.

2.0.0 ◆ CIRCUIT BREAKER RATINGS

The established voltage rating of a circuit breaker is based on its clearances or space, both through air and over surfaces between components of the electrical circuit and between the electrical components and ground. Circuit breaker voltage ratings indicate the maximum electrical system voltage on which they can be applied. Underwriters Laboratories, Inc. (UL) recognizes only the ratings listed in *Table 1* for molded-case circuit breakers.

A circuit breaker can be rated for either alternating current (AC) or direct current (DC) system applications or for both. Single-pole circuit breakers, rated at 120/240VAC or 125/250VDC, can be used singly and in pairs on three-wire circuits having a neutral connected to the mid-point of the load. Single-pole circuit breakers rated at

Table 1 Molded-Case Circuit Breaker Ratings Recognized by Underwriters Laboratories

For Alternating Current	For Direct Current
120V	125V
120/240V	—
240V	250V
277V	600V
277/480V	—
480V	—
600V	—

INSIDE TRACK

Tripped Circuit Breaker Indications

When tripped, the handle of a circuit breaker can position itself in one of several ways, depending on the design of the breaker. Some common handle positions (indications) of a tripped circuit breaker are:

- *Center trip* – The handle of the breaker moves to a center position on the breaker.
- *Center position flag* – The handle of the breaker moves to the center position on the breaker and displays a red flag.
- *Full off position* – The handle of the breaker moves to the full OFF position on the breaker.
- *OFF position outward* – A center button protrudes to display OFF.

INSIDE TRACK

Circuit Breakers

Circuit breakers are available in a wide variety of sizes and types to suit various applications.

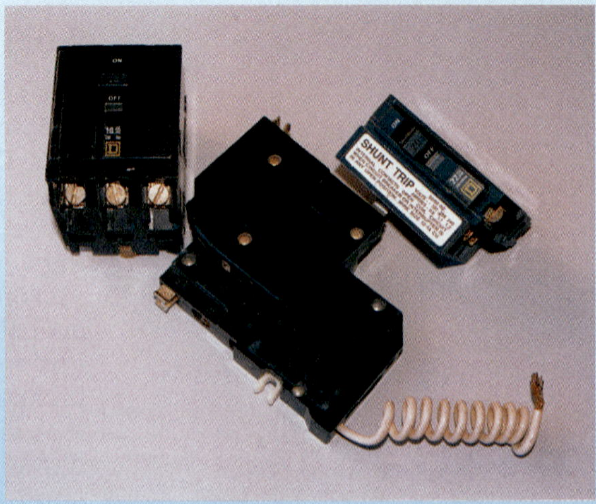

210P1001.EPS

INSIDE TRACK

60°C/75°C Ratings

Some circuit breakers are suitable for use with both 60°C and 75°C rated conductors. Others can be used with only 75°C rated conductors. Be sure to check the rating of the circuit breaker selected. For example, to use the 75°C ampacity rated conductors No. 1 and smaller, you must make sure to use a circuit breaker rated for 75°C. To comply with ***NEC Section 110.14(C),*** the termination at the load end of the same circuit must also be rated for 75°C.

120/240VAC or 125/250VDC can also be used in pairs on a two-wire circuit connected to the ungrounded conductors of a three-wire system. Two-pole circuit breakers rated at 120/240VAC or 125/250VDC can be used only on a three-wire, direct current, or single-phase, alternating current system having a grounded neutral. Circuit breaker voltage ratings must be equal to or greater than the voltage of the electrical system on which they are used.

Circuit breakers have two types of current ratings. The first—and the one that is used most often—is the continuous current rating. The second is the fault current interrupting capacity.

2.1.0 Current Rating

The rated continuous current of a device is the maximum current (in amperes) that it will carry continuously without exceeding the specified limits of observable temperature rise. The continuous current ratings of circuit breakers are established based on standard UL ampere ratings. As listed in *NEC Section 240.6(A),* these range from 15A to 6,000A. The ampere rating of a circuit breaker is located on the handle of the device (see *Figure 3*).

Generally, the circuit breaker current rating must be equal to or less than the load circuit conductor current-carrying capacity (ampacity).

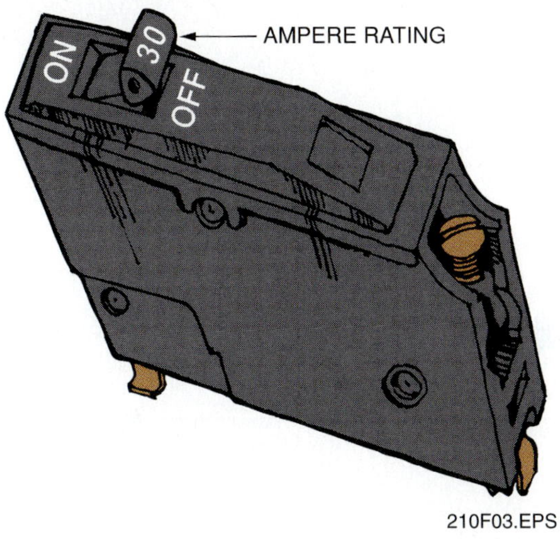

Figure 3 ◆ Circuit breaker current rating.

Most overcurrent protective devices are labeled with the following current ratings:

- Normal current rating
- Interrupting rating

2.2.0 Interrupting Capacity Rating

The amperage interrupting capacity (AIC) of a circuit breaker is the maximum short circuit current at which the breaker will safely interrupt the circuit. The AIC is at the rated voltage and frequency.

NEC Section 110.9 states that equipment intended to interrupt current at fault levels (fuses and circuit breakers) must have an interrupting rating sufficient for the nominal circuit voltage and the current that is available at the line terminals of the equipment.

Equipment intended to break current at other than fault levels must have an interrupting rating at nominal circuit voltage sufficient for the current that must be interrupted.

These *NEC*® statements mean that fuses and circuit breakers (and their related components) that are designed to break fault or operating currents (open the circuit) must have a rating sufficient to withstand such currents. This *NEC*® section emphasizes the difference between clearing fault level currents and clearing operating currents. Protective devices such as fuses and circuit breakers are designed to clear fault currents and, therefore, must have **short circuit** interrupting ratings that are sufficient for fault levels.

Equipment such as contactors and safety switches have interrupting ratings for currents at levels other than fault levels. Therefore, the interrupting rating of electrical equipment is now divided into two parts:

- Current at fault (short circuit) levels
- Current at operating levels

Most people are familiar with the normal current-carrying ampere rating of fuses and circuit breakers. If an overcurrent protective device is designed to open a circuit when the circuit load exceeds 20A for a given time period, as the current approaches 20A, the overcurrent protective device begins to overheat. If the current barely exceeds 20A, the circuit breaker will open normally or a **fuse link** will melt after a given period of time with little, if any, arcing. If 40A of current are instantaneously applied to the circuit, the overcurrent protective device will open faster, but again with very little arcing. However, if a fault current occurs that runs the amperage up to 5,000A, an explosion effect would occur within the protective device. One simple indication of this explosion effect is demonstrated by the blackened windows of blown **plug fuses.**

If this fault current exceeds the interrupting rating of a fuse or circuit breaker, the protective device can be damaged or destroyed; such current can also cause severe damage to equipment and injure personnel. Therefore, selecting overcurrent protective devices with the proper interrupting capacity is extremely important in all electrical systems.

Consider the following analogy using a dammed stream *(Figure 4)* as an example of interrupting

INSIDE TRACK

Circuit Breaker Current Interrupting Rating

NEC Section 240.83(C) requires that circuit breakers having a current interrupting rating other than 5,000A have the rating clearly shown on the circuit breaker. By definition, circuit breakers that have not been marked with an interrupting rating have a rating of 5,000A.

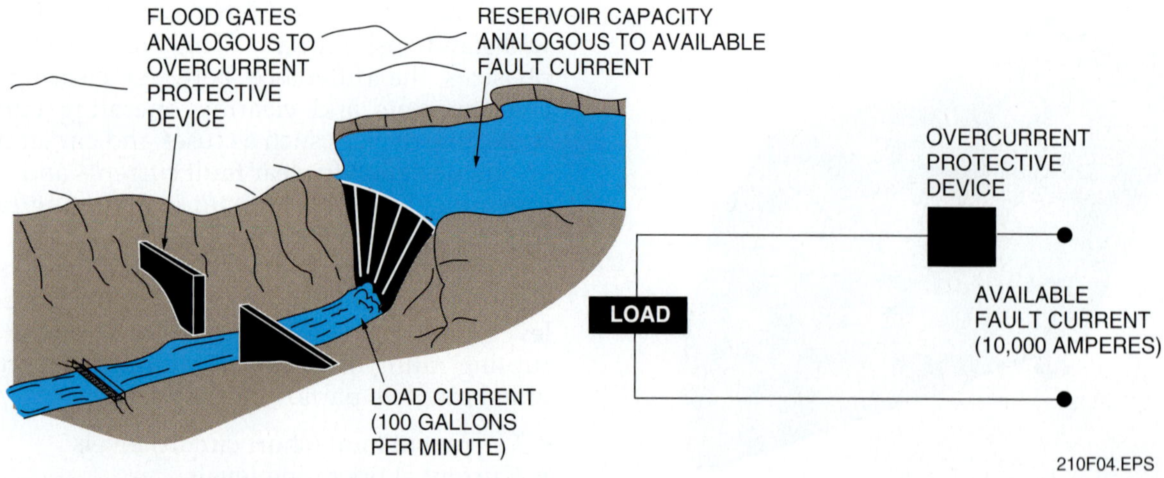

Figure 4 ◆ Normal current operation.

rating. Let the reservoir capacity be the available fault current in an electrical circuit; the flood gates (located downstream from the dam) be the overcurrent protective device in the circuit rated at 10,000 gallons per minute (10,000A); and the stream of water coming through the discharge pipes in the dam be the normal load current. Our drawing shows a normal flow of 100 gallons per minute (100A). Also, note the bridge downstream from the flood gates. This bridge will represent downstream circuit components or equipment connected to the circuit.

Figure 5 shows this same diagram with a fault in the dam—creating a water short circuit that allows 50,000 gallons per minute to flow (50,000 fault circuit amperes). Such a situation destroys the flood gates because of inadequate interrupting rating. The overcurrent protective device in the circuit will also be destroyed. With the flood gates damaged, this surge of water continues downstream, wrecking the bridge. Similarly, the downstream components may not be able to withstand the let-through current in an electrical circuit.

Figure 6 shows the same situation, but with adequate interrupting capacity. Note that the flood gates have adequately contained the surge of water and restricted the let-through current to an amount that can be withstood by the bridge or the components downstream.

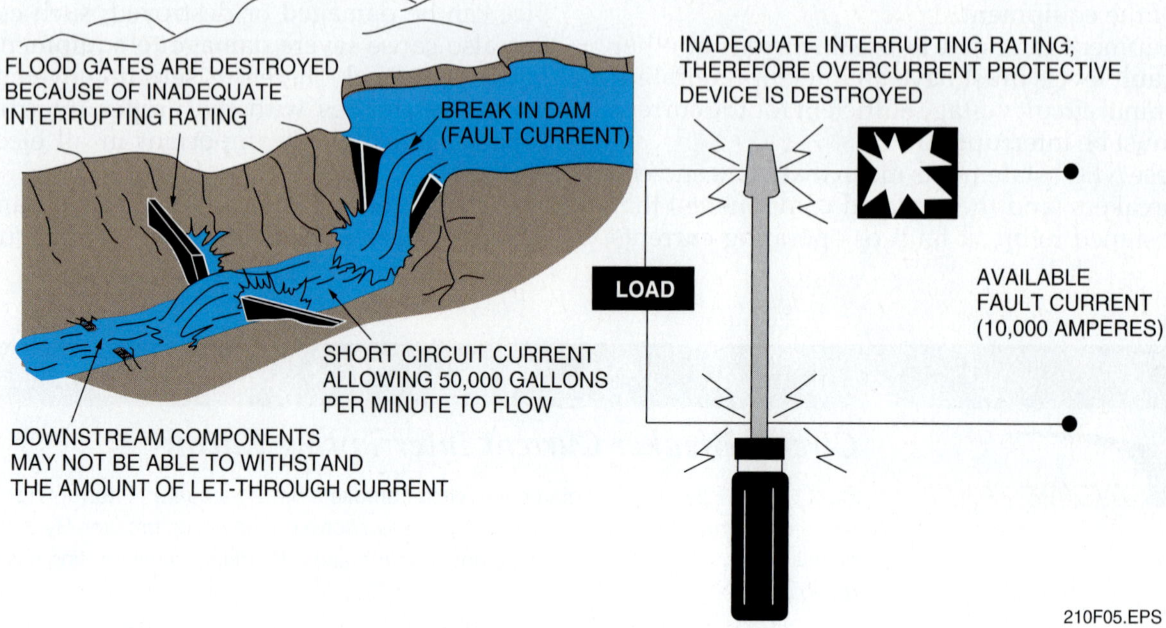

Figure 5 ◆ Short circuit operation with inadequate interrupting rating.

Current Interrupt Rating

THINK ABOUT IT

When replacing a circuit breaker, is it necessary to take into account the current interrupting rating of the replacement breaker? If so, why?

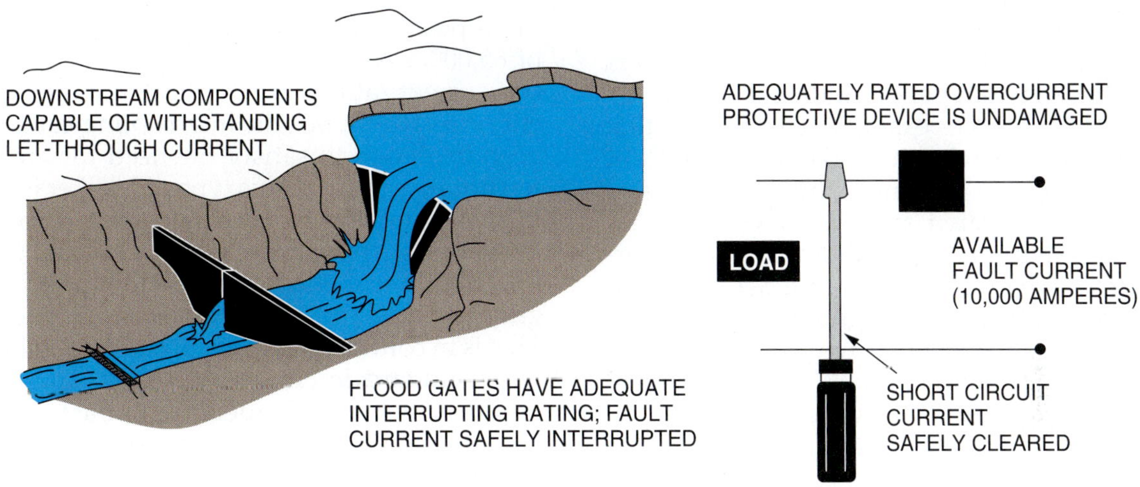

Figure 6 ◆ Short circuit operation with adequate interrupting rating.

There are several factors that must be considered when calculating the required interrupting capacity of an **overcurrent protection** device. *NEC Section 110.10* states that the overcurrent protective devices, the total impedance, the component short circuit current ratings, and other characteristics of the circuit to be protected shall be so selected and coordinated to permit the circuit protective devices used to clear a fault to do so without extensive damage to the electrical components of the circuit. This fault shall be assumed to be either between two or more of the circuit conductors, or between any circuit conductor and the grounding conductor or enclosing metal raceway.

The component short circuit rating is a current rating given to conductors, switches, circuit breakers, and other electrical components, which, if exceeded by fault currents, will result in extensive damage to the component. Short circuit damage can be the result of heat generated or the electromechanical force of a high-intensity magnetic field. The rating is expressed in terms of time intervals and/or current values.

The intent of the *NEC®* is that the design of a system must be such that short circuit currents cannot exceed the short circuit current ratings of the components selected as part of the system. Given specific system components and the level of available short circuit currents that could occur, overcurrent protection devices (mainly fuses and/or circuit breakers) must be used that will limit the let-through current to levels within the withstand ratings of the system components.

In most large commercial and industrial installations, it is necessary to calculate the available short circuit current at various points in a system to determine if the equipment meets the requirements of *NEC Sections 110.9 and 110.10.* There are a number of methods used to determine the short circuit requirements in an electrical system. Some give approximate values; others require extensive computations and are quite exacting. A simple, yet accurate method is the point-by-point method, which will be discussed in more detail later in this module.

An overcurrent protection device must be selected with an interrupting capacity that is equal to or greater than the available short circuit current at the point where the circuit breaker or fuse is applied in the system. The breaker interrupting capacity is based on tests to which the breaker is subjected. There are two such tests; one is set up by Underwriters Laboratories, Inc. (UL) and the other by the National Electrical Manufacturers Association (NEMA). The NEMA tests are self-certification tests, while the UL tests are certified by unbiased witnesses. UL tests have been limited to a maximum of 10,000A in the past, so the emphasis was placed on the NEMA tests with higher ratings. The UL tests now include the NEMA tests plus other ratings. Consequently, emphasis is now placed on the UL tests. It is important to remember that NEMA does not list devices. Two organizations that both test and list devices are UL and the Canadian Standards Association (CSA).

UL requires that molded-case circuit breakers open within a certain period of time during overcurrent conditions. For example, UL requires that a 240V molded-case circuit breaker trip at 300% of its rated continuous current within 50 seconds. If the breaker were rated at 600V, it would be required to trip within 70 seconds at 300% rated current. Currents between 7 and 15 times the rated current (depending on voltage rating and frame size) are handled as overcurrents by the thermal magnetic trip element. Currents above this level activate the magnetic trip element.

The interrupting capacity of a circuit breaker is based on its rated voltage. Where the circuit breaker can be used on more than one voltage, the interrupting capacity will be shown for each voltage level. For example, the LA-type circuit breaker has 42,000A symmetrical interrupting capacity at 240V, 30,000A symmetrical at 480V, and 22,000A symmetrical at 600V.

2.2.1 Standard Interrupting Capacity

Standard interrupting capacity circuit breakers can be identified by their black operating handles and black printed interrupting rating labels. The interrupting rating of a circuit breaker is as important in application as the voltage and current ratings and should be considered each time a breaker is applied. In residential applications, the available fault current is seldom higher than 10,000A or even near this value. The QO and A1 breakers (15A–150A) with the interrupting rating of 10,000A are used in these applications. Higher interrupting ratings are available when required.

2.2.2 High Interrupting Capacity

Where still higher interrupting capacity than the standard ratings discussed previously is required in the 15A–100A FA-type circuit breaker, the FH (H for high interrupting capacity) is available. The FH-type circuit breaker has an interrupting rating of 65,000A symmetrical at 240VAC. The continuous current ratings are duplicated in these breakers (15A–100A), but the interrupting capacity has been increased to satisfy the need for greater interrupting capacity. This type of breaker is applied in installations where higher fault currents are available (such as large industrial plants). FA and FH circuit breakers are quickly identified by their gray handles and red interrupting rating labels. This is in contrast to the black handles on breakers with standard interrupting ratings. These breakers are dimensionally identical to the lower interrupting capacity breakers, but they are built with a case material that will withstand the higher shocks from heat and interrupting forces. Circuit breakers with high interrupting capacity are available in the FH, KH, LH, MH, NH, and PH types.

Three ampere ratings (15A, 20A, and 30A) are also available in the QH high interrupting capacity breakers. These breakers satisfy conditions of lighting circuits supplied from these high available fault current circuit breakers.

2.2.3 Current-Limiting Circuit Breakers

The need for current limitation is the result of the increasingly higher available fault currents associated with the growth and interconnection of modern power systems. To meet this need, current-limiting circuit breakers have been developed. This type of breaker operates extremely fast to provide

INSIDE TRACK

Thermal, Magnetic, and Thermomagnetic Circuit Breakers

Thermal circuit breakers contain a spring-loaded bimetal element that pops the circuit breaker open when the element becomes overheated by excessive current. Magnetic circuit breakers use a magnetic field induced by current flow through a coil of wire to open its contacts and break the circuit. Some circuit breakers, called thermomagnetic breakers, operate on both principles. They are often used in motor protection applications.

downstream protection for other types of fuses with as little as 10,000 AIC on systems with 100,000A available fault current.

There are circuit breakers available for almost every need in any electrical system. For example, in circuits where the breaker must be highly sensitive to amperage changes (in an environment where the temperature changes are great), an ambient-compensating circuit breaker may be used. Its design permits it to compensate for temperature variation.

Figure 7 shows the internal parts of a current-limiting circuit breaker. This is basically a conventional, common-trip, thermal magnetic circuit breaker with an independent limiter section in series with each **pole.** For purposes of explanation, each limiter can be represented by a set of contacts shunted by a transformable resistor.

At currents below a threshold of about 1,000A, the current-limiting circuit breaker performs in a manner similar to conventional thermal magnetic circuit breakers. In the event of an overload or minor fault current, only the breaker section contacts open to interrupt the circuit. The limiter contacts remain closed. See *Figure 8*.

For 100A frame current-limiting circuit breakers, the threshold of current limitation occurs in the range of 1,000A to 2,000A. Above this point, the limiter contacts open first—starting the limiting action within ¼ of a millisecond of fault initiation. A substantial arc voltage is generated across the limiter contacts, as shown in *Figure 9*. Rapid generation of this arc voltage is essential to drastically limit the peak let-through current in the faulted circuit.

The rapid rising equivalent resistance of the arc across the limiter contacts forces the current to transfer to the alternate path provided by the limiter resistor. This special resistor has a high positive temperature coefficient of resistance. As current is transferred to it, the temperature of the resistor increases rapidly with a consequent rapid rise in resistance. The resistor dissipates and limits let-through energy and increases the short circuit power factor of the circuit nearly to unity, resulting in an easier, more reliable, transient-free interruption. See *Figure 10*.

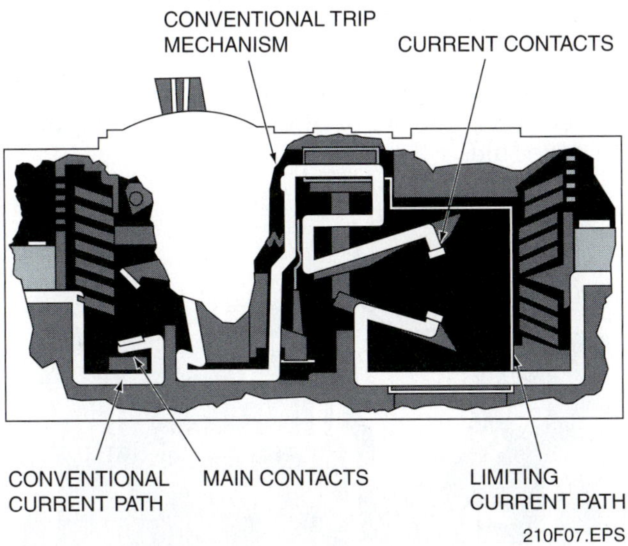

Figure 7 ◆ Current-limiting circuit breaker.

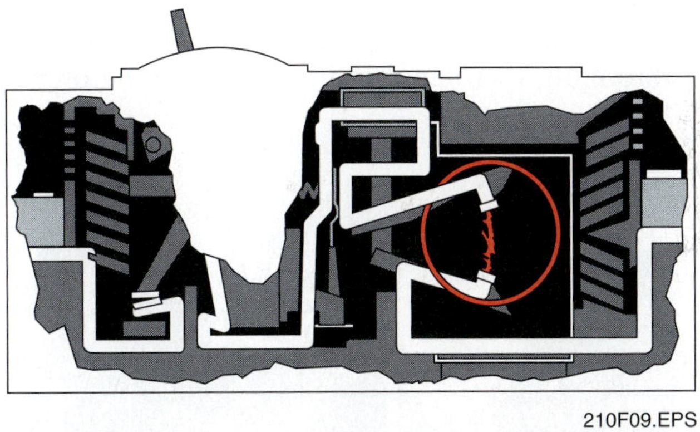

Figure 9 ◆ Limiting action begins.

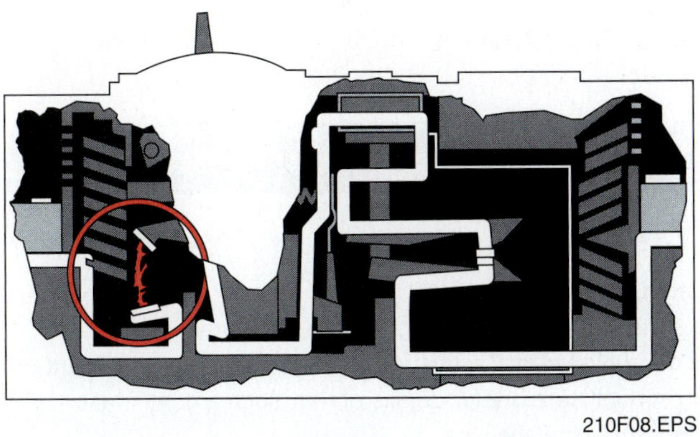

Figure 8 ◆ Conventional operation.

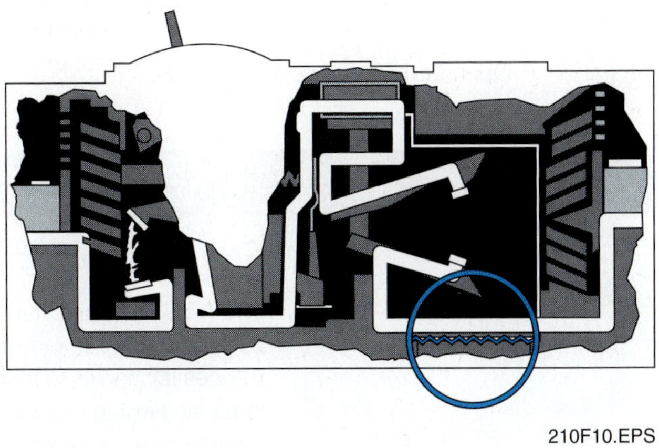

Figure 10 ◆ Limiter resistor.

During the extremely short period of operation of the limiter section, the breaker section tripping mechanism is activated and its contacts begin to open. After the arc between the limiter contacts has been extinguished and all the current is flowing through the limiter resistor, the breaker contacts interrupt the already drastically limited fault current. Because the fault current and system voltage are nearly in phase at this point, complete interruption of the circuit takes place near the first voltage zero of the AC wave. See *Figure 11*.

When a current-limiting circuit breaker trips to clear a faulted circuit, it can be reset in the same manner as a conventional molded-case circuit breaker. This is true whether tripping was due to low-level overload currents or major faults involving the breaker's current-limiting mode of operation. The limiter contacts are automatically closed after circuit interruption. Breaker contacts are reclosed by moving the handle to the extreme OFF position and then to ON (see *Figure 12*). There are no fusible elements in this type of circuit breaker.

Regular maintenance of circuit breakers and their enclosures is necessary to obtain the best service and performance.

Since every circuit breaker failure represents a potential hazard to other equipment in the system, it is difficult to calculate the risks involved in prolonging maintenance inspections. One of the chief causes of circuit breaker failure is high heat caused by loose connections at the load side of the breaker. Another cause of circuit breaker failure is defective loads, that is, electrical equipment that cycles too frequently, which in turn causes the circuit breaker to overheat. Therefore, in the majority of circuit breaker failures, heat is the biggest culprit.

2.3.0 Labeling Considerations

NEC Sections 110.22 and 240.86 require special marking for UL-recognized, series-rated systems (see *Figure 13*). For example, *NEC Section 110.22* states that where circuit breakers or fuses are applied in compliance with the series combination ratings marked on the equipment by the manufacturer, the equipment enclosure(s) shall be legibly marked in the field to indicate the equipment has been applied with a series combination rating. The marking shall be readily visible and state, "CAUTION—SERIES COMBINATION SYSTEM RATED ____ AMPERES. IDENTIFIED REPLACEMENT COMPONENTS REQUIRED."

Complete coverage of engineered electrical distribution systems giving complete details of these *NEC*® requirements are presented in Electrical Level Three and will not be covered in depth in this module.

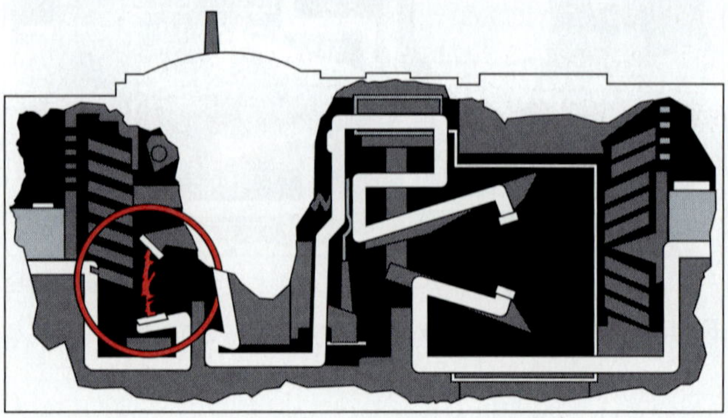

210F11.EPS

Figure 11 ◆ Operation of the breaker contacts.

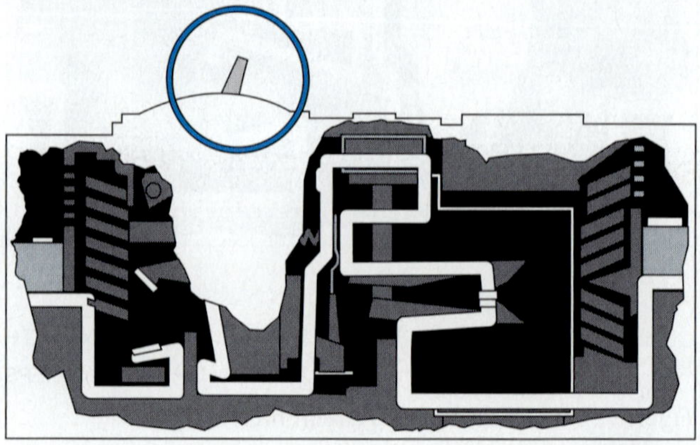

210F12.EPS

Figure 12 ◆ Resetting is done with the breaker handle.

INSIDE TRACK

Series-Rated Circuit Breakers

The *NEC*® recognizes series-rated circuit breakers and requires that the end use equipment be marked with the series combination rating. Such a series rating of devices is specific to individual manufacturers. For this reason, it is especially important to consult listing information or product literature for details of how such a combination rating was derived, especially for retrofit installations.

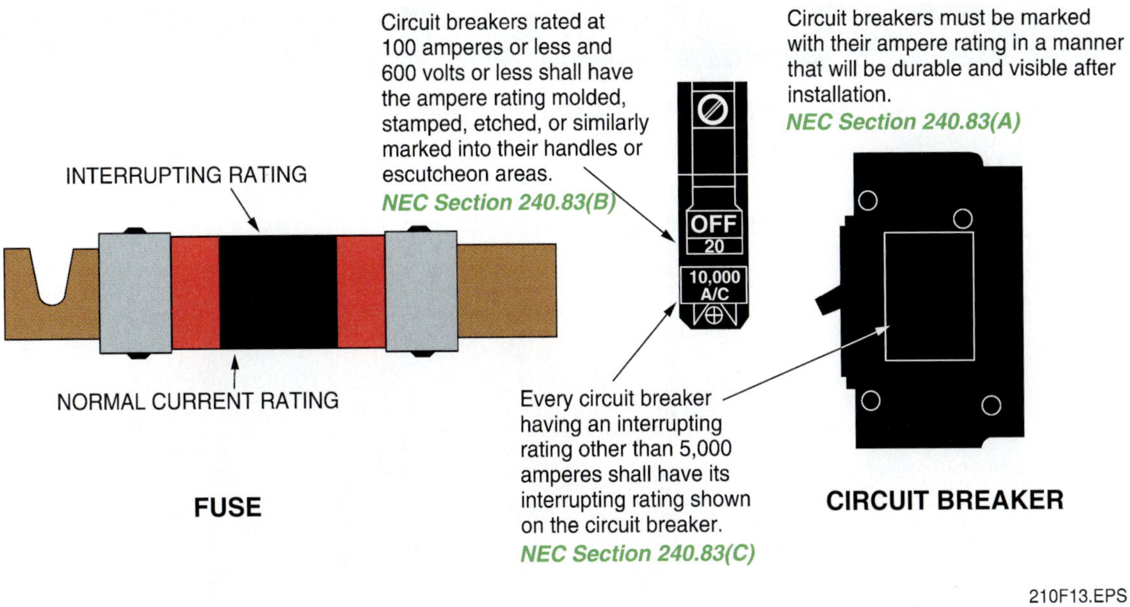

Figure 13 ◆ Most overcurrent protective devices are labeled with two current ratings.

2.4.0 *NEC*® Regulations

The following are critical *NEC*® sections with which you should become familiar. These are not the only *NEC*® requirements of importance when dealing with overcurrent protection devices, but they are considered by many to be the most important.

- *NEC Section 110.3(B), Installation and Use* – Listed or labeled equipment shall be used and installed in accordance with any instructions included in the listing or labeling.
- *NEC Section 110.9, Interrupting Rating* – Equipment intended to interrupt current at fault levels shall have an interrupting rating sufficient for the nominal circuit voltage and the current that is available at the terminals of the equipment. Equipment intended to interrupt current at other than fault levels shall have an interrupting rating at nominal circuit voltage sufficient for the current that must be interrupted.
- *NEC Section 110.10, Circuit Impedance and Other Characteristics* – The overcurrent protection devices, total impedance, component short circuit current ratings, and other characteristics of the circuit to be protected shall be so selected and coordinated as to permit the circuit protective devices that are used to clear a fault to do so without extensive damage to the electrical components of the circuit. The fault shall be assumed to be either between two or more of the circuit conductors, or between any circuit conductor and the grounding conductor or enclosing metal raceway.
- *NEC Section 240.1, Scope (FPN)* – Overcurrent protection for conductors and equipment is provided to open the circuit if the current reaches a value that will cause an excessively dangerous temperature in conductors or conductor insulation. See also *NEC Sections 110.9 and 110.10* for requirements for interrupting ratings and protection against fault currents.
- *NEC Section 240.2, Definition of Current-Limiting Overcurrent Protective Devices* – A current-limiting overcurrent protection device is a device which, when interrupting currents in its current-limiting range, will reduce the current flowing in the faulted circuit to a magnitude that is substantially less than that obtainable in the same circuit if the device were replaced with a solid conductor having comparable impedance.
- *NEC Section 240.83(C), Interrupting Rating* – Every circuit breaker having an interrupting rating other than 5,000A shall have its interrupting rating shown on the circuit breaker.
- *NEC Section 240.83(E), Voltage Marking* – Circuit breakers shall be marked with a voltage rating that is not less than the nominal system voltage that is indicative of their capability to interrupt fault currents between phases or from phase to ground.

THINK ABOUT IT

Installing Circuit Breakers

In addition to using the correct size circuit breaker (trip current and interrupt current), what are some other factors that should be taken into consideration when selecting and installing a circuit breaker in a panel?

210P1002.EPS

- *NEC Section 240.85, Applications* – A circuit breaker with a straight voltage rating, such as 240V or 480V, may be applied in a circuit in which the nominal voltage between any two conductors does not exceed the circuit breaker's voltage rating; except that a two-pole circuit breaker cannot be used for protecting a three-phase corner-grounded delta circuit unless it is marked single-phase/three-phase to indicate such suitability. A circuit breaker with a slash rating such as 120V/240V or 480Y/277V shall be permitted to be applied on a solidly grounded circuit where the nominal voltage of any conductor to ground does not exceed the lower of the two values of the circuit breaker's voltage rating, and the nominal voltage between any two conductors does not exceed the higher value of the circuit breaker's voltage rating.
- *NEC Section 250.4, General Requirements* – Systems and circuit conductors are grounded to limit voltages due to lightning, line surges, or unintentional contact with higher voltage lines, and to stabilize the voltage to ground during normal operation. System equipment and circuit conductors are solidly grounded to the system grounded conductor to facilitate overcurrent device operation in case of fault currents.

3.0.0 ◆ FAULT CURRENT CIRCUIT PROTECTION

Fault current protection is extremely important to anyone who works with and uses electrically operated equipment. A fault current exists when an unintended path is established between an ungrounded conductor and ground. This situation can occur not only from worn or defective electrical equipment, but also from accidental misuse of equipment that is in good working order.

Will a conventional overcurrent device (fuse or circuit breaker) detect a fault current and open the circuit before irreparable harm is done? Before answering this question, a study of the effects of current on the human body is in order.

The hand-to-hand body resistance of an adult lies between 1,000Ω and 4,000Ω, depending on moisture, muscular structure, and voltage. The average value is 2,100Ω at 240VAC and 2,000Ω at 120VAC.

Using Ohm's law, the current resulting from the above-listed average hand-to-hand resistance values is 120 milliamperes (mA) at 240VAC and 60mA at 120VAC. The effects of 60Hz alternating current on a normal healthy adult are as follows (note that the current is in milliamperes):

- *More than 5mA* – Generally painful shock.
- *More than 15mA* – Sufficient to cause freezing to the circuit for 50% of the population.
- *More than 30mA* – Breathing difficulty (possible suffocation).
- *50mA to 100mA* – Possible ventricular fibrillation or very rapid uncoordinated contractions of the ventricles of the heart resulting in loss of synchronization between the heartbeat and pulse beat. Once ventricular fibrillation occurs, it usually continues and death will happen within a few minutes unless special defibrillation equipment is used.
- *100mA to 200mA* – Certain ventricular fibrillation.
- *Over 200mA* – Severe burns and muscle contractions. The heart is more likely to stop than fibrillate.

Now, will a conventional overcurrent device open a circuit before irreparable harm is done? No! Here is why.

For example, the current that would flow from a defective electric drill through the metal housing and through the human body to ground would be 60mA, calculated using the above value for 120VAC. The 60mA current is less than one-half of 1% of the rating of 15A circuit breaker or fuse, and yet it approaches the current level, which may produce ventricular fibrillation. Obviously, the standard circuit breaker or fuse will not open the circuit under such low levels of current flow.

Ground fault circuit interrupters (GFCIs) are Class A protective devices built in accordance with *UL Standard No. 943* for ground fault circuit interrupters. UL defines a Class A device as one that will trip when a fault current to ground is 6mA or more. Also, Class A devices must not trip below 4mA.

Class A GFCIs provide a self-contained means of testing the fault current circuitry, as required by UL. To test, simply push the test button, and the device will respond with a trip indication. UL requires that the current generated by the test circuit shall not exceed 9mA. Also, UL requires the device to be functional at 85% of the rated voltage.

INSIDE TRACK

Listed HACR Circuit Breakers

HACR circuit breakers are UL listed for use in group motor applications such as heating, air conditioning, and refrigeration (HACR) applications. HACR breakers have a built-in time delay that allows a higher current than its rating to momentarily flow in the circuit. This compensates for the large starting current drawn by such loads. The equipment being protected must also be marked by the manufacturer as being suitable for protection by this type of breaker.

Knowing these facts, OSHA and other codes require the use of fault current circuit interrupters to be used on certain circuits, namely, all construction sites where temporary service is in use for electric power tools. Other required uses include:

- Circuits for electrically operated pool covers
- Power or lighting circuits for swimming pools, fountains, and similar locations
- Receptacles in both commercial and residential garages
- Receptacles in residential bathrooms
- Receptacles installed outdoors in public spaces
- Receptacles installed in residential crawl spaces or unfinished basements
- All kitchen countertop receptacles or any receptacles mounted within 6' of a wet bar sink
- Receptacles installed in bathhouses
- Receptacles installed in bathrooms of commercial or industrial buildings
- Receptacles installed on the roof of any building
- Dockside receptacles
- Branch circuits derived from autotransformers

There are two types of GFCI receptacles. The feed-through type will provide GFCI protection to the remainder of the outlets in a circuit, similar to a GFCI circuit breaker. The second, a nonfeed-through type, offers protection only at the point of installation. Both types are available in various styles.

When installed properly, GFCIs continuously monitor the current in the two conductor wires of the circuit—both the grounded and ungrounded conductors. The current in these two conductors should always be equal. If the GFCI senses a difference between them of more than 6mA, it assumes that the difference is due to fault current and automatically trips the circuit. Power is interrupted within $1/40$ of a second or less, which is fast enough to prevent injury to anyone in normal health. Note that the ungrounded and grounded conductors are constantly being monitored. If at any time the sensor detects a difference in current flow between these conductors, this difference will be amplified. See *Figure 14*. The amplified signal will then cause a circuit breaker to open to de-energize the circuit.

CASE HISTORY

GFCI Circuit Protection

Electricity to operate portable power tools at a construction site was supplied by a temporary service. A panel box on the pole had two duplex receptacles in waterproof boxes. One was a regular receptacle and the other a GFCI receptacle. The pole had not been inspected by the city and was not in compliance with code requirements. (It was not grounded.)

A worker used extension cords plugged into the regular receptacle to supply a power tool. The site was wet, humidity was high, and the worker was sweating. Reportedly, he was getting shocks whenever he used the tool. Unknown to the worker, the tool had a ground fault. As he was climbing down to the ground using a piece of metal floor truss as a makeshift ladder, he received a shock, causing him to fall from the floor truss into a puddle of water while still holding onto the tool. He was electrocuted.

The Bottom Line: This accident could have been prevented if the temporary electrical service at the construction site had complied with all local regulations and OSHA standards, including proper grounding. Also, OSHA requires that all 120V, single-phase 15A and 20A receptacles at construction sites that are not part of the permanent wiring be GFCI protected.

Another factor that contributed to this accident was the continued use of a power tool that should have been taken out of service immediately at the occurrence of the first shock. Also, the worker should have used an approved ladder instead of the floor truss to climb down to the ground.

INSIDE TRACK

Hurricane and Flood Damage Victims

The U.S. Consumer Product Safety Commission has issued a warning to hurricane and flood damage victims that circuit breakers, GFCIs, and fuses that have been submerged under water must be replaced to avoid electrocutions, explosions, and fires. This also applies to panels and electrical devices in general. This is because any flood water and silt trapped inside the devices may prevent them from performing properly. In addition, silt, sand, and salt can also allow tracking to occur between conductors, buses, and connection points.

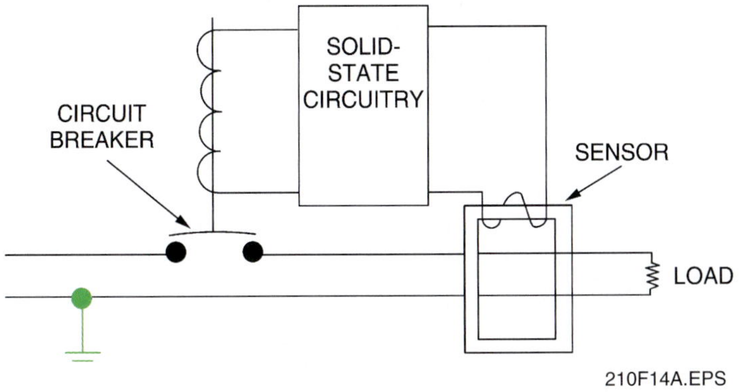

(A) DIAGRAM OF A GFCI CIRCUIT MONITORING A TWO-WIRE CIRCUIT

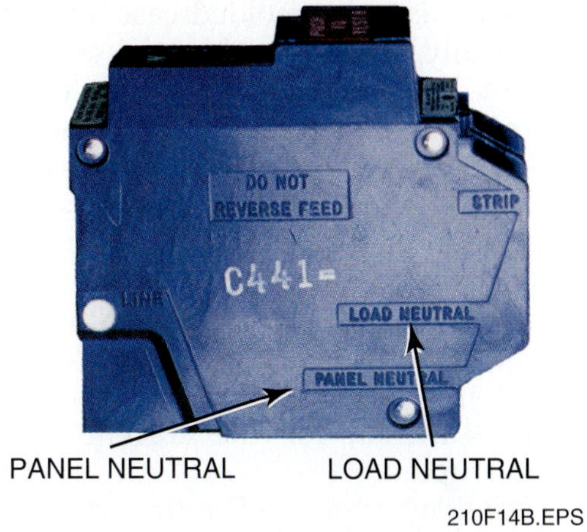

(B) GFCI CIRCUIT BREAKER

Figure 14 ◆ GFCI protection.

4.0.0 ◆ FUSES

A fuse is the simplest device for opening an electric circuit when excessive current flows due to an overload or such fault conditions as grounds and short circuits. A fusible link or links encapsulated in a tube and connected to contact terminals comprise the fundamental elements of the basic fuse. The electrical resistance of the link is so low that it simply acts as a conductor, and every fuse is intended to be connected in series with each phase conductor so that current flowing through the conductor or to any load must also pass through the fuse. The continuous current rating of the fuse in amps establishes the maximum amount of current the fuse will carry without opening. When the circuit current flow exceeds this value, an internal element (link) in the fuse melts due to the heat of the current flow and opens the circuit. Fuses are manufactured in a wide variety of types and sizes with different current ratings, different abilities to interrupt fault currents, various speeds of operation (either quick-opening or time-delay types), different internal and external constructions, and voltage ratings for both low-voltage (600V and below) and medium-voltage (over 600V) circuits.

- *Voltage rating* – Most low-voltage power distribution fuses have 250V or 600V ratings (other ratings are 125V and 300V). The voltage rating of a fuse must be at least equal to the circuit voltage. It can be higher but never lower. For instance, a 600V fuse can be used in a 208V circuit. The voltage rating of a fuse is a function of or depends upon its ability to open a circuit under an overcurrent condition. Specifically, the voltage rating determines the ability of the fuse to suppress the internal arcing that occurs after a fuse link melts and an arc is produced. If a fuse is used with a voltage rating lower than the circuit voltage, arc suppression will be impaired and, under some fault current conditions, the fuse may not safely clear the overcurrent.
- *Ampere rating* – Every fuse has a specific ampere rating. In selecting the ampacity of a fuse, consideration must be given to the type of load and code requirements. The ampere rating of a fuse should normally not exceed the current-carrying capacity of the circuit. For instance, if a conductor is rated to carry 20A, a 20A fuse is the largest that should be used in the conductor circuit. However, there are some specific circumstances under which the ampere rating is permitted to be greater than the current-carrying capacity of the circuit. A typical example is the motor circuit; **dual-element fuses** are generally permitted to be sized up to 175% and nontime-delay fuses up to 300% of the motor full-load amperes. Generally, the ampere rating of a fuse and switch combination should be selected at 125% of the load current. There are exceptions, such as when the fuse-switch combination is approved for continuous operation at 100% of its rating. A protective device must be able to withstand the destructive energy of a short circuit. If a fault current exceeds a level beyond the capability of the protective device, the device may actually rupture and cause severe damage. Thus, it is important when applying a fuse or circuit breaker to use one that can sustain the largest potential short circuit currents. The rating that defines the capacity of a protective device to maintain its integrity when reacting to fault currents is termed its interrupting rating. The interrupting rating of most

branch circuit, molded-case circuit breakers typically used in residential service entrance boxes is 10,000A. The rating is usually expressed as 10,000 AIC. Larger, more expensive circuit breakers may have ratings of 14,000 AIC or higher. In contrast, most modern, current-limiting fuses have an interrupting capacity of 200,000A and are commonly used to protect the lower-rated circuit breakers. *NEC Section 110.9* requires equipment intended to interrupt current at fault levels to have an interrupting rating sufficient for the current that must be interrupted.

- *Time-delay ratings* – The time-delay rating of a fuse is established by standard UL tests. All fuses have an inverse time current characteristic. That is, the fuse will open quickly on high currents and after a period of time delay on low overcurrents. Specific types of fuses are made to have specific time delays. The basic UL requirement for Class RK-1, RK-15, and J fuses that are marked time delay is that the fuse must carry a current equal to five times its continuous rating for a period of not less than ten seconds. UL has not developed time-delay tests for all fuse classes. Fuses are available for use where a time delay is needed along with current limitation on high-level short circuits. In all cases, the manufacturer's literature should be consulted to determine the degree of time delay in relation to the operating characteristics of the circuit being protected.

4.1.0 Plug Fuses

Plug fuses have a screw-shell base and are commonly used in dwellings for circuits that supply lighting, heating, and appliances. Plug fuses are supplied with standard screw bases (**Edison-base**) or Type S bases. An Edison-base fuse consists of a strip of fusible (capable of being melted) metal in a small porcelain or glass case, with the fuse strip or link visible through a window in the top of the fuse. The screw base corresponds to the base of a standard medium-base incandescent lamp. Edison-base fuses are permitted only as replacements in existing installations; all new work must use Type S fuses in accordance with *NEC Section 240.51(B),* which states that plug fuses of the Edison-base type may only be used for replacements in existing installations where there is no evidence of overfusing or tampering.

The chief disadvantage of the Edison-base plug fuse is that it is made in several ratings (from 0 to 30A), all with the same size base—permitting unsafe replacement of one rating by a higher rating. Type S fuses were developed to reduce the possibility of over-fusing a circuit (inserting a fuse with a rating greater than that required by the circuit). There are 15 classifications of Type S fuses, from 0 to 30A. Each Type S fuse has a base of a different size and a matching adapter. Once an adapter is screwed into a standard Edison-base fuseholder, it locks into place and is not readily removed without destroying the fuseholder. As a result, only a Type S fuse with a size the same as that of the adapter may be inserted. Two types of plug fuses are shown in *Figure 15;* a Type S adapter is also shown.

Plug fuses are also made in time-delay types that permit a longer period of overload flow before operation, such as on motor inrush current and other higher-than-normal currents. They are available in ratings up to 30A, both in Edison-base and Type S.

Their principal use is in motor circuits, where the starting inrush current to the motor is much higher than the running or continuous current. The time-delay fuse will not open on the inrush of

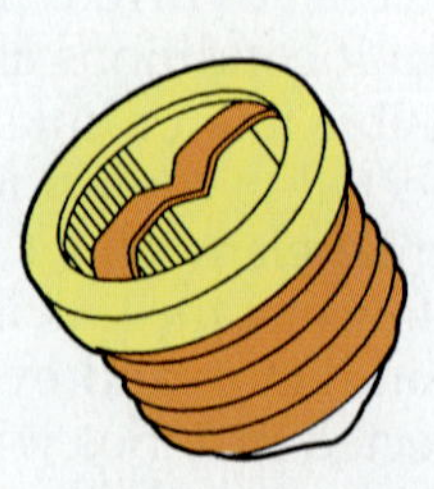

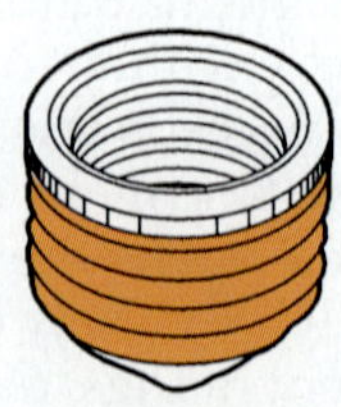

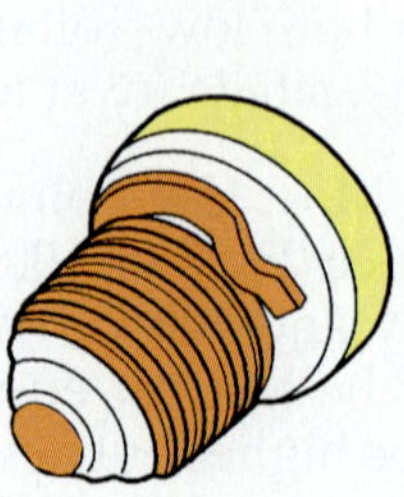

Figure 15 ◆ Standard Edison-base fuse on top; Type S with adapter shown on bottom.

high starting current. If, however, the high current persists, the fuse will open the circuit just as if a short circuit or heavy overload current had developed. All Type S fuses are time-delay fuses.

Plug fuses are normally permitted to be used in circuits of no more than 125V between phases, but they may also be used where the voltage between any ungrounded conductor and ground is not more than 150V. The screwshell of the fuseholder for plug fuses must be connected to the load side circuit conductor; the base contact is connected to the line side or conductor supply. A disconnecting means (switch) is not required on the supply side of a plug fuse.

The plug fuse is a **nonrenewable fuse,** or one-time fuse; that is, once it has opened the circuit because of a fault or overload, it cannot be used again or renewed. For safe and effective restoration of circuit operation, it must be replaced by a new fuse of the same rating and characteristics.

4.2.0 Cartridge Fuses

In most industrial and commercial applications, **cartridge fuses** are used because they have a wider range of types, sizes, and ratings than plug fuses. Older cartridge fuses were provided with a means to renew the fuse by unscrewing the end caps and replacing the links. These fuses presented safety hazards in that it was easy to tamper with them to override the safety feature. Their production has been discontinued. Two types of cartridge fuses are shown in *Figure 16*.

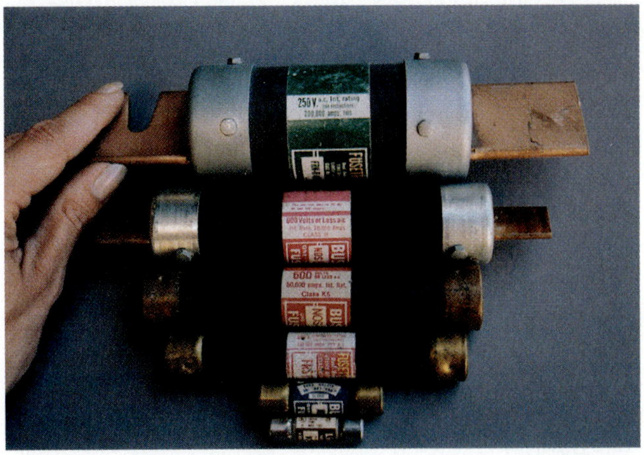

210F16.EPS

Figure 16 ◆ Cartridge fuses.

Single-element cartridge fuses – The basic component of this fuse is the link. Depending upon the ampere rating of the fuse, the single-element fuse may have one or more links. They are electrically connected to the end blades (or ferrules) and enclosed in a tube or cartridge surrounded by an arc-quenching filler material.

Under normal operation, when the fuse is operating at or near its ampere rating, it simply functions as a conductor. However, if an overload current occurs and persists for more than a short interval of time, the temperature of the link eventually reaches a level that causes a restricted segment of the link to melt; as a result, a gap is formed and an electric arc established. As the arc causes the link metal to burn back, the gap becomes progressively larger. The electrical resistance of the arc eventually reaches such a high level that the arc cannot be sustained and is extinguished; the fuse will have then completely cut off all current flow in the circuit. Suppression or quenching of the arc is accelerated by the filler material.

INSIDE TRACK

Plug Fuses

Plug fuses are used in the following applications:

- *W series* – Residential loads and nonmotor industrial circuits
- *TL series* – Normal-duty motor loads
- *T series* – When extra time delay or sensitivity to temperature is needed
- *S series* – Time-delay fuses with tamper-resistant dimensions to prevent substitution of incorrect fuses

The condition of plug fuses can be checked visually by looking into the fuse window to observe the condition of the fusible link. If the element is broken and/or the window is black or discolored, the fuse is blown and must be replaced. Note that the condition of the window gives an indication of why the fuse blew. If the element is broken but the window is clear, this indicates an overload occurred; if the window is black or discolored, this indicates a short circuit occurred.

Single-element fuses have a very high speed of response to overcurrents. They provide excellent short circuit component protection. However, temporary harmless overloads or surge currents may cause nuisance openings unless these fuses are oversized. They are best used, therefore, in circuits that are not subject to heavy transient surge currents and the temporary overload of circuits with inductive loads such as motors, transformers, and solenoids. Because single-element fuses have a high speed of response to short circuit currents, they are particularly suited for the protection of circuit breakers with low interrupting ratings.

Dual-element cartridge fuses – Unlike single-element fuses, the dual-element fuse can be applied in circuits subject to temporary motor overload and surge currents to provide both high performance short circuit and overload protection. Oversizing in order to prevent nuisance openings is not necessary. The dual-element fuse contains two distinct types of elements. Electrically, the two elements are series connected. The fuse links, similar to those used in the single-element fuse, perform the short circuit protection function; the overload element provides protection against low-level overcurrents or overloads and will hold an overload that is five times greater than the ampere rating of the fuse for a minimum time of ten seconds.

The overload section consists of a copper heat absorber and a spring-operated trigger assembly. The heat-absorber strip is permanently connected to the first short circuit link and to the second short circuit link on the opposite end of the fuse by the S-shaped connector of the trigger assembly. The connector electronically joins the one short circuit link to the heat absorber in the overload section of the fuse. These elements are joined by a calibrated fusing alloy. An overload current causes heating of the short circuit link connected to the trigger assembly. Transfer of heat from the short circuit link to the heat absorber in the mid-section of the fuse begins to raise the temperature of the heat absorber. If the overload is sustained, the temperature of the heat absorber eventually reaches a level that permits the trigger spring to fracture the calibrated fusing alloy and pull the connector free. The short circuit link is electrically disconnected from the heat absorber, the conducting path through the fuse is opened, and the overload current is interrupted. A critical aspect of the fusing alloy is that it retains its original characteristic after repeated temporary overloads without degradation.

Dual-element fuses may also be used in circuits other than motor branch circuits and feeders, such as lighting circuits and those feeding mixed lighting and power loads. The low-resistance construction of the fuses offers cooler operation of the equipment, which permits higher loading of fuses in switches and panel enclosures without heat damage and without nuisance openings from accumulated ambient heat.

4.3.0 Fuse Markings

It is a requirement of the *NEC®* that any cartridge fuses used for branch circuit or feeder protection must be plainly marked, either by printing on the fuse barrel or by a label attached to the barrel that shows the following:

- Ampere rating
- Voltage rating
- Interrupting rating (if other than 10,000A)
- Current limiting, where applicable
- The name or trademark of the manufacturer

4.4.0 UL Fuse Classes

Fuses are tested and listed by Underwriters Laboratories, Inc. in accordance with established standards of construction and performance. There are many varieties of miscellaneous fuses used for special purposes or for supplementary protection of individual types of electrical equipment. However, here we will chiefly be concerned with those fuses used for the protection of branch circuits and feeders on systems operating at 600V or below. Cartridge fuses in this category include UL Classes H, K, G, J, R, T, CC, and L.

5.0.0 ◆ OVERCURRENTS

An overcurrent is either an overload current or a short circuit current. The overload current is an excessive current relative to normal operating current, but one that is confined to the normal conductive paths provided by the conductors and other components and loads of the electrical system.

A short circuit *(Figure 17)* is probably the most common cause of electrical problems. It is an undesired current path that allows the electrical current to bypass the load on the circuit. Sometimes the short is between two wires due to faulty insulation, or it can occur between a wire and a grounded object, such as the metal frame of a motor.

5.1.0 Overloads

Overloads are most often between one and six times the normal current level. Usually, they are

210F17.EPS

Figure 17 ◆ Short circuit.

INSIDE TRACK

Current-Limiting Fuses

Current-limiting fuses have a much faster response to short circuit currents and are typically used to provide protection to mains, feeders, circuit breakers, and other components that lack an adequate interrupting rating.

INSIDE TRACK

Fuse Classes

Cartridge fuses are classified by Underwriters Laboratories, Inc., according to their maximum voltage and amperage ranges. Some typical values assigned for common fuse classifications are:

Class G	480V, 1 to 60A, interrupting rating of 100,000 rms symmetrical; listed as current limiting
Class H	250/600V, 1 to 600A, interrupting rating of 10,000 rms symmetrical
Class J	600V, 1 to 600A, interrupting ratings ranging from 200,000 to 300,000 rms symmetrical; listed as current limiting
Class K	250/600V, 1/10 to 600A, interrupting ratings ranging from 50,000 to 200,000 rms symmetrical
Class L	600V, 601 to 6,000A, interrupting ratings ranging from 200,000 to 300,000 rms symmetrical; listed as current limiting
Class R	250/600V, 1/10 to 600A, interrupting rating of 200,000 rms symmetrical; listed as current limiting
Class T	300/600V, 1 to 1200A, interrupting rating of 200,000 rms symmetrical; listed as current limiting
Class CC	600V, 1/10 to 30A, interrupting rating of 200,000 rms symmetrical; listed as current limiting

caused by harmless temporary surge currents that occur when motors are started up or transformers are energized. Since they are of brief duration, any temperature rise is trivial and has no harmful effect on the circuit components. Therefore, it is important that protective devices (fuses and circuit breakers) do not react to them.

Continuous overloads can result from motor defects (such as worn motor bearings), overloaded equipment, or too many loads on one circuit. Such sustained overloads are destructive and must be cut off by protective devices before they damage the electrical distribution system or affect the system loads. However, since they are of relatively low magnitude compared to short circuit currents, removal of the overload current within a few seconds will generally prevent equipment damage. A sustained overload current results in overheating of conductors and other components and will cause deterioration of insulation, which may eventually result in severe damage and short circuits if not interrupted.

6.0.0 ◆ GUIDE FOR SIZING FUSES

General guidelines for sizing fuses are given here for most circuits that will be encountered on conventional systems. Some specific applications may warrant other fuse sizing; in these cases, the load characteristics and appropriate *NEC*® sections should be considered. The selections shown here are not, in all cases, the maximum or minimum ampere ratings permitted by the *NEC*®. Demand factors as permitted by the *NEC*® are not included here.

6.1.0 Dual-Element Time-Delay Fuses

Application considerations for dual-element time-delay fuses include the following:

- *Main service* – Each ungrounded service-entrance conductor shall have a fuse in series with a rating not higher than the ampacity of the conductor, except as permitted in ***NEC Section 230.90.*** As specified in ***NEC Section 230.91,*** the service fuses must either be part of the service disconnecting means or located immediately adjacent to it.
- *Feeder circuit with no motor load* – The fuse size must be at least 125% of the continuous load plus 100% of the non-continuous load.
- *Feeder circuit with all motor loads* – Select the largest branch circuit overcurrent device and add 100% of the full-load current of all other motors connected on the same feeder. The *NEC*® will not permit you to size up; you must size down.
- *Feeder circuit with mixed loads* – Select the largest motor branch circuit fuse and add 100% of the full-load current of all other motors plus 125% of the continuous, non-motor load plus 100% of the non-continuous, non-motor load.
- *Branch circuit with no motor load* – The fuse size must be at least 125% of the continuous load plus 100% of the non-continuous load.

6.2.0 Nontime-Delay Fuses

Application considerations for nontime-delay fuses include the following:

- *Main service* – Service-entrance conductors shall have a short circuit protective device in each ungrounded conductor, either on the load side of or as an integral part of the service-entrance switch. The protective device shall be capable of detecting and interrupting all values of current in excess of its trip setting or melting point, which can occur at its location. A fuse rated in continuous amperes not to exceed three times the ampacity of the conductor, or a circuit breaker with a trip setting of not more than six times the ampacity of the conductors, shall be considered as providing the required short circuit protection.
- *Feeder circuit with no motor load* – The fuse size must be at least 125% of the continuous load plus 100% of the non-continuous load.
- *Feeder circuit with all motor loads* – Select the largest branch circuit overcurrent device and add 100% of the full-load current of all other motors connected on the same feeder. The *NEC*® will not permit you to size up; you must size down.
- *Feeder circuit with mixed loads* – Select the largest motor branch circuit fuse and add 100% of the full-load current of all other motors plus 125% of the continuous, non-motor load plus 100% of the non-continuous, non-motor load.
- *Branch circuit with no motor load* – The fuse size must be at least 125% of the continuous load plus 100% of the non-continuous load.

When sizing fuses for a given application, a schematic drawing of the system will help tremendously. Such drawings do not have to be detailed; a single-line schematic will suffice, such as the one shown in *Figure 18*.

Let us determine the feeder ampacity, motor overload protection, branch circuit short circuit and fault current protection, and feeder protection for one 25hp squirrel-cage induction motor (full starting voltage, nameplate current 31.6A, service factor 1.15, Code letter F), and two 30hp wound rotor induction motors (nameplate primary

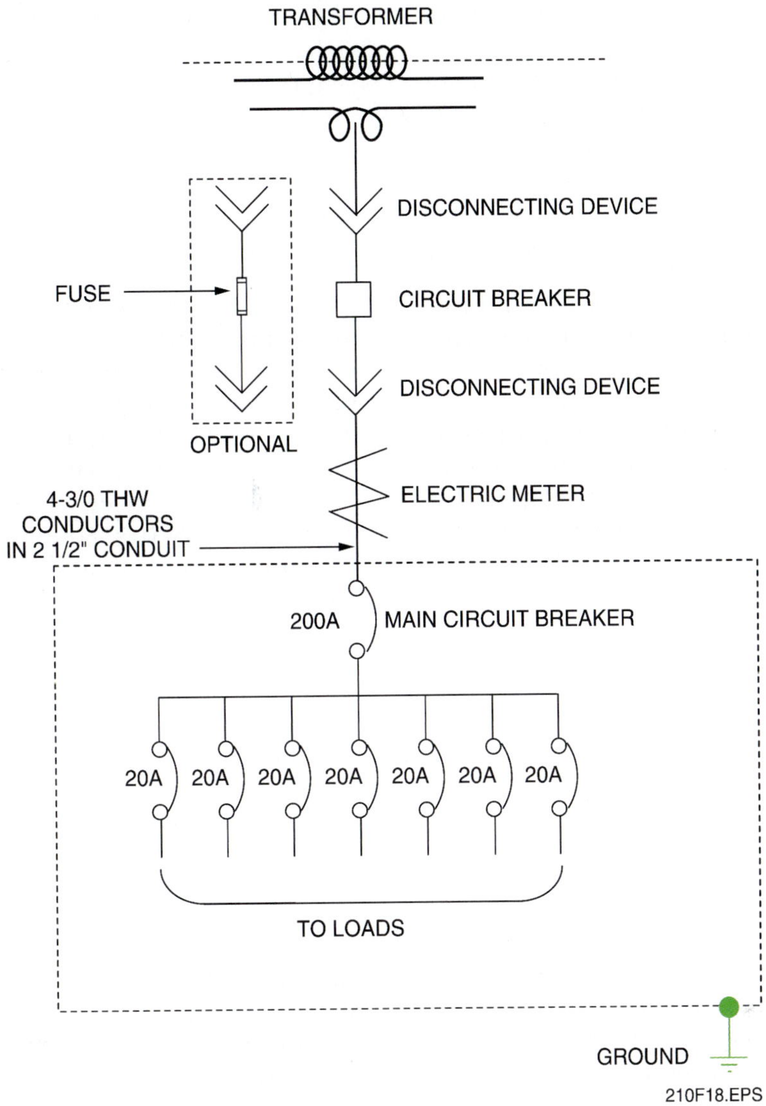

Figure 18 ◆ Typical single-line schematic diagram showing overcurrent protective devices.

current 36.4A, nameplate secondary current 65A, 40°C rise), on a 460V, three-phase, 60Hz electrical supply.

- *Feeder ampacity* – The full-load current value used to determine the ampacity of conductors for the 25hp motor is 34A [*NEC Section 430.6(A) and Table 430.250*]. A full-load current of 34A × 1.25 = 42.5A (*NEC Section 430.22*). The full-load current value used to determine the ampacity of primary conductors for each 30hp motor is 40A [*NEC Section 430.6(A) and Table 430.250*]. A full-load primary current of 40A × 1.25 = 50A (*NEC Section 430.22*). A full-load secondary current of 65A × 1.25 = 81.25A [*NEC Section 430.23(A)*]. The feeder ampacity will be (40A × 1.25) + 40A + 34A = 124A (*NEC Section 430.24*).
- *Overload* – Where protected by a separate overload device, the 25hp motor with a nameplate current of 31.6A must have overload protection of not over 39.5A [*NEC Sections 430.6(A) and 430.32(A)(1)*]. Where protected by a separate overload device, the 30hp motor with a nameplate current of 36.4A must have overload protection of not over 45.5A [*NEC Section 430.6(A)*]. If the overload protection is not sufficient to start the motor or carry the load, it may be increased according to *NEC Section 430.32(C)*. For a motor marked thermally protected, overload protection is provided by the thermal protector [*NEC Sections 430.7(A)(13) and 430.32(A)(2)*].
- *Branch circuit short circuit and fault current* – The branch circuit of the 25hp motor must have short circuit and fault current protection of not

over 300% for a nontime-delay fuse (*NEC Table 430.52*) or 3.00 × 34A = 102A. The next smallest standard size fuse is 100A. The fuse size may be increased to 110A [*NEC Section 430.52(C)(1), Exception No. 1 and NEC Section 240.6*]. Where the maximum value of branch circuit short circuit and fault current protection is not sufficient to start the motor, the value for a nontime-delay fuse may be increased to 400% [*NEC Section 430.52(C)(1), Exception No. 2(a)*]. If a time-delay fuse is to be used, see *NEC Section 430.52(C)(1), Exception No. 2(b).*

- *Feeder circuit* – The maximum rating of the feeder short circuit and fault current protection device is based on the sum of the largest branch circuit protective device (110A fuse) plus the sum of the full-load currents of the other motors, or 110A + 40A + 40A = 190A. The nearest standard fuse that does not exceed this value is 175A [*NEC Section 430.62(A)*].

7.0.0 ◆ SAFETY

Overcurrent protection offers one of the greatest means of protection against electrical faults to both equipment and personnel. However, there are certain precautionary measures that must be observed at all times:

- Make certain the switch or circuit breaker is open before making any inspections or changes to either the overcurrent device or the circuit it is protecting.
- Make certain that all components are compatible with each other and with the system on which they are used; that is, manufacturer, type, normal current ratings, interrupting capacity, voltage (especially voltage), etc.
- Use the proper tools for the job.
- When an overcurrent protection device opens a circuit, determine the reason before changing fuses or resetting the circuit breaker. A severe fault may not damage the circuit or equipment the first time, but repeated hits with high inrush current may do so the second or third time.
- Make sure that adjustable circuit breakers are calibrated correctly (refer to the manufacturer's instructions for calibration details). An improperly calibrated circuit breaker is considered by many to be the most dangerous of circuit breaker faults.
- Use the test breaker to trip devices one phase at a time.
- Never substitute a fuse or circuit breaker that is rated higher than the circuit conductors.
- When working on circuits protected by overcurrent protective devices, lock out the device so that it cannot be inadvertently closed. If locking out the device or switch is not possible, tagging is the next best means to help prevent someone from closing the device while the circuit is being worked on.

8.0.0 ◆ COORDINATION

Coordination is the name given to the time-current relationship among a number of overcurrent devices connected in series, such as fuses in a main feeder, subfeeder, and branch circuits. Safety is the prime consideration in the operation of fuses; however, coordination of the characteristics of fuses has become a very important factor in the large and complex electrical systems of the present time. Every fuse should be properly rated for continuous current and overloads and for the maximum short circuit current the electrical system could feed into a fault on the load side of the fuse—but this is not enough. It might still be possible for a fault on a feeder to open the main service fuse before the feeder fuse opens. Alternately, a branch circuit fault might open the feeder fuse before the branch circuit device opens. Such applications are said to be uncoordinated or nonselective—the fuse closest to the fault is not faster than the one farthest from the fault. When a fault on a feeder opens the main service fuse instead of the feeder fuse, the entire electrical system is taken out of service instead of just the one faulted feeder. Effective coordination minimizes the extent of electrical outage when a fault occurs. It therefore minimizes loss of production, interruption of critical continuous processes, loss of vital facilities, and possible panic.

Selective coordination is the selection of overcurrent devices with time/current characteristics that ensure the clearing of a fault or short circuit by the device nearest the fault on the line side of the fault. A fault on a branch circuit is cleared by the branch circuit device. The subfeeder, feeder, and main service overcurrent devices will not operate. Alternately, a fault on a feeder is opened by the feeder fuse without opening any other fuse on the supply side of the feeder. With selective coordination, only the faulted part of the system is taken out of service, which represents the condition of minimum outage.

With proper selective coordination, every device is rated for the maximum fault current it might be called upon to open. Coordination is achieved by studying the curve of current versus the time required for the operation of each device. The selection is then made so that the device nearest any load is faster than all devices closer to the supply, and each device going back to the service

Cartridge Fuse Removal

THINK ABOUT IT

After the ON-OFF handle of a general-purpose disconnect switch is set to OFF and the door has been opened, is there any danger involved in the removal or installation of the cartridge fuses?

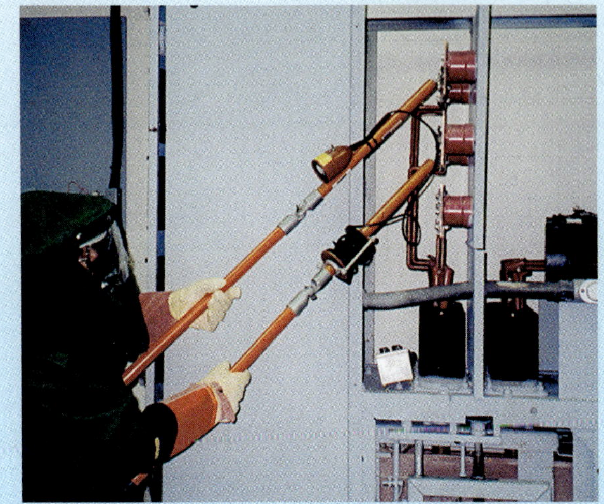

210P1003.EPS

entrance is faster than all devices closer to the supply. The main service fuses must have the longest opening time for any branch or feeder fault. The *NEC®* defines selective coordination as the "localization of an overcurrent condition to restrict outages to the circuit or equipment affected, accomplished by the choice of overcurrent protective devices and their ratings or settings."

9.0.0 ◆ SHORT CIRCUIT CALCULATIONS

Adequate interrupting capacity and protection of electrical components are two essential aspects required by *NEC Sections 110.9, 110.10, and 240.1.* The first step in ensuring that the system protective devices have the proper interrupting rating and provide component protection is to determine the available short circuit currents. Although several methods for determining short circuit currents are in use, the point-to-point method of short circuit calculation is normally the recommended method. The application of the point-to-point method permits the determination of available short circuit currents with a reasonable degree of accuracy at various points for either single-phase or three-phase electrical distribution systems. This method assumes unlimited primary short circuit current (infinite bus).

9.1.0 Basic Short Circuit Calculation Procedure

Step 1 Determine the transformer full-load amperes (I_{FLA}) from the nameplate, *Table 2*, or the following equation. ($E_{L\text{-}L}$ = line-to-line voltage.)

$$I_{FLA}(3\phi) = \frac{kVA \times 1{,}000}{E_{L\text{-}L} \times 1.73}$$

$$I_{FLA}(1\phi) = \frac{kVA \times 1{,}000}{E_{L\text{-}L}}$$

Step 2 Find the transformer multiplier using the following equation:

$$\text{Multiplier} = \frac{100}{\text{transformer } \%Z}$$

Step 3 Determine the transformer let-through short circuit current (I_{SCA}). Use *Table 3* or the following equation:

$$I_{SCA} = \text{transformer}_{FLA} \times \text{multiplier}$$

Table 2 Three-Phase Transformer—Full-Load Current Rating in Amperes

Voltage (Line-to-Line)	Transformer kVA Rating								
	150	167	225	300	500	750	1,000	1,500	2,000
208	417	464	625	834	1,388	2,080	2,776	4,164	5,552
220	394	439	592	788	1,315	1,970	2,630	3,940	5,260
240	362	402	542	722	1,203	1,804	2,406	3,609	4,812
440	197	219	296	394	657	985	1,315	1,970	2,630
460	189	209	284	378	630	945	1,260	1,890	2,520
480	181	201	271	361	601	902	1,203	1,804	2,406

Table 3 Short Circuit Currents Available from Various Size Transformers

Voltage and Phase	kVA	Full-Load Amps	% Impedance	Short Circuit Amps
120/240V Single phase	25	104	1.6	10,300
	37½	156	1.6	15,280
	50	209	1.7	19,050
	75	313	1.6	29,540
	100	417	1.6	38,540
	167	695	1.8	54,900
120/208V Three phase	150	417	2.0	20,850
	225	625	2.0	31,250
	300	834	2.0	41,700
	500	1,388	2.0	69,444
	750	2,080	3.5	59,426
	1,000	2,776	3.5	79,310
	1,500	4,164	3.5	118,965
	2,000	5,552	5.0	111,040
	2,500	6,950	5.0	139,000
277/480V Three phase	112½	135	1.0	13,500
	150	181	1.2	15,083
	225	271	1.2	22,583
	300	361	1.2	30,083
	500	601	1.3	46,230
	750	902	3.5	25,770
	1,000	1,203	3.5	34,370
	1,500	1,804	3.5	51,540
	2,000	2,406	5.0	48,120
	2,500	3,077	5.0	60,140

NOTE

The motor short circuit contribution, if significant, may be added to the transformer secondary short circuit current value as determined here. Proceed with the adjusted figure through Steps 4, 5, and 6 to follow. A practical estimate of motor short circuit contribution is to multiply the total load current in amperes by four.

Step 4 Calculate the f factor using the following equations along with *Table 4*.

For parallel runs, multiply the C value by the number of conductors per phase:

$$3\phi \text{ faults} = \frac{1.73 \times L \times I}{C \times E_{L\text{-}L}}$$

1ϕ line-to-line (L-L) faults on 1ϕ, center-tapped transformers:

$$f = \frac{2 \times L \times I}{C \times E_{L\text{-}L}}$$

Table 4 C Values for Conductors and Busways

AWG or kcmil	Copper, Three Single Conductors				Copper, Three-Conductor Cable			
	Steel Conduit		Nonmagnetic Conduit		Steel Conduit		Nonmagnetic Conduit	
	600V	5kV	600V	5kV	600V	5kV	600V	5kV
14	389	389	389	389	389	389	389	389
12	617	617	617	617	617	617	617	617
10	981	981	981	981	981	981	981	981
8	1,557	1,551	1,558	1,555	1,559	1,557	1,559	1,558
6	2,425	2,406	2,430	2,417	2,431	2,424	2,433	2,428
4	3,806	3,750	3,825	3,789	3,830	3,811	3,837	3,823
3	4,760	4,760	4,802	4,802	4,760	4,790	4,802	4,802
2	5,906	5,736	6,044	5,926	5,989	5,929	6,087	6,022
1	7,292	7,029	7,493	7,306	7,454	7,364	7,579	7,507
1/0	8,924	8,543	9,317	9,033	9,209	9,086	9,472	9,372
2/0	10,755	10,061	11,423	10,877	11,244	11,045	11,703	11,528
3/0	12,843	11,804	13,923	13,048	13,656	13,333	14,410	14,118
4/0	15,082	13,605	16,673	15,351	16,391	15,890	17,482	17,019
250	16,483	14,924	18,593	17,120	18,310	17,850	19,779	19,352
300	18,176	16,292	20,867	18,975	20,617	20,051	22,524	21,938
350	19,703	17,385	22,736	20,526	22,646	21,914	24,904	24,126
400	20,565	18,235	24,296	21,786	24,253	23,371	26,915	26,044
500	22,185	19,172	26,706	23,277	26,980	25,449	30,028	28,712
600	22,965	20,567	28,033	25,203	28,752	27,974	32,236	31,258
750	24,136	21,386	28,303	25,430	31,050	30,024	32,404	31,338
1,000	25,278	22,539	31,490	28,083	33,864	32,688	37,197	35,748

Note: The C values of other conductors/busways can be found in the manufacturer's literature for the device being sized.

1ϕ line-to-neutral (L-N) faults on 1ϕ, center-tapped transformers:

$$f = \frac{2 \times L \times I}{C \times E_{L\text{-}N}}$$

Where:

L = length (in feet) of the circuit to the fault

C = constant (from *Table 4*)

I = available short circuit current in amperes at the beginning of the circuit

Step 5 Calculate the multiplier (M) using the following equation, or use *Table 5:*

$$M = \frac{1}{1 + f}$$

Step 6 Compute the available short circuit current (symmetrical) at the fault using the following equation:

$$I_{SCA} \text{ at fault} = I_{SCA} \text{ at beginning of circuit} \times M$$

Table 5 Multipliers

f	M	f	M
0.01	0.99	1.50	0.40
0.02	0.98	1.75	0.36
0.03	0.97	2.00	0.33
0.04	0.96	2.50	0.29
0.06	0.94	3.50	0.22
0.07	0.93	4.00	0.20
0.08	0.93	5.00	0.17
0.09	0.92	6.00	0.14
0.10	0.91	7.00	0.13
0.15	0.87	8.00	0.11
0.20	0.83	9.00	0.10
0.25	0.80	10.00	0.09
0.30	0.77	15.00	0.06
0.35	0.74	20.00	0.05
0.40	0.71	30.00	0.03
0.50	0.67	40.00	0.02
0.60	0.63	50.00	0.02
0.70	0.59	60.00	0.02
0.80	0.55	70.00	0.01
0.90	0.53	80.00	0.01
1.00	0.50	90.00	0.01
1.20	0.45	100.00	0.01

> **NOTE**
>
> The line-to-neutral current is higher than the line-to-line fault current at the secondary terminals of a single-phase, center-tapped transformer. The short circuit current available (I) for Step 4 above should be adjusted at the transformer terminals as follows:
>
> I = 1.5 × L-L short circuit amperes at transformer terminals

At some distance from the terminals, depending on wire size, the L-N fault current is lower than the L-L fault current. The 1.5 multiplier is an approximation and will theoretically vary from 1.33 to 1.67. These figures are based on the change in turns ratio between the primary and the secondary, infinite source available, zero feet from the terminals of the transformer, and 1.2 × %X and 1.5 × %R for L-N vs. L-L resistance and reactance values. Begin L-N calculations at the transformer secondary terminals, then proceed point-to-point.

9.2.0 Practical Application

To demonstrate how the point-to-point method of short circuit calculation is used, find the total short circuit amperes at Fault #1 (I_{SCA1}) and Fault #2 (I_{SCA2}) in the circuit shown in *Figure 19*.

Step 1 $I_{FLA} = \frac{kVA \times 1{,}000}{E_{L\text{-}L} \times 1.73} = \frac{300 \times 1{,}000}{208 \times 1.73} = 834A$

Step 2 $\text{Multiplier} = \frac{100}{\text{transformer } \%Z} = \frac{100}{2} = 50$

Step 3 I_{SCA} at transformer secondary $= 834 \times 50 = 41{,}700A$

Step 4 $f_1 = \frac{1.73 \times L_1 \times I_{SCA}}{C_1 \times E_{L\text{-}L}}$

$$= \frac{1.73 \times 20 \times 41{,}700}{22{,}185 \times 208} = 0.313$$

Step 5 $M_1 = \frac{1}{1 + f_1} = \frac{1}{1 + 0.313} = 0.762$

Step 6 $I_{SCA1} = I_{SCA} \times M = I_{SCA1}$
$= 41{,}700 \times 0.762 = 31{,}775A$ at Fault #1

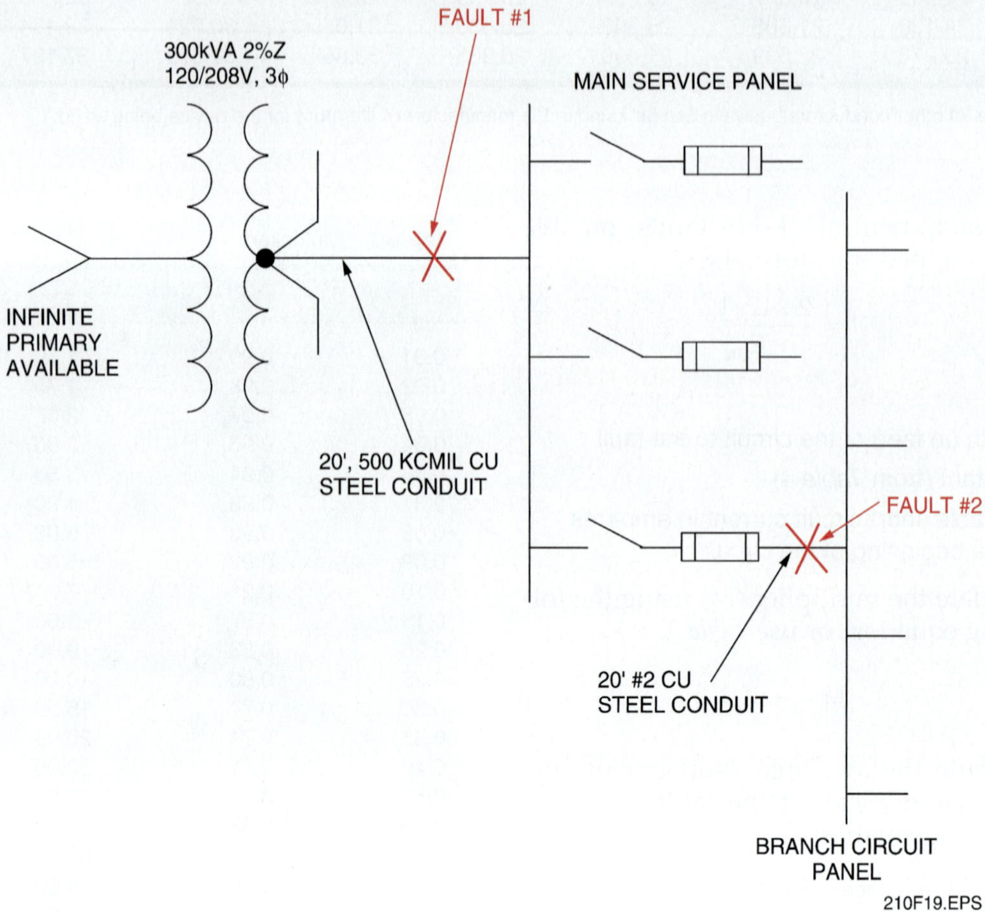

Figure 19 ◆ Sample one-line diagram for short circuit calculation.

To find the available short circuit current at Fault #2 (I_{SCA2}), use the current at Fault #1 as the available short circuit current.

Step 7
$$f_2 = \frac{1.73 \times L \times I_{SCA1}}{C_2 \times E_{L\text{-}L}}$$
$$= \frac{1.73 \times 20 \times 31{,}775}{5{,}906 \times 208} = 0.895$$

Step 8
$$M_2 = \frac{1}{1 + f_2} = \frac{1}{1 + 0.895} = 0.528$$

Step 9
$$I_{SCA2} = I_{SCA} \times M_2$$
$$= I_{SCA2} = 31{,}775 \times 0.528 = 16{,}777\text{A}$$
at Fault #2

9.3.0 Peak Let-Through Charts

Peak let-through charts *(Figure 20)* allow you to determine both the peak let-through current and the apparent prospective root-mean-square (rms) symmetrical let-through current. Such charts are commonly referred to as current limitation curves. These charts vary with the type of overcurrent protective device and are shown here only to familiarize you with their appearance. Their use will be explained in future modules.

10.0.0 ◆ TROUBLESHOOTING AND TESTING CIRCUIT BREAKERS AND FUSES

The main function of a circuit breaker or fuse is to protect the circuit wiring from damage caused by the heat generated by overloads and short circuits. While both perform similar functions, they do so in different ways and require slightly different procedures to troubleshoot them.

10.1.0 Circuit Breakers

A standard circuit breaker operates on both thermal and magnetic principles. The magnetic portion provides an instantaneous trip for currents of high levels such as those present in a short circuit. The thermal portion protects the circuit from excessive heat damage generated by lower level currents such as those caused by an overload.

The first step in troubleshooting is to verify the position of the breaker handle. When tripped, most breaker handles move to a position halfway between ON and OFF. Some have visual indicators, while others do not. To reset a circuit breaker, move the handle to the OFF position and then move it back to the ON position.

CAUTION

Whenever re-energizing a breaker, stand to one side and look away from the breaker. If the breaker trips immediately when reset, stop. Do not re-energize the circuit until the fault is located or the load is disconnected from the breaker.

An instant trip or failure to reset is a good sign that a short circuit is causing the magnetic portion to trip. This can be verified by turning the breaker off, checking to ensure there is no presence of voltage, and removing the load wires from the breaker. After removing the load wires, re-energize the breaker and verify the presence of voltage on all load terminals. If the breaker remains on, this may indicate a short circuit in the wiring or in a wiring device.

If the breaker stays on for a limited time, the problem is likely an overload condition, a high impedance or arcing fault, or a short circuit condition. Use the following procedure to troubleshoot the problem:

Step 1 Verify if voltage is present using a voltmeter. First verify the operation of the meter by testing it on a known voltage source. Then test for voltage by applying probe one to the load side of the circuit breaker and probe two to ground. Continue the test by moving probe two to other phases that may be present in the panel. If voltage is present, test for current using a clamp-on ammeter. Clamp the meter around the wire to be tested. In the case of multi-pole circuit breakers, this test will need to be performed on each wire. Verify that the current does not exceed the rating of the breaker.

NOTE

Thermal magnetic breakers are designed not to trip at 100% of the handle rating under normal operating conditions. At slightly over 100% (between 105% and 120%) of the handle rating, the breaker must trip between 2,000 and 10,000 seconds. At 150% of the handle rating, the same molded case breaker must trip between 200 and 900 seconds. This type of breaker will actually sustain a load in excess of its rating for a short time, depending upon the size of the load. If the current is less than the handle rating of the breaker but the breaker keeps tripping, it may be due to other factors such as unmeasured harmonic currents and/or excess heat or a defective breaker.

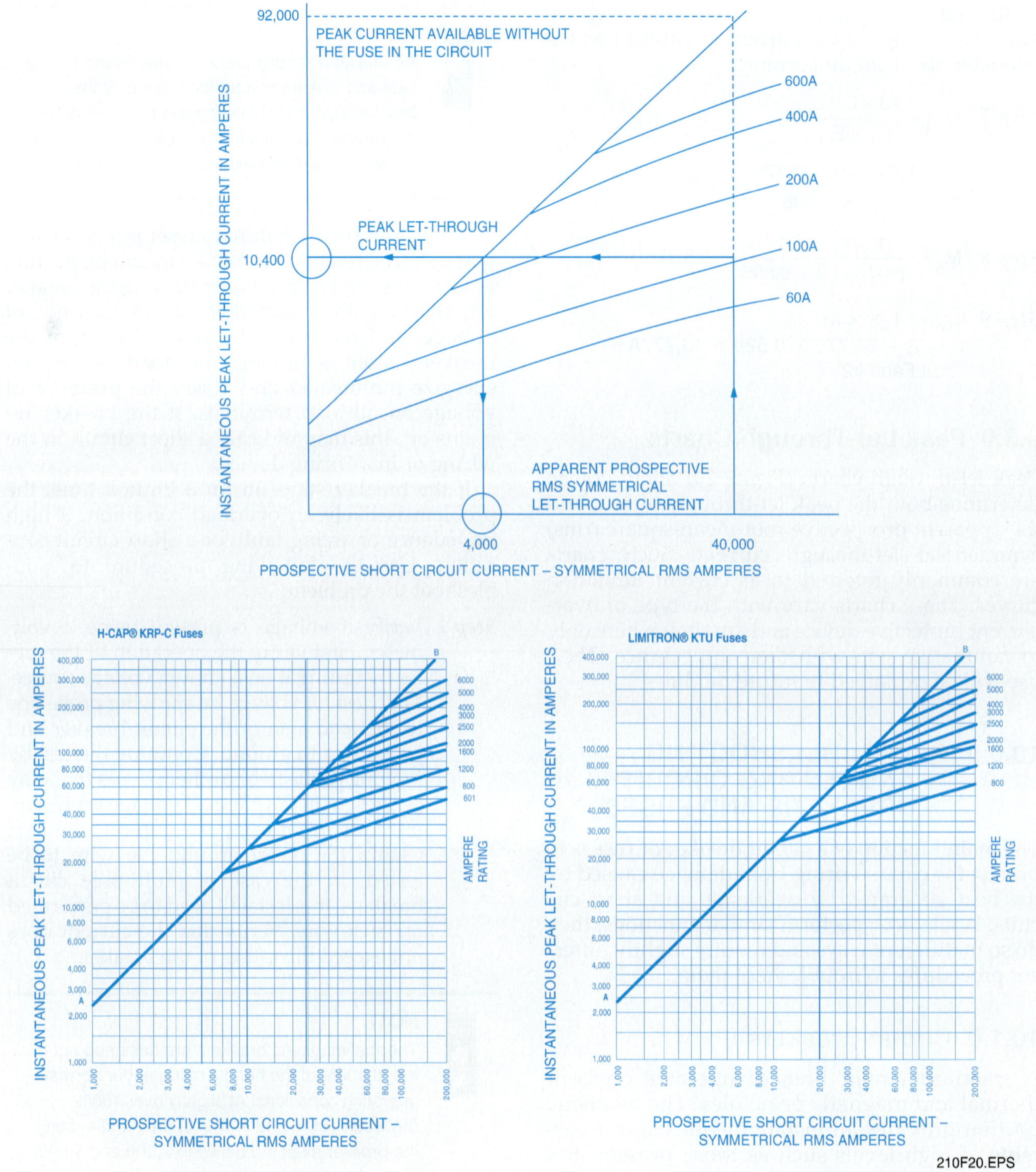

Figure 20 ◆ Peak let-through charts.

Step 2 Check for excess heat. If the breaker is remaining on for a limited amount of time and the current levels are within tolerance, the problem may be heat related. As mentioned earlier, this is a thermally operated device. It senses the heat produced due to current and trips if that heat is exceeded. Outside heat sources may also affect the operation. If the breaker is located in an area where there is a high ambient temperature, the breaker may trip at currents less than the handle rating of the breaker. Other factors that may cause this are loose or corroded connections, such as where the breaker is attached to the busbar in the panel, or loose connections where the load wire is attached to the breaker. If an infrared heat detector is available, it may be used to pinpoint the source of heat causing the breaker to trip or you may check by using your own senses. You may also check for heat in the panel by lightly placing your hand on top of the suspected breaker and the adjacent breakers. The breaker may be cool to slightly warm, but should not feel hot.

WARNING!
When testing for this condition, be aware of the voltages present and take the appropriate precautionary measures before proceeding.

Step 3 Listen for any crackling sounds that may indicate arcing. Electrical arcs generate ozone, which has a distinct sharp odor. Smell for signs of arcing.

Step 4 Remove the breaker to inspect the connection between the breaker plug-on jaw and the busbar. Follow the manufacturer's recommendations for removing the breaker. Inspect the breaker at the termination points. If excess heat has been present, the terminals may be discolored. Check the wire near the point of attachment for discoloration or charring of the insulation.

Step 5 Replace the breaker.

Breakers are usually stable, reliable, and long-lasting. However, excess current and tripping may damage the internal components and lead to premature wear or unreliable operation. If there is any doubt as to the operation or condition of a breaker, replace it immediately and inspect the connecting bus and nearby components for potential problems.

10.2.0 Fuses

There are two types of fuses: single-element fuses and dual-element fuses. A single-element fuse is designed to act immediately upon overcurrent, which may be either a short circuit or an overload. A dual-element fuse has two fuse elements: a quick-acting element similar to that of a single-element fuse and a thermal operation time-delay element. The quick-acting element is sized to react to short circuit conditions while allowing smaller overcurrents to pass. The time-delay element will protect against small overcurrents over a given time, similar to that of a thermal breaker.

WARNING!
Since fuses usually have live exposed parts within the fuseholder, extra care must be taken when troubleshooting.

To troubleshoot a fuse using an ohmmeter, proceed as follows:

Step 1 Disconnect the power to the fuse. This usually involves turning off a local disconnect switch. If the fuseholder is not adjacent to the power switch, follow appropriate lockout/tagout procedures.

Step 2 Verify that the voltage has been removed from all fuses associated with the circuit. Use a voltmeter to take readings from the top of the fuseholder to ground and other phases, and then repeat the procedure from the bottom of the fuseholder. Another check between the top and bottom of the fuseholder will verify that no voltage is present.

Step 3 After checking for voltage, remove the fuse. Using an ohmmeter or continuity tester, place one probe at one end of the fuse and the other probe on the opposite end. If the reading shows no resistance, indicating continuity, the fuse is good. Otherwise, the fuse is bad.

Step 4 To locate a blown fuse in an energized circuit, use a voltmeter to check for voltage first at the top of the fuse, checking to ground or another phase, and then at the bottom of the fuse. If voltage is present at the top but not the bottom, the fuse is bad. Disconnect the power and verify voltage is not present before replacing the fuse. If voltage-to-ground or another phase is present at both the top and bottom of the fuse and a bad fuse is still suspected, conduct further voltage checks from the bottom of the fuse to all phases (line-to-line in a 240V single-phase circuit or phase-to-phase and ground). If voltage is not obtained when checking phase-to-phase, a final check across the fuse from top to bottom should also be done. If voltage is present across the fuse, the fuse is bad.

NOTE
Fuses have varying time curves between types and classes and therefore vary in the amount of time they may sustain an overload. They are still affected by the same external problems that affect a circuit breaker, including unmeasured harmonic currents and excess heat.

Step 5 To check for excess heat, disconnect all power, verify that the voltage has been removed, and proceed to do the visual and other tests. In this case, a visual test is easy to perform. Look for signs of discoloration on the fuse, fuseholders, and terminals. If fuseholder clips are discolored from heat, they may need to be replaced. The fuse clip material will lose its spring temper, and squeezing it with pliers won't fix it.

Check the wire for insulation damage or discoloration that may result from excessive heat. Since the voltage is connected, the sound of arcing should not be present, but the smell may exist.

Step 6 Disconnect the power, verify that voltage has been removed, and replace the fuse. Note that voltage detected across a fuse (end-to-end) in a 240V or three-phase circuit is indicative of an open link.

Step 7 Once the new fuse is installed, verification should follow. After installing the new fuse, re-energize the circuit using the proper safety techniques. Verify the presence of voltage on all load terminals by checking phase-to-phase and phase-to-ground using a clamp-on ammeter to verify the current on the circuit.

WARNING!
Verify that all voltage has been removed prior to touching or replacing any fuse.

Fuses are stable devices that seldom fail due to problems within the fuses themselves. However, a fuse may be stressed over time from excessive currents and eventually fail. More often there is a problem within the circuit that causes the failure. Since fuses are good for one time only, it is sometimes more difficult to determine the problem that caused a fuse to fail. If replacement of one fuse does not correct the problem, and the visual inspections have been completed, it may be necessary to troubleshoot the circuit using known methods for that circuit. Do not continue to replace fuses hoping to eventually find one that holds. Do not increase the fuse size or change fuse types without investigating the impact on the protection of the equipment. Be aware of the limitations of single-element fuses and the limited use with motor circuits.

INSIDE TRACK

Other Types of Circuit Breakers

In addition to circuit breakers that protect against overloads and ground fault short circuit protection (overcurrents) some circuit breakers are designed with additional capabilities or other functions. The following are some examples of these breakers:

- *Shunt trip circuit breaker* – In addition to the normal operating devices that protect against overcurrents, this breaker has a built-in electric coil that causes it to open the breaker contacts when the coil is energized by an outside source. Typical sources may include fire suppression circuits, pushbuttons, or alarm circuits.
- *Arc fault circuit breaker* – In addition to the normal operating devices that protect against overcurrents, this breaker includes electronic circuits to monitor current flow within the breaker. If it detects a pattern of small continuous surges or spikes in that current flow, the breaker will operate and open the circuit. This pattern of surges or spikes is typical of currents in a short, high-resistance arcing circuit, such as a frayed electrical cord or a loose connection. The *NEC®* requires these types of breakers to be used on all residential bedroom circuits.
- *GFCI breaker* – In addition to the normal operating devices that protect against overcurrents, this breaker includes circuitry to monitor currents on both conductors in the circuit. If there is an Imbalance of those currents of 6mA or more, the breaker will operate and open the circuit. This imbalance would indicate that a current path other than that intended has formed, possibly through equipment or persons and is flowing to ground. GFCI breakers have different ranges to enable protection of people, equipment, or both.
- *Switched neutral breaker* – In addition to the normal operating devices that protect against overcurrents, this breaker disconnects the neutral or grounded conductor simultaneously with all ungrounded conductors. The use of breakers that switch grounded conductors is limited to installations such as fuel dispensing equipment as identified in ***NEC Section 514.11***.
- *Non-automatic breaker* – This breaker has no devices to protect the circuit against overcurrent. It is used as a means of manually disconnecting circuits by operating the handle. Its use would be similar to a non-fused disconnect switch. It is also known as a molded case switch.

THINK ABOUT IT

Putting It All Together

Examine the electrical service in your home. Does it use fuses or circuit breakers? What are the listed amperages? Are there any double breakers? What devices do they serve? Does your furnace use a separate fused disconnect? If the fuses/breakers aren't labeled, take some time to label them now. Can you visualize the internal circuits of your home based on the areas that are linked to the same breaker?

Review Questions

1. Which of the following is the most important consideration when working with overcurrent protection devices?
   a. The brand name
   b. The type of project
   c. The correct interrupting capacity
   d. The age of the device

2. On renewable cartridge fuses, which part of the fuse is normally renewed?
   a. The barrel
   b. The link
   c. The ferrule
   d. The ferrule ring

3. Which of the following best describes a dual-element fuse?
   a. A fuse having an overcurrent limit and a time delay before activation
   b. A fuse with two attachment points
   c. A fuse having two ferrules at each end
   d. A fuse with two barrels

4. When a common molded-case circuit breaker trips, in what position will the operating handle be?
   a. Halfway between the ON and OFF positions
   b. In the ON position
   c. In the OFF position
   d. In the RESET position

5. What is the highest voltage rating for circuit breakers used on DC systems that UL recognizes?
   a. 240V
   b. 277V
   c. 480V
   d. 600V

6. Where is the ampere rating normally marked on circuit breakers?
   a. On the side of the molded case
   b. On the handle
   c. Beneath the terminal lug
   d. On the bottom of the molded case

7. What is normally the highest short circuit current in residential applications?
   a. 10,000A
   b. 30,000A
   c. 50,000A
   d. 100,000A

8. Every circuit breaker having an interrupting rating other than _____ must have its interrupting rating shown on the circuit breaker.
   a. 1,000A
   b. 5,000A
   c. 10,000A
   d. 65,000A

9. Which of the following best describes where the *NEC*® allows the use of Edison-base fuses?
   a. On new installations only
   b. On circuits rated at over 40A
   c. On circuits rated under 40A
   d. For replacements in existing installations

10. What must be used in conjunction with Type S plug fuses when they are installed in a new fuse holder?
    a. A Type S adapter
    b. A copper penny
    c. A circuit breaker connected in series
    d. A setscrew holding device

Summary

Personal safety is the most important issue when completing electrical installations, and the protection of buildings and equipment comes second. Much of this protection is provided by overcurrent protection. This is why the *NEC*®, UL, NEMA, and other organizations put forth so much effort in regulating the type and methods of overcurrent protection.

To fully understand the purpose of overcurrent protection devices, you must first understand the meaning and importance of common electrical terms relating to overcurrent protection. Furthermore, you must understand the key *NEC*® requirements regarding overcurrent protection. When making electrical installations, always check the working drawings to make certain that all parts of the electrical system—that is, short circuit currents, fault currents, interrupting capacities, grounding electrode conductors, electrical grounding conductors, and the like—comply with the latest edition of the *NEC*®.

Also verify that all conductors (circuit, feeder, service, grounding electrode conductors, equipment grounding conductors, and bonding conductors) have adequate capacity to safely conduct any fault current that is likely to be imposed on them.

Determine let-through current values (peak and rms) when current-limiting overcurrent devices are used. Furthermore, apply current-limiting data to protect downstream electrical components that have interrupting ratings and/or withstand ratings less than the available fault current at any given point in the system.

Notes

Trade Terms Introduced in This Module

Cartridge fuse: A fuse enclosed in an insulating tube to confine the arc when the fuse blows.

Circuit breaker: A device that is designed to open a circuit automatically at a certain overcurrent. It can also be used to manually open and close the circuit.

Dual-element fuse: A fuse having two fuse characteristics; the usual combination is an overcurrent limit and a time delay before activation.

Edison-base: The standard screw base used for ordinary lamps and Edison-base plug fuses.

Fault current: The current that exists when an unintended path is established between an ungrounded conductor and ground.

Frame size: A method used to classify circuit breakers according to given current ranges.

Fuse: A protective device that opens a circuit when the fusible element is severed by heating due to a fault current or overcurrent passing through it.

Fuse link: The fusible part of a cartridge fuse.

Molded-case circuit breaker: A circuit breaker enclosed in an insulating housing.

Nonrenewable fuse: A fuse that must be replaced after it interrupts a circuit.

Overcurrent protection: De-energizing a circuit whenever the current exceeds a predetermined value; the usual devices are fuses, circuit breakers, or magnetic relays.

Plug fuse: A type of fuse that is held in position by a screw thread contact instead of spring clips, as is the case with a cartridge fuse.

Pole: That portion of a device associated exclusively with one electrically separated conducting path of the main circuit or device.

Short circuit: The current that exists when an unintended path is created between any two components in a circuit.

Additional Resources

This module is intended to present thorough resources for task training. The following reference works are suggested for further study. These are optional materials for continued education rather than for task training.

American Electrician's Handbook, Latest Edition. New York: Croft and Summers, McGraw-Hill.

National Electrical Code® Handbook, Latest Edition. Quincy, MA: National Fire Protection Association.

CONTREN® LEARNING SERIES — USER FEEDBACK

The NCCER makes every effort to keep these textbooks up-to-date and free of technical errors. We appreciate your help in this process. If you have an idea for improving this textbook, or if you find an error, a typographical mistake, or an inaccuracy in NCCER's *Contren®* textbooks, please write us, using this form or a photocopy. Be sure to include the exact module number, page number, a detailed description, and the correction, if applicable. Your input will be brought to the attention of the Technical Review Committee. Thank you for your assistance.

Instructors – If you found that additional materials were necessary in order to teach this module effectively, please let us know so that we may include them in the Equipment/Materials list in the Annotated Instructor's Guide.

Write: Product Development
National Center for Construction Education and Research
P.O. Box 141104, Gainesville, FL 32614-1104

Fax: 352-334-0932

E-mail: curriculum@nccer.org

Craft ______________________ Module Name ______________________

Copyright Date ____________ Module Number ____________ Page Number(s) ____________

Description

____

____

____

____

(Optional) Correction

____

____

____

(Optional) Your Name and Address

____

____

____

Contactors and Relays

26211-05

Texas City CHP Cogeneration Project

Texas City, Texas

Mega-Projects Over $100 Million Award Winner

S&B Engineers and Constructors

26211-05

Contactors and Relays

Topics to be presented in this module include:

Overview

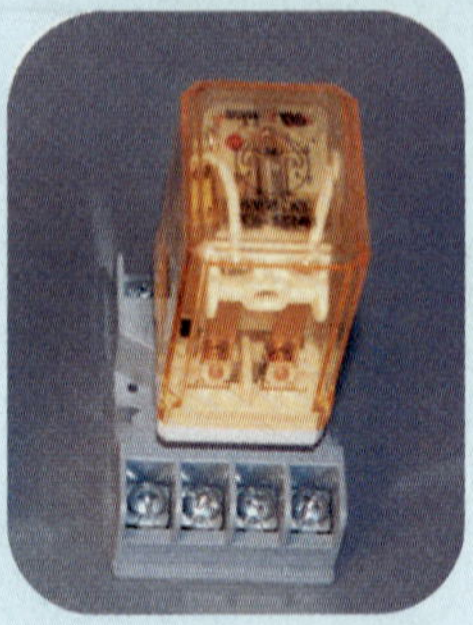

Contactors and relays are switching devices typically equipped with more than one set of contacts that are opened or closed by electromagnetic energy such as a coil. The coil usually gets its control current from an outside source such as a control circuit. Whenever current flows through the coil, the magnetic fields created by the current flow cause a mechanical action to occur that changes the state of the contacts. Since the circuits that are attached to the contacts are electrically isolated from the coil circuit, the contacts can control circuits with a much higher voltage and current flow than that of the control circuit.

Contactors can control more than one circuit at a time and are only limited by the number of contacts operated by the coil. A common multi-circuit application for contactors is the control of lighting. Contactors are regulated by their ratings. Depending on the type of contactor, typical ratings may include maximum horsepower, continuous current, types of lighting loads, maximum resistance heat loads, maximum transformer kVA, and ratings for switching capacitors.

Relays are contactors that are used to control circuits of lower voltage and amperage ratings. These devices are commonly applied in control circuits and are referred to as control relays. Control relays are available in a variety of configurations and functions, including timers and timing relays, solid-state relays, and overload relays.

Note: *National Electrical Code*® and *NEC*® are registered trademarks of the National Fire Protection Association, Inc., Quincy, MA 02269. All *National Electrical Code*® and *NEC*® references in this module refer to the 2005 edition of the *National Electrical Code*®.

Objectives

When you have completed this module, you will be able to do the following:

1. Describe the operating principles of contactors and relays.
2. Select contactors and relays for use in specific electrical systems.
3. Explain how mechanical contactors operate.
4. Explain how solid-state contactors operate.
5. Install contactors and relays according to the *NEC*® requirements.
6. Select and install contactors and relays for lighting control.
7. Read wiring diagrams involving contactors and relays.
8. Describe how overload relays operate.
9. Connect a simple control circuit.
10. Test control circuits.

Trade Terms

Ambient temperature
Break
Inductive load
Instantaneous
Latch coil
Make
Mechanically held relay
Normally closed (N.C.) contacts
Normally open (N.O.) contacts
Shading coil
Solid-state relay
Unlatch coil

Required Trainee Materials

1. Pencil and paper
2. Appropriate personal protective equipment
3. Copy of the latest edition of the *National Electrical Code*®

Prerequisites

Before you begin this module, it is recommended that you successfully complete the following: *Core Curriculum; Electrical Level One;* and *Electrical Level Two,* Modules 26201-05 through 26210-05.

This course map shows all of the modules in *Electrical Level Two.* The suggested training order begins at the bottom and proceeds up. Skill levels increase as you advance on the course map. The local Training Program Sponsor may adjust the training order.

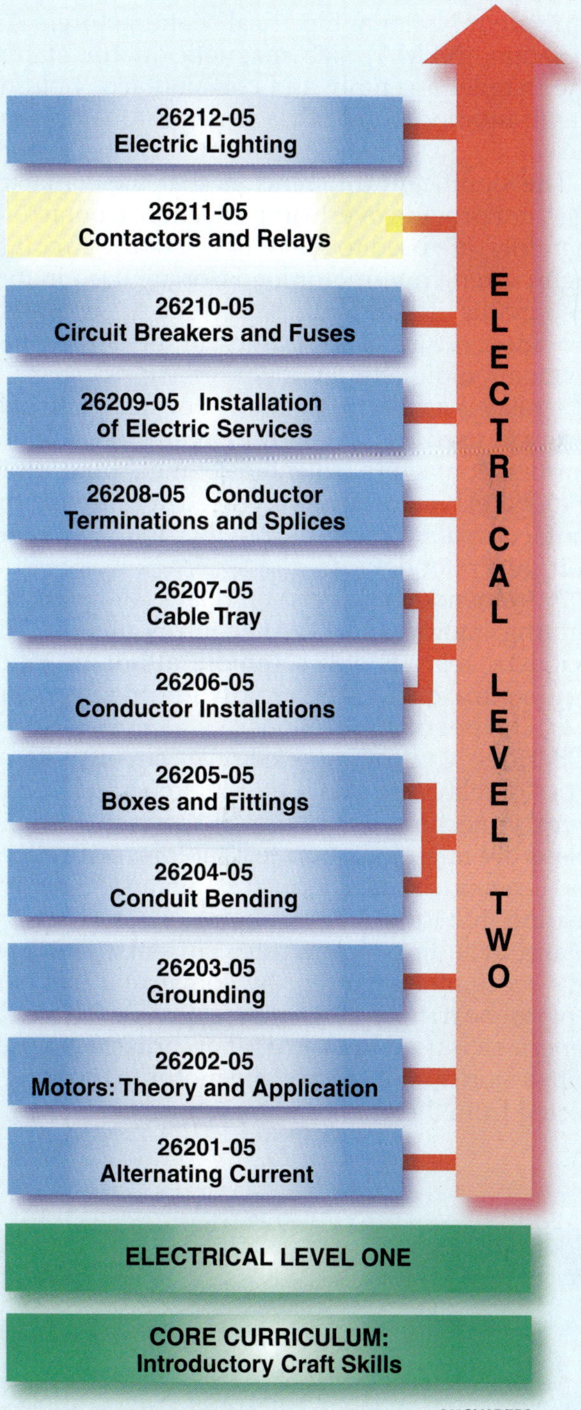

211CMAP.EPS

1.0.0 ◆ INTRODUCTION

The general classification of contactor covers a type of electromagnetic apparatus that is designed to handle relatively high currents. The conventional contactor is identical in appearance, construction, and current-carrying ability to the equivalent National Electrical Manufacturers Association (NEMA) size magnetic motor starter. The magnet assembly and coil, contacts, holding circuit interlock, and other structural features are the same.

The significant difference is that the contactor does not provide overload protection. Contactors, therefore, are used to **make** or **break** the circuit to high-current, non-motor loads, or are used in motor circuits if overload protection is separately provided. A typical application of the latter is in a reversing motor starter.

A relay is an electromagnetic device whose contacts are used in the control circuits of magnetic starters, contactors, solenoids, timers, and other relays. Relays are generally used to amplify the contact capability or multiply the switching functions of a pilot device.

A tremendous variety of relays are used for lighting, motor, and HVAC control circuits. The magnetic relay is still common, although solid-state control devices are rapidly finding their way into all types of applications.

Besides controlling electrical circuits, some types of relays are also used for circuit and equipment protection. For example, thermal overload relays are intended to protect motors, controllers, and branch circuit conductors against excessive heating due to prolonged motor overcurrents up to and including locked rotor currents. These relays sense motor current by converting this current to heat in a resistance element. The heat generated is used to open a set of contacts in series with a starter coil, causing the motor to be disconnected from the line.

There are also a great many electrical protective relays installed in modern power systems, including:

- Overcurrent (time delay and **instantaneous**)
- Over-/under-voltage
- Directional overcurrent
- Percentage differential

The operating characteristics of these protective relays are all similar and are covered in this module. However, the sensing, adjustment, and actuating functions are quite different and are beyond the scope of this module. For specific instructions, always consult the manufacturer's literature.

2.0.0 ◆ MAGNETIC CONTACTORS

The typical magnetic contactor (*Figure 1*) contains a coil of wire wound around an iron core or a laminated iron core (as shown) and an armature. The coil acts as an electromagnet, and when the proper

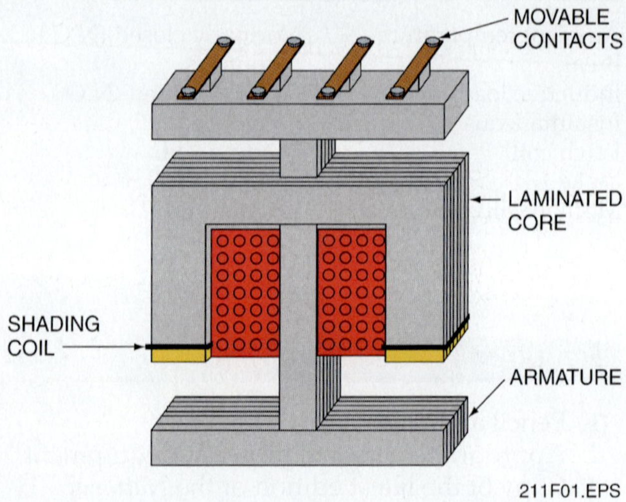

Figure 1 ◆ A magnetic contactor.

Relays

INSIDE TRACK

The term *relay* defines a device that accepts a signal and sends it on down the line, just as a runner in a relay race passes a baton to the next runner.

The invention of the relay is credited to Samuel Morse, one of the inventors of the telegraph. Morse developed the relay to boost signal strength. An incoming signal activated an electromagnet, which closed a battery circuit, thereby transmitting the signal to the next relay.

electric current is applied, it attracts or picks up the armature. Since the movable contacts are attached via an iron bar to the armature, when the armature moves upwards, the movable contacts make contact with a set of stationary contacts to close and complete the circuit.

When power is being applied to the coil, the device is said to be energized. The motion of the armature moves the contact points together or apart, according to their arrangement. When the power is disconnected from the coil, a spring returns the armature to its original position. When no power is being applied, the device is said to be in its de-energized or normal condition. Contacts that are together when the relay is de-energized are said to be **normally closed (N.C.) contacts.** Contacts that are apart when the relay is de-energized are said to be **normally open (N.O.) contacts.**

Notice that a **shading coil** has been added to the iron core in *Figure 1*. Shading coils are used with AC relays to prevent contact chatter and hum. They operate in the same way as in the shaded-pole motor. In general, the shading coil is used to provide a continuous magnetic flow to the armature when the voltage of the AC waveform is zero.

The relay operates on the solenoid principle; that is, it converts electrical energy to linear motion, as shown in *Figure 2*. In this example, a coil of wire is wound around an iron core. When current

THINK ABOUT IT

Coil Shading

Why does shading the coil prevent the contacts of a contactor from chattering?

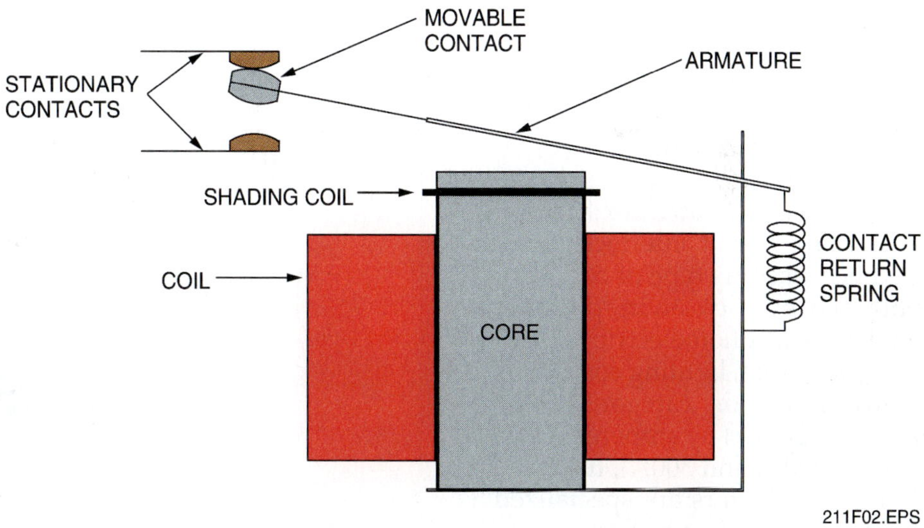

Figure 2 ◆ Contactor operating on the solenoid principle.

INSIDE TRACK

Troubleshooting Contactors

If the contacts of a contactor are defective, it can usually be spotted during a visual inspection. Another way to check is to place a voltmeter across the closed contacts with voltage applied. If the voltage drop is more than 5% of the rated value of the contactor, the contacts are defective.

flows through the coil, a magnetic field develops in the iron core. The magnetic field of the iron core attracts the movable arm, known as the armature. In its present position, the movable contact makes connection with a stationary contact. This contact is normally closed. When the armature is attracted to the iron core, the movable contact breaks the connection with the normally closed stationary contact and makes the connection with another stationary contact. This contact is normally open. In a wiring diagram, the movable contact would be the common, and the stationary contacts would be labeled normally open and normally closed.

2.1.0 Lighting Contactors

Filament-type lamps (e.g., tungsten, infrared, and quartz) have inrush currents of approximately 15 to 17 times their normal operating currents. To prevent welding of their contacts on the high initial current, standard motor control contactors must be derated if used to control this type of load.

Table 1 lists electrical ratings for AC magnetic contactors and starters. Referring to this table, a NEMA Size 1 contactor has a continuous current rating of 27A, but if it is used to switch certain lighting loads, it must be derated to 15A. The standard contactor, however, need not be derated for resistance heating or fluorescent lamp loads that do not impose as high an inrush current.

Lighting contactors differ from standard contactors in that the contacts are rated for high inrush currents. A holding circuit interlock is not normally provided because this type of contactor is frequently controlled by a two-wire pilot device, such as a time clock or photoelectric relay.

Unlike standard contactors, lighting contactors are not horsepower-rated or categorized by NEMA size, but are designated by ampere ratings (20A, 30A, 60A, 100A, 200A, and 300A). It should be noted that lighting contactors are specialized in their application and should not be used on motor loads.

A basic circuit using a lighting contactor and serving as a control for a high-amperage lamp is shown in *Figure 3*. Its purpose is to control the lamp from two different locations. The relay coil is indicated by the standard circular symbol and the relay contacts are indicated in the diagram by two parallel lines. The contacts are drawn separated from the coil to show their function. In reality, they are next to the coil in the relay itself.

Closing the N.O. pushbutton energizes the coil of relay R-1, and both sets of normally open R-1 contacts close. The R1-2 contacts provide a closed circuit to light the lamp. The R1-1 contacts act as holding or seal-in contacts for the relay coil, so that the coil remains energized when the N.O. pushbutton is released.

The lamp will remain lighted until either of the N.C. pushbuttons is momentarily pressed. Pressing either of these buttons opens the circuit to the coil. The coil will drop out or become de-energized and will open the contacts of the relay. The circuit through the lamp is now open and the lamp is turned off. The lamp remains off until one of the N.O. pushbuttons is pressed again. One set of N.O. and N.C. pushbuttons (on and off) is

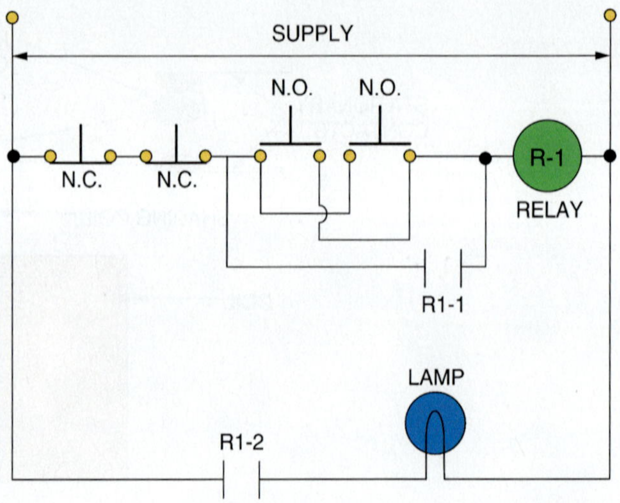

211F03.EPS

Figure 3 ◆ Pushbutton control circuit with a magnetic relay.

Replacing Contactors

INSIDE TRACK

Be careful when selecting replacement contactors because they are rated for different uses. Some are rated for inductive loads; others are rated for resistive loads; and still others are rated for both types of loads.

placed at one location where remote control of the lamp is desired, and another set is placed in another convenient location. The lamp can be turned on and off from either of these locations.

Another application of a magnetic relay is its use in an emergency lighting circuit. The lamps in such a circuit should be lighted instantly after the failure of the main power supply. A simple emergency lighting system controlling one lamp is shown in *Figure 4*. The main power is supplied through the power supply terminals to the relay coil. The relay has a set of normally open contacts and a set of normally closed contacts. As long as the main power is available, it energizes the relay coil, which in turn keeps the N.O. contacts closed. The power is therefore supplied to the transformer, whose primary is connected in series with the N.O. contacts. The secondary of the transformer keeps the storage battery charged through the rectifier. Since the coil is energized, the N.C. contacts are open and the emergency lamp is not lighted.

In the event of main power failure or low voltage in the line, the relay will drop out, the N.O. contacts will open, and the N.C. contacts will close. The lamp will be connected to the battery and give light as long as the relay coil remains de-energized. Such relay and circuit arrangements are frequently used in the emergency battery packs found in many commercial, industrial, and public buildings of all types.

INSIDE TRACK

Emergency Lighting Circuits

A typical emergency lighting circuit uses an automatic transfer switch. A potential transformer in the switch senses the absence of normal voltage and switches on the emergency generator. When it senses power from the generator, it closes the contactor to the main circuit.

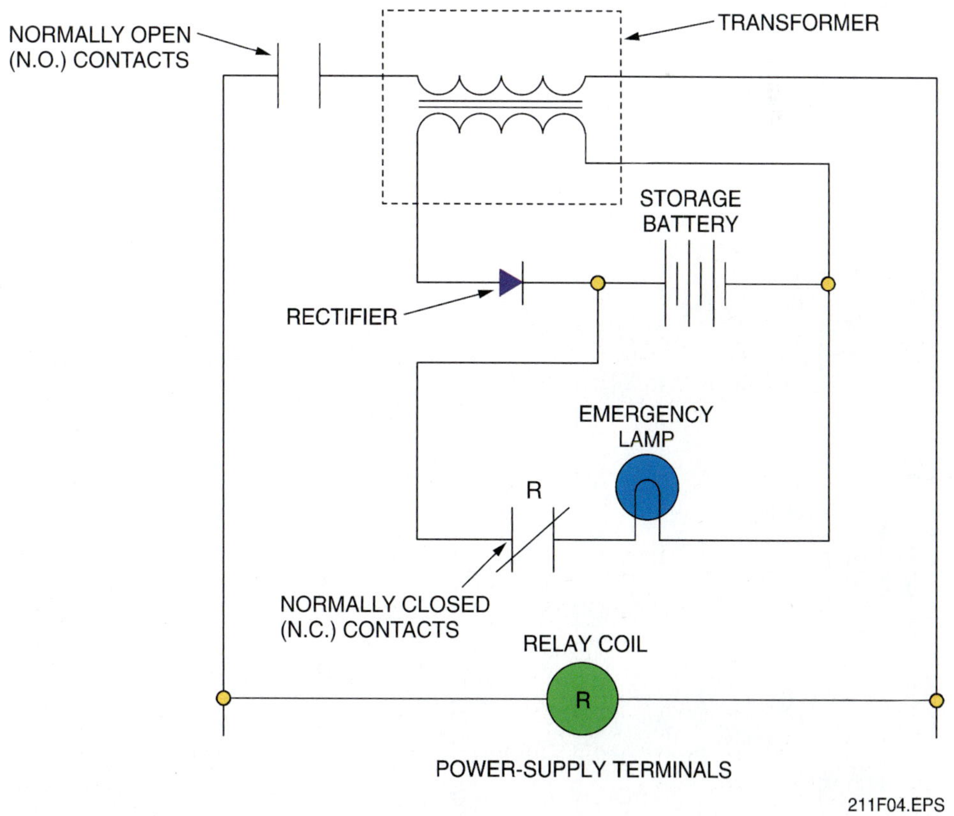

Figure 4 ◆ Emergency lighting circuit using a relay for operation.

Table 1 Electrical Ratings for AC Magnetic Contactors and Starters *(Reprinted with the permission of the National Electrical Manufacturers Association)*

NEMA Size	Volts	Maximum Horsepower Rating—Nonplugging and Nonjogging Duty		Maximum Horsepower Rating—Plugging and Jogging Duty		Continuous Current Rating, Amperes—600 Volt Max.	Service-Limit Current Rating, Amperes	Tungsten and Infrared Lamp Load, Amperes—250 Volts Max.	Resistance Heating Loads, kW—other than Infrared Lamp Loads		KVA Rating for Switching Transformer Primaries at 50 or 60 Cycles		3-Phase Rating for Switching Capacitors
		Single Phase	Poly-Phase	Single Phase	Poly-Phase				Single Phase	Poly-Phase	Single Phase	Poly-Phase	Kvar
	115	⅓	—	—	—	9	11	5	—	—	—	—	—
	200	—	1½	—	—	9	11	5	—	—	—	—	—
	230	1	1½	—	—	9	11	5	—	—	—	—	—
00	380	—	1½	—	—	9	11	—	—	—	—	—	—
	460	—	2	—	—	9	11	—	—	—	—	—	—
	575	—	2	—	—	9	11	—	—	—	—	—	—
	115	1	—	½	—	18	21	10	—	—	0.9	1.2	—
	200	—	3	—	1½	18	21	10	—	—	—	1.4	—
0	230	2	3	1	1½	18	21	10	—	—	1.4	1.7	—
	380	—	5	—	1½	18	21	—	—	—	—	2	—
	460	—	5	—	2	18	21	—	—	—	1.9	2.5	—
	575	—	5	—	2	18	21	—	—	—	1.9	2.5	—
	115	2	—	1	—	27	32	15	3	5	1.4	1.7	—
	200	—	7½	—	3	27	32	15	—	9.1	—	3.5	—
	230	3	7½	2	3	27	32	15	6	10	1.9	4.1	—
1	380	—	10	—	5	27	32	—	—	16.5	—	4.3	—
	460	—	10	—	5	27	32	—	12	20	3	5.3	—
	575	—	10	—	5	27	32	—	15	25	3	5.3	—
	115	3	—	1½	—	36	42	24	—	—	—	—	—
1P	230	5	—	3	—	36	42	24	—	—	—	—	—
	115	3	—	2	—	45	52	30	5	8.5	1	4.1	—
	200	—	10	—	7½	45	52	30	—	15.4	—	6.6	11.3
	230	7½	15	5	10	45	52	30	10	17	4.6	7.6	13

2	380	—	25	—	15	45	52	—	—	28	—	9.9	21
	460	—	25	—	15	45	52	—	20	34	5.7	12	26
	575	—	25	—	15	45	52	—	25	43	5.7	12	33
	115	7½	—	—	—	90	104	60	10	17	4.6	7.6	—
	200	—	25	—	15	90	104	60	—	31	—	13	23.4
	230	15	30	—	20	90	104	60	20	34	8.6	15	27
3	380	—	50	—	30	90	104	—	—	56	—	19	43.7
	460	—	50	—	30	90	104	—	40	68	14	23	53
	575	—	50	—	30	90	104	—	50	86	14	23	67
	200	—	40	—	25	135	156	120	—	45	—	20	34
	230	—	50	—	30	135	156	120	30	52	11	23	40
4	380	—	75	—	50	135	156	—	—	86.7	—	38	66
	460	—	100	—	60	135	156	—	60	105	22	46	80
	575	—	100	—	60	135	156	—	75	130	22	46	100
	200	—	75	—	60	270	311	240	—	91	—	40	69
	230	—	100	—	75	270	311	240	60	105	28	46	80
5	380	—	150	—	125	270	311	—	—	173	—	75	132
	460	—	200	—	150	270	311	—	120	210	40	91	160
	575	—	200	—	150	270	311	—	150	260	40	91	200
	200	—	150	—	125	540	621	480	—	182	—	79	139
	230	—	200	—	150	540	621	480	120	210	57	91	160
6	380	—	300	—	250	540	621	—	—	342	—	148	264
	460	—	400	—	300	540	621	—	240	415	86	180	320
	575	—	400	—	300	540	621	—	300	515	86	180	400
	230	—	300	—	—	810	932	720	180	315	—	—	240
7	460	—	600	—	—	810	932	—	360	625	—	—	480
	575	—	600	—	—	810	932	—	450	775	—	—	600
	230	—	450	—	—	1215	1400	1080	—	—	—	—	360
8	460	—	900	—	—	1215	1400	—	—	—	—	—	720
	575	—	900	—	—	1215	1400	—	—	—	—	—	900

The following example demonstrates how a relay and one single-pole, 15A toggle switch can be used to control 60A of lighting.

Figure 5 shows a diagram of a lighting contactor controlled by a single-pole switch. The contactor coil operates six sets of contacts, each of which is connected to a lighting circuit with lamps totaling 1,200W (10A). When the single-pole switch is turned to the ON position (contacts closed), the coil is energized and moves the armature, which in turn closes all of the contacts simultaneously. When these contacts are closed, this completes the circuit to all of the lamps. Thus, all of the lamps will come on at the same time.

Six single-pole switches could have been used in place of the contactor, but simultaneous switching of the lamps would not be possible.

2.2.0 Reversing Motor Starter

Reversing the direction of motor shaft rotation is often required. Three-phase squirrel cage motors can be reversed by reconnecting any two of the three line connections to the motor. By interwiring two contactors, an electromagnetic method of making the reconnection can be obtained.

As seen in the power circuit in *Figure 6*, the contacts (F) of the forward contactor, when closed, connect lines L1, L2, and L3 to the motor terminals T1, T2, and T3, respectively. As long as the forward contacts are closed, mechanical and electrical interlocks prevent the reverse contactor from being energized.

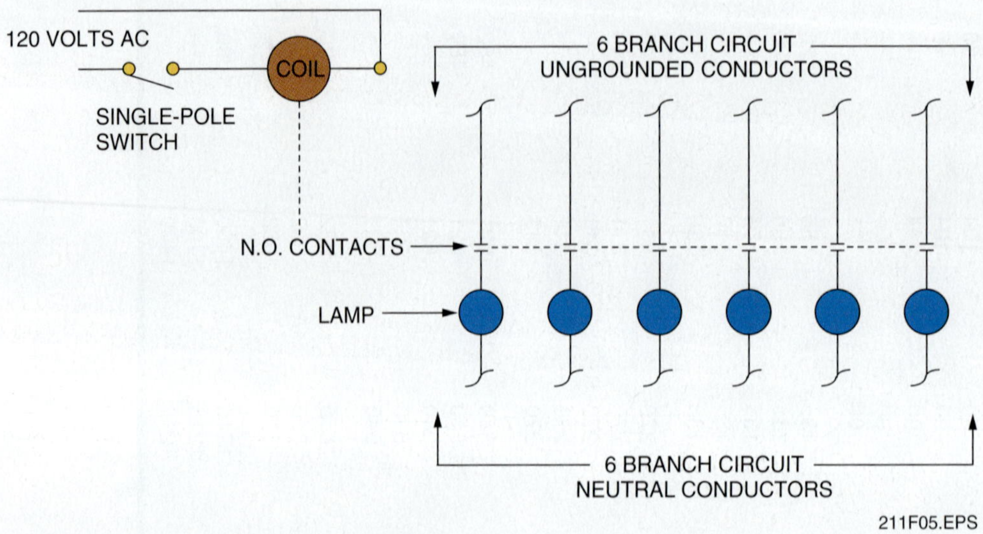

Figure 5 ◆ Several lighting circuits controlled by one contactor.

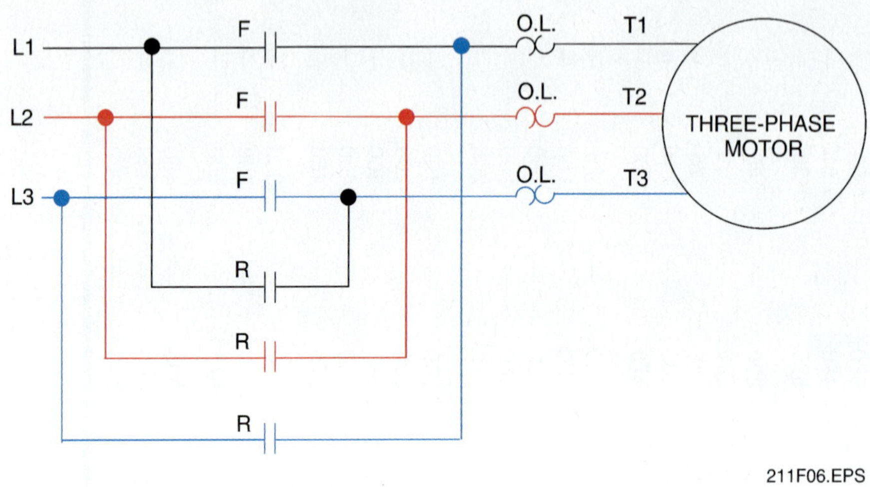

Figure 6 ◆ Three-pole reversing starting diagram.

INSIDE TRACK

Relays with Many Contacts

A common example of a relay that controls many circuits is the defrost control on a large refrigeration unit, such as a walk-in cooler. When a timing relay closes the defrost circuit, one set of N.C. contacts opens to shut the fans off, another set of N.C. contacts opens to turn the compressor off, and a set of N.O. contacts closes to turn on the heaters. A set of N.C. contacts then opens to activate a solenoid that mechanically impedes the refrigerant.

When the forward contactor is de-energized, the second contactor can be picked up, closing its contacts (R), which reconnect the lines to the motor. Note that by running through the reverse contacts, L1 is connected to motor terminal T3, and L3 is connected to motor terminal T1. The motor will now run in the opposite direction.

Whether operating through either the forward or reverse contactor, the power connections are run through an overload relay assembly, which provides motor overload protection. A magnetic reversing starter, therefore, consists of a starter and a set of contactors, suitably interwired, with electrical and mechanical interlocking to prevent the coil of both units from being energized at the same time.

Manual reversing starters (employing two manual starters) are also sometimes used. As in the magnetic version, the forward and reverse switching mechanisms are mechanically interlocked, but since coils are not used in manually operated equipment, electrical interlocks are not furnished.

2.3.0 Mechanically Held Contacts

In a conventional contactor, current flow through the coil creates a magnetic pull to seal in the armature and maintain the contacts in a switched position; that is, the N.O. contacts will be held closed while the N.C. contacts will be held open. Because the contactor action is dependent on the current flow through the coil, the contactor is described as electrically held. As soon as the coil is de-energized, the contacts will revert to their initial position.

Versions of **mechanically held relays** and contactors are also used in some applications. The action is accomplished through the use of two coils and a latching mechanism. Energizing one coil (called the **latch coil**) through a momentary signal causes the contacts to switch, and a mechanical latch holds the contacts in this position even though the initiating signal is removed. The coil is then de-energized. To restore the contacts to their initial position, a second coil (called the **unlatch coil**) is momentarily energized.

Mechanically held relays and contactors are used where the hum of an electrically held device would be objectionable, such as in auditoriums, hospitals, churches, etc.

3.0.0 ◆ RELAYS

A control relay is an electromagnetic device that is similar in operating characteristics to a contactor. The contactor, however, is generally employed to switch power circuits or relatively high-current loads.

Relays, with few exceptions, are used in control circuits, and consequently, their lower ratings (15A maximum at 600V) reflect the reduced current levels at which they operate.

Contactors generally have from one to five poles. Although both normally open and normally closed contacts are frequently encountered, the great majority of applications use normally open contacts. There is little, if any, conversion of contact operation on the job.

As compared to contactors, it is not uncommon to find relays used in applications requiring 10 or 12 poles per device, with various combinations of normally open and normally closed contacts. In addition, some relays have convertible contacts, permitting changes to be made in the field from N.O. to N.C. operation, or vice versa, without requiring modification kits or additional components.

3.1.0 Control Relays

A control relay is an electromagnetic device whose contacts are used in the control circuits of magnetic starters, solenoids, timers, and other relays. Relays are generally used to amplify the contact capability or multiply the switching functions of a pilot device.

The wiring diagrams in *Figures 7* and *8* demonstrate how a relay amplifies contact capacity. *Figure 7* represents a current amplification. The relay and starter coil voltages are the same, but the ampere rating of the temperature switch is too low to handle the current drawn by the starter coil (M). A relay is interposed between the temperature switch and the starter coil. The current drawn by the relay coil (CR) is within the rating of the

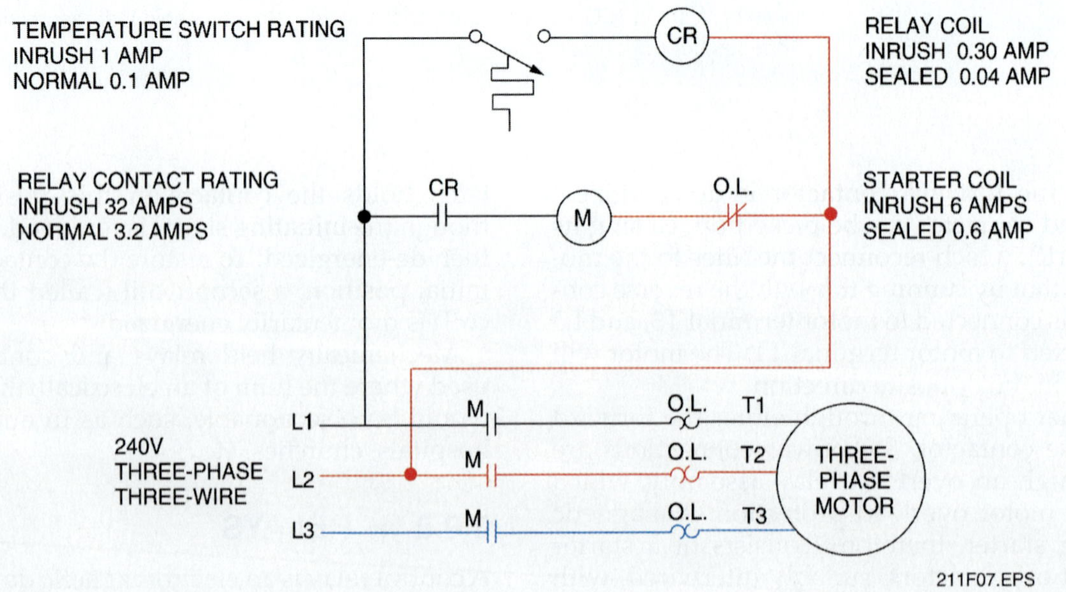

Figure 7 ◆ Relay used to amplify contact capacity.

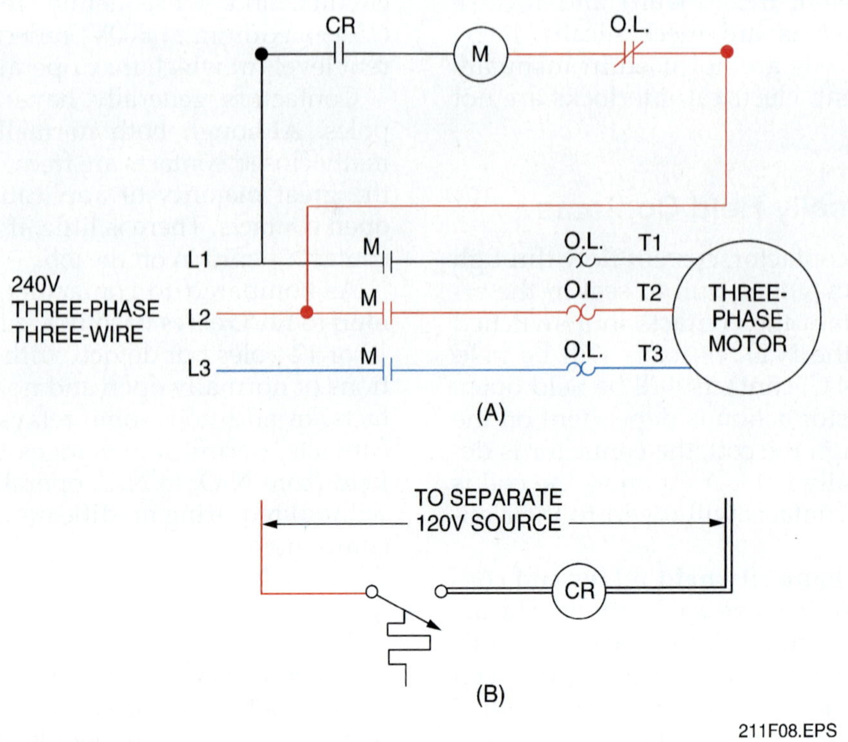

Figure 8 ◆ Relay used for voltage amplification.

temperature switch, and the relay contact (CR) has a rating that is adequate for the current drawn by the starter coil.

Figure 8 represents a voltage amplification. A condition may exist in which the voltage rating of the temperature switch is too low to permit its direct use in a starter control circuit operating at a higher voltage. In this application, the coil of the interposing relay and the pilot device are both wired to a low-voltage source of power that is compatible with the rating of the pilot device. The relay contact, with its higher voltage rating, is then used to control the operation of the starter.

Figure 9 represents another use of relays, which is to multiply the switching functions of a pilot device with a single or limited number of contacts. In this circuit, a single-pole pushbutton contact can, through the use of an interposing six-pole relay, control the operation of a number of different loads such as a pilot light, starter, contactor, solenoid, and timing relay.

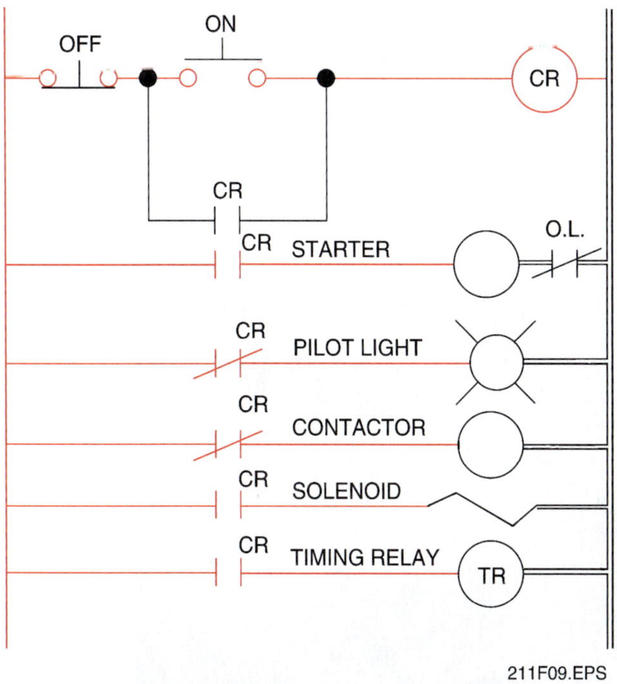

Figure 9 ◆ Relay used to multiply the switching functions of a pilot device.

Relays are commonly used in complex controllers to provide the logic required to set up and initiate the proper sequencing and control of a number of interrelated operations.

Relays differ in voltage ratings (up to 600V), number of contacts, contact convertibility, physical size, and attachments for accessory functions such as mechanical latching and timing.

In selecting a relay for a particular application, one of the first steps should be a determination of the control voltage at which the relay will operate. Once the voltage is known, the relays that have the necessary contact rating can be further reviewed and a selection made on the basis of the number of contacts and other characteristics required.

Figures 10 and *11* show an 8-pin plug-in relay with its base and wiring diagram.

3.2.0 Timers and Timing Relays

A pneumatic timer or timing relay is similar to a control relay, except that it contains contacts that are designed to operate at a specific time interval after the coil is energized or de-energized. A delay on energization is also referred to as an on delay. A delay on de-energization is called an off delay.

A timed function is useful in applications such as the lubricating system of a large industrial machine, in which a small oil pump must deliver lubricant to the bearings of the main motor for a set period of time before the main motor starts.

In pneumatic timers, the timing is accomplished by the transfer of air through a restricted orifice. The amount of restriction is controlled by an adjustable needle valve, permitting changes to be made in the timing period.

Timing relays are often motor-driven by a clock-type mechanism, but solid-state timing circuits will also be encountered.

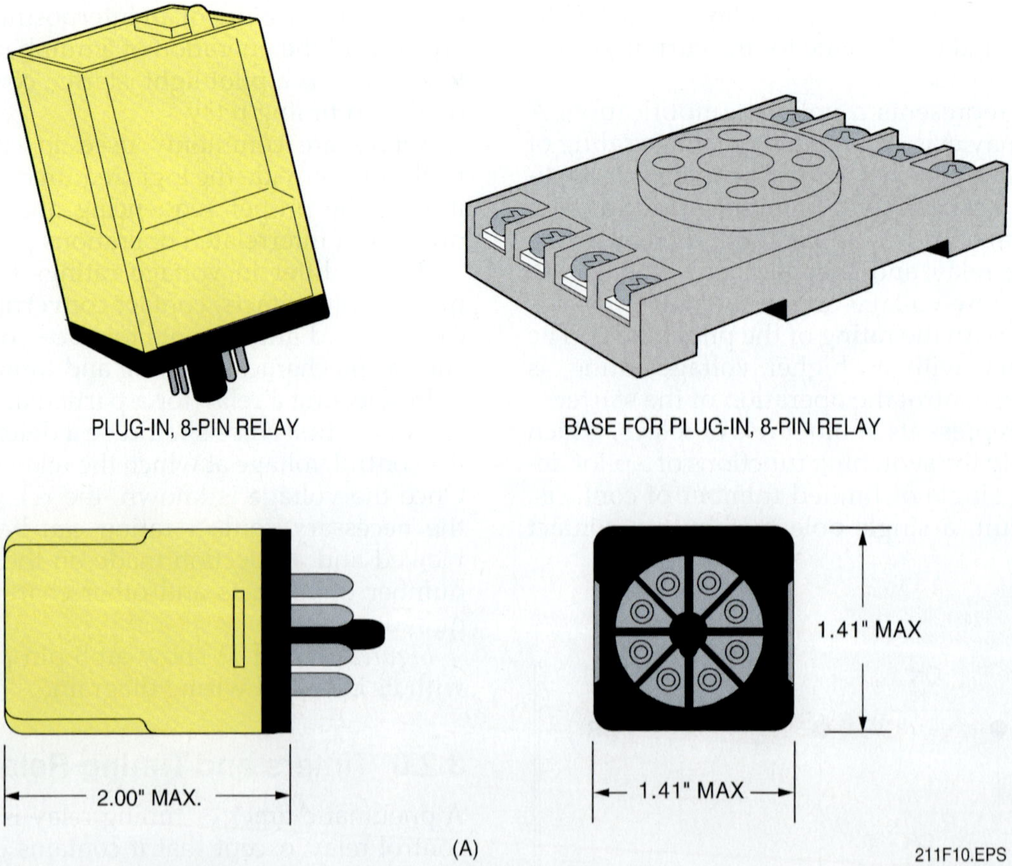

(B)

Figure 10 ◆ Plug-in relays.

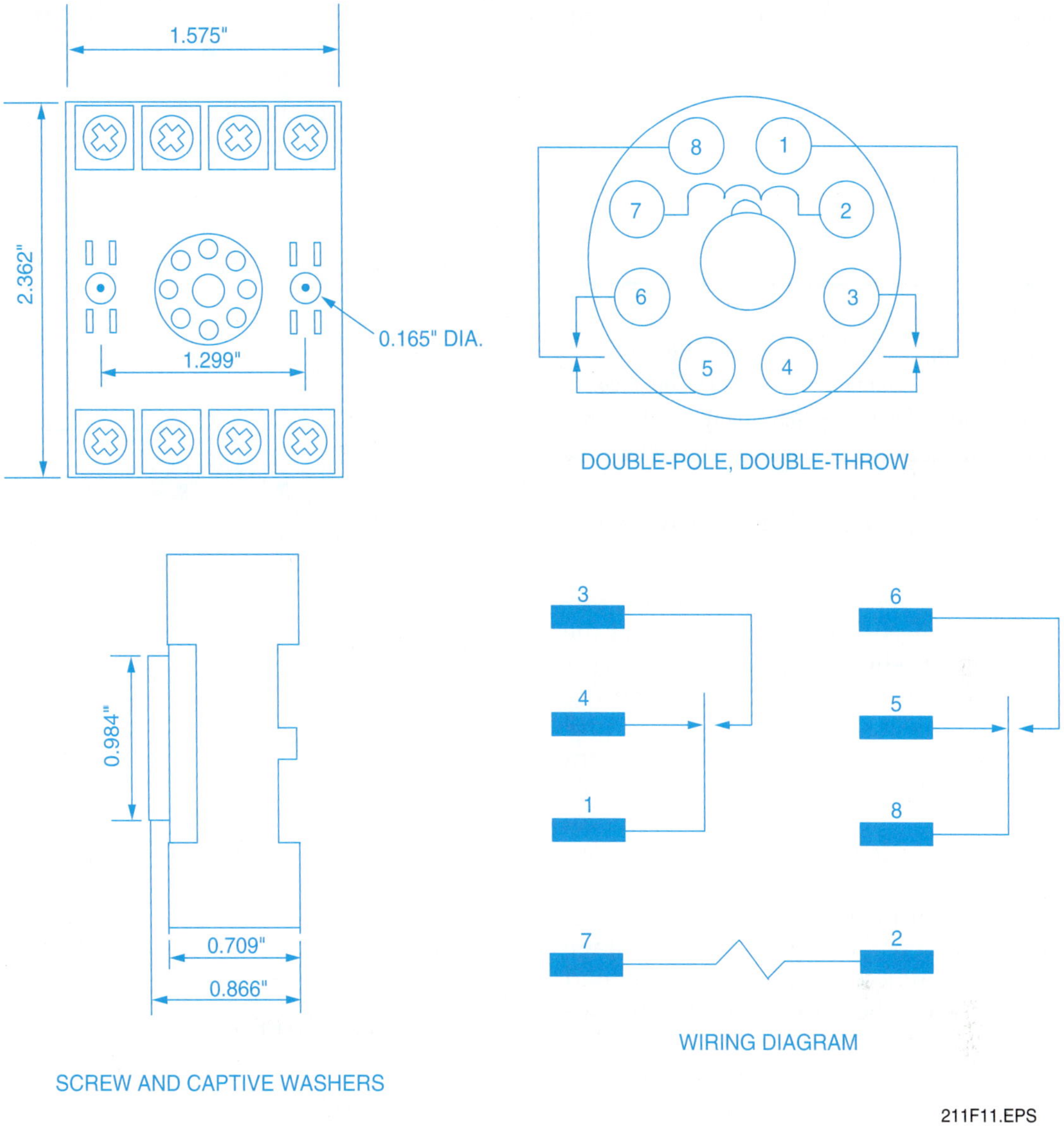

Figure 11 ◆ Wiring diagram for a plug-in relay.

4.0.0 ◆ SOLID-STATE RELAYS

Like magnetic relays, **solid-state relays** are also used in switching applications (*Figure 12*). They are available in different styles and ratings, including time-delay relays. One common use for a solid-state relay is the I/O track of a programmable controller, which is a controller that can be changed by reprogramming rather than requiring an extensive rewiring of the control system. This type of relay can be used to switch a 120V signal to a 5VDC signal.

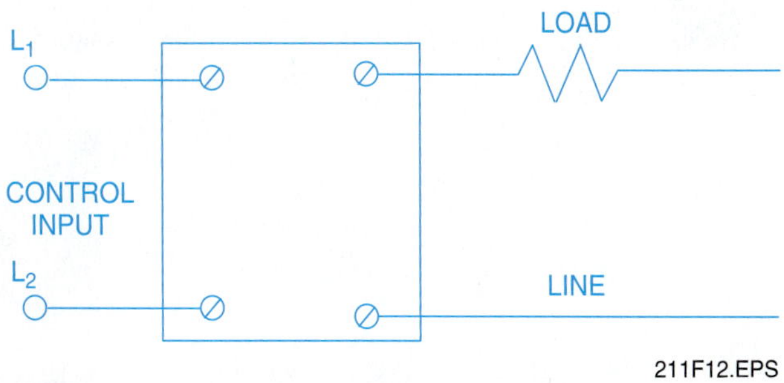

Figure 12 ◆ Basic solid-state relay.

Some of the advantages of solid-state relays include the following:

- No moving parts
- Resistant to shock and vibration
- Sealed against dirt and moisture
- Control input voltage isolated from relay-controlled circuit

> **NOTE**
> Solid-state relays are much more sensitive to loads than equivalent mechanical devices. Always follow the manufacturer's application instructions.

Figure 13 shows typical solid-state relay control circuits. These include:

- *Power transistors for DC loads* – In this DC circuit, a power transistor is used to connect the load to the line, as shown in *Figure 13(A)*. This is an opto-isolated relay with a light-emitting diode (LED) connected to the input or control voltage. This is a common circuit arrangement in solid-state relays, and relays that use this coupling method are referred to as opto-isolated because the load side of the relay is optically isolated from the control side of the relay. Since a light beam (rather than electricity) is used to control the relay, no voltage spikes or electrical noise produced on the load side of the relay can be transmitted to the control side of the relay. When the LED is activated by the input voltage, a photo-detector connected to the base of the transistor turns the

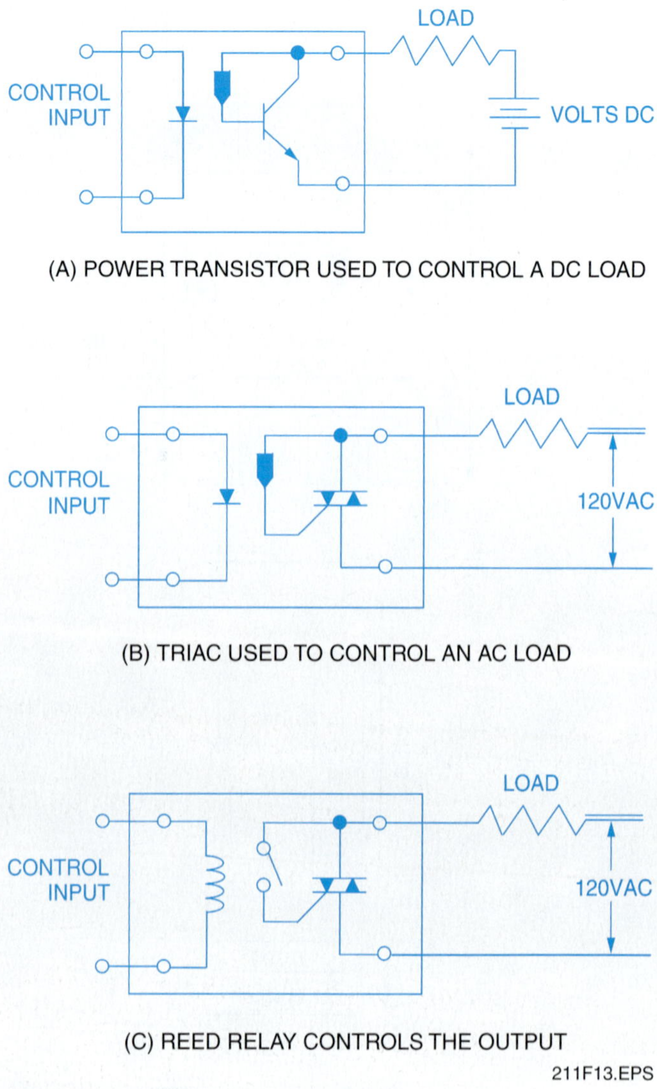

211F13.EPS

Figure 13 ◆ Solid-state relay circuits.

INSIDE TRACK

Plug-In Relays

Plug-in relays make troubleshooting and replacement of relays much easier. Instead of having to connect and disconnect wiring, a new relay can simply be snapped into place.

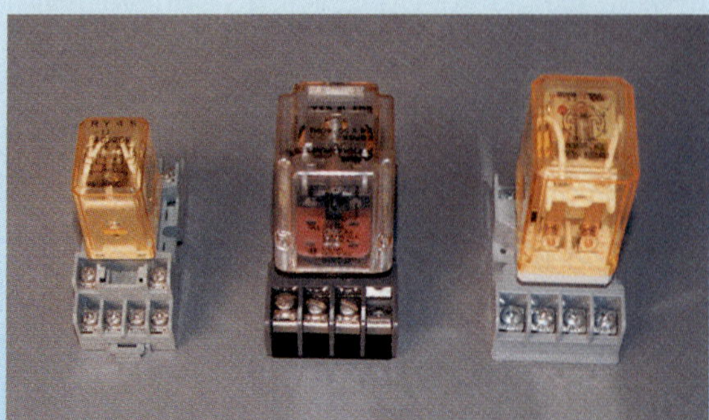

211P1101.EPS

INSIDE TRACK

Timing Relays

Some of the simplest examples of timing relays are the plug-in devices that turn a household lamp on and off or the clocks that control a night-rate water heater. These relays are usually mechanical switches. A more sophisticated example of a timing relay is the relay that energizes half the windings of a split-wound, three-phase motor a microsecond after initially energizing the motor.

Solid-State Relays

Several advantages of solid-state relays are longer life, higher reliability, higher speed switching, and high resistance to shock and vibration. The absence of mechanical contacts eliminates contact bounce, arcing when contacts open, and hazards from explosives and flammable gases.

transistor on and connects the load to the line via this optical coupling.

- *Triac-controlled AC relays* – In this AC circuit, a triac is connected to the load in the place of a power transistor. Refer to *Figure 13(B)*. This circuit also uses an LED as the control device. When the photo-detector detects the LED, it triggers the gate of the triac and connects the load to the line.
- *Reed-controlled AC relays* – Another method of controlling solid-state relays in AC circuits is to use a small reed relay to control the output, as shown in *Figure 13(C)*. In this circuit, small sets of reed contacts are connected to the gate of the triac and the control circuit is connected to the coil of the reed relay. When the control voltage sends a current through the coil, the magnetic field produced around the coil of the relay closes the reed contacts and activates the triac. Unlike opto-isolated relays, this type of solid-state relay uses a magnetic field to isolate the control circuit from the load circuit.

Solid-state relays normally have control voltages from 3V to 32V. Triac-controlled relays commonly have load voltage ratings of 120VAC to 240VAC, with current ratings between 5A and 25A.

Many solid-state relays incorporate a zero switching feature, which means that if the relay is programmed to turn off when the AC voltage is in the middle of a cycle, it will continue to conduct until the AC voltage returns to zero and turns off. Zero switching applications include some **inductive loads** such as transformers. In normal operation, the core material of a transformer can be left saturated on one end of the flux swing if power is removed from the primary winding when the AC voltage is at its positive or negative peak. This can cause inrush currents of up to 600% of the normal operating current when power is restored to the primary. If the circuit is equipped with a zero switching feature, the power will return to zero before the relay is de-energized.

5.0.0 ◆ OVERLOAD RELAYS

To protect a motor and its related circuits from accidental or prolonged overloads, either the starter or the motor should be equipped with automatic devices that will open the circuit in the event of an overload. This protection may be provided by fuses, circuit breakers, or overload relays.

Overcurrent protection must be provided in the line of every motor circuit, but additional protection should be provided in the form of magnetic overload relays. These are used in both manual and automatic starters.

Figure 14 shows a simplified bimetallic overload relay. The load current passes through a heater element in the overload relay. The heater element is replaceable and contains a resistance. When the current exceeds a certain value, the heat radiated from the element causes a bimetallic strip to spring away, interrupting the control circuit or other mechanical action. The heater element is sized to the full load current of the motor. A small overload may have a long delay, while a larger overload will have a shorter delay.

Most overload relays used on modern motor starters are thermally operated and usually consist of two strips of metal that are welded together. Each has a different degree of thermal expansion. If this bimetallic strip is heated, it will deflect sufficiently to trip two normally closed contacts,

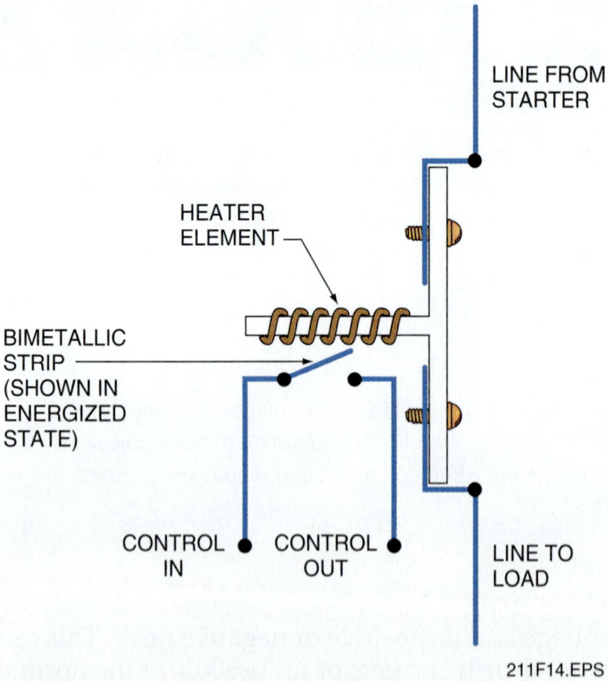

Figure 14 ◆ Simplified bimetallic overload relay.

which in turn will open the circuit connected to the holding coil of the magnetic contactor, causing the main contacts to open. An advantage of this protective device is that it provides a time delay, which prevents the circuit from being opened by momentary high starting currents and short overloads. At the same time, it protects the motor from prolonged overloading.

Conditions of motor overload may be caused by an overload on driven machinery, by a low line voltage, or by an open line in a poly-phase system, which results in single-phase operation. Under any conditions of overload, a motor draws excessive current that causes overheating. Since the motor winding insulation will deteriorate when subjected to overheating, there are established limits on motor operating temperatures. To protect a motor from overheating, overload relays are employed on a motor control to limit the amount of current drawn. This is known as overload protection or running protection.

The ideal overload protection for a motor is an element with current-sensing properties that are very similar to the heating curve of the motor it protects and will act to open the motor circuit when the full-load current is exceeded. The operation of the protective device should be such that the motor is allowed to carry harmless overloads but is quickly removed from the line when an overload has persisted for too long.

Fuses are not the best choice to provide overload protection. Their basic function is to protect against short circuits (overcurrent protection). Motors draw a high inrush current when starting, and conventional single-element fuses have no way of distinguishing between this temporary and harmless inrush current and a damaging overload. Such fuses, if chosen on the basis of motor full-load current, would blow every time the motor started. On the other hand, if a fuse were chosen so that it was large enough to pass the starting or inrush current, it would not protect the motor against small, sustained overloads that might occur later.

Dual-element or time-delay fuses can provide motor overload protection but have the disadvantage of being nonrenewable and must be replaced once tripped.

The overload relay is the heart of motor protection. It has inverse trip time characteristics, permitting it to hold in during the accelerating period (when inrush current is drawn) yet providing protection on small overloads above the full-load current when the motor is running. Unlike dual-element fuses, overload relays are renewable and can withstand repeated trip and reset cycles without needing replacement. They cannot, however, take the place of overcurrent protective equipment.

An overload relay consists of a current-sensing unit connected in the line to the motor, plus a mechanism that is actuated by the sensing unit and serves to directly or indirectly break the circuit. In a manual starter, an overload trips a mechanical latch, causing the starter contacts to open and disconnect the motor from the line. In magnetic starters, an overload opens a set of

INSIDE TRACK

Motor Protection

Some motors have a bimetal thermal device embedded in the stator winding. This device opens when the circuit overheats; when the device cools, the motor will resume operation. The electrician must determine whether the application of the motor warrants further overload protection.

contacts within the overload relay itself. These contacts are wired in series with the starter coil in the control circuit of the magnetic starter. Breaking the coil circuit causes the starter contacts to open, disconnecting the motor from the line.

Overload relays can be classified as being either thermal or magnetic. Magnetic overload relays react only to current excesses and are not affected by temperature. As the name implies, thermal overload relays rely on the rising temperatures caused by the overload current to trip the overload mechanism. Thermal overload relays can be further subdivided into two types: melting alloy (eutectic) and bimetallic.

The melting alloy assembly of a heater element and solder pot are shown in *Figure 15*. The motor load current passes through the heater element. When it exceeds the designated capacity of the heater, it melts an alloy solder pot. The ratchet wheel will then be allowed to turn in the molten pool, and a tripping action of the starter control circuit results, stopping the motor. A cooling-off period is required to allow the solder pot to harden before the overload relay assembly may be reset and the motor service restored.

Melting alloy thermal units are interchangeable and of a one-piece construction, which ensures a constant relationship between the heater element and solder pot and allows for factory calibration, making them virtually tamperproof in the field. These important features are not possible with any other type of overload relay construction. Many types of interchangeable thermal units are available to provide precise overload protection from any full-load current.

Bimetallic overload relays have two major advantages: first, the automatic reset feature is desirable when devices are mounted in locations that are not easily accessible for manual operation; and, second, these relays can easily be adjusted to trip within a range of 85% to 115% of the nominal trip rating of the heater unit. This feature is useful when the recommended heater size might result in unnecessary tripping, while the next larger size would not give adequate protection. The **ambient temperature** affects overload relays operating on the principle of heat.

Ambient-compensated bimetallic overload relays were designed for one particular situation, that is, when the motor is at a constant temperature and the controller is located separately and has a varying temperature. In this case, if a standard thermal overload relay were used, it would not trip consistently at the same level of motor current if the controller temperature changed. A thermal overload relay is always affected by the surrounding temperature. To compensate for temperature variations, an ambient-compensated overload relay is applied. Its trip point is not affected by temperature and it performs consistently at the same value of current.

Melting alloy and bimetallic overload relays are designed to approximate the heat actually generated in the motor. As the motor temperature increases, so does the temperature of the thermal unit. Motor and relay heating curves show this relationship. From such graphs (obtainable from the manufacturers of overload relays), you can see that, no matter how high the current drawn, the overload relay will provide protection, yet the relay will not trip out unnecessarily.

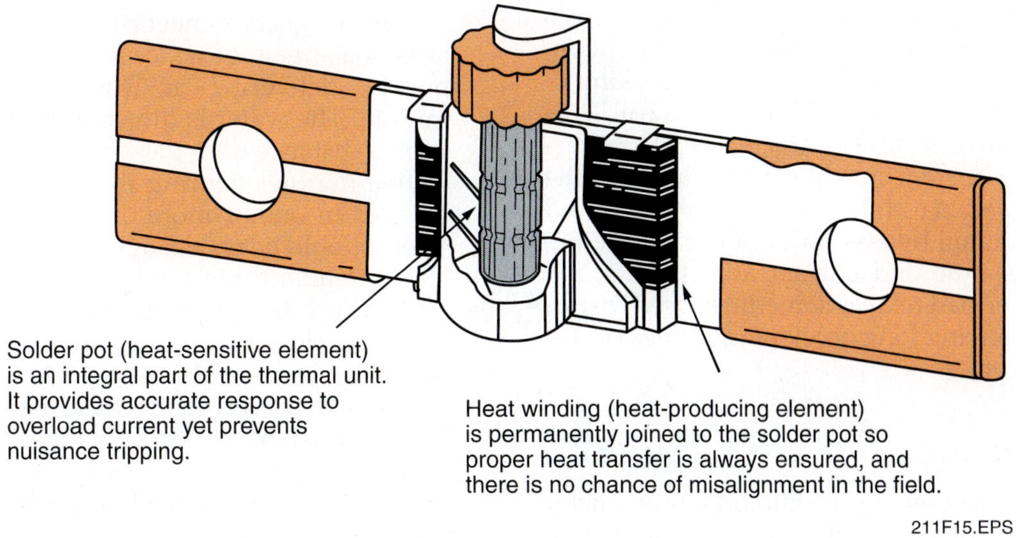

Figure 15 ◆ Melting alloy thermal unit.

When selecting thermal overload relays, the following must be considered:

- Motor full-load current
- Type of motor
- Difference in ambient temperature between the motor and controller

Motors of the same horsepower and speed do not all have the same full-load current, and the motor nameplate must always be referred to and checked to obtain the full-load amperes for a particular motor. Thermal unit selection tables are published on the basis of continuous-duty motors operating under normal conditions with a 1.15 service factor. These tables are shown in manufacturers' catalogs and also appear on the inside of the door or cover of the motor controller. These values are selected to properly protect the motor and allow it to develop its full horsepower, allowing for the service factor if the ambient temperature is the same at the motor as at the controller. If the temperatures are not the same, or if the motor service factor is less than 1.15, a special procedure is required to select the proper thermal unit.

Standard overload relay contacts are closed under normal conditions and open when the relay trips. An alarm signal is sometimes required to indicate when a motor has stopped due to an overload trip. Also, with some machines, particularly those associated with continuous processing, it may be required to signal an overload condition, rather than have the motor and process stop automatically. This is done by fitting the overload relay with a set of contacts that close when the relay trips, thus completing the alarm circuit. These contacts are appropriately called alarm contacts.

A magnetic overload relay has a movable magnetic core inside a coil that carries the motor current. The flux set up inside the coil pulls the core upward. When the core rises far enough, it trips a set of contacts on the top of the relay. The movement of the core is slowed by a piston working in an oil-filled dashpot mounted below the coil. This produces an inverse-time characteristic. The effective tripping current is adjusted by moving the core on a threaded rod. The tripping time is varied by uncovering oil bypass holes in the piston. Because of the time and current adjustments, the magnetic overload relay is sometimes used to protect motors having long accelerating times or unusual duty cycles.

6.0.0 ◆ PROTECTIVE ENCLOSURES

The correct selection and installation of an enclosure for a particular application can contribute considerably to extended service life and trouble-free operation. To shield live parts from accidental contact, some form of enclosure is always necessary. This function is usually filled by a general-purpose sheet steel cabinet. Frequently, however, dust, moisture, or explosive gases make it necessary to employ a special enclosure to protect the motor controller from corrosion or the surrounding equipment from explosion. In selecting and installing control apparatus, it is always necessary to carefully consider the conditions under which the apparatus must operate; there are many applications where a general-purpose enclosure does not afford protection.

Underwriters Laboratories, Inc., has defined the requirements for protective enclosures according to hazardous conditions, and the National Electrical Manufacturers Association (NEMA) has standardized enclosures from these requirements:

- *NEMA Type 1: General Purpose* – This enclosure is primarily intended to prevent accidental contact with the enclosed apparatus. It is suitable for general-purpose applications indoors where it is not exposed to unusual service conditions. A NEMA Type 1 enclosure serves as protection against dust and light and indirect splashing, but is not dust-tight.
- *NEMA Type 3: Dust-Tight, Raintight* – This enclosure is intended to provide suitable protection against specified weather hazards. A NEMA Type 3 enclosure is suitable for outdoor applications, such as construction work. It is also sleet-resistant.
- *NEMA Type 3R: Rainproof, Sleet Resistant* – This enclosure protects against interference in operation of the contained equipment due to rain and resists damage from exposure to sleet. It is designed with conduit hubs and external mounting as well as drainage provisions.
- *NEMA Type 4: Watertight* – A watertight enclosure is designed to meet a hose test, which consists of a stream of water from a hose with a 1" nozzle, delivering at least 65 gallons per minute. The water is directed on the enclosure from a distance of not less than 10' for a period of five minutes. During this period, it may be pointed in one or more directions, as desired. There should be no leakage of water into the enclosure under these conditions.
- *NEMA Type 4X: Watertight, Corrosion-Resistant* – These enclosures are generally constructed along the lines of NEMA Type 4 enclosures except that they are made of a material that is highly resistant to corrosion. For this reason, they are ideal in applications such as meat packing and chemical plants, where contaminants would ordinarily destroy a steel enclosure over a period of time.

- *NEMA Type 7: Hazardous Locations, Class I* – These enclosures are designed to meet the application requirements of the *NEC®* for Class I hazardous locations: *Class I locations are those in which flammable gases or vapors are or may be present in the air in sufficient quantities to produce explosive or ignitible mixtures.* In this type of equipment, the circuit interruption occurs in the air.
- *NEMA Type 9: Hazardous Locations, Class II* – These enclosures are designed to meet the application requirements of the *NEC®* for Class II hazardous locations: *Class II locations are those that are hazardous because of the presence of combustible dust.* The letter or letters following the type number indicate the particular group or groups of hazardous locations (as defined in the *NEC®*) for which the enclosure is designed. The designation is incomplete without a suffix letter or letters.
- *NEMA Type 12: Industrial Use* – This type of enclosure is designed for use in those industries where it is desired to exclude such materials as dust, lint, fibers and flyings, oil seepage, or coolant seepage. There are no conduit openings or knockouts in the enclosure, and mounting is by means of flanges or mounting feet.
- *NEMA Type 13: Oil-Tight, Dust-Tight* – NEMA Type 13 enclosures are generally made of cast iron, gasketed, or permit use in the same environments as NEMA Type 12 devices. The essential difference is that due to its cast housing, a conduit entry is provided as an integral part of the NEMA Type 13 enclosure, and mounting is by means of blind holes rather than mounting brackets.

7.0.0 ◆ LOW-VOLTAGE REMOTE CONTROL SWITCHING

In applications where lighting must be controlled from several points, where there is a complexity of lighting or power circuits, or where flexibility is desirable in certain systems, low-voltage remote control relay systems have been applied. Basically, these systems use special low-voltage components operated from a transformer to switch relays, which in turn control the standard line voltage circuits. Because the control wiring does not carry the line load directly, small, lightweight cable can be used. It can be installed wherever and however convenient—placed behind moldings, stapled to woodwork, buried in shallow plaster channels, or installed in holes bored in wall studs. A basic circuit of a remote control switching system is shown in *Figure 16*; another is shown in *Figure 17*. In both circuits, the relay permits positive control for on and off. It can be located near the load or installed in centrally located distribution panel boxes, depending on the application.

Because no line voltage flows through the control circuits and low voltage is used for all switch and relay wiring, it is possible to place the controls at a great distance from the source or load, thus offering many advantages.

INSIDE TRACK

Step-Down Transformers

Step-down transformers like the one shown here can provide low voltage for remote control switching circuits.

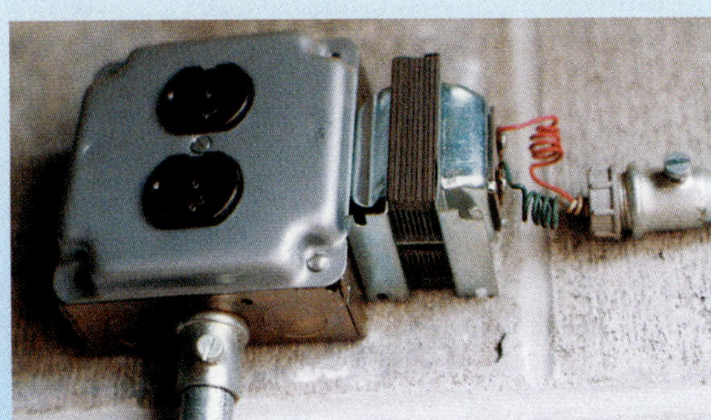

211P1102.EPS

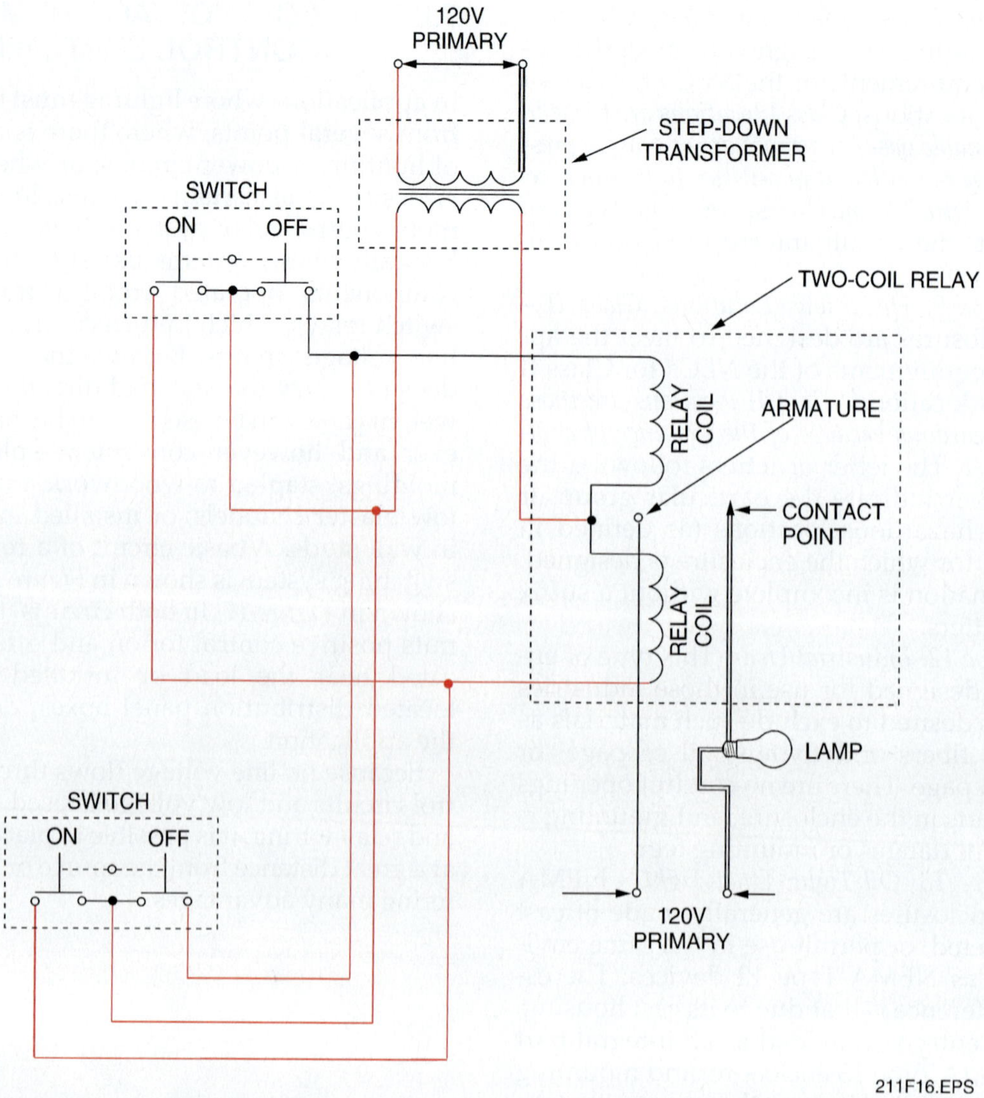

Figure 16 ◆ Basic remote control system.

7.1.0 Advantages of Remote Control Switching

Since branch circuits go directly to the loads in this type of system, no line voltage switch legs are required. This saves costly runs of larger conductor through all switches and saves installation time and costs if multipoint switches are used.

Under certain conditions, the *NEC*® allows low-voltage conductors to be run in open spaces above ceilings and through wall partitions without further protection. An outlet box is not required at the switch locations. When the electricians rough-in the wiring, they merely secure a plaster ring of the correct depth at each switch location and usually wrap the low-voltage cable around a nail driven behind the plate. The switch and its cover are then installed after the wall is finished.

Low-voltage switching is especially useful in rewiring existing buildings, because the small cables are as easy to run as telephone wires. They are easy to hide behind baseboards or even behind quarter-round molding. The cable can be run exposed without being very noticeable because of its small size. The small, flexible wires are easily fished in partitions.

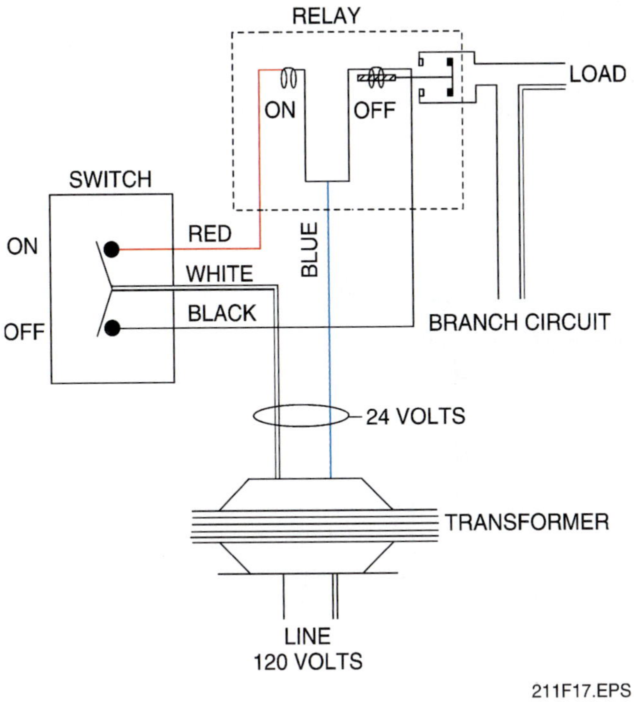

Figure 17 ◆ Basic circuit of a remote control switching system.

Figure 18 shows several types of remote control circuits. In the circuits shown, any number of on-off switches can be connected to provide control from many remote points. In a typical installation, outside lights could be controlled from any area within the building without the need to run full-size cable for three- or four-way switch legs through the building. By running a low-voltage line from each relay to a central point, such as an exit door, all relays can be operated from one spot using a selector switch; this allows quick, convenient control of all exterior and interior lighting and power circuits.

A motor-driven rotary switch is available that allows control of 25 (or more) separate circuits with the push of one button. Dimming can also be accomplished by using the motorized control unit together with a modular incandescent dimming system.

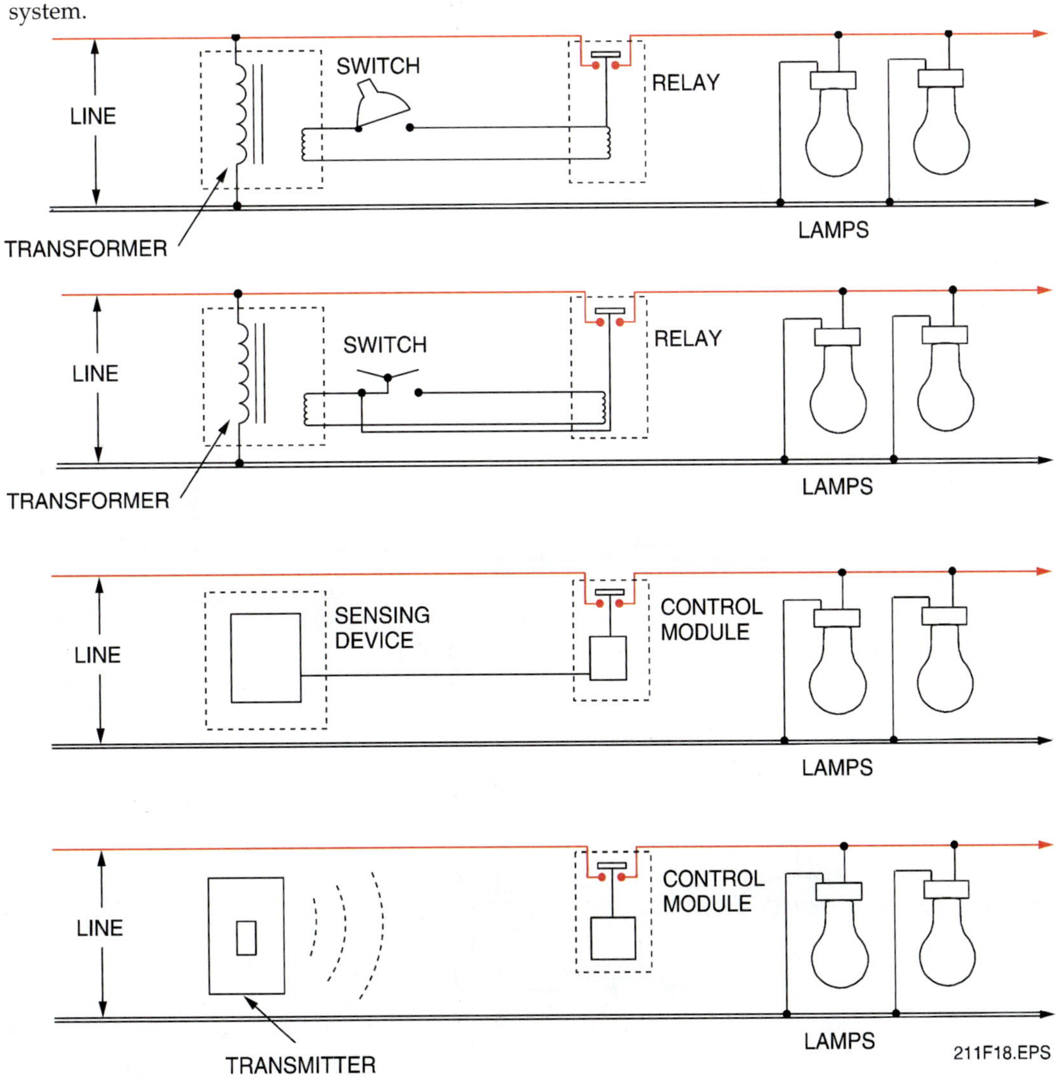

Figure 18 ◆ Several types of remote control circuits and related components.

7.2.0 System Components

The relay is the heart of a remote control switching system. It employs a split low-voltage coil to move the line voltage contact armature to the ON (latched) position. As illustrated in *Figure 19*, the ON coil moves the armature to the right when a 24VAC control signal is impressed across its leads. This is similar to a magnet attracting the handle of a standard single-pole switch to the ON position when energized. The armature (handle) latches in the ON position and will remain there until the OFF coil is energized, drawing the armature into the OFF position. *Figure 20* shows how a low-voltage relay is installed in an outlet box.

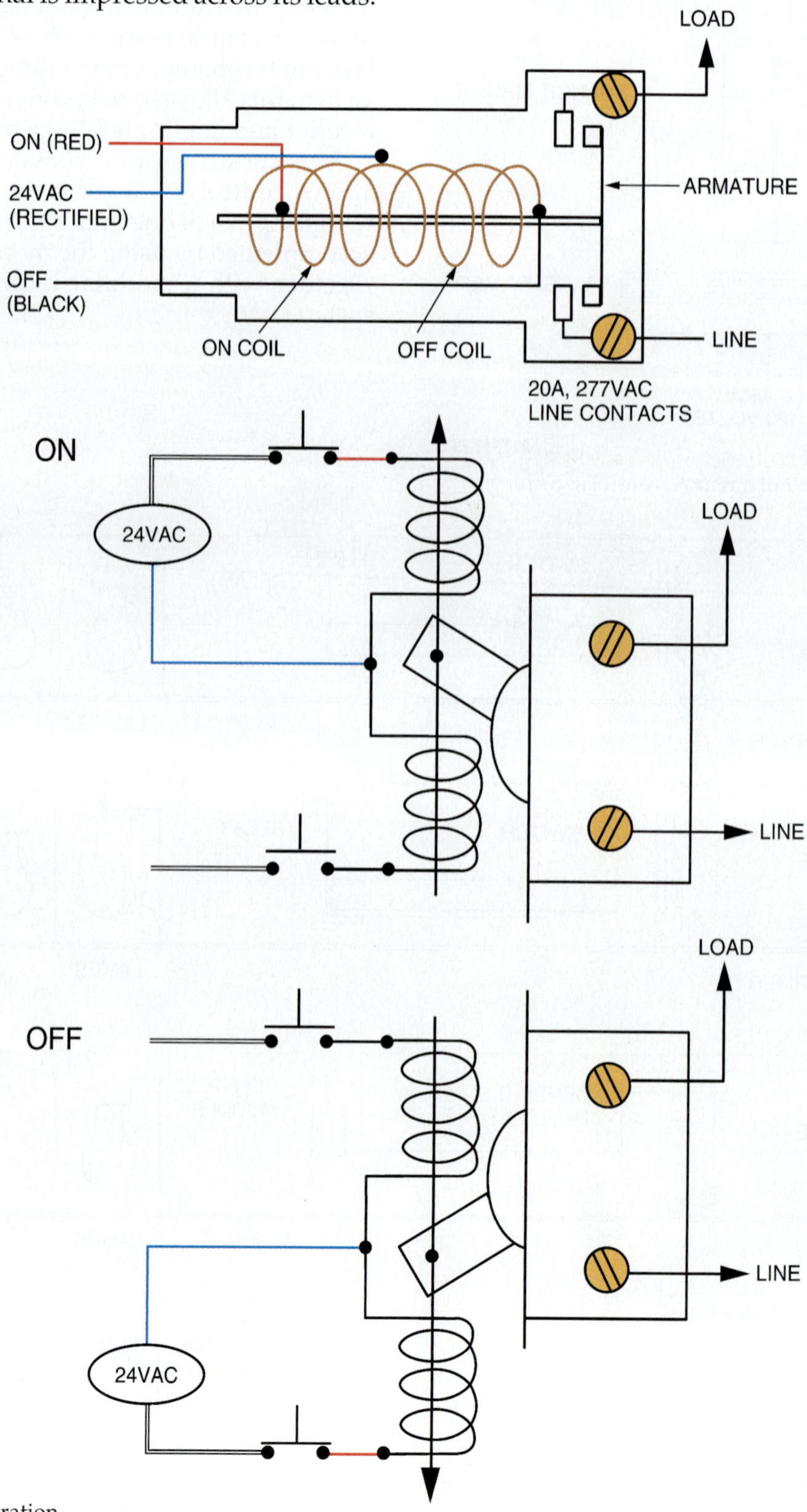

Figure 19 ◆ Relay operation.

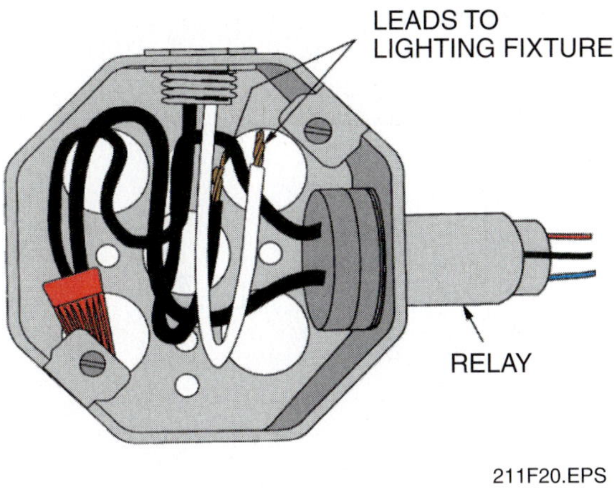

Figure 20 ◆ Relay installed in an outlet box.

7.2.1 Power Supplies

Transformers designed for use on remote control switching systems supply 24VAC power. This is rectified to 24VDC to operate the relays and pilot lights. The AC input voltage to the power supply can vary by manufacturer, with the more common voltages being 120V, 208V, 240V, and 277V. The type of voltage used depends on the system. *Figure 21* shows a control transformer along with connection details.

7.2.2 Switches

The standard low-voltage switch uses a rocker or two-button configuration to provide a momentary single-pole, double-throw action. Pushing the ON button completes the circuit to the ON coil of the relay, shifting the contact armature to the corresponding position. When the button is released, the relay remains in that position.

The pulse operation allows any number of switches to be wired in parallel, as shown in *Figure 22*. A group of relays could also be wired for common switch control by paralleling their control leads. Those relays would operate as a group.

Pilot switches include a lamp wired between the switch common (white) and a pilot terminal. The auxiliary contact in the relay provides power to drive this lamp when the relay is energized.

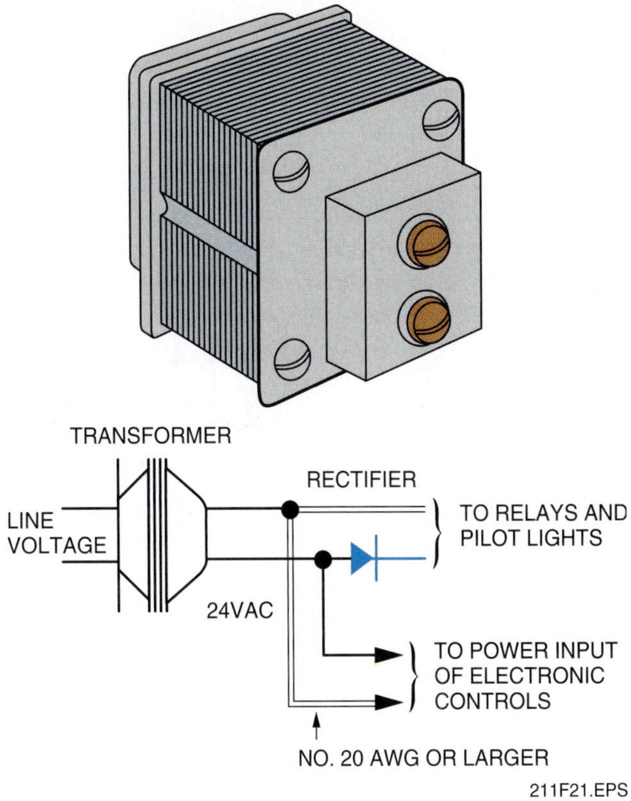

Figure 21 ◆ Control transformer with power-supply wiring diagram.

7.2.3 Master Sequencer

The master sequencer allows relays to be controlled as a group while still allowing individual switch control for each. When the master switch is turned ON, the sequencer pulses each of its ON relay outputs sequentially. A local switch can control an individual relay without affecting any other.

A second input channel allows time clocks, building automation system outputs, photocells, or other maintained contact devices to also control the sequencer. This provides simple automation coupled with local override of individual loads. See *Figure 23*.

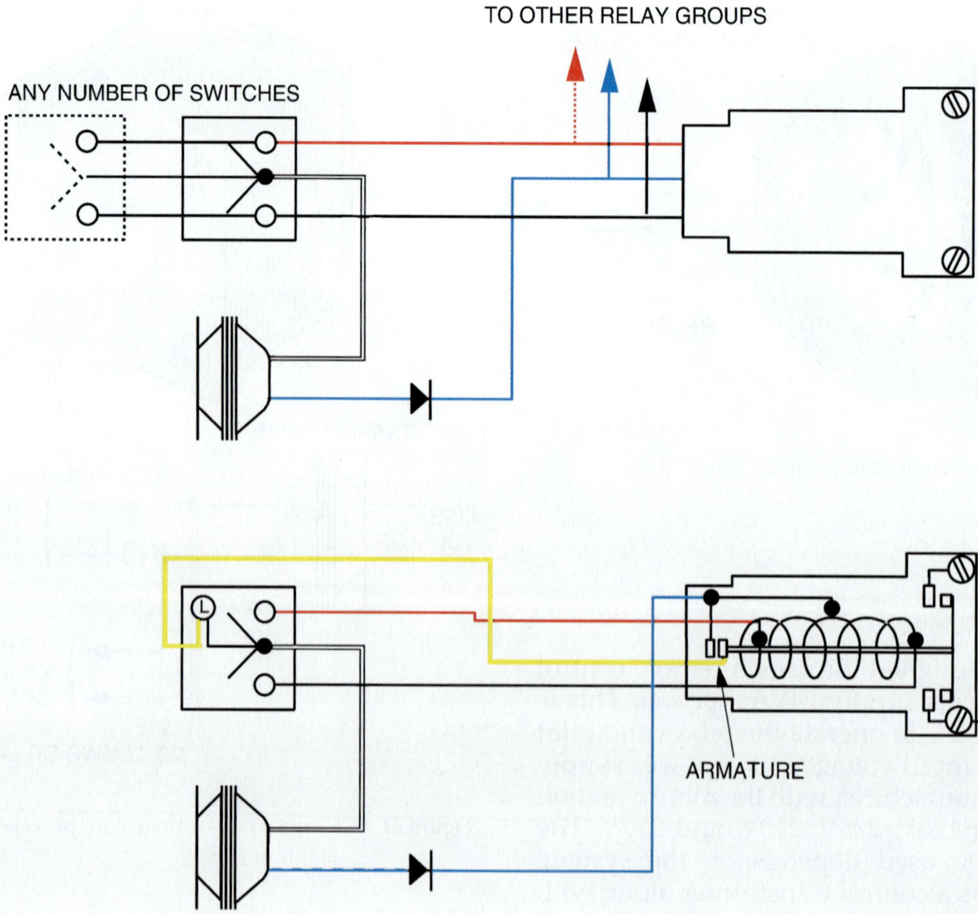

Figure 22 ◆ Switch operation.

211F22.EPS

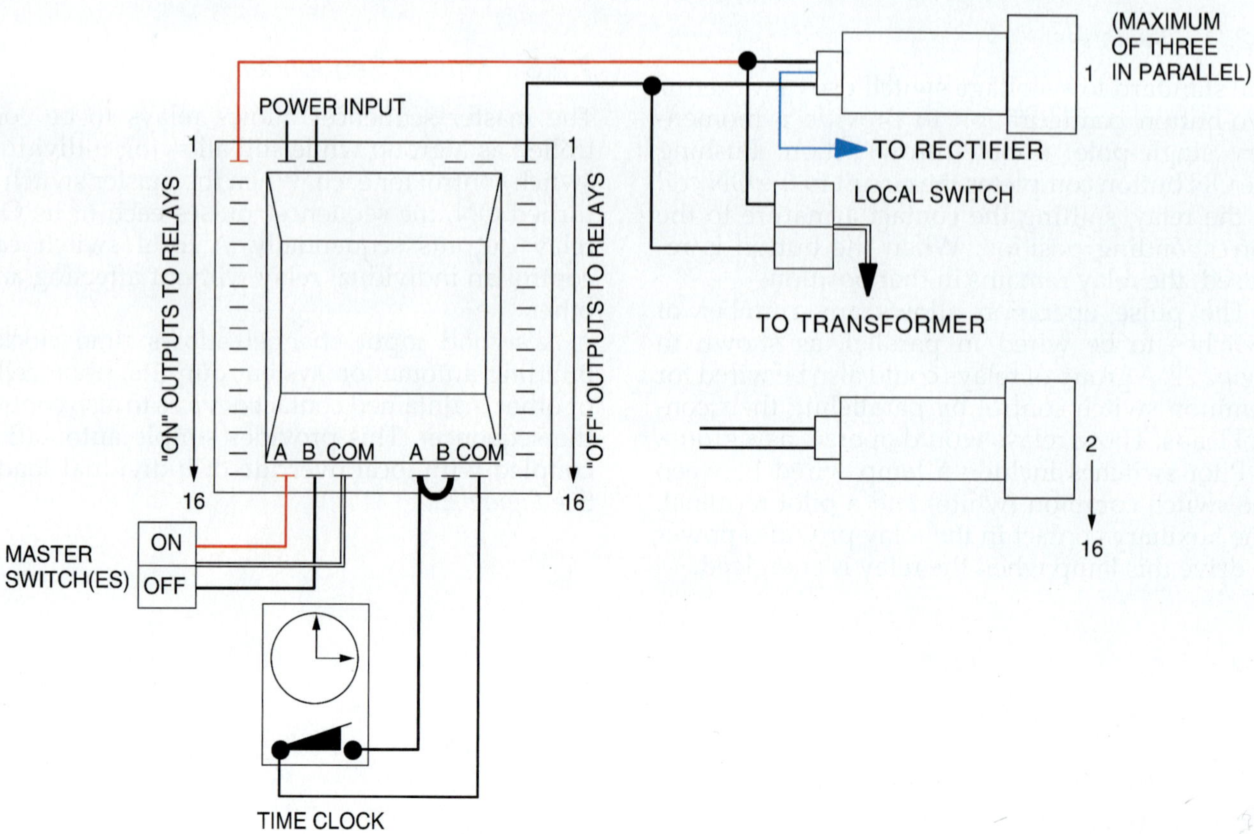

Figure 23 ◆ Master sequencer connections.

211F23.EPS

7.3.0 Basic Operation of Remote Control Switching

Figure 24 shows a single-switch control of a light or a group of lights in one area. This is the basic circuit of relay control and is similar to single-pole conventional switch circuits. The schematic wiring diagram of this circuit is shown in *Figure 25*.

The addition of a time clock and/or photoelectric cell allows applications in which any number of control circuits are turned on or off at desired intervals (e.g., the dusk-to-dawn control of outdoor lighting circuits).

Multipoint switching, or the switching of a single circuit from two or more switches, is shown in *Figure 26*, along with its related wiring diagram. In conventional wiring, multipoint switching requires costly three-way and four-way switches, plus the extension of switch legs and traveler wires through all switches. One of the greatest advantages of relay switching is the low cost of adding additional switch points.

Figure 27 shows many lighting circuits being controlled from one location. In conventional line voltage switches, this is accomplished by ganging the individual switches together. The same procedure can be employed with remote control switches. When more than six switches are required, it is usually desirable to install a master-selector switch that permits the control of individual circuits.

Master control of many individual and isolated circuits is possible. Where more than 12 circuits are to be controlled from a single location and the selection of individual circuits is not required, a motor master control unit automatically sweeps 25 circuits from either ON or OFF at the touch of a single master switch, which can be placed at one or more locations.

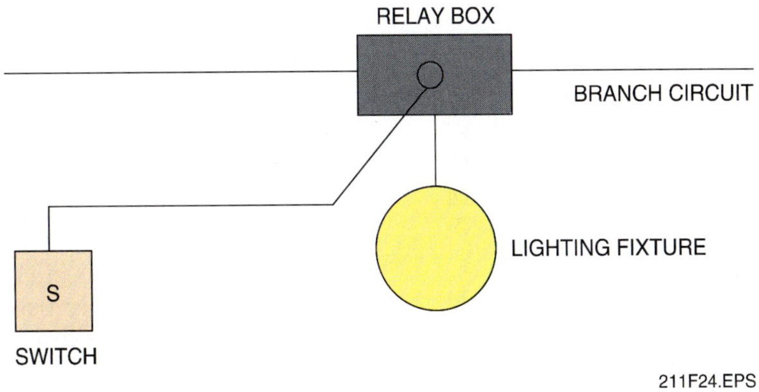

Figure 24 ◆ Single-switch control.

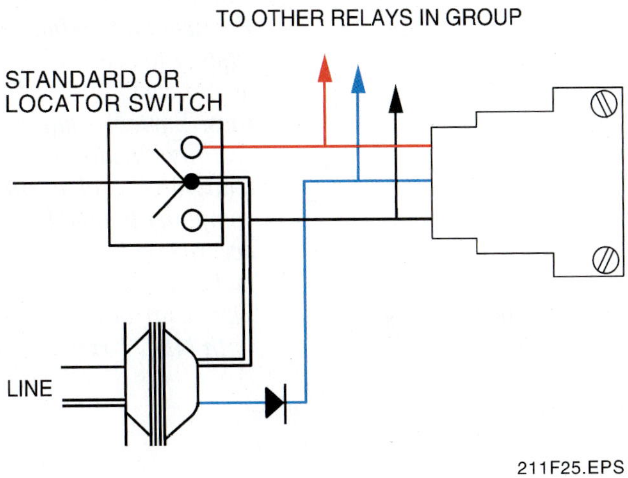

Figure 25 ◆ Schematic wiring diagram of single switches controlling each circuit.

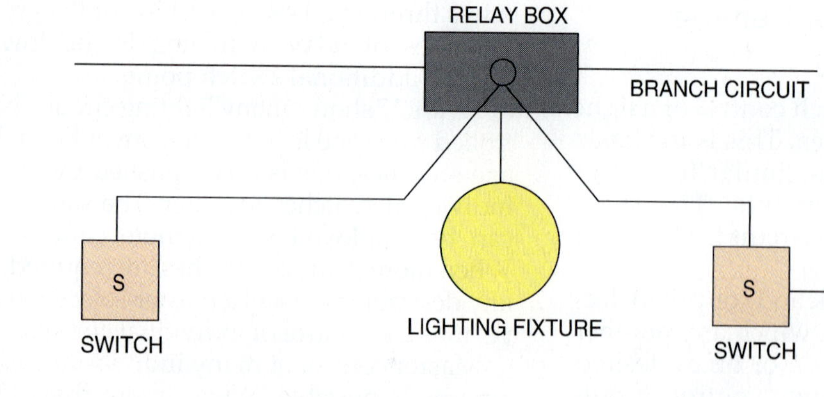

Figure 26 ◆ Diagram of multipoint switching of a single circuit.

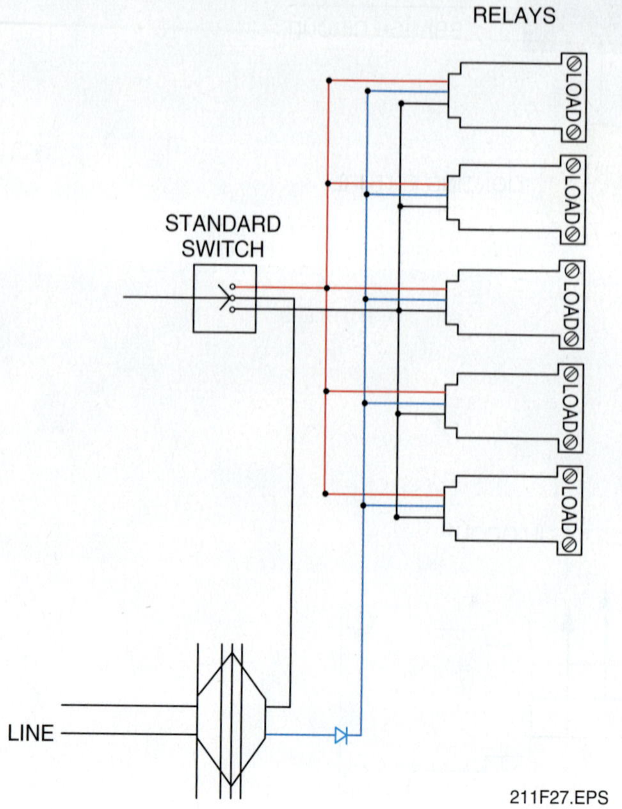

Figure 27 ◆ Switching circuits being controlled from one location.

7.4.0 Planning a Remote Control Switching System

Low-voltage lighting control systems can be installed in both residential and commercial installations. The systems installed in each type of location may differ. The manufacturer's instructions and recommendations should always be consulted and adhered to, as should the blueprint specifications. The first step in the installation procedure, whether the building is residential or commercial, is to develop a plan. When laying out a remote control switching system for a building, drawing symbols such as the ones shown in *Figure 28* are normally used to identify the various components. The drawing may be simple, requiring only the following:

- The location of each low-voltage control with a dashed line drawn from the switch to the outlet to be controlled
- A schematic wiring diagram showing all components
- The following note:

  Furnish and install complete remote control wiring system for control of lighting and other equipment as indicated on the drawings, diagrams, and schedules. System shall be complete with transformers, rectifiers, relays, switches, master-selector switches, wallplates, and wiring. All remote control wiring components shall be of the same type and installed in accordance with the recommendations of the manufacturer. Except where otherwise indicated, all remote control wiring shall be installed in accordance with NEC Article 725, Class 2.

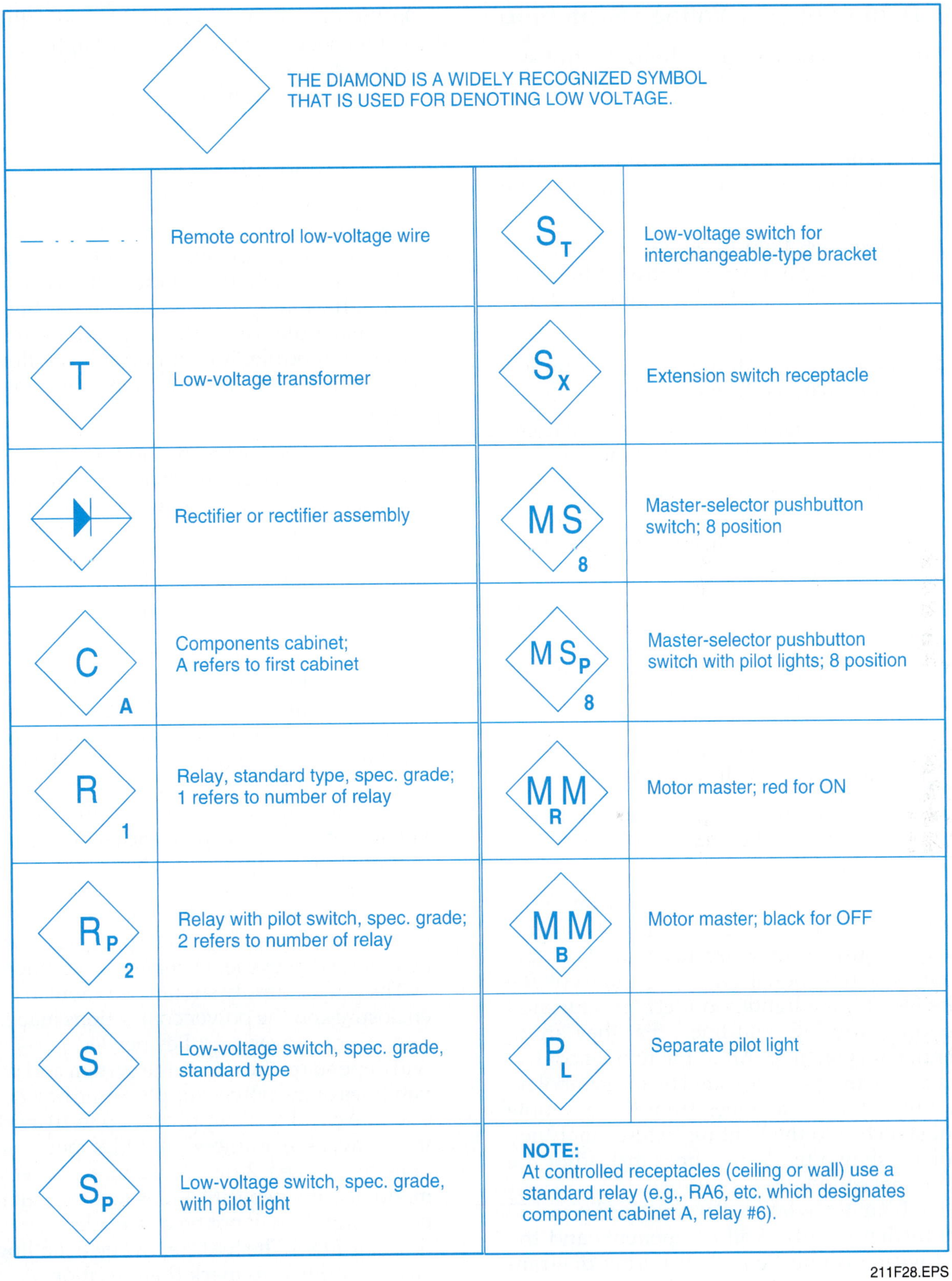

Figure 28 ◆ Recommended symbols for low-voltage drawings.

7.5.0 Design of Low-Voltage Switching

Many times the low-voltage lighting control system will be designed by architects, especially in new installations. However, in existing buildings, the electrician may be expected to design and install the system. If this situation occurs, work from the set of job blueprints, and use the following general procedure:

Step 1 Assign each light or receptacle a specific relay. This can be accomplished by using the letters R1 for relay 1, R2 for relay 2, and so on.

Step 2 Group together all the lights and outlets that are to be switched together.

Step 3 Determine the number of relays that can be controlled by each control module.

Step 4 Divide the number of relays by the number computed in Step 3. This is the number of control modules needed to control the relays.

Step 5 Determine the location of the switch points on the building blueprints.

Step 6 Indicate what lights and receptacles are being controlled by what relay. This step may be accomplished by indicating the group of lights on the blueprint. However, it is easier if a switch schedule is used.

Step 7 Review the blueprints or schedule to determine the number and size of the switching stations.

Figure 29 shows the floor plan of a building. Symbols are used to show the location of all lighting fixtures just as they are used to show the location of conventional switches. Line voltage circuits feeding the lighting outlets are indicated in the same way (by solid lines) that they are in conventional switching. All of the remote control switches, however, are indicated by the symbol S_L, instead of S, S_3, S_4, etc. Lines from these remote control switches to the lighting fixtures that they control are shown by dashed lines instead of the conventional solid line.

The schematic wiring diagram in *Figure 30* shows further details of all components and the related wiring connections. This wiring diagram aids the contractor or electrician in the installation of the system and leaves little doubt as to exactly what is required.

A brief note or specification, such as described earlier, completes the design of the low-voltage remote control system.

On the floor plan in *Figure 29*, notice that the door switches controlling the closet lights use conventional switches since low-voltage switches of this type are uncommon.

7.6.0 Installations

Before starting, consult both the *NEC®* and local codes for installation requirements. Once the layout plan has been developed and materials are acquired, the installation of the system can begin. The rough-in instructions presented here are only an example. As with the previous sections of this task module, it is important to follow the manufacturer's recommendations and blueprint specifications.

- *Enclosure* – The enclosure should be placed in a location that is easily accessible but hidden from sight. If at all possible, install a unit that is prewired. This will decrease installation time and also limit the possibility of mistakes. Note that even though the panel may be prewired, the supply cables will still have to be installed. This should be done according to the manufacturer's recommendations.

  Since heat will be produced by the enclosed control modules, vents are placed around the enclosure to facilitate cooling. The location chosen for the enclosure should not hinder the cooling capabilities for which the panels are designed. The manufacturer may require that the high-voltage portion of the enclosure panel be mounted in the upper right-hand corner. This is due to the *NEC®* requirement that is designed to keep low-voltage conductors separated from high-voltage conductors. The panel itself should be mounted at eye level between two studs.

  The power supply should be mounted on the enclosure and the power connections made. The number of power supplies needed per system will depend on the size of the system and the capabilities of the power supply. Some power supplies are capable of supplying up to 10 modules if the average number of LEDs per channel does not exceed 2 or 2.5. Be sure to check the manufacturer's specifications to ensure that the power supply will not be overloaded.
- *Switch points* – The first step when installing the switch points is to mark their location. As with ordinary switches, the switch points should be located as specified by the plans or specifications. Once the location has been determined, mount either a plaster ring or a nonmetallic box. A box is not needed in most frame construction installations.

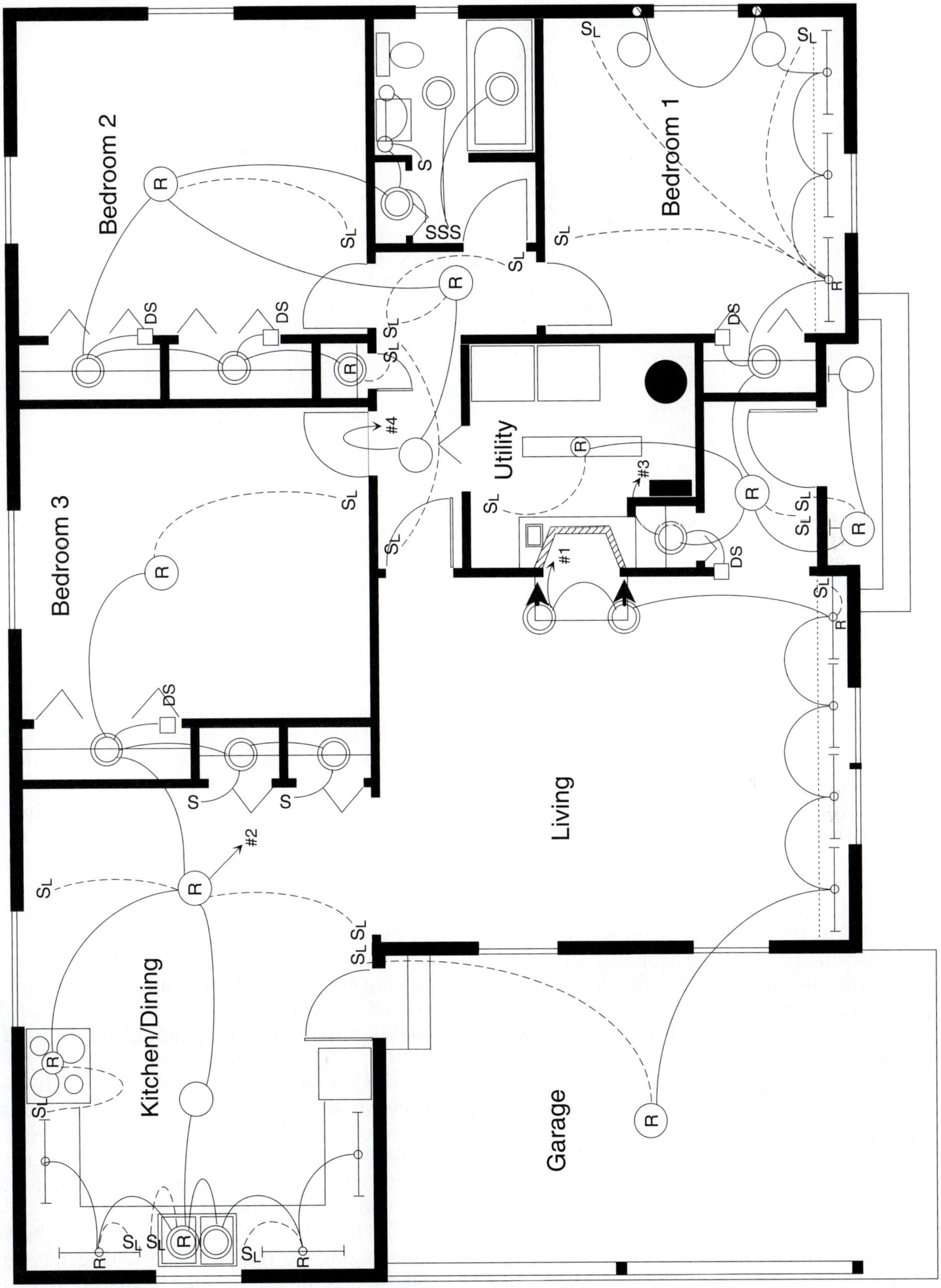

211F29.EPS

Figure 29 ◆ Floor plan of a residence using low-voltage switching.

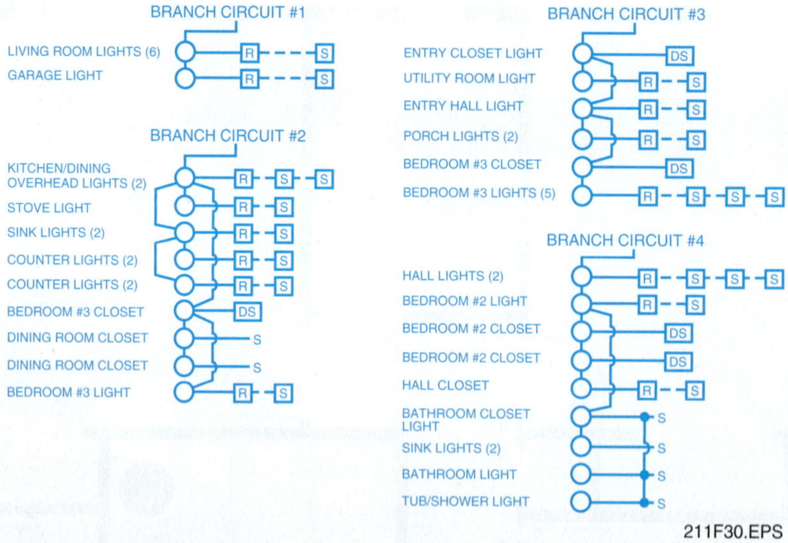

Figure 30 ◆ Schematic wiring diagram of lighting circuits.

A plaster ring or Romex® staple should then be positioned partway in the stud, behind the plaster ring or nonmetallic box. The preparation for running the low-voltage wiring is the same as for running 120V wiring. If the edge of the hole is closer than 1¼" from the nearest edge of the stud, a 1⁄16" steel nail plate or bushing must be used to protect the cable from future damage by nails or screws.

The wires are then pulled from the switch station to the panel enclosure. They should come from the left side if the manufacturer recommends that the enclosure be installed with the high-voltage section in the upper right-hand corner. There should be approximately 10" to 12" of extra wire left at each switch location and 12" to 18" left at the panel enclosure.

Multiconductor cable is frequently used due to its simpler installation. Run one wire (low-voltage lead) from the enclosure to the farthest switching point and anchor the wire to the plaster ring. This wire will supply the low-voltage power to each of the switches. Working back toward the enclosure, loop the wire down and anchor it to each switch location. Take care to leave a sufficient amount of wire for connecting the switch. The low-voltage lead is then connected to the appropriate power supply lead at the enclosure. The process is repeated for the DC common lead. When this wiring method is used, the DC positive and common lead will have the same color throughout the system and future connection mistakes will be minimized.

The next step is to select a multiple-switch circuit. Pull a different colored wire from the enclosure to the farthest switching location in this circuit. Working back toward the panel enclosure as before, loop the wire to the other switch locations in the multiple-switch circuit. Thus, there are two leads going to each switch with one of the leads common to all the switches in the system. However, if the switches contain a pilot light, two additional No. 18 wires must be run to the LED. The wires are connected to the appropriate pilot light transformer located in the panel enclosure.

If possible, use different colored wires for each run. This will limit confusion in the future when connecting the individual circuits to power. Wire schedules should be used to limit hookup mistakes and provide a clearer picture of the system.

Remember that the colors for the control wiring should not be the same as the colors for the DC power and common conductors. Additional marking methods should also be used for future reference. Note that if the wires are marked using adhesive tabs or punch cards, they should be placed in a plastic bag and shoved behind the ring or securely into the plastic box. This will prevent the markings from being covered by paint or plaster when the walls are finished.

The low-voltage wiring should also be kept apart from the high-voltage wiring whenever possible. If the cables must be placed in parallel runs, they should be separated by at least 6", otherwise the noise generated from the high-voltage conductors could seriously interfere with the operation of the low-voltage system.

- *Power and finish trim* – Connecting the circuits to power is now a fairly simple task. Each wire from the switch is connected to a relay with the other side of the relay connected to the common of the DC supply.

  When the walls of the building are finished, the switches are connected to the conductors located at each switch point location. Wall plates are then added. A common type of wall plate used with low-voltage lighting control systems requires no screws since it can be snapped into place.

8.0.0 ◆ RELAY TROUBLESHOOTING

Electromagnetic contactors and relays of various types are used to make and break circuits that control and protect the operation of such devices as electric motors, combustion and process controls, alarms, and annunciators. Successful operation of a relay is dependent on maintaining the proper interaction between its solenoid and its mechanical elements—springs, hinges, contacts, dashpots, and the like.

The operating coil is usually designed to operate the relay from 80% to 110% of rated voltage. At low voltage, the magnet may not be strong enough to pull in the armature against the armature spring and gravity, while at high voltages excessive coil temperatures may develop. Increasing the armature spring force or the magnet gap will require higher pull-in voltage or current values and will result in higher drop-out values.

When sufficient voltage is applied to the operating coil, the magnetic field builds up, and the armature is attracted and begins to close. The air gap is shortened, increasing the magnetic attraction and accelerating the closing action, so that the armature closes with a snap, closing normally open contacts and opening normally closed contacts. Binding at the hinges or excessive armature spring force may cause a contactor or relay that normally has snap action to make and break sluggishly. This condition is often encountered in relays that are rarely operated. Contactors and relays should be operated and observed from time to time to make sure their parts are working freely with proper clearances and spring actions.

Troubleshooting charts are provided in *Table 2*. Knowing these symptoms and their causes should help you to quickly spot and repair contactor and relay problems.

THINK ABOUT IT

Putting It All Together

Imagine you are designing your dream home. What systems would you like to be able to control from a central location? What types of devices would be required for this system?

Table 2 Contactor and Relay Troubleshooting Chart

Symptoms	Probable Cause	Action or Items to Check
Failure to pull in	Either no voltage or low voltage at coil terminals	Blown fuse, open line switch, break in wiring Line voltage below normal Overload relay open or set too low Tripping toggle (non-automatic breakers) fouled Control level or start button in off position Pull-in circuit open, shorted, or grounded Contacts in protective or controlling circuit open or one of their pigtail connections broken
	Operating coil open or grounded	Inspect and test coil
	Loose or disconnected coil lead wire	Inspect and correct
	Excessive magnet gap, improper alignment	Inspect and correct
	Armature obstructed or gumming deposits between armature and pole face	Inspect and correct
	Binding caused by deformed or gummy hinge	Replace if bent; degrease if gummed up
	Excessive armature spring force	Lessen spring force
	Normally closed contacts welded together	Replace contacts
Failure of equipment to start with contactor closed	One contact not closing	Replace set of contacts
	Contacts burned	Replace
	Contact pigtail connection broken	Replace
Failure to drop out	Operating coil is energized	Contacts in controlling or protective tripping circuits closed, shorted, or shunted Tripping devices defective, such as overload tripping toggles do not strike release, undervoltage relay plunger stuck or out of adjustment, defective stop button, or defective time-delay escape mechanism (closed air vent) Current supplied over an unintentional path due to grounds, defective insulation, pencil markings, moisture, or lacquer chipped off relay base
	Residual magnetism excessive due to armature closed tightly against pole face	Adjust or replace
	Armature obstructed or gumming deposits between armature and pole face	Clear and clean
	Binding caused by deformed or gummy hinge	Replace or degrease
	Contact pressure spring or armature spring too weak or improperly adjusted	Replace or adjust
	Improper mounting position (upside down)	Mount correctly
	Normally open contacts welded together	Replace contacts
Time-delay relays operate too fast	Air escapes too freely due to holes in bellows or open air vent	Escape mechanism faulty Worn dashpot plunger, dashpot oil too thin
	Non-magnetic shim in air gap too large or armature spring too strong	Adjust or replace
	Magnets out of adjustment	Adjust

Table 2 Contactor and Relay Troubleshooting Chart

Symptoms	Probable Cause	Action or Items to Check
Contacts pitted or discolored	Contacts overheated from overload	Reduce load and replace contacts
	Contacts not fitted properly	Refit
	Barriers broken from rough usage or breakers closing with too much force	Replace barriers; check for high voltage
	Wiping action of contacts on closing is insufficient	Adjust
	Excessive chatter or hum	Check causes listed below
	Exposure to weather, dripping water, salt air, or vibration	Use suitable NEMA enclosure
Excessive chatter or hum	Vibration of surrounding devices communicated to relay	Arrange differently
	Relay receiving contradictory signals	Correct
	Bouncing of controlling or protective contacts	Adjust
	Line voltage fluctuating or insufficient	Consider using buck-and-boost transformers
	Excessive coil circuit resistance	Lessen resistance
	Armature spring or contact pressure spring too strong	Lessen
	Excessive coil voltage drop on closing	Adjust
	Missing shading coil	Replace the contactor
	Improper antifreeze pin location	Locate in correct position
	Free movement of armature hindered due to deformed parts or dirt	Clean and replace appropriate parts
Excessive coil temperature	Excessive current or voltage	Reduce load or adjust taps on transformer
	Short circuit in coil	Replace coil

Review Questions

1. The function of a shading coil in a contactor is to _____.
   a. keep ultraviolet light off the contacts
   b. prevent contact chatter and hum
   c. act as a holding circuit to keep the contactor energized when power is removed
   d. help the contacts return to their normal position when the contactor becomes de-energized

2. What is the major difference between a contactor and a motor starter?
   a. Contactors are always larger than motor starters for the same current-carrying capacity.
   b. All contactors are suitable for full-load lighting loads; starters are not.
   c. A contactor always has a disconnect while a starter does not.
   d. The starter provides overload protection while a contactor does not.

3. Which of the following best describes a control relay?
   a. A device designed to handle relatively high currents
   b. A fuse with two attachment points
   c. An electromagnetic device used in control circuits
   d. A fuse with two barrels

4. Which of the following best describes a relay that is used for motor protection?
   a. Amplifying relay
   b. Thermal overload relay
   c. Solenoid
   d. Contactor

5. The highest voltage rating for magnetic relays is _____.
   a. 240V
   b. 277V
   c. 480V
   d. 600V

6. The approximate inrush current of filament-type lamps is _____ times the normal operating current.
   a. 15 to 17
   b. 20 to 30
   c. 30 to 40
   d. 50 to 60

7. What is the normal position of a stop pushbutton that is used in conjunction with contactors?
   a. Normally open
   b. Normally closed
   c. Normally closed 50% of the time and open 50% of the time
   d. Normally open 70% of the time and closed 30% of the time

8. The normal maximum amperage rating of control relays at 600V is _____.
   a. 15A
   b. 20A
   c. 0A
   d. 65A

9. When selecting overload relays for motors, which of the following must be considered?
   a. The motor manufacturer
   b. The cost of the motor
   c. The length of time the motor has been in use
   d. The motor full-load current

10. Which NEMA enclosure is used for general indoor applications?
    a. NEMA 1
    b. NEMA 3R
    c. NEMA 7
    d. NEMA 13

Summary

The general classification of contactor covers a type of electromagnetic apparatus that is designed to handle relatively high currents and is identical in appearance, construction, and current-carrying ability to the equivalent NEMA size magnetic starter. The magnet assembly and coil, contacts, holding circuit interlock, and other structural features are the same. The significant difference is that the contactor does not provide overload protection.

A relay is an electromagnetic device whose contacts are used in the control circuits of magnetic starters, contactors, solenoids, timers, and other relays. Relays are also used extensively in remote control switching, especially where an outlet or a group of outlets is to be controlled from several locations.

Notes

Trade Terms Introduced in This Module

Ambient temperature: The temperature of the medium (usually air) surrounding a device and into which the heat losses in the device are dissipated.

Break: The opening of closed contacts to interrupt an electrical circuit.

Inductive load: A load in which the voltage leads the current. Electromagnet coils such as those used in relays and starters are inductive loads.

Instantaneous: Contacts that make or break a circuit immediately upon coil energization or de-energization.

Latch coil: The coil of a latching relay that, when energized, causes the relay to assume the switched or latched position.

Make: The closing of open contacts to establish an electrical circuit.

Mechanically held relay: A latching relay in which the switched or latched position is maintained by the interference of a mechanical part that moves to a non-interfering position when the unlatch coil is energized, allowing the relay to return to the normal or unlatched position.

Normally closed (N.C.) contacts: A combination of a stationary contact and a movable contact, which are engaged when the coil is de-energized. In the case of a latching relay, the contacts that are engaged after the unlatch coil is energized.

Normally open (N.O.) contacts: A combination of a stationary contact and a movable contact, which are disengaged when the coil is de-energized. In the case of a latching relay, the contacts that are disengaged after the unlatch coil is energized.

Shading coil: A shorted turn surrounding a portion of the pole of an AC electromagnet, producing by mutual inductance a delay in the change of the magnetic field in that part, thereby tending to prevent chatter and reduce hum.

Solid-state relay: A semiconductor switch having no moving parts. These devices control power to the load and are analogous to relay contacts.

Unlatch coil: The coil of a latching relay that, when energized, causes the relay to assume the normal or unlatched position.

Additional Resources

This module is intended to present thorough resources for task training. The following reference works are suggested for further study. These are optional materials for continued education rather than for task training.

American Electrician's Handbook, Latest Edition. New York: Croft and Summers, McGraw-Hill.

National Electrical Code® Handbook, Latest Edition. Quincy, MA: National Fire Protection Association.

CONTREN® LEARNING SERIES — USER FEEDBACK

The NCCER makes every effort to keep these textbooks up-to-date and free of technical errors. We appreciate your help in this process. If you have an idea for improving this textbook, or if you find an error, a typographical mistake, or an inaccuracy in NCCER's *Contren®* textbooks, please write us, using this form or a photocopy. Be sure to include the exact module number, page number, a detailed description, and the correction, if applicable. Your input will be brought to the attention of the Technical Review Committee. Thank you for your assistance.

Instructors – If you found that additional materials were necessary in order to teach this module effectively, please let us know so that we may include them in the Equipment/Materials list in the Annotated Instructor's Guide.

Write: Product Development
National Center for Construction Education and Research
P.O. Box 141104, Gainesville, FL 32614-1104

Fax: 352-334-0932

E-mail: curriculum@nccer.org

Craft _______________ Module Name _______________

Copyright Date _______________ Module Number _______________ Page Number(s) _______________

Description

____

____

____

(Optional) Correction

____

____

(Optional) Your Name and Address

____

____

Electric Lighting
26212-05

Steven F. Udvar-Hazy Center
National Air and Space Museum
Chantilly, Virginia
Mega-Projects Over $100 Million Award Winner
Hensel Phelps Construction Co.

26212-05

Electric Lighting

Topics to be presented in this module include:

Overview

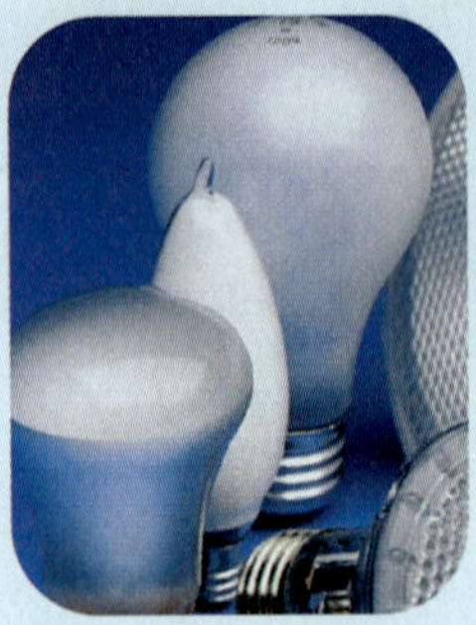

Electric lighting is divided into four primary groups: incandescent, halogen, fluorescent, and high-intensity lighting. Incandescent lighting was the first type of lighting developed, but is the least energy efficient. A variation of standard incandescent lighting is the tungsten-halogen incandescent lamp, which provides up to 20% greater efficiency than incandescent lamps, as well as longer lamp life and improved light quality.

Fluorescent lighting provides a bright, even lighting and is much more energy efficient than incandescent lighting. In addition, fluorescent lamps may last up to 20 times longer than standard incandescent lamps. Fluorescent lamps are available in a variety of lamp shapes, wattages, and color-rendering qualities.

High-intensity discharge lamps (HID) demonstrate a high energy efficiency and long life, but the color of their light is not generally appealing to the eye. Three primary types of HIDs are mercury vapor, metal halide, and sodium, with each type displaying its own unique color of light.

Lighting fixtures, now referred to as luminaires in the *NEC*®, come in many designs and mounting configurations. They may be designed for recess, surface, pendant, lay-in, suspended, or track mounting. Lighting must be installed in accordance with the manufacturer's procedures and the *NEC*®.

Note: *National Electrical Code*® and *NEC*® are registered trademarks of the National Fire Protection Association, Inc., Quincy, MA 02269. All *National Electrical Code*® and *NEC*® references in this module refer to the 2005 edition of the *National Electrical Code*®.

Objectives

When you have completed this module, you will be able to do the following:

1. Explain how the human eye works.
2. Describe the characteristics of light.
3. Recognize the different kinds of lamps and explain the advantages and disadvantages of each type:
   - Incandescent
   - Halogen
   - Fluorescent
   - High-intensity discharge (HID)
4. Properly select and install lamps into lighting fixtures.
5. Recognize and install various types of lighting fixtures:
   - Surface-mounted
   - Recessed
   - Suspended
   - Track-mounted

Trade Terms

Ballast
Color rendering index (CRI)
Glare
Incident light
Lumen (lm)
Luminaire (fixture)
Luminance
Reflected
Reflected light
Refracted/refraction
Starter
Troffer

Required Trainee Materials

1. Pencil and paper
2. Appropriate personal protective equipment
3. Copy of the latest edition of the *National Electrical Code*®

Prerequisites

Before you begin this module, it is recommended that you successfully complete *Core Curriculum; Electrical Level One;* and *Electrical Level Two,* Modules 26201-05 through 26211-05.

This course map shows all of the modules in *Electrical Level Two.* The suggested training order begins at the bottom and proceeds up. Skill levels increase as you advance on the course map. The local Training Program Sponsor may adjust the training order.

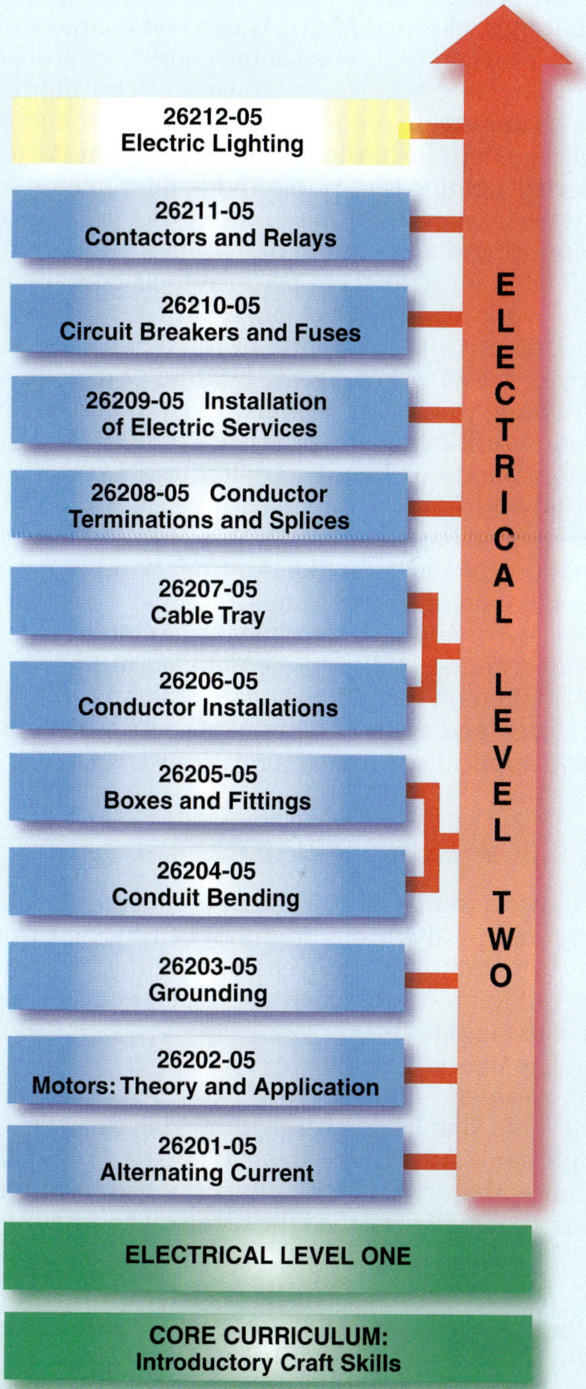

212CMAP.EPS

1.0.0 ◆ INTRODUCTION

Electric lighting is used extensively throughout residential structures, commercial businesses, industrial plants, and outdoor sites. Its use in residential structures can improve their appearance and that of the objects in them. In stores, hotels, and office buildings, electric lighting is used on a large scale to improve the efficiency of the employees and aid in selling merchandise. In industry, electric lighting helps to increase production, reduce errors, and increase safety. Regardless of the application, correct electric lighting serves to provide illumination for the performance of visual tasks with a maximum of comfort and a minimum of eyestrain and fatigue, allowing individuals to perform their daily living and work-related tasks more easily.

This module is the first of three modules that cover the subject of electric lighting. It introduces the basic principles of human vision and seeing and describes the characteristics of light. The different kinds of lamps and lighting fixtures are briefly introduced. The remainder of this module focuses on basic guidelines and procedural information for the receiving, storing, handling, and installation of electric lamps and lighting fixtures. The two remaining modules, presented later in your training, will provide detailed technical information on the various types of lamps, lighting fixtures, and related components, as well as practical applications involving their use.

2.0.0 ◆ HUMAN VISION

Human vision is a process that occurs partly in the eye and partly in the brain. Light **reflected** from objects stimulates the eye. This stimulation is conveyed to the brain, where it is registered as a conscious sensation.

The structure of the eye is similar in many ways to a camera, which consists of a lens system, a variable diaphragm, and film. The variable diaphragm is the iris of the eye (*Figure 1*), and the film is the retina. Light enters the eye through a transparent layer called the cornea. The amount of light that is allowed to strike the lens is controlled by the contraction and expansion of the iris. During low-light levels, the iris expands, and during high-light levels, it contracts.

Light passes through the pupil, which is the opening of the iris, and then through the lens, which is directly behind the iris. There, the light is focused to form an image that is passed through a transparent, jelly-like substance called the vitreous humor to the back wall or retina of the eye. The light on the retina stimulates nerve terminals called rods and cones. These nerve terminals are

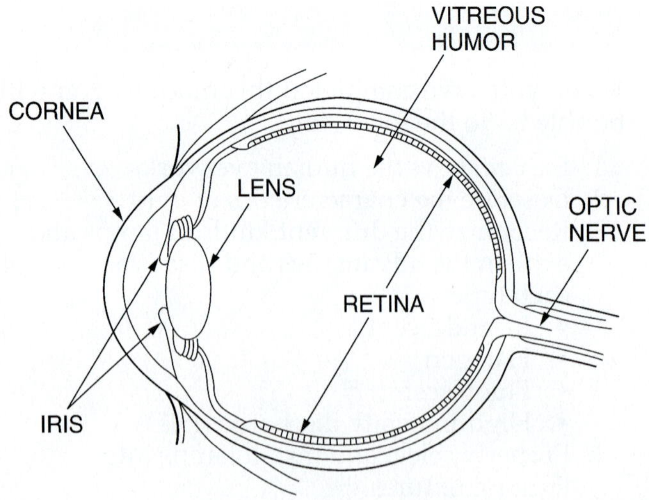

Figure 1 ◆ Structure of the human eye.

connected to the brain by the optic nerve, which furnishes the path by which the light impulses are transferred from the eye to the brain.

The field of vision covers a wide angle of about 200° horizontally and 120° vertically. In the central region of the field of vision, the eye is responsive to color and detail, whereas in the outer region, it is chiefly sensitive to motion. The limits of vision are determined mainly by the following factors:

- *Intensity threshold* – The lowest brightness level that can stimulate the eye.
- *Contrast* – The difference in the degree of **luminance** (brightness).
- *Visual angle* – When an object is made smaller or is placed at a greater distance from the eye, the angle formed by the light rays from the extremities of the object to the eye becomes smaller (*Figure 2*), and vice versa.
- *Time threshold* – The minimum time during which a light stimulus must be present in order to be effective. If the interval is too short, the rods and cones of the eye do not have time to respond to an image on the retina. The time threshold is also dependent on the size, brightness, and color of the object being observed.

Other characteristics of vision are:

- High contrast, sharp edges, and motion increase the sensitivity to detail.
- Straight lines are easier to see than curved lines. Horizontal or vertical lines are more easily seen than diagonal lines.
- The background surrounding an object changes its appearance. For example, a gray object appears lighter when it is placed upon a black background, but it appears darker when placed on a white background.

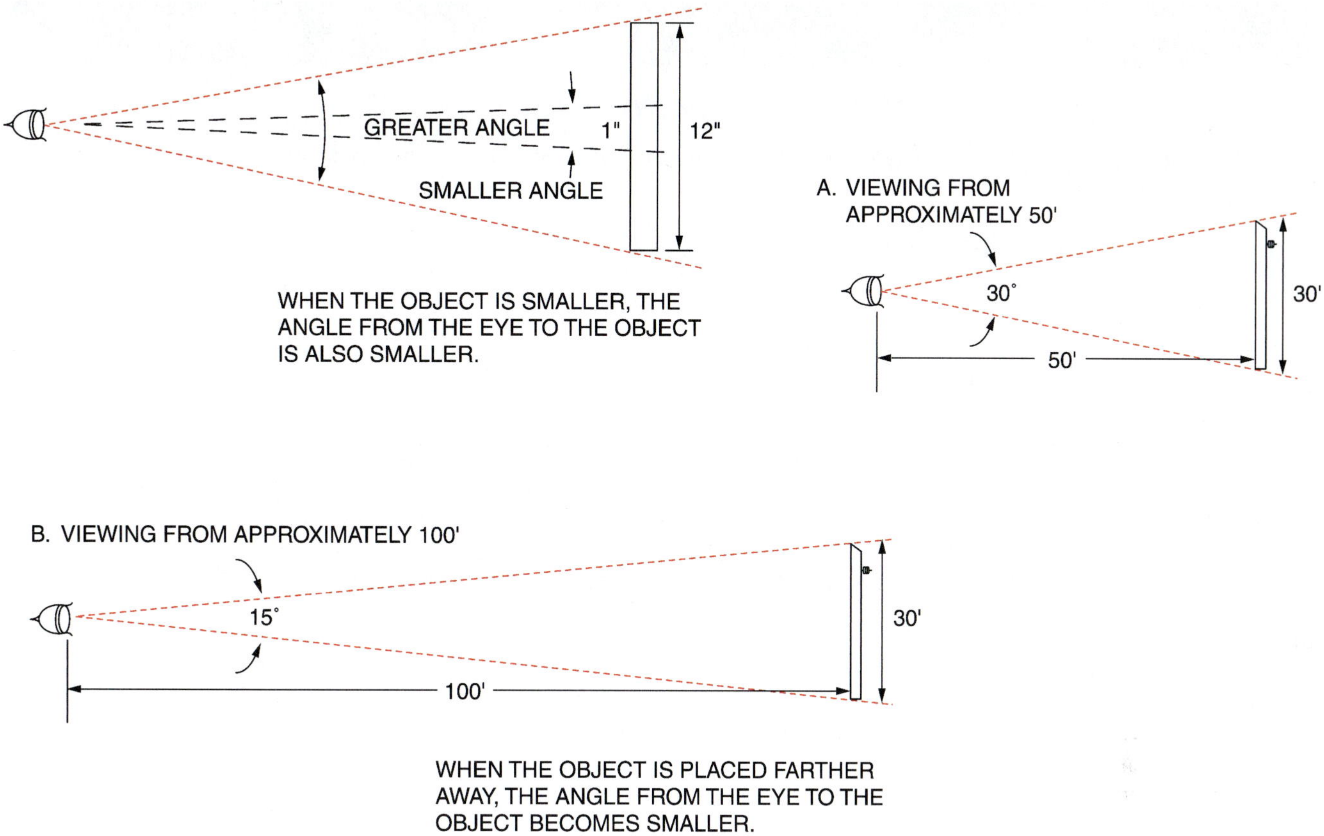

Figure 2 ◆ Visual angle.

3.0.0 ◆ LIGHT CHARACTERISTICS

When discussing light, we normally think of natural light from the sun or the light emitted from a light source, such as an electric light bulb. This type of light is called **incident light.** Another type of light is **reflected light,** which is light given off by an object when incident light strikes it. The difference between the two is that reflected light is dependent on the incident light. When light is not shining on an object, light will not be given off unless the object contains self-luminating properties.

3.1.0 Absorption, Reflection, and Refraction of Light

Three things can happen when light strikes objects or surfaces. Some of the light can be absorbed, some can be transmitted through the object media and **refracted** or diffused, and some can be reflected back from the surface (*Figure 3*). Absorption occurs when light rays pass through a transparent or translucent medium or meet a dense body, such as an opaque reflector surface. The amount of energy absorbed depends on the object's molecular construction, the wavelength

INSIDE TRACK

Rods and Cones

The retina contains millions of rods and cones. The rods are sensitive to the presence of light, even dim light. They detect black and white and function as night-vision receptors. Cones are sensitive to colors. They contain at least three different pigments that are responsible for our ability to see various colors. Cones function as bright-light (day-vision) color receptors.

THINK ABOUT IT

Visible Light Spectrum

Can you think of one example that sometimes occurs in nature that confirms the fact that white light is actually made up of different colors?

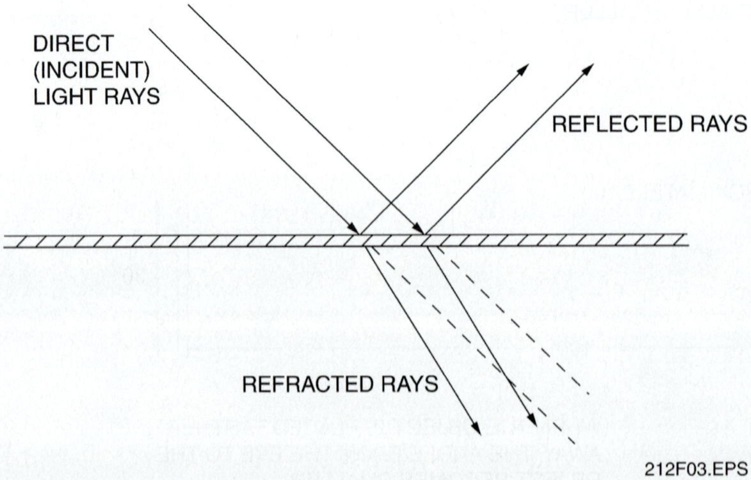

Figure 3 ◆ Simplified example of reflected and refracted light rays.

of the incident light striking it, and the angle at which the light strikes it. Light energy absorbed by an object is dissipated in the form of heat. This explains why walls, floors, metals, etc., feel warm or hot when exposed to direct sunlight or light from a strong artificial light source.

The energy not absorbed by an object when light strikes it can be transmitted through the object or reflected back from its surface, or both. If light rays strike the object perpendicular (at 90°) to its surface, the rays will be transmitted through it in a straight line and/or be reflected back from its surface in the same direction from which it came. If the light strikes the object's surface at an angle other than 90°, as shown in *Figure 3*, then the light transmitted through the object is bent in a different direction, called refraction, and/or the light is reflected back from the object in a different direction from that in which it came. It should be pointed out here that the resultant directions taken by reflected and/or refracted light waves can be very complex. This is because the redirection imparted to the waves depends on the type and density of the object's material, as well as whether its surface is smooth or rough, or if the material contains particles beneath its surface.

The larger the amount of light that is reflected by an object, the brighter the object will appear to the eye. For example, if all the direct light striking an object's surface is reflected from the surface, the object appears light; if it is entirely absorbed, the object appears dark. In addition, the more intense the light source, the brighter the object will become. For example, if you cast a shadow on a portion of an object and look at the difference in brightness between the two areas, the portion without the shadow will be brighter.

Some surfaces and materials reflect light better than others. Light-colored or highly-polished surfaces reflect more light than darker-colored or dull surfaces. This is because white and light-colored materials absorb less of the light energy than do dark or dull surfaces, thus leaving more energy available to be reflected. The color of the walls, ceilings, and floors and their reflecting ability are major considerations in interior lighting design. Typically, darker areas require the use of more artificial light. *Table 1* shows some typical examples of the percentages of light that will be reflected by common surface materials.

3.2.0 Light Colors

Any energy that travels by wave motion is considered radiant energy. Light is one of the many forms of radiant energy. Radio waves and X-rays are also forms of radiant energy. Visible light occupies only a small portion of the radiant energy spectrum. Light is made up of that portion of the spectrum encompassing wavelengths between 380 and 780 nanometers (*Figure 4*).

Table 1 Percentages of Light Reflected by Common Surface Materials

Surface Material	Percentage of Light Reflected
White plaster	90% to 92%
Mirrored glass	80% to 90%
White paint	75% to 90%
Metalized plastic	75% to 85%
Polished aluminum	75% to 80%
Stainless steel	55% to 65%
Limestone	35% to 65%
Marble (white)	45%
Concrete	40%
Dark red–glazed bricks	30%

When all the wavelengths of the light spectrum are presented to the eye in nearly equal proportions, white light is seen. This white light is made up of the wavelengths that create the different colors. This composition can be demonstrated by passing light through a prism. As shown in *Figure 5*, the light spectrum is broken up into its component wavelengths, with each representing a different color. The ability of the prism to separate the different wavelengths, and therefore colors, is because the white light is dispersed into its component wavelengths by refraction as it passes through the prism. The spectrum ranges from violet on the lower end to red on the upper end. In between fall blue, green, yellow, and orange. A total of six colors are visible when passing white light through a prism. Since the colors of the spectrum pass gradually from one to the other, the theoretical number of colors becomes infinite. It has been determined that about 125 colors can be identified over the visible spectrum.

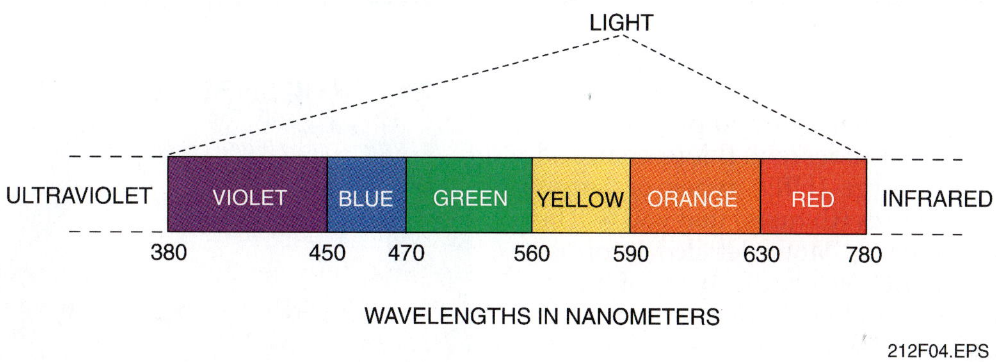

Figure 4 ◆ Visible light spectrum.

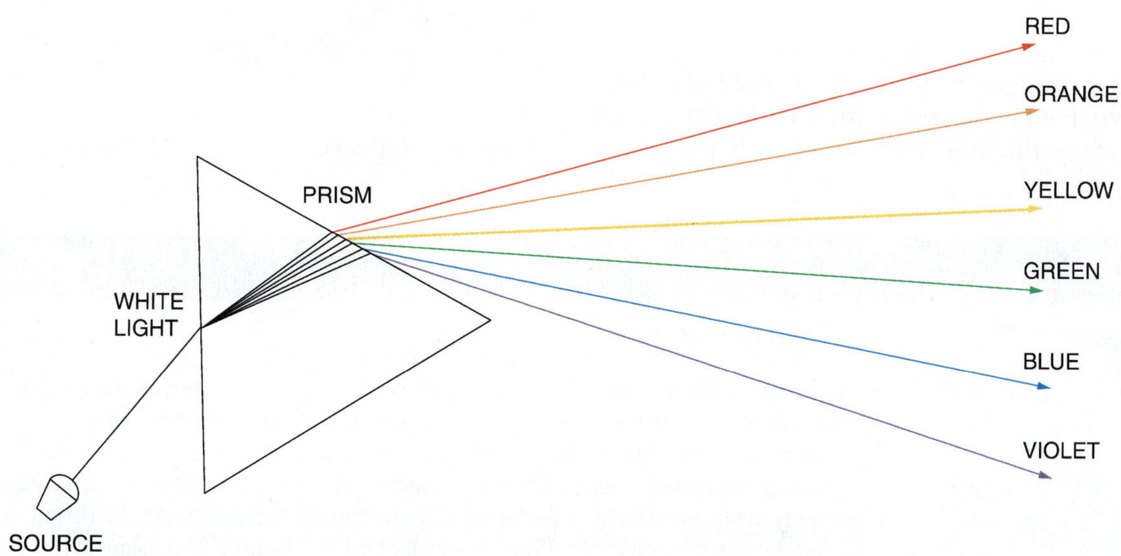

Figure 5 ◆ Light being separated into its component colors by refraction.

As previously described, the amount of light energy absorbed by an object depends on the wavelengths of the incident light striking the object. All objects absorb light of different wavelengths in different proportions. This is called selective absorption. Practically all colored objects owe their color to selective absorption in some part of the visible light spectrum, with resulting reflection and transmission in other selected parts of the spectrum. It is this characteristic that gives us our color distinction when viewing different objects, surfaces, etc. An object's appearance results from the way it reflects the particular light that is falling on it. For example, suppose an apple is red. Under white light, the apple appears red because it tends to reflect light in the red portion of the spectrum and absorb light of other wavelengths (colors).

4.0.0 ◆ ELECTRIC LAMPS AND LIGHTING FIXTURES

This section gives a brief description of the basic kinds of electric lamps and lighting fixtures widely installed in residential, commercial, and industrial lighting applications. The information given here is physical in nature and places an emphasis on the basic incandescent, fluorescent, and high-intensity discharge (HID) types of lamps and lighting **luminaires (fixtures)** that you will be handling and installing. More detailed information and theory about how specific types of incandescent, fluorescent, and HID lamps and fixtures operate is covered later in your training in the *Lamps, Ballasts, and Components* module.

4.1.0 Lamps

There are three main categories of lamps used for lighting: incandescent, fluorescent, and HID. Each category of lamp is made in a wide variety of shapes, sizes, finishes, and mounting bases.

NOTE

Legislation enacted as a result of the *National Energy Policy Act* (EPACT) passed in 1992 requires that lamps that are inefficient in converting electricity to light output be eliminated. This legislation also mandated that after April 1994, many common wattage lamp types could no longer be manufactured or imported into the U.S. It also required that high-efficiency lamps and lighting fixtures be used in all new construction and most retrofit construction.

4.1.1 Incandescent Lamps

Incandescent lamps (*Figure 6*) are used for general lighting and typically provide a warm and natural light. When compared to other types of lamps, incandescent lamps have the lowest efficacy (light output ÷ power input) and shortest life. The average rated life expectancy for a specific type of lamp is usually printed on the shipping carton or package. The number of hours shown on the

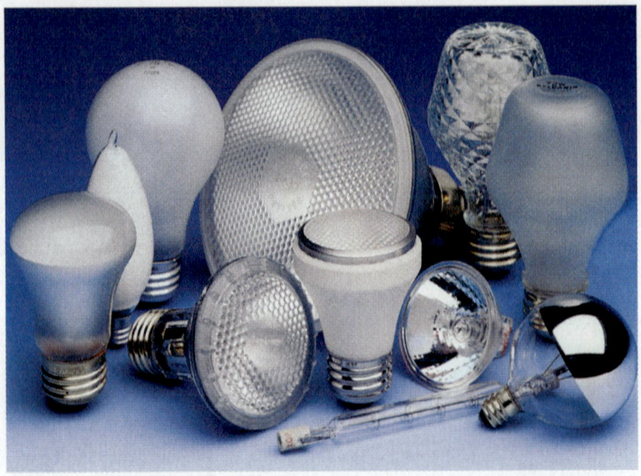

212F06.EPS

Figure 6 ◆ Typical incandescent and halogen lamps.

INSIDE TRACK

Retrofitting

For every dollar spent on the cost of lighting, 80% to 88% is spent on the electricity to run the lamp, while the remainder is spent on the purchase of lamps, lamp maintenance, etc. When doing retrofit work, there are many opportunities for replacing obsolete lamps with newer cost-efficient lamps. This will not only result in reduced energy costs, but usually provides equal or better-quality lighting. Another thing to consider is switching to more energy-efficient lighting fixtures. The initial expenditure will be recovered by the resultant energy savings over the long run.

package represents the number of hours at which 50% of the lamps have burned out. As shown in *Figure 6*, incandescent lamps are made in a wide variety of sizes and shapes. The most common lamp bases used with incandescent lamps are the screw-in or bayonet type, although other base configurations are used.

Figure 7 shows an example of a typical lamp manufacturer's catalog entry used to describe different kinds of incandescent lamps.

Approximate Beam Spread: *For reflector type lamps. The total angle of the directed beam (in degrees) to where the intensity of the beam falls to 50% of the maximum value.*

Filament Design: *Filaments are designated by a letter combination in which C is a coiled wire filament, CC is a coiled wire that is itself wound into a larger coil, and SR is a straight ribbon filament. Numbers represent the type of filament-support arrangement.*

CBCP (Center Beam Candlepower): *For reflector type lamps. Center Beam Candlepower is the intensity (candelas) at the center or maximum intensity of the beam.*

Ⓔ = Means this lamp meets Federal Minimum Efficiency Standards. To save energy costs, find the bulbs with the light output you need, then choose the one with the lowest watts.

Bulb: *Bulb shape followed by its size (the maximum diameter of the bulb expressed in eighths of an inch).*

Lamp Designation: *The lamp's identification code.*

Life: Hours: *Life (as defined by FTC lamp label rules) is rated average life (see notes 1-34).*

Color Temperature-Kelvins (K): *"Warmth" or "Coolness" of the lamp, measured in Kelvins (K). The higher the temperature, the cooler the appearance of the light.*

Volts: *Each lamp's voltage is listed.*

Energy Used Watts: *Energy Used (as defined by FTC lamp label rules). To find actual energy used (kWh) multiply power (watts shown) x time divided by 1000.*

Product Code: *It is important to use this five-digit code when ordering to ensure that you receive the exact product you require.*

LCL in. (mm): *Distance between the center of the filament and the Light Center Length reference plane, in inches and millimeters.*

Light Output- Lumens: *The lamp's rated output after the initial 2 hours of operation.*

Case Quantity: *Number of product units packed in a case.*

MOL in. (mm): *Maximum Overall Length in inches and millimeters.*

Additional Informaton: *Typical application and/or other important information.*

Base: *The type of base.*

Bulb	Base	Product Code	Ⓔ	Lamp Description	Volts	Case Qty.	Additional Information	Light Output Lumens	Energy Used Watts	Life Hours	Filament Design	MOL in. (mm)	LCL in. (mm)	Color Temp. K	CBCP	Appx. Beam Spread
100 WATTS																
A19	Med	41034		100A 48PK	120	144	Standard	1710	100	750	CC-8	4 7/16 (112.7)	3 1/8 (79.4)	-	-	-
		17933		100A	130	120	Inside Frost	1680	100	750	CC-8	4 7/16 (112.7)	3 1/8 (79.4)	-	-	-
							Ratings at 120 volts.	1275	89	1950						
		39321		100A/CL 24PK	120	120	Crystal Clear	1730	100	750	CC-8	4 7/16 (112.7)	3 1/8 (79.4)	-	-	-
		41036		100A/W 48PK	120	144	Soft White	1690	100	750	CC-8	4 7/16 (112.7)	3 1/8 (79.4)	-	-	-
100 WATTS STANDARD HALOGEN PAR LAMPS																
PAR38	Med Skirt	17951	Ⓔ	100PAR/H/SP	120	12	Spot (14,15,56,88,96)*	1400	100	2000	CC-8	5 5/16 (134.9)	- -	2900	17000	11
		17986	Ⓔ	100PAR/FL/S/HAL 6PK	120	6	Flood (14,15,56,88,96)*	1400	100	2000	CC-8	5 5/16 (134.9)	- -	2900	4800	27
		17947	Ⓔ	100PAR/H/FL	130	12	Flood (14,15,56,88,96)*	1400	100	2000	CC-8	5 5/16 (134.9)	- -	2900	4800	27
							Ratings at 120 volts.	1100	88	4000						

100 PAR / H / FL

Identifies the lamp's wattage. (100)

Identifies the lamp's shape. (PAR)

Identifies as a halogen lamp. (H)

Identifies as a flood lamp. (FL)

To save energy costs, find the bulbs with the light output you need, then choose the one with the lowest watts.
Ⓔ Means this lamp meets Federal Minimum Efficiency Standards. ( ) * All footnote references found at the end of this section. Reduced Wattage

212F07.EPS

Figure 7 ◆ Example of a lamp manufacturer's catalog data. *(Courtesy of General Electric Company.)*

As shown, all manufacturers describe the physical shape and size of their incandescent lamps using a designation consisting of a letter or letters followed by a number. The letter indicates the general shape of the lamp envelope. Some common shapes are standard (A) and tubular (T). The number following the letter indicates the diameter of bulb at its widest point in eighths of an inch.

For example, a lamp designation of A23 means that the lamp has a standard envelope and is $\frac{23}{8}$" or $2\frac{7}{8}$" in diameter. Other information typically given in a manufacturer's catalog data includes lamp watts, type of lamp base, product number, ordering number, volts, package quantity, a brief description of the lamp, class and filament type, light center length, maximum overall length, approximate hours of life, and **lumens (lm)** (the basic measurement of light).

Many people equate a lamp's wattage rating with the amount of light the lamp produces. This is a misconception. The wattage rating is a measure of the electrical energy consumed by a lamp to produce its rated light output, not the actual amount of light it produces. For example, a standard 100W lamp may produce about 1,600 lumens, while a typical 32W fluorescent lamp may produce about 2,600 lumens.

Lamps with standard, tubular, and similar envelope shapes provide lighting in all directions (omni-directional). Incandescent lamps, with shapes designated as R, ER, and PAR, are all reflector-type lamps. They direct their light out in front by reflecting it off their cone-shaped inside walls.

Tungsten halogen incandescent lamps are a refinement over the standard incandescent lamp. They provide up to 20% greater efficiency, longer service life, and improved light quality. Their light output contains more blue and less yellow than standard incandescent lamps, making their light appear whiter and brighter. Halogen lamps are commonly used to replace standard and PAR incandescent lamps. Their designs are constantly being upgraded to provide alternatives to popular general-service incandescent lamps, including those made obsolete by energy legislation. Halogen PAR-20 and PAR-30 lamps with long necks are common replacements for obsolete R lamps.

4.1.2 Fluorescent Lamps

Fluorescent lamps (*Figure 8*) provide bright, even lighting at a lower operating cost than incandescent lamps. They provide more light per watt of power used and last many times longer than incandescent lamps. Some last nearly 20 times as long as incandescent lamps. Fluorescent lamps are used mainly indoors because they are temperature sensitive.

Fluorescent lamps are available in various wattages, sizes, shapes, bases, color temperatures, and **color rendering indices.** Fluorescent lamps are made in straight, U-bent, and circular configurations. Straight lamps are made in varying lengths and diameters. Common fluorescent lamps are designated by the letter T, followed by the diameter of the tube expressed in eighths of an inch. They vary in diameter from T-5 ($\frac{5}{8}$") to PG (power groove)-17 ($2\frac{1}{8}$"). In overall length, straight fluorescent lamps range from 6" to 96". Higher wattages go with longer tubes. For example, a 20W straight T-12 tube is shorter than a 40W T-12 tube. T-8 straight fluorescent lamps are newer energy-efficient lamps that use about 20% less

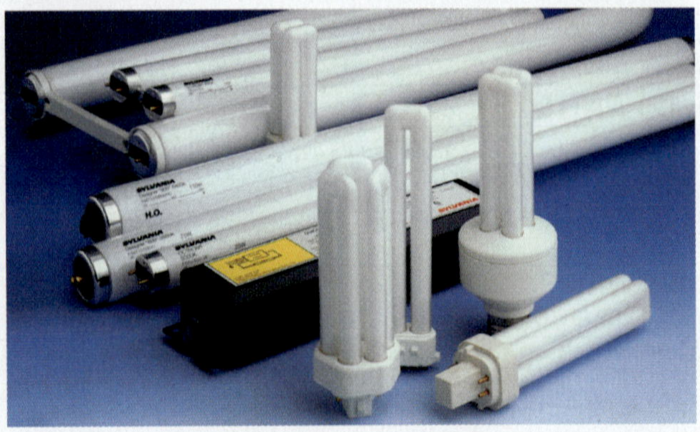

212F08.EPS

Figure 8 ◆ Typical fluorescent lamps and ballast.

INSIDE TRACK

Efficiency

Different lamp types have varying abilities to convert electrical power into visible light. The quantity of light emitted (measured in lumens) is divided by the input power (in watts) to determine a lamp's efficiency (efficacy). This number is expressed in lumens per watt (LPW) and is a measure of energy efficiency. For example, a 60W lamp that produces 800 lumens of light has an efficacy of 13.333 LPW.

Low-Voltage Halogen Lamps

INSIDE TRACK

Compact low-voltage halogen lamps that require less than 30V (typically 12V) rather than 120V line voltage are used for many lighting applications. These low-voltage lamps require a transformer to convert line voltage to the low voltage used by the lamp. The transformer may be incorporated into the lighting fixture itself or it may have to be installed as a separate unit.

These lamps are very sensitive to voltage variations. Even a small change in voltage can have a considerable impact on lamp life. For this reason, it is important to make sure that the fixture transformer rating matches the actual line voltage to ensure that the lamps operate at as close to their rated voltage as possible. Also, make sure that the fixture transformer you select is listed for use with the lighting product.

energy to produce the same light output as an older 40W T-12 lamp. Newer high-efficiency fluorescent lamps that meet the new minimum efficiency standards are normally marked with the letter E enclosed in a circle.

Some fluorescent lamps, called circline lamps, are made in the form of a complete circle with outside diameters ranging from 6½" to 16". U-shaped lamps having a 180° bend in the center are made to fit a specialized 2' long fixture. Another type of fluorescent lamp is the compact fluorescent lamp. These lamps were initially developed as energy-efficient direct replacements for conventional incandescent lamps, but are now also used in many new lighting fixtures designed specifically for their use. They are generally available in ranges from 5W to 28W, with light comparable to incandescent lamps of 15W to 100W. As shown in *Figure 8*, they consist of small, single and multiple U-shaped fluorescent tubes and a self-contained **ballast.** Some have replaceable tubes. A few of the newer versions are dimmable. Some compact fluorescent lamps use a tight-spiral tube and look much like a standard incandescent lamp in size and shape.

There are three electrical classes of fluorescent lamps: preheat, rapid start, and instant start. The lamp class depends on the type of circuit used in the fixture. Preheat fluorescent lamps are used with fluorescent lighting fixtures that have a replaceable **starter** that is separate from the ballast. A ballast is a circuit component that provides the required voltage surge to start up the lamp(s), then controls the subsequent flow of current through the lamp(s) during operation. The starter functions to preheat the lamp's electrodes before lighting. Preheat lamps and fixtures are nearly obsolete and are not used in new construction. Rapid-start fixtures have a ballast with a built-in starter, and instant-start models have no starter. Instant-start lamps are easily recognized because they have a single pin at each end of the lamp; preheat and rapid-start lamps have two pins at each end.

4.1.3 High-Intensity Discharge Lamps

High-intensity discharge (HID) lamps (*Figure 9*) produce light when electricity excites specific gases in pressurized bulbs. They provide long life

Fluorescent Lamp Life

INSIDE TRACK

The life of a fluorescent lamp is affected by the number of times the lamp is started. Frequent switching on and off of fluorescent lamps results in shorter lamp life, while continuous operation provides the longest lamp life. The typical fluorescent lamp ratings given by one major manufacturer of fluorescent lamps are based on three hours per start. For example, if a lamp is rated as having an average life of 12,000 hours, this is based on the lamp being turned on 4,000 times and being left on for a minimum of at least three hours.

Comparing Lamp Efficiency Ratings

INSIDE TRACK

The following listing shows how incandescent, halogen, fluorescent, and HID lamps compare in terms of efficiency and rated average life.

Lamp Type	Typical LPW	Rated Avg. Life (Hrs.)
Incandescent	5–22	750–2,000
Halogen	12–36	2,000–5,000
Fluorescent (compact)	27–80	9,000–10,000
Fluorescent	75–100	12,000–24,000+
Mercury	50–60	12,000–24,000+
Metal halide	80–115	10,000–20,000
High-pressure sodium	90–140	10,000–24,000+

Note that the typical LPW value given above is based on the initial lumen output produced by the lamp when it is new. Most lamps degrade in efficiency over time. For example, mercury vapor lamps have a long average life, but with a continual reduction in lumen output, especially after 10,000 hours.

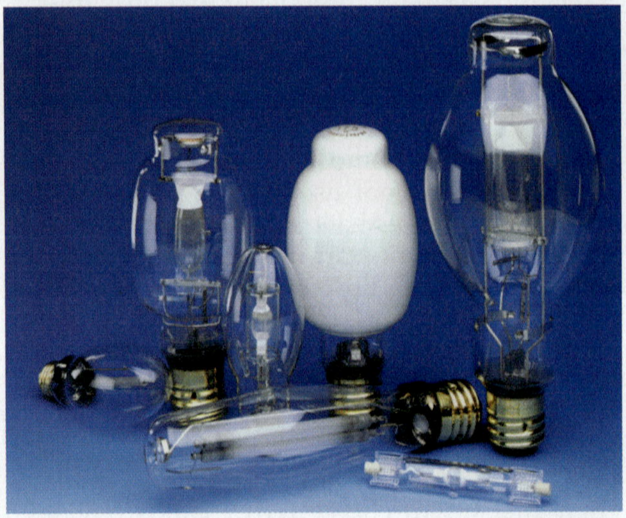

212F09.EPS

Figure 9 ◆ Typical high-intensity discharge lamps.

and high efficiency; however, the color of light emitted by some is rather unflattering. There are three main types of HID lamps: mercury vapor, metal halide, and sodium. The names refer to elements that are added to the gases in the lamp, which cause each type to have somewhat different color characteristics and efficiency. Like fluorescent lamps, HID lamps must be used in matched lighting fixtures with a ballast specifically designed for the lamp type and wattage.

Mercury vapor lamps are the oldest HID technology. These lamps produce light energy in both the visible and ultraviolet (UV) range and have specially-formulated glass outer jackets to filter the potentially harmful UV energy. Today, mercury vapor lamps are being replaced by other HID types that have better efficiency and color properties.

Metal halide lamps are the most energy-efficient source of white light. They have high efficacy, excellent color rendition, long service life, and good lumen maintenance (longevity). Because of these advantages, they are widely used in commercial interiors, especially where elevated ceilings require high light output.

High-pressure sodium HID lamps have extremely high efficacy and exceptionally long service life (typically up to 24,000 hours). However, they produce light that is concentrated in the yellow/orange portion of the spectrum and render colors poorly. This limits their use to outdoor and industrial applications where efficiency and long life are more important than color rendering.

Low-pressure sodium lamps produce light that is limited to a single wavelength in the yellow portion of the spectrum. These lamps are the most efficient of any lamp type and are used only where energy efficiency and long life are the only requirements. It should be pointed out here that technically speaking, low-pressure sodium lamps are not actually a type of HID lamp.

4.1.4 Comparison of Lamp Types

Table 2 provides a brief comparison of the major lamp types.

Table 2 Comparison of Major Lamp Types

Type of Lamp	Advantages	Disadvantages
Incandescent	• Low initial cost • Small size • Excellent color rendering index (CRI) • Variety of shapes • Dimmable • Wattage interchangeable	• Inefficient • Excessive heat output • High operating costs • Short service life • Glare potential
Halogen	• Small size • Increased efficiency* • Longer life* • Excellent CRI • Bright white light • Dimmable	• Excessive heat output • Glare potential
Fluorescent	• Highly efficient • Long service life • Choice of color temperatures and CRI • Low operating costs • Low heat output • Diffuse light source	• High initial cost • Temperature sensitive • Limited optical control • Requires matched fixture/ballast
Metal halide (HID)	• Highly efficient • Long service life • Low operating costs • Good color rendering	• High initial cost • Requires matched fixture/ballast • Long startup/restrike period • Glare potential
High-pressure sodium (HID)	• Long service life • Exceptionally efficient • Very low operating costs • High lumen maintenance	• High initial cost • Requires matched fixture/ballast • Long startup/restrike period • Poor color rendering • Glare potential

*Compared to standard incandescent lamps

4.1.5 Guidelines for Installing Lamps

Listed below are some general guidelines to follow when installing lamps in their related lighting fixtures.

- Make sure that the lamp and lighting fixture are compatible. Install incandescent lamps in incandescent lighting fixtures, fluorescent lamps in fluorescent fixtures, etc.
- Make sure that the lamp or lamps being installed in a lighting fixture have the correct voltage rating for use with the fixture. This is especially important when installing incandescent lamps because a small difference between the rating and the actual voltage has a great effect on a lamp's life and light (lumen) output.
- Make sure that you do not install a lamp, or combination of lamps, in a lighting fixture that exceeds the wattage rating for the lighting fixture; otherwise, the fixture can overheat and may cause a fire. In recessed lighting fixtures with built-in thermal protectors, it will cause the lamps to cycle on and off as the fixture overheats, then cools down.
- When installing fluorescent and HID lamps, make sure that the lamp type and wattage are matched to the lighting fixture and ballast being used. Always install the lamps specified by the lighting fixture manufacturer.
- When installing bi-pin fluorescent lamps, make sure to rotate them ¼ turn to seat and connect them.
- Do not install lamps that have scratched or otherwise damaged envelopes.
- Use a soft cloth or gloves to handle an unencapsulated halogen tubular lamp. This prevents oil from your hands from contacting the bulb. This is important because its quartz glass walls can withstand high operating temperatures, but may crack if etched with oil. Also make sure that no combustible materials can come in contact with the lamp.

- Unlike standard lamps, halogen and metal halide lamps may continue to light even if the exterior lamp glass is broken. If this occurs, replace the lamp immediately to prevent creating a hazard or personnel injury.
- In some applications, such as food processing plants, bakeries, dairies, schools, warehouses, etc., safety sleeves made to fit over fluorescent lamps should be used in order to retain broken glass and phosphors if the lamp ever breaks, falls, or bursts.

In general, energy legislation promotes replacing standard incandescent lamps with halogen and compact fluorescent lamps, replacing incandescent reflector lamps with halogen PARs, and using higher-efficiency fluorescent lamps, such as T-8s, instead of full wattage types. If it is necessary to relamp existing light fixtures designed for use with a discontinued or inefficient type of lamp, replace the lamp with an approved energy-efficient substitute. Lists of approved substitutes are available at all lighting and electrical distributors.

4.2.0 Lighting Fixtures

In the electrical trade, the terms lighting fixture and luminaire are used interchangeably (luminaire is an international term). A lighting fixture (luminaire) is defined as a complete lighting unit consisting of the lamp or lamps, together with the parts designed to distribute the light, position and protect the lamps, and connect the lamps to a power supply.

Lighting fixtures are classified in many ways, including: incandescent, fluorescent, and HID; indoor or outdoor; surface-mounted, recessed, or suspended; residential, commercial, or industrial. There are thousands of lighting fixtures manufactured in different styles for all of these applications. From an installation point of view, lighting fixtures can best be classified by the way in which they are mounted. These include:

- Surface-mounted
- Recessed
- Suspended
- Track-mounted

4.2.1 Surface-Mounted Lighting Fixtures

Surface-mounted lighting fixtures include those fixtures that are directly mounted on a ceiling or wall. They can be incandescent, fluorescent, or HID units. Some typical examples of surface-mounted indoor ceiling and wall lighting fixtures are shown in *Figures 10* and *11*, respectively. *Figure 12* shows an example of a wall-mounted outdoor lighting fixture.

4.2.2 Recessed Lighting Fixtures

As the name implies, recessed lighting fixtures are mounted into a ceiling so that the main body of the fixture is not visible. *Figure 13* shows examples of recessed lighting fixtures typical of those used with incandescent, metal halide HID, or compact fluorescent lamps. As shown, a typical unit for

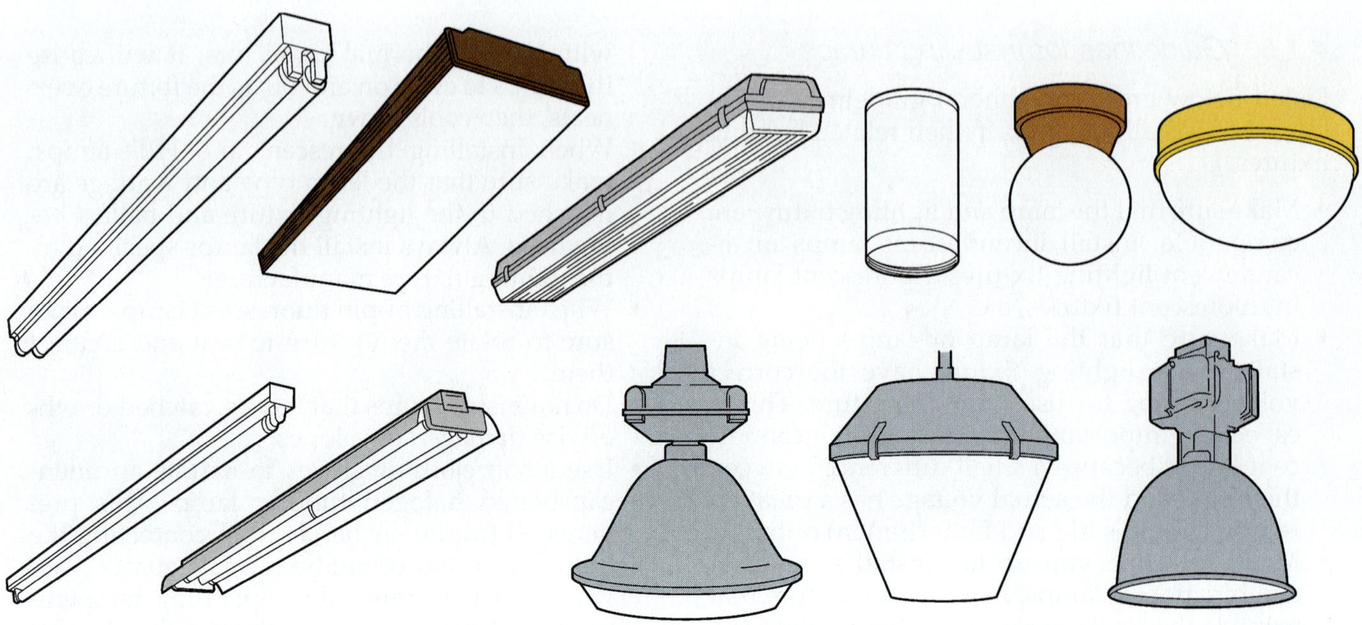

212F10.EPS

Figure 10 ◆ Surface-mounted ceiling light fixtures.

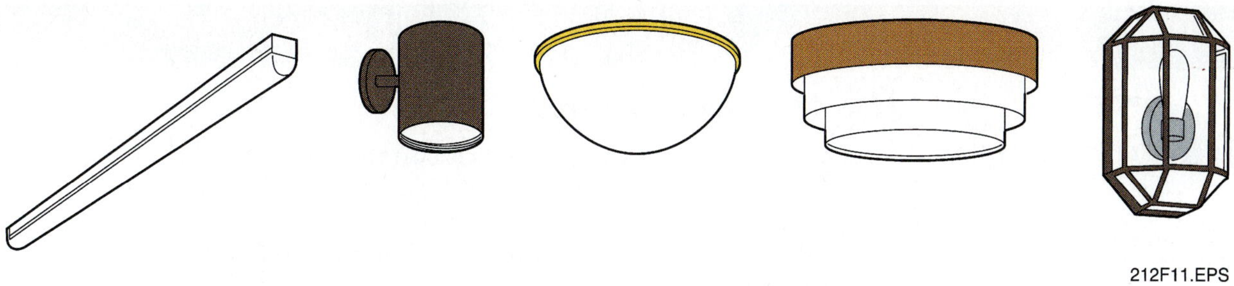

Figure 11 ◆ Surface-mounted indoor wall lighting fixtures.

Figure 12 ◆ Surface-mounted outdoor wall fixture.

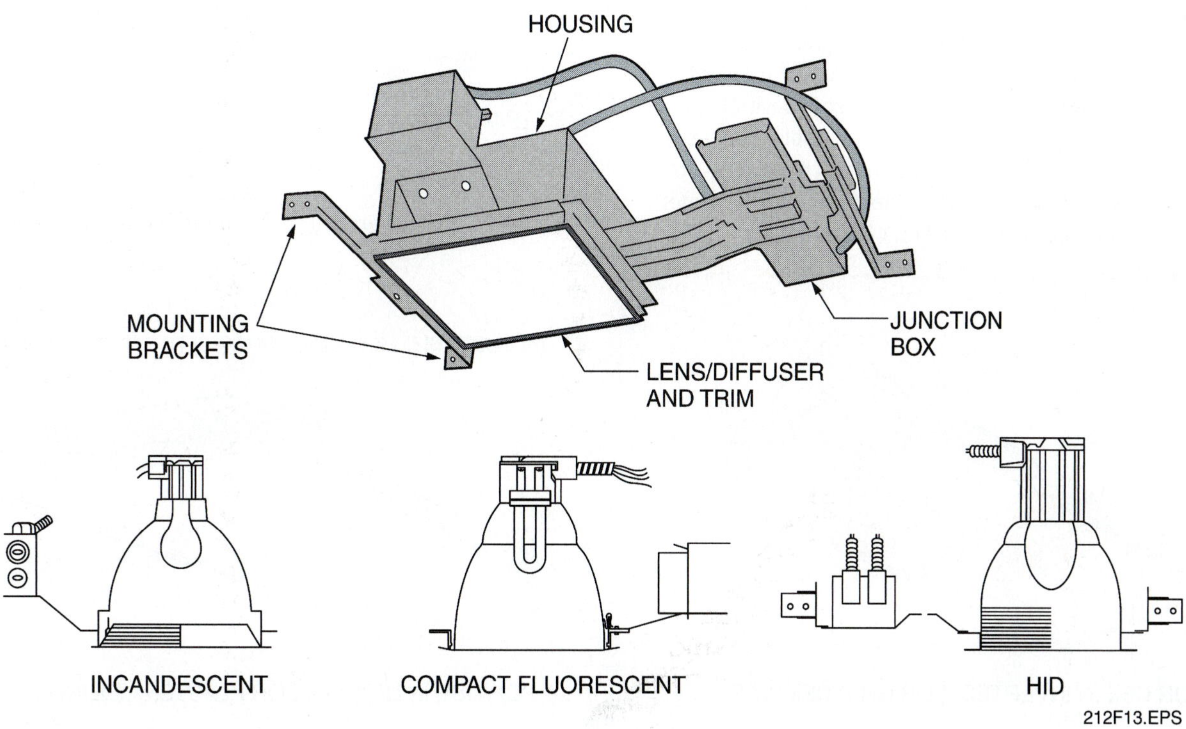

Figure 13 ◆ Typical recessed lighting fixtures.

Use of Ceiling Fixtures

INSIDE TRACK

For large residential rooms, consideration should be given to using more than one ceiling-mounted fixture to provide more efficient and even lighting throughout the space. A single fixture may be too bright in the center, leaving the walls and corners dark. A widely used alternative is to install perimeter recessed fixtures to light up the walls and corners of the room.

new work consists of a housing, junction box, and a mounting frame used to fasten it to the ceiling structure. These components of the lighting fixture are all located above the ceiling line. Only the related lens/diffuser and trim are viewable at the ceiling surface.

Other recessed fluorescent fixtures, called **troffers,** are square or rectangular units installed above the ceiling (*Figure 14*). These are typically 2' × 4' or 2' × 2' units, containing two or three T-8, T-10, or T-12 lamps. Some troffers may incorporate a mounting flange or are used with a mounting flange kit, allowing them to be installed in ceiling recesses over plaster, drywall, or concealed-frame ceiling systems. Other troffers, called grid-type troffers, are designed for installation in grid T-bar suspended ceilings (*Figure 15*).

CAUTION

Incandescent recessed lighting fixtures generate a considerable amount of heat within their enclosure and are a definite fire hazard if not installed properly. For this reason, several precautions and *NEC*® requirements must be observed when installing them. These precautions and requirements are covered later in the installation section of this module.

4.2.3 Suspended Lighting Fixtures

Suspended lighting fixtures are hung from a ceiling using a chain, aircraft wire, cord, or metal stem. Included in this group are chandeliers and pendant units (*Figure 16*).

4.2.4 Track-Mounted Lighting Fixtures

Track-mounted lighting systems (*Figure 17*) include track, an electrical feed box, and two or more fixtures that can be positioned along the track. Track lighting wiring is in the metal track channels. It can be single-circuit or multiple-

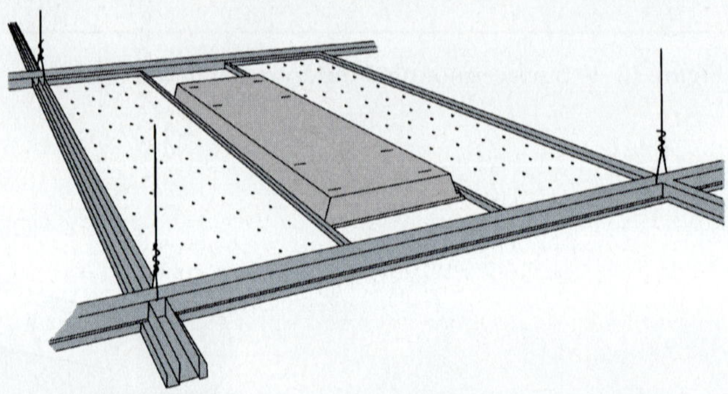

212F15.EPS

Figure 15 ◆ Fluorescent troffer mounted in a suspended ceiling.

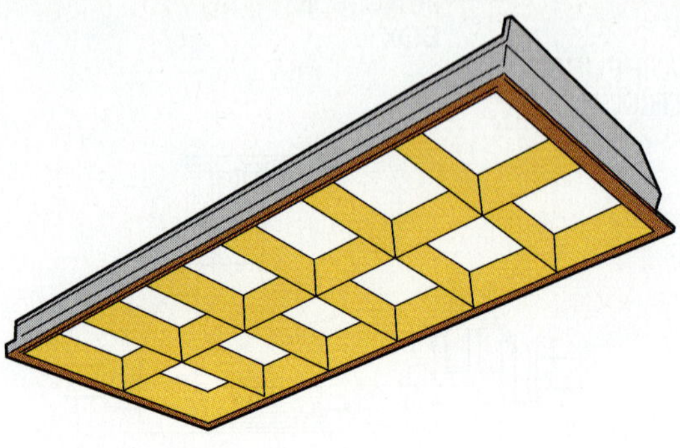

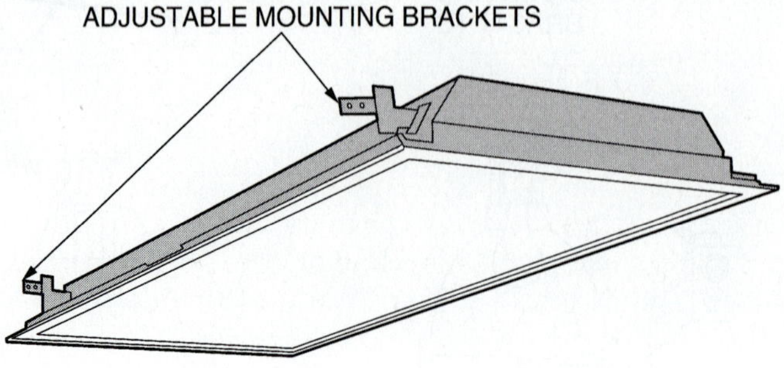

212F14.EPS

Figure 14 ◆ Typical recessed fluorescent troffers.

212F16.EPS

Figure 16 ◆ Suspended lighting fixtures.

212F17.EPS

Figure 17 ◆ Track lighting.

circuit wiring to allow two or more sets of lights to be independently controlled.

The tracks are normally mounted on or recessed in the ceiling, but can also be mounted horizontally or vertically on walls.

5.0.0 ◆ STORAGE, HANDLING, AND SECURITY OF LAMPS AND LIGHTING FIXTURES

Prior to installation, it is extremely important that lamps and lighting fixtures be handled and stored properly, both while in transit and at the job site. Failure to do so can result in broken and/or damaged lamps/lighting fixtures. This can result in expensive replacement costs and possible delays.

The proper care, handling, and storage of lamps/lighting fixtures begins by using some common sense. Manufacturers of lamps and lighting fixtures mark their shipping cartons, pallets, etc., with handling and storage instructions. The first thing you should always do is read and follow all such instructions. Obviously, lamps and lighting fixtures are fragile; therefore, pallets and/or cartons should not be dropped or otherwise handled roughly.

If you are responsible for receiving an order of lamps and/or lighting fixtures from a supplier or distributor, make sure that there is no obvious damage to the shipping cartons. Also make sure that the entire quantity that was ordered is delivered or that the delivery ticket shows what is missing.

If the lamps/lighting fixtures are not going to be installed immediately, store them in a dry place where they will be protected from loss or damage. If they must be stored in several locations, keep a record of what is stored at each of the locations.

THINK ABOUT IT

Tracking the Installation

When roughing-in lighting circuit wiring, how can you keep track of the work that has been accomplished?

6.0.0 ◆ PREPARATION OF LIGHTING FIXTURES FOR INSTALLATION

Before starting the actual installation of lighting fixtures, it is extremely important to review the lighting floor plan and related lighting fixture schedules to thoroughly familiarize yourself with the different types of lighting fixtures to be installed and their locations. Given this information, you can distribute the different kinds of lighting fixtures to the areas within the building where they are to be installed. *Figure 18* shows an example of a typical lighting floor plan and lighting fixture schedule. The *Appendix* shows a summary of several symbols commonly used on floor and building plans to identify different kinds of lighting fixtures. If available, shop drawings should be studied prior to installation, and if there are any questions, the fixture manufacturer or lighting designer should be consulted.

When opening lighting fixture shipping cartons, make sure to open the carton carefully so as not to damage the fixture. Do not rip or cut open the carton using knives with long blades, screwdrivers, or similar tools that can cut, scratch, gouge, or otherwise damage the fixture. When handling the fixture itself, wear gloves in order to prevent any grease, dirt, etc., on your hands from getting on its finished and/or decorative surfaces. Also, make sure not to lose or misplace the mounting hardware and insulation (if any) supplied with the fixture. Most lighting fixtures require a minimum amount of preparation before use. Depending on how they are purchased, some are shipped fully assembled, while others may require minor assembly, which should always be done according to the manufacturer's instructions.

Always read the labels on the lighting fixture and/or the installation instructions supplied with the fixture to familiarize yourself with its characteristics. Knowing this information can help prevent improper or inappropriate installation. *Figure 19* shows an example of one lighting manufacturer's installation instructions.

Typical information that can be found marked on lighting fixture labels and/or given in the installation instructions includes:

- Type(s) of lamps to use with the fixture and the maximum lamp wattage
- Approval for use in suspended ceilings, non-insulated ceilings (non-IC), or insulated ceilings (IC)
- Range of ambient temperatures for fixture operation
- Type of wire to use for fixture supply connections
- Thermal protection
- Suitability for wall or ceiling mount only
- Suitability for use in damp and/or wet locations
- Suitability for use as a raceway
- Suitability for installation in concrete
- Mechanical installation instructions
- Electrical connection instructions

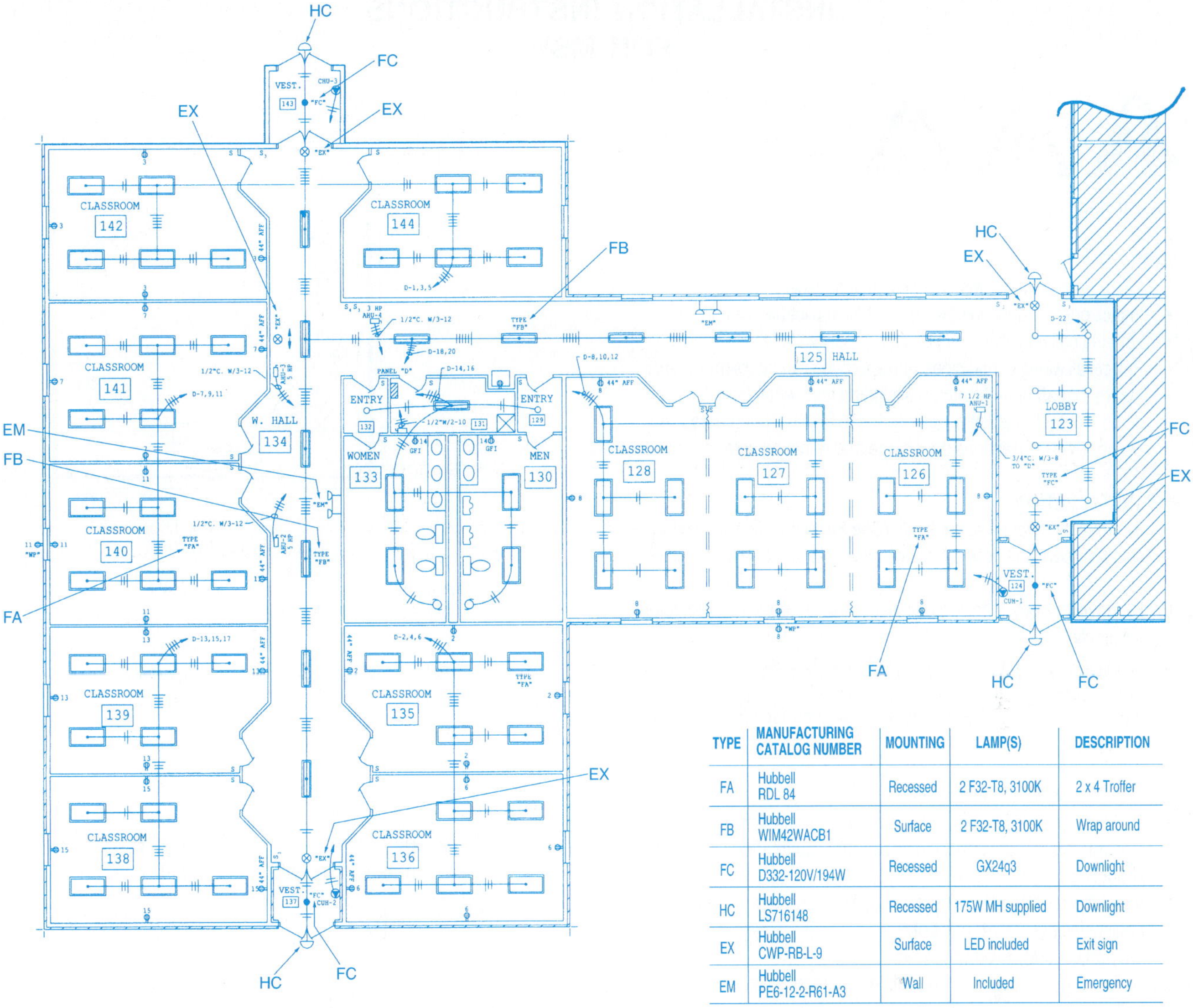

TYPE	MANUFACTURING CATALOG NUMBER	MOUNTING	LAMP(S)	DESCRIPTION
FA	Hubbell RDL 84	Recessed	2 F32-T8, 3100K	2 x 4 Troffer
FB	Hubbell WIM42WACB1	Surface	2 F32-T8, 3100K	Wrap around
FC	Hubbell D332-120V/194W	Recessed	GX24q3	Downlight
HC	Hubbell LS716148	Recessed	175W MH supplied	Downlight
EX	Hubbell CWP-RB-L-9	Surface	LED included	Exit sign
EM	Hubbell PE6-12-2-R61-A3	Wall	Included	Emergency

212F18.EPS

Figure 18 ◆ Typical lighting floor plan and lighting fixture schedule.

INSTALLATION INSTRUCTIONS FOR MSV

See warnings on carton.

WARNING

- Disconnect power before installation or servicing.
- Install, operate and maintain to meet all applicable codes.
- Labeled ballast voltage must match supply voltage.
- Protect all wiring connections with approved insulators (by others).
- Insure that **all** internal wiring, excluding lead wires from reflector, do not contact reflector assembly due to high lamp temperatures.
- Do not **overtighten** lens door hinge and retaining screws.

PRE-INSTALLATION

1. Remove lens door and set aside **(See Figure 1)**. To remove lens door, **loosen** lens door retaining screws (2) and one hinge screw (1).
2. Remove the two reflector retaining screws and lift out reflector assembly.
3. Remove shipping bolt located in arm mounting hole.
4. When a multi-tap ballast is supplied, the fixture is factory wired to the highest voltage. Verify that fixture wiring matches the supply voltage.

ARM INSTALLATION (See Figure 2)

NOTE: Guide electrical supply wires from fixture through the bearing plate, fixture, gasket, cover, positioning bracket, pole and bolt plate as each mounting step is completed.

1. Insert longer threaded end of hex rods (5) through lockwashers, positioning bracket (4) and pole.
2. Tighten securely to bolt plate (3).
3. Slide cover (6) over hex rods. Drain hole in cover must be on bottom.
4. Slide gasket (7) over threaded end of hex rods.
5. Place fixture onto hex rods.
6. Slide bearing plate (8) and lockwashers over end of hex rods on inside of fixture.
7. Install and tighten nuts securely.
8. Install supply wire (by others) thru the cord grip bracket in pole using appropriate supplied cord grip (9).
9. Make wiring connections.
10. Push wires and wirenuts into pole top and install cap (10).

FINAL ASSEMBLY

1. Re-install and secure reflector assembly using the reflector retaining screws. The reflector assembly is fully rotatable 360° by 90° increments.
2. Install proper lamp.
3. Re-install lens door and insure that all fasteners are securely tightened.

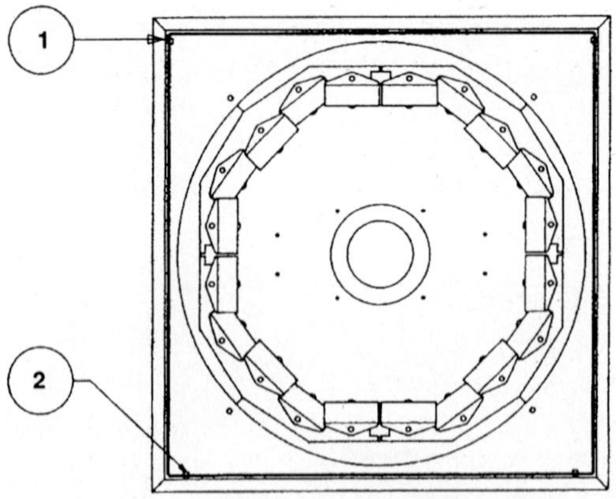

FIGURE 1
FIGURA 1

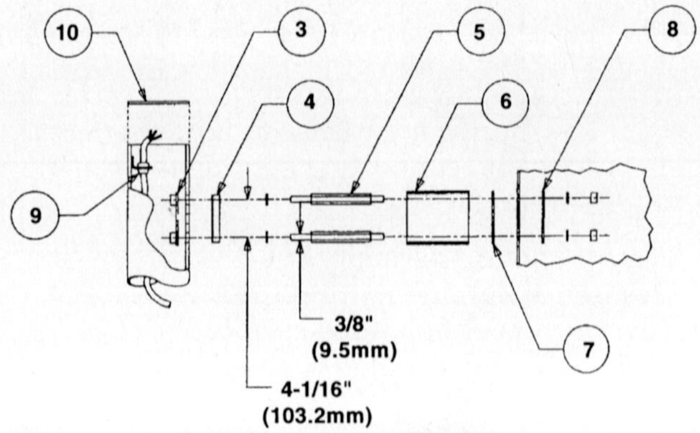

FIGURE 2
FIGURA 2

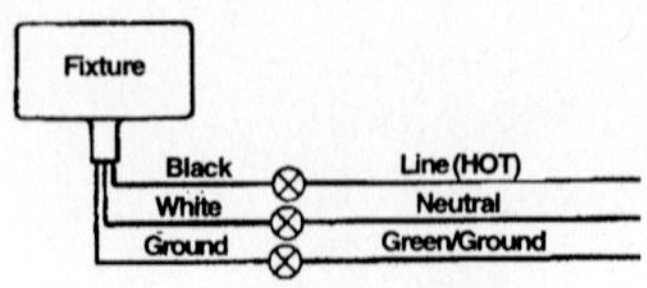

WIRING DIAGRAM

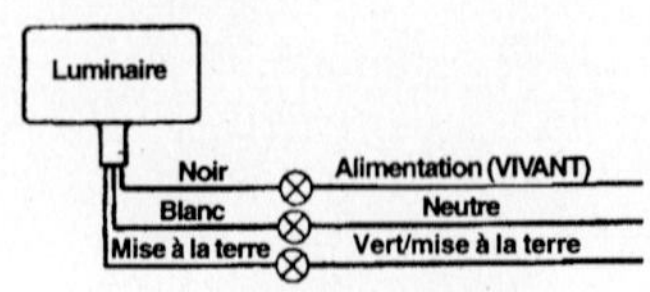

SCHÉMA DE CÂBLAGE

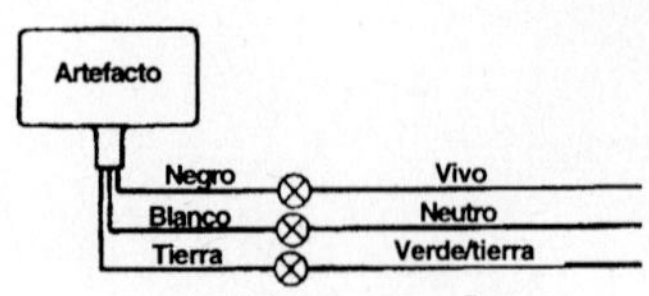

DIAGRAMA DE CABLEADO

Hubbell Lighting
® A Division of Hubbell Lighting, Inc.
2000 Electric Way
Christiansburg, VA 24073-2500
(540) 382-6111
FAX (540) 382-1526

268-1072-9901

Figure 19 ◆ Example of lighting fixture installation instructions.

7.0.0 ◆ LIGHTING FIXTURE INSTALLATION

This section gives guidelines and procedures for installing lighting fixtures and information on the various ways in which lighting fixtures can be supported from building structures and/or mounted to a branch circuit outlet box. General procedures for installing lighting branch circuit components and raceways were covered earlier in your training.

NEC Article 410 sets forth most of the requirements pertaining to the installation of lighting fixtures. You must meet all *NEC®* and/or local code requirements, including those for mounting, supporting, grounding, live-parts exposure, insulation clearances, supply conductor types, maximum lamp wattages, etc. Whenever you have questions or need clarification about the method to use, always refer to the appropriate codes.

It is important to understand how different types of lighting fixtures are safely and securely attached to a structure or outlet box. Code requirements for the installation of lighting fixtures, lighting outlet boxes, and related methods of support are covered in *NEC Sections 314.23(F), 314.27(A) and (B),* and *NEC Article 410, Parts III and IV.* The specific methods and hardware used for mounting and supporting lighting fixtures depend on several factors:

- Type of lighting fixture
- Location of lighting fixture
- Construction of mounting surface
- Weight of lighting fixture

7.1.0 Mechanical Installation of Surface-Mounted Lighting Fixtures

Surface-mounted lighting fixtures include those that are directly mounted on a ceiling or wall. They can be incandescent, fluorescent, or HID units. The general methods for mechanically installing ceiling and wall surface-mounted lighting fixtures are basically the same; only the location of the outlet box is different. Normally, boxes for indoor and outdoor wall lighting fixtures are installed at heights of 6'-0" and 6'-6" above the floor, respectively. A lighting fixture should be mounted to its outlet box using the mounting hardware supplied with the fixture and installed according to the manufacturer's instructions. If you are not thoroughly familiar with the lighting fixture to be installed, first read and be sure that you understand the related manufacturer's product literature and installation instructions supplied either with the fixture or printed on the shipping carton.

In buildings of all types, branch circuit wiring for surface-mounted ceiling and wall lighting fixture outlets normally terminates in octagonal or round boxes (*Figure 20*). Some may terminate in square or rectangular boxes. Both metal and nonmetallic types of boxes may be used. These boxes are designed to be covered by the mounting base of a lighting fixture with a similar shape. For concealed wiring in wooden frame construction, the outlet boxes are fastened directly to studs, joists, or other framing members using nails or screws. Some ceiling outlet boxes are mounted on an adjustable bar hanger so that the box can be installed in the space between two ceiling joists. The box is attached to the bar hanger by a fixture stud or two bolts. When properly installed, the front face of all concealed ceiling and wall boxes should be flush with the finished ceiling or wall. For exposed conduit installed on concrete or other masonry surfaces, the outlet boxes should be fastened to the ceiling or wall with screws and expansion shields, anchors, powder-actuated fasteners, or similar appropriate fasteners. On steel work, they should be bolted or clamped to the steel. Outlet boxes embedded in concrete are held securely in position by the hardened concrete.

The size and weight of the fixture determine what type of outlet box should be used with a fixture. When mounting lighting fixtures that weigh more than 50 pounds, *NEC Section 314.27(B)* requires either that the fixture be supported independently of the outlet box or that it be mounted to a listed outlet box designed to support a fixture

THINK ABOUT IT

Installing Surface-Mounted Fixtures in Closets

What precautions should you take when installing surface-mounted ceiling or wall fixtures in a clothes closet?

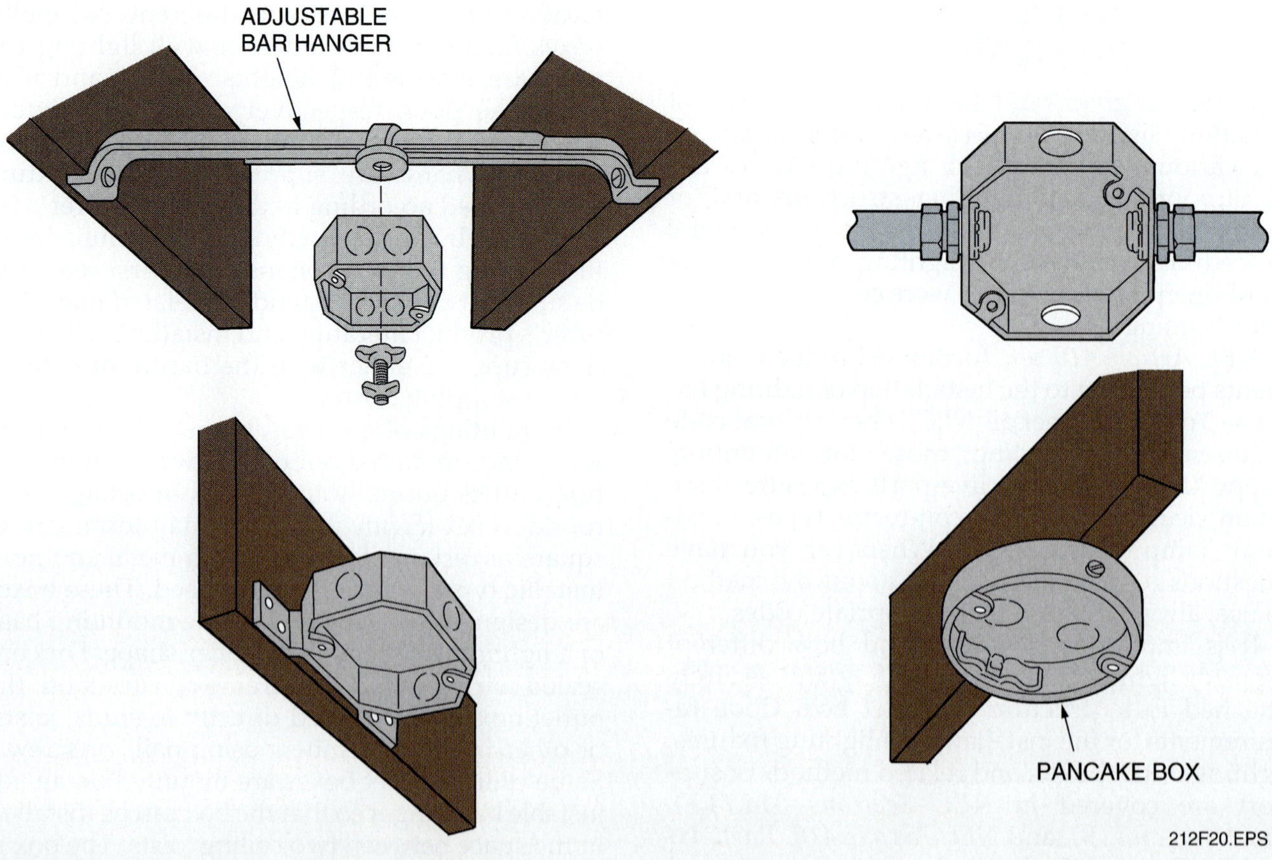

Figure 20 ◆ Typical outlet boxes used with surface-mounted lighting fixtures.

weighing more than 50 pounds. In addition to complying strictly with the code, good judgment must also be exercised. Remember, codes only give the minimum requirements for electrical installations, not necessarily those that make a quality installation.

NOTE

When a manufacturer supplies insulation with a surface-mounted lighting fixture, do not throw it away. It must be placed between the fixture and the mounting surface to insulate the mounting surface from the heat produced by the fixture. Make sure to install any such insulation as directed by the manufacturer.

7.1.1 Drum, Globe, Sconce, and Similar Lighting Fixtures

Figure 21 shows some common methods used to mount drum, globe, sconce, and similar types of indoor lighting fixtures to an outlet box. Usually, these types of lighting fixtures can be directly mounted to the outlet box with no additional support. As shown in *Figure 21(A)*, the common porcelain-type lamp holder fixture screws directly to the ears of the outlet box. Other ceiling fixtures and wall sconces are fastened to the outlet box using a mounting bar that is attached to a threaded stud in the box, as shown in *Figure 21(B)*. The mounting bar is positioned on the stud and fastened in place with a locknut. After the fixture wiring has been completed, the mounting holes in the base of the fixture are slipped over the machine screws installed in the mounting bar. Cap nuts are then installed on the machine screws and tightened until the fixture is drawn flush with the surface. Fixtures can be mounted to an outlet box without a box stud in a similar manner by screwing a mounting bar to the ears of the box, as shown in *Figure 21(C)*. Another common way some fixtures are installed is by screwing a reducing nut and nipple onto a box stud, as shown in *Figure 21(D)*. Following this, the hole in the center of the fixture base is placed over the nipple, then a cap nut is installed on the nipple and tightened until the fixture is drawn flush with the surface. A similar method is to screw a nipple into the threaded center hole of a mounting bar attached to the ears of the outlet box.

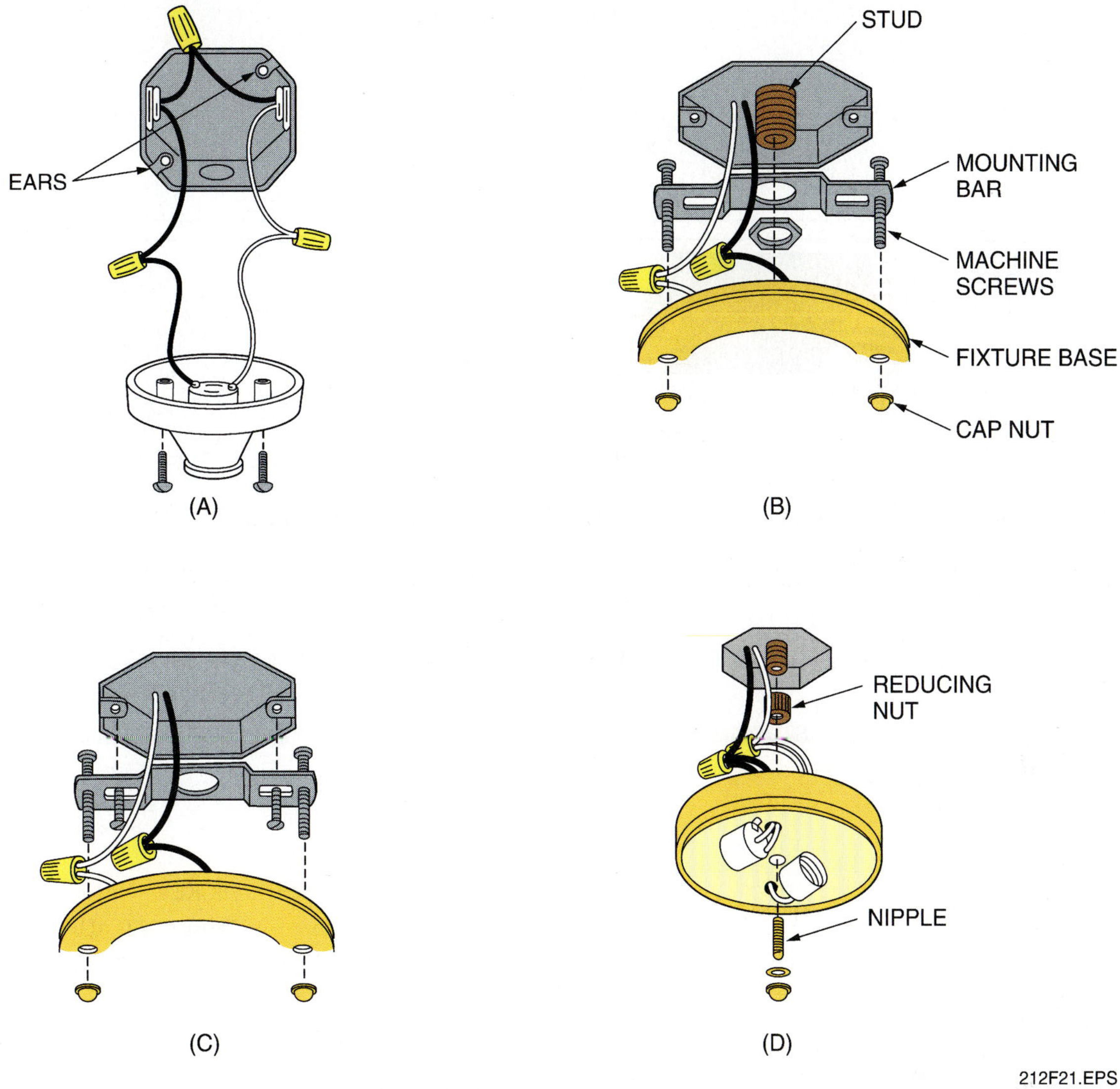

Figure 21 ◆ Common methods of installing surface-mounted lighting fixtures.

7.1.2 Fluorescent Lighting Fixtures

Some lightweight, surface-mounted fluorescent fixtures can be supported directly from an outlet box. Such types typically use a stud, reducing nut, and nipple in the box, with a mounting bar inside the housing, as shown in *Figure 22*. The assembly is held with a locknut. Heavier fluorescent fixtures are supported directly from the ceiling using toggle bolts, wood screws in joists, anchors in masonry, or other appropriate fasteners. These fixtures have an access opening that, when the fixture is in place, must be positioned directly under the outlet box. Many surface-mounted fluorescent fixtures have bumps or ridges on their mounting surfaces that are designed to hold the fixtures slightly off the ceiling in order to allow for the dissipation of heat.

7.1.3 Chandeliers and Pendants

Figure 23 shows some methods used for attaching chandeliers to an outlet box. Chandeliers are normally heavy, requiring that they be mounted on a box with a stud. As shown, a hickey (or a hickey together with a threaded adapter) is installed over the fixture chain and wires, then screwed onto the stud. The fixture base is then fastened to the hickey or adapter with a locknut or cap nut. Some are secured to a mounting bar by means of a nipple. Pendant lighting fixtures are typically supported by chains, metal stems, thin cables, or retractable cords. Most pendants are designed to be wired directly into a ceiling outlet box in a similar manner as that described for chandeliers.

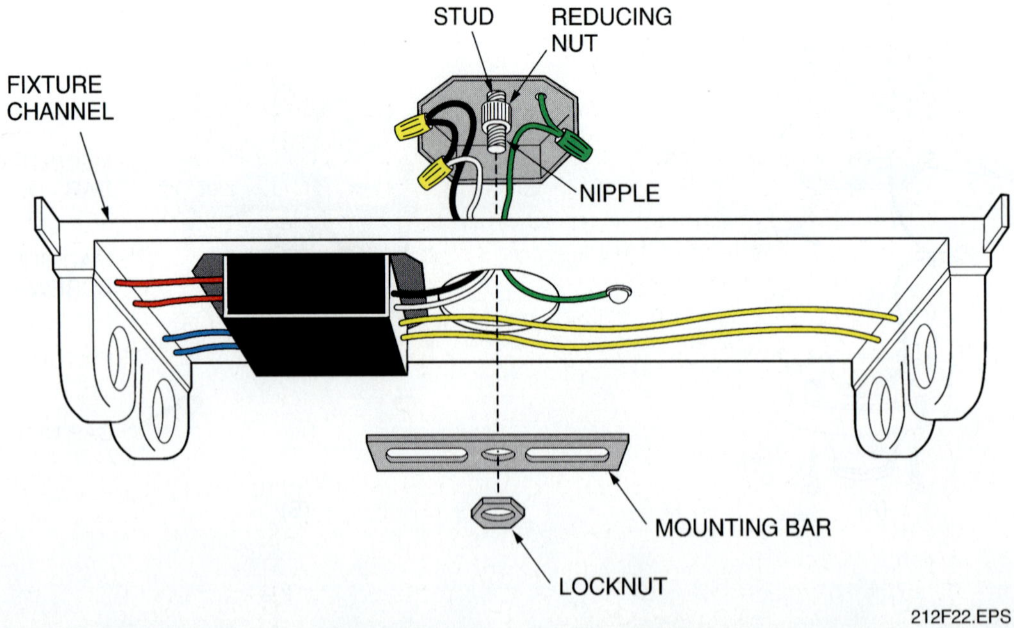

Figure 22 ◆ Typical method for mounting a smaller fluorescent fixture to an outlet box.

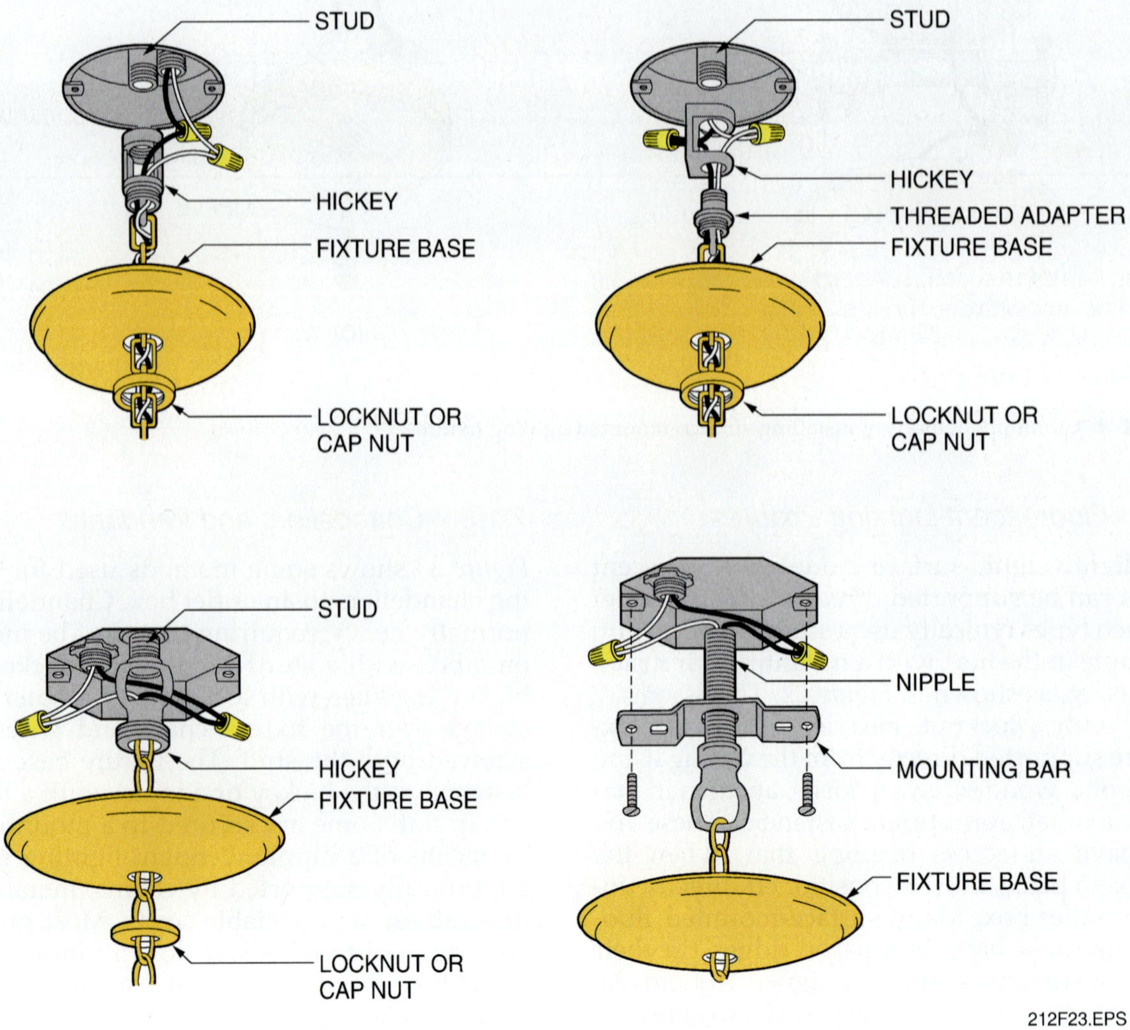

Figure 23 ◆ Methods for attaching chandeliers to an outlet box.

Installing Ceiling-Mounted Fluorescent Fixtures

INSIDE TRACK

While standing on a ladder to mount a fluorescent fixture on a ceiling, it is much easier to mark the mounting hole locations for the fixture on the ceiling using a cardboard template than it is to hold the actual fixture in place while marking the hole locations.

Use a portion of the cardboard carton that the fixture was shipped in to make a template of the fixture. Transfer the locations of all fixture mounting holes, and any other required holes, to the template. Cut out these holes, then use the template to locate and mark the mounting hole locations on the ceiling.

WARNING!

NEC Section 410.4(D) prohibits pendants and other hanging-type fixtures from being installed within a zone measured 3' horizontally and 8' vertically from the top of a bathtub rim or shower stall threshold.

7.1.4 Ceiling Fans/Fixtures

NEC Section 422.18 covers the support of ceiling-suspended (paddle) fans. It states that for fans weighing 35 pounds or less, the outlet box may not be used as the only support of the fan unless the box is listed for such use. Such boxes will be labeled as acceptable for fan support. The code also requires that fans weighing over 35 pounds be supported independently of the outlet box. This is typically done by using a box mounted on a ceiling fan joist hanger (*Figure 24*).

The specific method for mounting ceiling fans varies among ceiling fan manufacturers; therefore, the fan should be mounted to the outlet box per the manufacturer's directions. *Figure 24* shows an example of one mounting method. *NEC Section 410.4(D)* also prohibits ceiling-suspended (paddle) fans from being installed within a zone measured 3' horizontally and 8' vertically from the top of a bathtub rim or shower stall threshold.

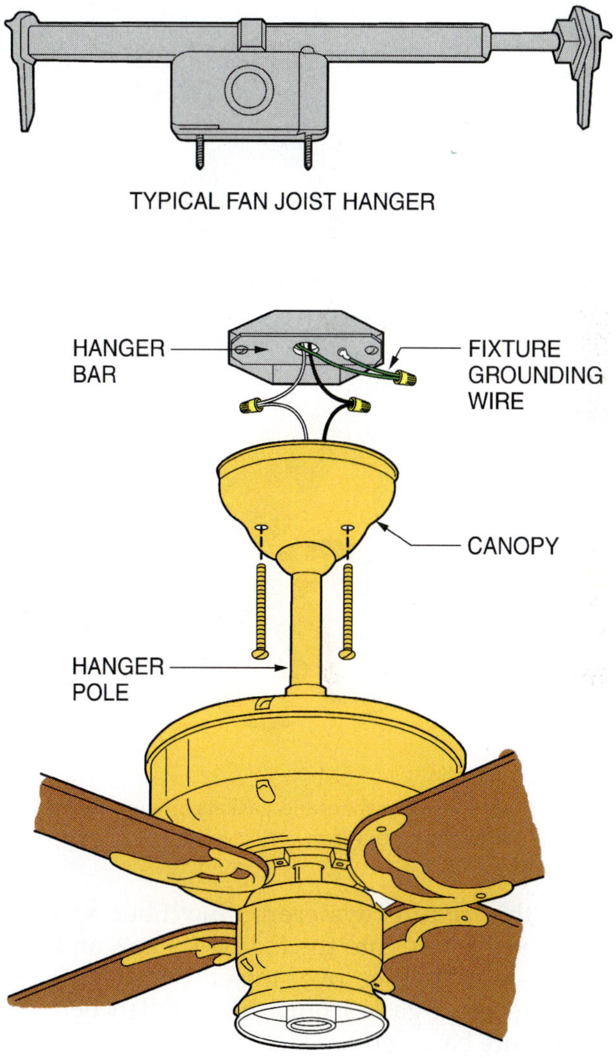

Figure 24 ◆ Typical ceiling fan/light mounting method.

7.1.5 Outdoor Lighting Fixtures

The methods for attaching outdoor surface-mounted lighting fixtures to outlet boxes are basically the same as those described for indoor surface-mounted fixtures. However, the installation must be done so that the final assembly is watertight.

All fixtures installed in wet locations must be marked as suitable for wet locations. All fixtures installed in damp locations must be marked as either suitable for wet locations or suitable for damp locations. When mounting a lighting fixture to an outlet box that is fully exposed to the weather, the outlet box must be a metal weatherproof-type box.

All washers, gaskets, etc. required to make a watertight assembly must be properly installed as directed by the manufacturer. *Figure 25* shows an example of a common outdoor spotlight fixture.

Installing Chandeliers and Pendants

INSIDE TRACK

It is important to make sure that a chandelier or pendant fixture is installed at the proper height. For example, you should allow for door clearances when installing the fixture in a foyer or entrance hallway. Such a fixture should also be hung at a height of 7' or more so that it allows safe passage for tall people.

If the fixture is being hung over a table, its width should be at least 12" less than the width of the table (6" from each edge) so that people do not hit their heads on the fixture when sitting down at or getting up from the table. It should be hung at a height of about 30" to 36" above the table, unless the ceiling is higher than 8'. If the ceiling height is greater than 8', the fixture may look better raised even higher than 36". To provide light for demanding tasks, pendants should be hung about 15" above the work surface.

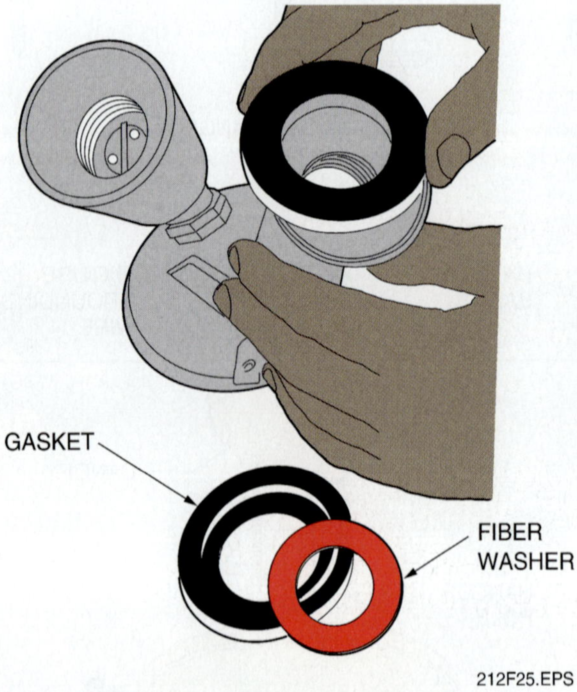

Figure 25 ◆ Install washers and gaskets on outdoor lighting fixtures.

With this fixture, waterproofing fiber washers must be inserted into the fixture sockets and rubber gaskets installed over the end of each socket. In addition, a gasket must be installed between the fixture base and the face of the outlet box.

7.2.0 Mechanical Installation of Recessed Lighting Fixtures

Recessed lighting fixtures are mounted into ceiling recesses so that the main body of the fixture is not visible.

Figure 26 lists some of the code requirements for recessed fixtures.

What's wrong with this picture?

212P1201.EPS

INSIDE TRACK

Methods of Hanging Suspended Lighting Fixtures

The mechanical methods used to hang and support specific suspended lighting fixtures can be quite different because of the wide variety of fixture types and the nature of the buildings or structures in which they are to be installed.

Generally, on large construction jobs, the specific method for hanging suspended fixtures is shown in the construction drawings. However, on many smaller jobs, the method used for suspending fixtures is frequently left up to the electrician. This can be a problem for newer and less experienced electricians. Fortunately, technical help is readily available.

To aid customers, manufacturers of fasteners and metal framing devices produce numerous catalogs and technical product support bulletins that show example solutions to common hanging/support problems. Other good sources of technical advice are your supervisor, other qualified electricians, and fastener/framing company sales engineers or representatives.

212P1202.EPS

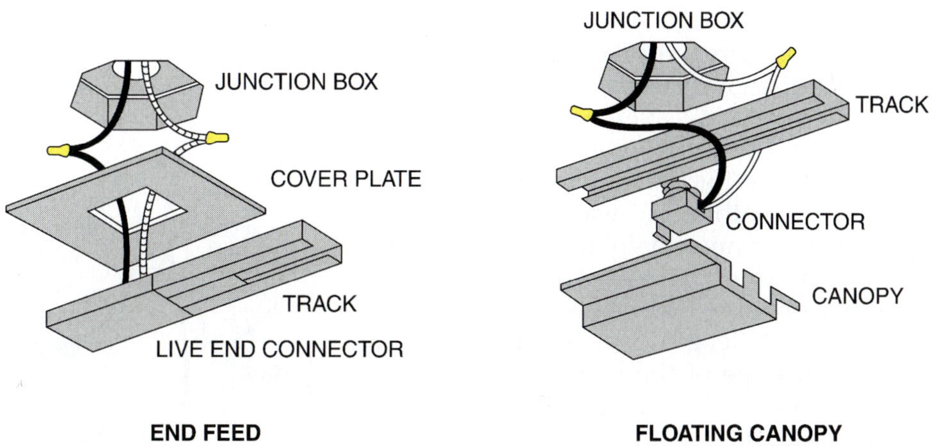

212F34.EPS

Figure 34 ◆ Typical end feed and floating canopy track connectors.

section of track is fitted solidly into the track connector, and then the track is snapped into the clips. Following this, setscrews in the sides of the clips are tightened to hold the track firmly in place. The remaining sections of track are installed in the same manner using the appropriate connectors and accessories. When it is necessary to cut a section of track to length, it can be cut using a hacksaw. Make sure to follow the manufacturer's instructions as to where and how to make the cut and how to close off the cut with a proper end cap.

7.5.0 Electrical Connection of Lighting Fixtures

Making the electrical connections between the lighting fixture and related outlet box wires is done at the appropriate point during the mechanical installation of the fixture. The specific way in which the supply wires in the outlet box are connected to the lighting fixture wires varies depending on the design of the branch circuit and the way in which the branch circuit cable or raceway wiring is routed. Therefore, you must refer to the lighting plan to determine the exact connections.

7.5.1 Conventional Hard-Wired Installations

The connection of the lighting fixture wiring to the outlet box supply wires should be done in accordance with the manufacturer's instructions. The general procedure is basically the same, no matter what type of fixture is being connected.

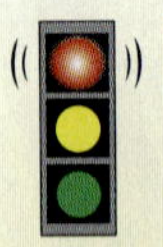

WARNING!

Before attempting to connect lighting fixtures to a branch circuit outlet box, make sure that the power is turned off to the branch circuit involved at the circuit breaker panel. Then, make sure to follow the prevailing lockout and tagout procedures.

When cutting outlet box wires in preparation for connecting the fixture, do not cut the branch circuit wires too short or too long. *NEC Section 300.14* requires that at least 6" of free conductor length be left at each outlet or junction box where it emerges from its cable sheath or raceway. Furthermore, for splices or the connection of fixtures or devices, each conductor's length shall extend at least 3" outside any box opening that is less than 8" in any dimension. The 3" length is generally measured from the outer edge of the box. Note that leaving the wires too long will make it difficult to position the wires in the box and will overcrowd the box, possibly resulting in overheating.

The lighting fixture's black wire(s) are connected to the outlet box hot wires (black, red, or marked hot) and the white wire(s) to white wires (*Figure 35*). If dealing with nonmetallic cable and metal boxes, the bare grounding wires should be spliced with one end of a grounding jumper. The other end of the grounding jumper should be attached to the box using either a grounding clip or the box grounding screw. If dealing with nonmetallic cable and a nonmetallic outlet box, the bare cable wires should be spliced together. In some cases, a green wire from the fixture and the bare wires will be connected to a grounding screw provided on the mounting bar supplied with the lighting fixture.

All wire splices should be made using the correct wire nuts sized for the number and gauges of wire you are splicing. If in doubt, most wire nut manufacturers mark their packages to show the maximum and minimum number of wires of different sizes that can be connected using a particular wire nut. Use this information as your guide. When properly installed and tightened, the wire nuts should cover all bare current-carrying conductors. Some wire nut manufacturers recommend twisting the wires together before installing the wire nut; others do not. Follow the manufacturer's directions for the wire nut you are using. The correct methods for stripping wires and using wire nuts were covered in the *Conductor Terminations and Splices* module.

The *NEC*® requires that all fixtures with exposed metal parts be grounded. If the fixture has a separate green grounding wire, it should be spliced together with the other grounding wires. Once the fixture box and/or metal mounting strap is grounded, the nipple or screws holding the fixture will ground the fixture.

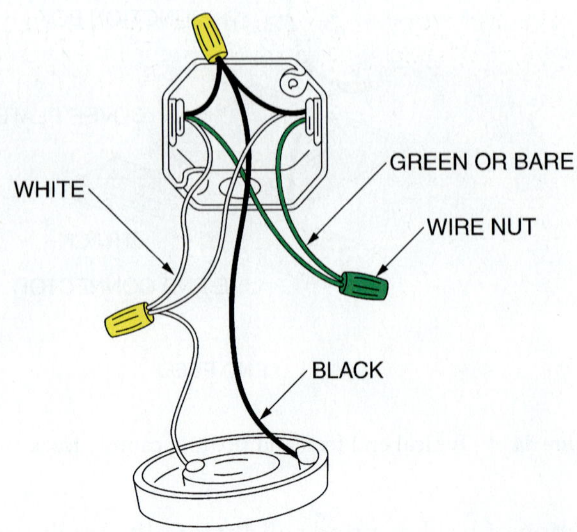

Figure 35 ◆ Example of basic lighting fixture connections.

7.5.2 Modular System Wiring

In some installations, lighting fixtures are connected using a modular wiring system (*Figure 36*). Modular wiring provides an alternative method to conventional hard wiring. Modular system components are interchangeable and can be reused. Component labels are color-coded by voltage, and the units are keyed to prevent mismatching voltages. Lighting branch circuit wiring is run from the building panel to an outlet or junction box that serves as the starting point for the modular wiring. There, a modular system feeder adapter is wired and mounted. From this point on, all the wiring to the lighting fixtures is modular using the appropriate cable adapters.

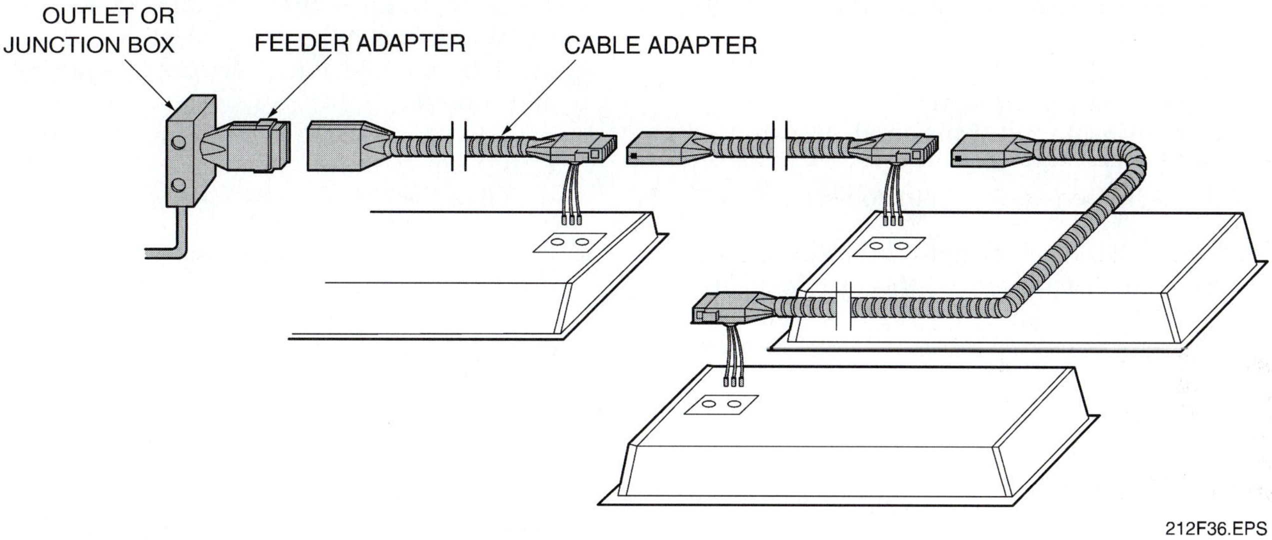

Figure 36 ◆ Basic modular wiring system.

Putting It All Together

THINK ABOUT IT

Examine the lighting system in your kitchen. What changes, if any, would help to maximize the lighting in the different work areas? What types of lighting could you use to make the other rooms in your home appear warmer and more inviting?

Review Questions

1. Light impulses are transferred from the eye to the brain by the part of the eye called the _____.
   a. rods
   b. cones
   c. pupil
   d. optic nerve

2. Light energy that is *not* absorbed by an object can be _____.
   a. reflected from the object
   b. transmitted through the object
   c. both reflected from and transmitted through the object
   d. dissipated as heat in the object

3. When all the wavelengths of the light spectrum are presented to the eye in nearly equal proportions, _____ light is seen.
   a. red
   b. white
   c. blue
   d. green

4. The tungsten halogen lamp is a type of _____ lamp.
   a. fluorescent
   b. incandescent
   c. high-intensity discharge (HID)
   d. high-pressure sodium

5. A(n) _____ lamp is *not* normally used outdoors.
   a. fluorescent
   b. incandescent
   c. halogen
   d. high-intensity discharge (HID)

6. The difference between a lighting fixture and a luminaire is that _____.
   a. luminaires have no ballast
   b. lighting fixtures have no ballast
   c. luminaires have a ballast and lighting fixtures do not
   d. There is no difference.

7. Most code requirements governing the installation of lighting fixtures, lampholders, and lamps are covered in _____.
   a. *NEC Article 411*
   b. *NEC Article 370*
   c. *NEC Article 410*
   d. *NEC Article 422*

8. Recessed fixtures that are labeled IC are suitable for installation in locations _____ insulation, while non-IC fixtures are not.
   a. in contact with
   b. 2" from
   c. 3" from
   d. All of the above.

9. Fixture whips used to connect from a junction box mounted above the ceiling to a fixture installed in a suspended ceiling cannot exceed _____ in length.
   a. 18"
   b. 2'
   c. 4'
   d. 6'

10. When cutting the branch circuit wires in an outlet box in preparation for connecting a light fixture, you should leave at least _____ of free conductor length after it emerges from the cable or raceway.
    a. 3"
    b. 5"
    c. 6"
    d. 8"

Summary

Electric lighting serves to provide illumination for the performance of visual tasks with a maximum of comfort and a minimum of eyestrain and fatigue, allowing individuals to perform their daily living and work-related tasks.

The proper installation of lighting fixtures requires that you become familiar with the different kinds of lighting fixtures and the lamps used with them. You must acquire the skills needed to mechanically and electrically install lighting fixtures for different applications according to the prevailing building and electrical codes.

Notes

Trade Terms Introduced in This Module

Ballast: A circuit component in fluorescent and HID lighting fixtures that provides the required voltage surge at startup and then controls the subsequent flow of current through the lamp during operation.

Color rendering index (CRI): A measurement of the way a light source reproduces color. The higher the index number (0–100), the closer colors are to how an object appears in full sunlight or incandescent light.

Glare: The sensation produced by luminance within the visual field that is significantly greater than the luminance to which the eyes are adapted.

Incident light: The light emitted from a self-luminous object such as the sun or an electric source.

Lumen (lm): The basic measurement of light. One lumen is defined as the amount of light cast upon one square foot of the inner surface of a hollow sphere with a one-foot radius with a light source of one candela in its center.

Luminaire (fixture): A complete lighting unit consisting of a lamp (or lamps) and ballasts (where applicable), together with the parts designed to distribute the light, position and protect lamps, and connect them to the power supply.

Luminance: The luminous intensity of any surface in a given direction per unit area of that surface as viewed from that direction. Measured in candela/m^2. The term luminance is commonly used to express brightness.

Reflected: The bouncing back of light waves or rays by a surface.

Reflected light: Light reflected from an object that does not have self-luminating properties but reflects light provided from another source.

Refracted/refraction: The bending of a light wave or ray as it passes obliquely from one medium to another of different density, or through layers of different density in the same medium.

Starter: A circuit component of certain fluorescent lighting fixtures that is used to heat the lamp electrodes before the lamps are lighted.

Troffer: A recessed lighting fixture installed with the opening flush with the ceiling.

Appendix

Summary of Lighting Fixture Symbols

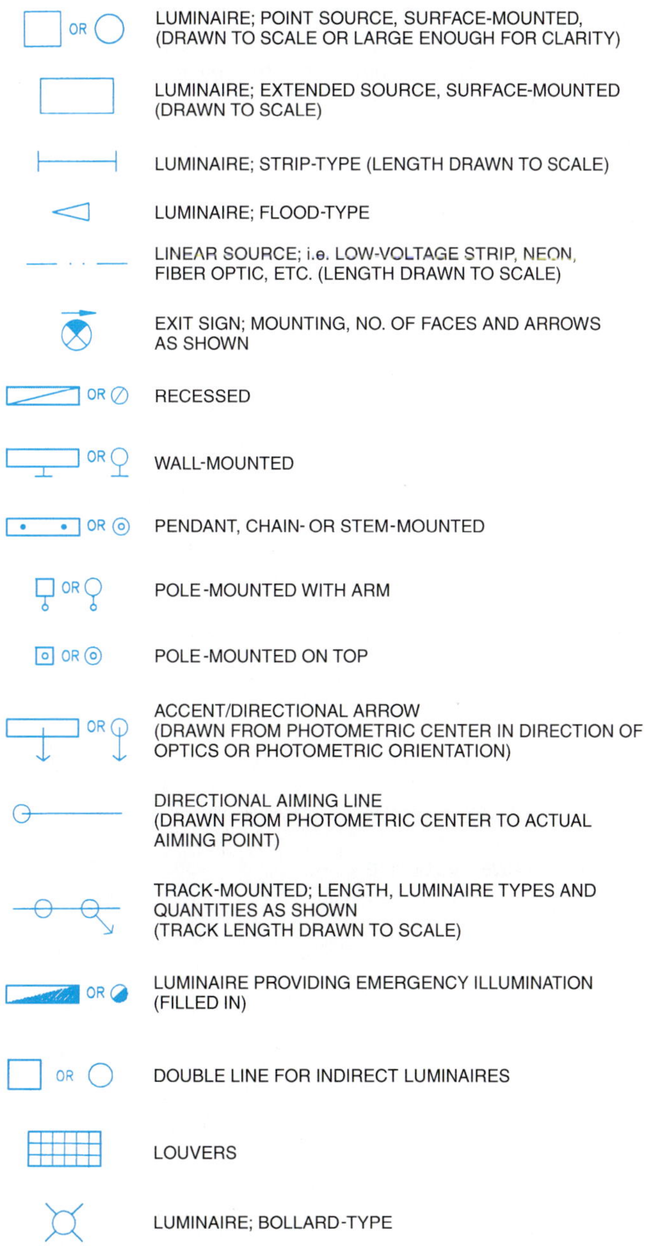

212UA01.EPS

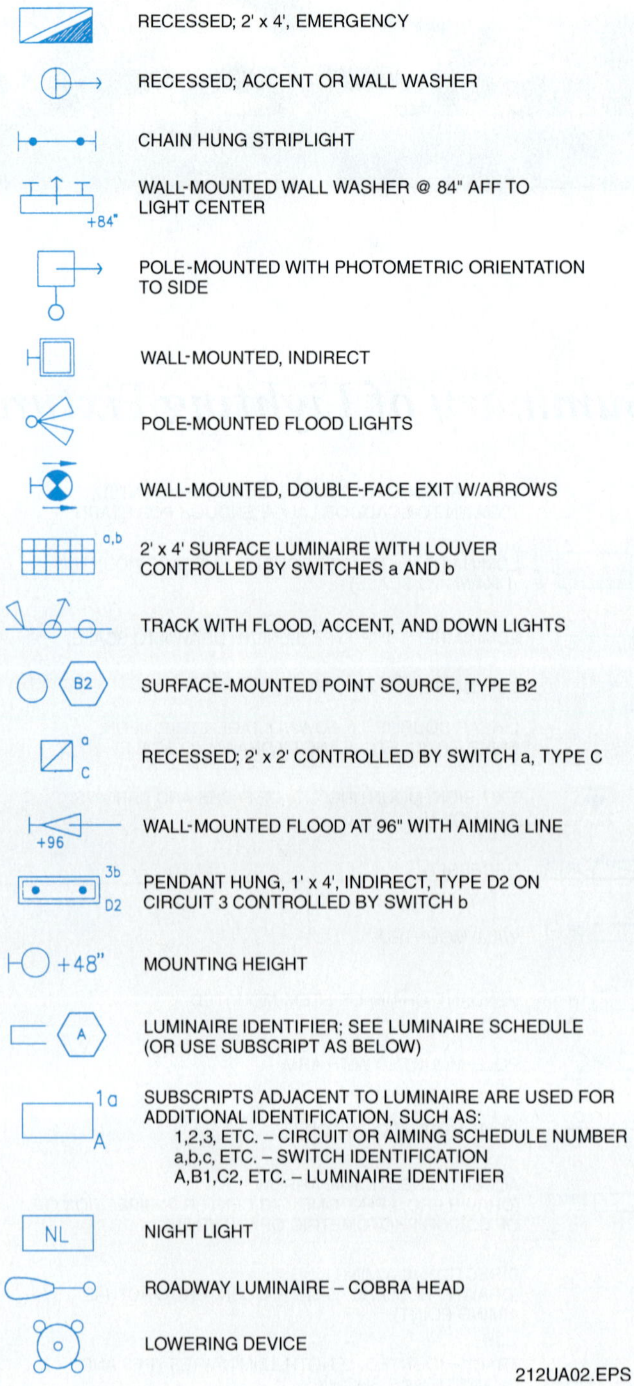
RECESSED; 2' x 4', EMERGENCY
RECESSED; ACCENT OR WALL WASHER
CHAIN HUNG STRIPLIGHT
+84"
WALL-MOUNTED WALL WASHER @ 84" AFF TO LIGHT CENTER
POLE-MOUNTED WITH PHOTOMETRIC ORIENTATION TO SIDE
WALL-MOUNTED, INDIRECT
POLE-MOUNTED FLOOD LIGHTS
WALL-MOUNTED, DOUBLE-FACE EXIT W/ARROWS
a,b
2' x 4' SURFACE LUMINAIRE WITH LOUVER CONTROLLED BY SWITCHES a AND b
TRACK WITH FLOOD, ACCENT, AND DOWN LIGHTS
B2
SURFACE-MOUNTED POINT SOURCE, TYPE B2
a
C
RECESSED; 2' x 2' CONTROLLED BY SWITCH a, TYPE C
+96
WALL-MOUNTED FLOOD AT 96" WITH AIMING LINE
3b
D2
PENDANT HUNG, 1' x 4', INDIRECT, TYPE D2 ON CIRCUIT 3 CONTROLLED BY SWITCH b
+48"
MOUNTING HEIGHT
A
LUMINAIRE IDENTIFIER; SEE LUMINAIRE SCHEDULE (OR USE SUBSCRIPT AS BELOW)
1a
A
SUBSCRIPTS ADJACENT TO LUMINAIRE ARE USED FOR ADDITIONAL IDENTIFICATION, SUCH AS:
1,2,3, ETC. – CIRCUIT OR AIMING SCHEDULE NUMBER
a,b,c, ETC. – SWITCH IDENTIFICATION
A,B1,C2, ETC. – LUMINAIRE IDENTIFIER
NL
NIGHT LIGHT
ROADWAY LUMINAIRE – COBRA HEAD
LOWERING DEVICE

212UA02.EPS

Additional Resources

This module is intended to present thorough resources for task training. The following reference works are suggested for further study. These are optional materials for continued education rather than for task training.

American Electrician's Handbook, Latest Edition. New York: Croft and Summers, McGraw-Hill.

Lighting Handbook, Latest Edition. New York: Illuminating Engineering Society of North America (IESNA).

National Electrical Code® Handbook, Latest Edition. Quincy, MA: National Fire Protection Association.

CONTREN® LEARNING SERIES — USER FEEDBACK

The NCCER makes every effort to keep these textbooks up-to-date and free of technical errors. We appreciate your help in this process. If you have an idea for improving this textbook, or if you find an error, a typographical mistake, or an inaccuracy in NCCER's *Contren®* textbooks, please write us, using this form or a photocopy. Be sure to include the exact module number, page number, a detailed description, and the correction, if applicable. Your input will be brought to the attention of the Technical Review Committee. Thank you for your assistance.

Instructors – If you found that additional materials were necessary in order to teach this module effectively, please let us know so that we may include them in the Equipment/Materials list in the Annotated Instructor's Guide.

Write: Product Development
National Center for Construction Education and Research
P.O. Box 141104, Gainesville, FL 32614-1104

Fax: 352-334-0932

E-mail: curriculum@nccer.org

Craft __________ Module Name __________

Copyright Date __________ Module Number __________ Page Number(s) __________

Description __________

(Optional) Correction __________

(Optional) Your Name and Address __________

Glossary

AL-CU: An abbreviation for aluminum and copper, commonly marked on terminals, lugs, and other electrical connectors to indicate that the device is suitable for use with either aluminum conductors or copper conductors.

Ambient temperature: The temperature of the medium (usually air) surrounding a device and into which the heat losses in the device are dissipated.

Approximate ram travel: The distance the ram of a hydraulic bender travels to make a bend. To simplify and speed bending operations, many benders are equipped with a scale that shows ram travel. Using a simple table (supplied with many benders), the degree of bend can easily be converted to inches of ram travel. This measurement, however, can only be approximated because of the variation in springback of the conduit being bent.

Armature: The rotating windings of a DC motor.

Auxiliary electrodes: Metallic electrodes pushed or driven into the earth to provide electrical contact for the purpose of performing measurements on grounding electrodes or ground grid systems.

Back-to-back bend: Any bend formed by two 90° bends with a straight section of conduit between the bends.

Ballast: A circuit component in fluorescent and HID lighting fixtures that provides the required voltage surge at startup and then controls the subsequent flow of current through the lamp during operation.

Barrier strip: A metal strip constructed to divide a section of cable tray so that certain kinds of cable may be separated from each other.

Basket grip: A flexible steel mesh grip that is used on the ends of cable and conductors for attaching the pulling rope. The more force exerted on the pull, the tighter the grip wraps around the cable.

Bending protractor: Made for use with benders mounted on a bending table and used to measure degrees; also has a scale for 18, 20, 21, and 22 shots when using it to make a large sweep bend.

Bending shot: The number of shots needed to produce a specific bend.

Bonding: The permanent joining of metallic parts to form an electrically conductive path that will assure electrical continuity and the capacity to safely conduct any current likely to be imposed on it.

Branch circuit: The circuit conductors between the final overcurrent device protecting the circuit and the outlet(s).

Break: The opening of closed contacts to interrupt an electrical circuit.

Brush: A conductor between the stationary and rotating parts of a machine. It is usually made of carbon.

Cable grip: A device used to secure ends of cables to a pulling rope during cable pulls.

Cable pulley: A device used to facilitate pulling conductor in cable tray where the tray changes direction. Several types are available (single, triple, etc.) to accommodate almost all pulling situations.

Cable roller: A device used to facilitate cable pulls (lower the pulling resistance) in cable trays. Several types are available, including straight rollers, radius rollers, etc.

Capacitance: The storage of electricity in a capacitor; capacitance produces an opposition to voltage change. The unit of measurement for capacitance is the farad (F) or microfarad (μF).

Capstan: The turning drum of a cable puller on which the rope is wrapped and pulled. An increase in the number of wraps increases the pulling force of the cable puller.

Cartridge fuse: A fuse enclosed in an insulating tube to confine the arc when the fuse blows.

Circuit breaker: A device designed to open and close a circuit by nonautomatic means and to open the circuit automatically on a predetermined overcurrent without injury to itself when properly applied within its rating.

Clevis: A device used in cable pulls to facilitate connecting the pulling rope to the cable grip.

Color rendering index (CRI): A measurement of the way a light source reproduces color. The higher the index number (0–100), the closer colors are to how an object appears in full sunlight or incandescent light.

Commutator: A device used on electric motors or generators to maintain a unidirectional current.

Concentric bending: The process of making 90° bends in parallel runs of conduit. This requires increasing the radius of each conduit from the inside of the bend toward the outside.

Conductor support: The act of providing support in vertical conduit runs to support the cables or conductors. The *NEC®* gives several methods in which cables may be supported, including wedges in the tops of conduits, supports to change the direction of cable in pull boxes, etc.

Conduit: Piping designed especially for pulling electrical conductors. Types include rigid, IMC, EMT, PVC, aluminum, and other materials.

Conduit body: A separate portion of a conduit or tubing system that provides access through a removable cover (or covers) to the interior of the system at a junction of two or more sections of the system or at a terminal point of the system. Boxes such as FS and FD or larger cast or sheet metal boxes are not classified as conduit bodies.

Conduit piston: A cylinder of foam rubber that fits inside the conduit and is then propelled by compressed air or vacuumed through the conduit run to pull a line, rope, or measuring tape. Also called a mouse.

Connection: That part of a circuit that has negligible impedance and joins components or devices.

Connector: A device used to physically and electrically connect two or more conductors; a device used to physically connect conduit or cable to an outlet box, cabinet, or other enclosure.

Continuous duty: Operation at a substantially constant load for an indefinitely long time.

Controller: A device that serves to govern, in some predetermined manner, the electric power delivered to the apparatus to which it is connected.

Cross: A four-way section of cable tray used when the tray assembly must branch off in four different directions.

Degree indicator: An instrument designed to indicate the exact degree of bend while it is being made.

Delta-connected: A three-phase transformer connection in which the terminals are connected in a triangular shape like the Greek letter delta (Δ).

Developed length: The amount of straight pipe needed to bend a given radius. Also, the actual length of the conduit that will be bent.

Direct rod suspension: A method used to support cable tray by means of threaded rods and hanger clamps. One end of the threaded rod is secured to an overhead structure, while the other end is connected to hanger clamps that are attached to the cable tray side rails.

Dropout: Cable leaving the tray assembly and travelling directly downward; that is, the cable is not routed into a conduit or channel.

Dropout plate: A metal plate used at the end of a cable tray section to ensure a greater cable bending radius as the cable leaves the tray assembly.

Dual-element fuse: A fuse having two fuse characteristics; the usual combination is an overcurrent limit and a time delay before activation.

Duty: Describes the length of operation. There are four designations for circuit duty: continuous, periodic, intermittent, and varying.

Edison-base: The standard screw base used for ordinary lamps and Edison-base plug fuses.

Effective grounding path: The path to ground from circuits, equipment, and metal enclosures for conductors, which must be permanent and continuous, have the capacity to safely conduct any fault current likely to be imposed on it, and have sufficiently low impedance to limit the voltage to ground and to facilitate the operation of the circuit protective devices.

Effectively grounded: Intentionally connected to the earth through a ground connection or connections of sufficiently low impedance and having sufficient current-carrying capacity to prevent the buildup of voltages that may result in undue hazards to personnel or connected equipment.

Elbow: A 90° bend in a conduit. Also, a section of cable tray used to change the direction of the tray assembly a full quarter turn (90°). Both vertical and horizontal elbows are common.

Equipment: A general term including material, fittings, devices, appliances, fixtures, apparatus, and the like used as a part of, or in connection with, an electrical installation.

Equipment bonding jumper: The connection between two or more portions of the equipment grounding conductor.

Equipment grounding conductor: The conductor used to connect the noncurrent-carrying metal parts of equipment, raceways, and other enclosures to the system grounded conductor, the grounding electrode conductor, or both, at the service equipment or at the source of a separately derived system.

Expansion joints: Plates used at intervals along a straight run of cable tray to allow space for thermal expansion or contraction of the tray.

Explosion-proof: Designed and constructed to withstand an internal explosion without creating an external explosion or fire.

Extension boom: Part of a cable-pulling apparatus that is usually used when pulling conductors through concealed conduit runs.

Extension bushing: Used to extend the end of the conduit in panels and other enclosures to provide a grip for the flexible adapter sheave.

Fault current: The current that exists when an unintended path is established between an ungrounded conductor and ground.

Field poles: The stationary portion of a DC motor that produces the magnetic field.

Fish line: Light cord used in conjunction with vacuum/blower power fishing systems that attaches to the conduit piston to be pushed or pulled through the conduit. Once through, a pulling rope is attached to one end and pulled back through the conduit for use in pulling conductors.

Fish tape: A flat iron wire or fiber cord used to pull conductors or a pulling rope through conduit.

Fittings: Devices used to assemble and/or change the direction of cable tray systems.

Frame size: A method used to classify circuit breakers according to given current ranges.

Frequency: The number of cycles an alternating electric current, sound wave, or vibrating object undergoes per second.

Fuse: A protective device that opens a circuit when the fusible element is severed by heating due to a fault current or overcurrent passing through it.

Fuse link: The fusible part of a cartridge fuse.

Gain: The amount of pipe saved by bending on a radius as opposed to right angles. Because conduit bends in a radius and not at right angles, the length of conduit needed for a bend will not equal the total determined length. Gain is the difference between the right angle distances A and B and the shorter distance C—the length of conduit actually needed for the bend.

Glare: The sensation produced by luminance within the visual field that is significantly greater than the luminance to which the eyes are adapted.

Ground: A conducting connection, whether intentional or accidental, between an electrical circuit or equipment and the earth, or to some conducting body that serves in place of the earth.

Ground current: Current in the earth or grounding connection.

Ground grids: System of grounding electrodes interconnected by bare cables buried in the earth to provide lower resistance than a single grounding electrode.

Ground mats: System of bare conductors, on or below the surface of the earth, connected to a ground or ground grid to provide protection from dangerous touch voltage.

Ground resistance: The ohmic resistance between a grounding electrode and a remote or reference grounding electrode that are spaced such that their mutual resistance is essentially zero.

Ground rod: A metal rod or pipe used as a grounding electrode.

Grounded: Connected to earth or to some conducting body that serves in place of the earth.

Grounded conductor: A system or circuit conductor that is intentionally grounded.

Grounding clip: A spring clip used to secure a bonding conductor to an outlet box.

Grounding conductor: A conductor used to connect equipment or the grounded circuit of a wiring system to a grounding electrode or electrodes.

Grounding connections: Connections used to establish a ground; they consist of a grounding conductor, a grounding electrode, and the earth surrounding the electrode.

Grounding electrode: A grounding electrode consists of (1) the nearest available effectively grounded structural metal member of the structure; or (2) the nearest available effectively grounded metal water pipe; or (3) other electrodes as specified in *NEC® Sections 250.50 and 250.52*.

Grounding electrode conductor: The conductor used to connect the grounding electrode to the equipment grounding conductor, the grounded conductor, or both, of the circuit at the service equipment or at the source of a separately derived system.

Handy box: Single-gang outlet box used for surface mounting to enclose receptacles or wall switches on concrete or concrete block construction of industrial and commercial buildings; nongangable; also made for recessed mounting; also known as a utility box.

Hertz (Hz): A unit of frequency; one hertz equals one cycle per second.

Horsepower: The rated output capacity of the motor. It is based on breakdown torque, which is the maximum torque a motor will develop without an abrupt drop in speed.

Hours: The duty cycle of a motor. Most fractional horsepower motors are marked continuous for around-the-clock operation at the nameplate rating in the rated ambient conditions. Motors marked one-half are for ½-hour ratings, and those marked one are for 1-hour ratings.

Impedance: The opposition to current flow in an AC circuit; impedance includes resistance (R), capacitive reactance (X_C), and inductive reactance (X_L). Impedance is measured in ohms.

Incident light: The light emitted from a self-luminous object such as the sun or an electric source.

Inductance: The creation of a voltage due to a time-varying current; also, the opposition to current change, causing current changes to lag behind voltage changes. The unit of measure for inductance is the henry (H).

Inductive load: A load in which the voltage leads the current. Electromagnet coils such as those used in relays and starters are inductive loads.

Inside diameter (I.D.): The inside diameter of conduit. All electrical conduit is measured in this manner. The outside dimensions, however, will vary with the type of conduit used.

Instantaneous: Contacts that make or break a circuit immediately upon coil energization or de-energization.

Insulating tape: Adhesive tape that has been manufactured from a nonconductive material and is used for covering wire joints and exposed parts.

Interlocked armor cable: Mechanically protected cable; usually a helical winding of metal tape formed so that each convolution locks mechanically upon the previous one (armor interlock).

Intermittent duty: Operation for alternate intervals of (1) load and no load; or (2) load and rest; or (3) load, no load, and rest.

Junction box: An enclosure where one or more raceways or cables enter, and in which electrical conductors can be, or are, spliced.

Kicks: Bends in a piece of conduit, usually less than 90°, made to change the direction of the conduit.

Ladder tray: A type of cable tray that consists of two parallel channels connected by rungs, similar in appearance to the common straight ladder.

Latch coil: The coil of a latching relay that, when energized, causes the relay to assume the switched or latched position.

Leg length: The distance from the end of the straight section of conduit to the bend, measured to the centerline or to the inside or outside of the bend or rise.

Load center: A specific type of panelboard designed for light-duty, single- or three-phase applications, such as residential or light commercial installations.

Lug: A device for terminating a conductor to facilitate the mechanical connection.

Lumen (lm): The basic measurement of light. One lumen is defined as the amount of light cast upon one square foot of the inner surface of a hollow sphere with a one-foot radius with a light source of one candela in its center.

Luminaire (fixture): A complete lighting unit consisting of a lamp (or lamps) and ballasts (where applicable), together with the parts designed to distribute the light, position and protect lamps, and connect them to the power supply.

Luminance: The luminous intensity of any surface in a given direction per unit area of that surface as viewed from that direction. Measured in candela/m^2. The term luminance is commonly used to express brightness.

Main bonding jumper: The connection between the grounded circuit conductor and the equipment grounding conductor at the service.

Make: The closing of open contacts to establish an electrical circuit.

Mechanically held relay: A latching relay in which the switched or latched position is maintained by the interference of a mechanical part that moves to a non-interfering position when the unlatch coil is energized, allowing the relay to return to the normal or unlatched position.

Micro: Prefix designating one-millionth of a unit. For example, one microfarad is one-millionth of a farad.

Molded-case circuit breaker: A circuit breaker enclosed in an insulating housing.

Neutral: A grounded conductor in an electrical system that does not carry current until the system is unbalanced.

Nineteen hundred box: A term commonly used to refer to any two-gang, four-inch-square outlet box.

Ninety-degree bend: A bend in a piece of conduit that changes its direction by 90°.

Nonrenewable fuse: A fuse that must be replaced after it interrupts a circuit.

Normally closed (N.C.) contacts: A combination of a stationary contact and a movable contact, which are engaged when the coil is de-energized. In the case of a latching relay, the contacts that are engaged after the unlatch coil is energized.

Normally open (N.O.) contacts: A combination of a stationary contact and a movable contact, which are disengaged when the coil is de-energized. In the case of a latching relay, the contacts that are disengaged after the unlatch coil is energized.

Offsets: Two equal bends made to avoid an obstruction blocking the run of the conduit.

One-shot shoe: A large bending shoe that is designed to make 90° bends in conduit.

Outlet box: A metallic or nonmetallic box installed in an electrical wiring system from which current is taken to supply some apparatus or device.

Outside diameter (OD): The size of any piece of conduit measured on the outside diameter.

Overcurrent: Any current in excess of the rated current of equipment or the ampacity of a conductor. It may result from an overload, short circuit, or ground fault.

Overcurrent protection: De-energizing a circuit whenever the current exceeds a predetermined value; the usual devices are fuses, circuit breakers, or magnetic relays.

Overload: Operation of equipment in excess of the normal, full-load rating, or of a conductor in excess of rated ampacity which, after a sufficient length of time, will cause damage or dangerous overheating. A fault, such as a short circuit or ground fault, is not an overload.

Peak voltage: The peak value of a sinusoidally varying (cyclical) voltage or current is equal to the root-mean-square (rms) value multiplied by the square root of two (1.414). AC voltages are usually expressed as rms values; that is, 120 volts, 208 volts, 240 volts, 277 volts, 480 volts, etc., are all rms values. The peak voltage, however, differs. For example, the peak value of 120 volts (rms) is actually $120 \times 1.414 = 169.71$ volts.

Periodic duty: Intermittent operation at a substantially constant load for a short and definitely specified time.

Pipe racks: Structural frames used to support the piping that interconnects equipment in outdoor industrial facilities.

Plug fuse: A type of fuse that is held in position by a screw thread contact instead of spring clips, as is the case with a cartridge fuse.

Pole: That portion of a device associated exclusively with one electrically separated conducting path of the main circuit or device.

Pressure connector: A connector applied using pressure to form a cold weld between the conductor and the connector.

Pull box: A sheet metal box-like enclosure used in conduit runs to facilitate the pulling of cables from point to point in long runs, or to provide for the installation of conduit support bushings needed to support the weight of long riser cables, or to provide for turns in multiple-conduit runs.

Radian: An angle at the center of a circle, subtending (opposite to) an arc of the circle that is equal in length to the radius.

Radius: The relative size of the bent portion of a pipe.

Raintight: Constructed or protected so that exposure to a beating rain will not result in the entrance of water under specified test conditions.

Reactance: The imaginary part of impedance. Also, the opposition to alternating current (AC) due to capacitance (X_C) and/or inductance (X_L).

Reducing connector: A connector used to join two different sized conductors.

Reflected: The bouncing back of light waves or rays by a surface.

Reflected light: Light reflected from an object that does not have self-luminating properties but reflects light provided from another source.

Refracted/refraction: The bending of a light wave or ray as it passes obliquely from one medium to another of different density, or through layers of different density in the same medium.

Resistivity: Resistance between opposite faces of a unit cube. Expressed in ohm-centimeters or ohms per cubic centimeter.

Resonance: A condition reached in an electrical circuit when the inductive reactance neutralizes the capacitance reactance, leaving ohmic resistance as the only opposition to the flow of current.

Revolutions per minute (rpm): The approximate full-load speed at the rated power line frequency. The speed of a motor is determined by the number of poles in the winding. A four-pole, 60Hz motor runs at an approximate speed of 1,725 rpm. A six-pole, 60Hz motor runs at an approximate speed of 1,140 rpm.

Rise: The length of the bent section of conduit measured from the bottom, centerline, or top of the straight section to the end of the bent section.

Root-mean-square (rms): The square root of the average of the square of the function taken throughout the period. The rms value of a sinusoidally varying voltage or current is the effective value of the voltage or current.

Rotation: For single-phase motors, the standard rotation, unless otherwise noted, is counterclockwise facing the lead or opposite shaft end. All motors can be reconnected at the terminal board for opposite rotation unless otherwise indicated.

Sealing compound: The material poured into an electrical fitting (seal-off) to seal and minimize the passage of vapors.

Segment bend: Any bend formed by a series of bends of a few degrees each, rather than a single one-shot bend.

Segmented bending shoe: A smaller type of shoe designed for bending segmented bends only (always less than 15°).

Self-inductance: A magnetic field induced in the conductor carrying the current.

Separately derived system: A premises wiring system whose power is derived from generator, transformer, or converter windings and that has no direct electrical connection, including a solidly connected grounded circuit conductor, to supply conductors originating in another system.

Service: The electric power delivered to the premises.

Service conductors: The conductors between the point of termination of the overhead service drop or underground service lateral and the main disconnecting device in the building or on the premises.

Service drop: The overhead conductors through which electrical service is supplied between the last power company pole and the point of their connection to the service-entrance conductors located at the building or other support used for the purpose.

Service entrance: The conductors between the point of termination of the overhead service drop or underground service lateral and the main disconnecting device in the building or on the premises.

Service equipment: The necessary equipment, usually consisting of a circuit breaker or switch and fuses and their accessories, located near the point of entrance of supply conductors to a building and intended to constitute the main control and cutoff means for the electric supply to the building.

Service lateral: The underground service conductors between the street main, including any risers at a pole or other structure or from transformers, and the first point of connection to the service-entrance conductors in a terminal box, meter, or other enclosure with adequate space, inside or outside the building wall.

Service raceway: The rigid conduit, IMC, or other raceway that encloses the service-entrance conductors.

Setscrew grip: A cable grip, usually with built-in clevis, in which the cable ends are inserted in holes and secured with one or more setscrews.

Shading coil: A shorted turn surrounding a portion of the pole of an AC electromagnet, producing by mutual inductance a delay in the change of the magnetic field in that part, thereby tending to prevent chatter and reduce hum.

Sheave: A pulley-like device used in cable pulls in both conduit and cable tray systems.

Short circuit: An often unintended low-resistance path through which current flows around, rather than through, a component or circuit; the current that exists when an unintended path is created between any two components in a circuit.

Soap: Slang for wire-pulling lubricant.

Solid-state relay: A semiconductor switch having no moving parts. These devices control power to the load and are analogous to relay contacts.

Splice: The electrical and mechanical connection between two pieces of cable.

Springback: The amount, measured in degrees, that a bent conduit tends to straighten after pressure is released on the bender ram. For example, a 90° bend, after pressure is released, will pull back about 2° to 88°.

Starter: A circuit component of certain fluorescent lighting fixtures that is used to heat the lamp electrodes before the lamps are lighted.

Step voltage: The potential difference between two points on the earth's surface separated by a distance of one pace, or about three feet.

Strand: A group of wires, usually stranded or braided.

Stub-up: Another name for the rise in a section of conduit.

Substation: An installation or area where an assembly of devices and apparatus are used to monitor, control, transform, or modify electrical power. Substations are normally installed at a point on the system where transmission voltages are reduced to distribution voltages.

Sweep bend: A 90° bend with a radius larger than that produced by a standard one-shot shoe.

Swivel plates: Devices used to make vertical offsets in cable tray.

Synchronous speed: When the speed of the rotor is equal to the speed of the stator. The speed is determined by multiplying 120 times the frequency divided by the number of poles.

System grounding: Intentional connection of one of the circuit conductors of an electrical system to ground potential.

Take-up (comeback): The amount that must be subtracted from the desired stub length to make the bend come out correctly using a point of reference on the bender or bending shoe.

Tap: A splice connection of a wire to another wire. Also called a tap-off.

Tee: A section of cable tray that branches off the main section in two other directions.

Terminal: A device used for connecting cables.

Termination: The connection of a cable.

Touch voltage: The potential difference between a grounded metallic structure and a point on the earth's surface equal to the normal maximum horizontal reach—approximately three feet.

Trapeze mounting: A method of supporting cable tray using metal channel, such as Unistrut®, Kindorf, etc., supported by two threaded rods, and giving the appearance of a swing or trapeze.

Tray cover: A flat piece of metal, fiberglass, or plastic designed to provide a solid covering that is needed in some locations where conductors in the tray system may be damaged.

Troffer: A recessed lighting fixture installed with the opening flush with the ceiling.

Trough: A type of cable tray consisting of two parallel channels (side rails) having a corrugated, ventilated bottom or a corrugated, solid bottom.

Ungrounded conductors: Conductors in an electrical system that are not intentionally grounded.

Unistrut®: A brand of metal channel used as the bottom bracket for hanging cable trays. Double Unistrut® adds strength and stability to the trays and also provides a means of securing future runs of conduit.

Unlatch coil: The coil of a latching relay that, when energized, causes the relay to assume the normal or unlatched position.

Wall mounting: A method of supporting cable tray systems using supports secured directly to the wall.

Waterproof: Constructed so that moisture will not interfere with successful operation.

Watertight: Constructed so that water will not enter the enclosure under specified test conditions.

Weatherproof: Constructed or protected so that exposure to the weather will not interfere with successful operation.

Wire nut: A form of wire connector that is tightened or loosened by screwing the device clockwise or counterclockwise, respectively. Some wire nuts are provided with two thin, flat wings extending from opposite sides to facilitate tightening the connectors.

Wye-connected: A three-phase transformer connection in which all three phases are connected at a central point, forming a Y configuration.

Figure Credits

Module 26201-05

Walter Johnson, 201P0101
Veronica Westfall, 201F13C

Module 26202-05

Reprinted with permission from NFPA 70-2005, the *National Electrical Code*®. Copyright 2004, National Fire Protection Association, Quincy, MA 02269, Table 6, Table 7, Table 8, Table 9
Reprinted with the permission of the National Electrical Manufacturers Association (NEMA), Table 2, Table 5
Walter Johnson, 202F27, 202P0201, 202P0202, 202P0203, 202F42, 202P0204, 202F47B, 202P0205, 202F57B, 202P0207
Veronica Westfall, 202P0206

Module 26203-05

Mike Powers, 203P0301
John Traister, 203F04, 203F05, 203F10, 203F11, 203F12, 203F13, 203F18, 203F22, 203F23
Veronica Westfall, 203P0302, 203F27B, 203F34

Module 26204-05

Tim Ely, 204F03B, 204F27
Reprinted with permission from NFPA 70-2005, the *National Electrical Code*®. Copyright 2004, National Fire Protection Association, Quincy, MA 02269, Table 1, Table 2
Greenlee Textron Inc., 204F15, 204F25, 204F28, 204F39, 204F40, 204F41, 204F42, 204F44
John Traister, 204F01A, 204F02, 204F09, 204F10
Veronica Westfall, 204P0401, 204F01B, 204F16, 204P0402, 204P0403, 204F26, 204F31, 204P0404, 204P0405, 204F52, 204F56

Module 26205-05

Walter Johnson, 205F25
John Traister, 205F09, 205F10, 205F11, 205F12, 205F20A, 205F20B, 205F21, 205F22, 205F28
Veronica Westfall, 205F01, 205F04, 205F06, 205F15, 205F16, 205F17, 205F18, 205F20C, 205P0501, 205P0502, 205P0503, 205P0504, 205F24, 205P0505, 205F30A, 205F30B, 205F31

Module 26206-05

Greenlee Textron Inc., 206F08, 206F12C, 208F13, 208F14A, 208F16, 208F17, 206F18, 206F19, 206F20, 206F27, 206F28, 206F29
Walter Johnson, 206F26
Veronica Westfall, 206F01, 206P0601, 206F02, 206F12A, 206F12B, 206F14B, 206P0602, 206P0603

Module 26207-05

John Traister, 207F01, 207F02, 207F03, 207F18
Veronica Westfall, 207P0701, 207P0702, 207P0703, 207P0704

Module 26208-05

Greenlee Textron Inc., 208F14
Reprinted with permission from NFPA 70-2005, the *National Electrical Code*®. Copyright 2004, National Fire Protection Association, Quincy, MA 02269, Table 4, Table 5
Veronica Westfall, 208F01, 208P0801, 208F05, 208F21A, 208P0802, 208P0803, 208P0804, 208P0805, 208P0806, 208F21A

Module 26209-05

Walter Johnson, 209F04

Reprinted with permission from NFPA 70-2005, the *National Electrical Code*®. Copyright 2004, National Fire Protection Association, Quincy, MA 02269, Table 1

John Traister, 209F01, 209F02, 209F03, 209F05, 209F06, 209F09, 209F10, 209F11, 209F12, 209F13, 209F14A, 209F15, 209F16, 209F17, 209F18, 209F19, 209F20, 209F21, 209F22, 209F23, 209F25, 209F26, 209F27, 209F28, 209F29, 209F30, 209F31, 209F32, 209F34, Table 3

Veronica Westfall, 209P0901, 209F14B, 209P0902, 209P0903, 209P0904, 209P0905, 209P0906, 209P0907, 209P0908, 209F24, 209P0909, 209F33A, 209F33B

Module 26210-05

Cooper Bussmann, 210F04, 210F05, 210F06, 210F20, Table 4

Tim Ely, 210P1003

Walter Johnson, 210F14B

John Traister, 210F01, 210F02, 210F03, 210F07, 210F08, 210F09, 210F10, 210F11, 210F12, 210F15

Veronica Westfall, 210P1001, 210P1002, 210F16

Module 26211-05

Walter Johnson, 211P1102

Reprinted with the permission of the National Electrical Manufacturers Association (NEMA), Table 1

Veronica Westfall, 211F10B, 211P1101

Module 26212-05

Courtesy of General Electric Company, 212F07

Courtesy of Hubbell Lighting, a Division of Hubbell Lighting, Inc., 212F19

Courtesy of Prescolite, a Division of Hubbell Lighting, Inc., 212F33

OSRAM SYLVANIA, 212F06, 212F08, 212F09

Veronica Westfall, 212P1201, 212P1202

Index

The designation "*f*" refers to figures.

C

D

E

F

G

H

I

N

O